MW00452049

TO MAJ (Y)

With best Wishes

Barend . Mahof

RADAR SYSTEMS ANALYSIS AND DESIGN USING MATLAB®

THIRD EDITION

RADAR SYSTEMS ANALYSIS AND DESIGN USING MATLAB®

THIRD EDITION

BASSEM R. MAHAFZA

deciBel Research Inc.
Huntsville, Alabama, USA

CRC Press
Taylor & Francis Group
Boca Raton London New York

CRC Press is an imprint of the
Taylor & Francis Group, an **informa** business

A CHAPMAN & HALL BOOK

MATLAB® is a trademark of The MathWorks, Inc. and is used with permission. The MathWorks does not warrant the accuracy of the text or exercises in this book. This book's use or discussion of MATLAB® software or related products does not constitute endorsement or sponsorship by The MathWorks of a particular pedagogical approach or particular use of the MATLAB® software.

CRC Press
Taylor & Francis Group
6000 Broken Sound Parkway NW, Suite 300
Boca Raton, FL 33487-2742

© 2013 by Taylor & Francis Group, LLC
CRC Press is an imprint of Taylor & Francis Group, an Informa business

No claim to original U.S. Government works

ISBN 13 : 978-1-4398-8495-9 (hbk)

This book contains information obtained from authentic and highly regarded sources. Reasonable efforts have been made to publish reliable data and information, but the author and publisher cannot assume responsibility for the validity of all materials or the consequences of their use. The authors and publishers have attempted to trace the copyright holders of all material reproduced in this publication and apologize to copyright holders if permission to publish in this form has not been obtained. If any copyright material has not been acknowledged please write and let us know so we may rectify in any future reprint.

Except as permitted under U.S. Copyright Law, no part of this book may be reprinted, reproduced, transmitted, or utilized in any form by any electronic, mechanical, or other means, now known or hereafter invented, including photocopying, microfilming, and recording, or in any information storage or retrieval system, without written permission from the publishers. Printed in Canada.

For permission to photocopy or use material electronically from this work, please access www.copyright.com (http://www.copyright.com/) or contact the Copyright Clearance Center, Inc. (CCC), 222 Rosewood Drive, Danvers, MA 01923, 978-750-8400. CCC is a not-for-profit organization that provides licenses and registration for a variety of users. For organizations that have been granted a photocopy license by the CCC, a separate system of payment has been arranged.

Trademark Notice: Product or corporate names may be trademarks or registered trademarks, and are used only for identification and explanation without intent to infringe.

Visit the Taylor & Francis Web site at
http://www.taylorandfrancis.com

and the CRC Press Web site at
http://www.crcpress.com

Book Dedication

To my four sons. May this book be my whisper through your minds and hearts, to guide you throughout your life, as you become the best you can be in this world.

Bassem R. Mahafza
Huntsville, Alabama
United States of America
November, 2012

Table of Contents

PART II: Radar Signals and Signal Processing

Chapter 3: Linear Systems and Complex Signal Representation, 93

Chapter 4: The Matched Filter Radar Receiver, 143

Chapter 9: Radar Clutter, 335

Chapter 10: Moving Target Indicator (MTI) and Pulse Doppler Radars, 361

PART IV: Radar Detection

Chapter 11: Random Variables and Random Processes, 403

Chapter 12: Single Pulse Detection, 419

PART V: Radar Special Topics

Chapter 14: Radar Cross Section (RCS), 485

Chapter 15: Phased Array Antennas, 541

Chapter 16: Adaptive Signal Processing, 607

Preface

In the year 2000 the first edition of *Radar Systems Analysis and Design Using MATLAB*®[1] was published. It was developed and organized based on my years of teaching graduate level courses on radar systems analysis and design including advanced topics in radar signal processing. At the time, the primary motivation behind the book was to introduce a college-suitable comprehensive textbook that provides hands-on experience with MATLAB companion software. This book very quickly turned into a bestseller, which prompted the publication of its second edition in the year 2005. The second edition continued in the same vein as its predecessor. It was updated, expanded, and reorganized to include advances in the field and to be more logical in sequence. New topics were introduced in the body of the text, and much of the MATLAB code was updated and improved upon to reflect the advancements of the latest MATLAB release.

Since the publication of the first edition, *Radar Systems Analysis and Design Using MATLAB* filled a void in the market by presenting a comprehensive and self-contained text on radar systems analysis and design. It was the first book on the market to provide companion MATLAB software to support the theoretical and mathematical discussion found within the pages of the text. These features were also supported with a detailed solutions manual of all end-of-chapter problems. This book quickly became the standard adopted by many books published on the subject; none of which, however, matched the clear presentation nor the transparency offered by this author, particularly when considering the end-of-chapter solutions manual and the complete and comprehensive set of MATLAB code, which was made available to all of the book audience without any restrictions. Users of this book were not only able to reproduce all plots found in the text, but they also had the ability to change the code by inputting their own parameters so that they could generate their own specific plots and outputs that met their own unique academic interest.

In addition to my academic tenure and experience in teaching the subject at the collegiate level, I have also taught numerous industry courses and conducted many seminars on the subject of radar systems. Based on this teaching experience, the following conclusion has become very evident to me: The need and the demand for a comprehensive textbook / reference book focused on all aspects of radar systems design and analysis remain very strong. Add to this the

---

1. All MATLAB® functions and programs provided in this book were developed using MATLAB R2011a version 7.12.0.635 with the Signal Processing Toolbox, on a PC with Windows XP Professional operating system. MATLAB® is a registered trademark of the The MathWorks, Inc. For product information, please contact: The MathWorks, Inc., 3 Apple Hill Drive, Natick, MA 01760-2098 USA. Web: *www.mathworks.com*.

fact that many college professors have adopted this book as the primary textbook for their courses on radar systems. Therefore, my desire to write this third edition was turned into reality and has materialized into this product.

It is my view that the third edition of *Radar Systems Analysis and Design Using MATLAB* is warranted for the following reasons: (1) bring the text to a more modern status to reflect the current state of the art; (2) incorporate into the new edition much of the feedback this author has received from professors using this book as a text and from other practicing engineers; (3) introduce several new topics that have not found much treatment by other authors, and even when they did, it was not on a level comparable to the comprehensive and exhaustive approach adopted by this author in the first two editions; (4) add many new end-of chapter problems; (5) restructure the presentation to be more convenient for users to adopt the text for either three graduate-level courses, or one senior-level and two graduate-level courses; and (6) take advantage of the new features offered by the latest MATLAB releases.

Note that all MATLAB code provided in this book was designed as an academic standalone tool and is not adequate for other purposes. The code was written in a way to assist the reader in gaining better understanding of the theory. The code was not developed, nor is it intended to be used as part of an open loop or a closed loop simulation of any kind. The MATLAB code found in this textbook can be downloaded from this book's web-page on the CRC Press web-site. Simply use your favorite web browser, go to *www.crcpress.com*, and search for keyword *"Mahafza"* to locate this book's web page.

Just like the first and second editions, this third edition provides easy-to-follow mathematical derivations of all equations and formulas present within the book, resulting in a user friendly coverage suitable for advanced as well as introductory level college courses. This third edition provides comprehensive up-to-date coverage of radar systems design and analysis issues. Users of this book will need only one book instead of several, to gain essential understanding of radar design, analysis, and signal processing. This edition contains numerous graphical plots and supporting artwork. The MATLAB code companion of this edition will help users evaluate the trade-offs between different radar parameters.

This book is composed of 18 chapters and is divided into 5 parts: Part I, Radar Principles, Part II, Radar Signals and Signal Processing, Part III, Special Radar Considerations, Part IV, Radar Detection, and Part V, Radar Special Topics. Part I comprises Chapters 1 and 2. Chapter 1, *Definitions and Nomenclature,* presents the basic radar definitions and establishes much of the nomenclature used throughout the text. In Chapter 2, *Basic Pulsed and Continuous Wave (CW) Radar Operations,* the radar equation is derived for both pulsed and CW radars, while other related material such as radar losses and noise are also discussed in details. The radar equation in the presence of electronic counter measures (ECM) is derived, as well as the bistatic radar equation.

Part II comprises Chapters 3 through 7. The main thrust of this part of the book is radar signals or waveforms and radar signal processing. Chapter 3, *Linear Systems and Complex Signal Representation*, contains a top-level discussion of elements of signal theory that are relevant to radar design and radar signal processing. It is assumed that the reader has sufficient and adequate background in signals and systems as well as in the Fourier transform and its associated properties. Lowpass and bandpass signals are discussed in the context of radar applications. Continuous as well as discrete systems are analyzed, and the sampling theorem is presented.

Chapter 4, *The Matched Filter Radar Receiver*, is focused on the matched filter. It presents the unique characteristic of the matched filter and develops a general formula for the output of the matched filter that is valid for any waveform. Chapter 5, *Ambiguity Function - Analog Waveforms*, and Chapter 6, *Ambiguity Function - Discrete Coded Waveforms,* analyze the output of the matched filter in the context of the ambiguity function. In Chapter 5 the most common analog radar waveforms are analyzed; this includes the single unmodulated pulse, Linear Frequency Modulation (LFM) pulse, unmodulated pulse train, LFM pulse train, stepped frequency waveforms, and nonlinear FM waveforms. Chapter 6 is concerned with discrete coded waveforms. In this chapter, unmodulated pulse-train codes are analyzed as well as binary codes, polyphase codes, and frequency codes. Chapter 7, *Pulse Compression*, contains details of radar signal processing using pulse compression. The correlation processor and stretch processor are presented. High range resolution processing using stepped frequency waveforms is also analyzed.

Part III comprises three chapters. Chapter 8, *Radar Wave propagation*, extends the free space analysis presented in the earlier chapters to include the effect of the atmosphere on radar performance. Topics such as refraction, diffraction, atmospheric attenuation, surface reflection, and multipath are discussed in a fair amount of detail. The subject of radar clutter is in Chapter 9, *Radar Clutter*. Area clutter as well as volume clutter are defined and the radar equation is re-derived to reflect the importance of clutter, where in this case, the signal to interference ratio becomes more critical than the signal to noise ratio. A step-by-step mathematical derivation of clutter RCS is presented, and the statistical models for the clutter backscatter coefficient is also presented. Chapter 10, *Moving Target Indicator (MTI) and Pulse Doppler Radars*, discusses how delay line cancelers can be used to mitigate the impact of clutter within the radar signal processor. PRF staggering is analyzed in the context of blind speeds and in the context of resolving range and Doppler ambiguities. Finally, pulsed Doppler radars are briefly analyzed.

In Part IV, radar detection is discussed and analyzed. The material presented in this part of the book requires a strong background in random variables and random processes. Therefore, Chapter 11, *Random Variables and Random Processes*, presents a review of the subject, and is written in such a way that it only highlights the major points of the subject. Users of this book are advised to use this chapter as a means for a quick top-level review of random variables and random processes. Instructors using this book as a text may assign Chapter 11 as a reading assignment to their students. Single pulse detection with known and unknown signal parameters is in Chapter 12, *Single Pulse Detection*. Chapter 13, *Detection of Fluctuating Targets*, extends the analysis of Chapter 12 to include target fluctuation where the Swerling target models are discussed. Detailed discussion of coherent and noncoherent integration in the context of a square law detector is in this chapter. An overview of CFAR, cumulative probability of detection, and M-out-of-N detection are also discussed.

Part V of this book addresses a few specialized topics in radar systems. In Chapter 14, *Radar Cross Section (RCS)*, the RCS dependency on aspect angle, frequency, and polarization are discussed. A target scattering matrix is developed. RCS formulas for many simple objects are presented. Complex object RCS is discussed, and RCS prediction methods are introduced. Chapter 15, *Phased Array Antennas*, starts by developing the general array formulation. Linear arrays and several planar array configurations such as rectangular, circular, rectangular with circular boundaries, and concentric circular arrays are discussed. Beam steering with and without using a finite number of bits is analyzed. Scan loss is also presented. A concept of a multiple input multiple output radar system developed by this author is discussed and analyzed. In Chapter 16, *Adaptive Signal Processing*, the concept behind conventual and adaptive beamforming is discussed. Adaptive signal processing using the least mean square algorithm is analyzed. Adaptive linear arrays and complex weights computation in the context of the least mean square algorithm are presented. Finally, this chapter discusses, space time adaptive processing.

Chapter 17, *Target Tracking*, discusses target tracking radar systems. The first part of this chapter covers the subject of single target tracking. Topics such as sequential lobing, conical scan, monopulse, and range tracking are discussed in detail. The second part of this chapter introduces multiple target tracking techniques. Fixed gain tracking filters such as the $\alpha\beta$ and the $\alpha\beta\gamma$ filters are presented in detail. The concept of the Kalman filter is introduced. Special cases of the Kalman filter are analyzed in depth and a MATLAB-based simulation of the Kalamn filter is developed. The last chapter of this book is Chapter 18, *Tactical Synthetic Aperture Radars*. The topics of this chapter include: SAR signal processing, SAR design considerations, and the SAR radar equation. Arrays operated in sequential mode are discussed in this chapter.

This book is written primarily as a graduate-level textbook, although parts of it can be used as a senior level course on radar systems. A companion solutions manual has been developed for use by professors that adopt this book as a text. This solutions manual is available through the publisher. Based on my own teaching experience, the following breakdown can be utilized by professors using this book as a text:

1. Option I: Chapters 1-4 (with omission of certain advanced sections) can be used as a senior-level course. Chapters 5-10 and the omitted sections in the previous course can be used as a first graduate level course. Finally, Chapters 11-18 can be used as a second advanced graduate-level course.
2. Option II: Chapters 1-4 can be used as an introductory graduate-level course. Chapters 5 10 can be used as a second graduate-level course, while Chapters 11-18 can be used as an advanced graduate course on the subject.

Bassem R. Mahafza
Huntsville, Alabama
United States of America
November, 2012

Part I

Radar Principles

Chapter 1

Definitions and Nomenclature

This chapter presents some basic radar definitions and establishes much of the nomenclature used throughout this text. The word radar is an abbreviation for *radio detection and ranging*. In most cases, radar systems use modulated waveforms and directive antennas to transmit electromagnetic energy into a specific volume in space to search for targets. Objects (targets) within a search volume will reflect portions of the incident energy (radar returns or echoes) in the direction of the radar. These echoes are then processed by the radar receiver to extract target information such as range, velocity, angular position, and other target identifying characteristics.

1.1. Radar Systems Classifications and Bands

Radars can be classified as ground-based, airborne, spaceborne, or ship-based radar systems. They can also be classified into numerous categories based on the specific radar characteristics, such as the frequency band, antenna type, and waveforms utilized. Radar systems using continuous waveforms, modulated or otherwise, are classified as Continuous Wave (CW) radars. Alternatively, radar systems using time-limited pulsed waveforms are classified as Pulsed Radars. Another radar systems classification is concerned with the mission and/or the functionality of the specific radar. This includes: weather, acquisition and search, tracking, track-while-scan, fire control, early warning, over-the-horizon, terrain following, and terrain avoidance radars. Phased array radars utilize phased array antennas, and are often called multifunction (multimode) radars. A phased array is a composite antenna formed from two or more basic radiators. Array antennas synthesize narrow directive beams that may be steered, mechanically or electronically. Electronic steering is achieved by controlling the phase of the electric current feeding the array elements, and thus the name phased arrays is adopted.

Historically, radars were first developed as military tools. It is for this primary reason the most common radar systems classification is the letter or band designation originally used by the military during and after World War II. This letter or band designation has also been adopted as an IEEE (Institute of Electrical and Electronics Engineers) standard. In recent years, NATO (North Atlantic Treaty Organization) has adopted a new band designation with easier abecedarian letters. Figure 1.1 shows the spectrum associated with these two letter or band radar classifications, while Table 1.1 presents the same information in a structured format.

3

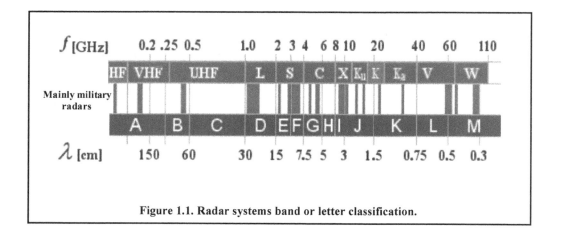

Figure 1.1. Radar systems band or letter classification.

Table 1.1. Radar systems band or letter classification.

Letter designation	Frequency range in GHz (IEEE Standard)	Frequency range in GHz (NATO or New band designation)
HF	0.003 - 0.03	A
VHF	0.03 - 0.3	A<0.25; B>0.25
UHF	0.3 - 1.0	B<0.5; C>0.5
L-band	1.0 - 2.0	D
S-band	2.0 - 4.0	E<3.0; F>3.0
C-band	4.0 - 8.0	G<6.0; H>6.0
X-band	8.0 - 12.5	I<10.0; J>10.0
Ku-band	12.5 - 18.0	J
K-band	18.0 - 26.5	J<20.0; K>20.0
Ka-band	26.5 - 40.0	K
V & W or Millimeter Wave (MMW)	Normally >34.0	L<60.0; M>60.0

High Frequency (HF) and Very High Frequency (VHF) Radars (A- and B-Bands): These radar bands below $300MHz$ represent the frontier of radio technology at the time during the World War II. However, in the modern radar era, these frequencies bands are used for early warning radars. These radars utilize the electromagnetic waves' reflection off the ionosphere to detect targets beyond the horizon, and so they are called Over-the-Horizon Radars (OTHR). Some examples include the United States (U.S.) Navy Relocatable over-the-horizon Radar (ROTHR) shown in Fig. 1.2, and the Russian Woodpecker radar shown in Fig. 1.3. By using these low HF and VHF frequency bands, one can use high-power transmitters. At these frequencies, the electromagnetic wave atmospheric attenuation is small and can be overcome by using high-power transmitters. Radar angular measurement accuracies are limited in these bands because lower frequencies require antennas with significant physical size, thus limiting the radar's angle accuracy and angle resolution. Other communication and broadcasting services typically use these frequency bands. Therefore, the available bandwidth for military radar systems is limited and highly contested throughout the world. Low-frequency systems can be used for Foliage Penetration (FoPen) applications, as well as in Ground Penetrating (GPen) applications.

Figure 1.2. U. S. Navy over-the-horizon Radar. Photograph obtained via the Internet *(http://www.fas.org/nuke/guide/usa/airdef/an-tps-71.htm)*.

Figure 1.3. Russian Woodpecker OTHR radar. Photograph obtained via the Internet *(http://passingstrangeness.wordpress.com/2010/04/23/the-russian-woodpecker/)*.

Ultra High Frequency (UHF) Radars (C-Band): UHF bands are used for very long range Early Warning Radars (EWR). Some examples include the Ballistic Missile Early Warning System (BMEWS) search-and-track monopulse radar that operates at $245\,MHz$ (see Fig. 1.4), the Perimeter and Acquisition Radar (PAR), which is a very long range multifunction phased array radar; and the early warning PAVE PAWS multifunction UHF phased array radar. This frequency band is also used for the detection and tracking of satellites and ballistic missiles over a long range. In recent years, ultra wideband (UWB) radar applications use all frequencies in the A- to C-Bands. UWB radars can be used in GPen applications as well as in see-through-the-wall applications.

Figure 1.4. Fylingdales BMEWS, United Kingdom. Photograph obtained via the Internet *(http://en.wikipedia.org/wiki/File:Radar_RAF_Fylingdales.jpg)*.

L-Band Radars (D-Band): Radars in the L-band are primarily ground-based and ship-based systems that are used in long range military and air traffic control search operations for up to 250 (~$500Km$) nautical miles. Therefore, due to earth curvature their maximum achievable range is limited when detecting low-altitude targets which can disappear very quickly below the horizon. The Air Traffic Management (ATM) long-range surveillance radars like the Air Route Surveillance Radar (ARSR), work in this frequency band. These radar systems are relatively large and demand sizable footprints. Historically, the designator L-Band was adopted since the "L" represent with large antenna or long range radars.

S-Band Radars (E- and F-Bands): Most ground- and ship-based medium range radars operate in the S-band. For example, the Airport Surveillance Radar (ASR) used for air traffic control, and the ship-based U.S. Navy AEGIS (Fig. 1.5) multifunction phased array are S-band radars, and the Airborne Warning and Control System (AWACS) shown in Fig. 1.6. The atmospheric attenuation in this band is higher than in the D-Band, and they are also more susceptible to weather conditions. Radar in this band usually need considerably high transmitting power as compared to the lower-frequency radars in order to achieve maximum detection range. Even with the considerable weather susceptibility, the National Weather Service Next Generation Doppler Weather Radar (NEXRAD) uses an S-band radar, because it can see beyond a severe storm. Special Airport Surveillance Radars (ASR) used at some civilian airports are also in this band where they can detect aircrafts for up to 60 nautical miles. The designator S-Band (contrary to L-Band) was adopted since the "S" represents the smaller antennas or shorter range radars.

Figure 1.5. U. S. Navy AEGIS. Photograph obtained via the Internet *(http://mostlymissiledefense.com/2012/08/03/ballistic-missile-defense-the-aegis-spy-1-radar-august-3-2012/)*.

Figure 1.6. U. S. Air Force AWACS. Photograph obtained via the Internet *(http://www.globalsecurity.org/military/systems/aircraft/e-3-pics.htm)*.

C-Band Radar (G-Band): Many of the mobile military battlefield surveillance, missile-control and ground surveillance radar systems operate in this band. Most weather radar systems are also C-band radars. Medium range search and fire control military radars and metric instrumentation radars are C-band systems. In this band, the size of the antenna allows for achieving excellent angular accuracies and resolution. Performance of systems operating in this band suffer severely from bad weather conditions and to counter that, they often employ antenna feeds with circular polarization.

X- and Ku-Band Radars (I- and J-Bands): In the X-band frequency range (8 to $12GHz$) the relationship between the wave length and size of the antenna is considerably better than in lower-frequency bands. Radar systems that require fine target detection capabilities and yet cannot tolerate the atmospheric attenuation of higher-frequency bands are typically X-Band. The X- and Ku-bands are relatively popular radar frequency bands for military applications like airborne radars, since the small antenna size provides good performance. Missile guidance systems use the Ku-Band (I- and J-Bands) because of the convenient antenna size where weight is a limiting requirement. Space borne or airborne imaging radars used in Synthetic Aperture Radar (SAR) for military electronic intelligence and civil geographic mapping typically use these frequency bands. Finally, these frequency bands are also widely used in maritime civil and military navigation radars.

K- and Ka- Band Radars (J- and K-Bands): These high-frequency bands suffer severe weather and atmospheric attenuation. Therefore, radars utilizing these frequency bands are limited to short range applications, such as police traffic radars, short range terrain avoidance, and terrain following radars. Alternatively, the achievable angular accuracies and range resolution are superior to other bands. In ATM applications these radars are often called Surface Movement Radar (SMR) or Airport Surface Detection Equipment (ASDE) radars.

Millimeter Wave (MMW) Radars (V- and W-Bands): Radars operating in this frequency band also suffer from severe high atmospheric attenuation. Radar applications are limited to very short range of up to a tens of meters. In the W-Band maximum attenuation occurs at about $75GHz$ and at about $96GHz$. Both of these frequencies are used in practice primarily in automotive industry where very small radars ($\sim 75\text{-}76GHz$) are used for parking assistants, blind spot and brake assists. Some radar systems operating at 96 to $98GHz$ are used as laboratory experimental or prototype systems.

1.2. Pulsed and Continuous Wave (CW) Radars

When the type of waveform is used as a classifier of radar systems, there are two types of radars; *pulsed and Continuous Wave* (CW) radar systems. Continuous wave radars are those that continuously emit electromagnetic energy, and use separate transmit and receive antennas. Unmodulated CW radars can accurately measure target radial velocity (Doppler shift) and angular position. Continuous wave waveforms can be viewed as pure sinewaves of the form $\cos 2\pi f_0 t$. Spectra of the radar echo from stationary targets and clutter will be concentrated around f_0. The center frequency for the echoes of a moving target will be shifted by f_d, the Doppler frequency. Thus, by measuring this frequency difference, CW radars can very accurately extract target radial velocity. Because of the continuous nature of CW emission, range measurement is not possible without some modifications to the radar operations and waveforms. Simply put, target range information cannot be extracted without utilizing some form of modulation. The primary use of CW radars is in target velocity search and track, and in missile guidance operations.

Pulsed radars use a train of pulsed waveforms (mainly with modulation). In this category, radar systems can be classified on the basis of the Pulse Repetition Frequency (PRF), as low PRF, medium PRF, and high PRF radars. Low PRF radars are primarily used for ranging where target velocity (Doppler shift) is not of interest. High PRF radars are mainly used to measure target velocity. Continuous wave as well as pulsed radars can measure both target range and radial velocity by utilizing different modulation schemes. The design, operation, and analysis of CW and pulsed radar systems are found in subsequent chapters of this book.

1.3. Range

Figure 1.7 shows a simplified pulsed radar block diagram. The time control box generates the synchronization timing signals required throughout the system. A modulated signal is generated and sent to the antenna by the modulator/transmitter block. Switching the antenna between the transmitting and receiving modes is controlled by the duplexer. The duplexer allows one antenna to be used to both transmit and receive. During transmission it directs the radar electromagnetic energy toward the antenna. Alternatively, on reception, it directs the received radar echoes to the receiver. The receiver amplifies the radar returns and prepares them for signal processing. Extraction of target information is performed by the signal processor block. The target's range, R, is computed by measuring the time delay, Δt; it takes a pulse to travel the two-way path between the radar and the target. Since electromagnetic waves travel at the speed of light, $c = 3 \times 10^8 m/s$, then

$$R = (c\Delta t)/2 \qquad \text{Eq. (1.1)}$$

where R is in meters and Δt is in seconds. The factor of $1/2$ is used to account for the two-way time delay.

In general, a pulsed radar transmits and receives a train of pulses, as illustrated by Fig. 1.8. The Inter Pulse Period (IPP) is T, and the pulse width is τ. The IPP is often referred to as the Pulse Repetition Interval (PRI). The inverse of the PRI is the PRF, which is denoted by f_r,

$$f_r = 1/PRI = 1/T. \qquad \text{Eq. (1.2)}$$

During each PRI the radar radiates energy only for τ seconds and listens for target returns for the rest of the PRI. The radar transmitting duty cycle (factor) d_t is defined as the ratio $d_t = \tau/T$. The radar average transmitted power is

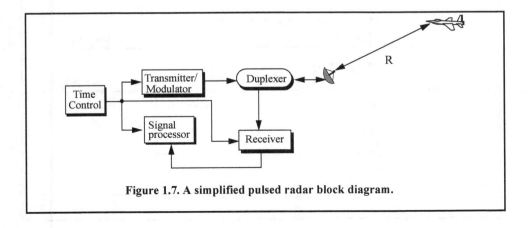

Figure 1.7. A simplified pulsed radar block diagram.

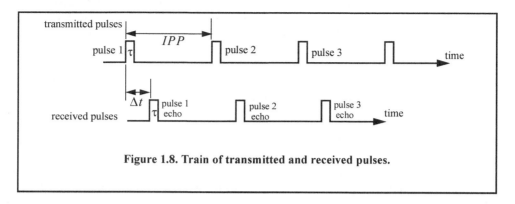

Figure 1.8. Train of transmitted and received pulses.

$$P_{av} = P_t \times d_t$$ **Eq. (1.3)**

where P_t denotes the radar peak transmitted power. The pulse energy is

$$E_p = P_t \tau = P_{av}T = P_{av}/f_r.$$ **Eq. (1.4)**

The range corresponding to the two-way time delay T is known as the radar unambiguous range, R_u. Consider the case shown in Fig. 1.9. Echo 1 represents the radar return from a target at range $R_1 = c\Delta t/2$ due to pulse 1. Echo 2 could be interpreted as the return from the same target due to pulse 2, or it may be the return from a faraway target at range R_2 due to pulse 1 again. In this case,

$$R_2 = \frac{c\Delta t}{2} \qquad or \qquad R_2 = \frac{c(T + \Delta t)}{2}.$$ **Eq. (1.5)**

Clearly, range ambiguity is associated with echo 2. Therefore, once a pulse is transmitted the radar must wait a sufficient length of time so that returns from targets at maximum range are back before the next pulse is emitted. It follows that the maximum unambiguous range must correspond to half of the PRI,

$$R_u = (cT)/2 = c/(2f_r).$$ **Eq. (1.6)**

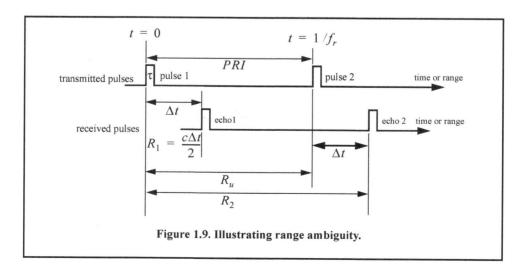

Figure 1.9. Illustrating range ambiguity.

Example:

A certain airborne pulsed radar has peak power $P_t = 10KW$, and uses two PRFs, $f_{r1} = 10KHz$ and $f_{r2} = 30KHz$. What are the required pulse widths for each PRF so that the average transmitted power is constant and is equal to $1500watts$? Compute the pulse energy in each case.

Solution:

Since P_{av} is constant, then both PRFs have the same duty cycle. More precisely,

$$d_t = \frac{1500}{10 \times 10^3} = 0.15.$$

The pulse repetition intervals are

$$T_1 = \frac{1}{10 \times 10^3} = 0.1ms$$

$$T_2 = \frac{1}{30 \times 10^3} = 0.0333ms.$$

It follows that

$$\tau_1 = 0.15 \times T_1 = 15\mu s$$

$$\tau_2 = 0.15 \times T_2 = 5\mu s$$

$$E_{p1} = P_t\tau_1 = 10 \times 10^3 \times 15 \times 10^{-6} = 0.15 Joules$$

$$E_{p2} = P_2\tau_2 = 10 \times 10^3 \times 5 \times 10^{-6} = 0.05 Joules.$$

MATLAB Function *"pulse_train.m"*

The MATLAB function *"pulse_train.m"* computes the duty cycle, average transmitted power, pulse energy, and the pulse repetition frequency; its syntax is as follows:

[dt, pav, ep, prf, ru] = pulse_train(tau, pri, p_peak)

where

Symbol	Description	Units	Status
tau	pulse width	seconds	input
pri	PRI	seconds	input
p_peak	peak power	watts	input
dt	duty cycle	none	output
pav	average transmitted power	watts	output
ep	pulse energy	joules	output
prf	PRF	Hz	output
ru	unambiguous range	Km	output

1.4. *Range Resolution*

Range resolution, denoted as ΔR, is a radar metric that describes its ability to detect targets in close proximity to each other as distinct objects. Radar systems are normally designed to operate between a minimum range R_{min} and maximum range R_{max}. The distance between R_{min} and R_{max} is divided into M range bins (gates), each of width ΔR,

$$M = \frac{R_{max} - R_{min}}{\Delta R}.$$ Eq. (1.7)

Targets separated by at least ΔR will be completely resolved in range, as illustrated in Fig. 1.10. Targets within the same range bin can be resolved in cross range (azimuth) utilizing signal processing techniques.

Consider two targets located at ranges R_1 and R_2, corresponding to time delays t_1 and t_2, respectively. Denote the difference between those two ranges as ΔR:

$$\Delta R = R_2 - R_1 = c\frac{(t_2 - t_1)}{2} = c\frac{\delta t}{2}.$$ Eq. (1.8)

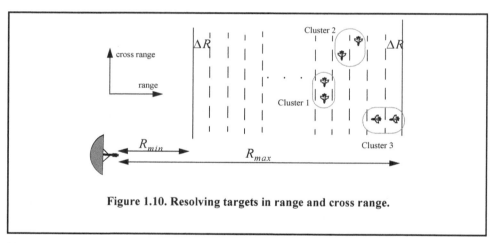

Figure 1.10. Resolving targets in range and cross range.

Now, try to answer the following question: What is the minimum time, δt, such that target 1 at R_1 and target 2 at R_2 will appear completely resolved in range (different range bins)? In other words, what is the minimum ΔR?

First, assume that the two targets are separated by $c\tau/4$, τ is the pulse width. In this case, when the pulse trailing edge strikes target 2, the leading edge would have traveled backward a distance $c\tau$, and the returned pulse would be composed of returns from both targets (i.e., unresolved return), as shown in Fig. 1.11a. However, if the two targets are at least $c\tau/2$ apart, then as the pulse trailing edge strikes the first target, the leading edge will start to return from target 2, and two distinct returned pulses will be produced, as illustrated by Fig. 1.11b. Thus, ΔR should be greater or equal to $c\tau/2$. And since the radar bandwidth B is equal to $1/\tau$, then

$$\Delta R = \frac{c\tau}{2} = \frac{c}{2B}.$$ Eq. (1.9)

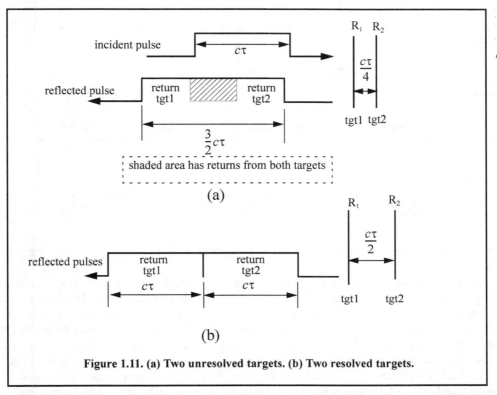

Figure 1.11. (a) Two unresolved targets. (b) Two resolved targets.

In general, radar users and designers alike seek to minimize ΔR in order to enhance the radar performance. As suggested by Eq. (1.9), in order to achieve fine range resolution one must minimize the pulse width. However, this will reduce the average transmitted power and increase the operating bandwidth. Achieving fine range resolution while maintaining adequate average transmitted power can be accomplished by using pulse compression techniques.

Example:

A radar system has an unambiguous range of 100Km, and a bandwidth 0.5MHz. Compute the required PRF, PRI, ΔR, and τ.

Solution:

$$PRF = \frac{c}{2R_u} = \frac{3 \times 10^8}{2 \times 10^5} = 1500\,Hz$$

$$PRI = \frac{1}{PRF} = \frac{1}{1500} = 0.6667\,ms$$

Using the function "range_resolution" yields

$$\Delta R = \frac{c}{2B} = \frac{3 \times 10^8}{2 \times 0.5 \times 10^6} = 300\,m$$

$$\tau = \frac{2\Delta R}{c} = \frac{2 \times 300}{3 \times 10^8} = 2\mu s \ .$$

MATLAB Function "range_resolution.m"

The MATLAB function *"range_resolution.m"* calculates range resolution; its syntax is as follows:

$$[delta\_R] = range\_resolution(var, indicator)$$

where

Symbol	Description	Units	Status
var	*bandwidth*	*Hz*	*input*
	OR	*OR*	
	pulsewidth	*seconds*	
delta_R	*range resolution*	*meters*	*output*

1.5. Doppler Frequency

Radars use Doppler frequency to extract target radial velocity (range rate), as well as to distinguish between moving and stationary targets or objects such as clutter. The Doppler phenomenon describes the shift in the center frequency of an incident waveform due to the target motion with respect to the source of radiation. Depending on the direction of the target's motion, this frequency shift may be positive or negative. A waveform incident on a target has equiphase wavefronts separated by λ, the wavelength. A closing target will cause the reflected equiphase wavefronts to compress and become closer to each other, resulting in a shorter wavelength of the reflected waveform. Alternatively, an opening or receding target (moving away from the radar) will cause the reflected equiphase wavefronts to expand, resulting in a longer wavelength of the reflected waveform. This is illustrated in Fig. 1.12.

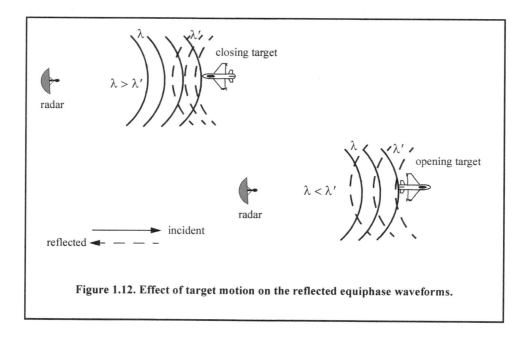

Figure 1.12. Effect of target motion on the reflected equiphase waveforms.

Consider a pulse of width τ (seconds) incident on a target that is moving toward the radar at velocity v, as shown in Fig. 1.13. Define d as the distance (in meters) that the target moves into the pulse during the interval Δt,

$$d = v\Delta t \qquad\qquad \textbf{Eq. (1.10)}$$

where Δt is equal to the time between the pulse leading edge striking the target and the trailing edge striking the target. Since the pulse is moving at the speed of light and the trailing edge has moved distance $c\tau - d$, then

$$c\tau = c\Delta t + v\Delta t \qquad\qquad \textbf{Eq. (1.11)}$$

$$c\tau' = c\Delta t - v\Delta t. \qquad\qquad \textbf{Eq. (1.12)}$$

Dividing Eq. (1.12) by Eq. (1.11) yields

$$\frac{c\tau'}{c\tau} = \frac{c\Delta t - v\Delta t}{c\Delta t + v\Delta t} \qquad\qquad \textbf{Eq. (1.13)}$$

which, after canceling the terms c and Δt from the left and right side of Eq. (1.13), respectively, one establishes the relationship between the incident and reflected pulses widths as

$$\tau' = \frac{c - v}{c + v}\tau. \qquad\qquad \textbf{Eq. (1.14)}$$

In practice, the factor $(c - v)/(c + v)$ is often referred to as the time dilation factor. Notice that if $v = 0$, then $\tau' = \tau$. In a similar fashion, one can compute τ' for an opening target. In this case,

$$\tau' = \frac{v + c}{c - v}\tau. \qquad\qquad \textbf{Eq. (1.15)}$$

To derive an expression for Doppler frequency, consider the illustration shown in Fig. 1.14. It takes Δt seconds for the leading edge of pulse 2 to travel a distance $(c/f_r) - d$ to strike the target. Over the same time interval, the leading edge of pulse 1 travels the same distance $c\Delta t$. More precisely,

$$d = v\Delta t \qquad\qquad \textbf{Eq. (1.16)}$$

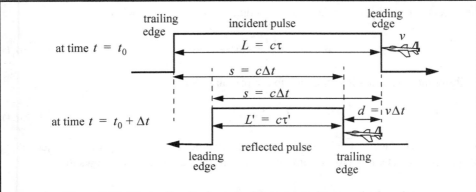

Figure 1.13. Illustrating the impact of target velocity on a single pulse.

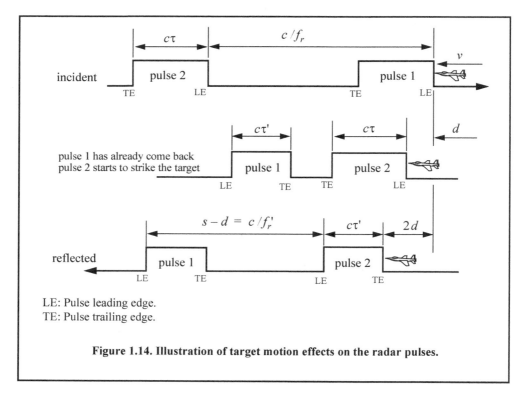

Figure 1.14. Illustration of target motion effects on the radar pulses.

LE: Pulse leading edge.
TE: Pulse trailing edge.

$$\frac{c}{f_r} - d = c\Delta t.$$

Eq. (1.17)

Solving for Δt yields

$$\Delta t = \frac{c/f_r}{c+v}$$

Eq. (1.18)

$$d = \frac{cv/f_r}{c+v}.$$

Eq. (1.19)

The reflected pulse spacing is now $s - d$ and the new PRF is f_r', where

$$s - d = \frac{c}{f_r'} = c\Delta t - \frac{cv/f_r}{c+v}$$

Eq. (1.20)

It follows that the new PRF is related to the original PRF by

$$f_r' = \frac{c+v}{c-v} \, f_r.$$

Eq. (1.21)

However, since the number of cycles does not change, the frequency of the reflected signal will go up by the same factor. Denoting the new frequency by f_0', it follows that

$$f_0' = \frac{c+v}{c-v} \, f_0$$

Eq. (1.22)

where f_0 is the carrier frequency of the incident signal. The Doppler frequency f_d is defined as the difference $f_0' - f_0$. More precisely,

$$f_d = f_0' - f_0 = \frac{c+v}{c-v} f_0 - f_0 = \frac{2v}{c-v} f_0,$$

Eq. (1.23)

but since $v \ll c$ and $c = \lambda f_0$, then

$$f_d \approx \frac{2v}{c} f_0 = \frac{2v}{\lambda}.$$

Eq. (1.24)

Eq. (1.24) indicates that the Doppler shift is proportional to the target velocity, and, thus, one can extract f_d from range rate and vice versa.

The result in Eq. (1.24) can also be derived using the following approach: Fig. 1.15 shows a closing target with velocity v. Let R_0 refer to the range at time t_0 (time reference); then the range to the target at any time t is

$$R(t) = R_0 - v(t - t_0).$$

Eq. (1.25)

The signal received by the radar is then given by

$$x_r(t) = x(t - \psi(t))$$

Eq. (1.26)

where $x(t)$ is the transmitted signal, and

$$\psi(t) = \frac{2}{c}(R_0 - vt + vt_0).$$

Eq. (1.27)

Substituting Eq. (1.27) into Eq. (1.26) and collecting terms yields

$$x_r(t) = x\left(\left(1 + \frac{2v}{c}\right)t - \psi_0\right)$$

Eq. (1.28)

where the constant phase ψ_0 is

$$\psi_0 = \frac{2R_0}{c} + \frac{2v}{c} t_0.$$

Eq. (1.29)

Define the compression or scaling factor γ by

$$\gamma = 1 + (2v/c)$$

Eq. (1.30)

Note that for a receding target the scaling factor becomes $\gamma = 1 - (2v/c)$. Utilizing Eq. (1.30), one can rewrite Eq. (1.28) as

$$x_r(t) = x(\gamma t - \psi_0).$$

Eq. (1.31)

Eq. (1.31) represents a time-compressed version of the return signal from a stationary target ($v = 0$). Hence, based on the scaling property of the Fourier transform, the spectrum of the received signal will be expanded in frequency to a factor of γ.

Consider the special case when

$$x(t) = y(t)\cos\omega_0 t$$

Eq. (1.32)

where ω_0 is the radar center frequency in radians per second. The received signal $x_r(t)$ is then given by

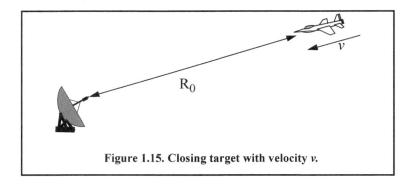

Figure 1.15. Closing target with velocity *v*.

$$x_r(t) = y(\gamma t - \psi_0)\cos(\gamma\omega_0 t - \psi_0).$$ **Eq. (1.33)**

The Fourier transform of Eq. (1.33) is

$$X_r(\omega) = \frac{1}{2\gamma}\left(Y\left(\frac{\omega}{\gamma} - \omega_0\right) + Y\left(\frac{\omega}{\gamma} + \omega_0\right)\right),$$ **Eq. (1.34)**

where for simplicity the effects of the constant phase ψ_0 have been ignored in Eq. (1.34). Therefore, the bandpass spectrum of the received signal is now centered at $\gamma\omega_0$ instead of ω_0. The difference between the two values corresponds to the amount of Doppler shift incurred due to the target motion,

$$\omega_d = \omega_0 - \gamma\omega_0 \Leftrightarrow f_d = f_0 - \gamma f_0.$$ **Eq. (1.35)**

ω_d and f_d are the Doppler frequency in radians per second and in Hz, respectively. Substituting the value of γ in Eq. (1.35) yields

$$f_d = \frac{2v}{c}\ f_0 = \frac{2v}{\lambda},$$ **Eq. (1.36)**

which is the same as Eq. (1.24). It can be shown that for a receding target, the Doppler shift is $f_d = -2v/\lambda$. This is illustrated in Fig. 1.16.

In both Eq. (1.36) and Eq. (1.24) the target radial velocity with respect to the radar is equal to v, but this is not always the case. In fact, the amount of Doppler frequency depends on the target velocity component in the direction of the radar (radial velocity). Fig. 1.17 shows three targets all having velocity v: target 1 has zero Doppler shift; target 2 has maximum Doppler frequency as defined in Eq. (1.36). The amount of Doppler frequency of target 3 is $f_d = 2v\cos\theta/\lambda$, where $v\cos\theta$ is the radial velocity, and θ is the total angle between the radar line of sight and the target. Thus, a more general expression for f_d that accounts for the total angle between the radar and the target is

$$f_d = \frac{2v}{\lambda}\cos\theta$$ **Eq. (1.37)**

and for an opening target

$$f_d = \frac{-2v}{\lambda}\cos\theta$$ **Eq. (1.38)**

where $\cos\theta = \cos\theta_e\ \cos\theta_a$. The angles θ_e and θ_a are, respectively, the elevation and azimuth angles; see Fig. 1.18.

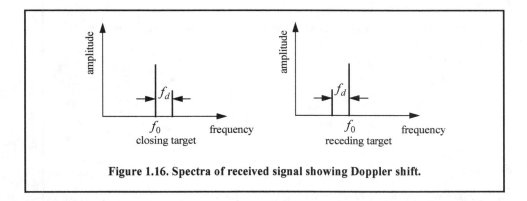

Figure 1.16. Spectra of received signal showing Doppler shift.

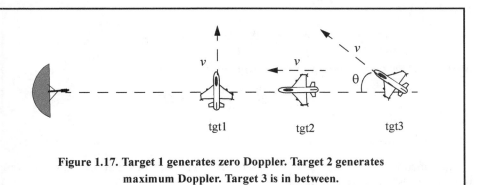

Figure 1.17. Target 1 generates zero Doppler. Target 2 generates maximum Doppler. Target 3 is in between.

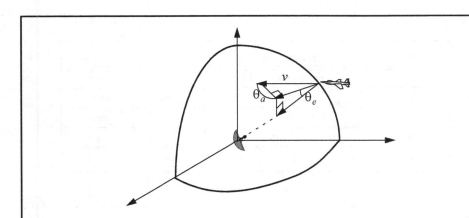

Figure 1.18. Radial velocity is proportional to the azimuth and elevation angles.

Example:

Compute the Doppler frequency measured by the radar shown in the figure below.

Solution:

The relative radial velocity between the radar and the target is $v_{radar} + v_{t\,\arg et}$. *Thus, using Eq.* *(1.36), we get*

$$f_d = 2\frac{(250 + 175)}{0.03} = 28.3\,KHz\,.$$

Similarly, if the target were opening the Doppler frequency is

$$f_d = 2\frac{250 - 175}{0.03} = 5\,KHz\,.$$

MATLAB Function "doppler_freq.m"

The function *"doppler_freq.m"* computes Doppler frequency and the associated time dilation factor; its syntax is as follows:

$$[fd, tdr] = doppler\_freq\ (freq,\ ang,\ tv,\ indicator)$$

where

Symbol	Description	Units	Status
freq	*radar operating frequency*	*Hz*	*input*
ang	*aspect angle*	*degrees*	*input*
tv	*target velocity*	*m/sec*	*input*
fd	*Doppler frequency*	*Hz*	*output*
tdr	*time dilation factor ratio* τ' / τ	*none*	*output*

1.6. Coherence

A radar is said to be coherent if the phase of any two transmitted pulses is consistent, i.e., there is a continuity in the signal phase from one pulse to the next, as illustrated in Fig. 1.19a. One can view coherence as the radar's ability to maintain an integer multiple of wavelengths between the equiphase wavefront from the end of one pulse to the equiphase wavefront at the beginning of the next pulse, as illustrated by Fig. 1.19b. Coherency can be achieved by using a STAble Local Oscillator (STALO). A radar is said to be coherent-on-receive or quasi-coherent if it stores in its memory a record of the phases of all transmitted pulses. In this case, the receiver phase reference is normally the phase of the most recent transmitted pulse.

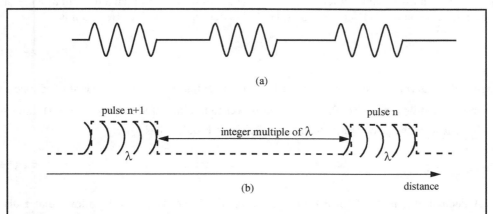

Figure 1.19. (a) Phase continuity between consecutive pulses. (b) Maintaining an integer multiple of wavelengths between the equiphase wavefronts of any two successive pulses guarantees coherency.

Coherence also refers to the radar's ability to accurately measure (extract) the received signal phase. Since Doppler represents a frequency shift in the received signal, then only coherent or coherent-on-receive radars can extract Doppler information. This is because the instantaneous frequency of a signal is proportional to the time derivative of the signal phase. More precisely,

$$f_i = \frac{1}{2\pi} \frac{d}{dt} \psi(t)$$ Eq. (1.39)

where f_i is the instantaneous frequency, and $\psi(t)$ is the signal phase.

For example, consider the following signal:

$$x(t) = \cos(\gamma \omega_0 t - \psi_0)$$ Eq. (1.40)

where the scaling factor γ is defined in Eq. (1.30), and ψ_0 is a constant phase. It follows that the instantaneous frequency of $x(t)$ is

$$f_i = \gamma f_0$$ Eq. (1.41)

where $\omega_0 = 2\pi f_0$. Substituting Eq. (1.30) into Eq. (1.41) yields

$$f_i = f_0\left(1 + \frac{2v}{c}\right) = f_0 + \frac{2v}{\lambda}$$ Eq. (1.42)

where the relation $c = \lambda f$ is utilized. Note that the second term of the most right-hand side of Eq. (1.42) is a Doppler shift.

1.7. Decibel Arithmetic

The decibel (dB) is a logarithmic unit of measurement that represents a ratio of a physical quantity (such as voltage, power, or antenna gain) to a specific reference quantity of the same type. The unit dB is named after Alexander Graham Bell, who originated the unit as a measure of power attenuation in telephone lines. By Bell's definition, a unit of Bell gain is

$$\log\left(\frac{P_0}{P_i}\right)$$

Eq. (1.43)

where the logarithm operation is base 10, P_0 is the output power of a standard telephone line (almost one mile long), and P_i is the input power to the line. If voltage (or current) ratios are used instead of the power ratio, then a unit Bell gain is defined as

$$\log\left(\frac{V_0}{V_i}\right)^2 \qquad or \qquad \log\left(\frac{I_0}{I_i}\right)^2.$$

Eq. (1.44)

A decibel, dB, is $1/10$ of a Bell (the prefix "deci" means 10^{-1}). It follows that a dB is defined as

$$10\log\left(\frac{P_0}{P_i}\right) = 10\log\left(\frac{V_0}{V_i}\right)^2 = 10\log\left(\frac{I_0}{I_i}\right)^2.$$

Eq. (1.45)

The inverse dB is computed from the relations

$$P_0/P_i = 10^{dB/10}$$
$$V_0/V_i = 10^{dB/20}$$
$$I_0/I_i = 10^{dB/20}$$

Eq. (1.46)

The decibel nomenclature is widely used by radar designers and users for several reasons, and perhaps, the most important one is that representing radar-related physical quantities using dBs drastically reduces the dynamic range that a designer or a user has to use. For example, an incoming radar signal may be as weak as $1 \times 10^{-9} V$, which can be expressed in dBs as $10\log(1 \times 10^{-9}) = -90dB$. Alternatively, a target may be located at range $R = 1000Km$, which can be expressed in dBs as $60dB$. Another advantage of using dB in radar design and analysis is to facilitate the arithmetic associated with calculating the different radar parameters. This is true since multiplication in base-10 arithmetic translates into addition in dB-arithmetic, and division translates into subtraction. For example,

$$\frac{250 \times 0.0001}{455} = [10\log(250) + 10\log(0.0001) - 10\log(455)]dB = -42.6dB.$$

Eq. (1.47)

In general,

$$10\log\left(\frac{A \times B}{C}\right) = 10\log A + 10\log B - 10\log C$$

Eq. (1.48)

$$10\log A^q = q \times 10\log A.$$

Eq. (1.49)

Other dB ratios that are often used in radar analysis include the dBsm (dB, squared meters). This definition is very important when referring to target Radar Cross Section (RCS), whose units are in squared meters. More precisely, a target whose RCS is $\sigma \ m^2$ can be expressed in dBsm as $10\log(\sigma \ m^2)$. For example, a $10m^2$ target is often referred to as a $10dBsm$ target, and a target with RCS $0.01m^2$ is equivalent to a $-20dBsm$.

Finally, the units dBm (dB, milliwatt) and dBW (dB, Watt) are power ratios of dBs with reference to one milliwatt and one Watt, respectively.

$$dBm = 10\log\left(\frac{P}{1\,mW}\right)$$ **Eq. (1.50)**

$$dBW = 10\log\left(\frac{P}{1\,W}\right)$$ **Eq. (1.51)**

To find dBm from dBW, add $30dB$, and to find dBW from dBm, subtract $30dB$. Other common dB units include dBz and dBi. dBz is used to measure weather radar reflectivity representing the amount of returned power received by the radar referenced to $mm^6\ m^{-3}$. The unit dBi (dB, isotropic) represents the forward gain of an antenna compared to an ideal isotropic antenna that emits energy equally in all directions.

Problems

1.1. (a) Calculate the maximum unambiguous range for a pulsed radar with PRF of $200Hz$ and $750Hz$. (b) What are the corresponding PRIs?

1.2. For the same radar in Problem 1.1, assume a duty cycle of 30% and peak power of $5KW$. Compute the average power and the amount of radiated energy during the first $20ms$.

1.3. A certain pulsed radar uses pulse width $\tau = 1\mu s$. Compute the corresponding range resolution.

1.4. An X-band radar uses PRF of $3KHz$. Compute the unambiguous range and the required bandwidth so that the range resolution is $30m$. What is the duty cycle?

1.5. Compute the Doppler shift associated with a closing target with velocity 100, 200, and 350 meters per second. In each case, compute the time dilation factor. Assume that $\lambda = 0.3m$.

1.6. Compute the round-trip delays, minimum PRIs, and corresponding PRFs for targets located $30Km$, $80Km$, and $150Km$ away from the radar.

1.7. Assume an S-band radar, what are the Doppler frequencies for the following target range rates: 50m/s; 200m/s; and 250m/s.

1.8. Repeat the previous problem for an X-Band radar ($9.5GHz$).

1.9. A certain L-band radar has center frequency $1.5GHz$, and PRF $f_r = 10KHz$. What is the maximum Doppler shift that can be measured by this radar?

1.10. Starting with a modified version of Eq. (1.25), derive an expression for the Doppler shift associated with a receding target.

1.11. In reference to Fig. 1.18, compute the Doppler frequency for $v = 150m/s$, $\theta_a = 30°$, and $\theta_e = 15°$. Assume that $\lambda = 0.1m$.

1.12. A pulsed radar system has a range resolution of $30cm$. Assuming sinusoid pulses at $45KHz$, determine the pulse width and the corresponding bandwidth.

1.13. (a) Develop an expression for the minimum PRF of a pulsed radar. (b) Compute $f_{r_{min}}$ for a closing target whose velocity is $400m/s$. (c) What is the unambiguous range? Assume that $\lambda = 0.2m$.

1.14. A certain radar is tasked with detecting and tracking the moon. Assume that the average distance to the moon is $3.844 \times 10^8 m$, and its average radar cross section is $6.64 \times 10^{11} m^2$. (a) Compute the delay to the moon. (b) What is required PRF so the range to the moon is unambiguous. (c) What is the moon's radar cross section in *dBsm*.

1.15. An L-band pulsed radar is designed to have an unambiguous range of $100 Km$ and range resolution $\Delta R \leq 100 m$. The maximum resolvable Doppler frequency corresponds to $v_{target} \leq 350 m / \text{sec}$. Compute the maximum required pulse width, the PRF, and the average transmitted power if $P_t = 500 W$.

1.16. A certain target has the following characteristics: its range away from the radar given in its corresponding x- y- and z- components is $\{25 Km, 32 Km, 12 Km\}$. The target velocity vector is $v_z = v_y = 0$, and $v_x = -250 m / s$. Compute the composite target range and range rate. If the radar's operating frequency is $9 GHz$, what is the corresponding Doppler frequency.

Appendix 1-A: Chapter 1 MATLAB Code Listings

The MATLAB code provided in this chapter was designed as an academic standalone tool and is not adequate for other purposes. The code was written in a way to assist the reader in gaining a better understanding of the theory. The code was not developed, nor is it intended to be used as part of an open-loop or a closed-loop simulation of any kind. The MATLAB code found in this textbook can be downloaded from this book's web page on the CRC Press website. Simply use your favorite web browser, go to *www.crcpress.com*, and search for keyword *"Mahafza"* to locate this book's web page.

MATLAB Function "pulse_train.m" Listing

```
function [dt, prf, pav, ep, ru] = pulse_train (tau, pri, p_peak)
% computes duty cycle, average transmitted power, pulse energy, and pulse repetition frequency
%% Inputs:
%   tau        == Pulse width in seconds
%   pri        == Pulse repetition interval in seconds
%   p_peak     == Peak power in Watts
%% Outputs:
%   dt         == Duty cycle - unitless
%   prf        == Pulse repetition frequency in Hz
%   pa         == Average power in Watts
%   ep         == Pulse energy in Joules
%   ru         == Unambiguous range in Km
%
c = 3e8; % speed of light
dt = tau / pri;
prf = 1. / pri;
pav = p_peak * dt;
ep = p_peak * tau;
ru = 1.e-3 * c * pri /2.0;
return
```

MATLAB Function "range_resolution.m" Listing

```
function [delta_R] = range_resolution (var)
% This function computes radar range resolution in meters
%%  Inputs:
%  var can be either
%   % var        == Bandwidth in Hz
%   % var        == Pulse width in seconds
% % Outputs:
%  % delta_R     == range resolution in meters
% Bandwidth may be equal to (1/pulse width)==> indicator = seconds
%
c = 3.e+8; % speed of light
indicator = input ('Enter 1 for var == Bandwidth, OR 2 for var == Pulse width \n');
switch (indicator)
   case 1
      delta_R = c / 2.0 / var; % del_r = c/2B
   case 2
      delta_R = c * var / 2.0; % del_r = c*tau/2
end
return
```

MATLAB Function "doppler_freq.m" Listing

```
function [fd, tdr] = doppler_freq (freq, ang, tv)
% This function computes Doppler frequency and time dilation factor ratio (tau_prime / tau)
% % Inputs:
    % freq         == radar operating frequency in Hz
    % ang          == target aspect angle in degrees
    % tv           == target velocity in m/sec
% % Outputs:
    % fd           == Doppler frequency in Hz
    % tdr          == time dilation factor; unitless
%
format long
indicator = input ('Enter 1 for closing target, OR 2 for opening target \n');
c = 3.0e+8;
ang_rad = ang * pi /180.;
lambda = c / freq;
switch (indicator)
    case 1
        fd = 2.0 * tv * cos(ang_rad) / lambda;
        tdr = (c - tv) / (c + tv);
    case 2
        fd = -2.0 * c * tv * cos(and_rad) / lambda;
        tdr = (c + tv) / (c -tv);
end
return
```

Chapter 2

Basic Pulsed and Continuous Wave (CW) Radar Operations

2.1. The Radar Range Equation

Consider a radar with an isotropic antenna (one that radiates energy equally in all directions). Since isotropic antennas have spherical radiation patterns, one can define the peak power density (power per unit area) at any point in space away from the radar as

$$P_D = \frac{Peak\ transmitted\ power}{area\ of\ a\ sphere} \qquad \frac{Watts}{m^2} .$$ **Eq. (2.1)**

The power density, in $Watts/m^2$, at range R away from the radar (assuming a lossless propagation medium) is

$$P_D = P_t/(4\pi R^2)$$ **Eq. (2.2)**

where P_t is the peak transmitted power and $4\pi R^2$ is the surface area of a sphere of radius R. Radar systems utilize directional antennas in order to increase the power density in a certain direction. Directional antennas are usually characterized by the antenna gain G and the antenna effective aperture A_e. They are related by

$$G = (4\pi A_e)/\lambda^2$$ **Eq. (2.3)**

where λ is the radar operating wavelength. The relationship between the antenna's effective aperture A_e and the physical aperture A is

$$A_e = \rho A$$ **Eq. (2.4)**
$$0 \le \rho \le 1$$

where ρ is referred to as the aperture efficiency, and good antennas require $\rho \rightarrow 1$. In this book, unless otherwise noted, A and A_e are used interchangeably to refer to the antenna's aperture, and will assume that antennas have the same gain in the transmitting and receiving modes. In practice, $\rho \approx 0.7$ is widely accepted.

The gain is also related to the antenna's azimuth and elevation antenna beamwidths by

$$G = K\frac{4\pi}{\theta_e \theta_a}$$ **Eq. (2.5)**

where $K \leq 1$ and depends on the physical aperture shape, and the angles θ_e and θ_a are, respectively, the antenna's elevation and azimuth beamwidths in radians. An excellent commonly used approximation of Eq. (2.5) is

$$G \approx \frac{26000}{\theta_e \theta_a}$$

Eq. (2.6)

where in this case the azimuth and elevation beamwidths are given in degrees.

The power density at a distance R away from a radar using a directive antenna of gain G is then given by

$$P_D = \frac{P_t G}{4\pi R^2}$$

Eq. (2.7)

When the radar radiated energy impinges upon a target, the induced surface currents on that target radiate electromagnetic energy in all directions. The amount of the radiated energy is proportional to the target size, orientation, physical shape, and material, which are all lumped together in one target-specific parameter called the Radar Cross Section (RCS) denoted symbolically by the Greek letter σ.

The radar cross section is defined as the ratio of the power reflected back to the radar to the power density incident on the target,

$$\sigma = \frac{P_r}{P_D} \ m^2$$

Eq. (2.8)

where P_r is the power reflected from the target. Thus, the total power delivered to the radar signal processor by its antenna is

$$P_{Dr} = \frac{P_t G \sigma}{(4\pi R^2)^2} \ A_e.$$

Eq. (2.9)

Substituting the value of A_e from Eq. (2.3) into Eq. (2.9) yields

$$P_{Dr} = \frac{P_t G^2 \lambda^2 \sigma}{(4\pi)^3 R^4}$$

Eq. (2.10)

Let S_{min} denote the minimum detectable signal power by the radar. It follows that the maximum radar range R_{max} is

$$R_{max} = \left(\frac{P_t G^2 \lambda^2 \sigma}{(4\pi)^3 S_{min}} \right)^{1/4}.$$

Eq. (2.11)

Eq. (2.11) suggests that in order to double the radar maximum range, one must increase the peak transmitted power P_t sixteen times; or equivalently, one must increase the effective aperture four times.

In practical situations the returned signals received by the radar will be corrupted with noise, which introduces unwanted voltages at all radar frequencies. Noise is random in nature and can be characterized by its Power Spectral Density (PSD) function. The noise power N is a function of the radar operating bandwidth, B. More precisely,

$$N = Noise\ PSD \times B.$$

Eq. (2.12)

The receiver input noise power is

$$N_i = kT_sB$$

Eq. (2.13)

where $k = 1.38 \times 10^{-23}\ Joule/degree\ Kelvin$ is Boltzmann's constant, and T_s is the total effective system noise temperature in degrees *Kelvin*. It is always desirable that the minimum detectable signal (S_{min}) be greater than the noise power. The fidelity of a radar receiver is normally described by a figure of merit referred to as the noise figure, F. The noise figure is defined as

$$F = \frac{(SNR)_i}{(SNR)_o} = \frac{S_i/N_i}{S_o/N_o}$$

Eq. (2.14)

where $(SNR)_i$ and $(SNR)_o$ are, respectively, the Signal to Noise Ratios (SNR) at the input and output of the receiver. The input signal power is S_i, and the input noise power immediately at the antenna terminal is N_i. The values S_o and N_o are, respectively, the output signal and noise powers.

The receiver effective noise temperature excluding the antenna is

$$T_e = T_o(F - 1)$$

Eq. (2.15)

where $T_0 = 290K$ and F is the receiver noise figure. It follows that the total effective system noise temperature T_s is given by

$$T_s = T_e + T_a = T_0(F-1) + T_a = T_oF - T_o + T_a$$

Eq. (2.16)

where T_a is the antenna temperature.

In many radar applications it is desirable to set the antenna temperature T_a to T_0 and thus, Eq. (2.16) is reduced to

$$T_s = T_oF.$$

Eq. (2.17)

Using Eq. (2.17) in Eq. (2.13) and substituting the result into Eq. (2.14) yields

$$S_i = kT_oBF(SNR)_o.$$

Eq. (2.18)

Thus, the minimum detectable signal power can be written as

$$S_{min} = kT_oBF(SNR)_{o_{min}}.$$

Eq. (2.19)

The radar detection threshold is set equal to the minimum output SNR, $(SNR)_{o_{min}}$. Substituting Eq. (2.19) in Eq. (2.11) gives

$$R_{max} = \left(\frac{P_tG^2\lambda^2\sigma}{(4\pi)^3 kT_oBF(SNR)_{o_{min}}}\right)^{1/4}$$

Eq. (2.20)

or equivalently,

$$(SNR)_{o_{min}} = \frac{P_tG^2\lambda^2\sigma}{(4\pi)^3 kT_oBFR_{max}^4}.$$

Eq. (2.21)

In general, radar losses denoted by L reduce the overall SNR, and hence

$$(SNR)_o = \frac{P_t G^2 \lambda^2 \sigma}{(4\pi)^3 k T_o B F L R^4}.$$ Eq. (2.22)

Although Eq. (2.22) is widely known and used as the Radar Range Equation, it is not quite correct unless the antenna temperature is equal to $290K$. In real-world cases, the antenna temperature may vary from a few degrees *Kelvin* to several thousand degrees. However, the actual error will be small if the radar receiver noise figure is large. In order to accurately account for the radar antenna temperature, one must use Eq. (2.17) in Eq. (2.22). Thus, the radar equation is now given by

$$(SNR)_o = \frac{P_t G^2 \lambda^2 \sigma}{(4\pi)^3 k T_s B L R^4}.$$ Eq. (2.23)

Example:

Assume a certain C-band radar with the following parameters: Peak power $P_t = 1.5MW$, operating frequency $f_0 = 5.6GHz$, antenna gain $G = 45dB$, effective temperature $T_o = 290K$, pulse width $\tau = 0.2\mu sec$. The radar threshold is $(SNR)_{min} = 20dB$. Assume target cross section $\sigma = 0.1m^2$. Compute the maximum range.

Solution:

The radar bandwidth is

$$B = \frac{1}{\tau} = \frac{1}{0.2 \times 10^{-6}} = 5MHz.$$

The wavelength is

$$\lambda = \frac{c}{f_0} = \frac{3 \times 10^8}{5.6 \times 10^9} = 0.054m.$$

From Eq. (2.20) one gets

$$(R^4)_{dB} = (P_t + G^2 + \lambda^2 + \sigma - (4\pi)^3 - k T_o B - F - (SNR)_{o_{min}})_{dB}$$

where, before summing, the dB calculations are carried out for each of the individual parameters on the right-hand side. One can now construct the following table with all parameters computed in dB:

P_t	λ^2	G^2	$k T_o B$	$(4\pi)^3$	F	$(SNR)_{o_{min}}$	σ
61.761	−25.421	90	−136.987	32.976	3	20	−10

It follows that

$$R^4 = 61.761 + 90 - 25.352 - 10 - 32.976 + 136.987 - 3 - 20 = 197.420dB$$

$$R^4 = 10^{(197.420 / 10)} = 55.208 \times 10^{18} m^4$$

$$R = \sqrt[4]{55.208 \times 10^{18}} = 86.199 Km.$$

Thus, the maximum detection range is 86.2 Km .

MATLAB Function "radar_eq.m"

The function *"radar_eq.m"* implements Eq. (2.22); its syntax is as follows:

[snr] = radar_eq (pt, freq, g, sigma, b, nf, loss, range)

where

Symbol	Description	Units	Status
pt	*peak power*	*Watts*	*input*
freq	*radar center frequency*	*Hz*	*input*
g	*antenna gain*	*dB*	*input*
sigma	*target cross section*	*m²*	*input*
b	*bandwidth*	*Hz*	*input*
nf	*noise figure*	*dB*	*input*
loss	*radar losses*	*dB*	*input*
range	*target range (can be single value or a vector)*	*Km*	*input*
snr	*SNR (single value or a vector, depending on the input range)*	*dB*	*output*

The function *"radar_eq.m"* is developed so that it can accept a single value for the input "range," or a vector containing many range values. Figure 2.1 shows typical plots generated using the function *"radar_eq.m,"* with the following inputs: Peak power $P_t = 1.5 MW$, operating frequency $f_0 = 5.6 GHz$, antenna gain $G = 45 dB$, radar losses $L = 6 dB$, noise figure $F = 3 dB$. The radar bandwidth is $B = 5 MHz$. The radar minimum and maximum detection range are $R_{min} = 25 Km$ and $R_{max} = 165 Km$. Figure 2.1 can be reproduced using MATLAB program *"Fig2_1.m"* listed in Appendix 2-A.

2.2. Low PRF Radar Equation

Consider a pulsed radar with pulse width τ, PRI T, and peak transmitted power P_t. The average transmitted power is $P_{av} = P_t d_t$, where $d_t = \tau / T$ is the transmission duty factor. One can define the receiving duty factor d_r as

$$d_r = \frac{T - \tau}{T} = 1 - \tau f_r. \qquad \text{Eq. (2.24)}$$

Thus, for low PRF radars ($T \gg \tau$) the receiving duty factor is $d_r \approx 1$.

Define the "time on target" T_i (the time that a target is illuminated by the beam) as

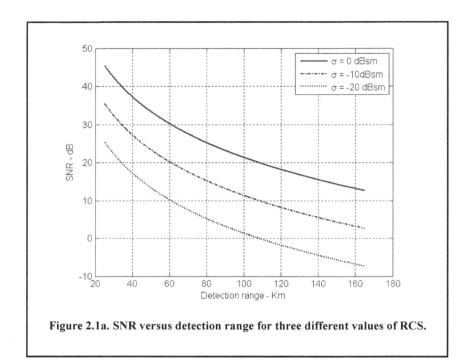

Figure 2.1a. SNR versus detection range for three different values of RCS.

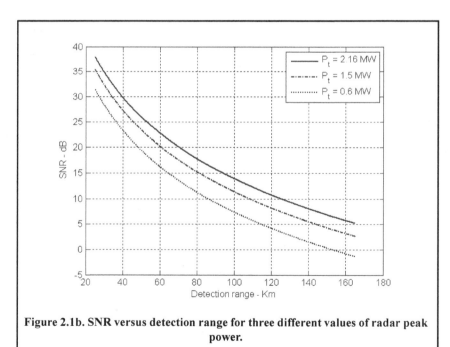

Figure 2.1b. SNR versus detection range for three different values of radar peak power.

$$T_i = n_p / f_r \Rightarrow n_p = T_i f_r \qquad \text{Eq. (2.25)}$$

where n_p is the total number of pulses that strike the target, and f_r is the radar PRF. Assuming low PRF, the single pulse radar equation is given by

$$(SNR)_1 = \frac{P_t G^2 \lambda^2 \sigma}{(4\pi)^3 R^4 k T_o BFL}, \qquad \text{Eq. (2.26)}$$

and for n_p coherently integrated pulses we get

$$(SNR)_{n_p} = \frac{P_t G^2 \lambda^2 \sigma \; n_p}{(4\pi)^3 R^4 k T_o BFL}. \qquad \text{Eq. (2.27)}$$

Now by using Eq. (2.25) and using $B = 1/\tau$, the low PRF radar equation can be written as

$$(SNR)_{n_p} = \frac{P_t G^2 \lambda^2 \sigma T_i f_r \tau}{(4\pi)^3 R^4 k T_o FL}. \qquad \text{Eq. (2.28)}$$

MATLAB Function "lprf_req.m"

The function *"lprf_req.m"* implements the low PRF radar equation given in Eq. (2.27). For a given set of input parameters, the function *"lprf_req.m"* computes $(SNR)_{np}$. Its syntax is as follows:

$$[snr] = lprf\_req(pt, g, freq, sigma, np, b, nf, loss, range)$$

where

Symbol	Description	Units	Status
pt	*peak power*	*W*	*input*
g	*antenna gain*	*dB*	*input*
freq	*frequency*	*Hz*	*input*
sigma	*target cross section*	*m²*	*input*
np	*number of pulses*	*none*	*input*
b	*bandwidth*	*Hz*	*input*
nf	*noise figure*	*dB*	*input*
loss	*radar losses*	*dB*	*input*
range	*target range (can be single value or a vector)*	*Km*	*input*
snr	*SNR (can be single value or a vector)*	*dB*	*output*

Figure 2.2 shows typical plots generated using the function *"lprf_req.m,"* with the following inputs: Peak power $P_t = 1.5MW$, operating frequency $f_0 = 5.6GHz$, antenna gain $G = 45dB$, radar losses $L = 6dB$, noise figure $F = 3dB$. The bandwidth is $B = 5MHz$. The target RCS is $\sigma = 0.1m^2$. Figure 2.2 can be reproduced using MATLAB program *"Fig2_2.m"* listed in Appendix 2-A.

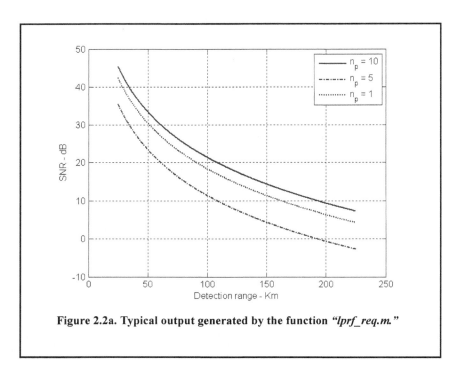

Figure 2.2a. Typical output generated by the function *"lprf_req.m."*

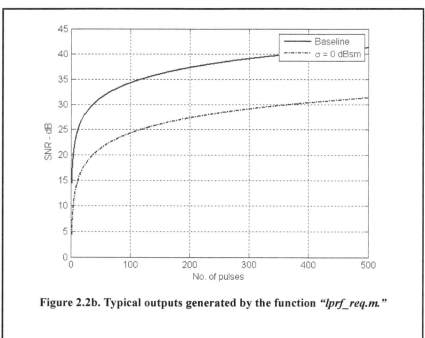

Figure 2.2b. Typical outputs generated by the function *"lprf_req.m."*

2.3. High PRF Radar Equation

In high PRF radars, the transmitted signal is assumed to be a periodic train of pulses, with pulse width of τ and period T. This pulse train can be represented using an exponential Fourier series, where the central power spectrum line (DC component) for this series contains most of the signal's power. Its value is $(\tau / T)^2$, and it is equal to the square of the transmit duty factor. Thus, the single pulse radar equation for a high PRF radar is

$$SNR = \frac{P_t G^2 \lambda^2 \sigma d_t^2}{(4\pi)^3 R^4 k T_o BFL d_r} \qquad \text{Eq. (2.29)}$$

where, in this case, one can no longer ignore the receive duty factor, since its value is comparable to the transmit duty factor. In fact, $d_r \approx d_t = \tau f_r$. Additionally, the operating radar bandwidth is now matched to the radar integration time (time-on-target), $B = 1 / T_i$. It follows that

$$SNR = \frac{P_t \tau f_r T_i G^2 \lambda^2 \sigma}{(4\pi)^3 R^4 k T_o FL} \qquad \text{Eq. (2.30)}$$

and finally,

$$SNR = \frac{P_{av} T_i G^2 \lambda^2 \sigma}{(4\pi)^3 R^4 k T_o FL} \qquad \text{Eq. (2.31)}$$

where P_{av} was substituted for $P_t \tau f_r$. Note that the product $P_{av} T_i$ is a "kind of energy" product, which indicates that high PRF radars can enhance detection performance by using relatively low power and longer integration time.

2.3.1 MATLAB Function "hprf_req.m"

The function *"hprf_req.m"* implements Eq. (2.30). Its syntax is as follows:

[snr] = hprf_req (pt, Ti, g, freq, sigma, dt, range, nf, loss)

where

Symbol	Description	Units	Status
pt	*peak power*	*W*	*input*
Ti	*time on target*	*seconds*	*input*
g	*antenna gain*	*dB*	*input*
freq	*frequency*	*Hz*	*input*
sigma	*target RCS*	m^2	*input*
dt	*duty cycle*	*none*	*input*
range	*target range (can be single value or a vector)*	*Km*	*input*
nf	*noise figure*	*dB*	*input*
loss	*radar losses*	*dB*	*input*
snr	*SNR (can be a single value or a vector)*	*dB*	*output*

Figure 2.3 shows typical outputs generated by the function *"hprf_req.m"*. This figure can be reproduced using MATLAB program *"Fig2_3.m"* listed in Appendix 2-A.

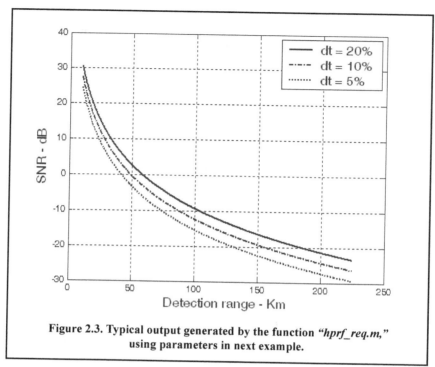

**Figure 2.3. Typical output generated by the function *"hprf_req.m,"*
using parameters in next example.**

Example:

Compute the single pulse SNR for a high PRF radar with the following parameters: peak power P_t = 100KW, antenna gain G = 20dB, operating frequency f_0 = 5.6GHz, losses L = 8dB, noise figure F = 5dB, dwell interval T_i = 2s, duty factor d_t = 0.3. The range of interest is R = 50Km. Assume target RCS σ = $0.01m^2$.

Solution:

From Eq. (2.31) we have

$$(SNR)_{dB} = (P_{av} + G^2 + \lambda^2 + \sigma + T_i - (4\pi)^3 - R^4 - kT_o - F - L)_{dB}$$

The following table gives all parameters in dB:

P_{av}	λ^2	T_i	kT_0	$(4\pi)^3$	R^4	σ
44.771	−25.421	3.01	−203.977	32.976	187.959	−20

$SNR = 44.771 + 40 - 25.421 - 20 + 3.01 - 32.976 + 203.977 - 187.959 - 5 - 8 = 12.4dB$

The same answer can be obtained by using the function "lprf_req.m" with the following syntax:

hprf_req (100e3, 2, 20, 5.6e9, 0.01, 0.3, 50e3, 5, 8)

2.4. Surveillance Radar Equation

The primary job for surveillance radars is to continuously scan a specified volume of space searching for targets of interest. Once detection is established, target information such as range, angular position, and possibly target velocity are extracted by the radar signal and data processors. Depending on the radar design and antenna, different search patterns can be adopted. A two-dimensional (2-D) fan beam search pattern is shown in Fig. 2.4a. In this case, the beamwidth is wide enough in elevation to cover the desired search volume along that coordinate; however, it has to be steered in azimuth. Figure 2.4b shows a stacked beam search pattern; here the beam has to be steered in azimuth and elevation. This latter kind of search pattern is normally employed by phased array radars.

Search volumes are normally specified by a search solid angle Ω in steradians, as illustrated in Fig. 2.5. Define the radar search volume extent for both azimuth and elevation as Θ_A and Θ_E. Consequently, the search volume is computed as

$$\Omega = (\Theta_A \Theta_E)/(57.296)^2 \quad steradians \qquad \text{Eq. (2.32)}$$

where both Θ_A and Θ_E are given in degrees. The radar antenna $3\,dB$ beamwidth can be expressed in terms of its azimuth and elevation beamwidths θ_a and θ_e, respectively. It follows that the antenna solid angle coverage is $\theta_a \theta_e$ and, thus, the number of antenna beam positions n_B required to cover a solid angle Ω is

$$n_B = \frac{\Omega}{(\theta_a \theta_e)/(57.296)^2}. \qquad \text{Eq. (2.33)}$$

In order to develop the search radar equation, start with Eq. (2.22), which is repeated here for convenience, as Eq. (2.34):

$$SNR = \frac{P_t G^2 \lambda^2 \sigma}{(4\pi)^3 kT_o BFLR^4}. \qquad \text{Eq. (2.34)}$$

Using the relations $\tau = 1/B$ and $P_t = P_{av}T/\tau$, where T is the PRI and τ is the pulse width, yields

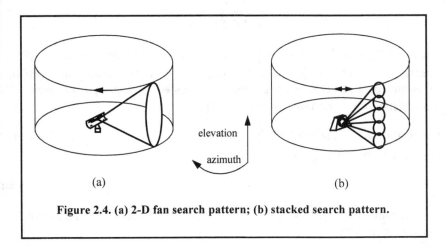

elevation

azimuth

(a)   (b)

Figure 2.4. (a) 2-D fan search pattern; (b) stacked search pattern.

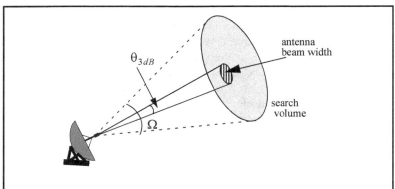

Figure 2.5. A cut in space showing the antenna beam width and the search volume.

$$SNR = \frac{T}{\tau} \frac{P_{av}G^2\lambda^2\sigma\tau}{(4\pi)^3 kT_o FLR^4}.$$

Eq. (2.35)

Define the time it takes the radar to scan a volume defined by the solid angle Ω as the scan time T_{sc}. The time on target can then be expressed in terms of T_{sc} as

$$T_i = \frac{T_{sc}}{n_B} = \frac{T_{sc}}{\Omega}\theta_a\theta_e.$$

Eq. (2.36)

Assume that during a single scan only one pulse per beam per PRI illuminates the target. It follows that $T_i = T$ and, thus, Eq. (2.35) can be written as

$$SNR = \frac{P_{av}G^2\lambda^2\sigma}{(4\pi)^3 kT_o FLR^4}\frac{T_{sc}}{\Omega}\theta_a\theta_e$$

Eq. (2.37)

Substituting Eqs. (2.3) and (2.5) into Eq. (2.37) and collecting terms yields the search radar equation (based on a single pulse per beam per PRI) as

$$SNR = \frac{P_{av}A_e\sigma}{4\pi kT_o FLR^4}\frac{T_{sc}}{\Omega}.$$

Eq. (2.38)

The quantity $P_{av}A$ in Eq. (2.38) is known as the power aperture product. In practice, the power aperture product is widely used to categorize the radar's ability to fulfill its search mission. Normally, a power aperture product is computed to meet a predetermined SNR and radar cross section for a given search volume defined by Ω.

As a special case, assume a radar using a circular aperture (antenna) with diameter D. The 3-dB antenna beamwidth θ_{3dB} is

$$\theta_{3dB} \approx \frac{\lambda}{D},$$

Eq. (2.39)

and when aperture tapering is used, $\theta_{3dB} \approx 1.25\lambda/D$. Substituting Eq. (2.39) into Eq. (2.33) and collecting terms yields

$$n_B = (D^2 / \lambda^2) \ \Omega .$$ **Eq. (2.40)**

In this case, the scan time T_{sc} is related to the time-on-target by

$$T_i = \frac{T_{sc}}{n_B} = \frac{T_{sc}\lambda^2}{D^2\Omega} .$$ **Eq. (2.41)**

Substitute Eq. (2.41) into Eq. (2.35) to get

$$SNR = \frac{P_{av}G^2\lambda^2\sigma}{(4\pi)^3 R^4 kT_oFL} \frac{T_{sc}\lambda^2}{D^2\Omega} ,$$ **Eq. (2.42)**

and by using Eq. (2.3) in Eq. (2.42) one can define the search radar equation for a circular aperture as

$$SNR = \frac{P_{av}A\sigma}{16R^4 kT_oLF} \frac{T_{sc}}{\Omega}$$ **Eq. (2.43)**

where the relation $A = \pi D^2 / 4$ (aperture area) was used.

MATLAB Function "power_aperture.m"

The function *"power_aperture.m"* implements the search radar equation given in Eq. (2.38); its syntax is as follows:

$PAP = power\_aperture \ (snr, \ tsc, \ sigma, \ range, \ nf, \ loss, \ az\_angle, \ el\_angle)$

where

Symbol	Description	Units	Status
snr	sensitivity snr	dB	input
tsc	scan time	seconds	input
sigma	target cross section	m^2	input
range	target range	Km	input
nf	noise figure	dB	input
loss	radar losses	dB	input
az_angle	search volume azimuth extent	degrees	input
el_angle	search volume elevation extent	degrees	input
PAP	power aperture product	dB	output

Plots of the power aperture product versus range and plots of the average power versus aperture area for three RCS choices are shown in Fig. 2.6, which can be reproduced using the MATLAB program *"Fig2_6.m"* listed in Appendix 2-A. In this case, the following radar parameters were used:

σ	T_{sc}	$\theta_e = \theta_a$	R	$nf \times loss$	snr
0.1 m^2	2.5 sec	2°	250 Km	13 dB	15 dB

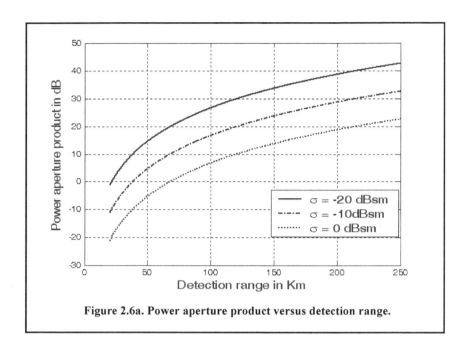

Figure 2.6a. Power aperture product versus detection range.

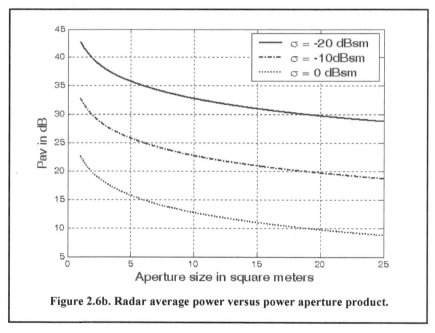

Figure 2.6b. Radar average power versus power aperture product.

Example:

Compute the power aperture product corresponding to the radar that has the following parameters: scan time $T_{sc} = 2s$, noise figure $F = 8dB$, losses $L = 6dB$, search volume $\Omega = 7.4$ steradians, range of interest is $R = 75Km$, and the required SNR is $20dB$. Assume that $\sigma = 3.162m^2$.

Solution:

Note that $\Omega = 7.4$ *steradians corresponds to a search sector that is three fourths of a hemisphere. Because of the three fourths of a hemisphere condition, one concludes that* $\theta_a = 180°$, *and using Eq. (2.32) yields* $\theta_e = 135°$. *Using the MATLAB function "power_aperture.m" with the following syntax:*

$$PAP = power\_aperture(20, 2, 3.162, 75e3, 8, 6, 180, 135)$$

one computes the power aperture product as 36.7 dB.

Example:

Compute the power aperture product for an X-band radar with the following parameters: signal-to-noise ratio $SNR = 15\,dB$; *losses* $L = 8\,dB$; *search volume* $\Omega = 2°$; *scan time* $T_{sc} = 2.5\,s$; *noise figure* $F = 5\,dB$. *Assume a* $-10\,dBsm$ *target cross section, and range* $R = 250\,Km$. *Also, compute the peak transmitted power corresponding to 30% duty factor, if the antenna gain is 45dB. Assume a circular aperture.*

Solution:

The angular coverage is $2°$ *in both azimuth and elevation. It follows that the solid angle coverage is*

$$\Omega = \frac{2 \times 2}{(57.23)^2} = -29.132\,dB.$$

The factor $360/2\pi = 57.23$ *converts angles into solid angles. From Eq. (2.43), one gets*

$$(SNR)_{dB} = (P_{av} + A + \sigma + T_{sc} - 16 - R^4 - kT_o - L - F - \Omega)_{dB}.$$

σ	T_{sc}	16	R^4	kT_o
$-10\,dB$	$3.979\,dB$	$12.041\,dB$	$215.918\,dB$	$-203.977\,dB$

It follows that

$$15 = P_{av} + A - 10 + 3.979 - 12.041 - 215.918 + 203.977 - 5 - 8 + 29.133.$$

Then the power aperture product is

$$P_{av} + A = 38.716\,dB.$$

Now, assume the radar wavelength to be $\lambda = 0.03\,m$, *then*

$$A = \frac{G\lambda^2}{4\pi} = 3.550\,dB$$

$$P_{av} = -A + 38.716 = 35.166\,dB$$

$$P_{av} = 10^{3.5166} = 3285.489\,W$$

$$P_t = \frac{P_{av}}{d_t} = \frac{3285.489}{0.3} = 10.9512\,KW.$$

2.5. Radar Equation with Jamming

Any deliberate electronic effort intended to disturb normal radar operations is usually referred to as an Electronic Countermeasure (ECM). This includes chaff, radar decoys, radar RCS alterations (e.g., radio frequency absorbing materials), and of course, radar jamming.

Jammers can be categorized into two general types: (1) barrage jammers and (2) deceptive jammers (repeaters). When strong jamming is present, detection capability is determined by receiver signal-to-noise plus interference ratio rather than SNR. In fact, in most cases, detection is established based on the signal-to-interference ratio alone.

Barrage jammers attempt to increase the noise level across the entire radar operating bandwidth, consequently lowering the receiver SNR, and, in turn, making it difficult to detect the desired targets. This is the reason why barrage jammers are often called maskers (since they mask the target returns). Barrage jammers can be deployed in the main beam or in the sidelobes of the radar antenna. If a barrage jammer is located in the radar main beam, it can take advantage of the antenna maximum gain to amplify the broadcasted noise signal. Alternatively, sidelobe barrage jammers must either use more power, or operate at a much shorter range than main-beam jammers. Main-beam barrage jammers can be deployed either onboard the attacking vehicle, or act as an escort to the target. Sidelobe jammers are often deployed to interfere with a specific radar, and since they do not stay close to the target, they have a wide variety of standoff deployment options.

Repeater jammers carry receiving devices onboard in order to analyze the radar's transmission, and then send back false target-like signals in order to confuse the radar. There are two common types of repeater jammers: spot noise repeaters and deceptive repeaters. The spot noise repeater measures the transmitted radar signal bandwidth and then jams only a specific range of frequencies. The deceptive repeater sends back altered signals that make the target appear in some false position (ghosts). These ghosts may appear at different ranges or angles than the actual target. Furthermore, there may be several ghosts created by a single jammer. By not having to jam the entire radar bandwidth, repeater jammers are able to make more efficient use of their jamming power. Radar frequency agility may be the only way possible to defeat spot noise repeaters.

In general, a jammer is characterized by its operating bandwidth B_J and Effective Radiated Power (ERP), which is proportional to the jammer transmitter power P_J. More precisely,

$$ERP = (P_J G_J) / L_J \qquad \textbf{Eq. (2.44)}$$

where G_J is the jammer antenna gain and L_J is the total jammer losses. The effect of a jammer on a radar is measured by the Signal-to-Jammer ratio (S/J).

2.5.1 Self-Screening Jammers (SSJ)

Self-screening jammers (SSJ), also known as self-protecting jammers and as main-beam jammers, are a class of ECM systems carried on the platform they are protecting. Escort jammers (carried on platforms that accompany the attacking vehicles) can also be treated as SSJs if they appear at the same range as that of the target(s).

Assume a radar with an antenna gain G, wavelength λ, aperture A_r, bandwidth B_r, receiver losses L, and peak power P_t. The single pulse power received by the radar from a target of RCS σ, at range R, is

$$S = \frac{P_t G^2 \lambda^2 \sigma \tau}{(4\pi)^3 R^4 L}$$

Eq. (2.45)

where τ is the radar pulse width. The power received by the radar from an SSJ jammer at the same range is

$$J = \frac{P_J G_J}{4\pi R^2} \frac{A_r}{B_J L_J}$$

Eq. (2.46)

where P_J, G_J, B_J, L_J are, respectively, the jammer's peak power, antenna gain, operating bandwidth, and losses. Using the relation

$$A_r = \lambda^2 G / 4\pi,$$

Eq. (2.47)

Eq. (2.46) can be written as

$$J = \frac{P_J G_J}{4\pi R^2} \frac{\lambda^2 G}{4\pi} \frac{1}{B_J L_J}.$$

Eq. (2.48)

Note that for jammers to be effective, they require $B_J > B_r$. This is needed in order to compensate for the fact that the jammer bandwidth is usually larger than the operating bandwidth of the radar. Jammers are normally designed to operate against a wide variety of radar systems with different bandwidths.

Substituting Eq. (2.44) into Eq. (2.48) yields

$$J = ERP \frac{\lambda^2 G}{(4\pi)^2 R^2} \frac{1}{B_J}.$$

Eq. (2.49)

Thus, the S/J ratio for an SSJ case is obtained from Eqs. (2.45) and (2.49) as,

$$\frac{S}{J} = \frac{P_t \tau G \sigma B_J}{(ERP)(4\pi) R^2 L},$$

Eq. (2.50)

and when pulse compression is used, with time-bandwidth-product G_{PC}, then Eq. (2.50) can be written as

$$\frac{S}{J} = \frac{P_t G \sigma B_J G_{PC}}{(ERP)(4\pi) R^2 B_r L}.$$

Eq. (2.51)

The jamming power reaches the radar on a one-way transmission basis, whereas the target echoes involve two-way transmission. Thus, the jamming power is generally greater than the target signal power. In other words, the ratio S/J is less than unity. However, as the target becomes closer to the radar, there will be a certain range such that the ratio S/J is equal to unity. This range is known as the cross-over range. The range window where the ratio S/J is sufficiently larger than unity is denoted as the detection range. In order to compute the cross-over range R_{co}, set S/J to unity in Eq. (2.51) and solve for range. It follows that

$$(R_{CO})_{SSJ} = \left(\frac{P_t G \sigma B_J}{4\pi B_r L (ERP)}\right)^{1/2}.$$

Eq. (2.52)

MATLAB Function "ssj_req.m"

The function *"ssj_req.m"* implements Eqs. (2.50) and (2.52). The syntax is as follows:

$$[BR\_range] = ssj\_req\ (pt,\ g,\ freq,\ sigma,\ br,\ loss,\ pj,\ bj,\ gj,\ lossj)$$

where

Symbol	Description	Units	Status
pt	radar peak power	W	input
g	radar antenna gain	dB	input
freq	radar operating frequency	Hz	input
sigma	target cross section	m^2	input
br	radar operating bandwidth	Hz	input
loss	radar losses	dB	input
pj	jammer peak power	W	input
bj	jammer bandwidth	Hz	input
gj	jammer antenna gain	dB	input
lossj	jammer losses	dB	input
BR_range	cross-over range	Km	output

This function generates data of relative S and J versus range normalized to the cross-over range, as illustrated in Fig. 2.7a. It also calculates the cross-over range as in Fig 2.7b. Figure 2.7b can be reproduced using MATLAB program *"Fig2_7b.m"* listed in Appendix 2-A. In this example, the following parameters were utilized: radar peak power $P_t = 50KW$, jammer peak power $P_J = 200W$, radar operating bandwidth $B_r = 667KHz$, jammer bandwidth $B_J = 50MHz$, radar and jammer losses $L = L_J = 0.10dB$, target cross section $\sigma = 10.m^2$, radar antenna gain $G = 35dB$, jammer antenna gain $G_J = 10dB$, and the radar operating frequency is $f = 5.6GHz$.

2.5.2. Burn-Through Range

If jamming is employed in the form of Gaussian noise, then the radar receiver has to deal with the jamming signal the same way it deals with noise power in the radar. Thus, detection, tracking, and other functions of the radar signal and data processors are no longer dependent on the SNR. In this case, the S/(J+N) ratio must be calculated. More precisely,

$$\frac{S}{J+N} = \frac{(P_t G \sigma A_r \tau)\ /((4\pi)^2 R^4 L)}{\left(\dfrac{(ERP)A_r}{4\pi R^2 B_J} + kT_0\right)}.$$

Eq. (2.53)

The S/(J+N) ratio should be used in place of the SNR when calculating the radar equation and when computing the probability of detection. Furthermore, S/(J+N) must also be used in place of the SNR when using coherent or noncoherent pulse integration. The range at which the radar can detect and perform proper measurements for a given S/(J+N) value is defined as the burn-through range. It is given by

$$R_{BT} = \left\{ \sqrt{\left(\frac{(ERP)A_r}{8\pi B_j kT_0}\right)^2 + \frac{P_t G \sigma A_r \tau}{(4\pi)^2 L \dfrac{S}{(J+N)} kT_0}} \; - \frac{(ERP)A_r}{8\pi B_j kT_0} \right\}^{\frac{1}{2}}. \qquad \textbf{Eq. (2.54)}$$

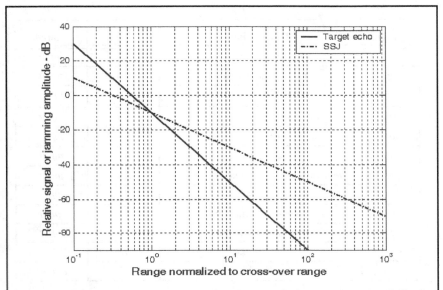

Figure 2.7a. Target and jammer echo signals using the input parameters defined on pp. 42.

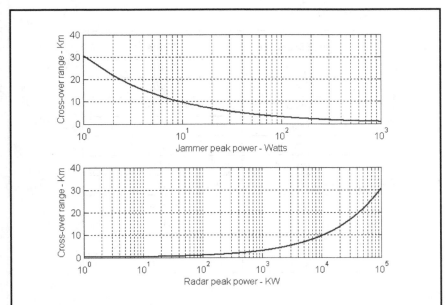

Figure 2.7b. Cross-over range versus jammer and radar peak powers corresponding to the example used in generating Fig. 2.7a.

MATLAB Function "sir.m"

The MATLAB function *"sir.m"* implements Eq. (2.53). The syntax is as follows:

[SIR] = sir (pt, g, sigma, freq, tau, loss, R, pj, bj, gj, lossj)

where

Symbol	Description	Units	Status
pt	radar peak power	W	input
g	radar antenna gain	dB	input
sigma	target cross section	m^2	input
freq	radar operating frequency	Hz	input
tau	radar pulse width	seconds	input
loss	radar losses	dB	input
R	range can be single value or a vector	Km	input
pj	jammer peak power	W	input
bj	jammer bandwidth	Hz	input
gj	jammer antenna gain	dB	input
lossj	jammer losses	dB	input
SIR	S/(J+N)	dB	output

The function *"sir.m"* generates data that can be used to plot the S/(J+N) versus detection range as shown in Fig. 2.8 using the input parameters defined in the table below. Figure 2.8 can be reproduced using the MATLAB program *"Fig2_8.m"* listed in Appendix 2-A.

Input Parameter	Value
pt	50KW
g	35dB
sigma	10 square meters
freq	5.6GHz
tau	50 micro-seconds
loss	5dB
R	linspace(10,400,5000) Km
pj	200Watts
bj	50MHz
gj	10dB
lossj	0.3dB

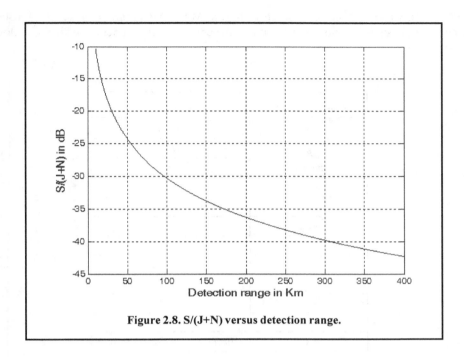

Figure 2.8. S/(J+N) versus detection range.

MATLAB Function "burn_thru.m"

The MATLAB function *"burn_thru.m"* implements Eqs. (2.54). It generates plots of the S/(J+N) versus detection range and plots of the burn-through range versus the jammer ERP. The syntax is as follows:

[Range] = burn_thru (pt, g, sigma, freq, tau, loss, pj, bj, gj, lossj, sir0, ERP)

where

Symbol	Description	Units	Status
pt	radar peak power	W	input
g	radar antenna gain	dB	input
sigma	target cross section	m^2	input
freq	radar operating frequency	Hz	input
tau	radar pulse width	seconds	input
loss	radar losses	dB	input
pj	jammer peak power	W	input
bj	jammer bandwidth	Hz	input
gj	jammer antenna gain	dB	input
lossj	jammer losses	dB	input
sir0	desired SIR	dB	input
ERP	desired ERP can be a vector	Watts	input
Range	burn-through range	Km	output

Figure 2.9, which can be reproduced using the MATLAB program *"Fig2_9.m"* listed in Appendix 2-A, shows some typical outputs generated by this function with the following inputs:

Input Parameter	Value
pt	*50KW*
g	*35dB*
sigma	*10 square meters*
freq	*5.6GHz*
tau	*0.5 Millie-seconds*
loss	*5dB*
pj	*200watts*
bj	*500MHz*
gj	*10dB*
lossj	*0.3dB*
sir0	*15dB*
ERP	*linspace(1, 1000, 1000) W*

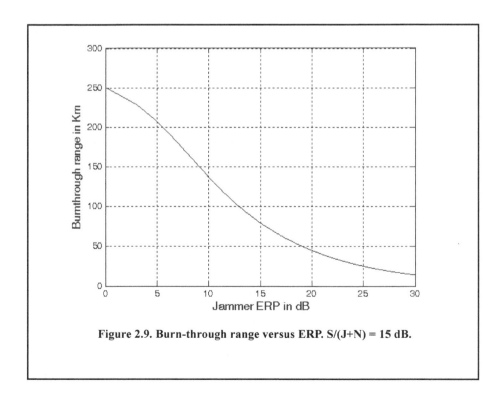

Figure 2.9. Burn-through range versus ERP. S/(J+N) = 15 dB.

2.5.3 Stand-Off Jammers (SOJ)

Stand-off jammers (SOJ) emit ECM signals from long ranges that are beyond the defense's lethal capability. The power received by the radar from an SOJ jammer at range R_J is

$$J = \frac{P_J G_J}{4\pi R_J^2} \frac{\lambda^2 G'}{4\pi} \frac{1}{B_J L_J} = \frac{ERP}{4\pi R_J^2} \frac{\lambda^2 G'}{4\pi} \frac{1}{B_J}$$

Eq. (2.55)

where all terms in Eq. (2.55) are the same as those for the SSJ case except for G'. The gain term G' represents the radar antenna gain in the direction of the jammer and is normally considered to be the sidelobe gain.

The SOJ radar equation is then computed as

$$\frac{S}{J} = \frac{P_t \tau G^2 R_J^2 \sigma B_J}{4\pi (ERP) G' R^4 L}$$

Eq. (2.56)

and when pulse compression is used, with time-bandwidth-product G_{PC}, then Eq. (2.56) can be written as

$$\frac{S}{J} = \frac{P_t G^2 R_J^2 \sigma B_J G_{PC}}{4\pi (ERP) G' R^4 B_r L}.$$

Eq. (2.57)

Again, the cross-over range is that corresponding to $S = J$; it is given by

$$(R_{CO})_{SOJ} = \left(\frac{P_t G^2 R_J^2 \sigma B_J G_{PC}}{4\pi (ERP) G' B_r L} \right)^{1/4}$$

Eq. (2.58)

MATLAB Function "soj_req.m"

The function *"soj_req.m"* implements Eqs. (2.57) and (2.58). The inputs to the program *"soj_req.m"* are the same as in the SSJ case, with two additional inputs: the radar antenna gain on the jammer G' and radar-to-jammer range R_J. Its syntax is as follows:

[BR_range] = soj_req (pt, g, sigma, b, freq, loss, range, pj, bj, gj, lossj, gprime, rangej)

Figure 2.10 shows plots generated using data generated by this function. In this case, the same input parameters as those in the SSJ case are used, with jammer peak power $P_J = 5000\,W$, jammer antenna gain $G_J = 30\,dB$, radar antenna gain on the jammer $G' = 10\,dB$, and radar-to-jammer range $R_J = 22.2\,Km$. Figure 2.10 can be reproduced using MATLAB program *"Fig2_10.m"* listed in Appendix 2-A. Again if the jamming is employed in the form of Gaussian noise, then the radar receiver has to deal with the jamming signal the same way it deals with noise power in the radar. In this case, the S/(J+N) is

$$\frac{S}{J+N} = \frac{\left(\dfrac{P_t G \sigma A_r \tau}{(4\pi)^2 R^4 L} \right)}{\left(\dfrac{(ERP) A_r G'}{4\pi R_J^2 B_J} + kT_0 \right)}.$$

Eq. (2.59)

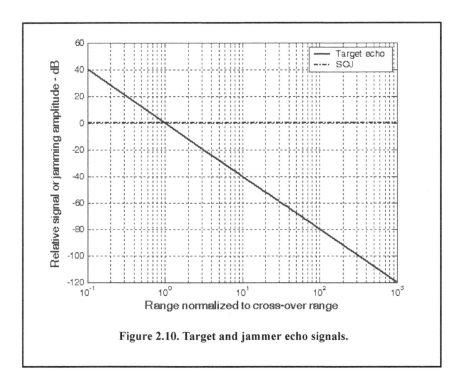

Figure 2.10. Target and jammer echo signals.

2.6. Range Reduction Factor

Consider a radar system whose detection range R in the absence of jamming is governed by

$$(SNR)_o = \frac{P_t G^2 \lambda^2 \sigma}{(4\pi)^3 k T_s B_r L R^4}.$$

Eq. (2.60)

The term Range Reduction Factor (RRF) refers to reduction in the radar detection range due to jamming. More precisely, in the presence of jamming, the effective radar detection range is

$$R_{dj} = R \times RRF.$$

Eq. (2.61)

In order to compute RRF, consider a radar characterized by Eq. (2.60), and a barrage jammer whose output power spectral density is J_o (i.e., Gaussian-like). Then the amount of jammer power in the radar receiver is

$$J = k T_J B_r$$

Eq. (2.62)

where T_J is the jammer effective temperature. It follows that the total jammer plus noise power in the radar receiver is given by

$$N_i + J = k T_s B_r + k T_J B_r.$$

Eq. (2.63)

In this case, the radar detection range is now limited by the receiver signal-to-noise plus interference ratio rather than SNR. More precisely,

$$\left(\frac{S}{J+N}\right) = \frac{P_t G^2 \lambda^2 \sigma}{(4\pi)^3 k(T_s + T_J)B_r L R^4}.$$

Eq. (2.64)

The amount of reduction in the signal-to-noise plus interference ratio because of the jammer effect is computed from the difference between Eqs. (2.60) and (2.64). It is expressed (in dB) by

$$\Upsilon = 10.0 \times \log\left(1 + \frac{T_J}{T_s}\right).$$

Eq. (2.65)

Consequently, the RRF is

$$RRF = 10^{\frac{-\Upsilon}{40}}.$$

Eq. (2.66)

2.7. Bistatic Radar Equation

Radar systems that use the same antenna for both transmitting and receiving are called monostatic radars. Bistatic radars use transmit and receive antennas that are placed at different locations. Under this definition CW radars, although they use separate transmit and receive antennas, are not considered bistatic radars unless the distance between the two antennas is considerable. Figure 2.11 shows the geometry associated with bistatic radars. The angle, β, is called the bistatic angle. A synchronization link between the transmitter and receiver is necessary in order to maximize the receiver's knowledge of the transmitted signal so that it can extract maximum target information.

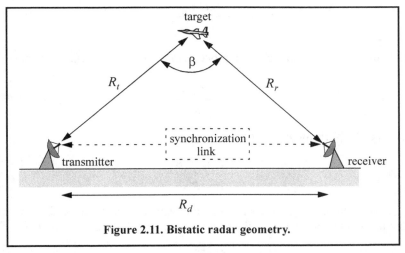

Figure 2.11. Bistatic radar geometry.

The synchronization link may provide the receiver with the following information: (1) the transmitted frequency in order to compute the Doppler shift, and (2) the transmit time or phase reference in order to measure the total scattered path ($R_t + R_r$). Frequency and phase reference synchronization can be maintained through line-of-sight communications between the transmitter and receiver. However, if this is not possible, the receiver may use a stable reference oscillator for synchronization.

One major distinction between monostatic and bistatic radar operations has to do with the measured bistatic target RCS, denoted by σ_B. In the case of a small bistatic angle, the bistatic RCS is similar to the monostatic RCS; but, as the bistatic angle approaches $180°$, the bistatic RCS becomes very large and can be approximated by

$$\sigma_{B_{max}} \approx (4\pi A_t^2)/\lambda^2 \qquad \text{Eq. (2.67)}$$

where λ is the wavelength and A_t is the target projected area.

The bistatic radar equation can be derived in a similar fashion to the monostatic radar equation. Referring to Fig. 2.11, the power density at the target is

$$P_D = (P_t G_t)/(4\pi R_t^2) \qquad \text{Eq. (2.68)}$$

where P_t is the peak transmitted power, G_t is the gain of the transmitting antenna, and R_t is the range from the radar transmitter to the target.

The effective power scattered off a target with bistatic RCS σ_B is

$$P' = P_D \sigma_B \qquad \text{Eq. (2.69)}$$

and the power density at the receiver antenna is

$$P_{refl} = \frac{P'}{4\pi R_r^2} = \frac{P_D \sigma_B}{4\pi R_r^2}. \qquad \text{Eq. (2.70)}$$

R_r is the range from the target to the receiver. Substituting Eq. (2.68) into Eq. (2.70) yields

$$P_{refl} = \frac{P_t G_t \sigma_B}{(4\pi)^2 R_t^2 R_r^2}. \qquad \text{Eq. (2.71)}$$

The total power delivered to the signal processor by a receiver antenna with aperture A_e is

$$P_{Dr} = \frac{P_t G_t \sigma_B A_e}{(4\pi)^2 R_t^2 R_r^2}. \qquad \text{Eq. (2.72)}$$

Substituting $(G_r \lambda^2/4\pi)$ for A_e yields

$$P_{Dr} = \frac{P_t G_t G_r \lambda^2 \sigma_B}{(4\pi)^3 R_t^2 R_r^2} \qquad \text{Eq. (2.73)}$$

where G_r is the gain of the receive antenna. Finally, when transmitter and receiver losses, L_t and L_r, are taken into consideration, the bistatic radar equation can be written as

$$P_{Dr} = \frac{P_t G_t G_r \lambda^2 \sigma_B}{(4\pi)^3 R_t^2 R_r^2 L_t L_r}. \qquad \text{Eq. (2.74)}$$

2.8. Radar Losses

As indicated by the radar equation, the receiver SNR is inversely proportional to the radar losses. Hence, any increase in radar losses causes a drop in the SNR, thus decreasing the probability of detection, since it is a function of the SNR. Often, the principal difference between a good radar design and a poor radar design is the radar losses. Radar losses include ohmic (resistance) losses and statistical losses. In this section, a brief summary of radar losses is presented.

2.8.1 Transmit and Receive Losses

Transmit and receive losses occur between the radar transmitter and antenna input port, and between the antenna output port and the receiver front end, respectively. Such losses are often called plumbing losses. Typically, plumbing losses are on the order of 1 to 2 dB.

2.8.2 Antenna Pattern Loss and Scan Loss

So far, when using the radar equation, maximum antenna gain was assumed. This is true only if the target is located along the antenna's boresight axis. However, as the radar scans across a target, the antenna gain in the direction of the target is less than maximum, as defined by the antenna's radiation pattern. The loss in the SNR due to not having maximum antenna gain on the target at all times is called the antenna pattern (shape) loss. Once an antenna has been selected for a given radar, the amount of antenna pattern loss can be mathematically computed.

For example, consider a $\sin x / x$ antenna radiation pattern as shown in Fig. 2.12. It follows that the average antenna gain over an angular region of $\pm \theta / 2$ about the boresight axis is

$$G_{av} \approx 1 - \left(\frac{\pi r}{\lambda}\right)^2 \frac{\theta^2}{36} \qquad \textbf{Eq. (2.75)}$$

where r is the aperture radius and λ is the wavelength. In practice, Gaussian antenna patterns are often adopted. In this case, if θ_{3dB} denotes the antenna 3dB beam width, then the antenna gain can be approximated by

$$G(\theta) = \exp\left(-\frac{2.776\theta^2}{\theta_{3dB}^2}\right). \qquad \textbf{Eq. (2.76)}$$

If the antenna scanning rate is so fast that the gain on receive is not the same as on transmit, additional scan loss has to be calculated and added to the beam shape loss. Scan loss can be computed in a similar fashion to beam shape loss. Phased array radars are often prime candidates for both beam shape and scan losses.

2.8.3 Atmospheric Loss

Detailed discussion of atmospheric loss and propagation effects will appear in a later chapter. Atmospheric attenuation is a function of the radar operating frequency, target range, and elevation angle. Atmospheric attenuation can be as high as a few dB.

2.8.4 Collapsing Loss

When the number of integrated returned noise pulses is larger than the target returned pulses, a drop in the SNR occurs. This is called collapsing loss. The collapsing loss factor is defined as

$$\rho_c = \frac{n + m}{n} \qquad \qquad \textbf{Eq. (2.77)}$$

where n is the number of pulses containing both signal and noise, while m is the number of pulses containing noise only. Radars detect targets in azimuth, range, and Doppler. When target returns are displayed in one coordinate, such as range, noise sources from azimuth cells adjacent to the actual target return converge in the target vicinity and cause a drop in the SNR. This is illustrated in Fig. 2.13.

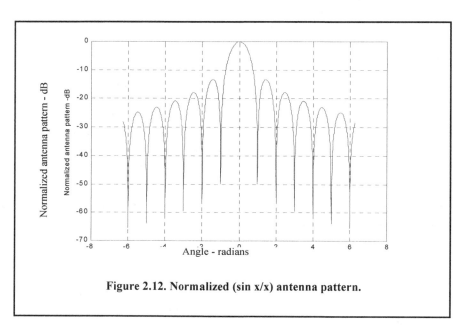

Figure 2.12. Normalized (sin x/x) antenna pattern.

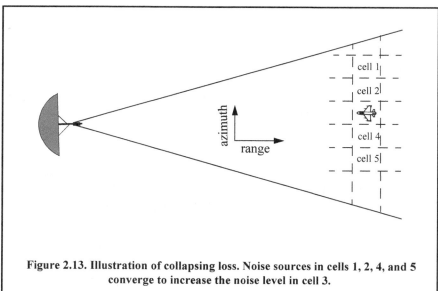

Figure 2.13. Illustration of collapsing loss. Noise sources in cells 1, 2, 4, and 5 converge to increase the noise level in cell 3.

2.8.5 Processing Loss

a. Detector Approximation

The output voltage signal of a radar receiver that utilizes a linear detector is

$$v(t) = \sqrt{v_I^2(t) + v_Q^2(t)}$$ Eq. (2.78)

where (v_I, v_Q) are the in-phase and quadrature components. For a radar using a square law detector, we have $v^2(t) = v_I^2(t) + v_Q^2(t)$.

Since in real hardware the operations of squares and square roots are time consuming, many algorithms have been developed for detector approximation. This approximation results in a loss of signal power, typically 0.5 to $1dB$.

b. Constant False Alarm Rate (CFAR) Loss

In many cases the radar detection threshold is constantly adjusted as a function of the receiver noise level in order to maintain a constant false alarm rate. For this purpose, Constant False Alarm Rate (CFAR) processors are utilized in order to keep the number of false alarms under control in a changing and unknown background of interference. CFAR processing can cause a loss in the SNR level on the order of $1dB$.

Three different types of CFAR processors are primarily used. They are adaptive threshold CFAR, nonparametric CFAR, and nonlinear receiver techniques. Adaptive CFAR assumes that the interference distribution is known and approximates the unknown parameters associated with these distributions. Nonparametric CFAR processors tend to accommodate unknown interference distributions. Nonlinear receiver techniques attempt to normalize the root-mean-square amplitude of the interference.

c. Quantization Loss

Finite word length (number of bits) and quantization noise cause an increase in the noise power density at the output of the Analog-to-Digital (A/D) converter. The A/D noise level is $q^2/12$, where q is the quantization level.

d. Range Gate Straddle

The radar receiver is normally mechanized as a series of contiguous range gates (bins). Each range bin is implemented as an integrator matched to the transmitted pulse width. Since the radar receiver acts as a filter that smears (smooths), the received target echoes. The smoothed target return envelope is normally straddled to cover more than one range gate.

Typically, three gates are affected; they are called the early, on, and late gates. If a point target is located exactly at the center of a range gate, then the early and late samples are equal. However, as the target starts to move into the next gate, the late sample becomes larger while the early sample gets smaller. In any case, the amplitudes of all three samples should always roughly add up to the same value. Fig. 2.14 illustrates the concept of range straddling. The envelope of the smoothed target echo is likely to be Gaussian shape. In practice, triangular shaped envelopes may be easier and faster to implement.

Since the target is likely to fall anywhere between two adjacent range bins, a loss in the SNR occurs (per range gate). More specifically, a target's returned energy is split between three range bins. Typically, straddle loss of about 2 to $3dB$ is not unusual.

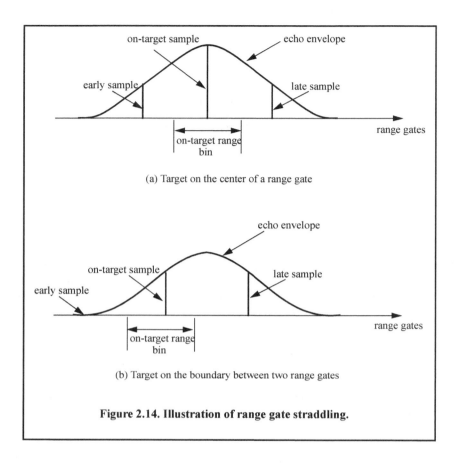

(a) Target on the center of a range gate

(b) Target on the boundary between two range gates

Figure 2.14. Illustration of range gate straddling.

Example:

Consider the smoothed target echo voltage shown below. Assume 1Ω resistance. Find the power loss due to range gate straddling over the interval $\{0, \tau\}$.

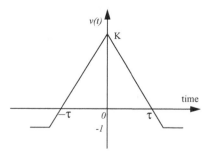

Solution:

The smoothed voltage can be written as

$$v(t) = \begin{cases} K + \left(\dfrac{K+1}{\tau}\right)t & ;t < 0 \\[3mm] K - \left(\dfrac{K+1}{\tau}\right)t & ;t \geq 0 \end{cases}.$$

The power loss due to straddle over the interval $\{0, \tau\}$ *is*

$$L_s = \frac{v^2}{K^2} = 1 - 2\left(\frac{K+1}{K\tau}\right)t + \left(\frac{K+1}{K\tau}\right)^2 t^2.$$

The average power loss is then

$$\bar{L}_s = \frac{2}{\tau} \int\limits_{0}^{\tau/2} \left(1 - 2\left(\frac{K+1}{K\tau}\right)t + \left(\frac{K+1}{K\tau}\right)^2 t^2\right) \, dt = 1 - \frac{K+1}{2K} + \frac{(K+1)^2}{12K^2}$$

and, for example, if $K = 15$, *then* $\bar{L}_s = 2.5\,dB$.

e. Doppler Filter Straddle

Doppler filter straddle is similar to range gate straddle. However, in this case the Doppler filter spectrum is spread (widened) due to weighting functions. Weighting functions are normally used to reduce the sidelobe levels. Since the target Doppler frequency can fall anywhere between two Doppler filters, signal loss occurs. This is illustrated in Fig. 2.15, where due to weighting, the cross-over frequency f_{co} is smaller than the filter cutoff frequency f_c, which normally corresponds to the $3\,dB$ power point.

f. Other Losses

Other losses may include equipment losses due to aging radar hardware, matched filter loss, and antenna efficiency loss. Tracking radars suffer from cross-over (squint) loss.

2.9. Noise Figure

Any signal other than the target returns in the radar receiver is considered to be noise. This includes interfering signals from outside the radar and thermal noise generated within the receiver itself. Thermal noise (thermal agitation of electrons) and shot noise (variation in carrier density of a semiconductor) are the two main internal noise sources within a radar receiver.

The power spectral density of thermal noise is given by

$$S_n(\omega) = \frac{|\omega|h}{\pi\left[\exp\left(\dfrac{|\omega|h}{2\pi kT}\right) - 1\right]} \qquad\qquad \textbf{Eq. (2.79)}$$

where $|\omega|$ is the absolute value of the frequency in radians per second, T is the temperature of the conducting medium in degrees *Kelvin*, k is Boltzman's constant, and h is Planck's constant ($h = 6.625 \times 10^{-34}$ *Joule s*). When the condition $|\omega| \ll 2\pi kT/h$ is true, it can be shown that Eq. (2.79) is approximated by

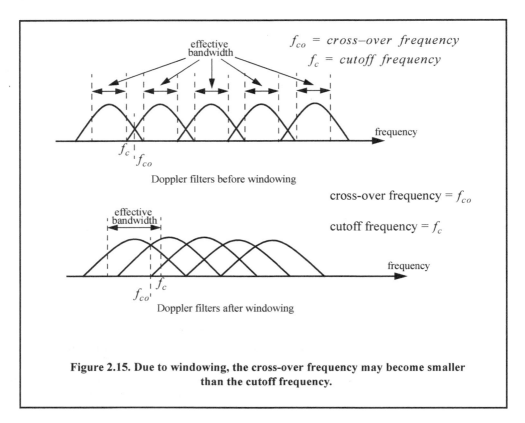

Figure 2.15. Due to windowing, the cross-over frequency may become smaller than the cutoff frequency.

$$S_n(\omega) \approx 2kT \qquad\qquad \text{Eq. (2.80)}$$

This approximation is widely accepted, since, in practice, radar systems operate at frequencies less than $100GHz$; and, for example, if $T = 290K$, then $2\pi kT/h \approx 6000GHz$.

The mean-square noise voltage (noise power) generated across a $1\,ohm$ resistance is then

$$\langle n^2 \rangle = \frac{1}{2\pi} \int\limits_{-2\pi B}^{2\pi B} 2kT \quad d\omega = 4kTB \qquad\qquad \text{Eq. (2.81)}$$

where B is the system bandwidth in hertz.

Any electrical system containing thermal noise and having input resistance R_{in} can be replaced by an equivalent noiseless system with a series combination of a noise equivalent voltage source and a noiseless input resistor R_{in} added at its input. This is illustrated in Fig. 2.16. The amount of noise power that can physically be extracted from $\langle n^2 \rangle$ is one fourth the value computed in Eq. (2.81). Consider a noisy system with power gain A_P, as shown in Fig. 2.17. The noise figure is defined by

$$F_{dB} = 10 \ \log \frac{total \ \ noise \ \ power \ \ out}{noise \ \ power \ \ out \ \ due \ \ to \ \ R_{in} \ \ alone}. \qquad\qquad \text{Eq. (2.82)}$$

More precisely,

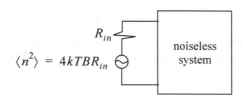

Figure 2.16. Noiseless system with an input noise voltage source.

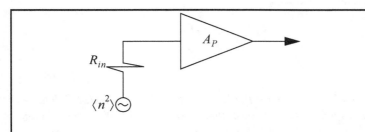

**Figure 2.17. Noisy amplifier replaced by its noiseless equivalent
and an input voltage source in series with a resistor.**

$$F_{dB} = 10 \ \log \frac{N_o}{N_i \ A_p} \qquad \text{Eq. (2.83)}$$

where N_o and N_i are, respectively, the noise power at the output and input of the system.

If we define the input and output signal power by S_i and S_o, respectively, then the power gain is

$$A_P = \frac{S_o}{S_i}. \qquad \text{Eq. (2.84)}$$

It follows that

$$F_{dB} = 10\log\left(\frac{S_i/Ni}{S_o/N_o}\right) = \left(\frac{S_i}{N_i}\right)_{dB} - \left(\frac{S_o}{N_o}\right)_{dB} \qquad \text{Eq. (2.85)}$$

where

$$\left(\frac{S_i}{N_i}\right)_{dB} > \left(\frac{S_o}{N_o}\right)_{dB}. \qquad \text{Eq. (2.86)}$$

Thus, the noise figure is the loss in the signal-to-noise ratio due to the added thermal noise of the amplifier $((SNR)_o = (SNR)_i - F \ \ in \ \ dB)$.

One can also express the noise figure in terms of the system's effective temperature T_e. Consider the amplifier shown in Fig. 2.17, and let its effective temperature be T_e. Assume the input noise temperature is T_o. Thus, the input noise power is

$$N_i = kT_oB \qquad \text{Eq. (2.87)}$$

and the output noise power is

$$N_o = kT_oB \; A_p + kT_eB \; A_p \qquad \text{Eq. (2.88)}$$

where the first term on the right-hand side of Eq. (2.88) corresponds to the input noise, and the latter term is due to thermal noise generated inside the system. It follows that the noise figure can be expressed as

$$F = \frac{(SNR)_i}{(SNR)_o} = \frac{S_i}{kT_oB} \; kBA_p \; \frac{T_o + T_e}{S_o} = 1 + \frac{T_e}{T_o} . \qquad \text{Eq. (2.89)}$$

Equivalently, we can write

$$T_e = (F-1)T_o . \qquad \text{Eq. (2.90)}$$

***Example*:**

An amplifier has a 4dB noise figure; the bandwidth is $B = 500KHz$. Calculate the input signal power that yields a unity SNR at the output. Assume $T_o = 290K$ and an input resistance of 1ohm.

Solution:

The input noise power is

$$kT_oB = 1.38 \times 10^{-23} \times 290 \times 500 \times 10^3 = 2.0 \times 10^{-15} W .$$

Assuming a voltage signal, then the input noise mean squared voltage is

$$\langle n_i^2 \rangle = kT_oB = 2.0 \times 10^{-15} \; v^2$$

$$F = 10^{0.4} = 2.51 .$$

From the noise figure definition we get

$$\frac{S_i}{N_i} = F\left(\frac{S_o}{N_o}\right) = F$$

$$\langle s_i^2 \rangle = F\langle n_i^2 \rangle = 2.51 \times 2.0 \times 10^{-15} = 5.02 \times 10^{-15} \; v^2 .$$

Finally,

$$\sqrt{\langle s_i^2 \rangle} = 70.852 nv .$$

Consider a cascaded system as in Fig. 2.18. Network 1 is defined by noise figure F_1, power gain G_1, bandwidth B, and temperature T_{e1}. Similarly, network 2 is defined by F_2, G_2, B, and T_{e2}. Assume the input noise has temperature T_0. The output signal power is

$$S_o = S_iG_1G_2 . \qquad \text{Eq. (2.91)}$$

The input and output noise powers are, respectively, given by

$$N_i = kT_oB \qquad \text{Eq. (2.92)}$$

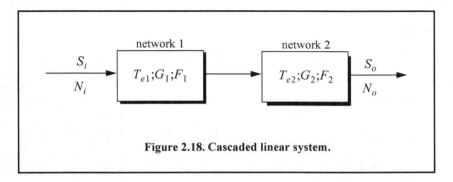

Figure 2.18. Cascaded linear system.

$$N_o = kT_0BG_1G_2 + kT_{e1}BG_1G_2 + kT_{e2}BG_2 \qquad \text{Eq. (2.93)}$$

where the three terms on the right-hand side of Eq. (2.93), respectively, correspond to the input noise power, thermal noise generated inside network 1, and thermal noise generated inside network 2.

Now, use the relation $T_e = (F-1)T_0$ along with Eq. (2.91) and Eq. (2.92) to express the overall output noise power as

$$N_o = F_1 N_i G_1 G_2 + (F_2 - 1)N_i G_2 . \qquad \text{Eq. (2.94)}$$

It follows that the overall noise figure for the cascaded system is

$$F = \frac{(S_i/N_i)}{(S_o/N_o)} = F_1 + \frac{F_2 - 1}{G_1} . \qquad \text{Eq. (2.95)}$$

In general, for an n-stage system we get

$$F = F_1 + \frac{F_2 - 1}{G_1} + \frac{F_3 - 1}{G_1 G_2} + \dots + \frac{F_n - 1}{G_1 G_2 G_3 \cdot \cdot \cdot G_{n-1}} . \qquad \text{Eq. (2.96)}$$

Also, the n-stage system effective temperatures can be computed as

$$T_e = T_{e1} + \frac{T_{e2}}{G_1} + \frac{T_{e3}}{G_1 G_2} + \dots + \frac{T_{en}}{G_1 G_2 G_3 \cdot \cdot \cdot G_{n-1}} . \qquad \text{Eq. (2.97)}$$

As suggested by Eq. (2.96) and Eq. (2.97), the overall noise figure is mainly dominated by the first stage. Thus, radar receivers employ low-noise power amplifiers in the first stage in order to minimize the overall receiver noise figure. However, for radar systems that are built for low RCS operations, every stage should be included in the analysis.

Example:

A radar receiver consists of an antenna with cable loss $L = 1dB = F_1$, an RF amplifier with $F_2 = 6dB$, and gain $G_2 = 20dB$, followed by a mixer whose noise figure is $F_3 = 10dB$ and conversion loss $L = 8dB$, and finally, an integrated circuit IF amplifier with $F_4 = 6dB$ and gain $G_4 = 60dB$. Find the overall noise figure.

Solution:

From Eq. (2.96)

$$F = F_1 + \frac{F_2 - 1}{G_1} + \frac{F_3 - 1}{G_1 G_2} + \frac{F_4 - 1}{G_1 G_2 G_3}.$$

G_1	G_2	G_3	G_4	F_1	F_2	F_3	F_4
$-1\,dB$	$20\,dB$	$-8\,dB$	$60\,dB$	$1\,dB$	$6\,dB$	$10\,dB$	$6\,dB$
0.7943	100	0.1585	10^6	1.2589	3.9811	10	3.9811

It follows that

$$F = 1.2589 + \frac{3.9811 - 1}{0.7943} + \frac{10 - 1}{100 \times 0.7943} + \frac{3.9811 - 1}{0.158 \times 100 \times 0.7943} = 5.3629$$

$$F = 10\log(5.3628) = 7.294\,dB.$$

2.10. Continuous Wave (CW) Radars

As mentioned earlier, in order to avoid interruption of the continuous radar energy emission, two antennas are used in CW radars, one for transmission and one for reception. Figure 2.19 shows a simplified CW radar block diagram. The appropriate values of the signal frequency at different locations are noted on the diagram. The individual Narrow Band Filters (NBF) must be as narrow as possible in bandwidth in order to allow accurate Doppler measurements and minimize the amount of noise power. In theory, the operating bandwidth of a CW radar is infinitesimal (since it corresponds to an infinite duration continuous sinewave). However, systems with infinitesimal bandwidths cannot physically exist, and thus, the bandwidth of CW radars is assumed to correspond to that of a gated CW waveform.

The NBF bank (Doppler filter bank) can be implemented using a Fast Fourier Transform (FFT). If the Doppler filter bank is implemented using an FFT of size N_{FFT}, and if the individual NBF bandwidth (FFT bin) is Δf, then the effective radar Doppler bandwidth is $N_{FFT}\Delta f/2$. The reason for the one-half factor is to account for both negative and positive Doppler shifts. The frequency resolution Δf is proportional to the inverse of the integration time.

Since range is computed from the radar echoes by measuring a two-way time delay, single frequency CW radars cannot measure target range. In order for CW radars to be able to measure target range, the transmit and receive waveforms must have some sort of timing marks. By comparing the timing marks at transmit and receive, CW radars can extract target range. The timing mark can be implemented by modulating the transmit waveform, and one commonly used technique is Linear Frequency Modulation (LFM). Before we discuss LFM signals, we will first introduce the CW radar equation and briefly address the general Frequency Modulated (FM) waveforms using sinusoidal modulating signals.

2.10.1 CW Radar Equation

As indicated by Fig. 2.19, the CW radar receiver declares detection at the output of a particular Doppler bin if that output value passes the detection threshold within the detector box. Since the NBF bank is implemented by an FFT, only finite length data sets can be processed at a time. The length of such blocks is normally referred to as the dwell interval, integration time, or coherent processing interval. The dwell interval determines the frequency resolution or the bandwidth of the individual NBFs. More precisely,

$$\Delta f = 1 / T_{Dwell}.$$ Eq. (2.98)

T_{Dwell} is the dwell interval. Therefore, once the maximum resolvable frequency by the NBF bank is chosen the size of the NBF bank is computed as

$$N_{FFT} = 2B / \Delta f.$$ Eq. (2.99)

B is the maximum resolvable frequency by the FFT. The factor 2 is needed to account for both positive and negative Doppler shifts. It follows that

$$T_{Dwell} = N_{FFT} / 2B.$$ Eq. (2.100)

The CW radar equation can now be derived. Consider the radar equation developed earlier in this chapter. That is

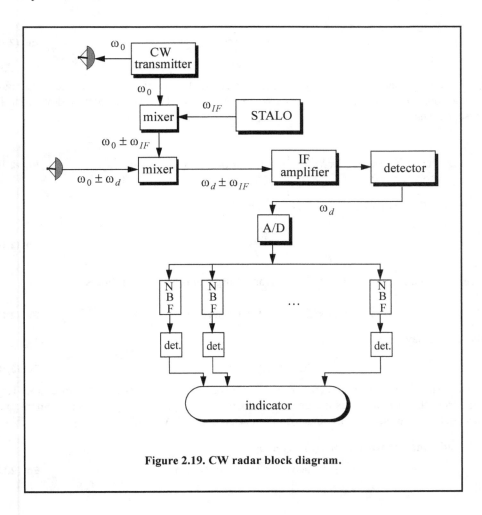

Figure 2.19. CW radar block diagram.

$$SNR = \frac{P_{av}TG^2\lambda^2\sigma}{(4\pi)^3 R^4 kT_o FL}$$

Eq. (2.101)

where $P_{av} = (\tau/T)P_t$, τ/T, and P_t is the peak transmitted power. In CW radars, the average transmitted power over the dwell interval P_{CW}, and T must be replaced by T_{Dwell}. Thus, the CW radar equation can be written as

$$SNR = \frac{P_{CW}T_{Dwell}G_t G_r \lambda^2\sigma}{(4\pi)^3 R^4 kT_o FLL_{win}}$$

Eq. (2.102)

where G_t and G_r are the transmit and receive antenna gains, respectively. The factor L_{win} is a loss term associated with the type of window (weighting) used in computing the FFT.

2.10.2 Frequency Modulation

The discussion presented in this section will be restricted to sinusoidal modulating signals. In this case, the general formula for an FM waveform can be expressed by

$$x(t) = A\cos\left(2\pi f_0 t + k_f \int_0^t \cos 2\pi f_m u\, du\right).$$

Eq. (2.103)

f_0 is the radar operating frequency (carrier frequency), $\cos 2\pi f_m t$ is the modulating signal, A is a constant, and $k_f = 2\pi\Delta f_{peak}$, where Δf_{peak} is the peak frequency deviation. The phase is given by

$$\psi(t) = 2\pi f_0 t + 2\pi\Delta f_{peak}\int_0^t \cos 2\pi f_m u\, du = 2\pi f_0 t + \beta\sin 2\pi f_m t$$

Eq. (2.104)

where β is the FM modulation index given by

$$\beta = \frac{\Delta f_{peak}}{f_m}.$$

Eq. (2.105)

Let $x_r(t)$ be the received radar signal from a target at range R. It follows that

$$x_r(t) = A_r\cos(2\pi f_0(t - \Delta t) + \beta\sin 2\pi f_m(t - \Delta t))$$

Eq. (2.106)

where the delay Δt is

$$\Delta t = (2R)/c.$$

Eq. (2.107)

c is the speed of light. CW radar receivers utilize phase detectors in order to extract target range from the instantaneous frequency, as illustrated in Fig. 2.20. A good measurement of the phase detector output $y(t)$ implies a good measurement of Δt, and hence range.

Consider the FM waveform $x(t)$ given by

$$x(t) = A\cos(2\pi f_0 t + \beta\sin 2\pi f_m t)$$

Eq. (2.108)

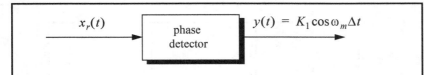

Figure 2.20. Extracting range from an FM signal return. K_1 is a constant.

which can be written as

$$x(t) = A Re\{e^{j2\pi f_0 t} e^{j\beta \sin 2\pi f_m t}\}$$

Eq. (2.109)

where $Re\{\ \}$ denotes the real part. Since the signal $\exp(j\beta \sin 2\pi f_m t)$ is periodic with period $T = 1/f_m$, it can be expressed using the complex exponential Fourier series as

$$e^{j\beta \sin 2\pi f_m t} = \left(\sum_{n = -\infty}^{\infty} C_n e^{jn2\pi f_m t} \right)$$

Eq. (2.110)

where the Fourier series coefficients C_n are given by

$$C_n = \frac{1}{2\pi} \int_{-\pi}^{\pi} e^{j\beta \sin 2\pi f_m t} e^{-jn2\pi f_m t} dt .$$

Eq. (2.111)

Make the change of variable $u = 2\pi f_m t$, and recognize that the Bessel function of the first kind of order n is

$$J_n(\beta) = \frac{1}{2\pi} \int_{-\pi}^{\pi} e^{j(\beta \sin u - nu)} du .$$

Eq. (2.112)

Thus, the Fourier series coefficients are $C_n = J_n(\beta)$, and consequently Eq. (2.110) can now be written as

$$e^{j\beta \sin 2\pi f_m t} = \left(\sum_{n = -\infty}^{\infty} J_n(\beta) e^{jn2\pi f_m t} \right) .$$

Eq. (2.113)

which is known as the Bessel-Jacobi equation. Figure 2.21 shows a plot of Bessel functions of the first kind for $n = 0, 1, 2, 3$.

The total power in the signal $s(t)$ is

$$P = \frac{1}{2}A^2 \sum_{n = -\infty}^{\infty} |J_n(\beta)|^2 = \frac{1}{2}A^2 .$$

Eq. (2.114)

Substituting Eq. (2.113) into Eq. (2.109) yields

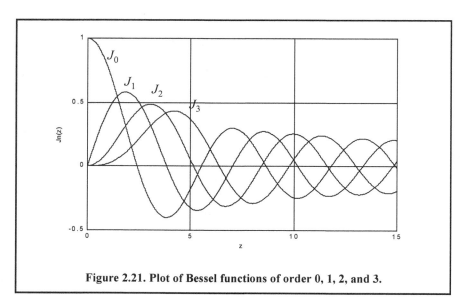

Figure 2.21. Plot of Bessel functions of order 0, 1, 2, and 3.

$$x(t) = A Re\left\{ e^{j2\pi f_0 t} \sum_{n=-\infty}^{\infty} J_n(\beta)e^{jn2\pi f_m t} \right\}.$$ **Eq. (2.115)**

Expanding Eq. (2.115) yields

$$x(t) = A \sum_{n=-\infty}^{\infty} J_n(\beta)\cos(2\pi f_0 + n2\pi f_m)t.$$ **Eq. (2.116)**

Finally, since $J_n(\beta) = J_{-n}(\beta)$ for n odd and $J_n(\beta) = -J_{-n}(\beta)$ for n even one can rewrite Eq. (2.116) as

$$\begin{aligned} x(t) = A\{ &J_0(\beta)\cos 2\pi f_0 t + J_1(\beta)[\cos(2\pi f_0 + 2\pi f_m)t - \cos(2\pi f_0 - 2\pi f_m)t] \\ &+ J_2(\beta)[\cos(2\pi f_0 + 4\pi f_m)t + \cos(2\pi f_0 - 4\pi f_m)t] \\ &+ J_3(\beta)[\cos(2\pi f_0 + 6\pi f_m)t - \cos(2\pi f_0 - 6\pi f_m)t] \\ &+ J_4(\beta)[\cos((2\pi f_0 + 8\pi f_m)t + \cos(2\pi f_0 - 8\pi f_m)t)] + \ldots \} \end{aligned}$$ **Eq. (2.117)**

which can be rewritten as

$$x(t) = A\left\{ J_0(\beta)\cos 2\pi f_0 t + \sum_{n=even}^{\infty} J_n(\beta)[\cos(2\pi f_0 + 2n\pi f_m)t + \right.$$ **Eq. (10.117b)**

$$\left. \cos(2\pi f_0 - 2n\pi f_m)t] + \sum_{q=odd}^{\infty} J_q(\beta)[\cos(2\pi f_0 + 2q\pi f_m)t - \cos(2\pi f_0 - 2q\pi f_m)t] \right\}$$

The spectrum of $x(t)$ is composed of pairs of spectral lines centered at f_0, as sketched in Fig. 2.22. The spacing between adjacent spectral lines is f_m. The central spectral line has an amplitude equal to $AJ_o(\beta)$, while the amplitude of the nth spectral line is $AJ_n(\beta)$.

As indicated by Eq. (2.117) the bandwidth of FM signals is infinite. However, the magnitudes of spectral lines of the higher orders are small, and thus the bandwidth can be approximated using Carson's rule,

$$B \approx 2(\beta + 1)f_m \qquad \text{Eq. (2.118)}$$

When β is small, only $J_0(\beta)$ and $J_1(\beta)$ have significant values. Thus, we may approximate Eq. (2.117) by

$$x(t) \approx A\{J_0(\beta)\cos 2\pi f_0 t + J_1(\beta)[\cos(2\pi f_0 + 2\pi f_m)t - \cos(2\pi f_0 - 2\pi f_m)t]\} . \qquad \text{Eq. (2.119)}$$

Finally, for small β, the Bessel functions can be approximated by

$$J_0(\beta) \approx 1 \qquad \text{Eq. (2.120)}$$

$$J_1(\beta) \approx \beta/2 . \qquad \text{Eq. (2.121)}$$

Thus, Eq. (2.119) may be approximated by

$$x(t) \approx A\left\{\cos 2\pi f_0 t + \frac{1}{2}\beta[\cos(2\pi f_0 + 2\pi f_m)t - \cos(2\pi f_0 - 2\pi f_m)t]\right\} . \qquad \text{Eq. (2.122)}$$

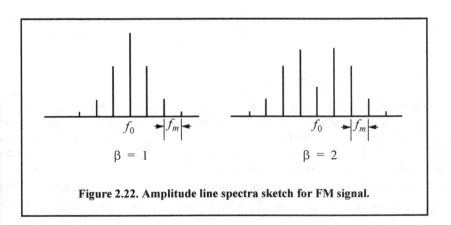

Figure 2.22. Amplitude line spectra sketch for FM signal.

Example:

If the modulation index is $\beta = 0.5$, *give an expression for the signal* $s(t)$.

Solution:

From Bessel function tables we get $J_0(0.5) = 0.9385$ *and* $J_1(0.5) = 0.2423$; *then using Eq. (2.119) we get*

$$x(t) \approx A\{(0.9385)\cos 2\pi f_0 t + (0.2423)[\cos(2\pi f_0 + 2\pi f_m)t - \cos(2\pi f_0 - 2\pi f_m)t]\} .$$

Example:

Consider an FM transmitter with output signal $s(t) = 100\cos(2000\pi t + \varphi(t))$. *The frequency deviation is* $4Hz$, *and the modulating waveform is* $x(t) = 10\cos 16\pi t$. *Determine the FM signal bandwidth. How many spectral lines will pass through a bandpass filter whose bandwidth is* $58Hz$ *centered at* $1000Hz$?

Solution:

The peak frequency deviation is $\Delta f_{peak} = 4 \times 10 = 40Hz$. *It follows that*

$$\beta = \frac{\Delta f_{peak}}{f_m} = \frac{40}{8} = 5.$$

Using Eq. (2.118) we get

$$B \approx 2(\beta + 1)f_m = 2 \times (5 + 1) \times 8 = 96Hz$$

However, only seven spectral lines pass through the bandpass filter as illustrated in the figure shown below.

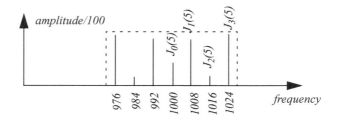

2.10.3 Linear Frequency Modulated CW Radar

Continuous Wave radars may use LFM waveforms so that both range and Doppler information can be measured. In practical CW radars, the LFM waveform cannot be continually changed in one direction, and thus, periodicity in the modulation is normally utilized. Figure 2.23 shows a sketch of a triangular LFM waveform. The modulation does not need to be triangular; it may be sinusoidal, sawtoothed, or some other form. The dashed line in Fig. 2.23 represents the return waveform from a stationary target at range R. The beat frequency f_b is also sketched in Fig. 2.23. It is defined as the difference (due to heterodyning) between the transmitted and received signals. The time delay Δt is a measure of target range; that is,

$$\Delta t = (2R)/c.$$ **Eq. (2.123)**

In practice, the modulating frequency f_m is selected such that

$$f_m = \frac{1}{2t_0}.$$ **Eq. (2.124)**

The rate of frequency change, $\dot{f}$, is

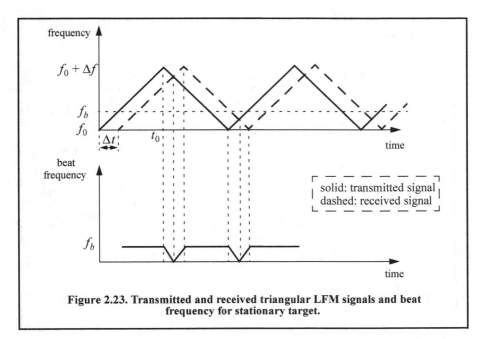

Figure 2.23. Transmitted and received triangular LFM signals and beat frequency for stationary target.

$$\dot{f} = \frac{\Delta f}{t_0} = \frac{\Delta f}{(1 / 2f_m)} = 2f_m \Delta f \qquad \text{Eq. (2.125)}$$

where Δf is the peak frequency deviation. The beat frequency f_b is given by

$$f_b = \Delta t \dot{f} = \frac{2R}{c} \dot{f}. \qquad \text{Eq. (2.126)}$$

Equation (2.126) can be rearranged as

$$\dot{f} = \frac{c}{2R} f_b. \qquad \text{Eq. (2.127)}$$

Equating Eqs. (2.125) and (2.127) and solving for f_b yields

$$f_b = (4Rf_m \Delta f) / c. \qquad \text{Eq. (2.128)}$$

Now consider the case when Doppler is present (i.e., non-stationary target). The corresponding triangular LFM transmitted and received waveforms are sketched in Fig. 2.24, along with the corresponding beat frequency. As previously noted the beat frequency is defined as

$$f_b = f_{received} - f_{transmitted}. \qquad \text{Eq. (2.129)}$$

When the target is not stationary, the received signal will contain a Doppler shift term in addition to the frequency shift due to the time delay Δt. In this case, the Doppler shift term subtracts from the beat frequency during the positive portion of the slope. Alternatively, the two terms add up during the negative portion of the slope. Denote the beat frequency during the positive (up) and negative (down) portions of the slope, respectively, as f_{bu} and f_{bd}. It follows that

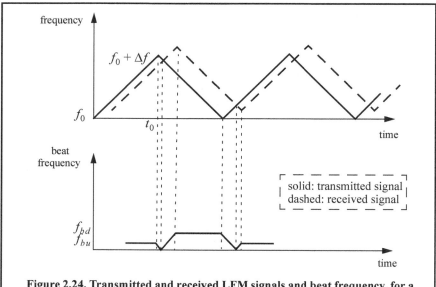

Figure 2.24. Transmitted and received LFM signals and beat frequency, for a moving target.

$$f_{bu} = \frac{2R}{c}\dot{f} - \frac{2\dot{R}}{\lambda}$$

Eq. (2.130)

where $\dot{R}$ is the range rate or the target radial velocity as seen by the radar. The first term of the right-hand side of Eq. (2.130) is due to the range delay defined by Eq. (2.123), while the second term is due to the target Doppler. Similarly,

$$f_{bd} = \frac{2R}{c}\dot{f} + \frac{2\dot{R}}{\lambda} .$$

Eq. (2.131)

Range is computed by adding Eq. (2.130) and Eq. (2.131). More precisely,

$$R = \frac{c}{4\dot{f}}(f_{bu} + f_{bd}) .$$

Eq. (2.132)

The range rate is computed by subtracting Eq. (2.131) from Eq. (2.130),

$$\dot{R} = \frac{\lambda}{4}(f_{bd} - f_{bu}) .$$

Eq. (2.133)

As indicated by Eqs. (2.132) and (2.133), CW radars utilizing triangular LFM can extract both range and range rate information. In practice, the maximum time delay Δt_{max} is normally selected as

$$\Delta t_{max} = 0.1 t_0 .$$

Eq. (2.134)

Thus, the maximum range is given by

$$R_{max} = \frac{0.1 c t_0}{2} = \frac{0.1c}{4 f_m}$$

Eq. (2.135)

and the maximum unambiguous range will correspond to a shift equal to $2t_0$.

2.10.4 Multiple Frequency CW Radar

Continuous wave radars do not have to use LFM waveforms in order to obtain good range measurements. Multiple frequency schemes allow CW radars to compute very adequate range measurements without using frequency modulation. In order to illustrate this concept, first consider a CW radar with the following waveform

$$x(t) = A\sin 2\pi f_0 t. \qquad \text{Eq. (2.136)}$$

The received signal from a target at range R is

$$x_r(t) = A_r \sin(2\pi f_0 t - \varphi) \qquad \text{Eq. (2.137)}$$

where the phase φ is equal to

$$\varphi = 2\pi f_0(2R/c). \qquad \text{Eq. (2.138)}$$

Solving for R we obtain

$$R = \frac{c\varphi}{4\pi f_0} = \frac{\lambda}{4\pi}\varphi. \qquad \text{Eq. (2.139)}$$

Clearly, the maximum unambiguous range occurs when φ is maximum, i.e., $\varphi = 2\pi$. Therefore, even for relatively large radar wavelengths, R is limited to impractical small values. Next, consider a radar with two CW signals, denoted by $x_1(t)$ and $x_2(t)$. More precisely,

$$x_1(t) = A_1 \sin 2\pi f_1 t \qquad \text{Eq. (2.140)}$$

$$x_2(t) = A_2 \sin 2\pi f_2 t. \qquad \text{Eq. (2.141)}$$

The received signals from a moving target are

$$x_{1r}(t) = A_{r_1}\sin(2\pi f_1 t - \varphi_1) \qquad \text{Eq. (2.142)}$$

$$x_{2r}(t) = A_{r_2}\sin(2\pi f_2 t - \varphi_2) \qquad \text{Eq. (2.143)}$$

where $\varphi_1 = (4\pi f_1 R)/c$ and $\varphi_2 = (4\pi f_2 R)/c$. After heterodyning (mixing) with the carrier frequency, the phase difference between the two received signals is

$$\varphi_2 - \varphi_1 = \Delta\varphi = \frac{4\pi R}{c}(f_2 - f_1) = \frac{4\pi R}{c}\Delta f. \qquad \text{Eq. (2.144)}$$

Again R is maximum when $\Delta\varphi = 2\pi$; it follows that the maximum unambiguous range is now

$$R = c/2\Delta f \qquad \text{Eq. (2.145)}$$

and since $\Delta f \ll c$, the range computed by Eq. (2.145) is much greater than that computed by Eq. (2.139), thus, indicating an increase in the unambiguous range when using more than one frequency.

2.11. MATLAB Program "range_calc.m"

The program *"range_calc.m"* solves the radar range equation of the form

$$R = \left(\frac{P_t \tau f_r T_i G_t G_r \lambda^2 \sigma}{(4\pi)^3 k T_0 F L (SNR)_o} \right)^{\frac{1}{4}}$$

Eq. (2.146)

where P_t is peak transmitted power, τ is pulse width, f_r is PRF, G_r and G_r are respectively the transmitting and receiving antenna gain, λ is wavelength, σ is target cross section, k is Boltzman's constant, T_0 is 290 *kelvin*, F is system noise figure, L is total system losses, and $(SNR)_o$ is the minimum SNR required for detection.

One can choose either CW or pulsed radars. In the case of CW radars, the term $P_t \tau f_r$ is replaced within the code by the average CW power P_{CW}. Additionally, the term T_i refers to the dwell interval. Alternatively, in the case of pulse radars T_i denotes the time on target. The plot inside Fig. 2.25 shows an example of the SNR versus the detection range for a pulse radar using the parameters shown in the figure. A MATLAB-based Graphical User Interface (GUI) (see Fig. 2.25) is utilized in inputting and editing all input parameters. The outputs include the maximum detection range versus minimum SNR plots.

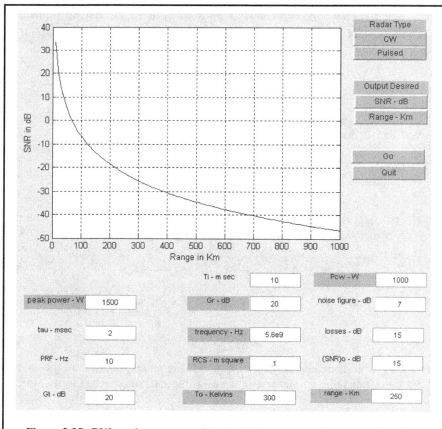

Figure 2.25. GUI work space associated with the program *"range_calc.m."*

Problems

2.1. Compute the aperture size for an X-band antenna at $f_0 = 9GHz$. Assume antenna gain $G = 10, 20, 30\, dB$.

2.2. An L-band radar ($1500MHz$) uses an antenna whose gain is $G = 30\, dB$. Compute the aperture size. If the radar duty cycle is $d_t = 0.2$ and the average power is $25KW$, compute the power density at range $R = 50Km$.

2.3. For the radar described in Problem 2.2, assume the minimum detectable signal is $5\, dBm$. Compute the radar maximum range for $\sigma = 1.0, 10.0, 20.0\, m^2$.

2.4. Consider an L-band radar with the following specifications: operating frequency $f_0 = 1500MHz$, bandwidth $B = 5MHz$, and antenna gain $G = 5000$. Compute the peak power, the pulse width, and the minimum detectable signal for this radar. Assume target RCS $\sigma = 10m^2$, the single pulse SNR is $15.4\, dB$, noise figure $F = 5\, dB$, temperature $T_0 = 290K$, and maximum range $R_{max} = 150Km$.

2.5. Repeat the example in Section 2.1 with $P_t = 1MW$, $G = 40\, dB$, and $\sigma = 0.5\, m^2$.

2.6. Show that the DC component is the dominant spectral line for high PRF waveforms.

2.7. Repeat the example in Section 2.3 with $L = 5\, dB$, $F = 10\, dB$, $T = 500K$, $T_i = 1.5\, s$, $d_t = 0.25$, and $R = 75Km$.

2.8. Consider a low PRF C-band radar operating at $f_0 = 5000MHz$. The antenna has a circular aperture with radius $2m$. The peak power is $P_t = 1MW$ and the pulse width is $\tau = 2\mu s$. The PRF is $f_r = 250Hz$, and the effective temperature is $T_0 = 600K$. Assume radar losses $L = 15\, dB$ and target RCS $\sigma = 10m^2$. (a) Calculate the radar's unambiguous range; (b) calculate the range R_0 that corresponds to $SNR = 0\, dB$; (c) calculate the SNR at $R = 0.75R_0$.

2.9. The atmospheric attenuation can be included in the radar equation as another loss term. Consider an X-band radar whose detection range at $20Km$ includes a $0.25\, dB/Km$ atmospheric loss. Calculate the corresponding detection range with no atmospheric attenuation.

2.10. Let the maximum unambiguous range for a low PRF radar be R_{max}. (a) Calculate the SNR at $(1/2)R_{max}$ and $(3/4)R_{max}$. (b) If a target with $\sigma = 10m^2$ exists at $R = (1/2)R_{max}$, what should the target RCS be at $R = (3/4)R_{max}$ so that the radar has the same signal strength from both targets.

2.11. A Millie-Meter Wave (MMW) radar has the following specifications: operating frequency $f_0 = 94GHz$, PRF $f_r = 15KHz$, pulse width $\tau = 0.05ms$, peak power $P_t = 10W$, noise figure $F = 5\, dB$, circular antenna with diameter $D = 0.254m$, antenna gain $G = 30\, dB$, target RCS $\sigma = 1m^2$, system losses $L = 8\, dB$, radar scan time $T_{sc} = 3s$, radar angular coverage $200°$, and atmospheric attenuation $3\, dB/Km$. Compute the following: (a) wavelength λ; (b) range resolution ΔR; (c) bandwidth B; (d) the SNR as a function of range; (e) the range for which $SNR = 15dB$; (f) antenna beam width; (g) antenna scan rate; (h) time on target; (i) the effective maximum range when atmospheric attenuation is considered.

2.12.   Repeat the second example in Section 2.4 with $\Omega = 4°$, $\sigma = 1m^2$, and $R = 400Km$.

2.13.   Using Eq. (2.53), compute (as a function of B_J/B) the cross-over range for the radar in Problem 2.11. Assume $P_J = 100W$, $G_J = 10dB$, and $L_J = 2dB$.

2.14.   Compute (as a function of B_J/B) the cross-over range for the radar in Problem 2.11. Assume $P_J = 200W$, $G_J = 15dB$, and $L_J = 2dB$. Assume $G' = 12dB$ and $R_J = 25Km$.

2.15.   A certain radar is subject to interference from an SSJ jammer. Assume the following parameters: radar peak power $P_t = 55KW$, radar antenna gain $G = 30dB$, radar pulse width $\tau = 2\mu s$, radar losses $L = 10dB$, jammer power $P_J = 150W$, jammer antenna gain $G_J = 12dB$, jammer bandwidth $B_J = 50MHz$, and jammer losses $L_J = 1dB$. Compute the cross-over range for a $5m^2$ target.

2.16.   A certain radar has losses of $6dB$ and a receiver noise figure of $8dB$. It has the requirement to detect targets within a search sector that is 360 degrees in azimuth and from 5 to 65 degrees in elevation. It must cover the search sector in 2 seconds. The RCS of the targets of interest is $5dBsm$ and the radar requires $20dB$ of signal-to-noise ratio to declare a detection. The required detection range of the radar is $75Km$. What is the average power aperture that the radar must have to satisfy the above search requirements

2.17.   Using Fig. 2.11 derive an expression for R_r. Assume 100% synchronization between the transmitter and receiver.

2.18.   A radar with antenna gain G is subject to a repeater jammer whose antenna gain is G_J. The repeater illuminates the radar with three fourths of the incident power on the jammer. (a) Find an expression for the ratio between the power received by the jammer and the power received by the radar; (b) what is this ratio when $G = G_J = 200$ and $R/\lambda = 10^5$?

2.19.   An X-band airborne radar transmitter and an air-to-air missile receiver act as a bistatic radar system. The transmitter guides the missile toward its target by continuously illuminating the target with a CW signal. The transmitter has the following specifications: peak power $P_t = 4KW$; antenna gain $G_t = 25dB$; operating frequency $f_0 = 9.5GHz$. The missile receiver has the following characteristics: aperture $A_r = 0.01m^2$; bandwidth $B = 750Hz$; noise figure $F = 7dB$; and losses $L_r = 2dB$. Assume that the bistatic RCS is $\sigma_B = 3m^2$. Assume $R_r = 35Km$; $R_t = 17Km$. Compute the SNR at the missile.

2.20.   Repeat the previous problem when there is $0.1dB/Km$ atmospheric attenuation.

2.21.   Consider an antenna with a $\sin x/x$ pattern. Let $x = (\pi r \sin\theta)/\lambda$, where r is the antenna radius, λ is the wavelength, and θ is the off-boresight angle. Derive Eq. (2.75). Hint: Assume small x, and expand $\sin x/x$ as an infinite series.

2.22.   Compute the amount of antenna pattern loss for a phased array antenna whose two-way pattern is approximated by $f(y) = [\exp(-2\ln 2(y/\theta_{3dB})^2)]^4$ where θ_{3dB} is the $3dB$ beam width. Assume circular symmetry.

2.23. A certain radar has a range gate size of $30m$. Due to range gate straddle, the envelope of a received pulse can be approximated by a triangular spread over three range bins. A target is detected in range bin 90. You need to find the exact target position with respect to the center of the range cell. (a) Develop an algorithm to determine the position of a target with respect to the center of the cell; (b) assuming that the early, on, and late measurements are, respectively, equal to $4/6$, $5/6$, and $1/6$, compute the exact target position.

2.24. Compute the amount of Doppler filter straddle loss for the filter defined by

$$H(f) = \frac{1}{1 + a^2 f^2}$$ Assume half-power frequency $f_{3dB} = 500Hz$ and cross-over frequency

$f_c = 350Hz$.

2.25. A radar has the following parameters: Peak power $P_t = 65KW$; total losses $L = 5dB$; operating frequency $f_o = 8GHz$; PRF $f_r = 4KHz$; duty cycle $d_t = 0.3$; circular antenna with diameter $D = 1m$; effective aperture is 0.7 of physical aperture; noise figure $F = 8dB$. (a) Derive the various parameters needed in the radar equation; (b) What is the unambiguous range? (c) Plot the SNR versus range ($1Km$ to the radar unambiguous range) for a $5dBsm$ target. (d) If the minimum SNR required for detection is $14dB$, what is the detection range for a $6dBsm$ target? What is the detection range if the SNR threshold requirement is raised to $18dB$?

2.26. A radar has the following parameters: Peak power $P_t = 50KW$; total losses $L = 5dB$; operating frequency $f_o = 5.6GHz$; noise figure $F = 10dB$ pulse width $\tau = 10\mu s$; PRF $f_r = 2KHz$; antenna beamwidth $\theta_{az} = 1°$ and $\theta_{el} = 5°$. (a) What is the antenna gain? (b) What is the effective aperture if the aperture efficiency is 60%? (c) Given a 14 dB threshold detection, what is the detection range for a target whose RCS is $\sigma = 1m^2$?

2.27. A certain radar has losses of $5dB$ and a receiver noise figure of $10dB$. This radar has a detection coverage requirement that extends over 3/4 of a hemisphere and must complete it in 3 seconds. The base line target RCS is $6dBsm$ and the minimum SNR is $15dB$. The radar detection range is less than $80Km$. What is the average power aperture product for this radar so that it can satisfy its mission?

2.28. A monostatic radar has the following parameters: Transmit power $100Kw$, transmit losses $2dB$, operating Frequency $7GHz$, PRF $2000Hz$, pulse width $10\mu sec$, antenna beamwidth $2°$ Az X $4°$ El, receive losses $3dB$, and receiver noise figure $12dB$. Assume that the radar uses pulses that employ $10MHz$ of linear frequency modulation and uses a processor that is matched to the transmitted pulse. (a) What is the antenna gain? (b) What is the effective aperture if the aperture efficiency is 50%? (c) What is the effective radiated power of the radar, in dBm? (d) Given a detection threshold of $13dB$, what is the detection range for a target with a radar cross-section of $6dBsm$?

2.29. A radar generates $100KW$ of power and has $1dB$ of loss between the power tube and the antenna. The radar is monostatic with a single antenna that has a gain of $38dB$. The radar is operating at $5GHz$. What is the power at the receive antenna output for the following targets:

(a) A $1m^2$ RCS target at a range of $30Km$. (b) A $10dBsm$ target at a range of $50\ Km$.

Assume that: the total radar losses of $1dB$.

2.30. A source with equivalent temperature $T_o = 290K$ is followed by three amplifiers with specifications shown in the table below.

Amplifier	F, dB	G, dB	T_e
1	You must compute	12	350
2	10	22	
3	15	35	

(a) Compute the noise figure for the three cascaded amplifiers. (b) Compute the effective temperature for the three cascaded amplifiers. (c) Compute the overall system noise figure.

2.31. A radar has the following receiver components. They are arranged in the order shown below

Receiver Stages			
Stage #	Component	Gain, dB	Noise Figure, dB
1	Waveguide	-2	2
2	RF Amp	28	5
3	1st Mixer	-3	15
4	IF Amp	100	30

(a) What is the receiver noise figure through the RF amp and referenced to the input of the waveguide (the first component after the antenna)? (b) What is the noise figure of the receiver through the IF amp and referenced to the input of the RF amp? (c) What is the effective noise temperature of the receiver through the IF amp and referenced to the input of the waveguide? (d) Suppose you want to determine how internal noise and sky noise contribute to noise power at various points in the receiver. Specifically, how does the noise power at the output of each component as a function of the effective noise temperature of the antenna, T_{ant}, and noise bandwidth, B. Derive four equations that will allow us to easily perform the computations. All of your equations should be of the form $P = B(K_1 T_{ant} + K_2)$ where K_1 and K_2 are constants. Provide a table with the four sets of values for K_1 and K_2.

2.32. Prove that

$$\sum_{n=-\infty}^{\infty} J_n(z) = 1.$$

2.33. Show that $J_{-n}(z) = (-1)^n J_n(z)$. Hint: You may utilize the relation

$$J_n(z) = \frac{1}{\pi} \int_0^{\pi} \cos(z\sin y - ny)\,dy$$

2.34. In a multiple-frequency CW radar, the transmitted waveform consists of two continuous sine waves of frequencies $f_1 = 105 KHz$ and $f_2 = 115 KHz$. Compute the maximum unambiguous detection range.

2.35. Consider a radar system using linear frequency modulation. Compute the range that corresponds to $\dot{f} = 20, 10 MHz$. Assume a beat frequency $f_b = 1200 Hz$.

2.36. A certain radar using linear frequency modulation has a modulation frequency $f_m = 300 Hz$, and frequency sweep $\Delta f = 50 MHz$. Calculate the average beat frequency differences that correspond to range increments of 10 and 15 meters.

2.37. A CW radar uses linear frequency modulation to determine both range and range rate. The radar wavelength is $\lambda = 3 cm$, and the frequency sweep is $\Delta f = 200 KHz$. Let $t_0 = 20 ms$. (a) Calculate the mean Doppler shift; (b) compute f_{bu} and f_{bd} corresponding to a target at range $R = 350 Km$, which is approaching the radar with radial velocity of $250 m/s$.

2.38. In Chapter 1 we developed an expression for the Doppler shift associated with a CW radar (i.e., $f_d = \pm 2v/\lambda$, where the plus sign is used for closing targets and the negative sign is used for receding targets). CW radars can use the system shown in Fig. P.2.34 to determine whether the target is closing or receding. Assuming that the emitted signal is $A \cos \omega_0 t$ and the received signal is $kA \cos((\omega_0 \pm \omega_d)t + \varphi)$, show that the direction of the target can be determined by checking the phase shift difference in the outputs $y_1(t)$ and $y_2(t)$.

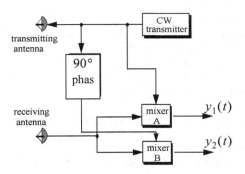

Figure P.2.34

Appendix 2-A: Chapter 2 MATLAB Code Listings

The MATLAB code provided in this chapter was designed as an academic standalone tool and is not adequate for other purposes. The code was written in a way to assist the reader in gaining a better understanding of the theory. The code was not developed, nor is it intended to be used as part of an open-loop or a closed-loop simulation of any kind. The MATLAB code found in this textbook can be downloaded from this book's web page on the CRC Press website. Simply use your favorite web browser, go to *www.crcpress.com*, and search for keyword *"Mahafza"* to locate this book's web page.

MATLAB Function "radar_eq.m" Listing

```
function [snr] = radar_eq(pt, freq, g, sigma, b, nf, loss, range)
% This function implements Eq. (2.22) of textbook
%% Inputs:
     % pt          == input peak power in Watts
     % freq        == radar operating frequency in Hz
     % g           == antenna gain in dB
     % sigma       == radar cross section in meter squared
     % b           == radar bandwidth in Hz
     % nf          == noise Figure in dB
  % loss          == total radar losses in dB
     % range       == range to target (single value or vector) in Km% % Outputs:
     % snr         == SNR in dB
%
c = 3.0e+8; % speed of light
lambda = c / freq; % wavelength
p_peak = 10*log10(pt); % convert peak power to dB
lambda_sqdb = 10*log10(lambda^2); % compute wavelength square in dB
sigmadb = 10*log10(sigma); % convert sigma to dB
four_pi_cub = 10*log10((4.0 * pi)^3); % (4pi)^3 in dB
k_db = 10*log10(1.38e-23); % Boltzman's constant in dB
to_db = 10*log10(290); % noise temp. in dB
b_db = 10*log10(b); % bandwidth in dB
range_pwr4_db = 10*log10(range.^4); % vector of target range^4 in dB
% Implement Equation (2.22)
num = p_peak + g + g + lambda_sqdb + sigmadb;
den = four_pi_cub + k_db + to_db + b_db + nf + loss + range_pwr4_db;
snr = num - den;
return
```

MATLAB Program "Fig2_1.m" Listing

```
% Use this program to reproduce Fig. 2.1 of text.
clc
close all
clear all
pt = 1.5e+6; % peak power in Watts
freq = 5.6e+9; % radar operating frequency in Hz
g = 45.0; % antenna gain in dB
sigma = 0.1; % radar cross section in m squared
b = 5.0e+6; % radar operating bandwidth in Hz
nf = 3.0; %noise figure in dB
loss = 6.0; % radar losses in dB
```

range = linspace(25e3,165e3,1000); % range to target from 25 Km 165 Km, 1000 points
snr1 = radar_eq(pt, freq, g, sigma, b, nf, loss, range);
snr2 = radar_eq(pt, freq, g, sigma/10, b, nf, loss, range);
snr3 = radar_eq(pt, freq, g, sigma*10, b, nf, loss, range);
% plot SNR versus range
figure(1)
rangekm = range ./ 1000;
plot(rangekm,snr3,'k',rangekm,snr1,'k -.',rangekm,snr2,'k:','linewidth',1.5)
grid
legend('\sigma = 0 dBsm', '\sigma = -10dBsm', '\sigma = -20 dBsm')
xlabel ('Detection range - Km');
ylabel ('SNR - dB');
snr1 = radar_eq(pt, freq, g, sigma, b, nf, loss, range);
snr2 = radar_eq(pt*.4, freq, g, sigma, b, nf, loss, range);
snr3 = radar_eq(pt*1.8, freq, g, sigma, b, nf, loss, range);
figure (2)
plot(rangekm,snr3,'k',rangekm,snr1,'k -.',rangekm,snr2,'k:','linewidth',1.5)
grid
legend('Pt = 2.16 MW','Pt = 1.5 MW','Pt = 0.6 MW')
xlabel ('Detection range - Km');
ylabel ('SNR - dB');

MATLAB Function "lprf_req.m" Listing

function [snr] = lprf_req(pt, g, freq, sigma, np, b, nf, loss, range)
% This program implements Eq. (2.27) of textbook
%% Inputs:
    % pt   == input peak power in Watts
    % freq   == radar operating frequency in Hz
    % g   == antenna gain in dB
    % sigma   == radar cross section in meter squared
    % b   == radar bandwidth in Hz
    % nf   == noise Figure in dB
    % np   == number of pulses
    % loss   == total radar losses in dB
    % range   == range to target (single value or vector) in Km
%% Outputs:
    % snr   == SNR in dB
%
c = 3.0e+8; % speed of light
lambda = c / freq; % wavelength
p_peak = 10*log10(pt); % convert peak power to dB
lambda_sqdb = 10*log10(lambda^2); % compute wavelength square in dB
sigmadb = 10*log10(sigma); % convert sigma to dB
four_pi_cub = 10*log10((4.0 * pi)^3); % (4pi)^3 in dB
k_db = 10*log10(1.38e-23); % Boltzman's constant in dB
to_db = 10*log10(290); % noise temp. in dB
b_db = 10*log10(b); % bandwidth in dB
np_db = 10.*log10(np); % number of pulses in dB
range_pwr4_db = 10*log10(range.^4); % vector of target range^4 in dB
% Implement Equation (1.68)
num = p_peak + g + g + lambda_sqdb + sigmadb + np_db;
den = four_pi_cub + k_db + to_db + b_db + nf + loss + range_pwr4_db;
snr = num - den;

return

MATLAB Program "Fig2_2.m" Listing

% Use this program to reproduce Fig. 2.2 of text.
clc
close all
clear all
pt = 1.5e+6; % peak power in Watts
freq = 5.6e+9; % radar operating frequency in Hz
g = 45.0; % antenna gain in dB
sigma = 0.1; % radar cross section in m squared
b = 5.0e+6; % radar operating bandwidth in Hz
nf = 3.0; %noise figure in dB
loss = 6.0; % radar losses in dB
np = 1;
range = linspace(25e3,225e3,1000); % range to target from 5 Km 225 Km, 1000 points
snr1 = lprf_req(pt, g, freq, sigma, np, b, nf, loss, range);
snr2 = lprf_req(pt, g, freq, sigma, 5*np, b, nf, loss, range);
snr3 = lprf_req(pt, g, freq, sigma, 10*np, b, nf, loss, range);
% plot SNR versus range
figure(1)
rangekm = range ./ 1000;
plot(rangekm,snr3,'k',rangekm,snr1,'k -.',rangekm,snr2,'k:','linewidth',1.5)
grid
legend('np = 10','np = 5','np = 1')
xlabel ('Detection range - Km');
ylabel ('SNR - dB');
np = linspace(1,500,500);
range = 150e3;
snr1 = lprf_req(pt, g, freq, sigma, np, b, nf, loss, range);
snr2 = lprf_req(pt, g, freq, 10*sigma, np, b, nf, loss, range);
figure (2)
plot(np,snr2,'k',np,snr1,'k -.','linewidth',1.5)
grid
legend('Baseline','\sigma = 0 dBsm')
xlabel ('No. of pulses');
ylabel ('SNR - dB');

MATLAB Function "hprf_req.m" Listing

function [snr] = hprf_req (pt, Ti, g, freq, sigma, dt, range, nf, loss)
% This program implements Eq. (2.31)of textbook
%% Inputs:
    % pt   == input peak power in Watts
    % freq   == radar operating frequency in Hz
    % g   == antenna gain in dB
    % sigma   == radar cross section in meter squared
    % Ti   == time on target in seconds
    % nf   == noise Figure in dB
    % dt   == duty cycle
    % loss   == total radar losses in dB
    % range   == range to target (single value or vector) in Km
%% Outputs:
    % snr   == SNR in dB

```
%
c = 3.0e+8; % speed of light
lambda = c / freq; % wavelength
pav = 10*log10(pt*dt); % compute average power in dB
Ti_db = 10*log10(Ti); % time on target in dB
lambda_sqdb = 10*log10(lambda^2); % compute wavelength square in dB
sigmadb = 10*log10(sigma); % convert sigma to dB
four_pi_cub = 10*log10((4.0 * pi)^3); % (4pi)^3 in dB
k_db = 10*log10(1.38e-23); % Boltzman's constant in dB
to_db = 10*log10(290); % noise temp. in dB
range_pwr4_db = 10*log10(range.^4); % vector of target range^4 in dB
% Implement Equation (1.72)
num = pav + Ti_db + g + g + lambda_sqdb + sigmadb;
den = four_pi_cub + k_db + to_db + nf + loss + range_pwr4_db;
snr = num - den;
return
```

MATLAB Program "Fig2_3.m" Listing

```
% Use this program to reproduce Fig. 2.3 of text.
clc
close all
clear all
pt = 10e03; % peak power in Watts
freq = 5.6e+9; % radar operating frequency in Hz
g = 20; % antenna gain in dB
sigma = 0.01; % radar cross section in m squared
b = 5.0e+6; % radar operating bandwidth in Hz
nf = 3.0; %noise figure in dB
loss = 8.0; % radar losses in dB
Ti = 2; % time on target in seconds
dt = .05; % 5% duty cycle
range = linspace(10e3,225e3,1000); % range to target from 10 Km 225 Km, 1000 points
snr1 = hprf_req (pt, Ti, g, freq, sigma, .05, range, nf, loss);
snr2 = hprf_req (pt, Ti, g, freq, sigma, .1, range, nf, loss);
snr3 = hprf_req (pt, Ti, g, freq, sigma, .2, range, nf, loss);
% plot SNR versus range
figure(1)
rangekm  = range ./ 1000;
plot(rangekm,snr3,'k',rangekm,snr2,'k -.',rangekm,snr1,'k:','linewidth',1.5)
grid on
legend('dt = 20%','dt = 10%','dt = 5%')
xlabel ('Detection range - Km');
ylabel ('SNR - dB');
```

MATLAB Function "power_aperture.m" Listing

```
function PAP = power_aperture(snr,tsc,sigma,range,nf,loss,az_angle,el_angle)
% This function implements Eq. (2.38) of textbook
%% Inputs:
    % snr        == SNR in dB
    % tsc        == scan time in seconds
    % sigma    == radar cross section in meter squared
    % range      == range to target in Km
```

```
   % nf          == noise Figure in dB
   % loss        == total radar losses in dB
   % az_angle    == azimuth search extent in degrees
   % el_angle    == elevation search extent in degrees
%% Outputs:
   % PAP         == power aperture product in dB
%
Tsc = 10*log10(tsc); % convert Tsc into dB
Sigma = 10*log10(sigma); % convert sigma to dB
four_pi = 10*log10(4.0 * pi); % (4pi) in dB
k_db = 10*log10(1.38e-23); % Boltzman's constant in dB
To = 10*log10(290); % noise temp. in dB
range_pwr4_db = 10*log10(range.^4); % target range^4 in dB
omega = (az_angle/57.296) * (el_angle / 57.296); % compute search volume in steraradians
Omega = 10*log10(omega); % search volume in dB
% implement Eq. (1.79)
PAP = snr + four_pi + k_db + To + nf + loss + range_pwr4_db + Omega - Sigma - Tsc;
return
```

MATLAB Program "Fig2_6.m" Listing

```
% Use this program to reproduce Fig. 2.6 of text.
clc
close all
clear all
tsc = 2.5; % Scan time i s2.5 seconds
sigma = 0.1; % radar cross section in m squared
te = 900.0; % effective noise temperature in Kelvins
snr = 15; % desired SNR in dB
nf = 6.0; %noise figure in dB
loss = 7.0; % radar losses in dB
az_angle = 2; % search volume azimuth extent in degrees
el_angle = 2; %serach volume elevation extent in degrees
range = linspace(20e3,250e3,1000); % range to target from 20 Km 250 Km, 1000 points
pap1 = power_aperture(snr,tsc,sigma/10,range,nf,loss,az_angle,el_angle);
pap2 = power_aperture(snr,tsc,sigma,range,nf,loss,az_angle,el_angle);
pap3 = power_aperture(snr,tsc,sigma*10,range,nf,loss,az_angle,el_angle);
% plot power aperture prodcut versus range
% generate Figure 2.6a
figure(1)
rangekm = range ./ 1000;
plot(rangekm,pap1,'k',rangekm,pap2,'k -.',rangekm,pap3,'k:','linewidth',1.5)
grid
legend('\sigma = -20 dBsm', '\sigma = -10dBsm', '\sigma = 0 dBsm')
xlabel ('Detection range in Km');
ylabel ('Power aperture product in dB');
% generate Figure 2.6b
lambda = 0.03; % wavelength in meters
G = 45; % antenna gain in dB
ae = linspace(1,25,1000);% aperture size 1 to 25 meter squared, 1000 points
Ae = 10*log10(ae);
range = 250e3; % rnage of interset is 250 Km
pap1 = power_aperture(snr,tsc,sigma/10,range,nf,loss,az_angle,el_angle);
pap2 = power_aperture(snr,tsc,sigma,range,nf,loss,az_angle,el_angle);
```

```
pap3 = power_aperture(snr,tsc,sigma*10,range,nf,loss,az_angle,el_angle);
Pav1 = pap1 - Ae;
Pav2 = pap2 - Ae;
Pav3 = pap3 - Ae;
figure(2)
plot(ae,Pav1,'k',ae,Pav2,'k -.',ae,Pav3,'k:','linewidth',1.5)
grid
xlabel('Aperture size in square meters')
ylabel('Pav in dB')
legend('\sigma = -20 dBsm','\sigma = -10dBsm','\sigma = 0dBsm')
```

MATLAB Program "ssj_req.m" Listing

```
function [BR_range] = ssj_req (pt, g, freq, sigma, br, loss, ...
  pj, bj, gj, lossj)
% This function implements Eq.s (2.50) and Eq. (2.52). It also generates
% plot 2.7a
% % Inputs
    % pt        == radar peak power in Watts
    % g         == radar antenna gain in dB
    % freq      == radar operating frequency in Hz
    % sigma     == target RCS in squared meters
    % br        == radar bandwidth in Hz
    % loss      == radar losses in dB
    % pj        == jammer power in Watts
    % bj        == jammer bandwidth in Hz
    % gj        == jammer antenna gain in dB
    % loosj     == jammer losses in dB
%%% Outputs
    % BR_range  == cross over range in Km
 %
c = 3.0e+8;
lambda = c / freq;
lambda_db = 10*log10(lambda^2);
if (loss == 0.0)
  loss = 0.000001;
end
if (lossj == 0.0)
  lossj =0.000001;
end
sigmadb =10*log10(sigma);
pt_db = 10*log10(pt);
b_db = 10*log10(br);
bj_db = 10*log10(bj);
pj_db = 10*log10(pj);
factor = 10*log10(4.0 *pi);
BR_range = sqrt((pt * (10^(g/10)) * sigma * bj * (10^(lossj/10))) / ...
   (4.0 * pi * pj * (10^(gj/10)) * br *  (10^(loss/10)))) / 1000.0
s_at_br = pt_db + 2.0 * g + lambda_db + sigmadb - 3.0 * factor - 4.* 10*log10(BR_range) - loss
index =0;
for ran_var = .1:10:10000
  index = index + 1;
  ran_db = 10*log10(ran_var * 1000.0);
  ssj(index) = pj_db + gj + lambda_db + g + b_db - 2.0 * factor - 2.0 * ran_db - bj_db - lossj + s_at_br ;
```

```
    s(index) = pt_db + 2.0 * g + lambda_db + sigmadb - 3.0 * factor - 4.* ran_db - loss + s_at_br ;
end
ranvar = .1:10:10000;
ranvar = ranvar ./ BR_range;
semilogx (ranvar,s,'k',ranvar,ssj,'k-.');
axis([.1 1000 -90 40])
xlabel ('Range normalized to cross-over range');
legend('Target echo','SSJ')
ylabel ('Relative signal or jamming amplitude - dB');
grid
```

MATLAB Program "Fig2_7b.m" Listing

```
% This program produces Fig 2.7 of text
clc;
clear all
close all
pt = 50.0e+3;    % peak power in Watts
g = 35.0;        % antenna gain in dB
freq = 5.6e+9;  % radar operating frequency in Hz
sigma = 10.0 ;   % radar cross section in m squared
b = 667.0e+3;    % radar operating bandwidth in Hz
loss = 0.1000;    % radar losses in dB
rangej = 50.0; % range to jammer in Km
pj = 200.0;     % jammer peak power in Watts
bj = 50.0e+6;  % jammer operating bandwidth in Hz
gj = 10.0;        % jammer antenna gain in dB
lossj = .10;    % jammer losses in dB
[BR_range] = ssj_req (pt, g, freq, sigma, b, loss, ...
   pj, bj, gj, lossj);
pj_var = 1:1:1000;
BR_pj = sqrt((pt * (10^(g/10)) * sigma * bj * (10^(lossj/10))) ...
   ./ (4.0 * pi .* pj_var * (10^(gj/10)) * b * (10^(loss/10)))) ./ 1000;
pt_var = 1000:100:10e6;
BR_pt = sqrt((pt_var * (10^(g/10)) * sigma * bj * (10^(lossj/10))) ...
   ./ (4.0 * pi .* pj * (10^(gj/10)) * b * (10^(loss/10)))) ./ 1000;
figure (2)
subplot (2,1,1)
semilogx (BR_pj,'k')
xlabel ('Jammer peak power - Watts');
ylabel ('Cross-over range - Km')
grid
subplot (2,1,2)
semilogx (BR_pt,'k')
xlabel ('Radar peak power - KW')
ylabel ('Cross-over range - Km')
grid
```

MATLAB Function "sir.m" Listing

```
function [SIR] = sir (pt, g, sigma, freq, tau, loss, R, pj, bj, gj, lossj);
% This function implements Eq. (2.53) of textbook
% % Inputs
    % pt         == radar peak power in Watts
    % g          == radar antenna gain in dB
```

```
    % freq     == radar operating frequency in Hz
    % tau      == radar pulse width in seconds
    % loss     == radar losses in dB
    % R        == target range in Km, can be single value or vector
    % pj       == jammer power in Watts
    % bj       == jammer bandwidth in Hz
    % gj       == jammer antenna gain in dB
    % loosj    == jammer losses in dB
%% Outputs
    % SIR      == S/(J+N) in dB
%
c = 3.0e+8;
k = 1.38e-23;
%R = linspace(rmin, rmax, 1000);
range = R .* 1000;
lambda = c / freq;
gj = 10^(gj/10);
G = 10^(g/10);
ERP1 = pj * gj / lossj;
ERP_db = 10*log10(ERP1);
Ar = lambda *lambda * G / 4 /pi;
num1 = pt * tau * G * sigma * Ar;
demo1 = 4^2 * pi^2 * loss .* range.^4;
demo2 = 4 * pi * bj .* range.^2;
num2 = ERP1 * Ar;
val11 = num1 ./ demo1;
val21 = num2 ./demo2;
sir = val11 ./ (val21 + k * 290);
SIR = 10*log10(sir);
end
```

MATLAB Program "Fig2_8.m" Listing

```
% This program generates Fig. 2.8 of text
clc
clear all
close all
R = linspace(10,400,5000);
[SIR] = sir (50e3, 35, 10, 5.6e9, 50e-6, 5, R, 200, 50e6, 10, .3);
figure (1)
plot (R, SIR,'k')
xlabel ('Detection range in Km');
ylabel ('S/(J+N) in dB')
grid
```

MATLAB Function "burn_thru.m" Listing

```
function [Range] = burn_thru (pt, g, sigma, freq, tau, loss, pj, bj, gj, lossj,sir0,ERP);
% This function implements Eq. (254) of textbook
% % Inputs
    % pt       == radar peak power in Watts
    % g        == radar antenna gain in dB
    % freq     == radar operating frequency in Hz
    % tau      == radar pulse width in seconds
    % loss     == radar losses in dB
```

```
        % pj       == jammer power in Watts
        % bj       == jammer bandwidth in Hz
        % gj       == jammer antenna gain in dB
        % loosj    == jammer losses in dB
        % sir0     == desired SIR in dB
        % ERP      == desired jammer ERP, single value or vector in Watts
%% Outputs
        % Range    == burn through range in Km
%
c = 3.0e+8;
k = 1.38e-23;
sir0 = 10^(sir0/10);
lambda = c / freq;
gj = 10^(gj/10);
G = 10^(g/10);
Ar = lambda *lambda * G / 4 /pi;
num32 = ERP .* Ar;
demo3 = 8 *pi * bj * k * 290;
demo4 = 4^2 * pi^2 * k * 290 * sir0;
val1 = (num32 ./ demo3).^2;
val2 = (pt * tau * G * sigma * Ar)/(4^2 * pi^2 * loss * sir0 * k * 290);
val3 = sqrt(val1 + val2);
val4 = (ERP .* Ar) ./ demo3;
Range = sqrt(val3 - val4) ./ 1000;
end
```

MATLAB Program "Fig2_9.m" Listing

```
% This program generates Fig. 2.9 of text
clc
clear all
close all
ERP = linspace(1,1000,1000);
[Range] = burn_thru (50e3, 35, 10, 5.6e9, 0.5e-3, 5, 200,500e6, 10, 0.3, 15,ERP);
figure (1)
plot (10*log10(ERP), Range,'k')
xlabel (' Jammer ERP in dB')
ylabel ('Burnthrough range in Km')
grid
```

MATLAB Function "soj_req.m" Listing

```
function [BR_range] = soj_req (pt, g, sigma, b, freq, loss, range, ...
   pj, bj,gj, lossj, gprime, rangej)
% This function implements Eqs. (257) and (2.58) of textbook
%% Inputs
        % pt        == radar peak power in Watts
        % g         == radar antenna gain in dB
        % sigma     == target RCS in sdBsm
        % freq      == radar operating frequency in Hz
        % tau       == radar pulse width in seconds
        % loss      == radar losses in dB
        % range     == range to target in Km
        % pj        == jammer power in Watts
        % bj        == jammer bandwidth in Hz
```

```
      % gj        == jammer antenna gain in dB
      % loosj     == jammer losses in dB
      % gprime  == jammer antenna gain
      % rangej     == range to jammer in Km
%% Outputs
      % BR_Range  == burn through range in Km
%
c = 3.0e+8;
lambda = c / freq;
lambda_db = 10*log10(lambda^2);
if (loss == 0.0)
   loss = 0.000001;
end
if (lossj == 0.0)
   lossj =0.000001;
end
sigmadb = 10*log10(sigma);
range_db = 10*log10(range * 1000.);
rangej_db = 10*log10(rangej * 1000.);
pt_db = 10*log10(pt);
b_db = 10*log10(b);
bj_db = 10*log10(bj);
pj_db = 10*log10(pj);
factor = 10*log10(4.0 *pi);
BR_range = ((pt * 10^(2.0*g/10) * sigma * bj * 10^(lossj/10) * ...
   (rangej)^2) / (4.0 * pi * pj * 10^(gj/10) * 10^(gprime/10) * ...
   b * 10^(loss/10)))^.25 / 1000.
end
```

MATLAB Program "Fig2_10.m" Listing

```
% This program generates Fig. 2.10 of text
clc
clear all
close all
pt = 5.0e+3; pt_db = 10*log10(pt);
g = 35.0;
freq = 5.6e+9;  lambda = 3e8 / freq;
lambda_db = 10*log10(lambda^2);
sigma = 10 ;
b = 667.0e+3;  b_db = 10*log10(b);
range = 20*1852;   range_db = 10*log10(range * 1000.);
gprime = 10.0;   sigmadb = 10*log10(sigma);
loss = 0.01;
rangej = 12*1852; rangej_db = 10*log10(rangej * 1000.);
pj = 5.0e+3;  pj_db = 10*log10(pj);
bj = 50.0e+6;  bj_db = 10*log10(bj);
gj = 30.0;
lossj =0.3;
factor = 10*log10(4.0 *pi);
[BR_range] = soj_req (pt, g, sigma, b, freq, loss, range, pj, bj,gj, lossj, gprime, rangej)
 soj_req (pt, g, sigma, b, freq, loss, range, pj, bj,gj, lossj, gprime, rangej)
s_at_br = pt_db + 2.0 * g + lambda_db + sigmadb - 3.0 * factor - 4.0 * 10*log10(BR_range) - loss
index =0;
```

```
for ran_var = .1:1:1000;
    index = index + 1;
    ran_db = 10*log10(ran_var * 1000.0);
    s(index) = pt_db + 2.0 * g + lambda_db + sigmadb - ...
        3.0 * factor - 4.0 * ran_db - loss + s_at_br;
    soj(index) = s_at_br - s_at_br;
end
 ranvar = .1:1:1000;
%ranvar = ranvar ./BR_range;
semilogx (ranvar,s,'k',ranvar,soj,'k-.','linewidth',1.5);
xlabel ('Range normalized to cross-over range');
legend('Target echo','SOJ')
ylabel ('Relative signal or jamming amplitude - dB');
grid
```

MATLAB Function "range_calc.m" Listing

```
function [output_par] = range_calc (pt, tau, fr, time_ti, gt, gr, freq, ...
    sigma, te, nf, loss, snro, pcw, range, radar_type, out_option)
c = 3.0e+8;
lambda = c / freq;
if (radar_type == 0)
    pav = pcw;
else
    % Compute the duty cycle
    dt = tau * 0.001 * fr;
    pav = pt * dt;
end
pav_db = 10.0 * log10(pav);
    lambda_sqdb = 10.0 * log10(lambda^2);
    sigmadb = 10.0 * log10(sigma);
    for_pi_cub = 10.0 * log10((4.0 * pi)^3);
                k_db = 10.0 * log10(1.38e-23);
    te_db = 10.0 * log10(te);
    ti_db = 10.0 * log10(time_ti);
    range_db = 10.0 * log10(range * 1000.0);
if (out_option == 0)
    %compute SNR
    snr_out = pav_db + gt + gr + lambda_sqdb + sigmadb + ti_db - ...
    for_pi_cub - k_db - te_db - nf - loss - 4.0 * range_db
    index = 0;
    for range_var = 10:10:1000
        index = index + 1;
        rangevar_db = 10.0 * log10(range_var * 1000.0);
        snr(index) = pav_db + gt + gr + lambda_sqdb + sigmadb + ti_db - ...
            for_pi_cub - k_db - te_db - nf - loss - 4.0 * rangevar_db;
    end
    var = 10:10:1000;
    plot(var,snr,'k')
    xlabel ('Range in Km');
    ylabel ('SNR in dB');
    grid
else
    range4 = pav_db + gt + gr + lambda_sqdb + sigmadb + ti_db - ...
```

```
   for_pi_cub - k_db - te_db - nf - loss - snro;
range = 10.0^(range4/40.) / 1000.0
index = 0;
for snr_var = -20:1:60
   index = index + 1;
   rangedb = pav_db + gt + gr + lambda_sqdb + sigmadb + ti_db - ...
     for_pi_cub - k_db - te_db - nf - loss - snr_var;
   range(index) = 10.0^(rangedb/40.) / 1000.0;
end
var = -20:1:60;
plot(var,range,'k')
xlabel ('Minimum SNR required for detection in dB');
ylabel ('Maximum detection range in Km');
 grid
end
return
```

Part II

Radar Signals and Signal Processing

Chapter 5:

Ambiguity Function - Analog Waveforms
> *Introduction*
> *Examples of the Ambiguity Function*
> *Stepped Frequency Waveforms*
> *Nonlinear FM*
> *Ambiguity Diagram Contours*
> *Interpretation of Range-Doppler Coupling in LFM Signals*
> *Problems*
> *Appendix 5-A: Chapter 5 MATLAB Code Listing*

Chapter 6:

Ambiguity Function - Discrete Coded Waveforms
> *Discrete Code Signal Representation*
> *Pulse-Train Codes*
> *Phase Coding*
> *Frequency Codes*
> *Ambiguity Plots for Discrete Coded Waveforms*
> *Problems*
> *Appendix 6-A: Chapter 6 MATLAB Code Listings*

Chapter 7:

Pulse Compression
> *Time-Bandwidth Product*
> *Radar Equation with Pulse Compression*
> *Basic Principal of Pulse Compression*
> *Correlation Processor*
> *Stretch Processor*
> *Problems*
> *Appendix 7-A: Chapter 7 MATLAB Code Listings*

Chapter 3

Linear Systems and Complex Signal Representation

In this chapter a top-level overview of elements of signal theory that are relevant to radar signal processing is presented. It is assumed that the reader has sufficient and adequate background in signals and systems as well as in Fourier transform and its associated properties.

3.1. Signal Classifications

In general, electrical signals can represent either current or voltage and may be classified into two main categories: energy signals and power signals. Energy signals can be deterministic or random, while power signals can be periodic or random. A signal is said to be random if it is a function of a random parameter (such as random phase or random amplitude). Additionally, signals may be divided into lowpass or bandpass signals. Signals that contain very low frequencies (close to DC) are called lowpass signals; otherwise they are referred to as bandpass signals. Through modulation, lowpass signals can be mapped into bandpass signals.

The average power P for the current or voltage signal $x(t)$ over the interval (t_1, t_2) across a 1Ω resistor is

$$P = \frac{1}{t_2 - t_1} \int_{t_1}^{t_2} |x(t)|^2 \ dt . \qquad \text{Eq. (3.1)}$$

The signal $x(t)$ is said to be a power signal over a very large interval $T = t_2 - t_1$, if and only if it has finite power and satisfies the relation:

$$0 < \lim_{T \to \infty} \frac{1}{T} \int_{-T/2}^{T/2} |x(t)|^2 \ dt \ < \infty . \qquad \text{Eq. (3.2)}$$

Using Parseval's theorem, the energy E dissipated by the current or voltage signal $x(t)$ across a 1Ω resistor, over the interval (t_1, t_2), is

$$E = \int_{t_1}^{t_2} |x(t)|^2 \ dt . \qquad \text{Eq. (3.3)}$$

The signal $x(t)$ is said to be an energy signal if and only if it has finite energy,

$$E = \int_{-\infty}^{\infty} |x(t)|^2 \ dt \quad < \infty .$$

Eq. (3.4)

A signal $x(t)$ is said to be periodic with period T if and only if

$$x(t) = x(t + nT) \qquad for \ all \ t$$

Eq. (3.5)

where n is an integer.

Example:

Classify each of the following signals as an energy signal, a power signal, or neither. All signals are defined over the interval $(-\infty < t < \infty)$: $x_1(t) = \cos t + \cos 2t$, $x_2(t) = \exp(-\alpha^2 t^2)$.

Solution:

$$P_{x_1} = \frac{1}{T} \int_{-T/2}^{T/2} (\cos t + \cos 2t)^2 dt = 1 \Rightarrow \ power \ signal .$$

Note that since the cosine function is periodic, the limit is not necessary.

$$E_{x_2} = \int_{-\infty}^{\infty} (e^{-\alpha^2 t^2})^2 dt = 2\int_{0}^{\infty} e^{-2\alpha^2 t^2} dt = 2\frac{\sqrt{\pi}}{2\sqrt{2}\alpha} = \frac{1}{\alpha}\sqrt{\frac{\pi}{2}} \ \Rightarrow \ energy \ signal$$

3.2. The Fourier Transform

The Fourier Transform (FT) of the signal $x(t)$ is

$$F\{x(t)\} = X(\omega) = \int_{-\infty}^{\infty} x(t)e^{-j\omega t} \ dt$$

Eq. (3.6)

$$F\{x(t)\} = X(f) = \int_{-\infty}^{\infty} x(t)e^{-j2\pi ft} \ dt$$

Eq. (3.7)

and the Inverse Fourier Transform (IFT) is

$$F^{-1}\{X(\omega)\} = x(t) = \frac{1}{2\pi} \int_{-\infty}^{\infty} X(\omega)e^{j\omega t} \ d\omega$$

Eq. (3.8)

$$F^{-1}\{X(f)\} = x(t) = \int_{-\infty}^{\infty} X(f)e^{j2\pi ft} \ df$$

Eq. (3.9)

where, in general, t represents time, while $\omega = 2\pi f$ and f represent frequency in radians per second and Hertz, respectively. In this book, we will use both notations for the transform, as appropriate (i.e., $X(\omega)$ or $X(f)$).

3.3. Systems Classification

Any system can mathematically be represented as a transformation (mapping) of an input signal into an output signal. This transformation or mapping relationship between the input signal $x(t)$ and the corresponding output signal $y(t)$ can be written as

$$y(t) = f[x(t); \ (-\infty < t < \infty)]. \qquad \text{Eq. (3.10)}$$

The relationship described in Eq. (3.10) can be linear or nonlinear, time invariant or time varying, causal or noncausal, and stable or nonstable systems. When the input signal is unit impulse (*Dirac delta function*) $\delta(t)$, the output signal is referred to as the system's impulse response $h(t)$.

3.3.1. Linear and Nonlinear Systems

A system is said to be linear if superposition holds true. More specifically, if

$$\begin{aligned} y_1(t) &= f[x_1(t)] \\ y_2(t) &= f[x_2(t)] \end{aligned} \qquad \text{Eq. (3.11)}$$

then for a linear system

$$f[ax_1(t) + bx_2(t)] = ay_1(t) + by_2(t) \qquad \text{Eq. (3.12)}$$

for any constants (a, b). If the relationship in Eq. (3.12) is not true, the system is said to be nonlinear.

3.3.2. Time Invariant and Time Varying Systems

A system is said to be time invariant (or shift invariant) if a time shift at its input produces the same time shift at its output. That is if

$$y(t) = f[x(t)] \qquad \text{Eq. (3.13)}$$

then

$$y(t - t_0) = f[x(t - t_0)]; -\infty < t_0 < \infty, \qquad \text{Eq. (3.14)}$$

If the above relationship is not true, the system is called a time varying system.

Any Linear Time Invariant (LTI) system can be described using the convolution integral between the input signal and the system's impulse response, as

$$y(t) = \int_{-\infty}^{\infty} x(t - u)h(u) \ du = x \otimes h \qquad \text{Eq. (3.15)}$$

where the operator $\otimes$ is used to symbolically describe the convolution integral. In the frequency domain, convolution translates into multiplication. That is

$$Y(f) = X(f)H(f).$$ **Eq. (3.16)**

$H(f)$ is the FT for $h(t)$ and it is referred to as the system transfer function.

3.3.3. Stable and Nonstable Systems

A system is said to be stable if every bounded input signal produces a bounded output signal. From Eq. (3.15)

$$|y(t)| = \left| \int_{-\infty}^{\infty} x(t-u)h(u) \ du \right| \le \int_{-\infty}^{\infty} |x(t-u)||h(u)| \ du.$$ **Eq. (3.17)**

If the input signal is bounded, then there is some finite constant K such that

$$|x(t)| \le K < \infty.$$ **Eq. (3.18)**

Therefore,

$$y(t) \le K \int_{-\infty}^{\infty} |h(u)| \ du$$ **Eq. (3.19)**

which can be finite if and only if

$$\int_{-\infty}^{\infty} |h(u)| \ du < \infty.$$ **Eq. (3.20)**

Thus, the requirement for stability is that the impulse response must be absolutely integrable. Otherwise, the system is said to be unstable.

3.3.4. Causal and Noncausal Systems

A causal (or physically realizable) system is one whose output signal does not begin before the input signal is applied. Thus, the following relationship is true when the system is causal:

$$y(t_0) = f[x(t); t \le t_0]; -\infty < t, t_0 < \infty.$$ **Eq. (3.21)**

A system that does not satisfy Eq. (3.21) is said to be noncausal which means it cannot exist in the real-world.

3.4. Signal Representation Using the Fourier Series

A set of functions $S = \{\varphi_n(t) \ ; \ n = 1, ..., N\}$ is said to be orthogonal over the interval (t_1, t_2) if and only if

$$\int_{t_1}^{t_2} \varphi_i^*(t)\varphi_j(t)dt = \int_{t_1}^{t_2} \varphi_i(t)\varphi_j^*(t)dt = \begin{cases} 0 & i \neq j \\ \lambda_i & i = j \end{cases}$$ **Eq. (3.22)**

where the asterisk indicates complex conjugation and λ_i are constants. If $\lambda_i = 1$ for all i, then the set S is said to be an orthonormal set. An electrical signal $x(t)$ can be expressed over the interval (t_1, t_2) as a weighted sum of a set of orthogonal functions as

$$x(t) \approx \sum_{n=1}^{N} X_n \varphi_n(t)$$

Eq. (3.23)

where X_n are, in general, complex constants and the orthogonal functions $\varphi_n(t)$ are called basis functions. If the integral-square error over the interval (t_1, t_2) is equal to zero as N approaches infinity, i.e.,

$$\lim_{N \to \infty} \int_{t_1}^{t_2} \left| x(t) - \sum_{n=1}^{N} X_n \varphi_n(t) \right|^2 dt = 0$$

Eq. (3.24)

then the set $S = \{\varphi_n(t)\}$ is said to be complete, and Eq. (3.23) becomes an equality. The constants X_n are computed as

$$X_n = \left(\int_{t_1}^{t_2} x(t) \varphi_n^*(t) dt \right) \bigg/ \left(\int_{t_1}^{t_2} |\varphi_n(t)|^2 dt \right).$$

Eq. (3.25)

Let the signal $x(t)$ be periodic with period T, and let the complete orthogonal set S be

$$S = \left\{ e^{\frac{j2\pi nt}{T}} \; ; \; n = -\infty, \infty \right\}.$$

Eq. (3.26)

Then the complex exponential Fourier series of $x(t)$ is

$$x(t) = \sum_{n=-\infty}^{\infty} X_n e^{\frac{j2\pi nt}{T}}.$$

Eq. (3.27)

Applying Eq. (3.25) yields

$$X_n = \frac{1}{T} \int_{-T/2}^{T/2} x(t) e^{\frac{-j2\pi nt}{T}} dt.$$

Eq. (3.28)

The FT of Eq. (3.27) is given by

$$X(\omega) = 2\pi \sum_{n=-\infty}^{\infty} X_n \delta\left(\omega - \frac{2\pi n}{T} \right)$$

Eq. (3.29)

where $\delta(\ )$ is delta function. When the signal $x(t)$ is real, we can compute its trigonometric Fourier series from Eq. (3.27) as

$$x(t) = a_0 + \sum_{n=1}^{\infty} a_n \cos\left(\frac{2\pi nt}{T} \right) + \sum_{n=1}^{\infty} b_n \sin\left(\frac{2\pi nt}{T} \right)$$

Eq. (3.30)

$$a_0 = X_0 \qquad\qquad\text{Eq. (3.31)}$$

$$a_n = \frac{1}{T} \int_{-T/2}^{T/2} x(t)\cos\left(\frac{2\pi nt}{T}\right) dt \qquad\qquad\text{Eq. (3.32)}$$

$$b_n = \frac{1}{T} \int_{-T/2}^{T/2} x(t)\sin\left(\frac{2\pi nt}{T}\right) dt . \qquad\qquad\text{Eq. (3.33)}$$

The coefficients a_n are all zeros when the signal $x(t)$ is an odd function of time. Alternatively, when the signal is an even function of time, then all b_n are equal to zero.

Consider the periodic energy signal defined in Eq. (3.30). The total energy associated with this signal is then given by

$$E = \frac{1}{T} \int_{t_0}^{t_0+T} |x(t)|^2 dt = \frac{a_0^2}{4} + \sum_{n=1}^{\infty}\left(\frac{a_n^2}{2} + \frac{b_n^2}{2}\right) . \qquad\qquad\text{Eq. (3.34)}$$

3.5. Convolution and Correlation Integrals

The convolution $\rho_{xh}(t)$ between the signals $x(t)$ and $h(t)$ is defined by

$$\rho_{xh}(t) = x(t) \otimes h(t) = \int_{-\infty}^{\infty} x(\tau)h(t-\tau)d\tau \qquad\qquad\text{Eq. (3.35)}$$

where τ is a dummy variable. Convolution is commutative, associative, and distributive. More precisely,

$$\begin{aligned}
x(t) \otimes h(t) &= h(t) \otimes x(t) \\
x(t) \otimes (h(t) \otimes g(t)) = (x(t) \otimes h(t)) \otimes g(t) &= x(t) \otimes (h(t) \otimes g(t))
\end{aligned} \qquad\text{Eq. (3.36)}$$

For the convolution integral to be finite at least one of the two signals must be an energy signal. The convolution between two signals can be computed using the FT:

$$\rho_{xh}(t) = F^{-1}\{X(\omega)H(\omega)\} . \qquad\qquad\text{Eq. (3.37)}$$

Consider an LTI system with impulse response $h(t)$ and input signal $x(t)$. It follows that the output signal $y(t)$ is equal to the convolution between the input signal and the system impulse response,

$$y(t) = \int_{-\infty}^{\infty} x(\tau)h(t-\tau)d\tau = \int_{-\infty}^{\infty} h(\tau)x(t-\tau)d\tau . \qquad\qquad\text{Eq. (3.38)}$$

The cross-correlation function between the signals $x(t)$ and $g(t)$ is

$$R_{xg}(t) = \int_{-\infty}^{\infty} x^*(\tau)g(t+\tau)d\tau = R^*_{gx}(-t) = \int_{-\infty}^{\infty} g^*(\tau)x(t+\tau)d\tau . \qquad\text{Eq. (3.39)}$$

Again, at least one of the two signals should be an energy signal for the correlation integral to be finite. The cross-correlation function measures the similarity between the two signals. The peak value of $R_{xg}(t)$ and its spread around this peak are an indication of how good this similarity is. This similarity is measured by a factor called *the correlation coefficient*, denoted by C_{xg}. For example, consider the signals $x(t)$ and $g(t)$, the correlation coefficient is

$$C_{xg} = \frac{\left| \int\limits_{-\infty}^{\infty} x(t)\ g^*(t)dt \right|^2}{\int\limits_{-\infty}^{\infty} |x(t)|^2 dt \int\limits_{-\infty}^{\infty} |g(t)|^2 dt} = C_{gx},$$

Eq. (3.40)

clearly the correlation coefficient is limited to $0 \le C_{xg} = C_{gx} \le 1$, with $C_{xg} = 0$ indicating no similarity while $C_{xg} = 1$ indicates 100% similarity between the signals $x(t)$ and $g(t)$.

The cross-correlation integral can be computed as

$$R_{xg}(t) = F^{-1}\{X^*(\omega)G(\omega)\},$$

Eq. (3.41)

When $x(t) = g(t)$, we get the autocorrelation integral,

$$R_x(t) = \int\limits_{-\infty}^{\infty} x^*(\tau)x(t+\tau)d\tau,$$

Eq. (3.42)

Note that the autocorrelation function is denoted by $R_x(t)$ rather than $R_{xx}(t)$. When the signals $x(t)$ and $g(t)$ are power signals, the correlation integral becomes infinite, and thus time averaging must be included. More precisely,

$$\bar{R}_{xg}(t) = \lim_{T \to \infty} \frac{1}{T} \int\limits_{-T/2}^{T/2} x^*(\tau)g(t+\tau)d\tau,$$

Eq. (3.43)

3.5.1. Energy and Power Spectrum Densities

Consider an energy signal $x(t)$. From Parseval's theorem, the total energy associated with this signal is

$$E = \int\limits_{-\infty}^{\infty} |x(t)|^2 dt = \frac{1}{2\pi} \int\limits_{-\infty}^{\infty} |X(\omega)|^2 d\omega,$$

Eq. (3.44)

When $x(t)$ is a voltage signal, the amount of energy dissipated by this signal when applied across a network of resistance R is

$$E = \frac{1}{R} \int\limits_{-\infty}^{\infty} |x(t)|^2 dt = \frac{1}{2\pi R} \int\limits_{-\infty}^{\infty} |X(\omega)|^2 d\omega,$$

Eq. (3.45)

Alternatively, when $x(t)$ is a current signal, we get

$$E = R \int_{-\infty}^{\infty} |x(t)|^2 dt = \frac{R}{2\pi} \int_{-\infty}^{\infty} |X(\omega)|^2 d\omega .$$ Eq. (3.46)

The quantity $\int |X(\omega)|^2 d\omega$ represents the amount of energy spread per unit frequency across a 1Ω resistor; therefore, the Energy Spectrum Density (ESD) function for the energy signal $x(t)$ is defined as

$$ESD = |X(\omega)|^2 .$$ Eq. (3.47)

The ESD at the output of an LTI system when $x(t)$ is at its input is

$$|Y(\omega)|^2 = |X(\omega)|^2 |H(\omega)|^2$$ Eq. (3.48)

where $H(\omega)$ is the FT of the system impulse response, $h(t)$. It follows that the energy present at the output of the system is

$$E_y = \frac{1}{2\pi} \int_{-\infty}^{\infty} |X(\omega)|^2 |H(\omega)|^2 d\omega .$$ Eq. (3.49)

Example:

The voltage signal $x(t) = e^{-5t}$; $t \geq 0$ is applied to the input of a lowpass LTI system. The system bandwidth is $5Hz$, and its input resistance is 5Ω. If $H(\omega) = 1$ over the interval $(-10\pi < \omega < 10\pi)$ and zero elsewhere, compute the energy at the output.

Solution:

From Eq. (2.49) one computes

$$E_y = \frac{1}{2\pi R} \int_{\omega = -10\pi}^{10\pi} |X(\omega)|^2 |H(\omega)|^2 d\omega .$$

Using Fourier transform tables and substituting $R = 5$ yields

$$E_y = \frac{1}{5\pi} \int_0^{10\pi} \frac{1}{\omega^2 + 25} d\omega .$$

Completing the integration yields

$$E_y = \frac{1}{25\pi} [\text{atanh}(2\pi) - \text{atanh}(0)] = 0.01799 \ Joules .$$

Note that an infinite bandwidth would give $E_y = 0.02$, only 11% larger.

The total power associated with a power signal $g(t)$ is

$$P = \lim_{T \to \infty} \frac{1}{T} \int_{-T/2}^{T/2} |g(t)|^2 dt .$$ Eq. (3.50)

The Power Spectrum Density (PSD) function for the signal $g(t)$ is $S_g(\omega)$, where

$$P = \lim_{T \to \infty} \frac{1}{T} \int_{-T/2}^{T/2} |g(t)|^2 dt = \frac{1}{2\pi} \int_{-\infty}^{\infty} S_g(\omega) d\omega .$$ Eq. (3.51)

It can be shown that

$$S_g(\omega) = \lim_{T \to \infty} \frac{|G(\omega)|^2}{T} .$$ Eq. (3.52)

Let the signals $x(t)$ and $g(t)$ be two periodic signals with period T. The complex exponential Fourier series expansions for those signals are, respectively, given by

$$x(t) = \sum_{n=-\infty}^{\infty} X_n e^{j\frac{2\pi n t}{T}}$$ Eq. (3.53)

$$g(t) = \sum_{m=-\infty}^{\infty} G_m e^{j\frac{2\pi m t}{T}} .$$ Eq. (3.54)

The power cross-correlation function $\bar{R}_{gx}(t)$ was given in Eq. (3.43) and is repeated here as Eq. (3.55),

$$\bar{R}_{gx}(t) = \frac{1}{T} \int_{-T/2}^{T/2} g^*(\tau) x(t + \tau) d\tau .$$ Eq. (3.55)

Note that since both signals are periodic the limit is no longer necessary in Eq. (3.55). Substituting Eqs. (3.53) and (2.54) into Eq. (3.55), collecting terms, and using the definition of orthogonality, yields

$$\bar{R}_{gx}(t) = \sum_{n=-\infty}^{\infty} G_n^* X_n e^{j\frac{2n\pi t}{T}} .$$ Eq. (3.56)

When $x(t) = g(t)$, Eq. (3.56) becomes the power autocorrelation function,

$$\bar{R}_x(t) = \sum_{n=-\infty}^{\infty} |X_n|^2 e^{j\frac{2n\pi t}{T}} = |X_0|^2 + 2 \sum_{n=1}^{\infty} |X_n|^2 e^{j\frac{2n\pi t}{T}} .$$ Eq. (3.57)

The power spectrum and cross-power spectrum density functions are then computed as the FT of Eqs. (3.57) and (3.56), respectively. More precisely,

$$\bar{S}_x(\omega) = 2\pi \sum_{n=-\infty}^{\infty} |X_n|^2 \delta\left(\omega - \frac{2n\pi}{T}\right)$$ Eq. (3.58a)

$$\bar{S}_{gx}(\omega) = 2\pi \sum_{n=-\infty}^{\infty} G_n^* X_n \delta\left(\omega - \frac{2n\pi}{T}\right) .$$ Eq. (3.58b)

The line (or discrete) power spectrum is defined as the plot of $|X_n|^2$ versus n, where the lines are $\Delta f = 1/T$ apart. The DC power is $|X_0|^2$, and the total power is $\displaystyle\sum_{n=-\infty}^{\infty} |X_n|^2$.

Consider a signal $x(t)$ and its FT $X(f)$. The corresponding autocorrelation function and power spectrum density are, respectively, $\bar{R}_x(t)$ and $\bar{S}_x(f)$. A few very useful relations that will be utilized often in this book include

$$x(0) = \int_{-\infty}^{\infty} X(f)\,df \qquad\qquad \text{Eq. (3.59)}$$

$$\int_{-\infty}^{\infty} x(t)\,dt = X(0) \qquad\qquad \text{Eq. (3.60)}$$

$$\bar{R}_x(0) = \int_{-\infty}^{\infty} |x(t)|^2\,dt = \int_{-\infty}^{\infty} |X(f)|^2\,df = \bar{S}_x(0) \qquad\qquad \text{Eq. (3.61)}$$

$$\int_{-\infty}^{\infty} |\bar{R}_x(t)|^2\,dt = \int_{-\infty}^{\infty} |X(f)|^4\,df. \qquad\qquad \text{Eq. (3.62)}$$

Note that Eq. (3.60) or Eq. (3.61) represents the total DC power (in the case of a power signal) or voltage (in the case of an energy signal). Equation (3.62) represents the signal's total power (for power signals) or total energy (for energy signals).

3.6. Bandpass Signals

Signals that contain significant frequency composition at a low frequency band including DC are called lowpass (LP) signals. Signals that have significant frequency composition around some frequency away from the origin are called bandpass (BP) signals. A real BP signal $x(t)$ can be represented mathematically by

$$x(t) = r(t)\cos(2\pi f_0 t + \phi_x(t)) \qquad\qquad \text{Eq. (3.63)}$$

where $r(t)$ is the amplitude modulation or envelope, $\phi_x(t)$ is the phase modulation, f_0 is the carrier frequency, and both $r(t)$ and $\phi_x(t)$ have frequency components significantly smaller than f_0. The frequency modulation is

$$f_m(t) = \frac{1}{2\pi}\frac{d}{dt}\phi_x(t) \qquad\qquad \text{Eq. (3.64)}$$

and the instantaneous frequency is

$$f_i(t) = \frac{1}{2\pi}\frac{d}{dt}(2\pi f_0 t + \phi_x(t)) = f_0 + f_m(t). \qquad\qquad \text{Eq. (3.65)}$$

If the signal bandwidth is B and f_0 is very large compared to B, then the signal $x(t)$ is referred to as a narrow bandpass signal.

Bandpass signals can also be represented by two lowpass signals known as the quadrature components; in this case Eq. (3.63) can be rewritten as

$$x(t) = x_I(t)\cos 2\pi f_0 t - x_Q(t)\sin 2\pi f_0 t \qquad \text{Eq. (3.66)}$$

where $x_I(t)$ and $x_Q(t)$ are real LP signals referred to as the quadrature components and are given, respectively, by

$$\begin{aligned} x_I(t) &= r(t)\cos\phi_x(t) \\ x_Q(t) &= r(t)\sin\phi_x(t) \end{aligned} \qquad \text{Eq. (3.67)}$$

3.6.1. The Analytic Signal (Pre-Envelope)

Given a real-valued signal $x(t)$, its Hilbert transform is

$$H\{x(t)\} = \hat{x}(t) = \frac{1}{\pi}\int_{-\infty}^{\infty}\frac{x(u)}{t-u}\,du \qquad \text{Eq. (3.68)}$$

Observation of Eq. (3.68) indicates that the Hilbert transform is computed as the convolution between the signals $x(t)$ and $h(t) = 1/(\pi t)$. More precisely,

$$\hat{x}(t) = x(t) \otimes \frac{1}{\pi t}. \qquad \text{Eq. (3.69)}$$

The Fourier transform of $h(t)$ is

$$FT\{h(t)\} = FT\left\{\frac{1}{\pi t}\right\} = H(\omega) = e^{-j\frac{\pi}{2}}\text{sgn}(\omega) \qquad \text{Eq. (3.70)}$$

where the function $\text{sgn}(\omega)$ is given by

$$\text{sgn}(\omega) = \frac{\omega}{|\omega|} = \begin{cases} 1 & ;\ \omega > 0 \\ 0; & \omega = 0 \\ -1 & ;\ \omega < 0 \end{cases}. \qquad \text{Eq. (3.71)}$$

Thus, the effect of the Hilbert transform is to introduce a phase shift of $\pi/2$ on the spectra of $x(t)$. It follows that,

$$FT\{\hat{x}(t)\} = \hat{X}(\omega) = X(\omega) - j\,\text{sgn}(\omega)X(\omega). \qquad \text{Eq. (3.72)}$$

The analytic signal $\psi(t)$ corresponding to the real signal $x(t)$ is obtained by canceling the negative frequency contents of $X(\omega)$. Then, by definition

$$\Psi(\omega) = \begin{cases} 2X(\omega) & ;\omega > 0 \\ X(\omega) & ;\omega = 0 \\ 0 & ;\omega < 0 \end{cases} \qquad \text{Eq. (3.73)}$$

or equivalently,

$$\Psi(\omega) = X(\omega)(1 + \text{sgn}(\omega)).$$

Eq. (3.74)

It follows that

$$\psi(t) = FT^{-1}\{\Psi(\omega)\} = x(t) + j\hat{x}(t).$$

Eq. (3.75)

The analytic signal is often referred to as the pre-envelope of $x(t)$ because the envelope of $x(t)$ can be obtained by simply taking the modulus of $\psi(t)$.

3.6.2. Pre-Envelope and Complex Envelope of Bandpass Signals

The Hilbert transform for the bandpass signal defined in Eq. (3.66) is

$$\hat{x}_{BP}(t) = x_I(t)\sin 2\pi f_0 t + x_Q(t)\cos 2\pi f_0 t.$$

Eq. (3.76)

The subscript BP is used to indicate that $x(t)$ is a bandpass signal. The corresponding bandpass analytic signal (pre-envelope) is then given by

$$\psi_{BP}(t) = x_{BP}(t) + j\hat{x}_{BP}(t)$$

Eq. (3.77)

using Eq. (3.66) and Eq. (3.76) into Eq. (3.77) and collecting terms yields

$$\psi_{BP}(t) = [x_I(t) + jx_Q(t)]e^{j2\pi f_0 t} = \tilde{x}_{BP}(t)e^{j2\pi f_0 t}.$$

Eq. (3.78)

The signal $\tilde{x}_{BP}(t) = x_I(t) + jx_Q(t)$ is the complex envelope of $x_{BP}(t)$. Thus, the envelope signal and associated phase deviation are given by

$$a(t) = |\tilde{x}_{BP}(t)| = |x_I(t) + jx_Q(t)| = |\psi_{BP}(t)|$$

Eq. (3.79)

$$\phi(t) = \arg(\tilde{x}_{BP}(t)) = \angle\tilde{x}_{BP}(t).$$

Eq. (3.80)

In the remainder of this text, unless it is indicated to be otherwise, all signals will be considered to be bandpass signals and consequently the subscript BP will not be used. More specifically, a bandpass signal $x(t)$ and its corresponding pre-envelope (analytic signal) and complex envelope will shown as

$$x(t) = x_I(t)\cos 2\pi f_0 t - x_Q(t)\sin 2\pi f_0 t$$

Eq. (3.81)

$$\psi(t) = x(t) + j\hat{x}(t) \equiv \tilde{x}(t)e^{j2\pi f_0 t}$$

Eq. (3.82)

$$\tilde{x}(t) = x_I(t) + jx_Q(t).$$

Eq. (3.83)

Obtaining the complex envelope for any bandpass signal requires extraction of the quadrature components. Figure 3.1 shows how the quadrature components can be extracted from a bandpass signal. First, the bandpass signal is split into two parts; one part is multiplied by $2\cos 2\pi f_0 t$ and the other is multiplied by $-2\sin 2\pi f_0 t$. From the figure, the two signals $z_1(t)$ and $z_2(t)$ are,

$$z_1(t) = 2x_I(t)(\cos 2\pi f_0 t)^2 - 2x_Q(t)\cos(2\pi f_0 t)\sin(2\pi f_0 t)$$

Eq. (3.84)

$$z_2(t) = -2x_I(t)\cos(2\pi f_0 t)\sin(2\pi f_0 t) + 2x_Q(t)(\sin 2\pi f_0 t)^2.$$

Eq. (3.85)

Utilizing the appropriate trigonometry identities and after lowpass filtering the quadrature components are extracted.

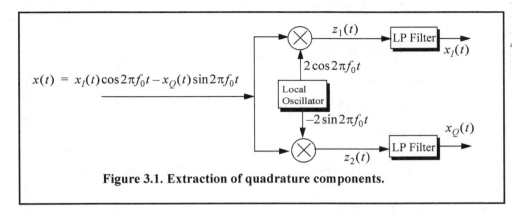

Figure 3.1. Extraction of quadrature components.

Example:

Extract the quadrature components, frequency modulation, instantaneous frequency, analytic signal, and complex envelope for the signals:

(a) $x(t) = Rect\left(\dfrac{t}{\tau}\right)\cos(2\pi f_0 t)$ *; (b)* $x(t) = Rect\left(\dfrac{t}{\tau}\right)\cos\left(2\pi f_0 t + \dfrac{\pi B}{\tau}t^2\right)$.

Solution:

(a) The quadrature components are extracted as described in Fig. 3.1. Define

$$z_1(t) = x(t) \times 2\cos(2\pi f_0 t), \quad z_2(t) = x(t) \times (-2)\sin(2\pi f_0 t),$$

then

$$z_1(t) = Rect\left(\frac{t}{\tau}\right)\cos(2\pi f_0 t) \times 2\cos(2\pi f_0 t) = Rect\left(\frac{t}{\tau}\right)\cos(0) + Rect\left(\frac{t}{\tau}\right)\cos(4\pi f_0 t)$$

$$z_2(t) = Rect\left(\frac{t}{\tau}\right)\cos(2\pi f_0 t) \times (-2)\sin(2\pi f_0 t) = Rect\left(\frac{t}{\tau}\right)\sin(0) - Rect\left(\frac{t}{\tau}\right)\sin(4\pi f_0 t).$$

Thus, the output of the LPFs are

$$x_I(t) = Rect\left(\frac{t}{\tau}\right) \quad ; \quad x_Q(t) = 0.$$

From Eq. (3.64) and Eq. (3.65) we get

$$f_m(t) = 0 \quad ; \quad f_i(t) = f_0.$$

Finally the complex envelope and the analytic signal are given by

$$\tilde{x}(t) = x_I(t) + jx_Q(t) = x_I(t) = Rect\left(\frac{t}{\tau}\right)$$

$$\psi(t) = \tilde{x}(t)e^{j2\pi f_0 t} = Rect\left(\frac{t}{\tau}\right)e^{j2\pi f_0 t}$$

(b)

$$z_1(t) = Rect\left(\frac{t}{\tau}\right)\cos\left(2\pi f_0 t + \frac{\pi B}{\tau}t^2\right) \times 2\cos(2\pi f_0 t)$$

which can be rewritten as

$$z_1(t) = Rect\left(\frac{t}{\tau}\right)\cos\left(\frac{\pi B}{\tau}t^2\right) + Rect\left(\frac{t}{\tau}\right)\cos\left(4\pi f_0 t + \frac{\pi B}{\tau}t^2\right)$$

and

$$z_2(t) = Rect\left(\frac{t}{\tau}\right)\cos\left(2\pi f_0 t + \frac{\pi B}{\tau}t^2\right) \times (-2)\sin(2\pi f_0 t),$$

which can be rewritten as

$$z_2(t) = Rect\left(\frac{t}{\tau}\right)\sin\left(\frac{\pi B}{\tau}t^2\right) - Rect\left(\frac{t}{\tau}\right)\sin\left(4\pi f_0 t + \frac{\pi B}{\tau}t^2\right).$$

Thus, the outputs of the LPFs are

$$x_I(t) = Rect\left(\frac{t}{\tau}\right)\cos\left(\frac{\pi B}{\tau}t^2\right) \quad ; \quad x_Q(t) = Rect\left(\frac{t}{\tau}\right)\sin\left(\frac{\pi B}{\tau}t^2\right).$$

From Eq. (3.64) and Eq.(3.65) we get

$$f_m(t) = \frac{B}{\tau}t \quad ; \quad f_i(t) = f_0 + \frac{B}{\tau}t.$$

The complex envelope is

$$\tilde{x}(t) = x_I(t) + jx_Q(t) = Rect\left(\frac{t}{\tau}\right)\cos\left(\frac{\pi B}{\tau}t^2\right) + jRect\left(\frac{t}{\tau}\right)\sin\left(\frac{\pi B}{\tau}t^2\right),$$

which can be written as

$$\tilde{x}(t) = Rect\left(\frac{t}{\tau}\right)e^{j\left(\frac{\pi B}{\tau}t^2\right)}.$$

Finally, the analytic signal is

$$\psi(t) = \tilde{x}(t)e^{j2\pi f_0 t} = Rect\left(\frac{t}{\tau}\right)e^{j\left(\frac{\pi B}{\tau}t^2\right)}e^{j2\pi f_0 t} = Rect\left(\frac{t}{\tau}\right)e^{j\left(2\pi f_0 t + \frac{\pi B}{\tau}t^2\right)}.$$

3.7. Spectra of a Few Common Radar Signals

The spectrum of a given signal describes the spread of its energy in the frequency domain. An energy signal (finite energy) can be characterized by its Energy Spectrum Density (ESD) function, while a power signal (finite power) is characterized by the Power Spectrum Density (PSD) function. The units of the ESD are Joules/Hertz and the PSD has units Watts/Hertz.

3.7.1. Continuous Wave Signal

Consider a Continuous Wave (CW) waveform given by

$$x_1(t) = \cos 2\pi f_0 t. \qquad\qquad \text{Eq. (3.86)}$$

The FT of $x_1(t)$ is

$$X_1(f) = \frac{1}{2}[\delta(f-f_0) + \delta(f+f_0)]. \qquad\qquad \text{Eq. (3.87)}$$

$\delta(\ )$ is the Dirac delta function. As indicated by the amplitude spectrum shown in Fig. 3.2, the signal $x_1(t)$ has infinitesimal bandwidth, located at $\pm f_0$.

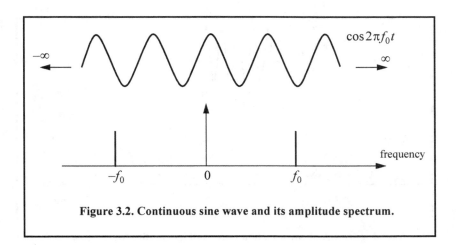

Figure 3.2. Continuous sine wave and its amplitude spectrum.

3.7.2. Finite Duration Pulse Signal

Consider the time-domain signal $x_2(t)$ given by

$$x_2(t) = x_1(t)Rect\left(\frac{t}{\tau_0}\right) = Rect\left(\frac{t}{\tau_0}\right)\cos 2\pi f_0 t \qquad \text{Eq. (3.88)}$$

$$Rect\left(\frac{t}{\tau_0}\right) = \begin{cases} 1 & -\dfrac{\tau_0}{2} \le t \le \dfrac{\tau_0}{2} \\ 0 & otherwise \end{cases} . \qquad \text{Eq. (3.89)}$$

The Fourier transform of the *Rect* function is

$$FT\left\{ Rect\left(\frac{t}{\tau_0}\right)\right\} = \tau_0 Sinc(f\tau_0) \qquad \text{Eq. (3.90)}$$

where

$$Sinc(u) = \frac{\sin(\pi u)}{\pi u} . \qquad \text{Eq. (3.91)}$$

It follows that the FT is

$$X_2(f) = X_1(f) \otimes \tau_0 Sinc(f\tau_0) = \frac{1}{2}[\delta(f-f_0) + \delta(f+f_0)] \otimes \tau_0 Sinc(f\tau_0), \qquad \text{Eq. (3.92)}$$

which can be written as

$$X_2(f) = \frac{\tau_0}{2}\{Sinc[(f-f_0)\tau_0] + Sinc[(f+f_0)\tau_0]\} . \qquad \text{Eq. (3.93)}$$

The amplitude spectrum of $x_2(t)$ is shown in Fig. 3.3. It is made up of two *Sinc* functions, as defined in Eq. (3.93), centered at $\pm f_0$.

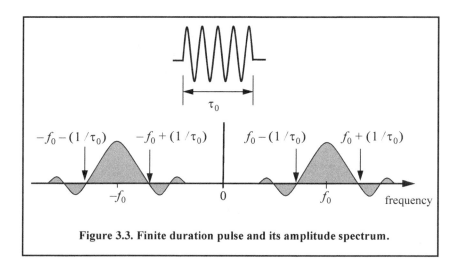

Figure 3.3. Finite duration pulse and its amplitude spectrum.

3.7.3. Periodic Pulse Signal

In this case, consider the coherent gated CW waveform $x_3(t)$ given by

$$x_3(t) = \sum_{n=-\infty}^{\infty} x_1(t) Rect\left(\frac{t-nT}{\tau_0}\right) = \cos 2\pi f_0 t \sum_{n=-\infty}^{\infty} Rect\left(\frac{t-nT}{\tau_0}\right).$$

Eq. (3.94)

The signal $x_3(t)$ is periodic, with period T (recall that $f_r = 1/T$ is the PRF), of course the condition $f_r \ll f_0$ is assumed. The FT of the signal $x_3(t)$ is

$$X_3(f) = X_1(f) \otimes FT\left\{ \sum_{n=-\infty}^{\infty} Rect\left(\frac{t-nT}{\tau_0}\right)\right\} =$$

Eq. (3.95)

$$\frac{1}{2}[\delta(f-f_0) + \delta(f+f_0)] \otimes FT\left\{ \sum_{n=-\infty}^{\infty} Rect\left(\frac{t-nT}{\tau_0}\right)\right\}$$

The complex exponential Fourier series of the summation inside Eq. (3.95) is

$$\sum_{n=-\infty}^{\infty} Rect\left(\frac{t-nT}{\tau_0}\right) = \sum_{n=-\infty}^{\infty} X_n e^{j\frac{nt}{T}}$$

Eq. (3.96)

where the Fourier series coefficients X_n are given by (see Eq. 3.28)

$$X_n = \frac{1}{T} FT\left\{ Rect\left(\frac{t}{\tau_0}\right)\right\}\bigg|_{f=\frac{n}{T}} = \frac{\tau_0}{T} Sinc(f\tau_0)\bigg|_{f=\frac{n}{T}} = \frac{\tau_0}{T} Sinc\left(\frac{n\tau_0}{T}\right).$$

Eq. (3.97)

It follows that

$$FT\left\{ \sum_{n=-\infty}^{\infty} X_n e^{j\frac{nt}{T}} \right\} = \left(\frac{\tau_0}{T}\right) \sum_{n=-\infty}^{\infty} Sinc(nf_r\tau_0)\delta(f - nf_r)$$ **Eq. (3.98)**

where the relation $f_r = 1/T$ was used. Substituting Eq. (3.98) into Eq. (3.95) yields the FT of $x_3(t)$. That is

$$X_3(f) = \frac{\tau_0}{2T}[\delta(f-f_0) + \delta(f+f_0)] \otimes \sum_{n=-\infty}^{\infty} Sinc(nf_r\tau_0)\delta(f - nf_r).$$ **Eq. (3.99)**

The amplitude spectrum of $x_3(t)$ has two parts centered at $\pm f_0$. The spectrum of the summation part is an infinite number of delta functions repeated every f_r, where the *nth* line is modulated in amplitude with the value corresponding to $Sinc(nf_r\tau_0)$. Therefore, the overall spectrum consists of an infinite number of lines separated by f_r and have $\sin u/u$ envelope that corresponds to X_n. This is illustrated in Fig. 3.4, for the positive portion of the spectrum only.

3.7.4. Finite Duration Pulse Train Signal

Define the function $x_4(t)$ as

$$x_4(t) = \cos(2\pi f_0 t) \sum_{n=0}^{N-1} Rect\left(\frac{t - nT}{\tau_0}\right) = \cos 2\pi f_0 t \times g(t)$$ **Eq. (3.100)**

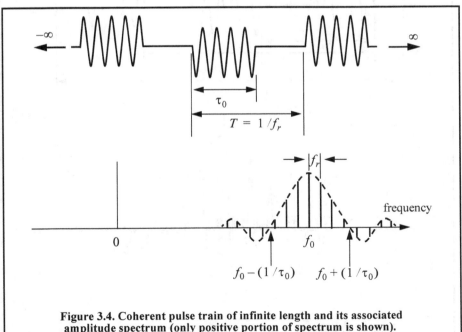

Figure 3.4. Coherent pulse train of infinite length and its associated amplitude spectrum (only positive portion of spectrum is shown).

where

$$g(t) = \sum_{n=0}^{N-1} Rect\left(\frac{t-nT}{\tau_0}\right).$$
Eq. (3.101)

The amplitude spectrum of the signal $x_4(t)$ is

$$X_4(f) = \frac{1}{2}G(f) \otimes [\delta(f-f_0) + \delta(f+f_0)]$$
Eq. (3.102)

where $G(f)$ is the FT of $g(t)$. This means that the amplitude spectrum of the signal $x_4(t)$ is equal to replicas of $G(f)$ centered at $\pm f_0$. Given this conclusion, one can then focus on computing $G(f)$.

The signal $g(t)$ can be written as (see top portion of Fig. 3.5)

$$g(t) = \sum_{n=-\infty}^{\infty} g_1(t)Rect\left(\frac{t-nT}{\tau_0}\right)$$
Eq. (3.103)

where

$$g_1(t) = Rect\left(\frac{t}{NT}\right).$$
Eq. (3.104)

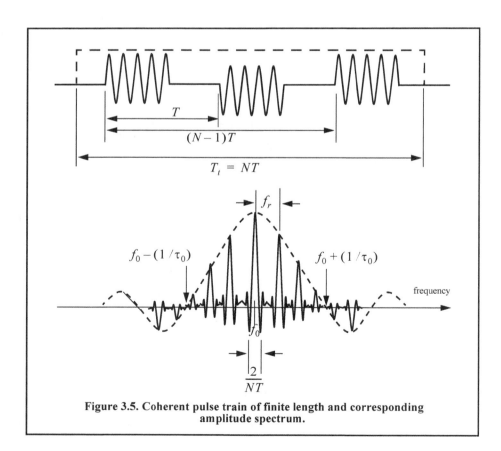

Figure 3.5. Coherent pulse train of finite length and corresponding amplitude spectrum.

It follows that the FT of Eq. (3.103) can be computed using analysis similar to that which led to Eq. (3.99). More precisely,

$$G(f) = \frac{\tau_0}{T} G_1(f) \otimes \sum_{n=-\infty}^{\infty} Sinc(nf_r\tau_0)\delta(f-nf_r) \qquad \text{Eq. (3.105)}$$

and the FT of $g_1(t)$ is

$$G_1(f) = FT\left\{ Rect\left(\frac{t}{T_t}\right) \right\} = T_t Sinc(fT_t). \qquad \text{Eq. (3.106)}$$

Using these results, the FT of $x_4(t)$ can be written as

$$X_4(f) = \frac{T_t \tau_0}{2T}\left(Sinc(fT_t) \otimes \sum_{n=-\infty}^{\infty} Sinc(nf_r\tau_0)\delta(f-nf_r) \right) \otimes [\delta(f-f_0) + \delta(f+f_0)]. \qquad \text{Eq. (3.107)}$$

Therefore, the overall spectrum of $x_4(t)$ consists of a two equal positive and negative portions, centered at $\pm f_0$. Each portion is made up of N $Sinc(fT_t)$ functions repeated every f_r with envelope corresponding to $Sinc(nf_r\tau_0)$. This is illustrated in Fig. 3.5; only the positive portion of the spectrum is shown.

3.7.5. Linear Frequency Modulation (LFM) Signal

Frequency or phase modulated signals can be used to achieve much wider operating bandwidths. Linear Frequency Modulation (LFM) is very commonly used in most modern radar systems. In this case, the frequency is swept linearly across the pulse width, either upward (up-chirp) or downward (down-chirp). Figure 3.6 shows a typical example of an LFM waveform. The pulse width is τ_0, and the bandwidth is B.

The LFM up-chirp instantaneous phase can be expressed by

$$\phi(t) = 2\pi\left(f_0 t + \frac{\mu}{2}t^2\right) \qquad -\frac{\tau_0}{2} \leq t \leq \frac{\tau_0}{2}, \qquad \text{Eq. (3.108)}$$

where f_0 is the radar center frequency, and $\mu = B/\tau_0$ is the LFM coefficient. Thus, the instantaneous frequency is

$$f(t) = \frac{1}{2\pi}\frac{d}{dt}\phi(t) = f_0 + \mu t \qquad -\frac{\tau_0}{2} \leq t \leq \frac{\tau_0}{2}. \qquad \text{Eq. (3.109)}$$

Similarly, the down-chirp instantaneous phase and frequency are given, respectively, by

$$\phi(t) = 2\pi\left(f_0 t - \frac{\mu}{2}t^2\right) \qquad -\frac{\tau_0}{2} \leq t \leq \frac{\tau_0}{2} \qquad \text{Eq. (3.110)}$$

$$f(t) = \frac{1}{2\pi}\frac{d}{dt}\phi(t) = f_0 - \mu t \qquad -\frac{\tau_0}{2} \leq t \leq \frac{\tau_0}{2}. \qquad \text{Eq. (3.111)}$$

A typical LFM waveform can be expressed by

$$x_1(t) = Rect\left(\frac{t}{\tau_0}\right)e^{j2\pi\left(f_0 t + \frac{\mu}{2}t^2\right)}$$

Eq. (3.112)

where $Rect(t/\tau_0)$ denotes a rectangular pulse of width τ_0. Remember that the signal $x_1(t)$ is the analytic signal for the LMF waveform. It follows that

$$x_1(t) = \tilde{x}(t)e^{j2\pi f_0 t}$$

Eq. (3.113)

$$\tilde{x}(t) = Rect\left(\frac{t}{\tau}\right)e^{j\pi\mu t^2}.$$

Eq. (3.114)

The spectrum of the signal $x_1(t)$ is determined from its complex envelope $\tilde{x}(t)$. The complex exponential term in Eq. (3.114) introduces a frequency shift about the center frequency f_o. Taking the FT of $\tilde{x}(t)$ yields

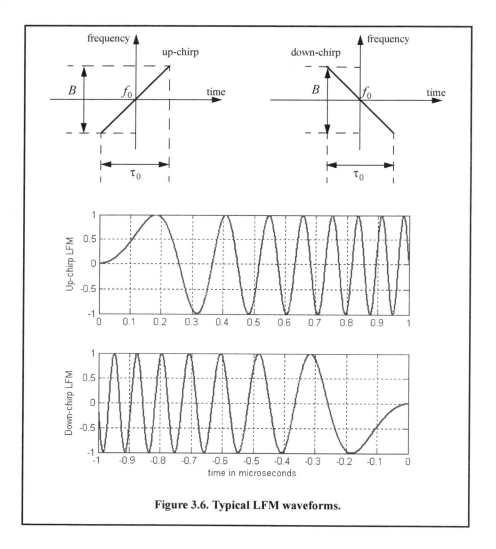

Figure 3.6. Typical LFM waveforms.

$$\tilde{X}(f) = \int_{-\infty}^{\infty} Rect\left(\frac{t}{\tau_0}\right) e^{j\pi\mu t^2} e^{-j2\pi ft} dt = \int_{-\frac{\tau_0}{2}}^{\frac{\tau_0}{2}} e^{j\pi\mu t^2} e^{-j2\pi ft} dt .$$

Eq. (3.115)

Let $\mu' = \pi\mu = \pi B/\tau_0$, and perform the change of variable

$$\left(z = \sqrt{\frac{2}{\pi}}\left(\sqrt{\mu'}t - \frac{\pi f}{\sqrt{\mu'}}\right)\right) \quad ; \quad \sqrt{\frac{\pi}{2\mu'}} dz = dt .$$

Eq. (3.116)

Thus, Eq. (3.115) can be written as

$$\tilde{X}(f) = \sqrt{\frac{\pi}{2\mu'}} e^{-j(\pi f)^2/\mu'} \int_{-z_1}^{z_2} e^{j\pi z^2/2} dz$$

Eq. (3.117)

$$\tilde{X}(f) = \sqrt{\frac{\pi}{2\mu'}} e^{-j(\pi f)^2/\mu'} \left\{ \int_0^{z_2} e^{j\pi z^2/2} dz - \int_0^{-z_1} e^{j\pi z^2/2} dz \right\}$$

Eq. (3.118)

$$z_1 = -\sqrt{\frac{2\mu'}{\pi}}\left(\frac{\tau_0}{2} + \frac{\pi f}{\mu'}\right) = \sqrt{\frac{B\tau_0}{2}}\left(1 + \frac{f}{B/2}\right)$$

Eq. (3.119)

$$z_2 = \sqrt{\frac{\mu'}{\pi}}\left(\frac{\tau_0}{2} - \frac{\omega}{\mu'}\right) = \sqrt{\frac{B\tau_0}{2}}\left(1 - \frac{f}{B/2}\right).$$

Eq. (3.120)

The Fresnel integrals, denoted by $C(z)$ and $S(z)$, are defined by

$$C(z) = \int_0^z \cos\left(\frac{\pi \upsilon^2}{2}\right) d\upsilon \text{ and } S(z) = \int_0^z \sin\left(\frac{\pi \upsilon^2}{2}\right) d\upsilon .$$

Eq. (3.121)

Fresnel integrals can be approximated by

$$C(z) \approx \frac{1}{2} + \frac{1}{\pi z}\sin\left(\frac{\pi}{2}z^2\right) \quad ; \quad z \gg 1$$

Eq. (3.122)

$$S(z) \approx \frac{1}{2} - \frac{1}{\pi z}\cos\left(\frac{\pi}{2}z^2\right) \quad ; \quad z \gg 1 .$$

Eq. (3.123)

Note that $C(-z) = -C(z)$ and $S(-z) = -S(z)$. Figure 3.7 shows a plot of both $C(z)$ and $S(z)$ for $0 \le z \le 4.0$. Using Eq. (3.121) into Eq. (3.118) and performing the integration yield

$$\tilde{X}(f) = \sqrt{\frac{\pi}{2\mu'}} e^{-j(\pi f)^2/(\mu')} \{[C(z_2) + C(z_1)] + j[S(z_2) + S(z_1)]\} .$$

Eq. (3.124)

Figure 3.8 shows typical plots for the LFM real part, imaginary part, and amplitude spectrum. The square-like spectrum shown in Fig. 3.8c is widely known as the Fresnel spectrum. Figure 3.8 can be reproduced using MATLAB program *"Fig3_8.m"* listed in Appendix 3-A.

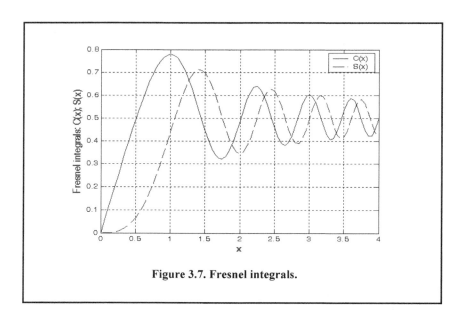

Figure 3.7. Fresnel integrals.

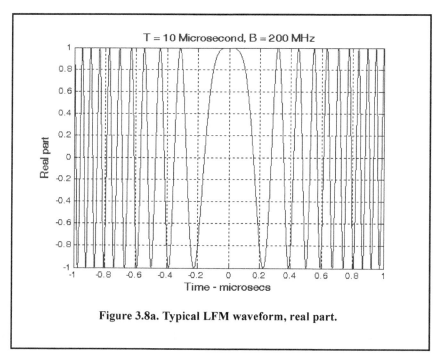

Figure 3.8a. Typical LFM waveform, real part.

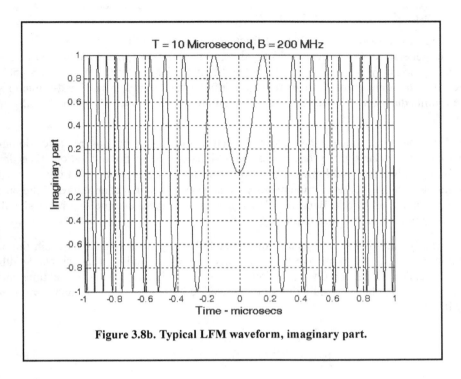

Figure 3.8b. Typical LFM waveform, imaginary part.

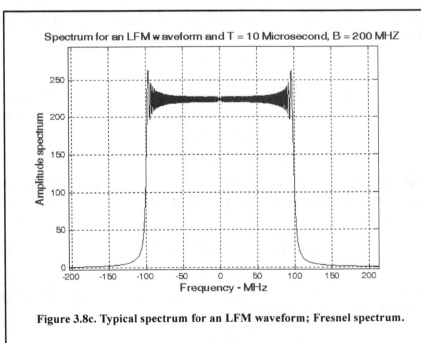

Figure 3.8c. Typical spectrum for an LFM waveform; Fresnel spectrum.

3.8. Signal Bandwidth and Duration

The signal bandwidth is the range of frequency over which the signal has a nonzero spectrum. In general, any signal can be defined using its duration (time domain) and bandwidth (frequency domain). A signal is said to be band-limited if it has finite bandwidth. Signals that have finite durations (time-limited) will have infinite bandwidths, while band-limited signals have infinite durations. The extreme case is a continuous sine-wave, whose bandwidth is infinitesimal.

Radar signal processing can be performed in either the time domain or frequency domain. In either case, the radar signal processor assumes signals to be of finite duration (time-limited) and finite bandwidth (band-limited). The trouble with this assumption is that time-limited and band-limited signals cannot simultaneously exist. That is, a signal cannot have finite duration and have finite bandwidth. Because of this, it is customary to assume that radar signals are essentially limited in time and frequency.

Essentially time-limited signals are considered to be very small outside a certain finite time duration. If the FT of a signal is very small outside a certain finite frequency bandwidth, the signal is called an essentially band-limited signal. A signal $x(t)$ over the time interval $\{T_1, T_2\}$ is said to be essentially time-limited relative to some very small signal level ε if and only if

$$\int_{T_1}^{T_2} |x(t)|^2 dt \geq (1-\varepsilon) \int_{-\infty}^{\infty} |x(t)|^2 dt \qquad \textbf{Eq. (3.125)}$$

where the interval $\tau_e = T_2 - T_1$ is called the effective duration. The effective duration is defined as

$$\tau_e = \frac{\left(\int_{-\infty}^{\infty} |x(t)|^2 dt\right)^2}{\int_{-\infty}^{\infty} |x(t)|^4 dt}. \qquad \textbf{Eq. (3.126)}$$

Similarly, a signal $x(t)$ over the frequency interval $\{B_1, B_2\}$ is said to be essentially band-limited relative to some small signal level η if and only if

$$\int_{B_1}^{B_2} |X(f)|^2 df \geq (1-\eta) \int_{-\infty}^{\infty} |X(f)|^2 df \qquad \textbf{Eq. (3.127)}$$

where $X(f)$ is the FT of $x(t)$ and the band $B_e = B_2 - B_1$ is called the effective bandwidth. The effective bandwidth is defined as

$$B_e = \left(\int_{-\infty}^{\infty} |X(f)|^2 df\right)^2 \bigg/ \left(\int_{-\infty}^{\infty} |X(f)|^4 df\right). \qquad \textbf{Eq. (3.128)}$$

Different, but equivalent, definitions for the effective bandwidth and effective duration can be found in the literature. In this book, the definitions cited in Burdic[1] are adopted. The quantity $B_e \tau_e$ is referred to as the time bandwidth product. In later chapters, it will be clear that large time bandwidth products are desirable in radar applications since they provide better pulse compression ratios (or compression gain).

Range resolution is defined as the reciprocal of the effective bandwidth. In Chapter 1, prior to introducing the concept of effective duration, the bandwidth was computed as the reciprocal of the pulse width, an approximation that is widely used and accepted, even though it is not quite 100% accurate. This is true since using one value or the other for the bandwidth does not make much difference in the overall calculation of the SNR when using the radar equation. Doppler resolution is computed as the reciprocal of the effective duration.

3.8.1. Effective Bandwidth and Duration Calculation

A few examples for computing the effective bandwidth and duration of most common radar signals are presented in this section.

Single Pulse

The single pulse was analyzed in the previous section. Consider the single pulse waveform given by

$$x(t) = Rect\left(\frac{t}{\tau_0}\right) \qquad ; \; \frac{-\tau_0}{2} < 0 < \frac{\tau_0}{2}.$$

Eq. (3.129)

The effective bandwidth for this signal can be computed using Eq. (3.128). For this purpose, the denominator of Eq. (3.128) is

$$\int_{-\infty}^{\infty} |X(f)|^4 df = \int_{-\infty}^{\infty} |R_x(\tau)|^4 d\tau = \int_{-\infty}^{\infty} |\tau_0 Sinc(f\tau_0)|^4 df = \frac{2\tau_0^3}{3}$$

Eq. (3.130)

and its numerator is computed utilizing Eq. (3.61) as

$$\left(\int_{-\infty}^{\infty} |X(f)|^2 df\right)^2 = |R_x(0)|^2 = \tau_0^2.$$

Eq. (3.131)

Note that this value represents the square of the signal total energy. Therefore, the effective bandwidth is

$$B_e = \frac{\left(\int_{-\infty}^{\infty} |X(f)|^2 df\right)^2}{\int_{-\infty}^{\infty} |X(f)|^4 df} = \frac{(\tau_0^2)}{\left(\frac{2\tau_0^3}{3}\right)} = \frac{3}{2\tau_0}.$$

Eq. (3.132)

1. Burdic, W. S., *Radar Signal Analysis,* Prentice-Hall, Englewood Cliffs, NJ, 1968.

The effective duration for the signal $x_2(t)$ is

$$\tau_e = \frac{\left(\int\limits_{-\infty}^{\infty} |x(t)|^2 dt\right)^2}{\int\limits_{-\infty}^{\infty} |x(t)|^4 dt} \qquad\qquad \text{Eq. (3.133)}$$

$$\tau_e = \frac{\left(\int\limits_{-\tau_0/2}^{\tau_0/2} (1)^2 dt\right)^2}{\int\limits_{-\tau_0/2}^{\tau_0/2} (1)^4 dt} = \frac{\tau_0^2}{\tau_0} = \tau_0 . \qquad\qquad \text{Eq. (3.134)}$$

Finally, the time bandwidth product for this signal is

$$B_e \tau_e = \frac{3}{2\tau_0} \tau_0 = \frac{3}{2} . \qquad\qquad \text{Eq. (3.135)}$$

Finite Duration Pulse Train Signal

The finite duration train signal was defined in the previous section; its complex envelope is given by

$$x(t) = Rect\left(\frac{t}{NT}\right) \sum_{n=-\infty}^{\infty} Rect\left(\frac{t-nT}{\tau_0}\right) . \qquad\qquad \text{Eq. (3.136)}$$

The corresponding FT is

$$X(f) = \frac{T_t \tau_0}{T} Sinc(fT_t) \otimes \sum_{n=-\infty}^{\infty} Sinc(nf_r\tau_0)\delta(f-nf_r) . \qquad\qquad \text{Eq. (3.137)}$$

The total energy for this signal is

$$\int\limits_{-\infty}^{\infty} |X(f)|^2 df = \frac{T_t \tau_0}{T} . \qquad\qquad \text{Eq. (3.138)}$$

It can be shown (see Problem 3.19) that

$$\int\limits_{-\infty}^{\infty} |R_x(t)|^2 dt = \int\limits_{-\infty}^{\infty} |X(f)|^4 df \approx \left(\frac{4}{3}\right)\left(\frac{T_t}{T}\right)^3 \left(\frac{2}{3}\right)(\tau_0)^3 . \qquad\qquad \text{Eq. (3.139)}$$

It follows that the effective bandwidth is

$$B_e \approx \frac{\left(\frac{T_t \tau_0}{T}\right)^2}{\left(\frac{4}{3}\right)\left(\frac{T_t}{T}\right)^3 \left(\frac{2}{3}\right)(\tau_0)^3} = \left(\frac{3T}{4T_t}\right)\left(\frac{3}{2\tau_0}\right).$$

Eq. (3.140)

The result of Eq. (3.140) clearly indicates that the effective bandwidth of the pulse train decreases as the length of the train is increased. This should intuitively make a lot of sense, since the bandwidth is inversely proportional to signal duration. Of course, when $T_t = T$ (i.e., single pulse case) Eq. (3.140) becomes identical to Eq. (3.132); note that in this case the factor $3/4$ would not have been present in Eq. (3.140).

The effective duration of this signal can be computed using Eq. (3.126). Again, the numerator of Eq. (3.126) represents the square of the total signal energy given in Eq. (3.44). In this case, the denominator of Eq. (3.126) is equal to unity (see Problem 3.20). Thus, the effective duration is

$$\tau_e = \frac{T_t \tau_0}{T}$$

Eq. (3.141)

and the time bandwidth product of this waveform is

$$B_e \tau_e \approx \left(\frac{3T}{4T_t}\right)\left(\frac{3}{2\tau_0}\right)\left(\frac{T_t \tau_0}{T}\right) = \frac{9}{8}.$$

Eq. (3.142)

LFM Signal

In this case, the LFM complex envelope can be written as

$$x(t) = Rect\left(\frac{t}{\tau_0}\right) e^{j\mu\pi t^2}$$

Eq. (3.143)

where $\mu = B/\tau_0$ and B is the LFM bandwidth. Make a change of variables $\mu' = \pi\mu$, then the modulus of the FT of this signal can be approximated as

$$|X(f)| \approx \sqrt{\frac{\pi}{\mu'}} \; Rect\left(\frac{\pi f}{\mu' \tau_0}\right).$$

Eq. (3.144)

The FT of the autocorrelation function is equal to the square of the modulus of the signal FT, i.e.,

$$FT\{R_x(\tau)\} = |X(f)|^2 = \frac{\pi}{\mu'} Rect\left(\frac{\pi f}{\mu' \tau_0}\right).$$

Eq. (3.145)

Therefore,

$$\left(\int_{-\infty}^{\infty} |X(f)|^2 \, df\right)^2 \approx \tau_0^2$$

Eq. (3.146)

also

$$\int_{-\infty}^{\infty} |X(f)|^4 df \approx \frac{\pi \tau_0}{\mu'}.$$ **Eq. (3.147)**

Then the effective bandwidth is

$$B_e \approx \frac{\tau_0^2}{\frac{\pi \tau_0}{\mu'}} = \frac{\mu' \tau_0}{\pi}.$$ **Eq. (3.148)**

The effective duration is

$$\tau_e = \frac{\left(\int_{-\infty}^{\infty} |x(t)|^2 dt\right)^2}{\int_{-\infty}^{\infty} |x(t)|^4 dt} = \frac{\left(\int_{-\tau_0/2}^{\tau_0/2} (1)^2 dt\right)^2}{\int_{-\tau_0/2}^{\tau_0/2} (1)^4 dt} = \frac{\tau_0^2}{\tau_0} = \tau_0.$$ **Eq. (3.149)**

And the time bandwidth product for LFM waveforms is computed as

$$B_e \tau_e \approx \frac{\mu' \tau_0}{\pi} \tau_0 = \frac{\mu' \tau_0^2}{\pi} = \frac{\pi \mu \tau_0^2}{\pi} = \frac{B \tau_0^2}{\tau_0} = B \tau_0.$$ **Eq. (3.150)**

3.9. Discrete Time Systems and Signals

Advances in computer hardware and in digital technologies completely revolutionized radar systems signal and data processing techniques. Virtually all modern radar systems use some form of a digital representation (signal samples) of their received signals for the purposes of signal and data processing. These samples of a time-limited signal are nothing more than a finite set of numbers (thought of as a vector) that represents discrete values of the continuous time domain signal. These samples are typically obtained by using Analog-to-Digital (A/D) conversion devices. Since in the digital world the radar receiver is now concerned with processing a set of finite numbers, its impulse response will also compose a set of finite numbers. Consequently, the radar receiver is now referred to as a discrete system. All input/output signal relationships are now carried out using discrete time samples. It must also be noted that just as in the case of continuous time-domain systems, the discrete systems of interest to radar applications must also be causal, stable, and linear time invariant.

Consider a continuous lowpass signal that is essentially time-limited with duration τ and band-limited with bandwidth B. This signal (as will be shown in the next section) can be completely represented by a set of $\{2\tau B\}$ samples. Since a finite set of discrete values (samples) is used to represent the signal, it is common to represent this signal by a finite dimensional vector of the same size. This vector is denoted by $\mathbf{x}$, or simply by the sequence $x[n]$,

$$\mathbf{x} \equiv x[n] = [x(0) \ x(1) \ \dots x(N-2) \ x(N-1)]^t$$ **Eq. (3.151)**

where the superscript t denotes transpose operation. The value N is at least $2\tau B$ for a real lowpass essentially limited signal $x(t)$ of duration τ and bandwidth B. If, however, the signal is complex, then N is at least τB and the components of the vector **x** are complex. The samples defined in Eq. (3.151) can be obtained from pulse-to-pulse samples at a fixed range (i.e., delay) of the radar echo signal. The PRF is denoted by f_r and the total observation interval is T_0; then N would be equal to $T_0 f_r$. Define the radar receiver transfer function as the discrete sequence $h[n]$ and the input signal sequence as $x[n]$; then the output sequence $y[n]$ is given by the convolution sum

$$y[n] = \sum_{m=0}^{M-1} h(m)x(n-m)$$

Eq. (3.152)

where $\{h[n] = [h(0)\ h(1)\ ...h(M-2)\ h(M-1)];\ M \leq N\}$.

3.9.1. Sampling Theorem

Lowpass Sampling Theorem

In general, it is required to determine the necessary condition such that a signal can be fully reconstructed from its samples by filtering, or data processing in general. The answer to this question lies in the sampling theorem, which may be stated as follows: let the signal $x(t)$ be real-valued, essentially band-limited by the bandwidth B; this signal can be fully reconstructed from its samples if the time interval between samples is no greater than $1/(2B)$. Figure 3.9 illustrates the sampling process concept. The sampling signal $p(t)$ is periodic with period T_s, which is called the sampling interval.

The Fourier series expansion of $p(t)$ and the sampled signal $x_s(t)$ expressed using this Fourier series definition are, respectively, given by

$$p(t) = \sum_{n=-\infty}^{\infty} P_n e^{j\frac{2\pi nt}{T_s}}$$

Eq. (3.153)

$$x_s(t) = p(t) \cdot x(t)$$

Eq. (3.154)

$$x_s(t) = \sum_{n=-\infty}^{\infty} x(t)P_n e^{j\frac{2\pi nt}{T_s}} .$$

Eq. (3.155)

Taking the FT of Eq. (3.155) yields

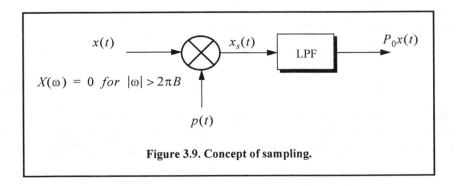

Figure 3.9. Concept of sampling.

$$X_s(\omega) = \sum_{n=-\infty}^{\infty} P_n X\left(\omega - \frac{2\pi n}{T_s}\right) = P_0 X(\omega) + \sum_{\substack{n=-\infty \\ n \neq 0}}^{\infty} P_n X\left(\omega - \frac{2\pi n}{T_s}\right) \qquad \text{Eq. (3.156)}$$

where $X(\omega)$ is the FT of $x(t)$. Therefore, we conclude that the spectral density, $X_s(\omega)$, consists of replicas of $X(\omega)$ spaced $(2\pi / T_s)$ apart and scaled by the Fourier series coefficients P_n. A lowpass filter (LPF) of bandwidth B can then be used to recover the original signal $x(t)$.

When the sampling rate is increased (i.e., T_s decreases), the replicas of $X(\omega)$ move farther apart. Alternatively, when the sampling rate is decreased (i.e., T_s increases), the replicas get closer to one another. The value of T_s such that the replicas are tangent to one another defines the minimum required sampling rate so that $x(t)$ can be recovered from its samples by using an LPF. It follows that

$$\frac{2\pi}{T_s} = 2\pi(2B) \Leftrightarrow T_s = \frac{1}{2B}. \qquad \text{Eq. (3.157)}$$

The sampling rate defined by Eq. (3.157) is known as the Nyquist sampling rate. When $T_s > (1/2B)$, the replicas of $X(\omega)$ overlap, and thus $x(t)$ cannot be recovered cleanly from its samples. This is known as aliasing. In practice, ideal LPF cannot be implemented; hence, practical systems tend to oversample in order to avoid aliasing.

Example:

Assume that the sampling signal $p(t)$ is given by $p(t) = \sum_{n=-\infty}^{\infty} \delta(t - nT_s)$.

Compute an expression for $X_s(\omega)$.

Solution:

The signal $p(t)$ is called the Comb function, with exponential Fourier series

$$p(t) = \sum_{n=-\infty}^{\infty} \frac{1}{T_s} e^{2\frac{\pi n t}{T_s}}.$$

It follows that

$$x_s(t) = \sum_{n=-\infty}^{\infty} x(t) \frac{1}{T_s} e^{2\frac{\pi n t}{T_s}}.$$

Taking the Fourier transform of this equation yields

$$X_s(\omega) = \frac{2\pi}{T_s} \sum_{n=-\infty}^{\infty} X\left(\omega - \frac{2\pi n}{T_s}\right).$$

It is desired to develop a general expression from which any lowpass signal can be recovered from its samples, provided that Eq. (3.157) is satisfied. In order to do that, let $x(t)$ and $x_s(t)$ be the desired lowpass signal and its corresponding samples, respectively. Then an expression for $x(t)$ in terms of its samples can be derived as follows: First, obtain $X(\omega)$ by filtering the signal $X_s(\omega)$ using an ideal LPF whose transfer function is

$$H(\omega) = T_s Rect\left(\frac{\omega}{4\pi B}\right).$$
Eq. (3.158)

Thus,

$$X(\omega) = H(\omega)X_s(\omega) = T_s Rect\left(\frac{\omega}{4\pi B}\right)X_s(\omega).$$
Eq. (3.159)

The signal $x(t)$ is now obtained from the inverse FT of Eq. (3.159) as

$$x(t) = FT^{-1}\{X(\omega)\} = FT^{-1}\left\{T_s Rect\left(\frac{\omega}{4\pi B}\right)X_s(\omega)\right\} = 2BT_s Sinc(2\pi Bt) \otimes x_s(t).$$
Eq. (3.160)

The sampled signal $x_s(t)$ can be represented using an ideal sampling signal

$$p(t) = \sum_n \delta(t - nT_s)$$
Eq. (3.161)

thus,

$$x_s(t) = \sum_n x(nT_s)\delta(t - nT_s).$$
Eq. (3.162)

Substituting Eq. (3.62) into Eq. (3.160) yields an expression for the signal $x(t)$ in terms of its samples

$$x(t) = 2BT_s \sum_n x(nT_s)\ Sinc(2\pi B(t - T_s))\ \ ; T_s \leq \frac{1}{2B}$$
Eq. (3.163)

Bandpass Sampling Theorem

It was established in Section 3.6 that any bandpass signal can be expressed using the quadrature components. It follows that it is sufficient to construct the bandpass signal $x(t)$ from samples of the quadrature components $\{x_I(t), x_Q(t)\}$. Let the signal $x(t)$ be essentially band-limited with bandwidth B, then each of the lowpass signals $x_I(t)$ and $x_Q(t)$ are also band-limited each with bandwidth $B/2$. Hence, if either of these lowpass signal sis sampled at a rate $f_s \leq 1/B$, then the Nyquist criterion is not violated. Assume that both quadrature components are sampled synchronously, that is

$$x_I(t) = BT_s \sum_{n=-\infty}^{\infty} x_I(nT_s)\ Sinc(\pi B(t - nT_s))$$
Eq. (3.164)

$$x_Q(t) = BT_s \sum_{n=-\infty}^{\infty} x_Q(nT_s)\ Sinc(\pi B(t - nT_s))$$
Eq. (3.165)

where if the Nyquist rate is satisfied, then $BT_s = 1$ (unity time bandwidth product). Substituting Eq. (3.164) and Eq. (3.165) into Eq. (3.66) yields

$$x(t) = BT_s \left\{ \sum_{n=-\infty}^{\infty} [x_I(nT_s)\cos 2\pi f_0 t - x_Q(nT_s)\sin 2\pi f_0 t] \ Sinc(\pi B(t - nT_s)) \right\} \quad \textbf{Eq. (3.166a)}$$

$$x(t) = Re\left\{ BT_s \sum_{n=-\infty}^{\infty} [x_I(nT_s) + jx_Q(nT_s)]e^{j2\pi f_0 t} Sinc(\pi B(t - nT_s)) \right\} \quad \textbf{Eq. (3.166b)}$$

where, of course, $T_s \leq 1/B$ is assumed. This leads to the conclusion that if the total period over which the signal $x(t)$ is sampled is T_0, then $2BT_0$ samples are required, BT_0 samples for $x_I(t)$ and BT_0 samples for $x_Q(t)$.

3.9.2. The Z-Transform

The Z-transform is a transformation that maps samples of a discrete time-domain sequence into a new domain known as the z-domain. It is defined as

$$Z\{x(n)\} = X(z) = \sum_{n=-\infty}^{\infty} x(n)z^{-n} \quad \textbf{Eq. (3.167)}$$

where $z = re^{j\omega}$, and for most cases, $r = 1$. It follows that Eq. (3.167) can be rewritten as

$$X(e^{j\omega}) = \sum_{n=-\infty}^{\infty} x(n)e^{-jn\omega}. \quad \textbf{Eq. (3.168)}$$

In the z-domain, the region over which $X(z)$ is finite is called the Region of Convergence (ROC).

Example:

Show that $Z\{nx(n)\} = -z\dfrac{d}{dz}X(z)$.

Solution:

Starting with the definition of the Z-transform,

$$X(z) = \sum_{n=-\infty}^{\infty} x(n)z^{-n}.$$

Taking the derivative, with respect to z, of the above equation yields

$$\frac{d}{dz}X(z) = \sum_{n=-\infty}^{\infty} x(n)(-n)z^{-n-1}$$

$$= (-z^{-1}) \sum_{n=-\infty}^{\infty} nx(n)z^{-n}.$$

It follows that

$$Z\{nx(n)\} = (-z)\frac{\mathrm{d}}{\mathrm{d}z}X(z).$$

A discrete LTI system has a transfer function $H(z)$ that describes how the system operates on its input sequence $x(n)$ in order to produce the output sequence $y(n)$. The output sequence $y(n)$ is computed from the discrete convolution between the sequences $x(n)$ and $h(n)$:

$$y(n) = \sum_{m=-\infty}^{\infty} x(m)h(n-m). \qquad \text{Eq. (3.169)}$$

However, since practical systems require the sequence $x(n)$ and $h(n)$ to be of finite length, we can rewrite Eq. (3.169) as

$$y(n) = \sum_{m=0}^{N} x(m)h(n-m). \qquad \text{Eq. (3.170)}$$

N denotes the input sequence length. The Z-transform of Eq. (3.170) is

$$Y(z) = X(z)H(z) \qquad \text{Eq. (3.171)}$$

and the discrete system transfer function is

$$H(z) = \frac{Y(z)}{X(z)}. \qquad \text{Eq. (3.172)}$$

Finally, the transfer function $H(z)$ can be written as

$$H(z)\big|_{z=e^{j\omega}} = \left|H(e^{j\omega})\right|e^{\angle H(e^{j\omega})} \qquad \text{Eq. (3.173)}$$

where $\left|H(e^{j\omega})\right|$ is the amplitude response, and $\angle H(e^{j\omega})$ is the phase response.

3.9.3. The Discrete Fourier Transform

The Discrete Fourier Transform (DFT) is a mathematical operation that transforms a discrete sequence, usually from the time domain into the frequency domain, in order to explicitly determine the spectral information for the sequence. The time-domain sequence can be real or complex. The DFT has finite length N and is periodic with period equal to N. The discrete Fourier transform pairs for the finite sequence $x(n)$ are defined by

$$X(k) = \sum_{n=0}^{N-1} x(n)e^{-j\frac{2\pi nk}{N}} \quad ; \ k = 0, ..., N-1 \qquad \text{Eq. (3.174)}$$

$$x(n) = \frac{1}{N} \sum_{k=0}^{N-1} X(k) e^{j\frac{2\pi nk}{N}} \qquad ; \ n = 0, \dots, N-1 .$$

Eq. (3.175)

The Fast Fourier Transform (FFT) is not a new kind of transform different from the DFT. Instead, it is an algorithm used to compute the DFT more efficiently. There are numerous FFT algorithms that can be found in the literature. In this book we will interchangeably use the DFT and the FFT to mean the same thing. Furthermore, we will assume a radix-2 FFT algorithm, where the FFT size is equal to $N = 2^m$ for some integer m.

3.9.4. Discrete Power Spectrum

Practical discrete systems utilize DFTs of finite length as a means of numerical approximation for the Fourier transform. The input signals must be truncated to a finite duration (denoted by T) before they are sampled. This is necessary so that a finite length sequence is generated prior to signal processing. Unfortunately, this truncation process may cause some serious problems.

To demonstrate this difficulty, consider the time-domain signal $x(t) = \sin 2\pi f_0 t$. The spectrum of $x(t)$ consists of two spectral lines at $\pm f_0$. Now, when $x(t)$ is truncated to length T seconds and sampled at a rate $T_s = T/N$, where N is the number of desired samples, we produce the sequence $\{x(n); \ n = 0, 1, \dots, N-1\}$.

The spectrum of $x(n)$ would still be composed of the same spectral lines if T is an integer multiple of T_s and if the DFT frequency resolution Δf is an integer multiple of f_0. Unfortunately, those two conditions are rarely met, and as a consequence, the spectrum of $x(n)$ spreads over several lines (normally the spread may extend up to three lines). This is known as spectral leakage. Since f_0 is normally unknown, this discontinuity caused by an arbitrary choice of T cannot be avoided. Windowing techniques can be used to mitigate the effect of this discontinuity by applying smaller weights to samples close to the edges.

A truncated sequence $x(n)$ can be viewed as one period of some periodic sequence with period N. The discrete Fourier series expansion of $x(n)$ is

$$x(n) = \sum_{k=0}^{N-1} X_k e^{j\frac{2\pi nk}{N}} .$$

Eq. (3.176)

It can be shown that the coefficients X_k are given by

$$X_k = \frac{1}{N} \sum_{n=0}^{N-1} x(n) e^{-j\frac{2\pi nk}{N}} = \frac{1}{N} X(k)$$

Eq. (3.177)

where $X(k)$ is the DFT of $x(n)$. Therefore, the Discrete Power Spectrum (DPS) for the band-limited sequence $x(n)$ is the plot of $|X_k|^2$ versus k, where the lines are Δf apart,

$$P_0 = \frac{1}{N^2} |X(0)|^2$$

Eq. (3.178)

$$P_k = \frac{1}{N^2}\{|X(k)|^2 + |X(N-k)|^2\} \qquad ; \ k = 1, 2, \ldots, \frac{N}{2} - 1 \qquad \text{Eq. (3.179)}$$

$$P_{N/2} = \frac{1}{N^2}|X(N/2)|^2 . \qquad \text{Eq. (3.180)}$$

Before proceeding to the next section, we will show how to select the FFT parameters. For this purpose, consider a band-limited signal $x(t)$ with bandwidth B. If the signal is not band-limited, an LPF can be used to eliminate frequencies greater than B. In order to satisfy the sampling theorem, one must choose a sampling frequency $f_s = 1/T_s$, such that

$$f_s \geq 2B . \qquad \text{Eq. (3.181)}$$

The truncated sequence duration T and the total number of samples N are related by

$$T = NT_s \qquad \text{Eq. (3.182)}$$

or equivalently,

$$f_s = N/T . \qquad \text{Eq. (3.183)}$$

It follows that

$$f_s = \frac{N}{T} \geq 2B \qquad \text{Eq. (3.184)}$$

and the frequency resolution is

$$\Delta f = \frac{1}{NT_s} = \frac{f_s}{N} = \frac{1}{T} \geq \frac{2B}{N} . \qquad \text{Eq. (3.185)}$$

3.9.5. Windowing Techniques

Truncation of the sequence $x(n)$ can be accomplished by computing the product

$$x_w(n) = x(n)w(n) \qquad \text{Eq. (3.186)}$$

where

$$w(n) = \left\{ \begin{array}{ll} f(n) & ; \ n = 0, 1, \ldots, \ N-1 \\ 0 & otherwise \end{array} \right\} \qquad \text{Eq. (3.187)}$$

where $f(n) \leq 1$. The finite sequence $w(n)$ is called a windowing sequence, or simply a window. The windowing process should not impact the phase response of the truncated sequence. Consequently, the sequence $w(n)$ must retain linear phase. This can be accomplished by making the window symmetrical with respect to its central point.

If $f(n) = 1$ for all n, we have what is known as the rectangular window. It leads to the Gibbs phenomenon, which manifests itself as an overshoot and a ripple before and after a discontinuity. Figure 3.10 shows the amplitude spectrum of a rectangular window. Note that the first sidelobe is at $-13.46\,dB$ below the main lobe. Windows that place smaller weights on the samples near the edges will have less overshoot at the discontinuity points (lower sidelobes); hence, they are more desirable than a rectangular window. However, reduction of the sidelobes is offset by a widening of the main lobe. Therefore, the proper choice of a windowing sequence is a continuous trade-off between sidelobe reduction and mainlobe widening. Table 3.1 gives a summary of some commonly used windows with the corresponding impact on main beam widening and peak reduction.

The multiplication process defined in Eq. (3.186) is equivalent to cyclic convolution in the frequency domain. It follows that $X_w(k)$ is a smeared (distorted) version of $X(k)$. To minimize this distortion, we would seek windows that have a narrow main lobe and small sidelobes. Additionally, using a window other than a rectangular window reduces the power by a factor P_w, where

$$P_w = \frac{1}{N}\sum_{n=0}^{N-1} w^2(n) = \sum_{k=0}^{N-1} |W(k)|^2.$$

Eq. (3.188)

It follows that the DPS for the sequence $x_w(n)$ is now given by

$$P_0^w = \frac{1}{P_w N^2}|X(0)|^2$$

Eq. (3.189)

$$P_k^w = \frac{1}{P_w N^2}\{|X(k)|^2 + |X(N-k)|^2\} \qquad ; \; k = 1, 2, ..., \frac{N}{2} - 1$$

Eq. (3.190)

$$P_{N/2}^w = \frac{1}{P_w N^2}|X(N/2)|^2$$

Eq. (3.191)

where P_w is defined in Eq. (3.188). Table 3.2 lists the mathematical expressions for some common windows. Figures 3.11 through 3.13 show the frequency domain characteristics for these windows. These plots can be reproduced using the following MATLAB code. Figures 3.11 through 3.13 can be reproduced using the MATLAB program *"Fig3_10_13.m"* listed in Appendix 3-A.

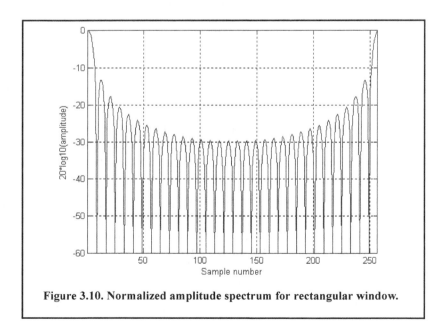

Figure 3.10. Normalized amplitude spectrum for rectangular window.

TABLE 3.1. Common Windows

Window	Null-to-Null Beamwidth Rectangular Window is the Reference	Peak Reduction
Rectangular	*1*	*1*
Hamming	*2*	*0.73*
Hanning	*2*	*0.664*
Blackman	*6*	*0.577*
Kaiser ($\beta = 6$)	*2.76*	*0.683*
Kaiser ($\beta = 3$)	*1.75*	*0.882*

TABLE 3.2. Some Common Windows. $n = 0, N-1$

Window	Expression	First Side-lobe	Main Lobe Width
Rectangular	$w(n) = 1$	$-13.46\,dB$	1
Hamming	$w(n) = 0.54 - 0.46\cos\left(\dfrac{2\pi n}{N-1}\right)$	$-41\,dB$	2
Hanning	$w(n) = 0.5\left[1 - \cos\left(\dfrac{2\pi n}{N-1}\right)\right]$	$-32\,dB$	2
Kaiser	$w(n) = \dfrac{I_0[\beta\sqrt{1 - (2n/N)^2}]}{I_0(\beta)}$ I_0 is the zero-order modified Bessel function of the first kind	$-46\,dB$ *for* $\beta = 2\pi$	$\sqrt{5}$ *for* $\beta = 2\pi$

3.9.6. Decimation and Interpolation

Decimation

Typically, radar systems use many signals for different functions, such as search, track, and discrimination, to name a few. All signals are assumed to be essentially limited; however, since these signals have different functions, they do not have the same time and bandwidth durations (τ, B). Earlier in this chapter, it was established that the number of samples required to sufficiently recover any signal from its samples is $N \geq 2\tau B$. Therefore, it is important to use an A/D with a high enough sampling rate to account for the largest possible number of samples required. As a result, it is often the case that some radar signals are sampled at a much higher rate than actually needed.

The process for decreasing the number of samples for a given sequence is called decimation. This is because the original data set has been reduced (decimated) in number. The process that increases the number of data samples is referred to as interpolation. The typical implementation for either operation is to alter the sampling rate, without violating the Nyquist sampling rate, of the input sequence. In decimation, the sampling rate is decreased by increasing the time steps between successive samples. More precisely, if the t_1 is the original sampling interval and t_2 is the decimated sampling interval, then

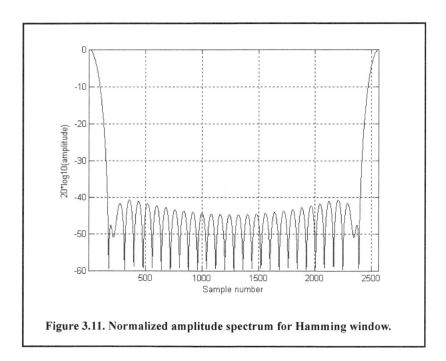

Figure 3.11. Normalized amplitude spectrum for Hamming window.

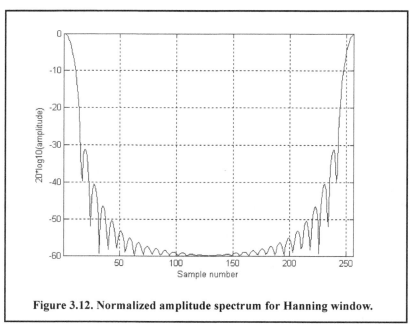

Figure 3.12. Normalized amplitude spectrum for Hanning window.

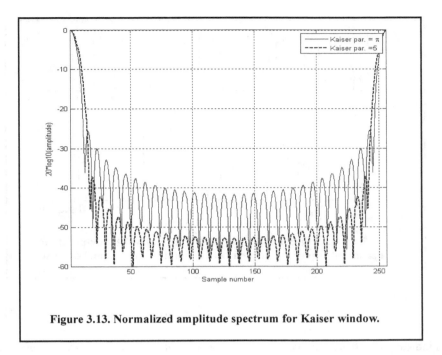

Figure 3.13. Normalized amplitude spectrum for Kaiser window.

$$t_2 = Dt_1 .$$ **Eq. (3.192)**

D is the decimation ratio and it is greater than unity. If D is an integer, then decimation effectively decreases the original sequence by discarding $(D-1)$ samples of D samples. This is illustrated in Fig. 3.14 for $D = 3$.

When D is not an integer, it is then necessary to first perform interpolation to determine new values for the new sequence. For example, if $D = 2.2$, then four out of every five samples in the decimated sequence are between samples in the original sequence and must be found by interpolation. This is illustrated in Fig. 3.15 for $D = 2.2$. In this example,

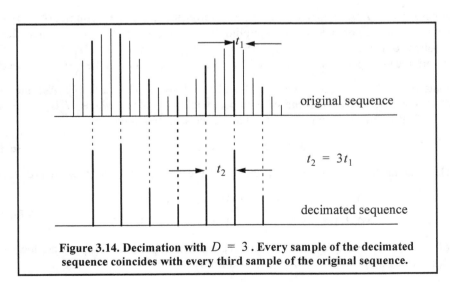

Figure 3.14. Decimation with $D = 3$. Every sample of the decimated sequence coincides with every third sample of the original sequence.

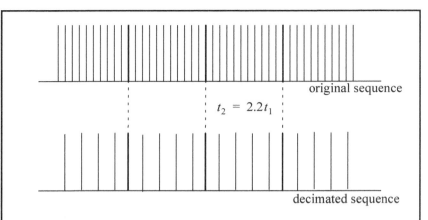

Figure 3.15. Decimation with $D = 2.2$. **Every fifth sample of the decimated sequence coincides with a sample in the original sequence.**

$$\left(t_2 = 2.2 t_1 = \frac{11}{5} t_1 \right) \Rightarrow 5 t_2 = 11 t_1 . \qquad \text{Eq. (3.193)}$$

which indicates that there are five samples in the decimated sequence for every eleven samples of the original sequence. Additionally, every fifth sample in the decimated sequence is equal to every eleventh sample of the original sequence.

Interpolation

Suppose that a signal $x(t)$ whose duration is T seconds has been sampled at a sampling rate t_1 to obtain a sequence

$$\mathbf{x} = x[n] = \{x(n t_1), n = 0, 1, ..., N_1 - 1\} \qquad \text{Eq. (3.194)}$$

in this case, $N_1 = T / t_1$. Suppose you want to interpolate between the samples of $x[n]$ to generate a new sequence of size N_2 and sampling interval t_2 , where $t_2 = t_1 / k$. This effectively corresponds to a new sampling frequency $f_{s2} = k f_{s1}$ where $f_{s1} = 1 / t_1$. A more efficient interpolation can be performed using the FFT, as will be described in the rest of this section.

Denote the FFT of the sequences $x_1[n]$ and $x_2[n]$ by $X_1[l]$ and $X_2[l]$. Assume that the signal $x(t)$ is essentially band-limited with bandwidth $B = M \Delta f$ where M is an integer and $\Delta f = 1 / T$. It follows that in order not to violate the sampling theorem

$$M \Delta f < f_{s1} / 2 < f_{s2} / 2 . \qquad \text{Eq. (3.195)}$$

It is clear that the coefficients of $X_1[l]$ and $X_2[l]$ are zero for all $|l| > M$. More precisely,

$$\begin{aligned} X_1[l] &= 0; \quad l = M + 1, M + 2, ..., N_1 - 3 \\ X_2[l] &= 0; \quad l = M + 1, M + 2, ..., N_2 - 3 \end{aligned} . \qquad \text{Eq. (3.196)}$$

Therefore, one can easily obtain the new sequence $X_2[l]$ from $X_1[l]$ by adding zeros in between the negative and positive frequencies from

$$N_1 - (2M + 1) \quad to \quad N_2 - (2M + 1)$$

Eq. (3.197)

and the sequence $x_2[n]$ is simply generated by computing the inverse DFT of the sequence $X_2[l]$. Interpolation can also be applied to the frequency domain sequence. For this purpose, one can simply zero pad the time-domain sequence to the desired size and then take the DFT of the newly interpolated sequence.

Problems

3.1. Classify each of the following signals as an energy signal, a power signal, or neither.

(a) $\exp(0.5t)$ $(t \geq 0)$,

(b) $\exp(-0.5t)$ $(t \geq 0)$,

(c) $\cos t + \cos 2t$ $(-\infty < t < \infty)$,

(d) $e^{-a|t|}$ $(a > 0)$.

3.2. A definition for the instantaneous frequency was given in Eq. (3.65). A more general definition is $f_i(t) = \dfrac{1}{2\pi} Im\left\{ \dfrac{d}{dt} \ln \psi(t) \right\}$ where Im {.}, indicates imaginary part and $\psi(t)$ is the analytic signal. Using this definition, calculate the instantaneous frequency for

$$x(t) = Rect\left(\frac{t}{\tau}\right) \cos\left(2\pi f_0 t + \frac{B}{2\tau} t^2\right).$$

3.3. Consider the two bandpass signals $x(t) = r_x(t)\cos(2\pi f_0 t + \phi_x(t))$ and $h(t) = r_h(t)\cos(2\pi f_0 t + \phi_h(t))$. Derive an expression for the complex envelope for the signal $s(t) = x(t) + h(t)$.

3.4. Consider the bandpass signal $x(t)$ whose complex envelope is equal to $\tilde{x}(t) = x_I(t) + jx_Q(t)$. Derive an expression for the autocorrelation function and the power spectrum density for $x(t)$ and $\tilde{x}(t)$. Assume that the signal $x(t)$ is the input to an LTI filter whose impulse response is $h(t)$; give an expression for the output's autocorrelation and power spectrum density.

3.5. Find the autocorrelation integral of the pulse train

$$y(t) = Rect(t/T) - Rect\left(\frac{t-T}{T}\right) + Rect\left(\frac{t-2T}{T}\right).$$

3.6. Compute the discrete convolution $y(n) = x(m) \bullet h(m)$ where $\{x(k), k = -1, 0, 1, 2\} = [-1.9, 0.5, 1.2, 1.5]$ $\{h(k), k = 0, 1, 2\} = [-2.1, 1.2, 0.8]$.

3.7. Define $\{x_I(n) = 1, -1, 1\}$ and $\{x_Q(n) = 1, 1, -1\}$. (a) Compute the discrete correlations: R_{x_I}, R_{x_Q}, $R_{x_I x_Q}$, and $R_{x_Q x_I}$. (b) A certain radar transmits the signal $s(t) = x_I(t)\cos 2\pi f_0 t - x_Q(t)\sin 2\pi f_0 t$. Assume that the autocorrelation $s(t)$ is equal to $y(t) = y_I(t)\cos 2\pi f_0 t - y_Q(t)\sin 2\pi f_0 t$. Compute and sketch $y_I(t)$ and $y_Q(t)$.

3.8. Compute the energy associated with the signal $x(t) = A Rect(t/\tau)$.

3.9. (a) Prove that $\varphi_1(t)$ and $\varphi_2(t)$, shown in the figure below, are orthogonal over the interval $(-2 \le t \le 2)$. (b) Express the signal $x(t) = t$ as a weighted sum of $\varphi_1(t)$ and $\varphi_2(t)$ over the same time interval.

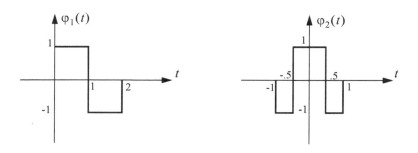

3.10. A periodic signal $x_p(t)$ is formed by repeating the pulse $x(t) = 2\Delta((t-3)/5)$ every 10 seconds. (a) What is the Fourier transform of $x(t)$? (b) Compute the complex Fourier series of $x_p(t)$. (c) Give an expression for the autocorrelation function $\bar{R}_{x_p}(t)$ and the power spectrum density $\bar{S}_{x_p}(\omega)$.

3.11. If the Fourier series is $x(t) = \sum_{n=-\infty}^{\infty} X_n e^{j2\pi nt/T}$, define $y(t) = x(t - t_0)$. Compute an expression for the complex Fourier series expansion of $y(t)$.

3.12. Derive Eq. (3.52).

3.13. Show that (a) $\bar{R}_x(-t) = \bar{R}_x^*(t)$, (b) If $x(t) = f(t) + m_1$ and $y(t) = g(t) + m_2$, show that $\bar{R}_{xy}(t) = m_1 m_2$, where the average values for $f(t)$ and $g(t)$ are zeroes.

3.14. What is the power spectral density for the signal $x(t) = A\cos(2\pi f_0 t + \theta_0)$?

2.15. A certain radar system uses linear frequency modulated waveforms of the form

$$x(t) = Rect\left(\frac{t}{\tau}\right)\cos\left(\omega_0 t + \mu\frac{t^2}{2}\right)$$

What are the quadrature components? Give an expression for both the modulation and instantaneous frequencies.

3.16. Consider the signal $x(t) = Rect(t/\tau)\cos(\omega_0 t - Bt^2/2\tau)$ and let $\tau = 15\mu s$ and $B = 10MHz$. What are the quadrature components?

3.17. Determine the quadrature components for the signal

$$h(t) = \delta(t) - \left(\frac{\omega_0}{\omega_d}\right)e^{-2t}\sin(\omega_0 t) \text{ for } t \ge 0.$$

3.18. If $x(t) = x_1(t) - 2x_1(t - 5) + x_1(t - 10)$, determine the autocorrelation functions $R_{x_1}(t)$ and $R_x(t)$ when $x_1(t) = \exp(-t^2/2)$.

3.19. Derive Eq. (3.139).

3.20. Prove that the effective duration of a finite pulse train is equal to $(T_t \tau_0) / T$, where τ_0 is the pulse width, T is the PRI, and T_t is as defined in Fig. 3.5.

3.21. Write an expression for the autocorrelation function $R_y(t)$, where

$$y(t) = \sum_{n=1}^{5} Y_n Rect\left(\frac{t-n5}{2}\right) \text{ and } \{Y_n\} = \{0.8, 1, 1, 1, 0.8\}.$$

Give an expression for the density function $S_y(\omega)$.

3.22. An LTI system has impulse response $h(t) = \begin{cases} \exp(-2t) & t \geq 0 \\ 0 & t < 0 \end{cases}$.

(a) Find the autocorrelation function $R_h(\tau)$. (b) Assume the input of this system is $x(t) = 3\cos(100t)$. What is the output?

3.23. Compute the Z-transform for

(a) $x_1(n) = \frac{1}{n!} u(n)$,

(b) $x_2(n) = \frac{1}{(-n)!} u(-n)$.

3.24. (a) Write an expression for the FT of $x(t) = Rect(t/3)$. (b) Assume that you want to compute the modulus of the FT using a DFT of size *512* with a sampling interval of *1* second. Evaluate the modulus at frequency $(80/512)Hz$. Compare your answer to the theoretical value and compute the error.

2.25. In Fig. 3.9, let

$$p(t) = \sum_{n=-\infty}^{\infty} ARect\left(\frac{t-nT}{\tau}\right)$$

Give an expression for $X_s(\omega)$.

3.26. Generate *512* samples of the signal $x(t) = 2.0e^{-5t}\sin(4\pi t)$, using a sampling interval equal to 0.002. Compute the resultant spectrum and then truncate the spectrum at $15Hz$. Generate the time-domain sequence for the truncated spectrum. Determine the sampling rate of the new sequence.

3.27. Assume that a time-domain sequence generated by using a sampling interval equal to *0.01* is given by $x(k) = \{0, 2, 5, 12, 5, 3, 3, -1, 1, 0\}$. Decimate this sequence so that the sampling interval is *0.02*.

3.28. Write a MATLAB program to decimate any sequence of finite length and demonstrate it using the previous problem.

3.29. You are given a sequence of samples $\{x(kT), k = -\infty, ..., \infty\}$ where the sampling interval T corresponds to twice the Nyquist rate. Give an expression to compute the samples of $x(t)$ at a new sampling rate corresponding to $T' = 0.7T$.

3.30. A certain band-limited signal has bandwidth $B = 20KHz$. Find the FFT size required so that the frequency resolution is $\Delta f = 50Hz$. Assume radix 2 FFT and a record length of 1 second.

3.31. Assume that a certain sequence is determined by its FFT. If the record length is $2ms$ and the sampling frequency is $f_s = 10KHz$, find N.

Appendix 3-A: Chapter 3 MATALAB Code Listings

The MATLAB code provided in this chapter was designed as an academic standalone tool and is not adequate for other purposes. The code was written in a way to assist the reader in gaining a better understanding of the theory. The code was not developed, nor is it intended to be used as part of an open-loop or a closed-loop simulation of any kind. The MATLAB code found in this textbook can be downloaded from this book's web page on the CRC Press website. Simply use your favorite web browser, go to *www.crcpress.com*, and search for keyword *"Mahafza"* to locate this book's web page.

MATLAB Program *"Fig3_6.m"* Listing

```
% Generates Figure 3.6 of text
close all
clear all
LFM_BW = 15e6;
tau = 1e-6;
ts = 1e-9; % 1000 samples per PW
beta = LFM_BW/tau;
t = 0: ts: +tau;
S = exp(j*pi*beta*(t.^2));
figure
subplot(2,1,1), plot(t*1e6,imag(S),'linewidth',1.5), grid
ylabel('Up-chirp LFM')
% The matched filter for S(t) is S*(-t)
t = -tau: ts: 0;
Smf = exp(-j*pi*beta*(t.^2));
subplot(2,1,2), plot(t*1e6,imag(Smf),'linewidth',1.5), grid
xlabel('time in microseconds')
ylabel('Down-chirp LFM')
```

MATLAB Program *"Fig3_8.m"* Listing

```
% use this program to reproduce Fig. 3.8 of text
clc
clear all
close all
%
nscat = 2; %two point scatterers
taup = 10e-6; % 100 microsecond uncompressed pulse
b = 40.0e6; % 50 MHz bandwdith
rrec = 50 ; % 50 meter processing window
scat_range = [15 25] ; % scattterers are 15 and 25 meters into window
scat_rcs = [1 2]; % RCS 1 m^2 and 2m^2
winid = 0; %no window used
%function [y] = matched_filter(nscat,taup,b,rrec,scat_range,scat_rcs,winid)
eps = 1.0e-16;
% time bandwidth product
time_B_product = b * taup;
if(time_B_product < 5 )
    fprintf('*********** Time Bandwidth product is TOO SMALL **************')
    fprintf('\n Change b and or taup')
    return
end
```

```
% speed of light
c = 3.e8;
% number of samples
n = fix(2 * taup * b);
% initialize input, output and replica vectors
x(nscat,1:n) = 0.;
y(1:n) = 0.;
replica(1:n) = 0.;
% determine proper window
if( winid == 0.)
   win(1:n) = 1.;
end
if(winid == 1.);
   win = hamming(n)';
end
if( winid == 2.)
   win = kaiser(n,pi)';
end
if(winid == 3.)
   win = chebwin(n,60)';
end
% check to ensure that scatterers are within recieve window
index = find(scat_range > rrec);
if (index ~= 0)
   'Error. Receive window is too large; or scatterers fall outside window'
  return
end
% calculate sampling interval
t = linspace(-taup/2,taup/2,n);
replica = exp(i * pi * (b/taup) .* t.^2);
figure(1)
plot(t,real(replica))
ylabel('Real (part) of replica')
xlabel('Time in seconds')
grid
figure(2)
plot(t,imag(replica))
ylabel('Imaginary (part) of replica')
xlabel('Time in seconds')
grid
figure(3)
sampling_interval = 1 / 2.5 /b;
freqlimit = 0.5/ sampling_interval;
freq = linspace(-freqlimit,freqlimit,n);
plot(freq,fftshift(abs(fft(replica))));
ylabel('Spectrum of replica')
xlabel('Frequency in Hz')
grid
for j = 1:1:nscat
   range = scat_range(j) ;;
   x(j,:) = scat_rcs(j) .* exp(i * pi * (b/taup) .* (t +(2*range/c)).^2) ;
   y = x(j,:)  + y;
end
```

MATLAB Program "Fig3_10_13.m" Listing

```
%Use this program to reproduce Figures 3.10 through 3.13 of textbook.
clear all; close all
eps = 0.001;
N = 32;
win_rect (1:N) = 1;
win_ham = hamming(N);
win_han = hanning(N);
win_kaiser = kaiser(N, pi);
win_kaiser2 = kaiser(N, 5);
Yrect = abs(fft(win_rect, 256));
Yrectn = Yrect ./ max(Yrect);
Yham = abs(fft(win_ham, 2562));
Yhamn = Yham ./ max(Yham);
Yhan = abs(fft(win_han, 256));
Yhann = Yhan ./ max(Yhan);
YK = abs(fft(win_kaiser, 256));
YKn = YK ./ max(YK);
YK2 = abs(fft(win_kaiser2, 256));
YKn2 = YK2 ./ max(YK2);
figure (1)
plot(20*log10(Yrectn+eps),'k')
xlabel('Sample number')
ylabel('20*log10(amplitude)')
axis tight; grid on
figure(2)
plot(20*log10(Yhamn + eps),'k')
xlabel('Sample number')
ylabel('20*log10(amplitude)')
grid on; axis tight
figure (3)
plot(20*log10(Yhann+eps),'k')
xlabel('Sample number'); ylabel('20*log10(amplitude)'); grid
axis tight
figure(4)
plot(20*log10(YKn+eps),'k')
grid on; hold on
plot(20*log10(YKn2+eps),'k--')
xlabel('Sample number'); ylabel('20*log10(amplitude)')
legend('Kaiser par. = \pi','Kaiser par. =5')
axis tight; hold off
```

Appendix 3-B: Fourier Transform Pairs

$x(t)$	$X(\omega)$
$ARect(t/\tau)$; rectangular pulse	$A\tau Sinc(\omega\tau/2)$
$A\Delta(t/\tau)$; triangular pulse	$A\dfrac{\tau}{2}Sinc^2(\tau\omega/4)$
$\dfrac{1}{\sqrt{2\pi}\sigma}\exp\left(-\dfrac{t^2}{2\sigma^2}\right)$; Gaussian pulse	$\exp\left(-\dfrac{\sigma^2\omega^2}{2}\right)$
$e^{-at}u(t)$	$1/(a+j\omega)$
$e^{-a|t|}$	$\dfrac{2a}{a^2+\omega^2}$
$e^{-at}\sin\omega_0 t \; u(t)$	$\dfrac{\omega_0}{\omega_0^2+(a+j\omega)^2}$
$e^{-at}\cos\omega_0 t \; u(t)$	$\dfrac{a+j\omega}{\omega_0^2+(a+j\omega)^2}$
$\delta(t)$	1
1	$2\pi\delta(\omega)$
$u(t)$	$\pi\delta(\omega)+\dfrac{1}{j\omega}$
$sgn(t)$	$\dfrac{2}{j\omega}$
$\cos\omega_0 t$	$\pi[\delta(\omega-\omega_0)+\delta(\omega+\omega_0)]$
$\sin\omega_0 t$	$j\pi[\delta(\omega+\omega_0)-\delta(\omega-\omega_0)]$
$u(t)\cos\omega_0 t$	$\dfrac{\pi}{2}[\delta(\omega-\omega_0)+\delta(\omega+\omega_0)]+\dfrac{j\omega}{\omega_0^2-\omega^2}$
$u(t)\sin\omega_0 t$	$\dfrac{\pi}{2j}[\delta(\omega+\omega_0)-\delta(\omega-\omega_0)]+\dfrac{\omega_0}{\omega_0^2-\omega^2}$
$|t|$	$\dfrac{-2}{\omega^2}$

Appendix 3-C: Z-Transform Pairs

$x(n); \; n \geq 0$	$X(z)$	ROC; $\lvert z \rvert > R$
$\delta(n)$	1	0
1	$\dfrac{z}{z-1}$	1
n	$\dfrac{z}{(z-1)^2}$	1
n^2	$\dfrac{z(z+1)}{(z-1)^3}$	1
a^n	$\dfrac{z}{z-a}$	$\lvert a \rvert$
na^n	$\dfrac{az}{(z-a)^2}$	$\lvert a \rvert$
$\dfrac{a^n}{n!}$	$e^{a/z}$	0
$(n+1)a^n$	$\dfrac{z^2}{(z-a)^2}$	$\lvert a \rvert$
$\sin n\omega T$	$\dfrac{z\sin\omega T}{z^2 - 2z\cos\omega T + 1}$	1
$\cos n\omega T$	$\dfrac{z(z-\cos\omega T)}{z^2 - 2z\cos\omega T + 1}$	1
$a^n \sin n\omega T$	$\dfrac{az\sin\omega T}{z^2 - 2az\cos\omega T + a^2}$	$\dfrac{1}{\lvert a \rvert}$
$a^n \cos n\omega T$	$\dfrac{z(z - a^2\cos\omega T)}{z^2 - 2az\cos\omega T + a^2}$	$\dfrac{1}{\lvert a \rvert}$
$\dfrac{n(n-1)}{2!}$	$\dfrac{z}{(z-1)^3}$	1
$\dfrac{n(n-1)(n-2)}{3!}$	$\dfrac{z}{(z-1)^4}$	1
$\dfrac{(n+1)(n+2)a^n}{2!}$	$\dfrac{z^3}{(z-a)^3}$	$\lvert a \rvert$
$\dfrac{(n+1)(n+2)\ldots(n+m)a^n}{m!}$	$\dfrac{z^{m+1}}{(z-a)^{m+1}}$	$\lvert a \rvert$

Chapter 4

The Matched Filter
Radar Receiver

4.1. The Matched Filter SNR

The topic of matched filtering is central to almost all radar systems. In this chapter the focus is the matched filter. The unique characteristic of the matched filter is that it produces the maximum achievable instantaneous SNR at its output when a signal plus noise (Gaussian noise is assumed in the analysis presented in this book) are present at its input. Maximizing the SNR is key in all radar applications, as was described in Chapter 2 in the context of the radar equation, and as will be discussed in a subsequent chapter in the context of target detection.

It is important to use a radar receiver which can be modeled as an LTI system that maximizes the signal's SNR at its output. For this purpose, the basic radar receiver of interest is often referred to as the matched filter receiver. The matched filter is an optimum filter in the sense of SNR because the SNR at its output is maximized at some delay t_0 that corresponds to the true target range R_0 (i.e., $t_0 = (2R_0)/c$). Figure 4.1 shows a simplified block diagram for the radar receiver of interest.

In order to derive the general expression for the transfer function and the impulse response of this optimum filter, adopt the following notation: $h(t)$ is the optimum filter impulse response, $H(f)$ is the optimum filter transfer function, $x(t)$ is the input signal, $X(f)$ is the FT of the input signal, $x_o(t)$ is the output signal, $X_o(f)$ is the FT of the output signal, $n_i(t)$ is the input noise signal, $N_i(f)$ is the input noise PSD (not necessarily white), $n_o(t)$ is the out noise signal, and $N_o(f)$ is the output noise PSD. As one would expect, the impulse response of this optimum filter will take on distinct forms depending on the noise characteristics, i.e., white versus non-white noise.

The optimum filter input or received signal (the words *input* and *received* will be used interchangeably in this book) can then be represented by

$$x_i(t) = x(t - t_0) + n_i(t) \qquad \text{Eq. (4.1)}$$

where t_0 is an unknown time delay proportional to the target range. The optimum filter output signal is

$$y(t) = x_o(t - t_0) + n_o(t) \qquad \text{Eq. (4.2)}$$

where

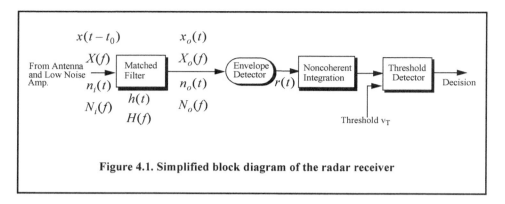

Figure 4.1. Simplified block diagram of the radar receiver

$$n_o(t) = n_i(t) \otimes h(t) \qquad\qquad \text{Eq. (4.3)}$$

$$x_o(t) = x(t - t_0) \otimes h(t). \qquad\qquad \text{Eq. (4.4)}$$

The operator ($\otimes$) indicates convolution. The FT of Eq. (4.4) is

$$X_o(f) = X(f)H(f)e^{j2\pi f t_0}. \qquad\qquad \text{Eq. (4.5)}$$

Integrating the right-hand side of Eq. (4.5) over all possible frequencies yields the signal output at time t_0, as

$$x_o(t_0) = \int_{-\infty}^{\infty} X(f)H(f)e^{j2\pi f t_0} \, df. \qquad\qquad \text{Eq. (4.6)}$$

From Parseval's theorem the modulus square of Eq. (4.6) is the total signal energy, E_x.

The total noise power at the output of the filter is calculated using Parseval's theorem as

$$N_o = \int_{-\infty}^{\infty} N_i(f)|H(f)|^2 \, df. \qquad\qquad \text{Eq. (4.7)}$$

Since the output signal power at time t_0 is equal to the modulus square of Eq. (4.6), then the instantaneous SNR at time t_0 is

$$SNR(t_0) = \frac{\left| \int_{-\infty}^{\infty} X(f)H(f)e^{j2\pi f t_0} \, df \right|^2}{\int_{-\infty}^{\infty} N_i(f)|H(f)|^2 \, df} = \frac{E_x}{\int_{-\infty}^{\infty} N_i(f)|H(f)|^2 \, df}. \qquad\qquad \text{Eq. (4.8)}$$

Equation (4.8) is the general form of the optimum SNR at the output of the matched filter. Of course, when the noise is white, a simpler formula will result.

Remember Schawrz's inequality, which has the form

$$\frac{\left|\int\limits_{-\infty}^{\infty} X_1(f) X_2(f) \; df\right|^2}{\int\limits_{-\infty}^{\infty} |X_1(f)|^2 \; df} \leq \int\limits_{-\infty}^{\infty} |X_2(f)|^2 \; df. \qquad \text{Eq. (4.9)}$$

The equal sign in Eq. (4.9) applies when $X_1(f) = KX_2^*(f)$ for some arbitrary constant K. Apply Schawrz's inequality to Eq. (4.8) with the following assumptions

$$X_1(f) = H(f)\sqrt{N_i(f)} \qquad \text{Eq. (4.10)}$$

$$X_2(f) = \frac{X(f)e^{j2\pi ft_0}}{\sqrt{N_i(f)}}. \qquad \text{Eq. (4.11)}$$

It follows that the SNR is maximized when

$$H(f) = K \frac{X^*(f)e^{-j2\pi ft_0}}{N_i(f)}. \qquad \text{Eq. (4.12)}$$

An alternative way of writing Eq. (4.12) is

$$X(f)H(f)e^{j2\pi ft_0} = \frac{K|X(f)|^2}{N_i(f)}. \qquad \text{Eq. (4.13)}$$

The optimum filter impulse response is computed using inverse FT integral

$$h(t) = \int\limits_{-\infty}^{\infty} K \frac{X^*(f)e^{-j2\pi ft_0}}{N_i(f)} \; e^{j2\pi ft} \; df. \qquad \text{Eq. (4.14)}$$

4.1.1. White Noise Case

A special case of great interest to radar systems is when the input noise is band-limited white noise with PSD given by

$$N_i(f) = \frac{\eta_0}{2}. \qquad \text{Eq. (4.15)}$$

η_0 is a constant. The transfer function for this optimum filter is then given by

$$H(f) = X^*(f)e^{-j2\pi ft_0} \qquad \text{Eq. (4.16)}$$

where the constant K was set equal to $\eta_0/2$. It follows that

$$h(t) = \int\limits_{-\infty}^{\infty} [X^*(f)e^{-j2\pi ft_0}] \; e^{j2\pi ft} \; df \qquad \text{Eq. (4.17)}$$

which can be written as

$$h(t) = x^*(t_0 - t) .$$ **Eq. (4.18)**

Observation of Eq. (4.18) indicates that the impulse response of the optimum filter is matched to the input signal, and thus, the term *matched filter* is used for this special case. Under these conditions, the maximum instantaneous SNR at the output of the matched filter is

$$SNR(t_0) = \frac{\left| \int_{-\infty}^{\infty} X(f)H(f)e^{j2\pi ft_0} \, df \right|^2}{\left(\frac{\eta_0}{2} \right)} .$$ **Eq. (4.19)**

Again, from Parseval's theorem the numerator in Eq. (4.19) is equal to the input signal energy, E_x; consequently one can write the output peak instantaneous SNR as

$$SNR(t_0) = \frac{2E_x}{\eta_0} .$$ **Eq. (4.20)**

Note that Eq. (4.20) is unitless since the units for η_0 are in watts per hertz (or joules). Finally, one can draw the conclusion that the peak instantaneous SNR depends only on the signal energy and input noise power, and is independent of the waveform utilized by the radar.

As indicated by Eq. (4.18), the impulse response $h(t)$ may not be causal if the value for t_0 is less than the signal duration. Thus, an additional time delay term $\tau_0 \geq T$ is added to ensure causality, where T is the signal duration. Thus, a realizable matched filter response is given by

$$h(t) = \begin{pmatrix} x^*(\tau_0 + t_0 - t) & ;t > 0, \tau_0 \geq T \\ 0 & ;t < 0 \end{pmatrix} .$$ **Eq. (4.21)**

The transfer function for this casual filter is

$$H(f) = \int_{-\infty}^{\infty} x^*(\tau_0 + t_0 - t)e^{-j2\pi ft} dt = \int_{\infty}^{-\infty} x^*(t + \tau_0 + t_0)e^{j2\pi ft} dt = X^*(f)e^{-j2\pi f(\tau_0 + t_0)} .$$ **Eq. (4.22)**

Substituting the right-hand side of Eq. (4.22) into Eq. (4.6) yields

$$x_o(\tau_0) = \int_{-\infty}^{\infty} X(f)X^*(f)e^{-j2\pi f(\tau_0 + t_0)} e^{j2\pi ft_0} \, df = \int_{-\infty}^{\infty} |X(f)|^2 e^{-j2\pi f\tau_0} \, df,$$ **Eq. (4.23)**

which has a maximum value when τ_0. This result leads to the following conclusion: The peak value of the matched filter output is obtained by sampling its output at times equal to the filter delay after the start of the input signal, and the minimum value for τ_0 is equal to the signal duration T.

Example:

Compute the maximum instantaneous SNR at the output of a linear filter whose impulse response is matched to the signal $x(t) = \exp(-t^2/2T)$.

Solution:

The signal energy is

$$E_x = \int\limits_{-\infty}^{\infty} |x(t)|^2 dt = \int\limits_{-\infty}^{\infty} e^{(-t^2)/T} dt = \sqrt{\pi T} \ joules \, .$$

It follows that the maximum instantaneous SNR is

$$SNR = \frac{\sqrt{\pi T}}{\dfrac{\eta_0}{2}} = \frac{2\sqrt{\pi T}}{\eta_0}$$

where $\eta_0/2$ is the input noise power spectrum density.

4.1.2. The Replica

Again, consider a radar system that uses a finite duration energy signal $x(t)$, and assume that a matched filter receiver is utilized. From Eq. (4.1), the input signal can be written as,

$$x_i(t) = x(t - t_0) + n_i(t) \, . \qquad \textbf{Eq. (4.24)}$$

The matched filter output $y(t)$ can be expressed by the convolution integral between the filter's impulse response and $x_i(t)$:

$$y(t) = \int\limits_{-\infty}^{\infty} x_i(u) h(t - u) du \, . \qquad \textbf{Eq. (4.25)}$$

Substituting Eq. (4.21) into Eq. (4.25) yields

$$y(t) = \int\limits_{-\infty}^{\infty} x_i(u) x^*(t - \tau_0 - t_0 + u) du = \bar{R}_{x_i x}(t - T_0) \qquad \textbf{Eq. (4.26)}$$

where $T_0 = \tau_0 + t_0$ and $\bar{R}_{x_i x}(t - T_0)$ is a cross-correlation between $x_i(t)$ and $x(T_0 - t)$. Therefore, the matched filter output can be computed from the cross-correlation between the radar received signal and a delayed replica of the transmitted waveform. If the input signal is the same as the transmitted signal, the output of the matched filter would be the autocorrelation function of the received (or transmitted) signal. In practice, replicas of the transmitted waveforms are normally computed and stored in memory for use by the radar signal processor when needed.

4.3. General Formula for the Output of the Matched Filter

Two cases are analyzed; the first is when a stationary target is present. The second case is concerned with a moving target whose velocity is constant. Assume the range to the target is

$$R(t) = R_0 - v(t - t_0) \qquad \textbf{Eq. (4.27)}$$

where v is the target radial velocity (i.e., the target velocity component on the radar line of sight.) The initial detection range R_0 is given by

$$t_0 = \frac{2R_0}{c}$$

Eq. (4.28)

where c is the speed of light and t_0 is the round trip delay it takes a certain radar pulse to travel from the radar to the target at range R_0 and back.

The general expression for the radar bandpass signal is

$$x(t) = x_I(t)\cos 2\pi f_0 t - x_Q(t)\sin 2\pi f_0 t$$

Eq. (4.29)

which can be written using its pre-envelope (analytic signal) as

$$x(t) = Re\{\psi(t)\} = Re\{\tilde{x}(t)e^{j2\pi f_0 t}\}$$

Eq. (4.30)

where $Re\{\ \}$ indicates "the real part of." Again, $\tilde{x}(t)$ is the complex envelope.

4.2.1. Stationary Target Case

In this case, the received radar return is given by

$$x_i(t) = x\left(t - \frac{2R_0}{c}\right) = x(t - t_0) = Re\{\tilde{x}(t - t_0)e^{j2\pi f_0(t - t_0)}\}.$$

Eq. (4.31)

It follows that the received (or input) analytic signal is,

$$\psi_i(t) = \{\tilde{x}(t - t_0)e^{-j2\pi f_0 t_0}\}e^{j2\pi f_0 t}$$

Eq. (4.32)

and by inspection the received (or input) complex envelope is,

$$\tilde{x}_i(t) = \tilde{x}(t - t_0)e^{-j2\pi f_0 t_0}.$$

Eq. (4.33)

Observation of Eq. (4.33) clearly indicates that the received complex envelope is more than just a delayed version of the transmitted complex envelope. It actually contains an additional phase shift φ_0 which represents the phase corresponding to the two-way optical length for the target range. That is,

$$\varphi_0 = -2\pi f_0 t_0 = -2\pi f_0 2\frac{R_0}{c} = -\frac{2\pi}{\lambda}2R_0$$

Eq. (4.34)

where λ is the radar wavelength and is equal to c/f_0. Since a very small change in range can produce significant change in this phase term, this phase is often treated as a random variable with uniform probability density function over the interval $\{0, 2\pi\}$. Furthermore, the radar signal processor will first attempt to remove (correct for) this phase term through a process known as phase unwrapping.

Substituting Eq. (4.33) into Eq. (4.25) provides the output of the matched filter. It is given by

$$y(t) = \int_{-\infty}^{\infty}\tilde{x}_i(u)h(t - u)du$$

Eq. (4.35)

where the impulse response $h(t)$ is in Eq. (4.18). It follows that

$$y(t) = \int_{-\infty}^{\infty} \tilde{x}(u-t_0)e^{-j2\pi f_0 t_0}\tilde{x}*(t-t_0+u)du.$$

Eq. (4.36)

Make the following change of variables:

$$z = u - t_0 \Rightarrow dz = du.$$

Eq. (4.37)

Therefore, the output of the matched filter when a stationary target is present is computed from Eq (4.36) as

$$y(t) = e^{-j2\pi f_0 t_0}\int_{-\infty}^{\infty}\tilde{x}(z)\tilde{x}*(t-z)dz = e^{-j2\pi f_0 t_0}\overline{R}_x(t).$$

Eq. (4.38)

$\overline{R}_x(t)$ is the autocorrelation function for the signal $\tilde{x}(t)$ (i.e., the transmitted waveform).

4.2.2. Moving Target Case

In this case, the received signal is not only delayed in time by t_0, but also has a Doppler frequency shift f_d corresponding to the target velocity, where

$$f_d = 2vf_0/c = 2v/(\lambda).$$

Eq. (4.39)

The pre-envelope of the received signal can be written as

$$\psi_i(t) = \psi\left(t - \frac{2R(t)}{c}\right) = \tilde{x}\left(t - \frac{2R(t)}{c}\right)e^{j2\pi f_0\left(t - \frac{2R(t)}{c}\right)}.$$

Eq. (4.40)

Substituting Eq. (4.27) into Eq. (4.40) yields

$$\psi_i(t) = \tilde{x}\left(t - \frac{2R_0}{c} + \frac{2vt}{c} - \frac{2vt_0}{c}\right)e^{j2\pi f_0\left(t - \frac{2R_0}{c} + \frac{2vt}{c} - \frac{2vt_0}{c}\right)}.$$

Eq. (4.41)

Collecting terms yields

$$\psi_i(t) = \tilde{x}\left(t\left(1 + \frac{2v}{c}\right) - t_0\left(1 + \frac{2v}{c}\right)\right)e^{j2\pi f_0\left(t - \frac{2R_0}{c} + \frac{2vt}{c} - \frac{2vt_0}{c}\right)}.$$

Eq. (4.42)

Define the scaling factor γ as

$$\gamma = 1 + \frac{2v}{c},$$

Eq. (4.43)

then Eq. (4.42) can be written as

$$\psi_i(t) = \tilde{x}(\gamma(t-t_0))e^{j2\pi f_0\left(t - \frac{2R_0}{c} + \frac{2vt}{c} - \frac{2vt_0}{c}\right)}.$$

Eq. (4.44)

Since $c \gg v$, the following approximation can be used

$$\tilde{x}(\gamma(t - t_0)) \approx \tilde{x}(t - t_0).$$ **Eq. (4.45)**

It follows that Eq. (4.44) can now be rewritten as

$$\psi_i(t) = \tilde{x}(t - t_0)e^{j2\pi f_0 t}e^{-j2\pi f_0 \frac{2R_0}{c}}e^{j2\pi f_0 \frac{2vt}{c}}e^{-j2\pi f_0 \frac{2vt_0}{c}}.$$ **Eq. (4.46)**

Recognizing that $f_d = (2vf_0)/c$ and $t_0 = (2R_0)/c$, the received pre-envelope signal is

$$\psi_i(t) = \tilde{x}(t - t_0)e^{j2\pi f_0 t}e^{-j2\pi f_0 t_0}e^{j2\pi f_d t}e^{-j2\pi f_d t_0} = \tilde{x}(t - t_0)e^{j2\pi(f_0 + f_d)(t - t_0)}$$ **Eq. (4.47)**

or

$$\psi_i(t) = \{\tilde{x}(t - t_0)e^{j2\pi f_d t}e^{-j2\pi(f_0 + f_d)t_0}\}e^{j2\pi f_0 t}.$$ **Eq. (4.48)**

Then by inspection the complex envelope of the received signal is

$$\tilde{x}_i(t) = \tilde{x}(t - t_0)e^{j2\pi f_d t}e^{-j2\pi(f_0 + f_d)t_0}.$$ **Eq. (4.49)**

Finally, it is concluded that the complex envelope of the received signal when the target is moving at a constant velocity v is a delayed (by t_0) version of the complex envelope signal of the stationary target case except that:

1. an additional phase shift term corresponding to the target's Doppler frequency is present, and

2. the phase shift term $(-2\pi f_d t_0)$ is present.

The output of the matched filter was defined in Eq. (4.25). Substituting Eq. (4.49) into Eq. (4.25) yields

$$y(t) = \int_{-\infty}^{\infty} \tilde{x}(u - t_0)e^{j2\pi f_d u}e^{-j2\pi(f_0 + f_d)t_0}\tilde{x}*(t - t_0 + u) \ du.$$ **Eq. (4.50)**

Applying the change of variables given in Eq. (4.37) and collecting terms provide

$$y(t) = e^{-j2\pi f_0 t_0} \int_{-\infty}^{\infty} \tilde{x}(z)\tilde{x}*(t - z)e^{j2\pi f_d z}e^{j2\pi f_d t_0}e^{-j2\pi f_d t_0} \ dz.$$ **Eq. (4.51)**

Observation of Eq. (4.51) shows that the output is a function of both t and f_d. Thus, it is more appropriate to rewrite the output of the matched filter as a two-dimensional function of both variables. That is,

$$y(t; f_d) = e^{-j2\pi f_0 t_0} \int_{-\infty}^{\infty} \tilde{x}(z)\tilde{x}*(t - z)e^{j2\pi f_d z} \ dz.$$ **Eq. (4.52)**

It is customary but not necessary to set $t_0 = 0$. Note that if the causal impulse response is used (i.e., Eq. (4.21)), the same analysis will hold true. However, in this case, the phase term is equal to $\exp(-j2\pi f_0 T_0)$, instead of $\exp(-j2\pi f_0 t_0)$, where $T_0 = \tau_0 + t_0$.

4.3. Waveform Resolution and Ambiguity

As indicated by Eq. (4.20), the radar sensitivity (in the case of white additive noise) depends only on the total energy of the received signal and is independent of the shape of the specific waveform. This leads to the following question: If the radar sensitivity is independent of the waveform, what is the best choice for the transmitted waveform? The answer depends on many factors; however, the most important consideration lies in the waveform's range and Doppler resolution characteristics, which can be determined from the output of the matched fitter.

As discussed in Chapter 1, range resolution implies separation between distinct targets in range. Alternatively, Doppler resolution implies separation between distinct targets in frequency. Thus, ambiguity and accuracy of this separation are closely associated terms.

4.3.1. Range Resolution

Consider radar returns from two stationary targets (zero Doppler) separated in range by distance ΔR. What is the smallest value of ΔR so that the returned signal is interpreted by the radar as two distinct targets? In order to answer this question, assume that the radar transmitted bandpass pulse is denoted by $x(t)$,

$$x(t) = r(t)\cos(2\pi f_0 t + \phi(t)) \qquad \text{Eq. (4.53)}$$

where f_0 is the carrier frequency, $r(t)$ is the amplitude modulation, and $\phi(t)$ is the phase modulation. The signal $x(t)$ can then be expressed as the real part of the pre-envelope signal $\psi(t)$, where

$$\psi(t) = r(t)e^{j(2\pi f_0 t - \phi(t))} = \tilde{x}(t)e^{2\pi f_0 t} \qquad \text{Eq. (4.54)}$$

and the complex envelope is

$$\tilde{x}(t) = r(t)e^{-j\phi(t)}. \qquad \text{Eq. (4.55)}$$

It follows that

$$x(t) = Re\{\psi(t)\}. \qquad \text{Eq. (4.56)}$$

The returns from two close targets are, respectively, given by

$$x_1(t) = \psi(t - \tau_0) \qquad \text{Eq. (4.57)}$$

$$x_2(t) = \psi(t - \tau_0 - \tau) \qquad \text{Eq. (4.58)}$$

where τ is the difference in delay between the two target returns. One can assume that the reference time is τ_0, and thus without any loss of generality, one may set $\tau_0 = 0$. It follows that the two targets are distinguishable by how large or small the delay τ can be.

In order to measure the difference in range between the two targets, consider the integral square error between $\psi(t)$ and $\psi(t - \tau)$. Denoting this error as ε_R^2, it follows that

$$\varepsilon_R^2 = \int_{-\infty}^{\infty} |\psi(t) - \psi(t - \tau)|^2 \, dt, \qquad \text{Eq. (4.59)}$$

which can be written as

$$\varepsilon_R^2 = \int_{-\infty}^{\infty} |\psi(t)|^2 \ dt + \int_{-\infty}^{\infty} |\psi(t-\tau)|^2 \ dt - \int_{-\infty}^{\infty} \{(\psi(t)\psi^*(t-\tau) + \psi^*(t)\psi(t-\tau)) \ dt\} . \quad \textbf{Eq. (4.60)}$$

Using Eq. (4.54) into Eq. (4.60) yields

$$\varepsilon_R^2 = 2\int_{-\infty}^{\infty} |\tilde{x}(t)|^2 \ dt - 2Re\left\{\int_{-\infty}^{\infty} \psi^*(t)\psi(t-\tau) \ dt\right\} = \qquad \textbf{Eq. (4.61)}$$

$$2\int_{-\infty}^{\infty} |\tilde{x}(t)|^2 \ dt - 2Re\left\{ e^{-j\omega_0\tau} \int_{-\infty}^{\infty} \tilde{x}^*(t)\tilde{x}(t-\tau) \ dt\right\}$$

This squared error is minimum when the second portion of Eq. (4.61) is positive and maximum. Note that the first term in the right-hand side of Eq. (4.61) represents the total signal energy, and is assumed to be constant. The second term is a varying function of τ with its fluctuation tied to the carrier frequency. The integral inside the rightmost side of this equation is defined as the range ambiguity function,

$$\chi_R(\tau) = \int_{-\infty}^{\infty} \tilde{x}^*(t)\tilde{x}(t-\tau) \ dt . \qquad \textbf{Eq. (4.62)}$$

This range ambiguity function is equivalent to the integral given in Eq. (4.38) with $t_0 = 0$. Comparison between Eq. (4.62) and Eq. (4.38) indicates that the output of the matched filter and the range ambiguity function have the same envelope (in this case the Doppler shift f_d is set to zero). This indicates that the matched filter, in addition to providing the maximum instantaneous SNR at its output, also preserves the signal range resolution properties. The value of $\chi_R(\tau)$ that minimizes the squared error in Eq. (4.61) occurs when $\tau = 0$.

Target resolvability in range is measured by the squared magnitude $|\chi_R(\tau)|^2$. It follows that if $|\chi_R(\tau)| = \chi_R(0)$ for some nonzero value of τ, then the two targets are indistinguishable. Alternatively, if $|\chi_R(\tau)| \neq \chi_R(0)$ for some nonzero value of τ, then the two targets may be distinguishable (resolvable). As a consequence, the most desirable shape for $\chi_R(\tau)$ is a very sharp peak (thumb tack shape) centered at $\tau = 0$ and falling very quickly away from the peak. The minimum range resolution corresponding to a time duration τ_e or effective bandwidth B_e is

$$\Delta R = \frac{c\tau_e}{2} = \frac{c}{2B_e} . \qquad \textbf{Eq. (4.63)}$$

The effective time duration and the effective bandwidth for any waveform were defined in Chapter 3 and are repeated here as Eq. (4.64) and Eq. (4.65), respectively

$$\tau_e = \left[\int_{-\infty}^{\infty} |\tilde{x}(t)|^2 dt\right]^2 / \int_{-\infty}^{\infty} |\tilde{x}(t)|^4 dt \qquad \textbf{Eq. (4.64)}$$

$$B_e = \frac{\left[\displaystyle\int_{-\infty}^{\infty} |\tilde{X}(f)|^2 \ df \right]^2}{\displaystyle\int_{-\infty}^{\infty} |\tilde{X}(f)|^4 \ df} .$$

Eq. (4.65)

4.3.2. Doppler Resolution

The Doppler shift corresponding to the target radial velocity is

$$f_d = \frac{2v}{\lambda} = \frac{2vf_0}{c}$$

Eq. (4.66)

where v is the target radial velocity, λ is the wavelength, f_0 is the frequency, and c is the speed of light.

The FT of the pre-envelope is

$$\Psi(f) = \int_{-\infty}^{\infty} \psi(t) e^{-j2\pi ft} \ dt .$$

Eq. (4.67)

Due to the Doppler shift associated with the target, the received signal spectrum will be shifted by f_d. In other words, the received spectrum can be represented by $\Psi(f - f_d)$. In order to distinguish between the two targets located at the same range but having different velocities, one may use the integral square error. More precisely,

$$\varepsilon_f^2 = \int_{-\infty}^{\infty} |\Psi(f) - \Psi(f - f_d)|^2 \ df .$$

Eq. (4.68)

Using similar analysis as that which led to Eq. (4.61), one should maximize

$$Re \left\{ \int_{-\infty}^{\infty} \Psi^*(f) \Psi(f - f_d) \ df \right\} .$$

Eq. (4.69)

Taking the FT of the pre-envelope (analytic signal) defined in Eq. (4.54) yields

$$\Psi(f) = \tilde{X}(2\pi f - 2\pi f_0) .$$

Eq. (4.70)

Thus,

$$\int_{-\infty}^{\infty} \tilde{X}^*(2\pi f) \tilde{X}(2\pi f - 2\pi f_d) \ df = \int_{-\infty}^{\infty} \tilde{X}^*(2\pi f - 2\pi f_0) \tilde{X}(2\pi f - 2\pi f_0 - 2\pi f_d) \ df .$$

Eq. (4.71)

The complex frequency correlation function is then defined as

$$\chi_f(f_d) = \int_{-\infty}^{\infty} \tilde{X}^*(2\pi f)\tilde{X}(2\pi f - 2\pi f_d) \ df = \int_{-\infty}^{\infty} |\tilde{x}(t)|^2 \ e^{j2\pi f_d t} \ dt \ . \qquad \text{Eq. (4.72)}$$

The velocity resolution (Doppler resolution) is by definition

$$\Delta v = (c\Delta f_d)/(2f_0) \qquad \text{Eq. (4.73)}$$

where Δf_d is the minimum resolvable Doppler difference between the Doppler frequencies corresponding to two moving targets, i.e., $\Delta f_d = f_{d1} - f_{d2}$, where f_{d1} and f_{d2} are the two individual Doppler frequencies for targets 1 and 2, respectively. The Doppler resolution Δf_d is equal to the inverse of the total effective duration of the waveform. Thus,

$$\Delta f_d = \frac{\displaystyle\int_{-\infty}^{\infty} |\chi_f(f_d)|^2 df_d}{\chi_f^2(0)} = \frac{\displaystyle\int_{-\infty}^{\infty} |\tilde{x}(t)|^4 dt}{\left[\displaystyle\int_{-\infty}^{\infty} |\tilde{x}(t)|^2 dt\right]^2} = \frac{1}{\tau_e} \ . \qquad \text{Eq. (4.74)}$$

4.3.3. Combined Range and Doppler Resolution

In this general case, one needs to use a two-dimensional function in the pair of variables (τ, f_d). For this purpose, assume that the pre-envelope of the transmitted waveform is

$$\psi(t) = \tilde{x}(t)e^{j2\pi f_0 t} \ . \qquad \text{Eq. (4.75)}$$

Then the delayed and Doppler-shifted signal is

$$\psi(t - \tau) = \tilde{x}(t - \tau)e^{j2\pi(f_0 - f_d)(t - \tau)} \ . \qquad \text{Eq. (4.76)}$$

Computing the integral square error between Eq. (4.75) and Eq. (4.76) yields

$$\varepsilon^2 = \int_{-\infty}^{\infty} |\psi(t) - \psi(t - \tau)|^2 dt \qquad \text{Eq. (4.77a)}$$

$$\varepsilon^2 = 2\int_{-\infty}^{\infty} |\psi(t)|^2 dt - 2Re\left\{\int_{-\infty}^{\infty} \psi^*(t) - \psi(t - \tau)dt\right\} \qquad \text{Eq. (4.77b)}$$

which can be written as

$$\varepsilon^2 = 2\int_{-\infty}^{\infty} |\tilde{x}(t)|^2 \ dt - 2Re\left\{e^{j2\pi(f_0 - f_d)\tau}\int_{-\infty}^{\infty} \tilde{x}(t)\tilde{x}^*(t - \tau)e^{j2\pi f_d t} dt\right\} \ . \qquad \text{Eq. (4.78)}$$

Again, in order to maximize this squared error for $\tau \neq 0$, one must minimize the last term of Eq. (4.78). Define the combined range and Doppler correlation function as

$$\chi(\tau, f_d) = \int_{-\infty}^{\infty} \tilde{x}(t)\tilde{x}*(t-\tau)e^{j2\pi f_d t} dt \,. \qquad \text{Eq. (4.79)}$$

In order to achieve the most range and Doppler resolution, the modulus square of this function must be minimized at $\tau \neq 0$ and $f_d \neq 0$. Note that except for a phase term, the output of the matched filter derived in Eq. (4.52) is identical to that given in Eq. (4.79). This means that the output of the matched filter exhibits maximum instantaneous SNR as well as the most achievable range and Doppler resolutions. The modulus square of Eq. (4.79) is often referred to as the ambiguity function:

$$\left|\chi(\tau, f_d)\right|^2 = \left|\int_{-\infty}^{\infty} \tilde{x}(t)\tilde{x}*(t-\tau)e^{j2\pi f_d t} dt\right|^2 \,. \qquad \text{Eq. (4.80)}$$

The ambiguity function is often used by radar designers and analysts to determine the *goodness* of a given radar waveform, where this *goodness* is measured by its range and Doppler resolutions. Remember that since the matched filter is used, maximum SNR is guaranteed.

4.4. Range and Doppler Uncertainty

The formula derived in Eq. (4.79) represents the output of the matched filter when the signal at its input comprises target returns only and has no noise components, an assumption that cannot be true in practical situations. In general, the input at the matched filter contains both target and noise returns. The noise signal is assumed to be an additive random process that is uncorrelated with the target and has a band-limited white spectrum. Referring to Eq. (4.79), a peak at the output of the matched filter at (τ_1, f_{d1}) represents a target whose delay (range) corresponds to τ_1 and Doppler frequency equal to f_{d1}. Therefore, measuring the targets' exact range and Doppler frequency is determined from measuring peak locations occurring in the two-dimensional space (τ, f_d). This last statement, however, is correct only if noise is not present at the input of the matched filter. When noise is present and because noise is random, it will generate ambiguity (uncertainty) about the exact location of the ambiguity function peaks in the (τ, f_d) space.

4.4.1. Range Uncertainty

Consider the received signal complex envelope (assuming stationary target); that is,

$$\tilde{x}_i(t) = \tilde{x}(t-t_0) + \tilde{n}(t) = \tilde{x}_r(t) + \tilde{n}(t) \qquad \text{Eq. (4.81)}$$

where $\tilde{x}_r(t)$ is the target return signal complex envelope, $\tilde{n}(t)$ is the noise signal complex envelope, and $t_0 = 2R/c$, where R is the target range. The integral squared error between the total received signal (target plus noise) and a shifted (delayed by τ) transmitted waveform is

$$\varepsilon^2 = \int_0^{T_{max}} \left|\tilde{x}(t-\tau) - \tilde{x}_i(t)\right|^2 dt \,. \qquad \text{Eq. (4.82)}$$

T_{max} corresponds to maximum range under consideration. Expanding this squared error yields

$$\varepsilon^2 = 2 \int_0^{T_{max}} |\tilde{x}(t)|^2 \, dt + 2 \int_0^{T_{max}} |\tilde{n}(t)|^2 \, dt - 2Re\left\{ \int_0^{T_{max}} \tilde{x}^*(t-\tau)\tilde{x}_i(t)dt \right\} \qquad \text{Eq. (4.83)}$$

which can be written as

$$\varepsilon^2 = E_x + E_n - 2Re\left\{ \int_0^{T_{max}} \tilde{x}^*(t-\tau)\tilde{x}_r(t)dt + \int_0^{T_{max}} \tilde{x}^*(t-\tau)\tilde{n}(t)dt \right\}. \qquad \text{Eq. (4.84)}$$

This expression is minimum at some τ that makes the integral term inside Eq. (4.88) maximum and positive. More precisely, the following correlation functions must be maximized

$$R_{x_r x}(\tau) = \int_0^{T_{max}} \tilde{x}^*(t-\tau)\tilde{x}_r(t)dt \qquad \text{Eq. (4.85)}$$

$$R_{nx}(\tau) = \int_0^{T_{max}} \tilde{x}^*(t-\tau)\tilde{n}(t)dt. \qquad \text{Eq. (4.86)}$$

Therefore, Eq. (4.84) can be written as

$$\varepsilon^2 = E - 2Re\{R_{x_r x}(\tau) + R_{nx}(\tau)\}. \qquad \text{Eq. (4.87)}$$

Expanding $\{R_{x_r x}(\tau)\}$ using Taylor series expansion about the point $\tau = t_0$ leads to

$$R_{x_r x}(\tau) = R_{x_r x}(t_0) + R'_{x_r x}(t_0)(\tau - t_0) + \frac{R''_{x_r x}(t_0)(\tau - t_0)^2}{2!} + \frac{R'''_{x_r x}(t_0)(\tau - t_0)^3}{3!} + \dots \quad \text{Eq. (4.88)}$$

where R', R'', and R''' respectively, indicate the first, second, and third derivatives of $R_{x_r x}$ with respect to τ. Remember that since the real part of the correlation function is an even function, then all of its odd number derivatives are equal to zero. Now, by approximating Eq. (4.88) using the first three terms (where the second and fourth terms are equal to zero) one gets

$$Re\{R_{x_r x}(\tau)\} \approx R_{x_r x}(t_0) + \frac{R''_{x_r x}(t_0)(\tau - t_0)^2}{2!}. \qquad \text{Eq. (4.89)}$$

There is some value τ_1 close to the exact target range, t_0, that will minimize the expression in Eq. (4.87). To find this minimum value, differentiate the quantity $Re\{R_{x_r x}(\tau) + R_{nx}(\tau)\}$ with respect to τ and set the result equal to zero to find τ_1. More specifically,

$$Re\left\{ \frac{d}{d\tau}R_{x_r x}(\tau) + \frac{d}{d\tau}R_{nx}(\tau) \right\} = Re\{R'_{x_r x}(\tau) + R'_{nx}(\tau)\} = 0. \qquad \text{Eq. (4.90)}$$

The derivative of the $Re\{R_{x_r x}(\tau)\}$ can be found from Eq. (4.89) as

$$Re\left\{\frac{d}{d\tau}R_{x,x}(\tau)\right\} = \frac{d}{d\tau}\left(R_{x,x}(t_0) + \frac{R''_{x,x}(t_0)(\tau - t_0)^2}{2!}\right) = R''_{x,x}(t_0)(\tau - t_0). \qquad \text{Eq. (4.91)}$$

Substituting the result of Eq. (4.91) into Eq. (4.90), collecting terms, and solving for τ_1, yield

$$(\tau_1 - t_0) = -\frac{Re\{R'_{nx}(\tau_1)\}}{R''_{x,x}(t_0)}. \qquad \text{Eq. (4.92)}$$

The value $(\tau_1 - t_0)$ represent the amount of target range error measurement. It is more meaningful, since noise is random, to compute this error in terms of the standard deviation of its rms value. Hence, the standard deviation for range measurement error is

$$\sigma_\tau = (\tau_1 - t_0)_{rms} = -\frac{Re\{R'_{nx}(\tau_1)\}_{rms}}{R''_{x,x}(t_0)}. \qquad \text{Eq. (4.93)}$$

By using the differentiation property of the Fourier transform and Parseval's theorem the denominator of Eq. (4.93) can be determined by

$$R''_{x,x}(t_0) = (2\pi)^2 \int_{-\infty}^{\infty} f^2 \; |X(f)|^2 df. \qquad \text{Eq. (4.94)}$$

Next, from relations developed in Chapter 3, one can write the FT of $R_{nx}(\tau)$ as

$$FT\{R_{nx}(\tau)\} = X^*(f)\frac{\eta_0}{2} \qquad \text{Eq. (4.95)}$$

where $\eta_0/2$ is the noise power spectrum density value (white noise). From the Fourier transform properties, the FT of the derivative of $R_{nx}(\tau)$ is

$$FT\{R'_{nx}(\tau)\} = (j2\pi f)\left(X^*(f)\frac{\eta_0}{2}\right) = (j2\pi f)S_{nx}(f). \qquad \text{Eq. (4.96)}$$

The rms value for $R'_{nx}(\tau)$ is by definition

$$\{R'_{nx}(\tau)\}_{rms} = \sqrt{\lim_{T_{max}} \frac{1}{T_{max}} \int_{0}^{T_{max}} R'_{nx}(\tau) \; d\tau}, \qquad \text{Eq. (4.97)}$$

which can be rewritten using Parseval's theorem as

$$\{R'_{nx}(\tau)\}_{rms} = \sqrt{\int_{0}^{T_{max}} |FT\{R'_{nx}(\tau)\}|^2 \; df}. \qquad \text{Eq. (4.98)}$$

Substituting Eq. (4.96) into Eq. (4.98) yields

$$\{R'_{nx}(\tau)\}_{rms} = \sqrt{\frac{\eta_0}{2}(2\pi)^2 \int_{0}^{T_{max}} f^2 \; |X(f)|^2 \; df}. \qquad \text{Eq. (4.99)}$$

Finally, the standard deviation for range measurement error can be written as

$$\sigma_\tau = \frac{\sqrt{\eta_0/2}}{\sqrt{(2\pi)^2 \int\limits_{-\infty}^{\infty} f^2 \, |X(f)|^2 df}} .$$

Eq. (4.100)

Define the bandwidth rms value, B_{rms}^2, as

$$B_{rms}^2 = \frac{(2\pi)^2 \int\limits_{-\infty}^{\infty} f^2 \, |X(f)|^2 df}{\int\limits_{-\infty}^{\infty} |X(f)|^2 df} .$$

Eq. (4.101)

It follows that Eq. (4.100) can now be written as

$$\sigma_\tau = \frac{\sqrt{\eta_0/2}}{B_{rms}\sqrt{\int\limits_{-\infty}^{\infty} |X(f)|^2 df}} = \frac{\sqrt{\eta_0/2}}{B_{rms}\sqrt{E_x}} = \frac{1}{B_{rms}\sqrt{2E_x/\eta_0}} ,$$

Eq. (4.102)

which leads to the conclusion that the uncertainty in range measurement is inversely proportional to the rms bandwidth and the square root of the ratio of signal energy to the noise power density (square root of the SNR).

4.4.2. Doppler Uncertainty

For this purpose, assume that the target range is completely known. In the next section the case where both target range and target Doppler are not known will be analyzed. Denote the signal transmitted by the radar as $x(t)$ and the received signal (target plus noise) as $x_r(t)$. The integral square difference between the two returns can be written as

$$\varepsilon^2 = \int\limits_{0}^{f_{max}} |X(f-f_c) - X_r(f)|^2 \, df$$

Eq. (4.103)

where $X(f)$ is the FT of $x(t)$, $X_r(f)$ is the FT of $x_r(t)$, and f_{max} is the maximum anticipated target Doppler. Again expand Eq. (4.103) to get

$$\varepsilon^2 = \int\limits_{0}^{f_{max}} |X(f)|^2 \, df + \int\limits_{0}^{f_{max}} |X_r(f)|^2 \, df - 2Re\left\{ \int\limits_{0}^{f_{max}} |X^*(f-f_c)X_r(f)|^2 \, df \right\} .$$

Eq. (4.104)

Minimizing the error squared in Eq. (4.104) requires maximizing the value

$$Re\left\{ \int_0^{f_{max}} |X^*(f-f_c)X_r(f)|^2 \ df \right\}.$$

Conducting similar analysis as that performed in the previous section, the duration rms, τ_{rms}^2, value can be defined as

$$\tau_{rms}^2 = \frac{(2\pi)^2 \int_{-\infty}^{\infty} t^2 \ |x(t)|^2 dt}{\int_{-\infty}^{\infty} |x(t)|^2 dt} . \qquad \text{Eq. (4.105)}$$

The standard deviation in the Doppler measurement can be derived as

$$\sigma_{f_d} = \frac{1}{\tau_{rms}\sqrt{2E_x/\eta_0}} . \qquad \text{Eq. (4.106)}$$

Comparison of Eq. (4.106) and Eq. (4.102) indicates that the error in estimating Doppler is inversely proportional to the signal duration, while the error in estimating range is inversely proportional to the signal bandwidth. Therefore, and as expected, larger bandwidths minimize the range measurement errors and longer integration periods minimize the Doppler measurement errors.

4.4.3. Range-Doppler Coupling

In the previous two sections, range estimate error and Doppler estimate error were derived by assuming that they are uncoupled estimates. In other words, range error was derived assuming a stationary target, while Doppler error was derived assuming a completely known target range. In this section a more general formula for the combined range and Doppler errors is derived.

The analytic signal for this case was derived in Section 4.2 and was given in Eq. (4.47), which is repeated here as Eq. (4.107) for easy reference:

$$\psi_i(t) = \tilde{x}(t-t_0)e^{j2\pi f_0 t}e^{-j2\pi f_0 t_0}e^{j2\pi f_d t}e^{-j2\pi f_d t_0} = \tilde{x}(t-t_0)e^{j2\pi(f_0+f_d)(t-t_0)} \qquad \text{Eq. (4.107)}$$

One can assume with any loss of generality that $t_0 = 0$, thus, Eq. (4.107) can be expressed as

$$\psi_i(t) = \tilde{x}_r(t)e^{j2\pi(f_0+f_d)t} = r(t)e^{j\varphi(t)}e^{j2\pi(f_0+f_d)t} \qquad \text{Eq. (4.108)}$$

where the complex envelope signal, $\tilde{x}_r(t)$, can be expressed as

$$\tilde{x}_r(t) = r(t)e^{j\varphi(t)} . \qquad \text{Eq. (4.109)}$$

Range Error Estimate

From the analysis performed in the previous section, the estimate for the range error is determined by maximizing the function

$$Re\{R_{x_r x}(\tau, f_d) + R_{nx}(\tau)\} \ .$$

Eq. (4.110)

It follows that for some fixed value f_{d1}, there is a value τ_1 close to $t_0 = 0$ that will maximize Eq. (4.110); that is,

$$Re\{R'_{x_r x}(\tau_1, f_{d1}) + R'_{nx}(\tau_1)\} = 0 \ .$$

Eq. (4.111)

Again, the Taylor series expansion of $R_{x_r x}$ about $\tau = 0$ is

$$R_{x_r x}(\tau, f_d) = Re\left\{ R_{x_r x}(0, f_{d1}) + R'_{x_r x}(0, f_{d1})(\tau) + \frac{R''_{x_r x}(0, f_{d1})\tau^2}{2!} + \dots \right\} \ .$$

Eq. (4.112)

Thus,

$$Re\left\{ \frac{d}{d\tau} R_{x_r x}(\tau, f_d) \right\} \approx Re\{R'_{x_r x}(0, f_{d1}) + R''_{x_r x}(0, f_{d1})\tau\} \ .$$

Eq. (4.113)

Substituting Eq. (4.113) into Eq. (4.111) and solving for τ_1 yields

$$\tau_1 = -\frac{Re\{R'_{nx}(\tau_1) + R'_{x_r x}(0, f_{d1})\}}{Re\{R''_{x_r x}(0, f_{d1})\}} \ .$$

Eq. (4.114)

The value of $R''_{x_r x}(0, f_{d1})$ is not much different from $R''_{x_r x}(0, 0)$; thus,

$$\tau_1 \approx -\frac{Re\{R'_{nx}(\tau_1) + R'_{x_r x}(0, f_{d1})\}}{R''_{x_r x}(0, 0)} \ .$$

Eq. (4.115)

To evaluate the term $R'_{x_r x}(0, f_{d1})$, start with the definition of $R_{x_r x}(\tau, f_d)$,

$$R_{x_r x}(\tau, f_d) = \int_{-\infty}^{\infty} r(t-\tau)e^{-j\varphi(t-\tau)} r(t)e^{j(\varphi(t) + 2\pi f_d t)} dt \ .$$

Eq. (4.116)

Compute the derivative of Eq. (4.116) with respect to τ

$$R'_{x_r x}(\tau, f_d) = -\int_{-\infty}^{\infty} \{r'(t-\tau)r(t) - j\varphi'(t-\tau)r(t-\tau)r(t)\} \times e^{j[\varphi(t) - \varphi(t-\tau) + 2\pi f_d t]} dt \ .$$ Eq. (4.117)

Evaluating Eq. (4.117) at $\tau = 0$ and $f_d = f_{d1}$ gives

$$R'_{x_r x}(0, f_{d1}) = -\int_{-\infty}^{\infty} \{r'(t)r(t) - j\varphi'(t)r^2(t)\} \times e^{j[2\pi f_{d1} t]} dt \ .$$

Eq. (4.118)

The exponential term in Eq. (4.118) can be approximated using small angle approximation as

$$e^{j[2\pi f_{d1} t]} = \cos(2\pi f_{d1} t) + j\sin(2\pi f_{d1} t) \approx 1 + j2\pi f_{d1} t \ .$$

Eq. (4.119)

Next, substitute Eq. (4.119) into Eq. (4.118), collect terms, and compute its real part to get

$$Re\{R'_{x,x}(0,f_{d1})\} = -\int_{-\infty}^{\infty} r'(t)r(t)dt - 2\pi f_{d1} \int_{-\infty}^{\infty} t\varphi'(t)r^2(t)dt. \qquad \text{Eq. (4.120)}$$

The first integral is evaluated (using FT properties and Parseval's theorem) as

$$\int_{-\infty}^{\infty} r'(t)r(t)dt = (j2\pi) \int_{-\infty}^{\infty} f_d |R(f)|^2 df. \qquad \text{Eq. (4.121)}$$

Remember that since the envelope function $r(t)$ is a real lowpass signal, its Fourier transform is an even function; thus, Eq. (4.121) is equal to zero. Using this result, Eq. (4.120) becomes

$$Re\{R'_{x,x}(0,f_{d1})\} = -2\pi f_{d1} \int_{-\infty}^{\infty} t\varphi'(t)r^2(t)dt. \qquad \text{Eq. (4.122)}$$

Substitute Eq. (4.122) into Eq. (4.115) to get

$$\tau_1 = -\frac{Re\{R'_{x,x}(\tau_1)\} - 2\pi f_{d1} \int_{-\infty}^{\infty} t\varphi'(t)r^2(t)dt}{R''_{x,x}(0,0)}. \qquad \text{Eq. (4.123)}$$

Equation (4.123) provides a measure for the degree of coupling between range and Doppler estimates. Clearly, if $\varphi(t) = 0 \Rightarrow \varphi'(t) = 0$, then there is zero coupling between the two estimates. Define the range-Doppler coupling constant as

$$\rho_{\tau RDC} = \frac{2\pi \int_{-\infty}^{\infty} t\varphi'(t)|\tilde{x}_r(t)|^2 dt}{\int_{-\infty}^{\infty} |\tilde{x}_r(t)|^2 dt}. \qquad \text{Eq. (4.124)}$$

Doppler Error Estimate

Applying similar analysis as that performed in the preceding section to the spectral cross correlation function yields an expression for the range-Doppler coupling term. It is given by

$$\rho_{f_d RDC} = \frac{2\pi \int_{-\infty}^{\infty} f\, \Phi'(f)|\tilde{X}_r(f)|^2 df}{\int_{-\infty}^{\infty} |\tilde{X}_r(f)|^2 df} \qquad \text{Eq. (4.125)}$$

where $\Phi(f)$ is the FT of $\varphi(t)$.

It can be shown that Eq. (4.124) and Eq. (4.125) are equal. Given this result, the subscripts τ and f_d in Eq. (4.124) and Eq. (4.125) are dropped and the range-Doppler term is simply referred to as ρ_{RDC}.

4.4.4. Range-Doppler Coupling in LFM Signals

Referring to Eq. (4.108) and Eq. (4.109), the phase for an LFM signal can be expressed as

$$\varphi(t) = \mu' t^2$$

Eq. (4.126)

where $\mu' = (\pi B)/\tau_0$, B is the LFM bandwidth, and τ_0 is the pulse width. Substituting Eq. (4.126) into Eq. (4.124) yields

$$\rho_{RDC} = \frac{4\pi\mu' \int\limits_{-\infty}^{\infty} t^2 |\tilde{x}_r(t)|^2 dt}{\int\limits_{-\infty}^{\infty} |\tilde{x}_r(t)|^2 dt} = \frac{\mu'}{\pi}\tau_e^2$$

Eq. (4.127)

where τ_e is the effective duration. Thus,

$$\sigma_\tau^2 = \frac{(\eta_0/2)}{B_e^2 2E_x} + \frac{f_{d1}^2 \rho_{RDC}^2}{B_e^4}.$$

Eq. (4.128)

Similarly,

$$\sigma_{f_d}^2 = \frac{(\eta_0/2)}{\tau_e^2 2E_x} + \frac{t_1^2 \rho_{RDC}^2}{\tau_e^4}$$

Eq. (4.129)

where f_{d1} and t_1 are constants. Since estimates of range or Doppler when noise is present cannot be 100% exact, it is better to replace these constants with their equivalent mean-squared errors. That is, let

$$f_{d1}^2 = \sigma_{fd}^2 \qquad , \qquad t_1^2 = \sigma_\tau^2$$

Eq. (4.130)

where σ_τ is as in Eq. (4.128) and σ_{fd} is in Eq. (4.129). Thus, Eq. (4.128) can be written as

$$\sigma_{\tau_{RDC}}^2 = \frac{(\eta_0/2)}{B_e^2 2E_x} + \frac{\rho_{RDC}^2}{B_e^4}\left(\frac{(\eta_0/2)}{\tau_e^2 2E_x} + \frac{\rho_{RDC}^2 \sigma_\tau^2}{\tau_e^4}\right),$$

Eq. (4.131)

which can be algebraically manipulated to get

$$\sigma_{\tau_{RDC}}^2 = \frac{(\eta_0/2)}{B_e^2 2E_x} \frac{1}{(1 - (\rho_{RDC}^2/B_e^2\tau_e^2))}.$$

Eq. (4.132)

Using similar analysis,

$$\sigma^2_{f_{dRDC}} = \frac{(\eta_0/2)}{\tau_e^2 2E_x} \frac{1}{(1-(\rho_{RDC}^2/B_e^2\tau_e^2))}.$$ **Eq. (4.133)**

These results lead to the conclusion that one can estimate target range and Doppler simultaneously only when the product of the rms bandwidth and rms duration is very large (i.e., very large time bandwidth products). This is the reason radars using LFM waveforms cannot estimate target Doppler accurately unless very large time bandwidth products are utilized. Often, the LFM waveforms are referred to as "Doppler insensitive" waveforms.

4.5. Target Parameter Estimation

Target parameters of interest to radar applications include, but are not limited to, target range (delay), amplitude, phase, Doppler, and angular location (azimuth and elevation). Target information (parameters) is typically embedded in the return signal's amplitude and phase. Different classes of waveforms are used by the radar signal and data processors to extract different target parameters more efficiently than others. Since radar echoes typically comprise signal plus additive noise, most if not all the target information is governed by the statistics of the input noise, whose statistical parameters most likely are not known but can be estimated. Thus, statistical estimates of the target parameters (amplitude, phase, delay, Doppler, etc.) are utilized instead of the actual corresponding measurements. The general form of the radar signal can be expressed in the following form

$$x(t) = Ar(t-t_0)\cos[2\pi(f_0+f_d)(t-t_0)+\phi(t-t_0)+\phi_0]$$ **Eq. (4.134)**

where A is the signal amplitude, $r(t)$ is the envelope lowpass signal, ϕ_0 is some constant phase, f_0 is the carrier frequency, and t_0 and f_d are the target delay and Doppler, respectively. The analysis in this section closely follows Melsa and Cohen[1].

4.5.1. What Is an Estimator?

In the case of radar systems, it always safe to assume, due to the central limit theorem, that the input noise is always Gaussian with mainly unknown parameters. Furthermore, one can assume that this noise is band-limited white noise. Consequently, the primary question that needs to be answered is as follows: Given that the probability density function of the observation is known (Gaussian in this case) and given a finite number of independent measurements, can one determine an estimate of a given parameter (such as range, Doppler, amplitude, or phase)?

Let $f_X(x;\theta)$ be the *pdf* of a random variable X with an unknown parameter θ. Define the values $\{x_1, x_2, ..., x_N\}$ as N observed independent values of the variable $\{X\}$. Define the function or estimator $\theta(x_1, x_2, ..., x_N)$ as an estimate of the unknown parameter θ. The bias of estimation is defined as

$$E[\hat{\theta}-\theta] = b$$ **Eq. (4.135)**

where $E[\ ]$ represents the "expected value of." The estimator $\hat{\theta}$ is referred to as an unbiased estimator if and only if

---

1. Melsa, J. L. Cohen, D. L., *Decision and Estimation Theory*, McGraw-Hill, New York, 1978.

$$E[\hat{\theta}] = \theta .$$

Eq. (4.136)

One of the most popular and common measures of the quality or effectiveness of an estimator is the Mean Square Deviation (MSD) referred to symbolically as $\Delta^2(\hat{\theta})$. For an unbiased estimator

$$\Delta^2(\hat{\theta}) = \sigma_{\hat{\theta}}^2$$

Eq. (4.137)

where $\sigma_{\hat{\theta}}^2$ is the estimator variance. It can be shown that the Cramer-Rao bound for this MSD is given by

$$\sigma^2(\hat{\theta}) \geq \sigma_{min}^2(\theta) = \frac{1}{N \int\limits_{-\infty}^{\infty} \left(\frac{\partial}{\partial \theta} \log \{f_X(x;\theta)\}\right)^2 f_X(x;\theta) \ dx}.$$

Eq. (4.138)

The efficiency of this unbiased estimator is defined by

$$\varepsilon(\hat{\theta}) = \frac{\sigma_{min}^2(\theta)}{\sigma^2(\hat{\theta})}.$$

Eq. (4.139)

When $\varepsilon(\hat{\theta}) = 1$, the unbiased estimator is called an efficient estimate.

Consider an essentially time-limited signal $x(t)$ with effective duration τ_e, and assume a band-limited white noise with PSD $\eta_0/2$. In this case, Eq. (4.139) is equivalent to

$$\sigma^2(\hat{\theta}_i) \geq \frac{1}{\frac{2}{\eta_0} \int\limits_0^{NT_r} \left(\frac{\partial}{\partial \theta_i} x(t)\right)^2 \ dt}$$

Eq. (4.140)

where $\hat{\theta}_i$ is the estimate for the ith parameter of interest and T_r is the pulse repetition interval for the pulsed sequence. In the next two sections, estimates of the target amplitude and phase are derived. It must be noted that since these estimates represent independent random variables, they are referred to as uncoupled estimates; that is, the computation of one estimate does not depend on a priori knowledge of the other estimates.

4.5.2. Amplitude Estimation

The signal amplitude A in Eq. (4.134) is the parameter of interest, in this case. Taking the partial derivative of Eq. (4.134) with respect to A and squaring the result yields

$$\left(\frac{\partial}{\partial A} x(t)\right)^2 = (r(t-t_0)\cos[2\pi(f_0+f_d)(t-t_0) + \phi(t-t_0) + \phi_0])^2$$

Eq. (4.141)

Thus,

$$\int\limits_0^{NT_r} \left(\frac{\partial}{\partial A} x(t)\right)^2 \ dt = \int\limits_0^{NT_r} (x(t))^2 \ dt = NE_x$$

Eq. (4.142)

where E_x is the signal energy (from Parseval's theorem). Substituting Eq. (4.142) into Eq. (4.140) and collecting terms yields the variance for the amplitude estimate as

$$\sigma_A^2 \geq \frac{1}{\dfrac{2}{\eta_0} NE_x} = \frac{1}{N \ SNR}.$$

Eq. (4.143)

In this case Eq. (4.20) used in Eq. (4.143) and SNR is the signal to noise ratio of the signal at the output of the matched filter. This clearly indicates that the signal amplitude estimate is improved as the SNR is increased.

4.5.3. Phase Estimation

In this case, it is desired to compute the best estimate for the signal phase ϕ_0. Again taking the partial derivative of the signal in Eq. (4.134) with respect to ϕ_0 and squaring the result yield

$$\left(\frac{\partial}{\partial\phi_0}x(t)\right)^2 = (-r(t-t_0)\sin[2\pi(f_0+f_d)(t-t_0)+\phi(t-t_0)+\phi_0])^2.$$

Eq. (4.144)

It follows that

$$\int_0^{NT_r}\left(\frac{\partial}{\partial\phi_0}x(t)\right)^2 \, dt = \int_0^{NT_r}(x(t))^2 \, dt = NE_x.$$

Eq. (4.145)

Thus, the variance of the phase estimate is

$$\sigma_{\phi_0}^2 \geq \frac{1}{\dfrac{2}{\eta_0}NE_x} = \frac{1}{N \ SNR}.$$

Eq. (4.146)

Problems

4.1. Compute the frequency response for the filter matched to the signal
(a) $x(t) = \exp\left(\dfrac{-t^2}{2T}\right)$;
(b) $x(t) = u(t)\exp(-\alpha t)$ where α is a positive constant.

4.2. Repeat the example in Section 4.1 using $x(t) = u(t)\exp(-\alpha t)$.

4.3. An closed form expression for the SNR at the output of the matched filter when the input noise is white was developed in Section 4.1.1. Derive an equivalent formula for the non-white noise case.

4.4. A radar system uses LFM waveforms. The received signal is of the form $s_r(t) = As(t-\tau)+n(t)$, where τ is a time delay that depends on range, $s(t) = Rect(t/\tau')\cos(2\pi f_0 t - \phi(t))$, and $\phi(t) = -\pi B t^2/\tau'$. Assume that the radar bandwidth is $B = 5MHz$, and the pulse width is $\tau' = 5\mu s$. (a) Give the quadrature components of the matched filter response that is matched to $s(t)$. (b) Write an expression for the output of the matched filter. (c) Compute the increase in SNR produced by the matched filter.

4.5. (a) Write an expression for the ambiguity function of an LFM waveform, where $\tau' = 6.4\mu s$ and the compression ratio is 32. (b) Give an expression for the matched filter impulse response.

4.6. (a) Write an expression for the ambiguity function of an LFM signal with bandwidth $B = 10MHz$, pulse width $\tau' = 1\mu s$, and wavelength $\lambda = 1cm$. (b) Plot the zero Doppler cut of the ambiguity function. (c) Assume a target moving toward the radar with radial velocity $v_r = 100m/s$. What is the Doppler shift associated with this target? (d) Plot the ambiguity function for the Doppler cut in part (c). (e) Assume that three pulses are transmitted with PRF $f_r = 2000Hz$. Repeat part (b).

4.7. (a) Give an expression for the ambiguity function for a pulse train consisting of 4 pulses, where the pulse width is $\tau' = 1\mu s$ and the pulse repetition interval is $T = 10\mu s$. Assume a wavelength of $\lambda = 1cm$. (b) Sketch the ambiguity function contour.

4.8. Hyperbolic frequency modulation (HFM) is better than LFM for high radial velocities. The HFM phase is

$$\phi_h(t) = \frac{\omega_0^2}{\mu_h}\ln\left(1 + \frac{\mu_h \alpha t}{\omega_0}\right)$$

where μ_h is an HFM coefficient and α is a constant. (a) Give an expression for the instantaneous frequency of an HFM pulse of duration τ'_h. (b) Show that HFM can be approximated by LFM. Express the LFM coefficient μ_l in terms of μ_h and in terms of B and τ'.

4.9. Consider a sonar system with range resolution $\Delta R = 4cm$. (a) A sinusoidal pulse at frequency $f_0 = 100KHz$ is transmitted. What is the pulse width, and what is the bandwidth? (b) By using an up-chirp LFM, centered at f_0, one can increase the pulse width for the same range resolution. If you want to increase the transmitted energy by a factor of 20, give an expression for the transmitted pulse. (c) Give an expression for the causal filter matched to the LFM pulse in part b.

4.10. A pulse train $y(t)$ is given by

$$y(t) = \sum_{n=0}^{2} w(n)x(t - n\tau')$$

where $x(t) = \exp(-t^2/2)$ is a single pulse of duration τ' and the weighting sequence is $\{w(n)\} = \{0.5, 1, 0.7\}$. Find and sketch the correlations R_x, R_w, and R_y.

4.11. Repeat the previous problem for $x(t) = \exp(-t^2/2)\cos 2\pi f_0 t$.

4.12. Show that

$$\int_{-\infty}^{\infty} t x^*(t) x'(t) \ dt = -\int_{-\infty}^{\infty} f X^*(f) X'(f) \ df$$

where $X(f)$, is the FT of $x(t)$ and $x'(t)$ is its derivative with respect to time. The function $X'(f)$ is the derivative of $X(f)$ with respect to frequency.

4.13. Using the range-Doppler coupling definition given in Eq. (4.125), develope an expression for the range-Doppler coupling for the following cases: (a) Linear FM pulse with a Gaussian envelope, and (b) parabolic FM signal.

Chapter 5

Ambiguity Function - Analog Waveforms

5.1. Introduction

The radar ambiguity function represents the modulus of the matched filter output, and it describes the interference caused by the range and/or Doppler shift of a target when compared to a reference target of equal RCS. The ambiguity function evaluated at $(\tau, f_d) = (0, 0)$ is equal to the matched filter output that is perfectly matched to the signal reflected from the target of interest. In other words, returns from the nominal target are located at the origin of the ambiguity function. Thus, the ambiguity function at nonzero τ and f_d represents returns from some range and Doppler different from those for the nominal target.

The formula for the output of the matched filter was derived in Chapter 4, and it is, assuming a moving target with Doppler frequency f_d,

$$\chi(\tau, f_d) = \int_{-\infty}^{\infty} \tilde{x}(t)\tilde{x}^*(t-\tau)e^{j2\pi f_d t}\,dt\,. \qquad \text{Eq. (5.1)}$$

The modulus square of Eq. (5.1) is referred to as the ambiguity function. That is,

$$|\chi(\tau, f_d)|^2 = \left| \int_{-\infty}^{\infty} \tilde{x}(t)\tilde{x}^*(t-\tau)e^{j2\pi f_d t}\,dt \right|^2\,. \qquad \text{Eq. (5.2)}$$

The radar ambiguity function is normally used by radar designers as a means of studying different waveforms. It can provide insight about how different radar waveforms may be suitable for the various radar applications. It is also used to determine the range and Doppler resolutions for a specific radar waveform. The three-dimensional (3-D) plot of the ambiguity function versus frequency and time delay is called the radar ambiguity diagram.

Denote E_x as the energy of the signal $\tilde{x}(t)$,

$$E_x = \int_{-\infty}^{\infty} |\tilde{x}(t)|^2\,dt\,. \qquad \text{Eq. (5.3)}$$

The following list includes the properties for the radar ambiguity function:

1) The maximum value for the ambiguity function occurs at $(\tau, f_d) = (0, 0)$ and is equal to $4E_x^2$,

$$max\{|\chi(\tau;f_d)|^2\} = |\chi(0;0)|^2 = (2E_x)^2 \qquad \text{Eq. (5.4)}$$

$$|\chi(\tau;f_d)|^2 \leq |\chi(0;0)|^2. \qquad \text{Eq. (5.5)}$$

2) The ambiguity function is symmetric,

$$|\chi(\tau;f_d)|^2 = |\chi(-\tau;-f_d)|^2. \qquad \text{Eq. (5.6)}$$

3) The total volume under the ambiguity function is constant,

$$\int\int |\chi(\tau;f_d)|^2 \ d\tau \ df_d = (2E_x)^2. \qquad \text{Eq. (5.7)}$$

4) If the function $X(f)$ is the Fourier transform of the signal $x(t)$, then by using Parseval's theorem we get

$$|\chi(\tau;f_d)|^2 = \left| \int X^*(f)X(f-f_d)e^{-j2\pi f\tau} df \right|^2. \qquad \text{Eq. (5.8)}$$

5) Suppose that $|\chi(\tau;f_d)|^2$ is the ambiguity function for the signal $\tilde{x}(t)$. Adding a quadratic phase modulation term to $\tilde{x}(t)$ yields

$$\tilde{x}_1(t) = \tilde{x}(t)e^{j\pi\mu t^2} \qquad \text{Eq. (5.9)}$$

where μ is a constant. It follows that the ambiguity function for the signal $\tilde{x}_1(t)$ is given by

$$|\chi_1(\tau;f_d)|^2 = |\chi(\tau;(f_d+\mu\tau))|^2. \qquad \text{Eq. (5.10)}$$

5.2. Examples of the Ambiguity Function

The ideal radar ambiguity function is represented by a spike of infinitesimal width that peaks at the origin and is zero everywhere else, as illustrated in Fig. 5.1. An ideal ambiguity function provides perfect resolution between neighboring targets regardless of how close they may be to each other. Unfortunately, an ideal ambiguity function cannot physically exist because the ambiguity function must have a finite peak value equal to $(2E_x)^2$ and a finite volume also equal to $(2E_x)^2$. Clearly, the ideal ambiguity function cannot meet those conditions.

5.2.1. Single Pulse Ambiguity Function

The complex envelope of a single pulse is $\tilde{x}(t)$ defined by

$$\tilde{x}(t) = \frac{1}{\sqrt{\tau_0}} Rect\left(\frac{t}{\tau_0}\right). \qquad \text{Eq. (5.11)}$$

From Eq. (5.1) we have

$$\chi(\tau;f_d) = \int_{-\infty}^{\infty} \tilde{x}(t)\tilde{x}^*(t-\tau)e^{j2\pi f_d t} dt. \qquad \text{Eq. (5.12)}$$

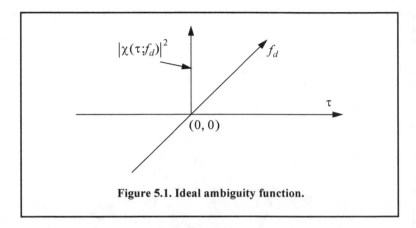

Figure 5.1. Ideal ambiguity function.

Substituting Eq. (5.11) into Eq. (5.12) and performing the integration yields

$$|\chi(\tau;f_d)|^2 = \left|\left(1 - \frac{|\tau|}{\tau_0}\right)\frac{\sin(\pi f_d(\tau_0 - |\tau|))}{\pi f_d(\tau_0 - |\tau|)}\right|^2 \qquad |\tau| \le \tau_0. \qquad \textbf{Eq. (5.13)}$$

MATLAB Function "single_pulse_ambg.m"

The MATLAB function *"single_pulse_ambg.m"* implements Eq. (5.13). The syntax is as follows:

single_pulse_ambg [taup]

where *taup* is the pulse width. Figures 5.2 a and b show 3-D and contour plots of single pulse ambiguity functions. This figure can be reproduced using MATLAB program *"Fig5_2.m"* listed in Appendix 5-A. The ambiguity function cut along the time-delay axis τ is obtained by setting $f_d = 0$. More precisely,

$$|\chi(\tau;0)| = \left(1 - \frac{|\tau|}{\tau_0}\right)^2 \qquad |\tau| \le \tau_0. \qquad \textbf{Eq. (5.14)}$$

Note that the time autocorrelation function of the signal $\tilde{x}(t)$ is equal to $\chi(\tau;0)$. Similarly, the cut along the Doppler axis is

$$|\chi(0;f_d)|^2 = \left|\frac{\sin \pi \tau_0 f_d}{\pi \tau_0 f_d}\right|^2. \qquad \textbf{Eq. (5.15)}$$

Figures 5.3 and 5.4, respectively, show the plots of the uncertainty function cuts defined by Eqs. (5.14) and (5.15). Since the zero Doppler cut along the time-delay axis extends between $-\tau_0$ and τ_0, close targets will be unambiguous if they are at least τ_0 seconds apart.

The zero time cut along the Doppler frequency axis has a $(\sin x / x)^2$ shape. It extends from $-\infty$ to ∞. The first null occurs at $f_d = \pm 1/\tau_0$. Hence, it is possible to detect two targets that are shifted by $1/\tau_0$, without any ambiguity. Thus, a single pulse range and Doppler resolutions are limited by the pulse width τ_0. Fine range resolution requires that a very short pulse be used. Unfortunately, using very short pulses requires very large operating bandwidths and may limit the radar average transmitted power to impractical values.

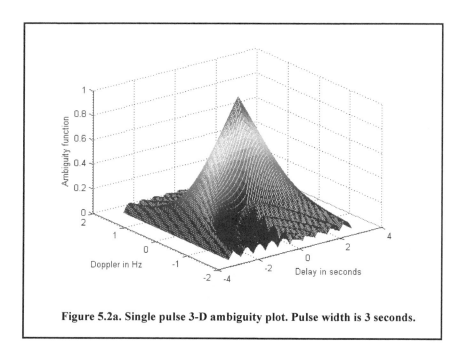

Figure 5.2a. Single pulse 3-D ambiguity plot. Pulse width is 3 seconds.

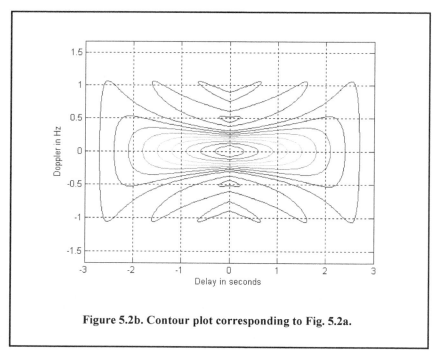

Figure 5.2b. Contour plot corresponding to Fig. 5.2a.

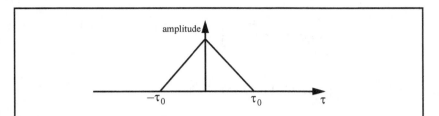

Figure 5.3. Zero Doppler ambiguity function cut along the time-delay axis.

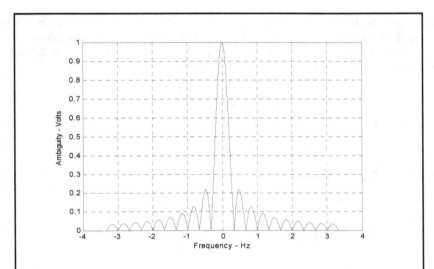

Figure 5.4. Ambiguity function of a single frequency pulse (zero delay). The pulse width is 3 seconds.

5.2.2. LFM Ambiguity Function

Consider the LFM complex envelope signal defined by

$$\tilde{x}(t) = \frac{1}{\sqrt{\tau_0}} Rect\left(\frac{t}{\tau_0}\right) e^{j\pi\mu t^2}.$$

Eq. (5.16)

In order to compute the ambiguity function for the LFM complex envelope, we will first consider the case when $0 \leq \tau \leq \tau_0$. In this case the integration limits are from $-\tau_0/2$ to $(\tau_0/2) - \tau$. Substituting Eq. (5.16) into Eq. (5.1) yields

$$\chi(\tau;f_d) = \frac{1}{\tau_0} \int_{-\infty}^{\infty} Rect\left(\frac{t}{\tau_0}\right) Rect\left(\frac{t-\tau}{\tau_0}\right) e^{j\pi\mu t^2} e^{-j\pi\mu(t-\tau)^2} e^{j2\pi f_d t} dt.$$

Eq. (5.17)

It follows that

$$\chi(\tau;f_d) = \frac{e^{-j\pi\mu\tau^2}}{\tau_0} \int\limits_{\frac{-\tau_0}{2}}^{\frac{\tau_0}{2}-\tau} e^{j2\pi(\mu\tau+f_d)t} \, dt \, . \qquad\qquad \textbf{Eq. (5.18)}$$

Finishing the integration process in Eq. (5.18) yields

$$\chi(\tau;f_d) = e^{j\pi\tau f_d}\left(1-\frac{\tau}{\tau_0}\right)\frac{\sin\left(\pi\tau_0(\mu\tau+f_d)\left(1-\frac{\tau}{\tau_0}\right)\right)}{\pi\tau_0(\mu\tau+f_d)\left(1-\frac{\tau}{\tau_0}\right)} \qquad 0\le\tau\le\tau_0. \qquad \textbf{Eq. (5.19)}$$

Similar analysis for the case when $-\tau_0 \le \tau \le 0$ can be carried out, where, in this case, the integration limits are from $(-\tau_0/2)-\tau$ to $\tau_0/2$. The same result can be obtained by using the symmetry property of the ambiguity function ($|\chi(-\tau,-f_d)| = |\chi(\tau,f_d)|$). It follows that an expression for $\chi(\tau;f_d)$ that is valid for any τ is given by

$$\chi(\tau;f_d) = e^{j\pi\tau f_d}\left(1-\frac{|\tau|}{\tau_0}\right)\frac{\sin\left(\pi\tau_0(\mu\tau+f_d)\left(1-\frac{|\tau|}{\tau_0}\right)\right)}{\pi\tau_0(\mu\tau+f_d)\left(1-\frac{|\tau|}{\tau_0}\right)} \qquad |\tau|\le\tau_0 \qquad \textbf{Eq. (5.20)}$$

and the LFM ambiguity function is

$$|\chi(\tau;f_d)|^2 = \left|\left(1-\frac{|\tau|}{\tau_0}\right)\frac{\sin\left(\pi\tau_0(\mu\tau+f_d)\left(1-\frac{|\tau|}{\tau_0}\right)\right)}{\pi\tau_0(\mu\tau+f_d)\left(1-\frac{|\tau|}{\tau_0}\right)}\right|^2 \qquad |\tau|\le\tau_0. \qquad \textbf{Eq. (5.21)}$$

Again the time autocorrelation function is equal to $\chi(\tau,0)$. The reader can verify that the ambiguity function for a down-chirp LFM waveform is given by

$$|\chi(\tau;f_d)|^2 = \left|\left(1-\frac{|\tau|}{\tau_0}\right)\frac{\sin\left(\pi\tau_0(\mu\tau-f_d)\left(1-\frac{|\tau|}{\tau_0}\right)\right)}{\pi\tau_0(\mu\tau-f_d)\left(1-\frac{|\tau|}{\tau_0}\right)}\right|^2 \qquad |\tau|\le\tau_0. \qquad \textbf{Eq. (5.22)}$$

Incidentally, either Eq. (5.21) or (5.22) can be obtained from Eq. (5.13) by applying property 5 from Section 5.1.

Figures 5.5 a and b show 3-D and contour plots for the LFM uncertainty and ambiguity functions for $\tau_0 = 1$ second and $B = 5Hz$ for a down-chirp pulse. This figure can be reproduced using MATLAB program "*Fig5_5.m*," listed in Appendix 5-A.

The up-chirp ambiguity function cut along the time-delay axis τ is

$$|\chi(\tau;0)|^2 = \left|\left(1-\frac{|\tau|}{\tau_0}\right)\frac{\sin\left(\pi\mu\tau\tau_0\left(1-\frac{|\tau|}{\tau_0}\right)\right)}{\pi\mu\tau\tau_0\left(1-\frac{|\tau|}{\tau_0}\right)}\right|^2 \qquad |\tau|\le\tau_0. \qquad \textbf{Eq. (5.23)}$$

MATLAB Function "lfm_ambg.m" Listing

The function *"lfm_ambg.m"* implements Eq. (5.21). The syntax is as follows:

$$lfm\_ambg \; [taup, \; b, \; up\_down]$$

where

Symbol	Description	Units	Status
taup	pulse width	seconds	input
b	bandwidth	Hz	input
up_down	up_down = 1 for up-chirp	none	input
	up_down = -1 for down-chirp		

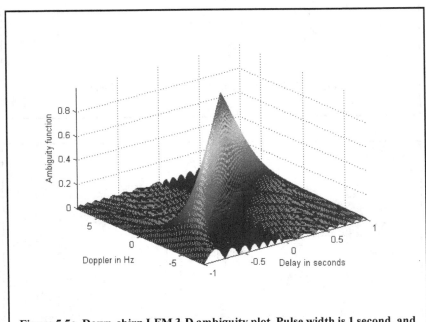

Figure 5.5a. Down-chirp LFM 3-D ambiguity plot. Pulse width is 1 second, and bandwidth is 5*Hz*.

Note that the LFM ambiguity function cut along the Doppler frequency axis is similar to that of the single pulse. This should not be surprising since the pulse shape has not changed (only frequency modulation was added). However, the cut along the time-delay axis changes significantly. It is now much narrower compared to the unmodulated pulse cut. In this case, the first null occurs at

$$\tau_{n1} \approx 1/B. \qquad\qquad \textbf{Eq. (5.24)}$$

Figure 5.6 shows a plot for a cut in the uncertainty function corresponding to Eq. (5.23). This figure can be reproduced using MATLAB program *"Fig5_6.m,"* listed in Appendix 5-A.

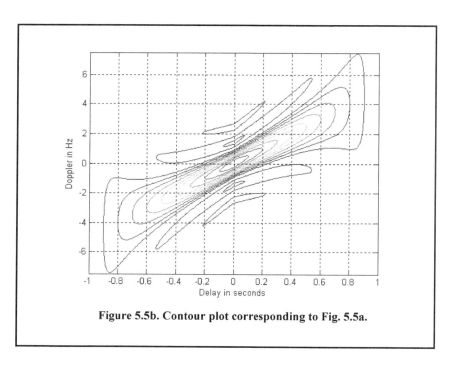

Figure 5.5b. Contour plot corresponding to Fig. 5.5a.

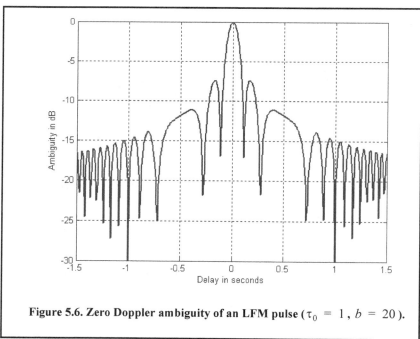

Figure 5.6. Zero Doppler ambiguity of an LFM pulse ($\tau_0 = 1$, $b = 20$).

Equation (5.24) indicates that the effective pulse width (compressed pulse width) of the matched filter output is completely determined by the radar bandwidth. It follows that the LFM ambiguity function cut along the time-delay axis is narrower than that of the unmodulated pulse by a factor

$$\xi = \frac{\tau_0}{(1/B)} = \tau_0 B \qquad \textbf{Eq. (5.25)}$$

ξ is referred to as the compression ratio (also called the time-bandwidth product and compression gain). All three names can be used interchangeably to mean the same thing. As indicated by Eq. (5.25), the compression ratio also increases as the radar bandwidth is increased.

Example:

Compute the range resolution before and after pulse compression corresponding to an LFM waveform with the following specifications: Bandwidth $B = 1GHz$ and pulse width $\tau_0 = 10ms$.

Solution:

The range resolution before pulse compression is

$$\Delta R_{uncomp} = \frac{c\tau_0}{2} = \frac{3 \times 10^8 \times 10 \times 10^{-3}}{2} = 1.5 \times 10^6 \ meters.$$

Using Eq. (5.24) yields

$$\tau_{n1} = \frac{1}{1 \times 10^9} = 1 \ ns$$

$$\Delta R_{comp} = \frac{c\tau_{n1}}{2} = \frac{3 \times 10^8 \times 1 \times 10^{-9}}{2} = 15 \ cm.$$

5.2.3. Coherent Pulse Train Ambiguity Function

Figure 5.7 shows a plot of a coherent pulse train. The pulse width is denoted as τ_0 and the PRI is T. The number of pulses in the train is N; hence, the train's length is $(N-1)T$ seconds. A normalized individual pulse $\tilde{x}(t)$ is defined by

$$\tilde{x}_1(t) = \frac{1}{\sqrt{\tau_0}} Rect\left(\frac{t}{\tau_0}\right). \qquad \textbf{Eq. (5.26)}$$

When coherency is maintained between the consecutive pulses, then an expression for the normalized train is

$$\tilde{x}(t) = \frac{1}{\sqrt{N}} \sum_{i=0}^{N-1} \tilde{x}_1(t - iT). \qquad \textbf{Eq. (5.27)}$$

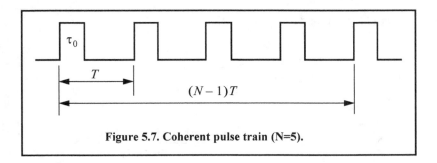

Figure 5.7. Coherent pulse train (N=5).

The output of the matched filter is

$$\chi(\tau;f_d) = \int_{-\infty}^{\infty} \tilde{x}(t)\tilde{x}*(t-\tau)e^{j2\pi f_d t}\,dt.$$

Eq. (5.28)

Substituting Eq. (5.27) into Eq. (5.28) and interchanging the summations and integration yield

$$\chi(\tau;f_d) = \frac{1}{N}\sum_{i=0}^{N-1}\sum_{j=0}^{N-1}\int_{-\infty}^{\infty}\tilde{x}_1(t-iT)\ \tilde{x}_1*(t-jT-\tau)e^{j2\pi f_d t}\,dt.$$

Eq. (5.29)

Making the change of variable $t_1 = t - iT$ yields

$$\chi(\tau;f_d) = \frac{1}{N}\sum_{i=0}^{N-1}e^{j2\pi f_d iT}\sum_{j=0}^{N-1}\int_{-\infty}^{\infty}\tilde{x}_1(t_1)\ \tilde{x}_1*(t_1 - [\tau-(i-j)T])e^{j2\pi f_d t_1}\,dt_1.$$

Eq. (5.30)

The integral inside Eq. (5.30) represents the output of the matched filter for a single pulse, and is denoted by χ_1. It follows that

$$\chi(\tau;f_d) = \frac{1}{N}\sum_{i=0}^{N-1}e^{j2\pi f_d iT}\sum_{j=0}^{N-1}\chi_1[\tau-(i-j)T;f_d].$$

Eq. (5.31)

When the relation $q = i-j$ is used, then the following relation is true:

$$\sum_{i=0}^{N}\sum_{m=0}^{N} = \left.\sum_{q=-(N-1)}^{0}\sum_{i=0}^{N-1-|q|}\right|_{for\ j=i-q} + \left.\sum_{q=1}^{N-1}\sum_{j=0}^{N-1-|q|}\right|_{for\ i=j+q}.$$

Eq. (5.32)

Substituting Eq. (5.32) into Eq. (5.31) gives

$$\chi(\tau;f_d) = \frac{1}{N}\sum_{q=-(N-1)}^{0}\left\{\chi_1(\tau-qT;f_d)\sum_{i=0}^{N-1-|q|}e^{j2\pi f_d iT}\right\}$$

$$+\frac{1}{N}\sum_{q=1}^{N-1}\left\{e^{j2\pi f_d qT}\chi_1(\tau-qT;f_d)\sum_{j=0}^{N-1-|q|}e^{j2\pi f_d jT}\right\}$$

Eq. (5.33)

Setting $z = \exp(j2\pi f_d T)$, and using the relation

$$\sum_{j=0}^{N-1-|q|}z^j = \frac{1-z^{N-|q|}}{1-z}$$

Eq. (5.34)

yields

$$\sum_{i=0}^{N-1-|q|}e^{j2\pi f_d iT} = e^{[j\pi f_d(N-1-|q|T)]}\ \frac{\sin[\pi f_d(N-1-|q|)T]}{\sin(\pi f_d T)}.$$

Eq. (5.35)

Using Eq. (5.35) in Eq. (5.31) yields two complementary sums for positive and negative q. Both sums can be combined as

$$\chi(\tau;f_d) = \frac{1}{N} \sum_{q=-(N-1)}^{N-1} \chi_1(\tau - qT;f_d) e^{[j\pi f_d(N-1+q)T]} \frac{\sin[\pi f_d(N-|q|)T]}{\sin(\pi f_d T)} . \qquad \text{Eq. (5.36)}$$

The second part of the right-hand side of Eq. (5.36) is the impact of the train on the ambiguity function, while the first part is primarily responsible for its shape details (according to the pulse type being used).

Finally, the ambiguity function associated with the coherent pulse train is computed as the modulus square of Eq. (5.36). For $\tau_0 < T/2$, the ambiguity function reduces to

$$|\chi(\tau;f_d)| = \frac{1}{N} \sum_{q=-(N-1)}^{N-1} |\chi_1(\tau - qT;f_d)| \left| \frac{\sin[\pi f_d(N-|q|)T]}{\sin(\pi f_d T)} \right| \;;|\tau| \leq NT. \qquad \text{Eq. (5.37)}$$

Within the region $|\tau| \leq \tau_0 \Rightarrow q = 0$, Eq. (5.37) can be written as

$$|\chi(\tau;f_d)| = |\chi_1(\tau;f_d)| \left| \frac{\sin[\pi f_d NT]}{N\sin(\pi f_d T)} \right| \;;|\tau| \leq \tau_0. \qquad \text{Eq. (5.38)}$$

Thus, the ambiguity function for a coherent pulse train is the superposition of the individual pulse's ambiguity functions. The ambiguity function cuts along the time-delay and the Doppler axes are, respectively, given by

$$|\chi(\tau;0)|^2 = \left| \sum_{q=-(N-1)}^{N-1} \left(1 - \frac{|q|}{N}\right)\left(1 - \frac{|\tau - q|T}{\tau_0}\right) \right|^2 \;;\; |\tau - qT| < \tau_0 \qquad \text{Eq. (5.39)}$$

$$|\chi(0;f_d)|^2 = \left| \frac{1}{N} \frac{\sin(\pi f_d \tau_0)}{\pi f_d \tau_0} \frac{\sin(\pi f_d NT)}{\sin(\pi f_d T)} \right|^2 . \qquad \text{Eq. (5.40)}$$

MATLAB Function "tarin_ambg.m"

The function *"train_ambg.m"* implements Eq. (5.37). The syntax is as follows:

train_ambg [taup, n, pri]

where

Symbol	Description	Units	Status
taup	*pulse width*	*seconds*	*input*
n	*number of pulses in train*	*none*	*input*
pri	*pulse repetition interval*	*seconds*	*input*

Figures 5.8a and 5.8b show the 3-D ambiguity plot and the corresponding contour plot for $N = 5$, $\tau_0 = 0.4$, and $T = 1$. This plot can be reproduced using MATLAB program *"Fig5_8.m,"* listed in Appendix 5-A. Figures 5.8c and 5.8d, respectively, show sketches of the zero Doppler and zero delay cuts in the ambiguity function. The ambiguity function peaks

along the frequency axis are located at multiple integers of the frequency $f = 1/T$. Alternatively, the peaks are at multiple integers of T along the delay axis. The width of the ambiguity function peaks along the delay axis are $2\tau_0$. The peak width along the Doppler axis is $1/(N-1)T$.

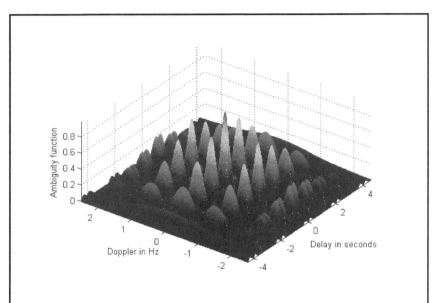

Figure 5.8a. Three-dimensional ambiguity plot for a five-pulse equal amplitude coherent train. Pulse width is 0.4 seconds; and PRI is 1 second, N=5.

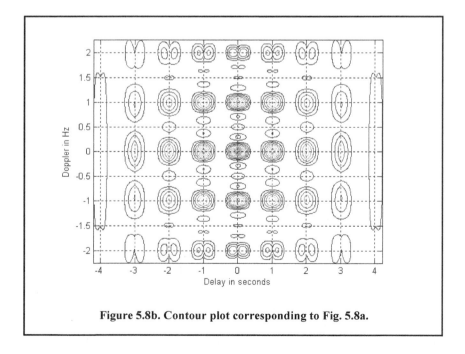

Figure 5.8b. Contour plot corresponding to Fig. 5.8a.

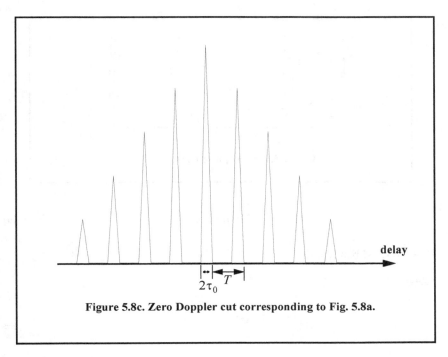

Figure 5.8c. Zero Doppler cut corresponding to Fig. 5.8a.

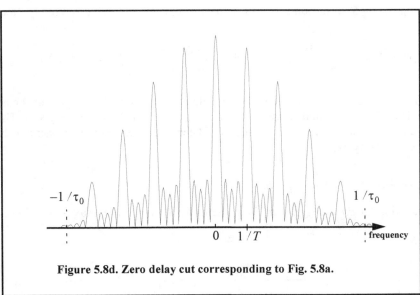

Figure 5.8d. Zero delay cut corresponding to Fig. 5.8a.

5.2.4. Pulse Train Ambiguity Function with LFM

In this case, the signal is as given in the previous section except for the LFM modulation within each pulse. This is illustrated in Fig. 5.9. Again, let the pulse width be denoted by τ_0 and the PRI by T. The number of pulses in the train is N; hence, the train's length is $(N-1)T$ seconds. A normalized individual pulse $\tilde{x}_1(t)$ is defined by

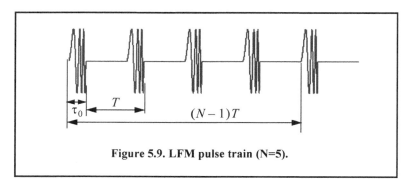

Figure 5.9. LFM pulse train (N=5).

$$\tilde{x}_1(t) = \frac{1}{\sqrt{\tau_0}} Rect\left(\frac{t}{\tau_0}\right) e^{j\pi\frac{B}{\tau_0}t^2}$$ Eq. (5.41)

where B is the LFM bandwidth.

The signal is now given by

$$\tilde{x}(t) = \frac{1}{\sqrt{N}} \sum_{i=0}^{N-1} \tilde{x}_1(t - iT).$$ Eq. (5.42)

Utilizing property 5 of Section 5.1 and Eq. (5.37) yields the following ambiguity function

$$|\chi(\tau;f_d)| = \sum_{q=-(N-1)}^{N-1} \left|\chi_1\left(\tau - qT;f_d + \frac{B}{\tau_0}\tau\right)\right| \left|\frac{\sin[\pi f_d(N-|q|)T]}{N\sin(\pi f_d T)}\right| \quad ; |\tau| \le NT$$ Eq. (5.43)

where χ_1 is the ambiguity function of the single pulse. Note that the shape of the ambiguity function is unchanged from the case of the unmodulated train along the delay axis. This should be expected since only a phase modulation has been added, which will impact the shape only along the frequency axis.

MATLAB Function "train_ambg_lfm.m"

The function *"train_ambg_lfm.m"* implements Eq. (5.43). The syntax is as follows:

x = train_ambg_lfm(taup, n, pri, bw)

where

Symbol	Description	Units	Status
taup	*pulse width*	*seconds*	*input*
n	*number of pulses in train*	*none*	*input*
pri	*pulse repetition interval*	*seconds*	*input*
bw	*the LFM bandwidth*	*Hz*	*input*
x	*array of bimodality function*	*none*	*output*

Note that this function will generate identical results to the function *"train_ambg.m"* when the value of *"bw"* is set to zero. In this case, Eqs. (4.43) and (4.35) are identical. Figures 5.10 a and b show the ambiguity plot and its associated contour plot for the same example listed in the previous section except, in this case, LFM modulation is added and $N = 3$ pulses. This figure can be reproduced using MATLAB program *"Fig5_10.m,"* listed in Appendix 5-A.

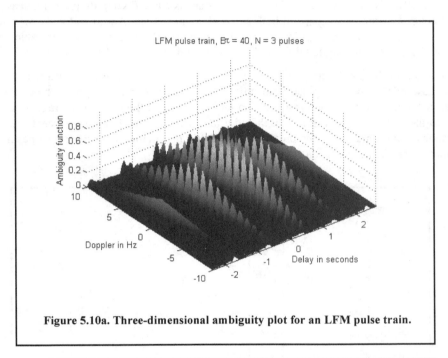

Figure 5.10a. Three-dimensional ambiguity plot for an LFM pulse train.

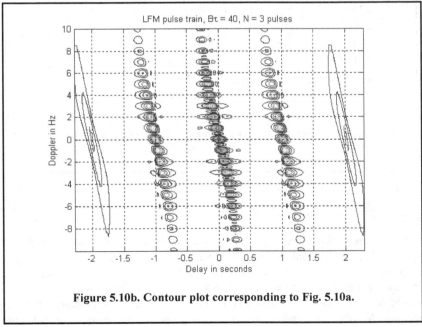

Figure 5.10b. Contour plot corresponding to Fig. 5.10a.

Understanding the difference between the ambiguity diagrams for a coherent pulse train and an LFM pulse train can be done with the help of Fig. 5.11a and Fig. 5.11b. In both figures a train of three pulses is used; in both cases the pulse width is $\tau_0 = 0.4\,sec$ and the period is $T = 1\,sec$. In the case of the LFM pulse train, each pulse has LFM modulation with $B\tau_0 = 20$. Locations of the ambiguity peaks along the delay and Doppler axes are the same in both cases. This is true because peaks along the delay axis are T seconds apart and peaks along the Doppler axis are $1/T$ apart; in both cases T is unchanged. Additionally, the width of the ambiguity peaks along the Doppler axis are the same in both cases, because this value depends only on the pulse train length, which is the same in both cases (i.e., $(N-1)T$).

The width of the ambiguity peaks along the delay axis are significantly different, however. In the case of the coherent pulse train, this width is approximately equal to twice the pulse width. Alternatively, this value is much smaller in the case of the LFM pulse train. This clearly leads to the expected conclusion that the addition of LFM modulation significantly enhances the range resolution. Finally, the presence of the LFM modulation introduces a slope change in the ambiguity diagram; again a result that is also expected.

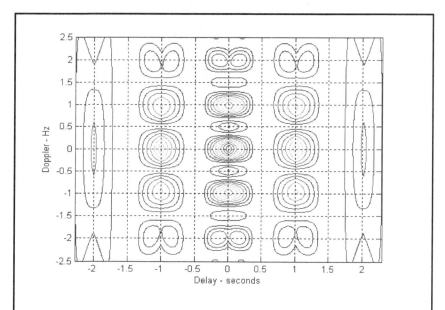

Figure 5.11a. Contour plot for the ambiguity function of a coherent pulse train.
$$N = 3; \tau_0 = 0.4; \quad T = 1$$

5.3. Stepped Frequency Waveforms

Stepped Frequency Waveforms (SFW) is a class of radar waveforms that are used in extremely wide bandwidth applications where very large time bandwidth product (or compression ratio as defined in Eq. (5.25) is required. One may think of SFW as a special case of an extremely wide bandwidth LFM waveform. For this purpose, consider an LFM signal whose bandwidth is B_i and whose pulse width is T_i, and refer to it as the primary LFM. Divide this long pulse into N subpulses, each of width τ_0, to generate a sequence of pulses whose PRI is denoted by T. It follows that $T_i = (n-1)T$. One reason SFW is favored over an extremely wideband LFM is that it may be very difficult to maintain the LFM slope when the time bandwidth product is large. By using SFW, the same equivalent bandwidth can be achieved; however, phase errors are minimized since the LFM is chirped over a much shorter duration.

Define the beginning frequency for each subpulse as that value measured from the primary LFM at the leading edge of each subpulse, as illustrated in Fig. 5.12. That is

$$f_i = f_0 + i\Delta f; \quad i = 0, N-1 \qquad \text{Eq. (5.44)}$$

where Δf is the frequency step from one subpulse to another. The set of n subpulses is often referred to as a burst. Each subpulse can have its own LFM modulation. To this end, assume that the subpulse LFM modulation corresponds to an LFM slope of $\mu = B/\tau_0$.

The complex envelope of a single subpulse with LFM modulation is

$$\tilde{x}_1 = \frac{1}{\sqrt{\tau_0}} Rect\left(\frac{t}{\tau_0}\right) e^{j\pi\mu t^2}. \qquad \text{Eq. (5.45)}$$

Of course if the subpulses do not have any LFM modulation, then the same equation holds true by setting $\mu = 0$. The overall complex envelope of the whole burst is

$$\tilde{x}(t) = \frac{1}{\sqrt{N}} \sum_{i=0}^{N-1} \tilde{x}_1(t - iT). \qquad \text{Eq. (5.46)}$$

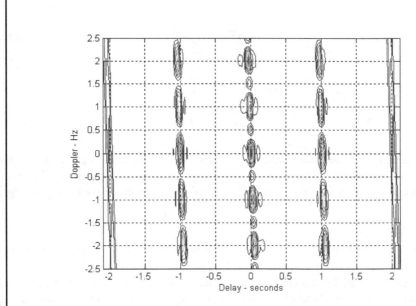

Figure 5.11b. Contour plot for the ambiguity function of a coherent pulse train.
$N = 3; \quad B\tau_0 = 20; \quad T = 1$

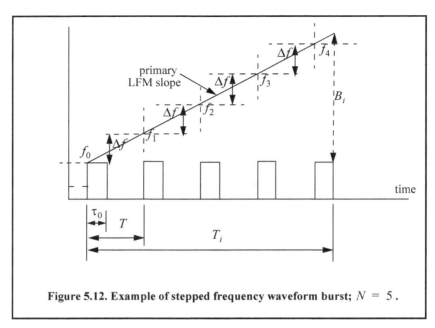

Figure 5.12. Example of stepped frequency waveform burst; $N = 5$.

The ambiguity function of the matched filter corresponding to Eq. (5.46) can be obtained from that of the coherent pulse train developed in Section 5.2.3 along with property 5 of the ambiguity function. The details are fairly straightforward and are left to the reader as an exercise. The result is (see Problem 5.2)

$$|\chi(\tau;f_d)| = \sum_{q=-(N-1)}^{N-1} \left|\chi_1\left(\tau - qT;\left(f_d + \frac{B}{\tau_0}\tau\right)\right)\right| \times \qquad \text{Eq. (5.47)}$$

$$\left|\frac{\sin\left[\pi\left(f_d + \frac{\Delta f}{T}\tau\right)(N - |q|)T\right]}{N\sin\left(\pi\left(f_d + \frac{\Delta f}{T}\tau\right)T\right)}\right| \; ;|\tau| \leq NT$$

where χ_1 is the ambiguity function of the single pulse. Unlike the case in Eq. (5.43), the second part of the right-hand side of Eq. (5.47) is now modified according to property 5 of Section 5.1. This is true since each subpulse has its own beginning frequency derived from the primary LFM slope.

5.4. Nonlinear FM

As clearly shown by Fig. 5.6, the output of the matched filter corresponding to an LFM pulse has sidelobe levels similar to those of the $|\sin(x)/x|^2$ signal, that is, $13.4 dB$ below the main beam peak. In many radar applications, these sidelobe levels are considered too high and may present serious problems for detection particularly in the presence of nearby interfering targets or other noise sources. Therefore, in most radar applications, sidelobe reduction of the output of the matched filter is always required. This sidelobe reduction can be accomplished using windowing techniques as described in Chapter 3. However, windowing techniques reduce the sidelobe levels at the expense of reducing of the SNR and widening the main beam (i.e., loss of resolution), which are considered to be undesirable features in many radar applications.

These effects can be mitigated by using non-linear FM (NLFM) instead of LFM waveforms. In this case, the LFM waveform spectrum is shaped according to a specific predetermined frequency function. Effectively, in NLFM, the rate of change of the LFM waveform phase is varied so that less time is spent on the edges of the bandwidth, as illustrated in Fig. 5.13. The concept of NLFM can be better analyzed and understood in the context of the stationary phase.

5.4.1. The Concept of Stationary Phase

Consider the following bandpass signal

$$x(t) = x_I(t)\cos(2\pi f_0 t + \phi(t)) - x_Q(t)\sin(2\pi f_0 t + \phi(t)),$$ Eq. (5.48)

where $\phi(t)$ is the frequency modulation. The corresponding analytic signal (pre-envelope) is

$$\psi(t) = \tilde{x}(t)e^{j2\pi f_0 t} = r(t)e^{j\phi(t)}e^{j2\pi f_0 t}$$ Eq. (5.49)

where $\tilde{x}(t)$ is the complex envelope and is given by

$$\tilde{x}(t) = r(t)e^{j\phi(t)},$$ Eq. (5.50)

The lowpass signal $r(t)$ represents the envelope of the transmitted signal; it is given by

$$r(t) = \sqrt{x_I^2(t) + x_Q^2(t)},$$ Eq. (5.51)

It follows that the FT of the signal $\tilde{x}(t)$ can then be written as

$$X(\omega) = \int_{-\infty}^{\infty} r(t)e^{j(-\omega t + \phi(t))}\,dt,$$ Eq. (5.52)

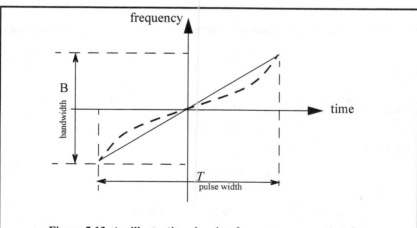

Figure 5.13. An illustration showing frequency versus time for an LFM waveform (solid line) and a NLFM (dashed line).

$$X(\omega) = |X(\omega)|e^{j\Phi(\omega)} \qquad \textbf{Eq. (5.53)}$$

where $|X(\omega)|$ is the modulus of the FT and $\Phi(\omega)$ is the corresponding phase frequency response. It is clear that the integrand is an oscillating function of time varying at a rate of

$$\frac{d}{dt}[\omega t - \phi(t)] . \qquad \textbf{Eq. (5.54)}$$

Most contribution to the FT spectrum occurs when this rate of change is minimal. More specifically, it occurs when

$$\frac{d}{dt}[\omega t - \phi(t)] = 0 \Rightarrow \omega - \phi'(t) = 0 . \qquad \textbf{Eq. (5.55)}$$

The expression in Eq. (5.55) is parametric since it relates two independent variables. Thus, for each value ω_n there is only one specific $\phi'(t_n)$ that satisfies Eq. (5.55). Thus, the time when this phase term is stationary will be different for different values of ω_n. Expanding the phase term in Eq. (5.55) about an incremental value t_n using Taylor series expansion yields

$$\omega_n t - \phi(t) = \omega_n t_n - \phi(t_n) + (\omega_n - \phi'(t_n))(t - t_n) - \frac{\phi''(t_n)}{2!}(t - t_n)^2 + \dots \qquad \textbf{Eq. (5.56)}$$

An acceptable approximation of Eq. (5.56) is obtained by using the first three terms, provided that the difference $(t - t_n)$ is very small. Now, using the right-hand side of Eq. (5.55) into Eq. (5.56) and terminating the expansion at the first three terms yields

$$\omega_n t - \phi(t) = \omega_n t_n - \phi(t_n) - \frac{\phi''(t_n)}{2!}(t - t_n)^2 . \qquad \textbf{Eq. (5.57)}$$

By substituting Eq. (5.57) into Eq. (5.52) and using the fact that $r(t)$ is relatively constant (slow varying) when compared to the rate at which the carrier signal is varying, gives

$$X(\omega_n) = r(t_n) \int_{t_n^-}^{t_n^+} e^{-j\left(\omega_n t_n - \phi(t_n) - \frac{\phi''(t_n)}{2}(t - t_n)^2\right)} dt \qquad \textbf{Eq. (5.58)}$$

where t_n^+ and t_n^- represent infinitesimal changes about t_n. Equation (5.58) can be written as

$$X(\omega_n) = r(t_n)e^{j(-\omega_n t_n - \phi(t_n))} \int_{t_n^-}^{t_n^+} e^{j\left(\frac{\phi''(t_n)}{2}(t - t_n)^2\right)} dt . \qquad \textbf{Eq. (5.59)}$$

Consider the changes of variables

$$t - t_n = \lambda \Rightarrow dt = d\lambda \qquad \textbf{Eq. (5.60)}$$

$$\sqrt{\phi''(t_n)}\lambda = \sqrt{\pi} \; y \Rightarrow d\lambda = \frac{\sqrt{\pi}}{\sqrt{\phi''(t_n)}}dy . \qquad \textbf{Eq. (5.61)}$$

Using these changes of variables leads to

$$X(\omega_n) = \frac{2\sqrt{\pi} \ r(t_n)}{\sqrt{\phi''(t_n)}} e^{j(-\omega_n t_n - \phi(t_n))} \int\limits_{0}^{y_0} e^{j\left(\frac{\pi y^2}{2}\right)} \, dy$$ Eq. (5.62)

where

$$y_0 = \sqrt{\frac{|\phi''(t_n)|}{\pi}}.$$ Eq. (5.63)

The integral in Eq. (5.62) is of the form of a Fresnel integral, which has an upper limit approximated by

$$\frac{\exp\left(j\frac{\pi}{4}\right)}{\sqrt{2}}.$$ Eq. (5.64)

Substituting Eq. (5.64) into Eq. (5.62) yields

$$X(\omega_n) = \frac{\sqrt{2\pi} \ r(t_n)}{\sqrt{\phi''(t_n)}} e^{j\left(-\omega_n t_n - \phi(t_n) + \frac{\pi}{4}\right)}.$$ Eq. (5.65)

Thus, for all possible values of ω

$$|X(\omega_t)|^2 \approx 2\pi \frac{r^2(t)}{|\phi''(t)|} \Rightarrow |X(\omega)| = \frac{\sqrt{2\pi}}{\sqrt{|\phi''(t)|}} \ r(t).$$ Eq. (5.66)

The subscript t was used to indicate the dependency of ω on time.

Using a similar approach that led to Eq. (5.66), an expression for $\tilde{x}(t_n)$ can be obtained. From Eq. (5.53), the signal $\tilde{x}(t)$

$$\tilde{x}(t) = \frac{1}{2\pi} \int\limits_{-\infty}^{\infty} |X(\omega)| \ e^{j(\Phi(\omega) + \omega t)} \, d\omega.$$ Eq. (5.67)

The phase term $\Phi(\omega)$ is (using Eq. (5.65))

$$\Phi(\omega) = -\omega t - \phi(t) + \frac{\pi}{4}.$$ Eq. (5.68)

Differentiating with respect to ω yields

$$\frac{d}{d\omega}\Phi(\omega) = -t - \left(\frac{dt}{d\omega}\right)\left[\omega - \frac{d}{dt}\phi(t)\right] = \Phi'(\omega).$$ Eq. (5.69)

Using the stationary phase relation in Eq. (5.55) (i.e., $\omega - \phi'(t) = 0$) yields

$$\Phi'(\omega) = -t$$ Eq. (5.70)

and

$$\Phi''(\omega) = -\frac{dt}{d\omega}.$$ Eq. (5.71)

Define the signal group time-delay function as

$$T_g(\omega) = -\Phi'(\omega),$$ Eq. (5.72)

then the signal instantaneous frequency is the inverse of the $T_g(\omega)$. Figure 5.14 shows a drawing illustrating this inverse relationship between the NLFM frequency modulation and the corresponding group time-delay function.

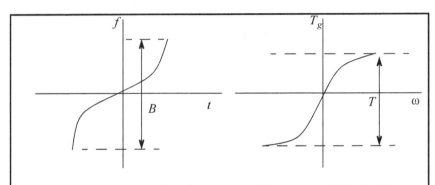

Figure 5.14. Matched filter time delay and frequency modulation for a NLFM waveform.

Comparison of Eq. (5.67) and Eq. (5.52) indicates that both equations have similar form. Thus, if one substitutes $X(\omega)/2\pi$ for $r(t)$, $\Phi(\omega)$ for $\phi(t)$, ω for t, and $-t$ for ω in Eq. (5.52), a similar expression to that in Eq. (5.65) can be derived. That is,

$$|\tilde{x}(t_\omega)|^2 \approx \frac{1}{2\pi}\frac{|X(\omega)|^2}{|\Phi''(\omega)|}.$$ Eq. (5.73)

The subscript ω was used to indicate the dependency of t on frequency. However, from Eq. (5.60)

$$|\tilde{x}(t)|^2 = \left|r(t)e^{j\phi(t)}\right|^2 = r^2(t).$$ Eq. (5.74)

It follows that Eq. (5.73) can be rewritten as

$$r^2(t_\omega) \approx \frac{1}{2\pi}\frac{|X(\omega)|^2}{|\Phi''(\omega)|} \Rightarrow r(t) = \frac{|X(\omega)|}{\sqrt{2\pi|\Phi''(\omega)|}}.$$ Eq. (5.75)

Substituting Eq. (5.71) into Eq. (5.75) yields a general relationship for any t

$$r^2(t)\ dt = \frac{1}{2\pi}|X(\omega)|^2 d\omega,$$ Eq. (5.76)

Clearly, the functions $r(t)$, $\phi(t)$, $X(\omega)$, and $\Phi(\omega)$ are related to each other as Fourier transform pairs, as given by

$$r(t)e^{j\phi(t)} = \frac{1}{2\pi}\int_{-\infty}^{\infty}|X(\omega)|\ e^{j(\Phi(\omega)+\omega t)}\ d\omega$$ Eq. (5.77)

$$|X(\omega)| \ e^{j\Phi(\omega)} = \int\limits_{-\infty}^{\infty} r(t) \ e^{-j(\omega t - \phi(t))} \ d\omega \ . \qquad \text{Eq. (5.78)}$$

They are also related using Parseval's theorem by

$$\int\limits_{-\infty}^{t} r^2(\zeta) \ d\zeta = \frac{1}{2\pi} \int\limits_{\omega} |X(\lambda)|^2 \ d\lambda \qquad \text{Eq. (5.79)}$$

or

$$\int\limits_{-\infty}^{t} r^2(\zeta) \ d\zeta = \frac{1}{2\pi} \int\limits_{-\infty}^{\omega} |X(\lambda)|^2 \ d\lambda \ . \qquad \text{Eq. (5.80)}$$

The formula for the output of the matched filter was derived earlier and is repeated here as Eq. (5.81)

$$\chi(\tau, f_d) = \int\limits_{-\infty}^{\infty} \tilde{x}(t)\tilde{x}^*(t-\tau)e^{j2\pi f_d t} dt \ . \qquad \text{Eq. (5.81)}$$

Substituting the right-hand side of Eq. (5.50) into Eq. (5.81) yields

$$\chi(\tau, f_d) = \int\limits_{-\infty}^{\infty} r(t)r^*(t-\tau)e^{j2\pi f_d t} dt \ . \qquad \text{Eq. (5.82)}$$

It follows that the zero Doppler and zero delay cuts of the ambiguity function can be written as

$$\chi(\tau, 0) = \frac{1}{2\pi} \int\limits_{-\infty}^{\infty} |X(\omega)|^2 \ e^{j\omega\tau} d\omega \qquad \text{Eq. (5.83)}$$

$$\chi(0, f_d) = \int\limits_{-\infty}^{\infty} |r(t)|^2 \ e^{j2\pi f_d t} dt \ . \qquad \text{Eq. (5.84)}$$

These two equations imply that the shape of the ambiguity function cuts are controlled by selecting different functions X and r (related as defined in Eq. (5.76)). In other words, the ambiguity function main beam and its delay axis sidelobes can be controlled (shaped) by the specific choices of these two functions; hence, the term *spectrum shaping* is used. Using this concept of spectrum shaping, one can control the frequency modulation of an LFM (see Fig. 5.13) to produce an ambiguity function with the desired sidelobe levels.

5.4.2. Frequency Modulated Waveform Spectrum Shaping

One class of FM waveforms which takes advantage of the stationary phase principles to control (shape) the spectrum is

$$|X(\omega;n)|^2 = \left(\cos \pi \left(\frac{\pi \omega}{B_n} \right) \right)^n \qquad ; \ |\omega| \le \frac{B_n}{2} \qquad\qquad \text{Eq. (5.85)}$$

where the value n is an integer greater than zero. It can be easily shown using direct integration and by utilizing Eq. (5.85) that

$$n = 1 \Rightarrow T_{g1}(\omega) = \frac{T}{2} \sin \left(\frac{\pi \omega}{B_1} \right) \qquad\qquad \text{Eq. (5.86)}$$

$$n = 2 \Rightarrow T_{g2}(\omega) = T \left[\frac{\omega}{B_2} + \frac{1}{2\pi} \sin \left(\frac{2\pi \omega}{B_2} \right) \right] \qquad\qquad \text{Eq. (5.87)}$$

$$n = 3 \Rightarrow T_{g3}(\omega) = \frac{T}{4} \left\{ \sin \left(\frac{\pi \omega}{B_3} \right) \left[\left(\cos \frac{\pi \omega}{B_3} \right)^2 + 2 \right] \right\} \qquad\qquad \text{Eq. (5.88)}$$

$$n = 4 \Rightarrow T_{g4}(\omega) = T \left\{ \frac{\omega}{B_4} + \frac{1}{2\pi} \sin \frac{2\pi \omega}{B_4} + \frac{2}{3\pi} \left(\cos \frac{\pi \omega}{B_4} \right)^3 \sin \frac{\pi \omega}{B_4} \right\} \qquad\qquad \text{Eq. (5.89)}$$

Figure 5.15 shows a plot for Eq. (5.86) through Eq. (5.89). These plots assume $T = 1$ and the x-axis is normalized, with respect to B. This figure can be reproduced using the MATLAB program "*Fig5_15.m,*" listed in Appendix 5-A.

The Doppler mismatch (i.e., a peak of the ambiguity function at a delay value other than zero) is proportional to the amount of Doppler frequency f_d. Hence, an error in measuring target range is always expected when LFM waveforms are used. To achieve sidelobe levels for the output of the matched filter that do not exceed a predetermined level, use this class of NLFM waveforms

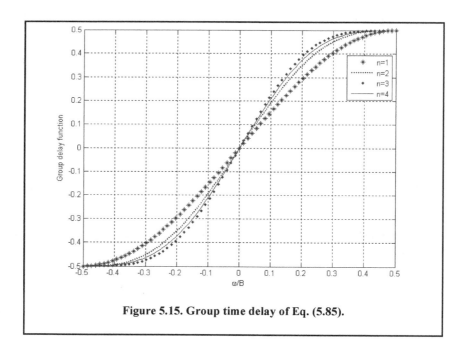

Figure 5.15. Group time delay of Eq. (5.85).

$$|X(\omega;n;k)|^2 = k + (1 - k)\left(\cos\pi\left(\frac{\pi\omega}{B_n}\right)\right)^n \qquad ;|\omega| \le \frac{B_n}{2}. \qquad \textbf{Eq. (5.90)}$$

For example, using the combination $n = 2$, $k = 0.08$ yields sidelobe levels less than $-40\,dB$.

5.5. *Ambiguity Diagram Contours*

Plots of the ambiguity function are called ambiguity diagrams. For a given waveform, the corresponding ambiguity diagram is normally used to determine the waveform properties such as the target resolution capability, measurements (time and frequency) accuracy, and its response to clutter. The ambiguity diagram contours are cuts in the 3-D ambiguity plot at some value, Q, such that $Q < |\chi(0, 0)|^2$. The resulting plots are ellipses (see Problem 5.11). The width of a given ellipse along the delay axis is proportional to the signal effective duration, τ_e, defined in Chapter 2. Alternatively, the width of an ellipse along the Doppler axis is proportional to the signal effective bandwidth, B_e.

Figure 5.16 shows a sketch of typical ambiguity contour plots associated with a single unmodulated pulse. As illustrated in Fig. 5.16, narrow pulses provide better range accuracy than long pulses. Alternatively, the Doppler accuracy is better for a wider pulse than it is for a short one. This trade-off between range and Doppler measurements comes from the uncertainty associated with the time-bandwidth product of a single sinusoidal pulse, where the product of uncertainty in time (range) and uncertainty in frequency (Doppler) cannot be much smaller than unity (see Problem 5.12). Figure 5.17 shows the ambiguity contour plot associated with an LFM waveform. The slope is an indication of the LFM modulation. The values σ_τ, σ_{f_d}, $\sigma_{\tau RDC}$, and $\sigma_{f_d RDC}$ were derived in Chapter 4 and were respectively given in Eq. (4.107), Eq. (4.111), Eq. (4.136), and Eq. (4.137).

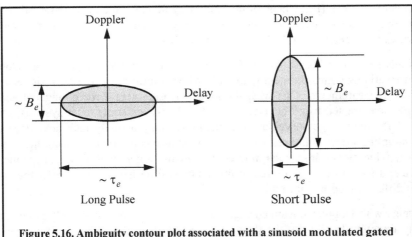

Figure 5.16. Ambiguity contour plot associated with a sinusoid modulated gated CW pulse.

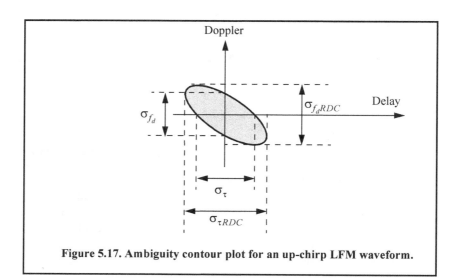

Figure 5.17. Ambiguity contour plot for an up-chirp LFM waveform.

5.6. Interpretation of Range-Doppler Coupling in LFM Signals

An expression of the range-Doppler for LFM signals was derived in Chapter 4. Range-Doppler coupling affects the radar's ability to compute target range and Doppler estimates. An interpretation of this term in the context of the ambiguity function can be explained further with the help of Eq. (5.20). Observation of this equation indicates that the ambiguity function for the LFM pulse has a peak value, not at $\tau = 0$, but rather at

$$(B/\tau_0)\tau + f_d = 0 \Rightarrow \tau = -f_d(\tau_0/B) \qquad \text{Eq. (5.91)}$$

This Doppler mismatch (i.e., a peak of the ambiguity function at a delay value other than zero) is proportional to the amount of Doppler frequency f_d. Hence, an error in measuring target range is always expected when LFM waveforms are used.

Most radar systems using LFM waveforms will correct for the effect of range-Doppler coupling by repeating the measurement with an LFM waveform of the opposite slope and averaging the two measurements. This way, the range measurement error is negated and the true target range is extracted from the averaged value. However, some radar systems, particularly those used for long-range surveillance applications, may actually take advantage of range-Doppler coupling effect; and here is how it works: Typically, radars during the search mode utilize very wide range bins which may contain many targets with different distinct Doppler frequencies. It follows that the output of the matched filter has several targets that have equal delay but different Doppler mismatches.

All targets with Doppler mismatches greater than $1/\tau_0$ are significantly attenuated by the ambiguity function (because of the sharp decaying slope of the ambiguity function along the Doppler axis), and thus will most likely go undetected along the Doppler axis. The combined target complex within that range bin is then detected by the LFM as if all targets had a Doppler mismatch corresponding to the target whose Doppler mismatch is less than or equal to $1/\tau_0$. Thus, all targets within that wide range bin are detected as one narrowband target. Because of this range-Doppler coupling, LFM waveforms are often referred to as Doppler intolerant (insensitive) waveforms.

Problems

5.1. From Eq. (5.15) one can deduce that the transfer function of the matched filter is given by $H(f) = \sin((\pi\tau_0 f)/(\pi\tau_0 f))$. Show that

$$\int_{-\frac{1}{2\tau_0}}^{\frac{1}{2\tau_0}} H(f)\ df = \frac{1}{2\tau_0}$$

5.2. Prove Eq. (5.5) through Eq. (5.10).

5.3. Derive an expression for the ambiguity function of a Gaussian pulse defined by

$$x(t) = \frac{1}{\sqrt{\sigma}\ ^{1/4}\!\sqrt{\pi}} \exp\left[\frac{-t^2}{2\sigma^2}\right] \qquad ;0 < t < T$$

where T is the pulsewidth and σ is a constant.

5.4. Write a MATLAB program that computes and plots the 3-D and the contour plots for the results in Problem 5.3.

5.5. Derive an expression for the ambiguity function of a V-LFM waveform, illustrated in figure below. In this case, the overall complex envelope is

$$\tilde{x}(t) = \tilde{x}_1(t) + \tilde{x}_2(t) \qquad ;-T < t < T$$

where

$$\tilde{x}_1(t) = \frac{1}{\sqrt{2T}} \exp[-\mu t^2] \qquad ;-T < t < 0$$

and

$$\tilde{x}_2(t) = \frac{1}{\sqrt{2T}} \exp[\mu t^2] \qquad ;0 < t < T$$

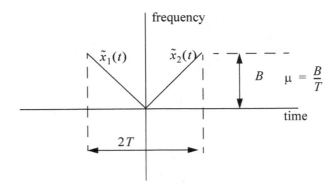

5.6. Using the stationary phase concept, find the instantaneous frequency for the waveform whose envelope and complex spectrum are, respectively, given by

$$r(t) = \frac{1}{\sqrt{T}} \exp\left[-\left(\frac{2t}{T}\right)^2\right] \qquad ; 0 < t < T$$

and

$$|X(f)| = \frac{1}{\sqrt{B}} \exp\left[-\left(\frac{2f}{B}\right)^2\right].$$

5.7. Using the stationary phase concept, find the instantaneous frequency for the waveform whose envelope and complex spectrum are respectively given by

$$r(t) = \frac{1}{\sqrt{\tau_0}} Rect\left(\frac{t}{\tau_0}\right) \qquad ; \ 0 < t < \tau_0$$

and

$$|X(\omega)| = \frac{2}{\sqrt{B}} \frac{1}{\sqrt{1 + (2\omega/B)^2}}.$$

5.8. Write a MATLAB program to compute the ambiguity function for the parabolic FM waveform. Your code must be able to produce 3-D and contour plots of the resulting ambiguity function.

5.9. Write a detailed MATLAB code to compute the ambiguity function for an SFW waveform. Your code must be able to produce 3-D and contour plots of the resulting ambiguity function.

5.10. Prove that cuts in the ambiguity function are always defined by an ellipse. Hint: Approximate the ambiguity function using a Taylor series expansion about the values $(\tau, f_d) = (0, 0)$; use only the first three terms in the Taylor series expansion.

5.11. The radar uncertainty principle establishes a lower bound for the time bandwidth product. More specifically, if the radar effective duration is τ_e and its effective bandwidth is B_e, show that $B_e^2 \tau_e^2 (1 - \rho_{RDC}^2) \geq \pi^2$, where ρ_{RDC} is the range-Doppler coupling coefficient defined in Chapter 4.

Appendix 5-A: Chapter 5 MATLAB Code Listings

The MATLAB code provided in this chapter was designed as an academic standalone tool and is not adequate for other purposes. The code was written in a way to assist the reader in gaining a better understanding of the theory. The code was not developed, nor is it intended to be used as part of an open-loop or a closed-loop simulation of any kind. The MATLAB code found in this textbook can be downloaded from this book's web page on the CRC Press website. Simply use your favorite web browser, go to *www.crcpress.com*, and search for keyword *"Mahafza"* to locate this book's web page.

MATLAB Function "single_pulse_ambg.m" Listing

```
function [x] = single_pulse_ambg (taup)
% Computes the ambiguity of a single pulse
% % Inputs
    % taup        == pulsewidth in seconds
%%Outputs
    % x           == ambiguity surface array
eps = 0.000001;
i = 0;
del = 2*taup/150;
for tau = -taup:del:taup
  i = i + 1;
  j = 0;
  fd = linspace(-5/taup,5/taup,151);
  val1 = 1. - abs(tau) / taup;
  val2 = pi * taup .* (1.0 - abs(tau) / taup) .* fd;
  x(:,i) = abs( val1 .* sin(val2+eps)./(val2+eps));
end
```

MATLAB Program "Fig5_2.m" Listing

```
% Use this program to reproduce Fig. 5.2 of text
close all;
clear all;
eps = 0.000001;
taup = 3;
[x] = single_pulse_ambg (taup);
taux = linspace(-taup,taup, size(x,1));
fdy = linspace(-5/taup+eps,5/taup-eps, size(x,1));
mesh(taux,fdy,x);
xlabel ('Delay in seconds');
ylabel ('Doppler in Hz');
zlabel ('Ambiguity function')
figure(2)
contour(taux,fdy,x);
xlabel ('Delay in seconds');
ylabel ('Doppler in Hz'); grid
```

MATLAB Program "Fig5_4.m" Listing

```
% Use this program to reproduce Fig 5.4 of text
close all
```

```
clear all
eps = 0.0001;
taup = 3.;
fd = -10./taup:.001:10./taup;
uncer = abs( sinc(taup .* fd));
figure(2)
plot (fd, uncer,'k','linewidth',1);
xlabel ('Frequency - Hz')
ylabel ('Ambiguity - Volts')
grid
```

MATLAB Function "lfm_ambg.m" Listing

```
function x = lfm_ambg(taup, b, up_down)
% Implements Eq. (5.21) of textbook
%% Inputs
   % taup                == pulsewidth in seconds
   % b                   == bandwidth in Hz
   % up_down             == 1 to indicate an up-chirp LFM
   % up_down             == -1 to indicate an down-chirp LFM
%% Output
   % x                   == ambiguity matrix
eps = 0.000001;
i = 0;
mu = up_down * b / taup;
del = 2*taup/200;
for tau = -1.*taup:del:taup
  i = i + 1;
  j = 0;
  fd = linspace(-1.5*b,1.5*b,201);
  val1 = 1. - abs(tau) / taup;
  val2 = pi * taup * (1.0 - abs(tau) / taup);
  val3 = (fd + mu * tau);
  val = val2 * val3;
  x(:,i) = abs( val1 .* (sin(val+eps)./(val+eps))).^2;
  end
end
```

MATLAB Program "Fig5_5.m" Listing

```
% Use this program to reproduce Fig. 5.5 of text
close all
clear all
eps = 0.0001;
taup = 1.;
b = 5.;
up_down = -1.;
x = lfm_ambg(taup, b, up_down);
taux = linspace(-1.*taup,taup,size(x,1));
fdy = linspace(-1.5*b,1.5*b,size(x,1));
figure(1)
mesh(taux,fdy,sqrt(x))
xlabel ('Delay in seconds')
ylabel ('Doppler in Hz')
```

```
zlabel ('Ambiguity function')
axis tight
figure(2)
contour(taux,fdy,sqrt(x))
xlabel ('Delay in seconds')
ylabel ('Doppler in Hz')
```

MATLAB Program "Fig5_6.m" Listing

```
% Use this program to reproduce Fig. 5.6 of text
close all
clear all
eps = 0.001;
taup = 1;
b =20.;
up_down = 1.;
taux = -1.5*taup:.01:1.5*taup;
fd = 0.;
mu = up_down * b / 2. / taup;
ii = 0.;
for tau = -1.5*taup:.01:1.5*taup
   ii = ii + 1;
   val1 = 1. - abs(tau) / taup;
   val2 = pi * taup * (1.0 - abs(tau) / taup);
   val3 = (fd + mu * tau);
   val = val2 * val3;
   x(ii) = abs( val1 * (sin(val+eps)/(val+eps)));
end
figure(1)
plot(taux,10*log10(x+eps))
grid
xlabel ('Delay in seconds')
ylabel ('Ambiguity in dB')
axis tight
```

MATLAB Function "train_ambg.m" Listing

```
function x = train_ambg(taup, n, pri)
% This function implements Eq. (5.37) of textbook
% % Inputs
   % taup        == pulse width in seconds
   % n           == number of pulses in train
   % pri         == pulse repetition interval in seconds
%% Outputs
   % x           == ambiguity matrix
if (taup >= pri/2)
   'ERROR. Pulse width must be less than the PRI/2.'
   return
end
eps = 1.0e-6;
bw = 1/taup;
q = -(n-1):1:n-1;
offset = 0:0.031:pri;
[Q, S] = meshgrid(q, offset);
```

```matlab
Q = reshape(Q, 1, length(q)*length(offset));
S = reshape(S, 1, length(q)*length(offset));
tau = (-taup * ones(1,length(S))) + S   ;
fd = -bw:0.011:bw;
[T, F] = meshgrid(tau, fd);
Q = repmat(Q, length(fd), 1);
S = repmat(S, length(fd), 1);
N = n * ones(size(T));
val1 = 1.0-(abs(T))/taup;
val2 = pi*taup*F.*val1;
val3 = abs(val1.*sin(val2+eps)./(val2+eps));
val4 = abs(sin(pi*F.*(N-abs(Q))*pri+eps)./sin(pi*F*pri+eps));
x = val3.*val4./N;
[rows, cols] = size(x);
x = reshape(x, 1, rows*cols);
T = reshape(T, 1, rows*cols);
indx = find(abs(T) > taup);
x(indx) = 0.0;
x = reshape(x, rows, cols);
return
```

MATLAB Program "Fig5_8.m" Listing

```matlab
% Use this program to reproduce Fig. 5.8 of text
clear all
close all
taup = .4;
pri = 1;
n = 5;
x = train_ambg(taup, n, pri);
figure(1)
time = linspace(-(n-1)*pri-taup, n*pri-taup, size(x,2));
doppler = linspace(-1/taup, 1/taup, size(x,1));
%mesh(time, doppler, x);
mesh(time, doppler, x); %shading interp;
xlabel('Delay in seconds');
ylabel('Doppler in Hz');
zlabel('Ambiguity function');
axis tight;
figure(2)
contour(time, doppler, (x));
%surf(time, doppler, x); shading interp; view(0,90);
xlabel('Delay in seconds');
ylabel('Doppler in Hz');
grid;
axis tight;
```

MATLAB Program "Fig5_9.m" Listing

```matlab
% Use this program to reproduce Fig. 5.9 of textbook
close all
clear all
LFM_BW = 20;
time = linspace(0,1,3000);
```

```
S = zeros(1,3000);
tau = .3;
index = find(time<=tau);
ts = tau / 3000; % 1000 samples per PW
beta = LFM_BW/tau;
S(index) = exp(j*pi*beta*(time(index).^2));
SS = repmat(S,1,5);
figure
timet = linspace(0,5,5*3000);
plot(timet,imag(SS),'linewidth',1.5), grid
ylabel('Up chirp LFM')
```

MATLAB Function "train_ambg_lfm.m" Listing

```
function x = train_ambg_lfm(taup, n, pri, bw)
% This function implemenst Eq. (5.43) of textbook
%% Inputs
  % taup       == pulsewidth in seconds
  % n          == number of pulses in train
  % pri        == pulse repetition interval in seconds
  % bw         == the LFM bandwidth in Hz
%%Outputs
  % x          == array of bimodality function
if (taup >= pri/2)
    'ERROR. Pulse width must be less than the PRI/2.'
    return
end
eps = 1.0e-6;
 q = -(n-1):1:n-1;
offset = 0:0.033:pri;
[Q, S] = meshgrid(q, offset);
Q = reshape(Q, 1, length(q)*length(offset));
S = reshape(S, 1, length(q)*length(offset));
tau = (-taup * ones(1,length(S))) + S ;
fd = -bw:0.033:bw;
[T, F] = meshgrid(tau, fd);
Q = repmat(Q, length(fd), 1);
S = repmat(S, length(fd), 1);
N = n * ones(size(T));
val1 = 1.0-(abs(T))/taup;
val2 = pi*taup*(F+T*(bw/taup)).*val1;
val3 = abs(val1.*sin(val2+eps)./(val2+eps));
val4 = abs(sin(pi*F.*(N-abs(Q))*pri+eps)./sin(pi*F*pri+eps));
x = val3.*val4./N;
[rows, cols] = size(x);
x = reshape(x, 1, rows*cols);
T = reshape(T, 1, rows*cols);
indx = find(abs(T) > taup);
x(indx) = 0.0;
x = reshape(x, rows, cols);
return
```

MATLAB Program "Fig5_10.m" Listing

```
% Use this program to reproduce Fig. 5.10 of the textbook.
clear all
close all
taup = 0.4;
pri = 1;
n = 3;
bw = 10;
x = train_ambg_lfm(taup, n, pri, bw);
figure(1)
time = linspace(-(n-1)*pri-taup, n*pri-taup, size(x,2));
doppler = linspace(-bw,bw, size(x,1));
%mesh(time, doppler, x);
surf(time, doppler, x); shading interp;
xlabel('Delay in seconds');
ylabel('Doppler in Hz');
zlabel('Ambiguity function');
axis tight;
title('LFM pulse train, B\tau = 40, N = 3 pulses')
figure(2)
contour(time, doppler, (x));
%surf(time, doppler, x); shading interp; view(0,90);
xlabel('Delay in seconds');
ylabel('Doppler in Hz');
grid;
axis tight;
title('LFM pulse train, B\tau = 40, N = 3 pulses')
```

MATLAB Program "Fig5_15.m" Listing

```
% Use this program to reproduce Fig. 5.15
clear all;
close all;
delw = linspace(-.5,.5,75);
T1 = .5 .* sin(pi.*delw);
T2 = delw + (1/2/pi) .* sin(2*pi.*delw);
T3 = .25 .* (sin(pi.*delw)) .* ((cos(pi.*delw)).^2 + 2);
T4 = delw + (1/2/pi) .* sin(2*pi.*delw) + (2/3/pi) .* (cos(pi.*delw)).^3 .* sin(delw);
figure (1)
plot(delw,T1,'k*',delw,T2,'k:',delw,T3,'k.',delw,T4,'k');
grid
ylabel('Group delay function'); xlabel('\omega/B')
legend('n=1','n=2','n=3','n=4')
```

Chapter 6

Ambiguity Function -
Discrete Coded Waveforms

The concepts of resolution and ambiguity were introduced in Chapter 4. The relationship between the waveform resolution (range and Doppler) and its corresponding ambiguity function was discussed and analyzed. It was determined that the *goodness* of a given waveform is based on its range and Doppler resolutions, which can be analyzed in the context of the ambiguity function. For this purpose, a few common analog radar waveforms were analyzed in Chapter 5. In this chapter, another type of radar waveform based on discrete codes is analyzed. This topic has been and continues to be a major research thrust area for many radar scientists, designers, and engineers. Discrete coded waveforms are more effective in improving range characteristics than Doppler (velocity) characteristics. Furthermore, in some radar applications, discrete coded waveforms are heavily favored because of their inherent anti-jamming capabilities. In this chapter, a quick overview of discrete coded waveforms is presented. Three classes of discrete codes are analyzed. They are unmodulated pulse-train codes (uniform and staggered), phase-modulated (binary or polyphase) codes, and frequency modulated codes.

6.1. Discrete Code Signal Representation

The general form for a discrete coded signal can be written as

$$x(t) = e^{j\omega_0 t} \sum_{n=1}^{N} u_n(t) = e^{j\omega_0 t} \sum_{n=1}^{N} P_n(t) e^{j(\omega_n t + \theta_n)} \qquad \text{Eq. (6.1)}$$

where ω_0 is the carrier frequency in radians, (ω_n, θ_n) are constants, N is the code length (number of bits in the code), and the signal $P_n(t)$ is given by

$$P_n(t) = a_n Rect\left(\frac{t}{\tau_0}\right). \qquad \text{Eq. (6.2)}$$

The constant a_n is either (1) or (0), and

$$Rect\left(\frac{t}{\tau_0}\right) = \begin{cases} 1 & ; \quad 0 < t < \tau_0 \\ 0 & ; \quad elsewhere \end{cases}. \qquad \text{Eq. (6.3)}$$

Using this notation, the discrete code can be described through the sequence

$$U[n] = \{u_n, n = 1, 2, \ldots, N\} \qquad \text{Eq. (6.4)}$$

which, in general, is a complex sequence depending on the values of ω_n and θ_n. The sequence $U[n]$ is called the code, and for convenience it will be denoted by U.

In general, the output of the matched filter is

$$\chi(\tau, f_d) = \int_{-\infty}^{\infty} x^*(t)x(t+\tau)e^{-j2\pi f_d t} dt, \qquad \text{Eq. (6.5)}$$

Substituting Eq. (6.1) into Eq. (6.5) yields

$$\chi(\tau, f_d) = \sum_{n=1}^{N} \sum_{k=1}^{N} \int_{-\infty}^{\infty} u_n^*(t)u_k(t+\tau)e^{-j2\pi f_d t} dt, \qquad \text{Eq. (6.6)}$$

Depending on the choice of combination for a_n, ω_n, and θ_n, different class of codes can be generated. To this end, pulse-train codes are generated when

$$\theta_n = \omega_n = 0 \quad ; \; and \; a_n = 1, or \; 0, \qquad \text{Eq. (6.7)}$$

Binary phase codes and polyphase codes are generated when

$$\omega_n = 0 \quad ; \; and \; a_n = 1, \qquad \text{Eq. (6.8)}$$

Finally, frequency codes are generated when

$$\theta_n = 0 \quad ; \; and \; a_n = 1, or \; 0, \qquad \text{Eq. (6.9)}$$

6.2. Pulse-Train Codes

The idea behind this class of code is to divide a relatively long pulse of length T_P into N subpulses, each being a rectangular pulse with pulse width τ_0 and amplitude of 1 or 0. It follows that the code U is the sequence of 1s and 0s. More precisely, the signal representing this class of code can be written as

$$x(t) = e^{j\omega_0 t} \sum_{n=1}^{N} P_n(t) = e^{j\omega_0 t} \sum_{n=1}^{N} a_n Rect\left(\frac{t}{\tau_0}\right) \qquad \text{Eq. (6.10)}$$

One way to generate a train-pulse class code can be by setting

$$a_n = \begin{cases} 1 & n-1 = 0 \; modulu \; q \\ 0 & n-1 \neq 0 \; modulu \; q \end{cases} \qquad \text{Eq. (6.11)}$$

where q is a positive integer that divides evenly into $N-1$. That is,

$$M-1 = (N-1)/q \qquad \text{Eq. (6.12)}$$

where M is the number of 1s in the code. For example, when $N = 21$ and $q = 5$, then $M = 5$, and the resulting code is

$${U} = {10000 \quad 10000 \quad 10000 \quad 10000 \quad 1}. \qquad \text{Eq. (6.13)}$$

This is illustrated in Fig. 6.1. In previous chapters this code would have been represented by the following continuous time domain signal

$$x_1(t) = e^{j\omega_0 t} \sum_{m=0}^{4} Rect\left(\frac{t - mT}{\tau_0}\right) \qquad \text{Eq. (6.14)}$$

where the period is $T = 5\tau_0$. Using this analogy yields

$$\frac{T_p}{M-1} \equiv T \qquad \text{Eq. (6.15)}$$

and Eq. (6.10) can now be written as

$$x(t) = e^{j\omega_0 t} \sum_{m=1}^{M-1} Rect\left(\frac{t - m\left(\dfrac{T_p}{M-1}\right)}{\tau_0}\right). \qquad \text{Eq. (6.16)}$$

In Chapter 5 an expression for the ambiguity function for a coherent train of pulses was derived. Comparison of Eq. (6.16) and Eq. (5.27) show that the two equations are equivalent when the condition in Eq. (6.15) is true except for some constants. It follows that the ambiguity function for the signal defined in Eq. (6.16) is

$$|\chi(\tau;f_d)| = \sum_{k=-M}^{M} \left| \frac{\sin\left[\pi f_d \left([M - |k|] \dfrac{T_p}{M-1}\right)\right]}{\sin\left(\pi f_d \dfrac{T_p}{M-1}\right)} \right| \left| \frac{\sin\left[\pi f_d \left(\tau_0 - \left|\tau - \dfrac{kT_p}{M-1}\right|\right)\right]}{\pi f_d} \right|. \qquad \text{Eq. (6.17)}$$

The zero Doppler and zero delay cuts of the ambiguity function are derived from Eq. (6.17). They are given by

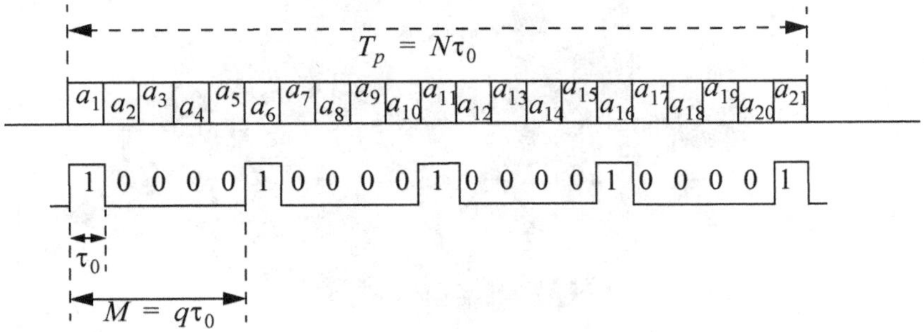

Figure 6.1. Generating a pulse-train code of length $N = 21$ bits.

$$|\chi(\tau;0)| = M\tau_0 \sum_{k=-M}^{M} \left[1 - \frac{|k|}{M}\right]\left(1 - \frac{\left|\tau - \frac{kT_p}{M-1}\right|}{\tau_0}\right)$$

Eq. (6.18)

$$|\chi(0;f_d)| = \sum_{k=-M}^{M} \left|\frac{\sin\left[\pi M f_d\left(\frac{T_p}{M-1}\right)\right]}{\sin\left(\pi f_d \frac{T_p}{M-1}\right)}\right| \left|\frac{\sin(\pi f_d \tau_0)}{\pi f_d \tau_0}\right|.$$

Eq. (6.19)

Figure 6.2a shows the three-dimensional ambiguity plot for the code shown in Fig. 6.1, while Fig. 6.2b shows the corresponding contour plot. This figure can be reproduced using MATLAB program "*Fig6_2.m,*" listed in Appendix 6-A.

A cartoon showing contour cuts of the ambiguity function for a pulse-train code is shown in Fig. 6.2c. Clearly, the width of the ambiguity function main lobe (i.e., resolution) is directly tied to the code length. As one would expect, longer codes will produce a narrower main lobe and thus have better resolution than shorter ones. Further observation of Fig. 6.2 shows that this ambiguity function has a strong grating lobe structure along with high sidelobe levels. The presence of such strong lobing structure limits the effectiveness of the code and will cause detection ambiguities. These lobes are a direct result from the uniform equal spacing between the 1s within a code (i.e., periodicity of the code). These lobes can be significantly reduced by getting rid of the periodic structure of the code, i.e., placing the pulses at nonuniform spacing. This is called code staggering (PRF staggering).

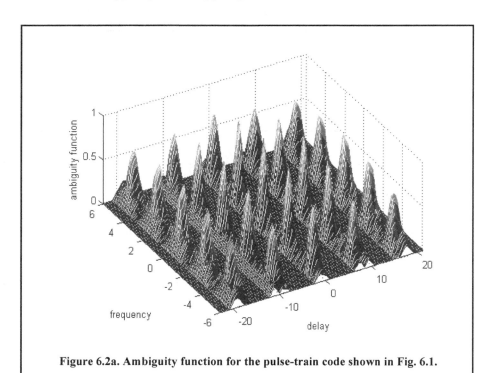

Figure 6.2a. Ambiguity function for the pulse-train code shown in Fig. 6.1.

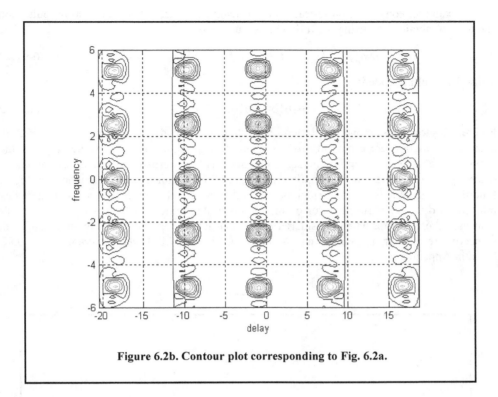

Figure 6.2b. Contour plot corresponding to Fig. 6.2a.

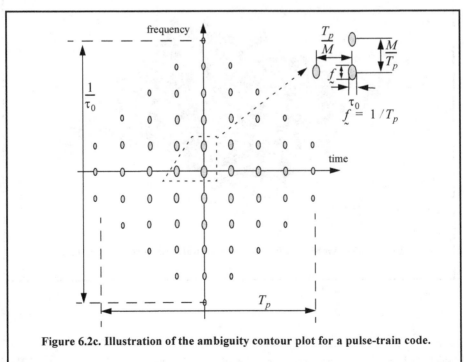

Figure 6.2c. Illustration of the ambiguity contour plot for a pulse-train code.

For example, consider a pulse-train code of length $N = 21$. A staggered train-pulse code can then be obtained by using the following sequence a_n

$$\{a_n\} = 1 \qquad n = 1, 4, 6, 12, 15, 21 .$$ **Eq. (6.20)**

Thus, the resulting code is

$$\{U\} = \{100101000001001000001\} .$$ **Eq. (6.21)**

Figure 6.3 shows the ambiguity plot corresponding to this code. As indicated by Fig. 6.3, the ambiguity function corresponding to a staggered pulse-train code approaches a thumbtack shape. The choice of the optimum staggered code has been researched extensively by numerous people. Resnick[1] defined the optimum staggered pulse-train code as that whose ambiguity function has absolutely uniform sidelobe levels that are equal to unity. Other researchers have introduced different definitions for optimum staggering, none of which is necessarily better than the others, except when considered for the particular application being analyzed by the respective researcher. Figure 6.3 can be reproduced using MATLAB program *"Fig6_3.m,"* listed in Appendix 6-A.

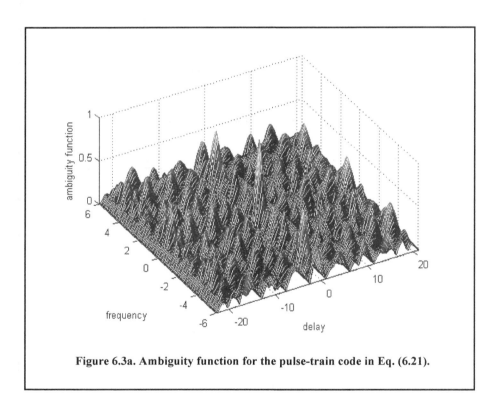

Figure 6.3a. Ambiguity function for the pulse-train code in Eq. (6.21).

---

1. Resnick, J. B., *High Resolution Waveforms Suitable for a Multiple Target Environment*, MS thesis, MIT, Cambridge, MA, June 1962.

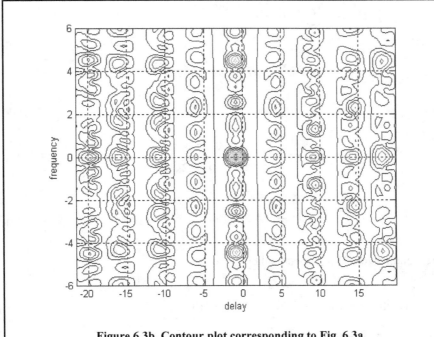

frequency

delay

Figure 6.3b. Contour plot corresponding to Fig. 6.3a.

6.3. Phase Coding

The signal corresponding to this class of code is obtained from Eq. (6.1) by letting $\omega_n = 0$. It follows that

$$x(t) = e^{j\omega_0 t} \sum_{n=1}^{N} u_n(t) = e^{j\omega_0 t} \sum_{n=1}^{N} P_n(t) e^{j\theta_n} .$$

Eq. (6.22)

Two subclasses of phase codes are analyzed. They are binary phase codes and polyphase codes.

6.3.1. Binary Phase Codes

In this case, the phase θ_n is set equal to either (0) or (π), and hence, the term *binary* is used. For this purpose, define the coefficient D_n as

$$D_n = e^{j\theta_n} = \pm 1 .$$

Eq. (6.23)

The ambiguity function for this class of code is derived by substituting Eq. (6.22) into Eq. (6.5). The resulting ambiguity function is given by

$$\chi(\tau;f_d) = \begin{cases} \chi_0(\tau',f_d)\displaystyle\sum_{n=1}^{N-k} D_n D_{n+k} e^{-j2\pi f_d(n-1)\tau_0} + \\[2em] \chi_0(\tau_0-\tau',f_d)\displaystyle\sum_{n=1}^{N-(k+1)} D_n D_{n+k+1} e^{-j2\pi f_d n\tau_0} \end{cases} \qquad 0<\tau<N\tau_0 \qquad \textbf{Eq. (6.24)}$$

where

$$\tau = k\tau_0 + \tau' \qquad \begin{cases} 0<\tau'<\tau_0 \\ k = 0, 1, 2, ..., N \end{cases} \qquad \textbf{Eq. (6.25)}$$

$$\chi_0(\tau',f_d) = \int_0^{\tau_0-\tau'} \exp(-j2\pi f_d t)dt \qquad 0<\tau'<\tau_0. \qquad \textbf{Eq. (6.26)}$$

The corresponding zero Doppler cut is then given by

$$\chi(\tau;0) = \tau_0\left(1-\frac{|\tau'|}{\tau_0}\right)\sum_{n=1}^{N-|k|} D_n D_{n+k} + |\tau'|\sum_{n=1}^{N-|k+1|} D_n D_{n+k+1}, \qquad \textbf{Eq. (6.27)}$$

and when $\tau' = 0$ then

$$\chi(k;0) = \tau_0\sum_{n=1}^{N-|k|} D_n D_{n+k}. \qquad \textbf{Eq. (6.28)}$$

Barker Code

Barker code is one of the most commonly known codes from the binary phase code class. In this case, a long pulse of width T_p is divided into N smaller pulses; each is of width $\tau_0 = T_p/N$. Then, the phase of each subpulse is chosen as either 0 or π radians relative to some code. It is customary to characterize a subpulse that has 0 phase (amplitude of +1 volt) as either "1" or "+." Alternatively, a subpulse with phase equal to π (amplitude of -1 volt) is characterized by either "0" or "-." Barker code is optimum in accordance with the definition set by Resnick. Figure 6.4 illustrates this concept for a Barker code of length seven. A Barker code of length N is denoted as B_N. There are only seven known Barker codes that share this unique property; they are listed in Table 6.1. Note that B_2 and B_4 have complementary forms that have the same characteristics.

In general, the autocorrelation function (which is an approximation for the matched filter output) for a B_N Barker code will be $2N\tau_0$ wide. The main lobe is $2\tau_0$ wide; the peak value is equal to N. There are $(N-1)/2$ sidelobes on either side of the main lobe; this is illustrated in Fig. 6.5 for a B_{13}. Notice that the main lobe is equal to 13, while all sidelobes are unity.

The most sidelobe reduction offered by a Barker code is $-22.3\,dB$, which may not be sufficient for the desired radar application. However, Barker codes can be combined to generate much longer codes. In this case, a B_M code can be used within a B_N code (M within N) to generate a code of length MN. The compression ratio for the combined B_{MN} code is equal to MN. As an example, a combined B_{54} is given by

$$B_{54} = \{11101, 11101, 00010, 11101\} \qquad \text{Eq. (6.29)}$$

and is illustrated in Fig. 6.6. Unfortunately, the sidelobes of a combined Barker code autocorrelation function are no longer equal to unity. Some sidelobes of a combined Barker code autocorrelation function can be reduced to zero if the matched filter is followed by a linear transversal filter with impulse response given by

$$h(t) = \sum_{k = -N}^{N} \beta_k \delta(t - 2k\tau_0) \qquad \text{Eq. (6.30)}$$

where N is the filter's order, the coefficients β_k $(\beta_k = \beta_{-k})$ are to be determined, $\delta(\ )$ is the delta function, and τ_0 is the Barker code subpulse width. A filter of order N produces N zero sidelobes on either side of the main lobe. The main lobe amplitude and width do not change, as illustrated in Fig. 6.7.

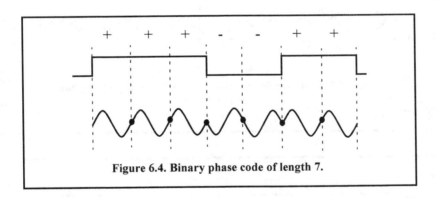

Figure 6.4. Binary phase code of length 7.

TABLE 6.1. Barker codes

Code Symbol	Code Length	Code Elements	Side Lode Reduction (dB)
B_2	2	+-	6.0
		++	
B_3	3	++-	9.5
B_4	4	++-+	12.0
		+++-	
B_5	5	+++-+	14.0
B_7	7	+++--+-	16.9
B_{11}	11	+++---+--+-	20.8
B_{13}	13	+++++--++-+-+	22.3

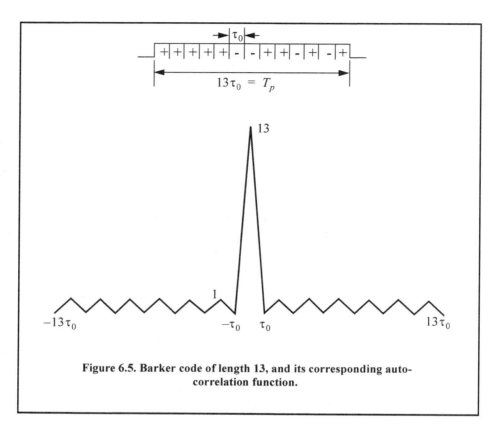

Figure 6.5. Barker code of length 13, and its corresponding auto-correlation function.

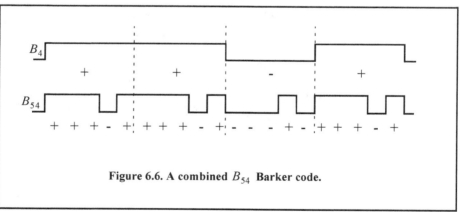

Figure 6.6. A combined B_{54} Barker code.

In order to illustrate this approach, consider the case where the input to the matched filter is B_{11}, and assume $N = 4$. The autocorrelation for a B_{11} is

$$R_{11} = \{-1, 0, -1, 0, -1, 0, -1, 0, -1, 0, 11,$$
$$0, -1, 0, -1, 0, -1, 0, -1, 0, -1\} \qquad\qquad \text{Eq. (6.31)}$$

The output of the transversal filter is the discrete convolution between its impulse response and the sequence R_{11}. At this point we need to compute the coefficients β_k that guarantee the desired filter output (i.e., unchanged main lobe and four zero sidelobe levels).

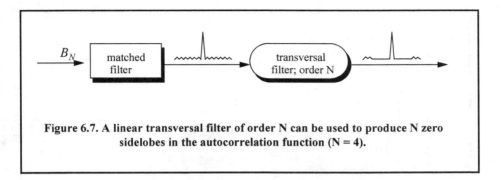

Figure 6.7. A linear transversal filter of order N can be used to produce N zero sidelobes in the autocorrelation function (N = 4).

Performing the discrete convolution as defined in Eq. (6.30) and collecting equal terms ($\beta_k = \beta_{-k}$) yield the following set of five linearly independent equations:

$$\begin{bmatrix} 11 & -2 & -2 & -2 & -2 \\ -1 & 10 & -2 & -2 & -1 \\ -1 & -2 & 10 & -2 & -1 \\ -1 & -2 & -1 & 11 & -1 \\ -1 & -1 & -1 & -1 & 11 \end{bmatrix} \begin{bmatrix} \beta_0 \\ \beta_1 \\ \beta_2 \\ \beta_3 \\ \beta_4 \end{bmatrix} = \begin{bmatrix} 11 \\ 0 \\ 0 \\ 0 \\ 0 \end{bmatrix}. \qquad \text{Eq. (6.32)}$$

Solving Eq. (6.32) yields

$$\begin{bmatrix} \beta_0 \\ \beta_1 \\ \beta_2 \\ \beta_3 \\ \beta_4 \end{bmatrix} = \begin{bmatrix} 1.1342 \\ 0.2046 \\ 0.2046 \\ 0.1731 \\ 0.1560 \end{bmatrix}. \qquad \text{Eq. (6.33)}$$

Note that setting the first equation equal to 11 and all other equations to 0 and then solving for β_k guarantees that the main peak remains unchanged, and that the next four sidelobes are zeros. So far we have assumed that coded pulses have rectangular shapes. Using other pulses of other shapes, such as Gaussian, may produce better sidelobe reduction and a larger compression ratio.

Figure 6.8 shows the output of this function when B_{13} is used as an input. Figure 6.9 is similar to Fig. 6.8, except in this case B_7 is used as an input. Figure 6.10 shows the ambiguity function, the zero Doppler cut, and the contour plot for the combined Barker code defined in Fig. 6.6.

Figures 6.8 through 6.10 can be reproduced using the MATALB program *"Fig6_8_10.m,"* listed in Appendix 6-A.

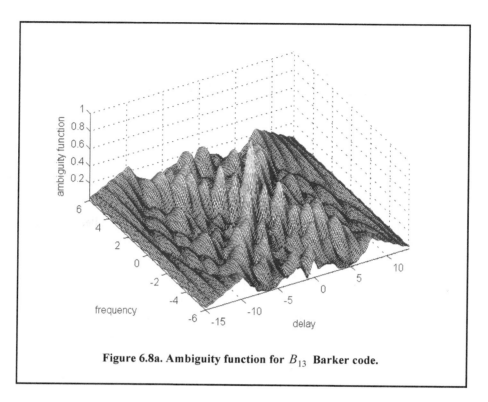

Figure 6.8a. Ambiguity function for B_{13} Barker code.

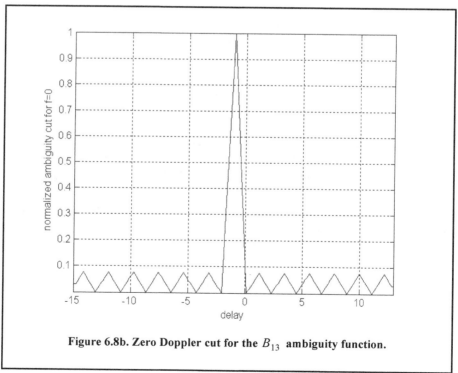

Figure 6.8b. Zero Doppler cut for the B_{13} ambiguity function.

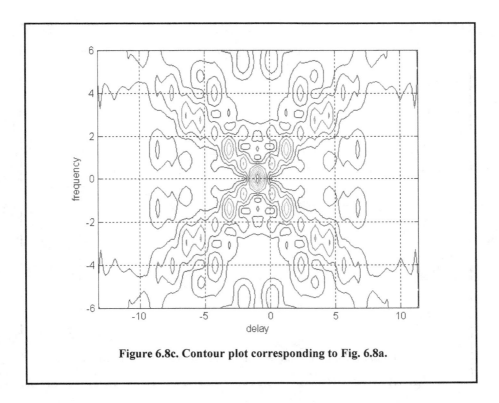

Figure 6.8c. Contour plot corresponding to Fig. 6.8a.

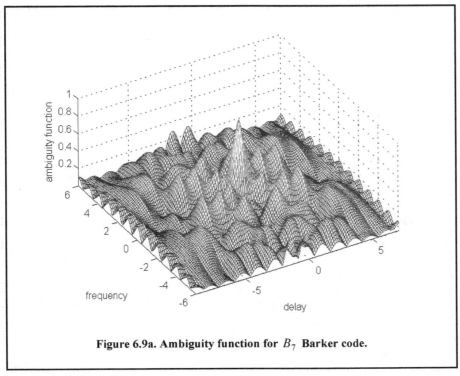

Figure 6.9a. Ambiguity function for B_7 Barker code.

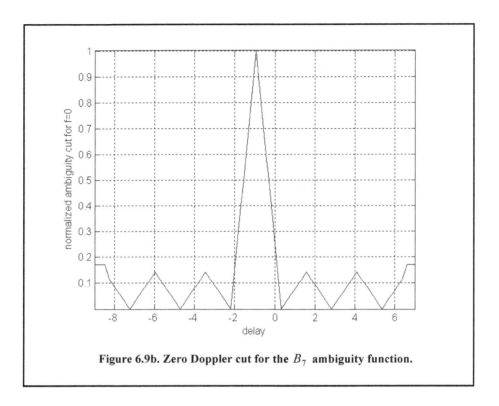

Figure 6.9b. Zero Doppler cut for the B_7 ambiguity function.

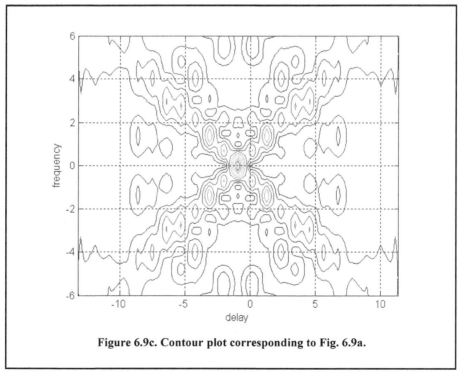

Figure 6.9c. Contour plot corresponding to Fig. 6.9a.

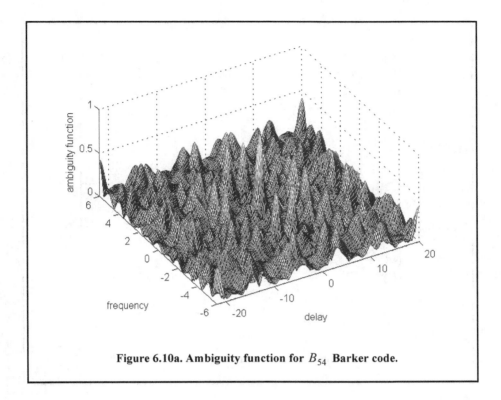

Figure 6.10a. Ambiguity function for B_{54} Barker code.

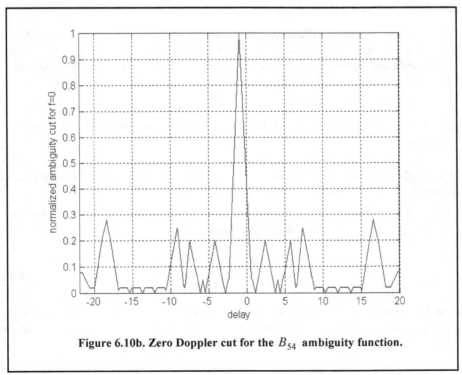

Figure 6.10b. Zero Doppler cut for the B_{54} ambiguity function.

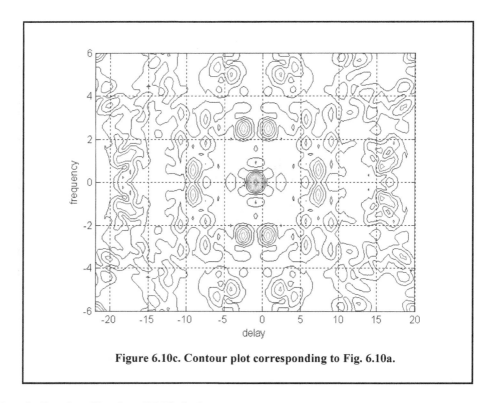

Figure 6.10c. Contour plot corresponding to Fig. 6.10a.

Pseudo-Random Number (PRN) Codes

Pseudo-Random Number (PRN) codes are also known as Maximal Length Sequences (MLS) codes. These codes are called pseudo-random because the statistics associated with their occurrence are similar to those associated with the coin-toss sequences. Maximum length sequences are periodic. The MLS codes have the following distinctive properties:

1. The number of ones per period is one more than the number of minus ones.
2. Half the runs (consecutive states of the same kind) are of length one and one fourth are of length two.
3. Every maximal length sequence has the "shift and add" property. This means that, if a maximal length sequence is added (modulo 2) to a shifted version of itself, then the resulting sequence is a shifted version of the original sequence.
4. Every n-tuple of the code appears once and only once in one period of the sequence.
5. The correlation function is periodic and is given by

$$\phi(n) = \begin{cases} L & n = 0, \pm L, \pm 2L, \dots \\ -1 & elsewhere \end{cases}.$$

Eq. (6.34)

Figure 6.11 shows a typical sketch for an MLS autocorrelation function. Clearly these codes have the advantage that the compression ratio becomes very large as the period is increased. Additionally, adjacent peaks (grating lobes) become farther apart.

Linear Shift Register Generators

There are numerous ways to generate MLS codes. The most common is to use linear shift registers. When the binary sequence generated using a shift register implementation is periodic and has maximal length, it is referred to as an MLS binary sequence with period L, where

$$L = 2^n - 1.$$
Eq. (6.35)

n is the number of stages in the shift register generator. A linear shift register generator basically consists of a shift register with modulo-two adders added to it. The adders can be connected to various stages of the register, as illustrated in Fig. 6.12 for $n = 4$ (i.e., $L = 15$). Note that the shift register initial state cannot be 0.

The feedback connections associated with a shift register generator determine whether the output sequence will be maximal. For a given size shift register, only a few feedback connections lead to maximal sequence outputs. In order to illustrate this concept, consider the two 5-stage shift register generators shown in Fig. 6.13. The shift register generator shown in Fig. 6.13 a generates a maximal length sequence, as clearly depicted by its state diagram. However, the shift register generator shown in Fig. 6.13 b produces three non-maximal length sequences (depending on the initial state).

Given an n-stage shift register generator, one would be interested in knowing how many feedback connections will yield maximal length sequences. Zierler[1] showed that the number of maximal length sequences possible for a given n-stage linear shift register generator is

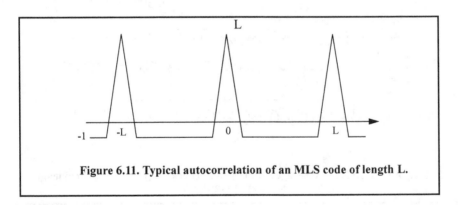

Figure 6.11. Typical autocorrelation of an MLS code of length L.

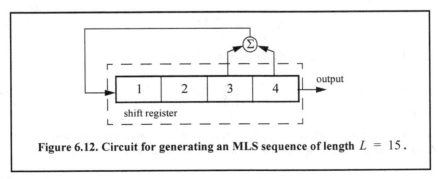

Figure 6.12. Circuit for generating an MLS sequence of length $L = 15$.

---

1. Zierler, N., *Several Binary-Sequence Generators*, MIT Technical Report No. 95, Sept. 1955.

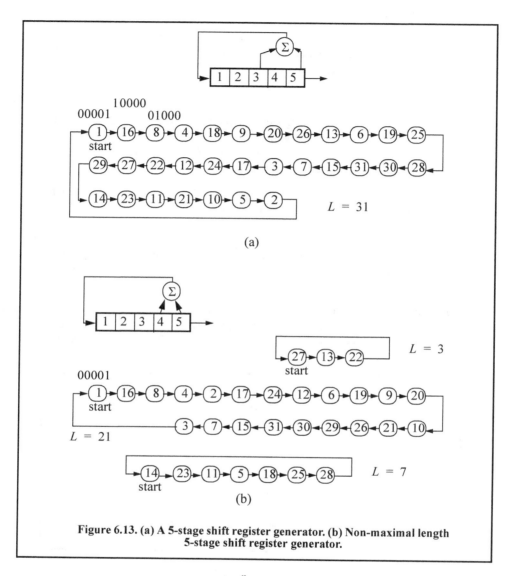

Figure 6.13. (a) A 5-stage shift register generator. (b) Non-maximal length
5-stage shift register generator.

$$N_L = \frac{\varphi(2^n - 1)}{n}.$$

Eq. (6.36)

φ is the Euler's totient (Euler's phi) function and is defined by

$$\varphi(k) = k \prod_i \frac{(p_i - 1)}{p_i}$$

Eq. (6.37)

where p_i are the prime factors of k. Note that when p_i has multiples, only one of them is used. Also note that when k is a prime number, the Euler's phi function is

$$\varphi(k) = k - 1.$$

Eq. (6.38)

For example, a 3-stage shift register generator will produce

$$N_L = \frac{\varphi(2^3 - 1)}{3} = \frac{\varphi(7)}{3} = \frac{7-1}{3} = 2, \qquad \text{Eq. (6.39)}$$

and a 6-stage shift register,

$$N_L = \frac{\varphi(2^6 - 1)}{6} = \frac{\varphi(63)}{6} = \frac{63}{6} \times \frac{(3-1)}{3} \times \frac{(7-1)}{7} = 6, \qquad \text{Eq. (6.40)}$$

Maximal Length Sequence Characteristic Polynomial

Consider an n-stage maximal length linear shift register whose feedback connections correspond to *n, k, m, etc.* This maximal length shift register can be described using its characteristic polynomial defined by

$$x^n + x^k + x^m + \ldots + 1 \qquad \text{Eq. (6.41)}$$

where the additions are modulo 2. Therefore, if the characteristic polynomial for an n-stage shift register is known, one can easily determine the register feedback connections and consequently deduce the corresponding maximal length sequence. For example, consider a 6-stage shift register whose characteristic polynomial is

$$x^6 + x^5 + 1. \qquad \text{Eq. (6.42)}$$

It follows that the shift register which generates a maximal length sequence is shown in Fig. 6.14.

One of the most important issues associated with generating a maximal length sequence using a linear shift register is determining the characteristic polynomial. This has been and continues to be a subject of research for many radar engineers and designers. It has been shown that polynomials which are both irreducible (not factorable) and primitive will produce maximal length shift register generators.

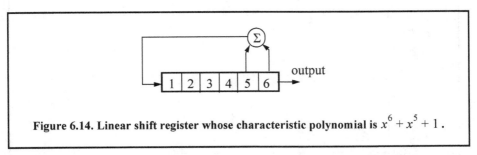

Figure 6.14. Linear shift register whose characteristic polynomial is $x^6 + x^5 + 1$.

A polynomial of degree n is irreducible if it is not divisible by any polynomial of degree less than *n*. It follows that all irreducible polynomials must have an odd number of terms. Consequently, only linear shift register generators with an even number of feedback connections can produce maximal length sequences. An irreducible polynomial is primitive if and only if it divides $x^n - 1$ for no value of *n* less than $2^n - 1$. Figure 6.15 shows the output of this function for

u31 = [1 -1 -1 -1 -1 1 -1 1 -1 1 1 1 -1 1 1 -1 -1 -1 1 1 1 1 1 -1 -1 1 1 -1 1 -1 -1].

Figure 6.16 is similar to Fig. 6.15, except in this case the input maximal length sequence is

u15=[1 -1 -1 -1 1 1 1 1 -1 1 -1 1 1 -1 -1].

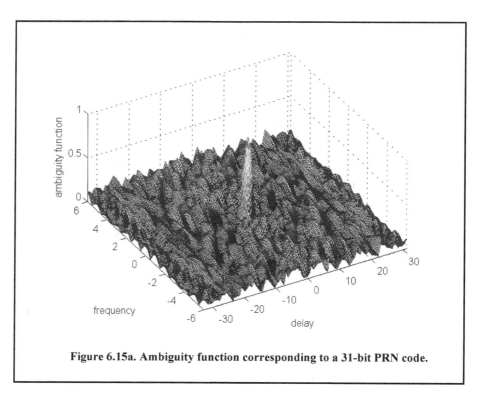

Figure 6.15a. Ambiguity function corresponding to a 31-bit PRN code.

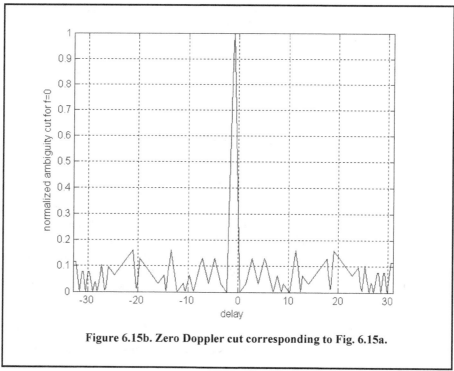

Figure 6.15b. Zero Doppler cut corresponding to Fig. 6.15a.

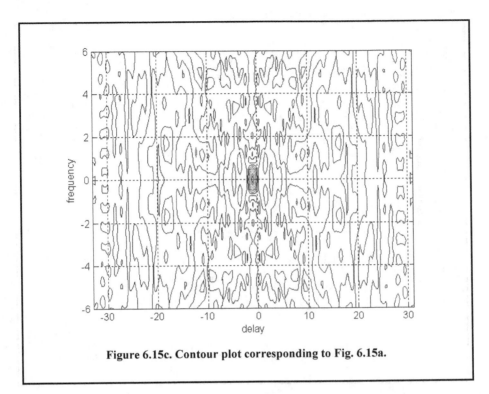

Figure 6.15c. Contour plot corresponding to Fig. 6.15a.

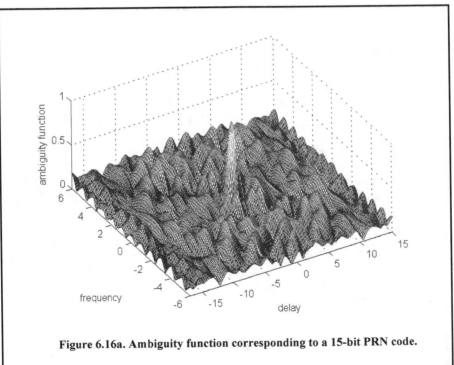

Figure 6.16a. Ambiguity function corresponding to a 15-bit PRN code.

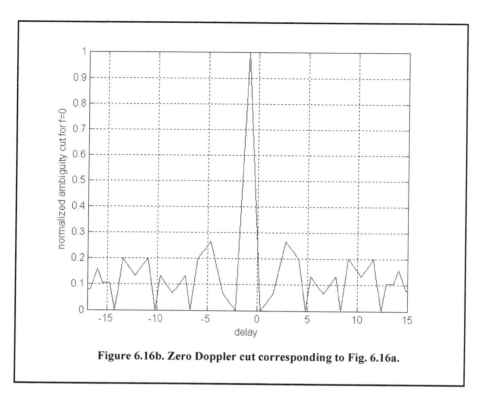

Figure 6.16b. Zero Doppler cut corresponding to Fig. 6.16a.

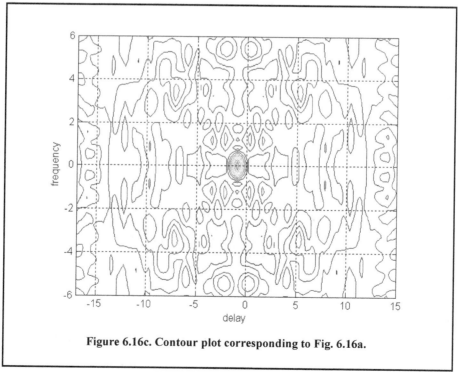

Figure 6.16c. Contour plot corresponding to Fig. 6.16a.

6.3.2. Polyphase Codes

The signal corresponding to polyphase codes is that given in Eq. (6.22) and the corresponding ambiguity function was given in Eq. (6.24). The only exception is that the phase θ_n is no longer restricted to $(0, \pi)$. Hence, the coefficient D_n are no longer equal to ± 1 but can be complex depending on the value of θ_n. Polyphase Barker codes have been investigated by many scientists, and much is well documented in the literature. In this chapter the discussion will be limited to Frank codes.

Frank Codes

In this case, a single pulse of width T_p is divided into N equal groups; each group is subsequently divided into other N subpulses, each of width τ_0. Therefore, the total number of subpulses within each pulse is N^2, and the compression ratio is $\xi = N^2$. As previously, the phase within each subpulse is held constant with respect to some CW reference signal.

A Frank code of N^2 subpulses is referred to as an N-phase Frank code. The first step in computing a Frank code is to divide $360°$ by N and define the result as the fundamental phase increment $\Delta\varphi$. More precisely,

$$\Delta\varphi = 360° / N. \qquad \text{Eq. (6.43)}$$

Note that the size of the fundamental phase increment decreases as the number of groups is increased, and because of phase stability, this may degrade the performance of very long Frank codes. For N-phase Frank code the phase of each subpulse is computed from

$$\begin{pmatrix} 0 & 0 & 0 & 0 & \dots & 0 \\ 0 & 1 & 2 & 3 & \dots & N-1 \\ 0 & 2 & 4 & 6 & \dots & 2(N-1) \\ \dots & \dots & \dots & \dots & \dots & \dots \\ \dots & \dots & \dots & \dots & \dots & \dots \\ 0 & (N-1) & 2(N-1) & 3(N-1) & \dots & (N-1)^2 \end{pmatrix} \Delta\varphi \qquad \text{Eq. (6.44)}$$

where each row represents a group, and a column represents the subpulses for that group. For example, a 4-phase Frank code has $N = 4$, and the fundamental phase increment is $\Delta\varphi = (360° / 4) = 90°$. It follows that

$$\begin{pmatrix} 0 & 0 & 0 & 0 \\ 0 & 90° & 180° & 270° \\ 0 & 180° & 0 & 180° \\ 0 & 270° & 180° & 90° \end{pmatrix} \Rightarrow \begin{pmatrix} 1 & 1 & 1 & 1 \\ 1 & j & -1 & -j \\ 1 & -1 & 1 & -1 \\ 1 & -j & -1 & j \end{pmatrix}. \qquad \text{Eq. (6.45)}$$

Therefore, a Frank code of 16 elements is given by

$$F_{16} = \{1\ 1\ 1\ 1\ 1\ j\ -1\ -j\ 1\ -1\ 1\ -1\ 1\ -j\ -1\ j\}. \qquad \text{Eq. (6.46)}$$

A plot of the ambiguity function for F_{16} is shown in Fig. 6.17. Note the thumbtack shape of the ambiguity function. This plot can be reproduced using MATLAB program *"Fig6_17.m,"* listed in Appendix 6-A. The phase increments within each row represent a step-wise approximation of an up-chirp LFM waveform. The phase increments for subsequent rows increase linearly versus time. Thus, the corresponding LFM chirp slopes also increase linearly for subsequent rows. This is illustrated in Fig. 6.18, for F_{16}.

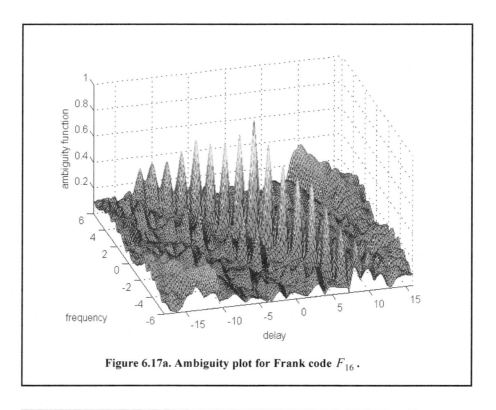

Figure 6.17a. Ambiguity plot for Frank code F_{16} .

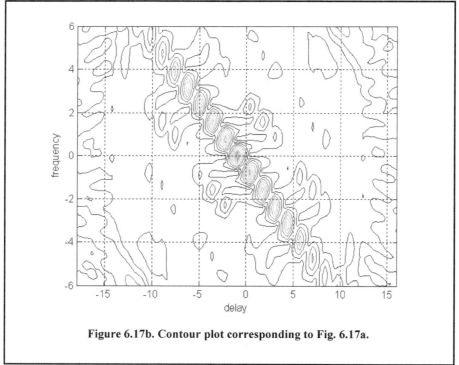

Figure 6.17b. Contour plot corresponding to Fig. 6.17a.

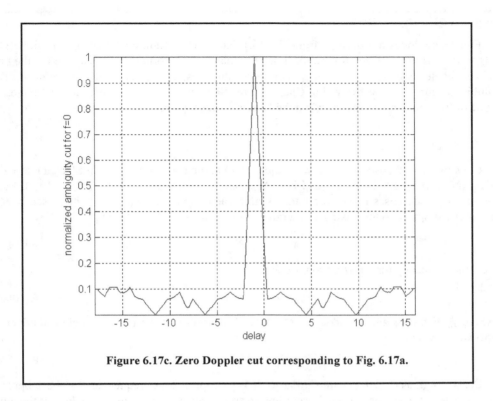

Figure 6.17c. Zero Doppler cut corresponding to Fig. 6.17a.

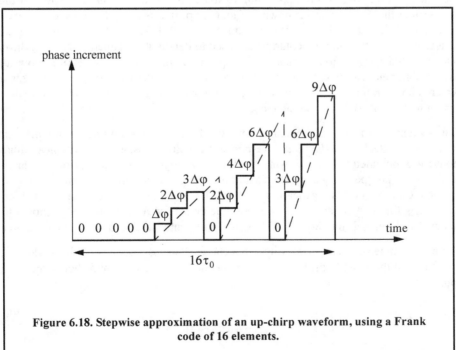

Figure 6.18. Stepwise approximation of an up-chirp waveform, using a Frank code of 16 elements.

6.4. Frequency Codes

Frequency codes are derived from Eq. (6.1) under the condition stated in Eq. (6.9) (i.e., $\theta_n = 0$; *and* $a_n = 1$, *or* 0). The Stepped Frequency Waveform (SFW) discussed in the previous chapter is considered to be a code under this class of discrete coded waveforms. The ambiguity function was derived in Chapter 5 for SFW. In this chapter the focus is on another type of frequency codes that is called the Costas frequency code.

6.4.1. Costas Codes

Construction of Costas codes can be understood in the context of SFW. In SFW, a relatively long pulse of length T_p is divided into N subpulses, each of width τ_0 ($T_p = N\tau_0$). Each group of N subpulses is called a burst. Within each burst the frequency is increased by Δf from one subpulse to the next. The overall burst bandwidth is $N\Delta f$. More precisely,

$$\tau_0 = T_p / N \qquad\qquad \text{Eq. (6.47)}$$

and the frequency for the *ith* subpulse is

$$f_i = f_0 + i\Delta f; \quad i = 1, N \qquad\qquad \text{Eq. (6.48)}$$

where f_0 is a constant frequency and $f_0 \gg \Delta f$. It follows that the time-bandwidth product of this waveform is

$$\Delta f T_p = N^2 . \qquad\qquad \text{Eq. (6.49)}$$

Costas[1] signals (or codes) are similar to SFW, except that the frequencies for the subpulses are selected in a random fashion, according to some predetermined rule or logic. For this purpose, consider the $N \times N$ matrix shown in Fig. 6.19 b. In this case, the rows are indexed from $i = 1, 2, ..., N$ and the columns are indexed from $j = 0, 1, 2, ..., (N-1)$. The rows are used to denote the subpulses and the columns are used to denote the frequency. A *dot* indicates the frequency value assigned to the associated subpulse. In this fashion, Fig. 6.19 a shows the frequency assignment associated with an SFW. Alternatively, the frequency assignments in Fig. 6.19b are chosen randomly. For a matrix of size $N \times N$, there are a total of $N!$ possible ways of assigning the dots (i.e., $N!$ possible codes).

The sequences of dot assignments for which the corresponding ambiguity function approaches an ideal or a *thumbtack* response are called Costas codes. A near thumbtack response was obtained by Costas using the following logic: There is only one frequency per time slot (row) and per frequency slot (column). Therefore, for an $N \times N$ matrix, the number of possible Costas codes is drastically less than $N!$. For example, there are $N_c = 4$ possible Costas codes for $N = 3$, and $N_c = 40$ possible codes for $N = 5$. It can be shown that the code density, defined as the ratio $N_c / N!$, gets significantly smaller as N becomes larger.

There are numerous analytical ways to generate Costas codes. In this section we will describe two of these methods. First, let q be an odd prime number, and choose the number of subpulses as

$$N = q - 1 . \qquad\qquad \text{Eq. (6.50)}$$

---

1. Costas, J. P., A Study of a Class of Detection Waveforms Having Nearly Ideal Range-Doppler Ambiguity Properties, *Proc. IEEE* 72, 1984, pp. 996-1009.

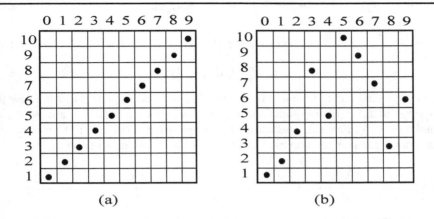

Figure 6.19. Frequency assignment for a burst of N subpulses. (a) SFW (stepped LFM); (b) Costas code of length Nc = 10.

Define γ as the primitive root of q. A primitive root of q (an odd prime number) is defined as γ such that the powers $\gamma, \gamma^2, \gamma^3, ..., \gamma^{q-1}$ modulo q generate every integer from 1 to $q-1$.

In the first method, for an $N \times N$ matrix, label the rows and columns, respectively, as

$$i = 0, 1, 2, ..., (q-2)$$
$$j = 1, 2, 3, ..., (q-1)$$

Eq. (6.51)

Place a dot in the location (i, j) corresponding to f_i if and only if

$$i = (\gamma)^j \ (modulo \ q).$$

Eq. (6.52)

In the next method, Costas code is first obtained from the logic described above; then by deleting the first row and first column from the matrix a new code is generated. This method produces a Costas code of length $N = q - 2$.

Define the normalized complex envelope of the Costas signal as

$$x(t) = \frac{1}{\sqrt{N\tau_0}} \sum_{l=0}^{N-1} x_l(t - l\tau_0)$$

Eq. (6.53)

$$x_l(t) = \begin{pmatrix} \exp(j2\pi f_l t) & 0 \le t \le \tau_0 \\ 0 & elsewhere \end{pmatrix}.$$

Eq. (6.54)

Costas showed that the output of the matched filter is

$$\chi(\tau, f_d) = \frac{1}{N} \sum_{l=0}^{N-1} \exp(j2\pi l f_d \tau) \left\{ \Phi_{ll}(\tau, f_d) + \sum_{\substack{q=0 \\ q \neq l}}^{N-1} \Phi_{lq}(\tau - (l-q)\tau_0, f_d) \right\}$$

Eq. (6.55)

$$\Phi_{lq}(\tau, f_d) = \left(\tau_0 - \frac{|\tau|}{\tau_0}\right)\frac{\sin\alpha}{\alpha}\ \exp(-j\beta - j2\pi f_q\tau) \qquad , \ |\tau| \le \tau_1 \qquad \text{Eq. (6.56)}$$

$$\alpha = \pi(f_l - f_q - f_d)(\tau_0 - |\tau|) \qquad\qquad\qquad \text{Eq. (6.57)}$$

$$\beta = \pi(f_l - f_q - f_d)(\tau_0 + |\tau|). \qquad\qquad\qquad \text{Eq. (6.58)}$$

Three-dimensional plots of the ambiguity function of Costas signals show the near thumb-tack response of the ambiguity function. All side-lobes, except for a few around the origin, have amplitude $1/N$. Few sidelobes close to the origin have amplitude $2/N$, which is typical of Costas codes. The compression ratio of a Costas code is approximately N.

6.5. Ambiguity Plots for Discrete Coded Waveforms

Plots of the ambiguity function for a given code and the corresponding cuts along zero delay and zero Doppler provide a strong indication about the code's characteristics in range and Doppler. Earlier, it was stated that the *goodness* of a given code is measured by its range and Doppler resolution characteristics. Therefore, plotting the ambiguity function of a given code is a key part of the design and analysis of radar waveforms. Unfortunately, some of the formulas for the ambiguity function are rather complicated and fairly difficult to code by the nonexpert programmer. In this section, a numerical technique for plotting the ambiguity function of any code is presented. This technique takes advantage of the computation power of MATLAB by exploiting one of the properties of the ambiguity function. Three-dimensional plots are built successively from cuts of the ambiguity function as different Doppler mismatches.

For this purpose, consider the ambiguity function property given in Eq. (5.8) and repeated here as Eq. (6.59)

$$|\chi(\tau; f_d)|^2 = \left| \int X^*(f) X(f - f_d) e^{-j2\pi f\tau} df \right|^2 \qquad \text{Eq. (6.59)}$$

where $X(f)$ is the Fourier transform of the signal $x(t)$. Using Eq. (6.59), one can compute the ambiguity function by first computing the FT of the signal under consideration, delaying it by some value f_d, and then taking the inverse FT. When the signal under consideration is a discrete coded waveform then the Fast Fourier transform is utilized. From this one can compute plots of the ambiguity function using the following technique:

1. Determine the code U under consideration. Note that U may have complex values in accordance with the class of code being considered.
2. Extend the length of the code to the next power of 2 by zero padding (see Chapter 2 for details on interpolation).
3. For better display utilize an FFT whose size is 8 times or higher than the power integer of 2 computed in step 2.
4. Compute the FFT of the extended sequence.
5. Generate vectors of frequency mismatches and delay cuts.
6. Calculate the value of $X(f - f_d)$ using vector notation.
7. Compute and store the vector resulting from the point-by-point multiplication $X^*(f)X(f - f_d)$.

8. Compute the inverse FFT of the product in step 7 for each delay value and store in a two-dimensional (2-D) array.
9. Plot the amplitude square of the resulting 2-D array to generate the ambiguity plot for the specific code under consideration.

An implementation of this algorithm is in MATLAB function *"ambiguity_code.m,"* listed in Appendix 6-A.

Problems

6.1. Show that the zero Doppler cut for the ambiguity function of an arbitrary phase coded pulse with a pulse width τ_p is given by $Y(f) = |\operatorname{sin} c(f\tau_p)|^2$.

6.2. Consider the 7-bit Barker code, designated by the sequence $x(n)$. (a) Compute and plot the autocorrelation of this code. (b) A radar uses binary phase-coded pulses of the form $s(t) = r(t)\cos(2\pi f_0 t)$, where $r(t) = x(0), \ for \ 0 < t < \Delta t$,

$r(t) = x(n), \ for \ n\Delta t < t < (n+1)\Delta t$, and $r(t) = 0, \ for \ t > 7\Delta t$. Assume $\Delta t = 0.5\,\mu s$. (a) Give an expression for the autocorrelation of the signal $s(t)$, and for the output of the matched filter when the input is $s(t - 10\Delta t)$. (b) Compute the time bandwidth product, the increase in the peak SNR, and the compression ratio.

6.3. (a) Perform the discrete convolution between the sequence R_{11} defined in Eq. (6.31), and the transversal filter impulse response; and (b) sketch the corresponding transversal filter output.

6.4. Repeat the previous problem for $N = 13$ and $k = 6$. Use a Barker code of length 13.

6.5. Develop a Barker code of length 35. Consider both B_{75} and B_{57}.

6.6. The smallest positive primitive root of $q = 11$ is $\gamma = 2$; for $N = 10$, generate the corresponding Costas matrix.

6.7. Compute the discrete autocorrelation for an F_{16} Frank code.

6.8. Generate a Frank code of length 8, i.e., F_8.

6.9. Using the MATLAB program developed in this chapter, plot the matched filter output for a 3-, 4-, and 5-bit Barker code.

Appendix 6-A: Chapter 6 MATLAB Code Listings

The MATLAB code provided in this chapter was designed as an academic standalone tool and is not adequate for other purposes. The code was written in a way to assist the reader in gaining a better understanding of the theory. The code was not developed, nor is it intended to be used as part of an open-loop or a closed-loop simulation of any kind. The MATLAB code found in this textbook can be downloaded from this book's web page on the CRC Press website. Simply use your favorite web browser, go to *www.crcpress.com*, and search for keyword *"Mahafza"* to locate this book's web page.

MATLAB Program "Fig6_2.m" Listing

```
% Use to reproduce Fig 6.2 of textbook
clc
close all
clear all
uinput = [1 0 0 0 1 0 0 0 0 1 0 0 0 0 1 0 0 0 0 1];
[ambig] = ambiguity_code(uinput);
freq = linspace(-6,6, size(ambig,1));
N = size(uinput,2);
% set code length to tau
tau = N;
code = uinput;
samp_num = size(code,2) * 10;
% compute the next power integer of 2 for FFT purposes
n = ceil(log(samp_num) / log(2));
% compute FFT size, nfft
nfft = 2^n;
% set a dummy array in preparation for interpolation
delay = linspace(-N-2,N,nfft);
plot_figuiures_chap6 ( ambig, delay, freq)
```

MATLAB Function "plot_figures_chap6.m" Listing

```
function plot_figures_chap6( ambig, delay, freq)
% This function is used to plot figures in Chapter 6
%
mesh(delay,freq,(ambig ./ max(max(ambig))))
view (-30,55);
axis tight
ylabel('frequency')
xlabel('delay')
zlabel('ambiguity function')
figure(2)
Nhalf = (size(ambig,1)-1)/2
plot(delay,ambig(Nhalf+1,:)/(max(max(ambig))),'k')
xlabel('delay')
ylabel('normalized ambiguity cut for f=0')
grid
axis tight
figure(3)
contour(delay,freq,(ambig ./ max(max(ambig))))
axis tight
```

```
ylabel('frequency')
xlabel('delay')
grid
end
```

MATLAB Program "Fig6_3.m" Listing

```
% Use to reproduce Fig 6.3 of textbook
clc
close all
clear all
uinput = [1 0 0 1 0 1 0 0 0 0 0 1 0 0 1 0 0 0 0 0 1];
[ambig] = ambiguity_code(uinput);
freq = linspace(-6,6, size(ambig,1));
N = size(uinput,2);
% set code length to tau
tau = N;
code = uinput;
samp_num = size(code,2) * 10;
% compute the next power integer of 2 for FFT purposes
n = ceil(log(samp_num) / log(2));
% compute FFT size, nfft
nfft = 2^n;
% set a dummy array in preparation for interpolation
delay = linspace(-N-2,N,nfft);
plot_figures_chap6 ( ambig, delay, freq)
```

MATLAB Program "Fig6_8_10.m" Listing

```
% Use to reproduce Figs 6.8 trhough 6.10 of textbook
clc
close all
clear all
% Figure 8
uinput = [1 1 1 1 1 -1 -1 1 1 -1 1 -1 1];
[ambig] = ambiguity_code(uinput);
freq = linspace(-6,6, size(ambig,1));
N = size(uinput,2);
% set code length to tau
tau = N;
code = uinput;
samp_num = size(code,2) * 10;
% compute the next power integer of 2 for FFT purposes
n = ceil(log(samp_num) / log(2));
% compute FFT size, nfft
nfft = 2^n;
% set a dummy array in preparation for interpolation
delay = linspace(-N-2,N,nfft);
plot_figures_chap6 ( ambig, delay, freq)
%
uinput = [1 1 1 -1 -1 1 1 -1 ];
[ambig] = ambiguity_code(uinput);
freq = linspace(-6,6, size(ambig,1));
N = size(uinput,2);
```

```
% set code length to tau
tau = N;
code = uinput;
samp_num = size(code,2) * 10;
% compute the next power integer of 2 for FFT purposes
n = ceil(log(samp_num) / log(2));
% compute FFT size, nfft
nfft = 2^n;
% set a dummy array in preparation for interpolation
delay = linspace(-N-2,N,nfft);
plot_figures_chap6 ( ambig, delay, freq)
%
uinput = [1 1 1 -1 1 1 1 1 -1 1 -1 -1 -1 1 -1 1 1 1 -1 1];
[ambig] = ambiguity_code(uinput);
freq = linspace(-6,6, size(ambig,1));
N = size(uinput,2);
% set code length to tau
tau = N;
code = uinput;
samp_num = size(code,2) * 10;
% compute the next power integer of 2 for FFT purposes
n = ceil(log(samp_num) / log(2));
% compute FFT size, nfft
nfft = 2^n;
% set a dummy array in preparation for interpolation
delay = linspace(-N-2,N,nfft);
plot_figures_chap6 ( ambig, delay, freq)
```

MATLAB Program "Fig6_15_16.m" Listing

```
% Use to reproduce Figs 6.15 and 6.16 of textbook
clc
close all
clear all
% Figure 15
uinput = [1 -1 -1 -1 -1 1 1 -1 1 -1 1 1 1 -1 1 1 1 -1 -1 -1 1 1 1 1 1 1 -1 -1 1 1 1 -1 1 1 -1 -1];
[ambig] = ambiguity_code(uinput);
freq = linspace(-6,6, size(ambig,1));
N = size(uinput,2);
% set code length to tau
tau = N;
code = uinput;
samp_num = size(code,2) * 10;
% compute the next power integer of 2 for FFT purposes
n = ceil(log(samp_num) / log(2));
% compute FFT size, nfft
nfft = 2^n;
% set a dummy array in preparation for interpolation
delay = linspace(-N-2,N,nfft);
plot_figures_chap6 ( ambig, delay, freq)
%Figure 6.16
uinput = [1 -1 -1 -1 1 1 1 1 -1 1 -1 1 1 -1 -1];
[ambig] = ambiguity_code(uinput);
freq = linspace(-6,6, size(ambig,1));
```

```
N = size(uinput,2);
% set code length to tau
tau = N;
code = uinput;
samp_num = size(code,2) * 10;
% compute the next power integer of 2 for FFT purposes
n = ceil(log(samp_num) / log(2));
% compute FFT size, nfft
nfft = 2^n;
% set a dummy array in preparation for interpolation
delay = linspace(-N-2,N,nfft);
plot_figures_chap6 ( ambig, delay, freq)
```

MATLAB Program "Fig6_17.m" Listing

```
% Use to reproduce Figs 6.17 text
clc
close all
clear all
uinput = [1 1 1 1 1 i -1 -i 1 -1 1 -1 1 -i -1 i];
[ambig] = ambiguity_code(uinput);
freq = linspace(-6,6, size(ambig,1));
N = size(uinput,2);
% set code length to tau
tau = N;
code = uinput;
samp_num = size(code,2) * 10;
% compute the next power integer of 2 for FFT purposes
n = ceil(log(samp_num) / log(2));
% compute FFT size, nfft
nfft = 2^n;
% set a dummy array in preparation for interpolation
delay = linspace(-N-2,N,nfft);
plot_figures_chap6 ( ambig, delay, freq)
```

MATLAB Function "ambiguity_code.m" Listing

```
function [ambig] = ambiguity_code(uinput)
% Compute and plot the ambiguity function for any give code u
% Compute the ambiguity function by utilizing the FFT
% through combining multiple range cuts
N = size(uinput,2);
tau = N;
code = uinput;
samp_num = size(code,2) * 10;
n = ceil(log(samp_num) / log(2));
nfft = 2^n;
u(1:nfft) = 0;
j = 0;
for index = 1:10:samp_num
   index;
   j = j+1;
   u(index:index+10-1) = code(j);
end
```

```
% set-up the array v
v = u;
delay = linspace(0,5*tau,nfft);
freq_del = 12 / tau /100;
j = 0;
vfft = fft(v,nfft);
for freq = -6/tau:freq_del:6/tau;
    j = j+1;
    exf = exp(sqrt(-1) * 2. * pi * freq .* delay);
    u_times_exf = u .* exf;
    ufft = fft(u_times_exf,nfft);
    prod = ufft .* conj(vfft);
    ambig(j,:) = fftshift(abs(ifft(prod))');
end
```

Chapter 7

Pulse Compression

Range resolution for a given radar can be significantly improved by using very short pulses. Unfortunately, utilizing short pulses decreases the average transmitted power, hence reducing the SNR. Since the average transmitted power is directly linked to the receiver SNR, it is often desirable to increase the pulse width (i.e., the average transmitted power) while simultaneously maintaining adequate range resolution. This can be made possible by using pulse compression techniques and the matched filter receiver. Pulse compression allows us to achieve the average transmitted power of a relatively long pulse, while obtaining the range resolution corresponding to a short pulse. In this chapter, two pulse compression techniques are discussed. The first technique is known as correlation processing, which is predominantly used for narrowband and some medium-band radar operations. The second technique is called stretch processing and is normally used for extremely wideband radar operations.

7.1. Time-Bandwidth Product

Consider a radar system that employs a matched filter receiver. Let the matched filter receiver bandwidth be denoted as B. Then the noise power available within the matched filter bandwidth is given by

$$N_i = 2B(\eta_0/2) \qquad \text{Eq. (7.1)}$$

where the factor of two is used to account for both negative and positive frequency bands, as illustrated in Fig. 7.1. The average input signal power over a pulse duration τ_0 is

$$S_i = E_x/\tau_0. \qquad \text{Eq. (7.2)}$$

E_x is the signal energy. Consequently, the matched filter input SNR is given by

$$(SNR)_i = S_i/N_i = E/(\eta_0 B \tau_0). \qquad \text{Eq. (7.3)}$$

The output peak instantaneous SNR to the input SNR ratio, at a specific time t_0, is

$$\frac{SNR(t_0)}{(SNR)_i} = 2B\tau_0. \qquad \text{Eq. (7.4)}$$

The quantity $B\tau_0$ is referred to as the time-bandwidth product for a given waveform or its corresponding matched filter. The factor $B\tau_0$ by which the output SNR is increased over that at the input is called the matched filter gain, or simply the compression gain.

In general, the time-bandwidth product of an unmodulated pulse approaches unity. The time-bandwidth product of a pulse can be made much greater than unity by using frequency or phase modulation. If the radar receiver transfer function is perfectly matched to that of the input waveform, then the compression gain is equal to $B\tau_0$. Clearly, the compression gain becomes smaller than $B\tau_0$ as the spectrum of the matched filter deviates from that of the input signal.

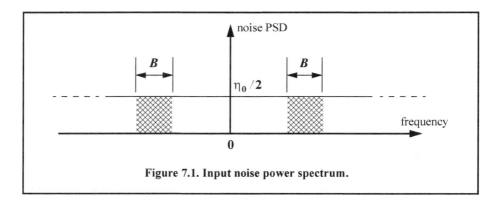

Figure 7.1. Input noise power spectrum.

7.2. Radar Equation with Pulse Compression

The radar equation for a pulsed radar can be written as

$$SNR = \frac{P_t \tau_0 G^2 \lambda^2 \sigma}{(4\pi)^3 R^4 k T_0 F L}$$

Eq. (7.5)

where P_t is peak power, τ_0 is pulse width, G is antenna gain, σ is target RCS, R is range, k is Boltzmann's constant, T_0 is 290 degrees *Kelvin*, F is noise figure, and L is total radar losses.

Pulse compression radars transmit relatively long pulses (with modulation) and process the radar echo into very short pulses (compressed). One can view the transmitted pulse as being composed of a series of very short subpulses (duty is 100%), where the width of each subpulse is equal to the desired compressed pulse width. Denote the compressed pulse width as τ_c. Thus, for an individual subpulse, Eq. (7.5) can be written as

$$(SNR)_{\tau_c} = \frac{P_t \tau_c G^2 \lambda^2 \sigma}{(4\pi)^3 R^4 k T_0 F L} .$$

Eq. (7.6)

The SNR for the uncompressed pulse is then derived from Eq. (7.6) as

$$SNR = \frac{P_t (\tau_0 = n_p \tau_c) G^2 \lambda^2 \sigma}{(4\pi)^3 R^4 k T_0 F L}$$

Eq. (7.7)

where n_p is the number of subpulses. Equation (7.7) is denoted as the radar equation with pulse compression.

Observation of Eq. (7.5) and Eq. (7.7) indicates the following (note that both equations have the same form): For a given set of radar parameters, and as long as the transmitted pulse remains unchanged, the SNR is also unchanged regardless of the signal bandwidth. More precisely, when pulse compression is used, the detection range is maintained while the range resolution is drastically improved by keeping the pulse width unchanged and by increasing the bandwidth. Remember that range resolution is proportional to the inverse of the signal bandwidth:

$$\Delta R = c/2B.$$

Eq. (7.8)

7.3. Basic Principle of Pulse Compression

For this purpose, consider a long pulse with LFM modulation and assume a matched filter receiver. The output of the matched filter (along the delay axis, i.e., range) is an order of magnitude narrower than that at its input. More precisely, the matched filter output is compressed by a factor $\xi = B\tau_0$, where τ_0 is the pulse width and B is the bandwidth. Thus, by using long pulses and wideband LFM modulation, large compression ratios can be achieved.

Figure 7.2 illustrates the ideal LFM pulse compression process. Part (a) shows the envelope of a pulse, part (b) shows the frequency modulation (in this case it is an upchirp LFM) with bandwidth $B = f_2 - f_1$. Part (c) shows the matched filter time-delay characteristic while part (d) shows the compressed pulse envelope. Finally, part (e) shows the matched filter input/output waveforms.

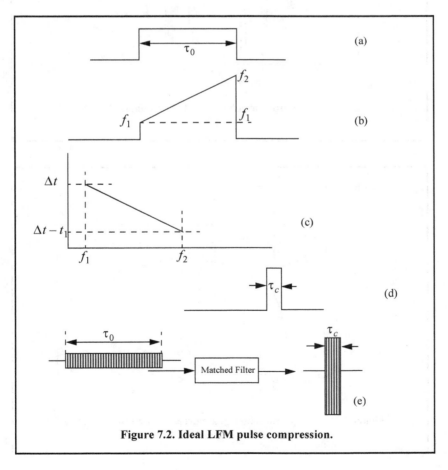

Figure 7.2. Ideal LFM pulse compression.

Figure 7.3 illustrates the advantage of pulse compression using a more realistic LFM waveform. In this example, two targets with RCS, $\sigma_1 = 1m^2$ and $\sigma_2 = 0.5m^2$, are detected. The two targets are not separated enough in time to be resolved. Figure 7.3a shows the composite echo signal from those targets. Clearly, the target returns overlap, and thus they are not resolved. However, after pulse compression, the two pulses are completely separated and are resolved as two distinct targets. In fact, when using LFM, returns from neighboring targets are resolved as long as they are separated in time by τ_c, the compressed pulse width.

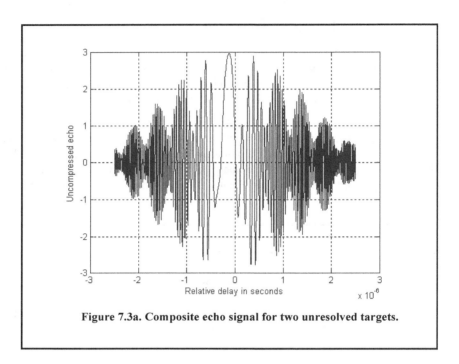

Figure 7.3a. Composite echo signal for two unresolved targets.

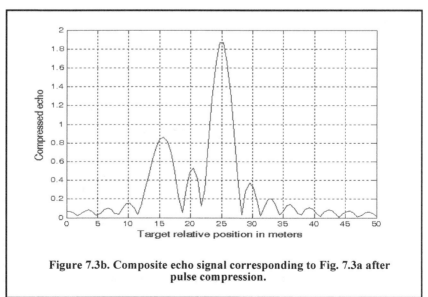

Figure 7.3b. Composite echo signal corresponding to Fig. 7.3a after pulse compression.

7.4. Correlation Processor

Radar operations (search, track, etc.) are usually carried out over a specified range window, referred to as the receive window, and defined by the difference between the radar maximum and minimum range. Returns from all targets within the receive window are collected and passed through matched filter circuitry to perform pulse compression. One implementation of such analog processors is the Surface Acoustic Wave (SAW) devices. Because of the recent advances in digital computer development, the correlation processor is often performed digitally using the FFT. This digital implementation is called Fast Convolution Processing (FCP) and can be implemented at the base band. The fast convolution process is illustrated in Fig. 7.4.

Since the matched filter is a linear time invariant system, its output can be described mathematically by the convolution between its input and its impulse response,

$$y(t) = x(t) \otimes h(t) \qquad\qquad \text{Eq. (7.9)}$$

where $x(t)$ is the input signal, $h(t)$ is the matched filter impulse response (replica), and the ($\otimes$) operator symbolically represents convolution. From the Fourier transform properties,

$$FFT\{x(t) \otimes h(t)\} = X(f) \cdot H(f), \qquad\qquad \text{Eq. (7.10)}$$

and when both signals are sampled properly, the compressed signal $y(t)$ can be computed from

$$y = FFT^{-1}\{X \cdot H\} \qquad\qquad \text{Eq. (7.11)}$$

where FFT^{-1} is the inverse FFT. When using pulse compression, it is desirable to use modulation schemes that can accomplish a maximum pulse compression ratio and can significantly reduce the sidelobe levels of the compressed waveform. For the LFM case, the first sidelobe is approximately $13.4 dB$ below the main peak, and for most radar applications this may not be sufficient. In practice, high sidelobe levels are not preferable because noise and/or jammers located at the sidelobes may interfere with target returns in the main lobe.

Weighting functions (windows) can be used on the compressed pulse spectrum in order to reduce the sidelobe levels. The cost associated with such an approach is a loss in the main lobe resolution, and a reduction in the peak value (i.e., loss in the SNR). Weighting the time domain transmitted or received signal instead of the compressed pulse spectrum will theoretically achieve the same goal. However, this approach is rarely used, since amplitude modulating the transmitted waveform introduces extra burdens on the transmitter.

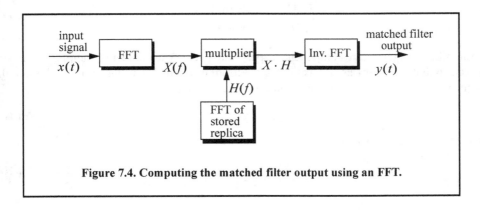

Figure 7.4. Computing the matched filter output using an FFT.

Consider a radar system that utilizes a correlation processor receiver (i.e., matched filter). The receive window in meters is defined by

$$R_{rec} = R_{max} - R_{min}$$
Eq. (7.12)

where R_{max} and R_{min}, respectively, define the maximum and minimum range over which the radar performs detection. Typically, R_{rec} is limited to the extent of the target complex. The normalized complex transmitted signal has the form

$$x(t) = \exp\left(j2\pi\left(f_0 t + \frac{\mu}{2}t^2\right)\right) \qquad 0 \leq t \leq \tau_0.$$
Eq. (7.13)

τ_0 is the pulse width, $\mu = B/\tau_0$, and B is the bandwidth.

The radar echo signal is similar to the transmitted one with the exception of a time delay and an amplitude change that correspond to the target RCS. Consider a target at range R_1. The echo received by the radar from this target is

$$x_r(t) = a_1 \exp\left(j2\pi\left(f_0(t - t_1) + \frac{\mu}{2}(t - t_1)^2\right)\right)$$
Eq. (7.14)

where a_1 is proportional to target RCS, antenna gain, and range attenuation. The time delay t_1 is given by

$$t_1 = 2R_1/c.$$
Eq. (7.15)

The first step of the processing consists of removing the frequency f_0. This is accomplished by mixing $x_r(t)$ with a reference signal whose phase is $2\pi f_0 t$. The phase of the resultant signal, after lowpass filtering, is then given by

$$\phi(t) = 2\pi\left(-f_0 t_1 + \frac{\mu}{2}(t - t_1)^2\right)$$
Eq. (7.16)

and the instantaneous frequency is

$$f_i(t) = \frac{1}{2\pi}\frac{d}{dt}\phi(t) = \mu(t - t_1) = \frac{B}{\tau_0}\left(t - \frac{2R_1}{c}\right).$$
Eq. (7.17)

The quadrature components are

$$\begin{pmatrix} x_I(t) \\ x_Q(t) \end{pmatrix} = \begin{pmatrix} \cos\phi(t) \\ \sin\phi(t) \end{pmatrix}.$$
Eq. (7.18)

Sampling the quadrature components is performed next. The number of samples, N, must be chosen so that foldover (ambiguity) in the spectrum is avoided. For this purpose, the sampling frequency, f_s (based on the Nyquist sampling rate), must be

$$f_s \geq 2B$$
Eq. (7.19)

and the sampling interval is

$$\Delta t \leq 1/2B.$$
Eq. (7.20)

Using Eq. (7.17) it can be shown that (the proof is left as an exercise) the frequency resolution of the FFT is

$$\Delta f = 1/\tau_0. \qquad \text{Eq. (7.21)}$$

The minimum required number of samples is

$$N = \frac{1}{\Delta f \Delta t} = \frac{\tau_0}{\Delta t}. \qquad \text{Eq. (7.22)}$$

Equating Eqs. (7.20) and (7.22) yields

$$N \geq 2B\tau_0. \qquad \text{Eq. (7.23)}$$

Consequently, a total of $2B\tau_0$ real samples, or $B\tau_0$ complex samples, is sufficient to completely describe an LFM waveform of duration τ_0 and bandwidth B. For example, an LFM signal of duration $\tau_0 = 20$ μs and bandwidth $B = 5$ MHz requires 200 real samples to determine the input signal (100 samples for the I-channel and 100 samples for the Q-channel).

For better implementation of the FFT, N is extended to the next power of two, by zero padding. Thus, the total number of samples, for some positive integer n, is

$$N_{FFT} = 2^n \geq N. \qquad \text{Eq. (7.24)}$$

The final steps of the FCP processing include (1) taking the FFT of the sampled sequence, (2) multiplying the frequency domain sequence of the signal with the FFT of the matched filter impulse response, and (3) performing the inverse FFT of the composite frequency domain sequence in order to generate the time domain compressed pulse. Of course, weighting, antenna gain, and range attenuation compensation must also be performed.

Assume that M targets at ranges R_1, R_2, and so forth are within the receive window. From superposition, the phase of the down-converted signal is

$$\phi(t) = \sum_{m=1}^{M} 2\pi\left(-f_0 t_m + \frac{\mu}{2}(t - t_m)^2\right). \qquad \text{Eq. (7.25)}$$

The times $\{t_m = (2R_m/c); \; m = 1, 2, ..., M\}$ represent the two-way time delays, where t_1 coincides with the start of the receive window.

MATLAB Function "matched_filter.m"

The function *"matched_filter.m"* performs fast convolution processing. The user can access this function either by a MATLAB function call or by executing the MATLAB program *"matched_filter_gui.m,"* which utilizes a MATLAB-based GUI. The work space associated with this program is shown in Fig. 7.5. The outputs for this function include plots of the compressed and uncompressed signals as well as the replica used in the pulse compression process. This function utilizes the function *"power_integer_2.m."* Its syntax is as follows:

[y] = matched_filter(nscat, rrec, taup, b, scat_range, scat_rcs, win)

where

Symbol	Description	Units	Status
nscat	*number of point scatterers within the received window*	*none*	*input*
rrec	*receive window size*	*m*	*input*
taup	*uncompressed pulse width*	*seconds*	*input*
b	*chirp bandwidth*	*Hz*	*input*
scat_range	*scatterers' relative range (within the receive window)*	*m*	*input*
scat_rcs	*vector of scatterers' RCS*	*m²*	*input*
win	*0 = no window; 1 = Hamming; 2 = Kaiser with parameter pi; and 3 = Chebychev sidelobes at -60dB*	*none*	*input*
y	*normalized compressed output*	*volts*	*output*

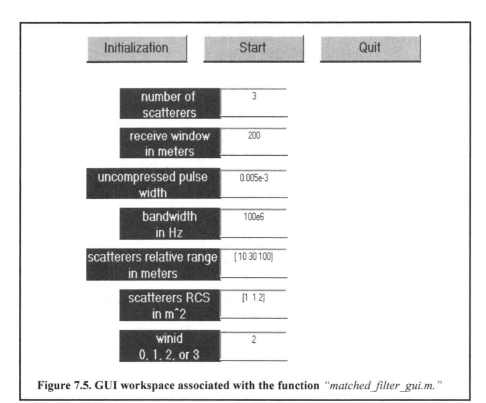

Figure 7.5. GUI workspace associated with the function *"matched_filter_gui.m."*

As an example, consider the case where

# Targets	R_{rec}	Pulse Width	Bandwidth	Targets Range	Target RCS	Window Type
3	*200m*	*0.005ms*	*100e6 Hz*	*[30 70 120] m*	*[1 1 1]m²*	*Hamming*

Note that the compressed pulsed range resolution is $\Delta R = 1.5m$. Figure 7.6a and Fig. 7.6b shows the real part and the amplitude spectrum of the replica used for this example. Figure 7.7a shows the uncompressed echo, while Fig. 7.7b shows the compressed MF output. Note that the scatterer amplitude attenuation is also a function of the inverse of the scatterer's range within the receive window. Figure 7.7c is similar to Fig. 7.7b except in this case the first and second scatterers are less than 1.5 meters apart (they are at 70 and 71 meters).

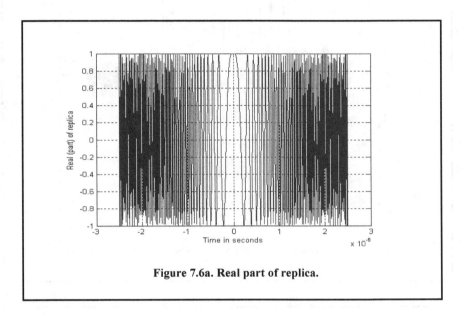

Figure 7.6a. Real part of replica.

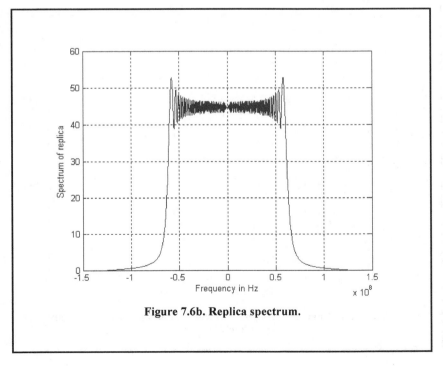

Figure 7.6b. Replica spectrum.

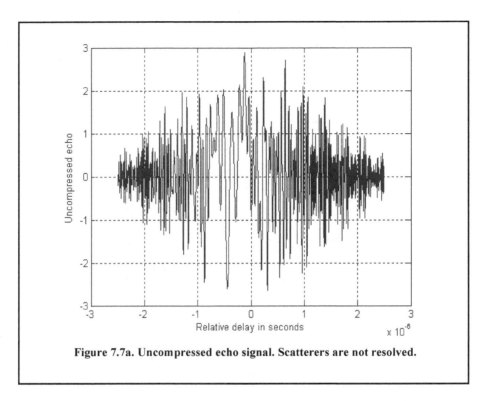

Figure 7.7a. Uncompressed echo signal. Scatterers are not resolved.

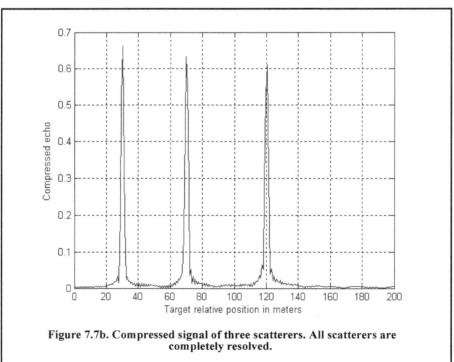

Figure 7.7b. Compressed signal of three scatterers. All scatterers are completely resolved.

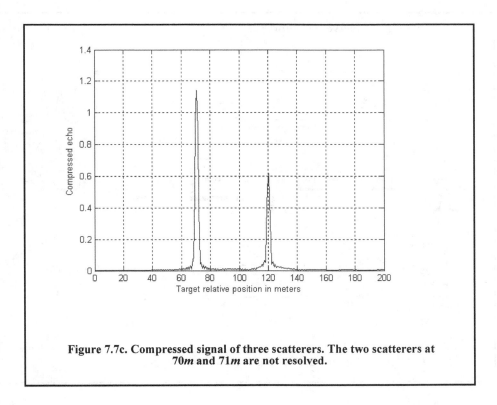

Figure 7.7c. Compressed signal of three scatterers. The two scatterers at 70m and 71m are not resolved.

7.5. Stretch Processor

Stretch processing, also known as *active correlation,* is normally used to process extremely high-bandwidth LFM waveforms. This processing technique consists of the following steps: First, the radar returns are mixed with a replica (reference signal) of the transmitted waveform. This is followed by Low Pass Filtering (LPF) and coherent detection. Next, Analog-to-Digital (A/D) conversion is performed; and finally, a bank of Narrow-Band Filters (NBFs) is used in order to extract the tones that are proportional to target range, since stretch processing effectively converts time delay into frequency. All returns from the same range bin produce the same constant frequency.

7.5.1. Single LFM Pulse

Figure 7.8 shows a block diagram for a stretch processing receiver. The reference signal is an LFM waveform that has the same LFM slope as the transmitted LFM signal. It exists over the duration of the radar "receive-window," which is computed from the difference between the radar maximum and minimum range. Denote the start frequency of the reference chirp as f_r. Consider the case when the radar receives returns from a few close (in time or range) targets, as illustrated in Fig. 7.8. Mixing with the reference signal and performing lowpass filtering are effectively equivalent to subtracting the return frequency chirp from the reference signal. Thus, the LPF output consists of constant tones corresponding to the targets' positions. The normalized transmitted signal can be expressed by

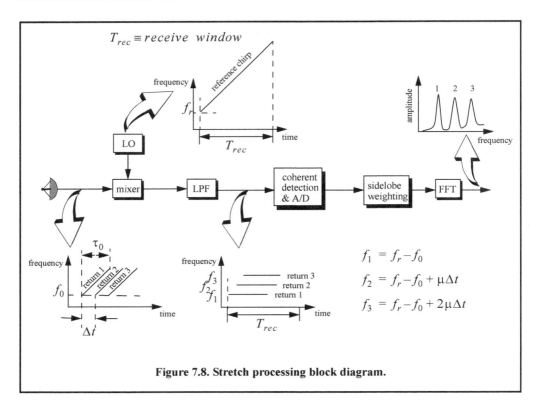

Figure 7.8. Stretch processing block diagram.

$$x(t) = \cos\left(2\pi\left(f_0 t + \frac{\mu}{2}t^2\right)\right) \qquad 0 \le t \le \tau_0 \qquad\qquad \textbf{Eq. (7.26)}$$

where $\mu = B/\tau_0$ is the LFM coefficient and f_0 is the chirp start frequency. Assume a point scatterer at range R_1. The signal received by the radar is

$$x_r(t) = a\cos\left[2\pi\left(f_0(t - t_1) + \frac{\mu}{2}(t - t_1)^2\right)\right] \qquad\qquad \textbf{Eq. (7.27)}$$

where a is proportional to target RCS, antenna gain, and range attenuation. The time delay t_1 is

$$t_1 = 2R_1/c. \qquad\qquad \textbf{Eq. (7.28)}$$

The reference signal is

$$x_{ref}(t) = 2\cos\left(2\pi\left(f_r t + \frac{\mu}{2}t^2\right)\right) \qquad 0 \le t \le T_{rec}. \qquad\qquad \textbf{Eq. (7.29)}$$

The receive window in seconds is

$$T_{rec} = \frac{2(R_{max} - R_{min})}{c} = \frac{2R_{rec}}{c}. \qquad\qquad \textbf{Eq. (7.30)}$$

It is customary to let $f_r = f_0$. The output of the mixer is the product of the received and reference signals. After lowpass filtering, the signal is

$$x_0(t) = a\cos(2\pi f_0 t_1 + 2\pi\mu t_1 t - \pi\mu(t_1)^2).$$ Eq. (7.31)

Substituting Eq. (7.28) into Eq. (7.31) and collecting terms yields

$$x_0(t) = a\ \cos\left[\left(\frac{4\pi BR_1}{c\tau_0}\right)t + \frac{2R_1}{c}\left(2\pi f_0 - \frac{2\pi BR_1}{c\tau_0}\right)\right],$$ Eq. (7.32)

and since $\tau_0 \gg 2R_1/c$, Eq. (7.32) is approximated by

$$x_0(t) \approx a\ \cos\left[\left(\frac{4\pi BR_1}{c\tau_0}\right)t + \frac{4\pi R_1}{c}f_0\right].$$ Eq. (7.33)

The instantaneous frequency is

$$f_{inst} = \frac{1}{2\pi}\frac{\mathrm{d}}{\mathrm{d}t}\left(\frac{4\pi BR_1}{c\tau_0}t + \frac{4\pi R_1}{c}f_0\right) = \frac{2BR_1}{c\tau_0},$$ Eq. (7.34)

which clearly indicates that target range is proportional to the instantaneous frequency. Therefore, proper sampling of the LPF output and taking the FFT of the sampled sequence lead to the following conclusion: a peak at some frequency f_1 indicates the presence of a target at range

$$R_1 = f_1 c\tau_0/2B.$$ Eq. (7.35)

Assume M close targets at ranges R_1, R_2, and so forth ($R_1 < R_2 < ... < R_M$). From superposition, the total signal is

$$x_r(t) = \sum_{m=1}^{M} a_m(t)\cos\left[2\pi\left(f_0(t-t_m) + \frac{\mu}{2}(t-t_m)^2\right)\right]$$ Eq. (7.36)

where $\{a_m(t); m = 1, 2, ..., M\}$ are proportional to the targets' cross sections, antenna gain, and range. The times $\{t_m = (2R_m/c); m = 1, 2, ..., M\}$ represent the two-way time delays, where t_1 coincides with the start of the receive window. Using Eq. (7.32), the overall signal at the output of the LPF can then be described by

$$x_o(t) = \sum_{m=1}^{M} a_m\cos\left[\left(\frac{4\pi BR_m}{c\tau_0}\right)t + \frac{2R_m}{c}\left(2\pi f_0 - \frac{2\pi BR_m}{c\tau_0}\right)\right].$$ Eq. (7.37)

Hence, target returns appear as constant frequency tones that can be resolved using the FFT. Consequently, determining the proper sampling rate and FFT size is very critical. The rest of this section presents a methodology for computing the proper FFT parameters required for stretch processing.

Assume a radar system using a stretch processor receiver. The pulse width is τ_0 and the chirp bandwidth is B. Since stretch processing is normally used in extreme bandwidth cases (i.e., very large B), the receive window over which radar returns will be processed is typically limited to from a few meters to possibly less than 100 meters. The compressed pulse range resolution is computed from Eq. (7.8). Declare the FFT size to be N and its frequency resolution to be Δf. The frequency resolution can be computed using the following procedure: Consider two adjacent point scatterers at ranges R_1 and R_2. The minimum frequency separation, Δf, between those scatterers so that they are resolved can be computed from Eq. (7.34). More precisely,

$$\Delta f = f_2 - f_1 = \frac{2B}{c\tau_0}(R_2 - R_1) = \frac{2B}{c\tau_0}\Delta R. \qquad \text{Eq. (7.38)}$$

Substituting Eq. (7.8) into Eq. (7.38) yields

$$\Delta f = \frac{2B}{c\tau_0}\frac{c}{2B} = \frac{1}{\tau_0}. \qquad \text{Eq. (7.39)}$$

The maximum frequency resolvable by the FFT is limited to the region $\pm N\Delta f/2$. Thus, the maximum resolvable frequency is

$$\frac{N\Delta f}{2} > \frac{2B(R_{max} - R_{min})}{c\tau_0} = \frac{2BR_{rec}}{c\tau_0}. \qquad \text{Eq. (7.40)}$$

Using Eqs. (7.30) and (7.39) into Eq. (7.40) and collecting terms yields

$$N > 2BT_{rec}. \qquad \text{Eq. (7.41)}$$

For better implementation of the FFT, choose an FFT of size

$$N_{FFT} \geq N = 2^n \qquad \text{Eq. (7.42)}$$

where n is a nonzero positive integer. The sampling interval is then given by

$$\Delta f = \frac{1}{T_s N_{FFT}} \Rightarrow T_s = \frac{1}{\Delta f N_{FFT}}. \qquad \text{Eq. (7.43)}$$

MATLAB Function "stretch.m"

The function *"stretch.m"* presents a digital implementation of the stretch processing described in this section. The user can access this function either by a MATLAB function call or by executing the MATLAB program *"stretch_gui.m,"* which utilizes MATLAB-based GUI and is shown in Fig. 7.9.

The outputs of this function are the complex array y containing pulsed compressed signal samples. The syntax for this function is as follows:

[y] = stretch (nscat, taup, f0, b, scat_range, rrec, scat_rcs, win)

where

Symbol	Description	Units	Status
nscat	*number of point scatterers within the receive window*	*none*	*input*
taup	*uncompressed pulse width*	*seconds*	*input*
f0	*chirp start frequency*	*Hz*	*input*
b	*chirp bandwidth*	*Hz*	*input*
scat_range	*vector of scatterers' range*	*m*	*input*

Symbol	Description	Units	Status
rrec	*range receive window*	*m*	*input*
scat_rcs	*vector of scatterers' RCS*	*m²*	*input*
win	*0 = no window; 1 = Hamming; 2 = Kaiser with parameter pi; 3 = Chebychev sidelobes at -60dB*	*none*	*input*
y	*compressed output*	*volts*	*output*

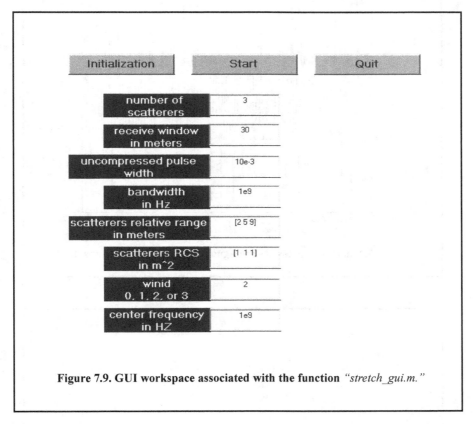

Figure 7.9. GUI workspace associated with the function *"stretch_gui.m."*

As an example, consider the case where

# Targets	3
Pulse Width	*10ms*
Center Frequency	*5.6GHz*
Bandwidth	*1GHz*
Receive Window	*30m*
Relative Target's Range	*[2 5 10]m*
Target's RCS	*[1, 1, 2]m²*
Window	*2 (Kaiser)*

Note that the compressed pulse range resolution, without using a window, is $\Delta R = 0.15m$. Figure 7.10a and Fig. 7.10b, respectively, show the uncompressed and compressed echo signals corresponding to this example. Figure 7.11 is similar to Fig. 7.10 except in this case two of the scatterers are less than 15 cm apart (i.e., unresolved targets at $R_{relative} = [3, 3.1]m$).

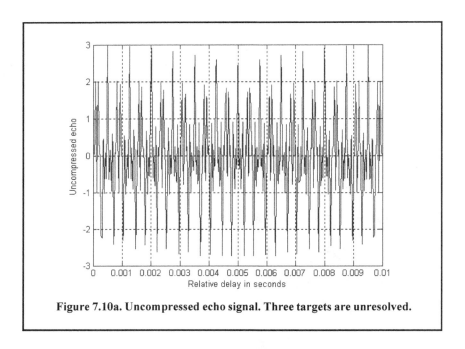

Figure 7.10a. Uncompressed echo signal. Three targets are unresolved.

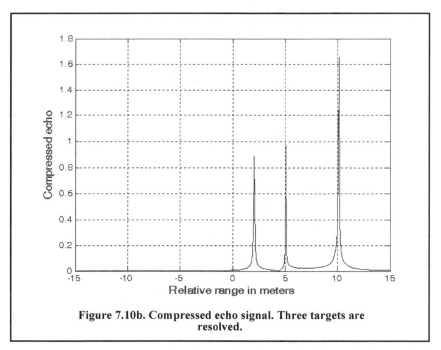

Figure 7.10b. Compressed echo signal. Three targets are resolved.

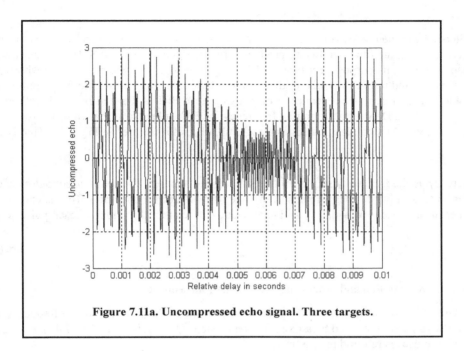

Figure 7.11a. Uncompressed echo signal. Three targets.

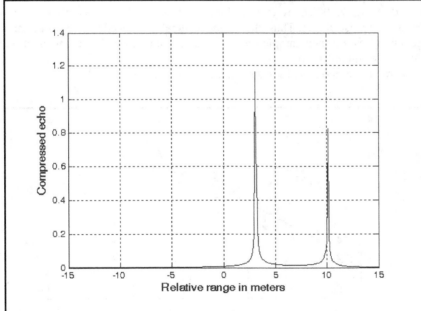

Figure 7.11b. Compressed echo signal of three targets; the two targets at *3m* and *3.1m* are not resolved.

7.5.2. Stepped Frequency Waveforms

Stepped Frequency Waveforms (SFW) are used in extremely wideband radar applications where a very large time-bandwidth product is required. Generation of SFW was discussed in Chapter 5. For this purpose, consider an LFM signal whose bandwidth is B_i and whose pulse width is T_i, and refer to it as the primary LFM. Divide this long pulse into N subpulses, each of width τ_0, to generate a sequence of pulses whose PRI is denoted by T. It follows that $T_i = (n-1)T$. Define the beginning frequency for each subpulse as that value measured from the primary LFM at the leading edge of each subpulse, as illustrated in Fig. 7.12. That is

$$f_i = f_0 + i\Delta f; \quad i = 0, N-1 \qquad \textbf{Eq. (7.44)}$$

where Δf is the frequency step from one subpulse to another. The set of n subpulses is often referred to as a burst. Each subpulse can have its own LFM modulation. To this end, assume that each subpulse is of width τ_0 and bandwidth B, then the LFM slope of each pulse is

$$\mu = \frac{B}{\tau_0}. \qquad \textbf{Eq. (7.45)}$$

The SFW operation and processing involve the following steps:

1. A series of N narrowband LFM pulses is transmitted. The chirp beginning frequency from pulse to pulse is stepped by a fixed frequency step Δf, as defined in Eq. (7.44). Each group of N pulses is referred to as a burst.
2. The LFM slope (quadratic phase term) is first removed from the received signal, as described in Fig. 7.10. The reference slope must be equal to the combined primary LFM and single subpulse slopes. Thus, the received signal is reduced to a series of subpulses.
3. These subpulses are then sampled at a rate that coincides with the center of each pulse, sampling rate equivalent to $(1/T)$.
4. The quadrature components for each burst are collected and stored.

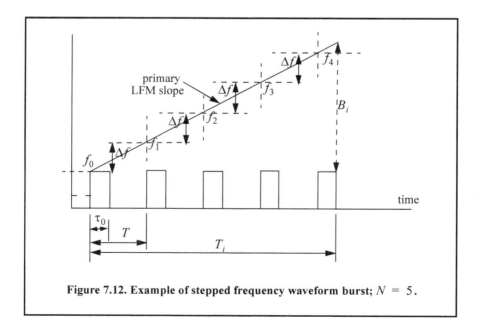

Figure 7.12. Example of stepped frequency waveform burst; $N = 5$.

5. Spectral weighting (to reduce the range sidelobe levels) is applied to the quadrature components. Corrections for target velocity, phase, and amplitude variations are applied.
6. The IDFT of the weighted quadrature components of each burst is calculated to synthesize a range profile for that burst. The process is repeated for M bursts to obtain consecutive high resolution range profiles.

Within a burst, the transmitted waveform for the ith step can be described as

$$x_i(t) = \left(C_i \frac{1}{\sqrt{\tau_0}} Rect\left(\frac{t}{\tau_0}\right) e^{j2\pi\left(f_i t + \frac{\mu}{2}t^2\right)} \quad ; \quad \begin{array}{c} iT \le t \le iT + \tau_0 \\ elsewhere \end{array} \right)$$

$$\qquad \qquad \qquad \qquad 0$$

$$\text{Eq. (7.46)}$$

where C_i are constants. The received signal from a target located at range R_0 is then given by

$$x_{ri}(t) = C_i' e^{j2\pi\left[f_i(t - \Delta(t)) - \frac{\mu}{2}(t - \Delta(t))^2\right]} \quad , \quad iT + \Delta(t) \le t \le iT + \tau_0 + \Delta(t) \qquad \text{Eq. (7.47)}$$

where C_i' are constant and the round-trip delay $\Delta(t)$ is given by

$$\Delta(t) = \frac{R_0 - vt}{c/2} \qquad \text{Eq. (7.48)}$$

where c is the speed of light and v is the target radial velocity.

In order to remove the quadratic phase term, mixing is first performed with the reference signal given by

$$y_i(t) = e^{j2\pi\left(f_i t + \frac{\mu}{2}t^2\right)} \quad ; \quad iT \le t \le iT + \tau_0 . \qquad \text{Eq. (7.49)}$$

Next lowpass filtering is performed to extract the quadrature components. More precisely, the quadrature components are given by

$$\begin{pmatrix} x_I(t) \\ x_Q(t) \end{pmatrix} = \begin{pmatrix} A_i \cos\phi_i(t) \\ A_i \sin\phi_i(t) \end{pmatrix} \qquad \text{Eq. (7.50)}$$

where A_i are constants, and

$$\phi_i(t) = -2\pi f_i\left(\frac{2R_0}{c} - \frac{2vt}{c}\right) \qquad \text{Eq. (7.51)}$$

where now $f_i = \Delta f$. For each pulse, the quadrature components are then sampled at

$$t_i = iT + \frac{\tau_r}{2} + \frac{2R_0}{c} . \qquad \text{Eq. (7.52)}$$

τ_r is the time delay associated with the range that corresponds to the start of the range profile.

The quadrature components can then be expressed in complex form as

$$X_i = A_i e^{j\phi_i} . \qquad \text{Eq. (7.53)}$$

Equation (7.53) represents samples of the target reflectivity, due to a single burst, in the frequency domain. This information can then be transformed into a series of range delay reflectivity (i.e., range profile) values by using the IDFT. It follows that

$$H_l = \frac{1}{N} \sum_{i=0}^{N-1} X_i \; \exp\left(j\frac{2\pi l i}{N}\right) \qquad ; \; 0 \le l \le N-1 \,. \qquad \text{Eq. (7.54)}$$

Substituting Eq. (7.51) and Eq. (7.53) into (7.54) and collecting terms yields

$$H_l = \frac{1}{N} \sum_{i=0}^{N-1} A_i \; \exp\left\{ j\left(\frac{2\pi l i}{N} - 2\pi f_i \left(\frac{2R_0}{c} - \frac{2vt_i}{c}\right)\right) \right\} . \qquad \text{Eq. (7.55)}$$

By normalizing with respect to N and by assuming that $A_i = 1$ and that the target is stationary (i.e., $v = 0$), then Eq. (7.55) can be written as

$$H_l = \sum_{i=0}^{N-1} \exp\left\{ j\left(\frac{2\pi l i}{N} - 2\pi f_i \frac{2R_0}{c}\right) \right\} . \qquad \text{Eq. (7.56)}$$

Using $f_i = i\Delta f$ inside Eq. (7.56) yields

$$H_l = \sum_{i=0}^{N-1} \exp\left\{ j\frac{2\pi i}{N}\left(-\frac{2NR_0\Delta f}{c} + l\right) \right\} , \qquad \text{Eq. (7.57)}$$

which can be simplified to

$$H_l = \frac{\sin\pi\zeta}{\sin\frac{\pi\zeta}{N}} \; \exp\left(j\frac{N-1}{2} \frac{2\pi\zeta}{N}\right) \qquad \text{Eq. (7.58)}$$

where

$$\zeta = \frac{-2NR_0\Delta f}{c} + l \,. \qquad \text{Eq. (7.59)}$$

Finally, the synthesized range profile is

$$|H_l| = \left|\frac{\sin\pi\zeta}{\sin\frac{\pi\zeta}{N}}\right| . \qquad \text{Eq. (7.60)}$$

Range Resolution and Range Ambiguity in SFW

As usual, range resolution is determined from the overall system bandwidth. Assuming an SFW with N steps and step size Δf, the corresponding range resolution is equal to

$$\Delta R = \frac{c}{2N\Delta f} . \qquad \text{Eq. (7.61)}$$

Range ambiguity associated with an SFW can be determined by examining the phase term that corresponds to a point scatterer located at range R_0. More precisely,

$$\phi_i(t) = 2\pi f_i \frac{2R_0}{c}.$$

Eq. (7.62)

It follows that

$$\frac{\Delta\phi}{\Delta f} = \frac{4\pi(f_{i+1} - f_i)R_0}{(f_{i+1} - f_i)} \frac{1}{c} = \frac{4\pi R_0}{c},$$

Eq. (7.63)

or equivalently,

$$R_0 = \frac{\Delta\phi}{\Delta f} \frac{c}{4\pi}.$$

Eq. (7.64)

It is clear from Eq. (7.64) that range ambiguity exists for $\Delta\phi = \Delta\phi + 2N\pi$. Therefore,

$$R_0 = \frac{\Delta\phi + 2N\pi}{\Delta f} \frac{c}{4\pi} = R_0 + N\left(\frac{c}{2\Delta f}\right)$$

Eq. (7.65)

and the unambiguous range window is

$$R_u = \frac{c}{2\Delta f}.$$

Eq. (7.66)

A range profile synthesized using a particular SFW represents the relative range reflectivity for all scatterers within the unambiguous range window, with respect to the absolute range that corresponds to the burst time delay. Additionally, if a specific target extent is larger than R_u, then all scatterers falling outside the unambiguous range window will fold over and appear in the synthesized profile. This foldover problem is identical to the spectral foldover that occurs when using a Fast Fourier Transform (FFT) to resolve certain signal frequency contents. For example, consider an FFT with frequency resolution $\Delta f = 50 Hz$ and size $NFFT = 64$. In this case, this FFT can resolve frequency tones between $-1600 Hz$ and $1600 Hz$. When this FFT is used to resolve the frequency content of a sine-wave tone equal to $1800 Hz$, foldover occurs and a spectral line at the fourth FFT bin (i.e., $200 Hz$) appears. Therefore, in order to avoid foldover in the synthesized range profile, the frequency step Δf must be

$$\Delta f \leq c/2E$$

Eq. (7.67)

where E is the target extent in meters.

Additionally, the pulse width must be large enough to contain the whole target extent. Thus,

$$\Delta f \leq 1/\tau_0$$

Eq. (7.68)

and in practice,

$$\Delta f \leq 1/2\tau_0.$$

Eq. (7.69)

This is necessary in order to reduce the amount of contamination of the synthesized range profile caused by the clutter surrounding the target under consideration.

MATLAB Function "SFW.m"

The function *"SFW.m"* computes and plots the range profile for a specific SFW. This function utilizes an Inverse Fast Fourier Transform (IFFT) of a size equal to twice the number of steps. A Hamming window of the same size is also assumed. The syntax is as follows:

$$[hl] = SFW\ (nscat,\ scat\_range,\ scat\_rcs,\ n,\ deltaf,\ prf,\ v,\ r0,\ winid)$$

where

Symbol	Description	Units	Status
nscat	*number of scatterers that make up the target*	*none*	*input*
scat_range	*vector containing range to individual scatterers*	*m*	*input*
scat_rcs	*vector containing RCS of individual scatterers*	m^2	*input*
n	*number of steps*	*none*	*input*
deltaf	*frequency step*	*Hz*	*input*
prf	*PRF of SFW*	*Hz*	*input*
v	*target velocity*	*m/sec*	*input*
r0	*profile starting range*	*meters*	*input*
winid	*number>0 for Hamming window* *number < 0 for no window*	*none*	*input*
hl	*range profile*	*dB*	*output*

For example, assume that the range profile starts at $R_0 = 900m$ and that

# Targets	Pulse Width	N	Δf	1/T	v
3	$100\mu\sec$	64	$10MHz$	$100KHz$	0.0

In this case,

$$\Delta R = \frac{3 \times 10^8}{2 \times 64 \times 10 \times 10^6} = 0.235m\ , \text{ and } R_u = \frac{3 \times 10^8}{2 \times 10 \times 10^6} = 15m\ .$$

Thus, scatterers that are more than 0.235 meters apart will appear as distinct peaks in the synthesized range profile. Assume two cases; in the first case, *[scat_range] = [908, 910, 912]* meters, and in the second case, *[scat_range] = [908, 910, 910.2]* meters. In both cases, let *[scat_rcs] = [100, 10, 1]* meters squared. Figure 7.13 shows the synthesized range profiles generated using the function *"SWF.m"* and the first case when the Hamming window is not used. Figure 7.14 is similar to Fig. 7.13, except in this case the Hamming window is used. Figure 7.15 shows the synthesized range profile that corresponds to the second case (Hamming window is used). Note that all three scatterers were resolved in Fig. 7.13 and Fig. 7.14; however, the last two scatterers are not resolved in Fig. 7.15, because they are separated by less than ΔR.

Next, consider another case where *[scat_range] = [908, 912, 916]* meters. Figure 7.16 shows the corresponding range profile. In this case, foldover occurs, and the last scatterer appears at the lower portion of the synthesized range profile. Also, consider the case where *[scat_range] = [908, 910, 923]* meters. Figure 7.17 shows the corresponding range profile. In this case, ambiguity is associated with the first and third scatterers since they are separated by $15m$. Both appear at the same range bin.

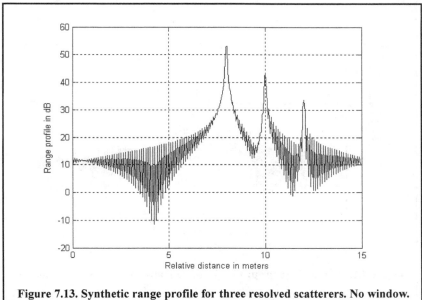

Figure 7.13. Synthetic range profile for three resolved scatterers. No window.

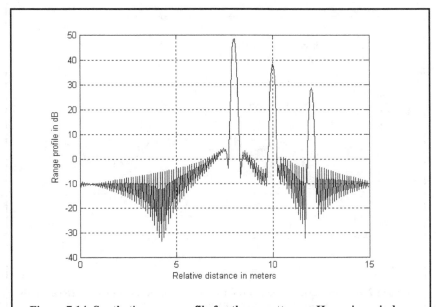

Figure 7.14. Synthetic range profile for three scatterers. Hamming window.

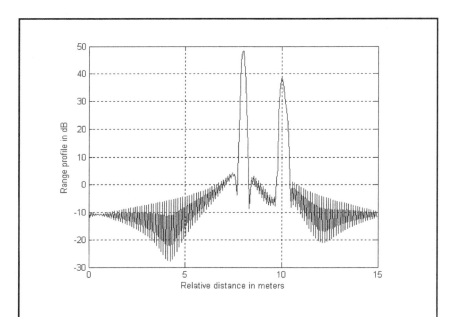

Figure 7.15. Synthetic range profile for three scatterers. Two are unresolved.

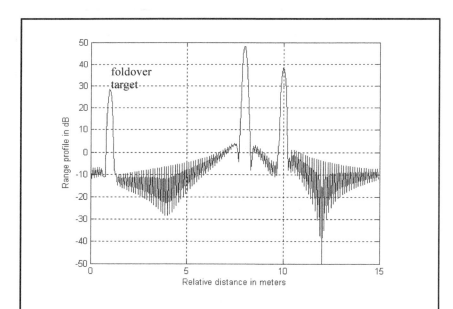

Figure 7.16. Synthetic range profile for three scatterers. Third scatterer folds over.

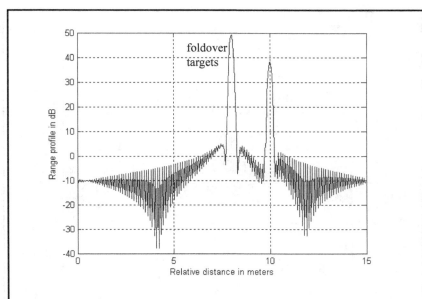

Figure 7.17. Synthetic range profile for three scatterers. The first and third scatterers appear in the same FFT bin.

7.5.3. Effect of Target Velocity

The range profile defined in Eq. (7.60) is obtained by assuming that the target under examination is stationary. The effect of target velocity on the synthesized range profile can be determined by starting with Eq. (7.55) and assuming that $v \neq 0$. Performing similar analysis as that of the stationary target case yields a range profile given by

$$H_l = \sum_{i=0}^{N-1} A_i \exp\left\{ j\frac{2\pi li}{N} - j2\pi f_i \left[\frac{2R}{c} - \frac{2v}{c}\left(iT + \frac{\tau_r}{2} + \frac{2R}{c} \right) \right] \right\}.$$ **Eq. (7.70)**

The additional phase term present in Eq. (7.70) distorts the synthesized range profile. In order to illustrate this distortion, consider the SFW described in the previous section, and assume the three scatterers of the first case. Also, assume that $v = 200m/s$. Figure 7.18 shows the synthesized range profile for this case. Comparisons of Figs. 7.13 and 7.18 clearly show the distortion effects caused by the uncompensated target velocity. Figure 7.19 is similar to Fig. 7.18 except in this case, $v = -200m/s$. Note in either case, the targets have moved from their expected positions (to the left or right) by $Disp = 2 \times n \times v / PRF$ *(1.28 m)*.

This distortion can be eliminated by multiplying the complex received data at each pulse by the phase term

$$\Phi = \exp\left(-j2\pi f_i \left[\frac{2\hat{v}}{c}\left(iT + \frac{\tau_r}{2} + \frac{2\hat{R}}{c} \right) \right] \right).$$ **Eq. (7.71)**

$\hat{v}$ and $\hat{R}$ are, respectively, estimates of the target velocity and range.

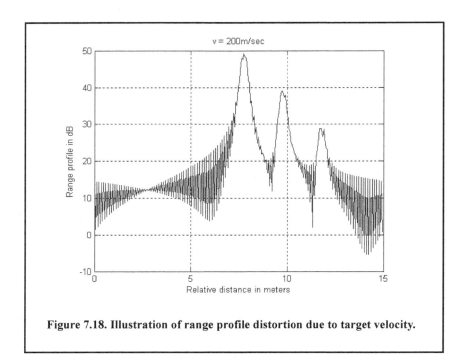

Figure 7.18. Illustration of range profile distortion due to target velocity.

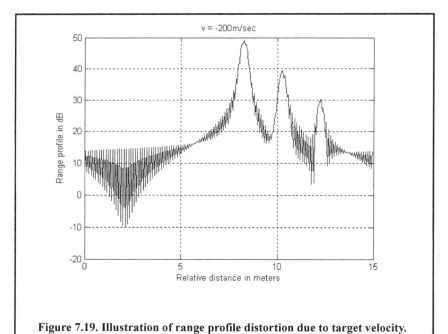

Figure 7.19. Illustration of range profile distortion due to target velocity.

This process of modifying the phase of the quadrature components is often referred to as "phase rotation." In practice, when good estimates of $\hat{v}$ and $\hat{R}$ are not available, then the effects of target velocity are reduced by using frequency hopping between the consecutive pulses within the SFW. In this case, the frequency of each individual pulse is chosen according to a predetermined code. Waveforms of this type are often called Frequency Coded Waveforms (FCW). Costas waveforms or signals are a good example of this type of waveform.

Figure 7.20 shows a synthesized range profile for a moving target whose RCS is $\sigma = 10m^2$ and $v = 10m/s$. The initial target range is at $R = 912m$. All other parameters are as before. This figure can be reproduced using the following MATLAB code.

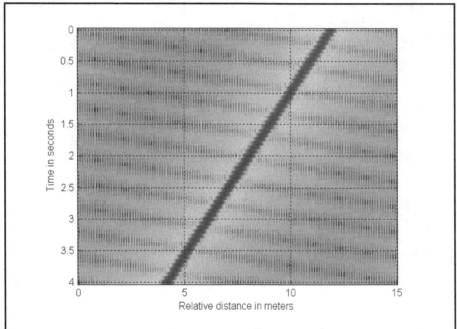

Figure 7.20. Synthesized range profile for a moving target (4 seconds long).

Problems

7.1. Starting with Eq. (7.17) derive Eq. (7.21).

7.2. Using MATLAB, generate a baseband (complex-valued) LFM waveform having a time duration of $10\mu s$ and bandwidth of $200MHz$ using a sampling step of $1ns$. Plot the real part, imaginary part, and the modulus of the FFT of this waveform.

7.3. Compress the waveform developed in Problem 7.3 using the *"xcorr"* function. Generate the magnitude-squared signal using the MATLAB command *"y.*conj(y)."* Plot the resulting compressed pulse and verify that the half power points correspond to the inverse bandwidth (i.e., $5ns$, or 5 samples).

7.4. The Synthetic Aperture Radar (SAR) ambiguity function can be approximated by

$$\chi = \frac{\sin kx}{x} \frac{\sin Nry}{\sin ry}$$

where x is the variable for the range-compressed axis, y is the azimuth-compressed axis, and k and r are related to the SAR range and azimuth resolutions. (a) Generate the x-axis from $-40m$ to $40m$ using a sampling interval of $0.1m$. Assume $k = 1$. Plot the magnitude of this range profile. (b) Generate the y-axis from $-40m$ to $40m$ using a sampling interval of $0.1m$. Assume $r = 0.00015$ and $N = 1000$. Plot the magnitude of this azimuth profile. (c) Use the findings in (a) and (b) to generate the two-dimensional ambiguity surface plot.

7.5. Derive Eq. (7.60).

7.6. Reproduce Fig. 7.19 for $v = 10, 50, 100, -150, 250 \ m/s$. Compare your outputs. What are your conclusions?

7.7. Using MATLAB, generate the waterfall plot corresponding to Fig.7.20.

Appendix 7-A: Chapter 7 MATLAB Code Listings

The MATLAB code provided in this chapter was designed as an academic standalone tool and is not adequate for other purposes. The code was written in a way to assist the reader in gaining a better understanding of the theory. The code was not developed, nor is it intended to be used as part of an open-loop or a closed-loop simulation of any kind. The MATLAB code found in this textbook can be downloaded from this book's web page on the CRC Press website. Simply use your favorite web browser, go to *www.crcpress.com,* and search for keyword *"Mahafza"* to locate this book's web page.

MATLAB Program "Fig7_3.m" Listing

```
% use this program to reproduce Fig. 7.3 of text
clc
clear all
close all
nscat = 2; %two point scatterers
taup = 10e-6; % 100 microsecond uncompressed pulse
b = 50.0e6; % 50 MHz bandwidth
rrec = 50 ; % 50 meter processing window
scat_range = [15 25] ; % scatterers are 15 and 25 meters into window
scat_rcs = [1 2]; % RCS 1 m^2 and 2m^2
winid = 0; %no window used
[y] = matched_filter(nscat,taup,b,rrec,scat_range,scat_rcs,winid);
```

MATLAB Function "matched_filter.m" Listing

```
function [y] = matched_filter(nscat,taup,b,rrec,scat_range,scat_rcs,winid)
% This function implements the matched filter processor
%% Inputs
    % nscat          == number of point scatterers within the received window
    % rrec           == receive window size in m
    % taup           == uncompressed pulse width in seconds
    % b              == chirp bandwidth in Hz
    % scat_range     == scatterers' relative range in m
    % scat_rcs       == vector of scatterers' RCS in meter squared
    % win            == 0 = no window; 1 = Hamming; 2 = Kaiser with parameter pi; ...
                        and 3 = Chebychev side-lobes at -60dB
%% Output
    % y              == normalized compressed output
%
eps = 1.0e-16;
% time bandwidth product
time_B_product = b * taup;
if(time_B_product < 5 )
    fprintf('*********** Time Bandwidth product is TOO SMALL **************')
    fprintf('\n Change b and or taup')
    return
end
%
% speed of light
c = 3.e8;
```

```
% number of samples
n = fix(5 * taup * b);
% initialize input, output, and replica vectors
x(nscat,1:n) = 0.;
y(1:n) = 0.;
replica(1:n) = 0.;
% determine proper window
if( winid == 0.)
   win(1:n) = 1.;
end
if(winid == 1.);
   win = hamming(n)';
end
if( winid == 2.)
   win = kaiser(n,pi)';
end
if(winid == 3.)
   win = chebwin(n,60)';
end
% check to ensure that scatterers are within recieve window
index = find(scat_range > rrec);
if (index ~= 0)
   'Error. Receive window is too large; or scatterers fall outside window'
   return
end
%
% calculate sampling interval
t = linspace(-taup/2,taup/2,n);
replica = exp(i * pi * (b/taup) .* t.^2);
figure(1)
subplot(2,1,1)
plot(t,real(replica))
ylabel('Real (part) of replica')
xlabel('Time in seconds')
grid
subplot(2,1,2)
sampling_interval = taup / n;
freqlimit = 0.5/ sampling_interval;
freq = linspace(-freqlimit,freqlimit,n);
plot(freq,fftshift(abs(fft(replica))));
ylabel('Spectrum of replica')
xlabel('Frequency in Hz')
grid
for j = 1:1:nscat
   range = scat_range(j) ;
   x(j,:) = scat_rcs(j) .* exp(i * pi * (b/taup) .* (t +(2*range/c)).^2) ;
   y = x(j,:)  + y;
end
%
figure(2)
```

```
y = y .* win;
plot(t,real(y),'k')
xlabel ('Relative delay in seconds')
ylabel ('Uncompressed echo')
grid
out =xcorr(replica, y);
out = out ./ n;
s = taup * c /2;
Npoints = ceil(rrec * n /s);
dist =linspace(0, rrec, Npoints);
delr = c/2/b;
figure(3)
plot(dist,abs(out(n:n+Npoints-1)),'k')
xlabel ('Target relative position in meters')
ylabel ('Compressed echo')
grid
return
```

MATLAB Function "power_integer_2.m" Listing

```
function n = power_integer_2 (x)
m = 0.;
for j = 1:30
  m = m + 1.;
  delta = x - 2.^m;
  if(delta < 0.)
    n = m;
    return
  else
  end
end
return
```

MATLAB Function "stretch.m" Listing

```
function [y] = stretch(nscat, taup, f0, b, scat_range, rrec, scat_rcs, winid)
% This function implements the stretch processor
%% Inputs
  % nscat          == number of point scatterers within the receive window
  % taup           == uncompressed pulse width in seconds
  % f0             == chirp start frequency in Hz
  % b              == chirp bandwidth in Hz
  % scat_range     == vector of scatterers' range in m
  % rrec           == range receive window in m
  % scat_rcs       == vector of scatterers' RCS in m^2
  % win            == 0 = no window; 1 = Hamming; 2 = Kaiser with parameter pi; ...
                      3 = Chebychev side-lobes at -60dB
%% Outputs
  % y              == compressed output in volts
%
eps = 1.0e-16;
```

```
htau = taup / 2.;
c = 3.e8;
trec = 2. * rrec / c;
n = fix(2. * trec * b);
m = power_integer_2(n);
nfft = 2.^m;
x(nscat,1:n) = 0.;
y(1:n) = 0.;
if( winid == 0.)
  win(1:n) = 1.;
  win =win';
else
  if(winid == 1.)
    win = hamming(n);
  else
    if( winid == 2.)
      win = kaiser(n,pi);
    else
      if(winid == 3.)
        win = chebwin(n,60);
      end
    end
  end
end
deltar = c / 2. / b;
max_rrec = deltar * nfft / 2.;
maxr = max(scat_range);
if(rrec > max_rrec | maxr >= rrec )
  'Error. Receive window is too large; or scatterers fall outside window'
  return
end
t = linspace(0,taup,n);
for j = 1:1:nscat
  range = scat_range(j);% + rmin;
  psi1 = 4. * pi * range * f0 / c - ...
    4. * pi * b * range * range / c / c/ taup;
  psi2 = (2*4. * pi * b * range / c / taup) .* t;
  x(j,:) = scat_rcs(j) .* exp(i * psi1 + i .* psi2);
  y = y + x(j,:);
end
%
figure(1)
plot(t,real(y), 'k')
xlabel ('Relative delay in seconds')
ylabel ('Uncompressed echo')
grid
ywin = y .* win';
yfft = fft(y,n) ./ n;
out= fftshift(abs(yfft));
figure(2)
```

```
delinc = rrec/ n;
%dist = linspace(-delinc-rrec/2,rrec/2,n);
dist = linspace((-rrec/2), rrec/2,n);
plot(dist,out,'k')
xlabel ('Relative range in meters')
ylabel ('Compressed echo')
axis auto
grid
```

MATLAB Function "SFW.m" Listing

```
function [hl] = SFW (nscat, scat_range, scat_rcs, n, deltaf, prf, v, rnote, winid)
% Range or Time domain Profile
% Range_Profile returns the Range or Time domain plot of a simulated
% HRR SFWF returning from a predetermined number of targets with a predetermined
% RCS for each target.
%% Inputs
    % nscat           == number of scatterers that make up the target
    % scat_range      == vector containing range to individual scatterers m
    % scat_rcs        == vector containing RCS of individual scatterers m^2
    % n               == number of steps
    % deltaf          == frequency step in Hz
    % prf             == PRF of SFW in Hz
    % v               == target velocity m/sec
    % r0              == profile starting range im m
    % winid           == number>0 for Hamming window; umber < 0 no window
%% Output
    % hl              == range profile    dB
%
c=3.0e8;  % speed of light (m/s)
num_pulses  = n;
SNR_dB = 40;
nfft = 256;
% carrier_freq = 9.5e9; %Hz (10GHz)
freq_step   = deltaf; %Hz (10MHz)
V = v;  % radial velocity (m/s)  -- (+)=towards radar (-)=away
PRI = 1. / prf; % (s)
if (nfft > 2*num_pulses)
   num_pulses = nfft/2;
else
end
%
Inphase = zeros((2*num_pulses),1);
Quadrature = zeros((2*num_pulses),1);
Inphase_tgt   = zeros(num_pulses,1);
Quadrature_tgt = zeros(num_pulses,1);
IQ_freq_domain = zeros((2*num_pulses),1);
Weighted_I_freq_domain = zeros((num_pulses),1);
Weighted_Q_freq_domain = zeros((num_pulses),1);
Weighted_IQ_time_domain = zeros((2*num_pulses),1);
Weighted_IQ_freq_domain = zeros((2*num_pulses),1);
abs_Weighted_IQ_time_domain = zeros((2*num_pulses),1);
dB_abs_Weighted_IQ_time_domain = zeros((2*num_pulses),1);
```

```
taur = 2. * rnote / c;
for jscat = 1:nscat
  ii = 0;
  for i = 1:num_pulses
    ii = ii+1;
    rec_freq = ((i-1)*freq_step);
    Inphase_tgt(ii) = Inphase_tgt(ii) + sqrt(scat_rcs(jscat)) * cos(-2*pi*rec_freq*...
      (2.*scat_range(jscat)/c - 2*(V/c)*((i-1)*PRI + taur/2 + 2*scat_range(jscat)/c)));
    Quadrature_tgt(ii) = Quadrature_tgt(ii) + sqrt(scat_rcs(jscat))*sin(-2*pi*rec_freq*...
      (2*scat_range(jscat)/c - 2*(V/c)*((i-1)*PRI + taur/2 + 2*scat_range(jscat)/c)));
  end
end
if(winid >= 0)
  window(1:num_pulses) = hamming(num_pulses);
else
  window(1:num_pulses) = 1;
end
Inphase = Inphase_tgt;
Quadrature = Quadrature_tgt;
Weighted_I_freq_domain(1:num_pulses) = Inphase(1:num_pulses).* window';
Weighted_Q_freq_domain(1:num_pulses) = Quadrature(1:num_pulses).* window';
Weighted_IQ_freq_domain(1:num_pulses)= Weighted_I_freq_domain + ...
  Weighted_Q_freq_domain*j;
Weighted_IQ_freq_domain(num_pulses:2*num_pulses)=0.+0.i;
Weighted_IQ_time_domain = (ifft(Weighted_IQ_freq_domain));
abs_Weighted_IQ_time_domain = (abs(Weighted_IQ_time_domain));
dB_abs_Weighted_IQ_time_domain =
20.0*log10(abs_Weighted_IQ_time_domain)+SNR_dB;
% calculate the unambiguous range window size
Ru = c /2/deltaf;
hl = dB_abs_Weighted_IQ_time_domain;
 numb = 2*num_pulses;
delx_meter = Ru / numb;
xmeter = 0:delx_meter:Ru-delx_meter;
plot(xmeter, dB_abs_Weighted_IQ_time_domain,'k')
xlabel ('Relative distance in meters')
ylabel ('Range profile in dB')
grid
```

MATLAB Program "Fig7_20.m" Listing

```
% Use this program to reproduce Fig 7.20 of text
clc;
clear all;
close all;
nscat = 1;
scat_range = 912;
scat_rcs = 10;
n =64;
deltaf = 10e6;
prf = 10e3;
v = 10;
rnote = 900,
winid = 1;
```

```
count = 0;
for time = 0:.05:3
   count = count +1;
   hl = SFW (nscat, scat_range, scat_rcs, n, deltaf, prf, v, rnote, winid);
   array(count,:) = transpose(hl);
   hl(1:end) = 0;
   scat_range = scat_range - 2 * n * v / prf;
end
figure (1)
 numb = 2*256;% this number matches that used in hrr_profile.
 delx_meter = 15 / numb;
 xmeter = 0:delx_meter:15-delx_meter;
 imagesc(xmeter, 0:0.05:4,array)
ylabel ('Time in seconds')
xlabel('Relative distance in meters')
```

Part III

Special Radar Considerations

Chapter 10:

Chapter 8

Radar Wave Propagation

8.1. The Earth's Impact on the Radar Equation

So far in this book, all analysis presented implicitly assumed that the radar electromagnetic waves travel as if they were in free space. Simply put, all analysis presented did not account for the effects of the earth's atmosphere nor the effects of the earth's surface. Despite the fact that *"free space analysis"* may be adequate to provide a general understanding of radar systems, it is only an approximation. In order to accurately predict radar performance, one must modify free space analysis to include the effects of the earth and its atmosphere. These modifications should account for ground reflections from the surface of the earth, diffraction of electromagnetic waves, bending or refraction of radar waves due to the earth's atmosphere, Doppler errors, rotation of the polarization plane, time delays, dispersion effects, and attenuation or absorption of radar energy by the gases constituting the atmosphere.

The earth's impact on the radar equation manifests itself by introducing an additional power term in the radar equation. This term is referred to as the *pattern propagation factor* and is symbolically denoted by F_p. The propagation factor can actually introduce constructive as well as destructive interference onto the SNR depending on the radar frequency and the geometry under consideration. In general, the pattern propagation factor is defined as

$$F_p = |E/E_0| \qquad \text{Eq. (8.1)}$$

where E is the electric field in the medium and E_0 is the free space electric field. In this case, the radar equation is now given by

$$SNR = \frac{P_t G^2 \lambda^2 \sigma}{(4\pi)^3 k T_0 BFLR^4} F_p^4. \qquad \text{Eq. (8.2)}$$

8.2. Earth's Atmosphere

The earth's atmosphere compromises several layers, as illustrated in Fig. 8.1. The first layer, which extends in altitude to about $30Km$, is known as the troposphere. Electromagnetic waves refract (bend downward) as they travel in the troposphere. The troposphere refractive effect is related to its dielectric constant, which is a function of the pressure, temperature, water vapor, and gaseous content. Additionally, due to gases and water vapor in the atmosphere, radar energy suffers a loss. This loss is known as the atmospheric attenuation. Atmospheric attenuation increases significantly in the presence of rain, fog, dust, and clouds. The region above the troposphere (altitude from 30 to 85Km) behaves like free space, and thus little refraction occurs in this region. This region is known as the interference zone.

The ionosphere extends from about 85Km to about 1000Km. It has very low gas density compared to the troposphere. It contains a significant amount of ionized free electrons. The ionization is primarily caused by the sun's ultraviolet and X-rays. This presence of free electrons in the ionosphere affects electromagnetic wave propagation in different ways. These effects include refraction, absorption, noise emission, and polarization rotation. The degree of degradation depends heavily on the frequency of the incident waves. For example, frequencies lower than about 4 to 6MHz are completely reflected from the lower region of the ionosphere. Frequencies higher than 30MHz may penetrate the ionosphere with some level of attenuation. In general, as the frequency is increased, most of the ionosphere's effects become less prominent. The region below the horizon, close to the earth's surface, is called the diffraction region. Diffraction is a term used to describe the bending of radar waves around physical objects. In this region, two types of diffraction are common.

In free space, electromagnetic waves travel in straight lines. However, in the presence of the earth's atmosphere, they bend (refract), as illustrated in Fig. 8.2. Refraction is a term used to describe the deviation of radar wave propagation from straight lines. The deviation from straight line propagation is caused by the variation of the index of refraction. The index of refraction is defined as

$$n = c/v \qquad\qquad \text{Eq. (8.3)}$$

where c is the velocity of electromagnetic waves in free space and v is the wave group velocity in the medium. In the troposphere, the index of refraction decreases uniformly with altitude, while in the ionosphere the index of refraction is minimum at the level of maximum electron density. Alternatively, the interference zone acts like free space and in it the index of refraction is unity.

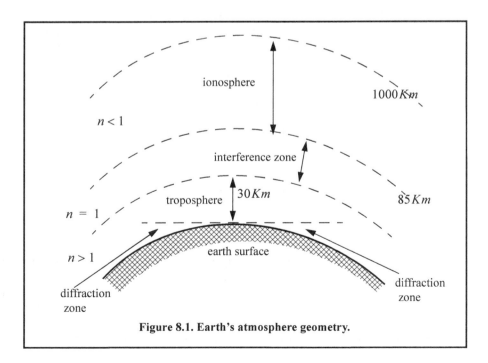

Figure 8.1. Earth's atmosphere geometry.

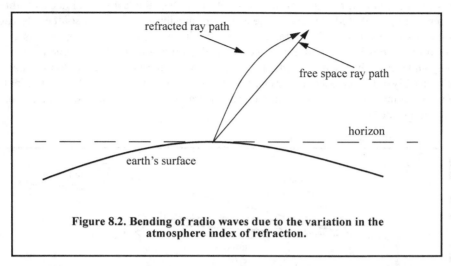

Figure 8.2. Bending of radio waves due to the variation in the atmosphere index of refraction.

In order to effectively study the effects of the atmosphere on the propagation of radar waves, it is necessary to have accurate knowledge of the height variation of the index of refraction in the troposphere and the ionosphere. The index of refraction is a function of the geographic location on the earth, weather, time of day or night, and the season of the year. Therefore, analyzing the atmospheric propagation effects under all parametric conditions becomes an overwhelming task. Typically, this problem is simplified by analyzing atmospheric models that are representative of an average of atmospheric conditions.

In most applications, including radars, one can assume a *well-mixed atmosphere* condition, where the index of refraction decreases in a smooth monotonic fashion with height. The rate of change of the earth's index of refraction n with altitude h is normally referred to as the refractivity gradient, dn/dh. As a result of the negative rate of change in dn/dh, electromagnetic waves travel at slightly higher velocities in the upper troposphere than in the lower part. As a result of this, waves traveling horizontally in the troposphere gradually bend downward. In general, since the rate of change in the refractivity index is very slight, waves do not curve downward appreciably unless they travel very long distances through the atmosphere.

Refraction affects radar waves in two different ways depending on height. For targets that have altitudes typically above 100 meters, the effect of refraction is illustrated in Fig. 8.3. In this case, refraction imposes limitations on the radar's capability to measure target position, and introduces an error in measuring the elevation angle. In a well-mixed atmosphere and very low altitudes (less than $100m$), the refractivity gradient close to the earth's surface is almost constant. However, temperature changes and humidity lapses close to the earth's surface may cause serious changes in the refractivity profile. When the refractivity index becomes large enough, electromagnetic waves bend around the curve of the earth. Consequently, the radar's range to the horizon is extended. This phenomenon is called ducting, and is illustrated in Fig. 8.4. Ducting can be serious over the sea surface, particularly during a hot summer.

8.3. Atmospheric Models

The amount of bending electromagnetic waves experience due to refraction has a lot to do with the medium propagation index of refraction n, defined in Eq. (8.3). Because the index of refraction is not constant as one rises in altitude, it is necessary to analyze the formulas for the index of refraction as a function of height or altitude. Over the last several decades, this topic has been a subject of study by many scientists and physicists; thus, open source references on the subject are abundant in the literature. However, due to differences in notation used as well as the application being studied, it is rather difficult to sift through all available information in a timely and productive manner, particularly for the non-experts in the field. In this chapter, the subject is analyzed in the context of radar wave propagation in the atmosphere. In order to simplify the presentation of the theory, the index of refraction is first analyzed in the troposphere, then the ionosphere.

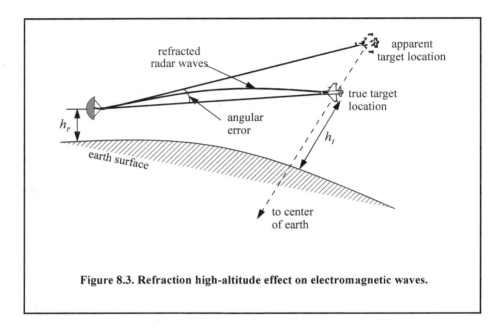

Figure 8.3. Refraction high-altitude effect on electromagnetic waves.

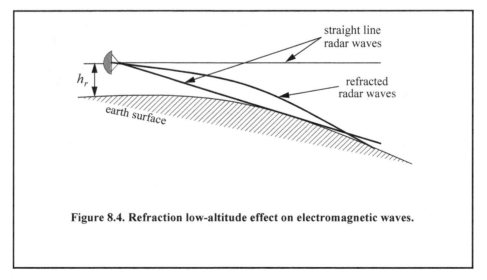

Figure 8.4. Refraction low-altitude effect on electromagnetic waves.

8.3.1. Index of Refraction in the Troposphere

As mentioned earlier, the index of refraction is a function of water vapor, air temperature, and air pressure in the medium, which all vary as a function of height. Because the rate of change of the index of refraction as a function of height is so small, it is very common to introduce a new quantity referred to as *refractivity* N, where

$$N = (n-1) \cdot 1 \times 10^6.$$

Eq. (8.4)

Using this notation, refractivity in the troposphere is given by

$$N = \frac{K_1}{T}\left(P + \frac{K_2 P_w}{T}\right)$$

Eq. (8.5)

where T is the air temperature of the medium in degrees *Kelvin*, P is the total air pressure in *millibars*, P_w is the partial pressure of water vapor in *millibars*, and K_1, K_2 are constants. The first term of Eq. (8.5) (i.e., $(K_1 P)/T$) applies to all frequencies, while the second term (i.e., $(K_1 K_2 P_w)/T^2$) is applicable to radio frequencies only. Experts in the field differ on the exact values for K_1, K_2 based on their relevant applications. However, for most radar applications K_1 can be assumed to be $77.6°$ *Kelvin/millibar* and K_2 is $4810°$ *Kelvin*. Therefore, Eq. (8.5) can now be written as,

$$N = \frac{77.6}{T}\left(P + \frac{4810 P_w}{T}\right).$$

Eq. (8.6)

The lowest values of N occur in dry areas where both P and P_w are low. In the United States, the surface value of N, denoted by N_0, varies between 285 and 345 in the winter, and from 275 to 385 in the summer. Note that Eq. (8.6) is valid for heights up to $h \le 50 Km$.

If the values for T, P, and P_w are known everywhere and at all times, then N can be computed everywhere. However, knowing these variables everywhere and at all times is a very daunting task. Therefore, approximations are made for N, where the assumption that pressure and water vapor tend to decrease with height in a well-mixed atmosphere is taken into consideration. On average, the refractivity will decrease exponentially from N_0 in accordance with the following relation,

$$N = N_0 e^{-c_e \cdot h}$$

Eq. (8.7)

where h is the altitude in Km and c_e is a constant (in Km^{-1}) related to refractivity by

$$c_e = -\frac{\left(\frac{d}{dh}N\right)\Big|_{h=0}}{N_0}.$$

Eq. (8.8)

In general, c_e can be computed from Eq. (8.7) using two different altitudes, for example,

$$c_e = -\ln\left(\frac{N|_{1Km}}{N_0}\right).$$

Eq. (8.9)

The International Telecommunication Union (ITU) has established that for an average atmosphere, $N_0 = 315$ and $c_e = 0.1360 Km^{-1}$, while in the United States these average values are given by $N_0 = 313$ and $c_e = 0.1439 Km^{-1}$. Table 8.1 lists a few values for these variables.

Table 8.1. Published Values for the Parameters in Eq. (8.7).

N_0	c_e (h in Km)	c_e (h in *feet*)
200	0.1184	3.609×10^{-5}
250	0.1256	3.829×10^{-5}
301	0.1396	4.256×10^{-5}
313	0.1439	4.385×10^{-5}
350	0.1593	4.857×10^{-5}
400	0.1867	5.691×10^{-5}
450	0.2233	6.805×10^{-5}

8.3.2. Index of Refraction in the Ionosphere

Unlike the troposphere, refraction in the ionosphere occurs because of the high electron density (ionization) inside the ionosphere and not due to water vapor or other variables. The average electron density as a function of height is given by the Chapman function as

$$\rho_e = \rho_{max} \cdot e^{\frac{1 - z - e^{-z}}{2}} \qquad \text{Eq. (8.10)}$$

where ρ_e is the electron density in electrons per cubic meters, ρ_{max} is the maximum electron density along the propagation path, and z is the normalized altitude or normalized height. The normalized height is given by

$$z = \frac{h - h_m}{H} \qquad \text{Eq. (8.11)}$$

where h_m is the height of maximum electron density and the height scale H is given by

$$H = \frac{kT}{mg} \qquad \text{Eq. (8.12)}$$

where k is Boltzmann's constant, T is the temperature in degrees *Kelvin*, m is the mean molecular mass of an air particle, and g is the gravitational constant. Table 8.2 shows some representative values for H, h_m and the corresponding values for ρ_{max}.

Table 8.2. Representative Values for H, h_m and ρ_{max}.

$h_m - Km$	$H - Km$	$\rho_{max} - electron/cm^3$
100	10	1.5×10^5
200	35	3.0×10^5
300	70	12.5×10^5

Electrons in the ionosphere travel in spiral paths along the earth's magnetic field lines at an angular rate ϖ_p given by

$$\varpi_p^2 = \frac{\rho_e Q}{m\varepsilon_0}$$

Eq. (8.13)

where Q is the charge of an electron ($1.6022 \times 10^{-19} Columbs$) and ε_0 is the permittivity of free space ($8.8542 \times 10^{-12} Columbs/m$). The index of refraction is given by

$$n = \sqrt{1 - \left(\frac{\varpi_p}{\omega}\right)^2}$$

Eq. (8.14)

where $\omega = 2\pi f$ is the radar wave frequency in radians and f is the frequency in hertz. Substituting Eq. (813) into Eq. (8.14) and collecting terms yields

$$n = \sqrt{1 - \frac{80.6\rho_e}{f^2}} \approx 1 - \frac{40.3\rho_e}{f^2}.$$

Eq. (8.15)

Note that Eq. (8.15) is valid for $h > 50 Km$ and the refractivity is given by

$$N \approx -\frac{40.3\rho_e \times 10^6}{f^2}.$$

Eq. (8.16)

8.3.3. Mathematical Model for Computing Refraction

Consider the geometry shown in Fig 8.5. The different variables shown in this figure are defined as follows: R is the range to the target in free space, R_a is the actual refracted range to the target, r_0 is the earth's radius and is equal to $6375\ Km$, r is the distance from the center of earth to the target, h is the target height above the earth's surface, β_f is the elevation angle of the free space range ray, β_0 is the elevation angle of the actual refracted range ray, β is the target elevation angle, the rest of the variables are as defined in the figure. From the geometry, ds and dr are related by the relationships

$$(ds)^2 = (dr)^2 + r^2(d\theta)^2$$

Eq. (8.17)

$$\sin\beta = \frac{dr}{ds}.$$

Eq. (8.18)

Hence,

$$\cos\beta = \sqrt{1 - \left(\frac{dr}{ds}\right)^2}.$$

Eq. (8.19)

From Eq. (8.3), the time it takes a radar wave to travel from point r_1 to r_2 is given by

$$t = \frac{1}{c}\int_{r_1}^{r_2} n\ dr.$$

Eq. (8.20)

In radar applications, this time represents the time difference between the time it takes the wave to travel from its source to the target using the refracted and the free space rays. From the law of sines,

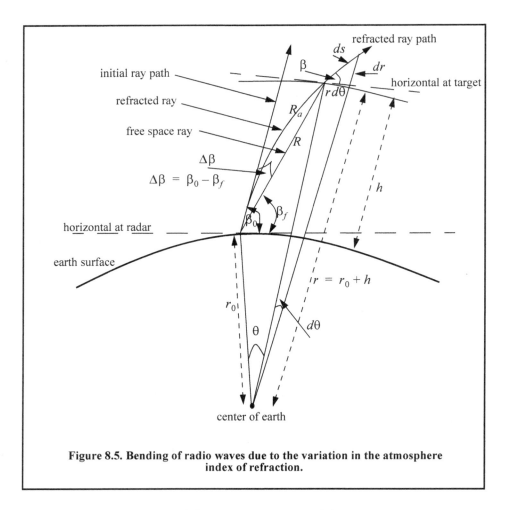

Figure 8.5. Bending of radio waves due to the variation in the atmosphere index of refraction.

$$\sin\left(\frac{\pi}{2} + \beta_f\right) = \frac{r_0 + h}{R}\sin\theta \Rightarrow \beta_f = a\cos\left(\frac{r_0 + h}{R}\sin\theta\right) \qquad \text{Eq. (8.21)}$$

and the free space range using the law of cosines is given by

$$R = \sqrt{r_0^2 + (r_0 + h)^2 - 2r_0(r_0 + h)\cos\theta}. \qquad \text{Eq. (8.22)}$$

Clearly the range error due to refraction is the difference between the apparent range R_a and the free space range R, which is defined in Eq. (8.22). More precisely,

$$\delta R = R_a - R. \qquad \text{Eq. (8.23)}$$

Calculating the error in Eq. (8.23) can be a cumbersome task; it requires minimizing the integral defined in Eq. (8.20) using Fermat's principle. This process is well documented in the literature and only the results are shown here. One can easily show (see Problem 8.3) that

$$\sin\beta = \sqrt{1 - \left(\frac{n_0 r_0 \cos\beta_0}{nr}\right)^2} \qquad \text{Eq. (8.24)}$$

where n_0 and n are, respectively, the medium indices of refraction at the radar and at the target. From Eq. (8.20) the apparent range is

$$R_a = \int_{r_0}^{r} n \, dr \, . \qquad \text{Eq. (8.25)}$$

Substituting Eqs. (8.18) and (8.24) into Eq. (8.25) and collecting terms yields

$$R_a\big|_{troposphere} = \frac{1}{n_0 r_0 \cos\beta_0} \int_{r_0}^{r} \frac{n^2 r \, dr}{\sqrt{\left(\dfrac{nr}{n_0 r_0 \cos\beta_0}\right)^2 - 1}} \qquad \text{Eq. (8.26)}$$

$$R_a\big|_{ionosphere} = \frac{1}{n_0 r_0 \cos\beta_0} \int_{r_0}^{r} \frac{r \, dr}{\sqrt{\left(\dfrac{nr}{n_0 r_0 \cos\beta_0}\right)^2 - 1}} \, . \qquad \text{Eq. (8.27)}$$

Eq. (8.26) is used to calculate R_a in the troposphere while Eq. (8.27) is used in the ionosphere. Recall that Eq. (8.4) should be used for n in Eq. (8.26) while Eq. (8.15) should be used for n in Eq. (8.27).

8.3.4. Stratified Atmospheric Refraction Model

In this section, an excellent approximation method for calculating the range measurement errors and the time-delay errors experienced by radar waves due to refraction is presented. This method is referred to as the *stratified atmospheric model,* and is capable of producing very accurate theoretical estimates of the propagation errors. The basic assumption for this approach is that the atmosphere is stratified into M spherical layers, each is of thickness $\{h_m; \; m = 1, ..., M\}$ and a constant refractive index $\{n_m; \; m = 1, ..., M\}$, as illustrated in Fig. 8.6. In this figure, β_o is the apparent elevation angle and β_{oM} is the true elevation angle. The free space path is denoted by R_{oM}, while the refracted path comprises the sum of $\{R_1, R_2, ..., R_M\}$. From the figure,

$$r_m = r_o + \sum_{j=1}^{m} h_j \qquad ; m = 1, 2, ..., M \qquad \text{Eq. (8.28)}$$

where r_o is the actual radius of the earth.

Using the law of sines, the angle of incidence α_o is given by

$$\frac{\sin\alpha_1}{r_0} = \frac{\sin(\pi/2 + \beta_o)}{r_1} \, . \qquad \text{Eq. (8.29)}$$

Using Snell's law for spherically symmetrical surfaces, the angle β_{m+1} that the ray makes with the horizon in layer *(m+1)* is given by

$$n_m r_m \cos\beta_m = n_{(m+1)} r_{(m+1)} \cos\beta_{(m+1)} \qquad , m = 0, 1, ..., M-1 \, . \qquad \text{Eq. (8.30)}$$

Consequently,

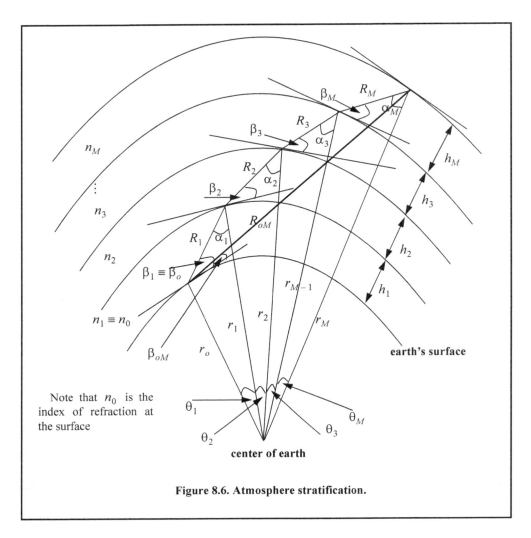

Figure 8.6. Atmosphere stratification.

$$\beta_{(m+1)} = \mathrm{acos}\left[\frac{n_m r_m}{n_{(m+1)} r_{(m+1)}} \cos\beta_m\right] \qquad ,m = 0, 1, ..., M-1.\qquad \text{Eq. (8.31)}$$

Recall from Fig. 8.6 that β_0 and β_1 are defined to be one in the same, and so are n_0 and n_1. From Eq. (8.29), one can write the general expression for the angle of incidence. More precisely,

$$\alpha_m = \mathrm{asin}\left[\frac{r_{(m-1)}}{r_m} \cos\beta_m\right] \qquad ;m = 1, 2, ..., M.\qquad \text{Eq. (8.32)}$$

Applying the law of sines of the direct path R_{0m} yields

$$\beta_{om} = \mathrm{acos}\left\{\frac{r_m}{R_{om}} \sin\left(\sum_{j=1}^{m}\theta_j\right)\right\} \qquad ;m = 1, 2, ..., M\qquad \text{Eq. (8.33)}$$

where

$$R_{om}^2 = r_o^2 + r_m^2 - 2r_o r_m \cos\left(\sum_{j=1}^m \theta_j\right) \qquad ;m = 1, 2, ..., M \qquad \text{Eq. (8.34)}$$

$$\theta_m = \frac{\pi}{2} - \beta_m - \alpha_m \qquad ;m = 1, 2, ..., M. \qquad \text{Eq. (8.35)}$$

The refraction angle error is measured as the difference between the apparent and true elevation angles. Thus, it is given by

$$\Delta\beta_m = \beta_o - \beta_{om}. \qquad \text{Eq. (8.36)}$$

In this notation, $\beta_{01} = \beta_0$; thus, when $m = 1$, then

$$R_{o1} = R_1; \ and \ \Delta\beta_1 = 0. \qquad \text{Eq. (8.37)}$$

Furthermore, when $\beta_o = 90°$,

$$R_{oM} = \sum_{m=1}^M h_m. \qquad \text{Eq. (8.38)}$$

Now, in order to determine the time-delay error due to refraction, refer again to Fig. 8.6. The time it takes an electromagnetic wave to travel through a given layer, $\{R_m; \ m = 1, 2, ..., M\}$, is defined as $\{t_m; \ m = 1, 2, ..., M\}$ where

$$t_m = R_m / v_{\varphi_m} \qquad \text{Eq. (8.39)}$$

and where v_{φ_m} is the phase velocity in the mth layer and is defined by

$$v_{\varphi_m} = c / n_m. \qquad \text{Eq. (8.40)}$$

It follows that the total time of travel of the refracted wave in a stratified atmosphere is

$$t_T = \frac{1}{c}\sum_{j=1}^M n_j R_j. \qquad \text{Eq. (8.41)}$$

The free space travel time of an unrefracted wave is denoted by t_{oM},

$$t_{oM} = R_{oM} / c. \qquad \text{Eq. (8.42)}$$

Therefore, the range error resulting from refraction at the mth is δR_m and is given by

$$\delta R_m = \sum_{j=1}^m n_j R_j - R_{om} \qquad ;m = 1, 2, ..., M. \qquad \text{Eq. (8.43)}$$

By using the law of cosines, one computes R_m as

$$R_m^2 = r_{(m-1)}^2 + r_m^2 - 2r_m r_{(m-1)}\cos\theta_m \qquad ;m = 1, 2, ..., M. \qquad \text{Eq. (8.44)}$$

The results stated in Eqs. (8.41) and (8.43) are valid only in the troposphere. In the ionosphere, which is a dispersive medium, the index of refraction is also a function of frequency. In this case, the group velocity must be used when estimating the range errors of radar measurements. The group velocity is

$$v = nc.$$
Eq. (8.45)

Thus, the total time of travel in the medium is now given by

$$t_T = \frac{1}{c} \sum_{j=1}^{M} \frac{R_j}{n_j}.$$
Eq. (8.46)

Finally, the range error at the mth in the ionosphere is

$$\delta R_m = \sum_{j=1}^{m} \frac{R_j}{n_j} - R_{om} \qquad ;m = 1, 2, ..., M.$$
Eq. (8.47)

MATLAB Function "refraction.m"

The MATLAB function *"refraction.m"* computes the apparent range, range error, and the time delay due to refraction. It implements the analysis presented in the previous two sections. Its syntax is as follows:

[deltaR, Rm, Rt] = refraction(Rmax, el, H, No, Ce, pmax, hm, f)

where

Symbol	Description	Units	Status
Rmax	maximum down range	Km	input
el	initial radar ray elevation angle	degrees	input
No	surface refractivity	none	input
Ce	constant	Km^{-1}	input
pmax	maximum electron density	C	input
hm	height at which maximum electron density occurs	Km	input
f	radar operating center frequency	Hz	input
deltaR	array of range measurement error	Km	output
Rm	stratified range (apparent range)	Km	output
Rt	time delay incurred	sec	output

Figure 8.7 shows a plot for the total range error incurred versus range due to refraction at $f = 9.5\,GHz$ for a few elevation angles. This figure can be reproduced using MATLAB program *"Fig8_7.m,"* listed in Appendix 8-A.

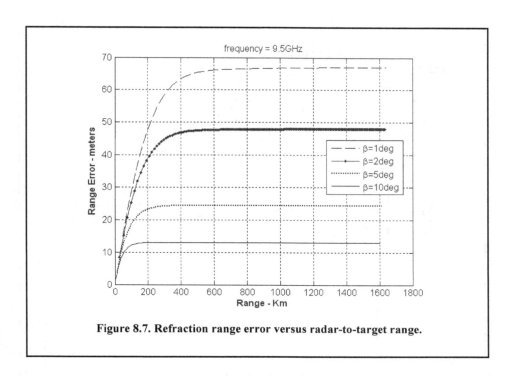

Figure 8.7. Refraction range error versus radar-to-target range.

8.4. Four-Third Earth Model

A very common way of dealing with refraction is to replace the actual earth with an imaginary earth whose effective radius is $r_e = kr_0$, where r_0 is the actual earth radius, and k is

$$k = \frac{1}{1 + r_0(dn/dh)}.$$ Eq. (8.48)

When the refractivity gradient is assumed to be constant with altitude and is equal to 39×10^{-9} per meter, then $k = 4/3$. Using an effective earth radius $r_e = (4/3)r_0$ produces what is known as the *four-third earth model.* In general, choosing

$$r_e = r_0(1 + 6.37 \times 10^{-3}(dn/dh))$$ Eq. (8.49)

produces a propagation model where waves travel in straight lines. Selecting the correct value for k depends heavily on the region's meteorological conditions. At low altitudes (typically less than $10Km$) when using the 4/3 earth model, one can assume that radar waves (beams) travel in straight lines and do not refract. This is illustrated in Fig. 8.8.

8.4.1. Target Height Equation

Using ray tracing (geometric optics), an integral-relating range-to-target height with the elevation angle as a parameter can be derived and calculated. However, such computations are complex and numerically intensive. Thus, in practice, radar systems deal with refraction in two different ways, depending on height. For altitudes higher than $3Km$, actual target heights are estimated from look-up tables or from charts of target height versus range for different elevation angles.

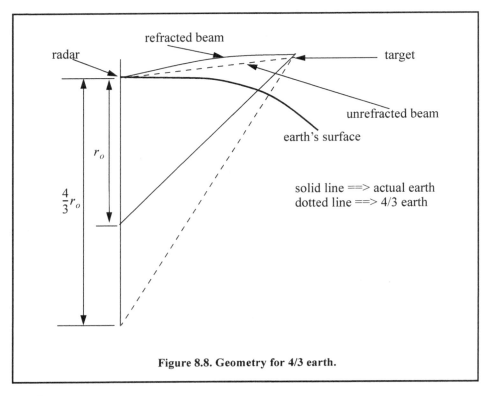

Figure 8.8. Geometry for 4/3 earth.

Blake[1] derives the *height-finding equation* for the 4/3 earth (see Fig. 8.9); it is

$$h = h_r + 6076R\sin\theta + 0.6625R^2(\cos\theta)^2$$

Eq. (8.50)

where h and h_r are in feet and R is nautical miles.

The distance to the horizon for a radar located at height h_r can be calculated with the help of Fig. 8.10. For the right-angle triangle OBA we get

$$r_h = \sqrt{(r_0 + h_r)^2 - r_0^2}$$

Eq. (8.51)

where r_h is the distance to the horizon. By expanding Eq. (8.51) and collecting terms, one can derive the expression for the distance to the horizon as

$$r_h^2 = 2r_0h_r + h_r^2.$$

Eq. (8.52)

Finally, since $r_0 \gg h_r$ Eq. (8.52) is approximated by

$$r_h \approx \sqrt{2r_0h_r},$$

Eq. (8.53)

and when refraction is accounted for, Eq. (8.53) becomes

$$r_h \approx \sqrt{2r_e h_r}.$$

Eq. (8.54)

---

1. Blake, L. V., *Radar Range-Performance Analysis*, Artech House, 1986.

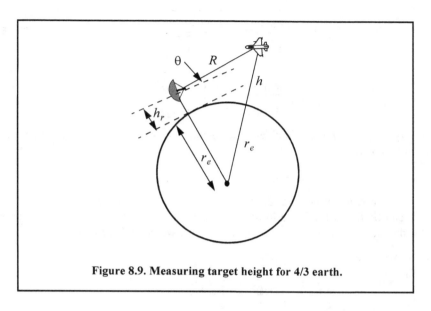

Figure 8.9. Measuring target height for 4/3 earth.

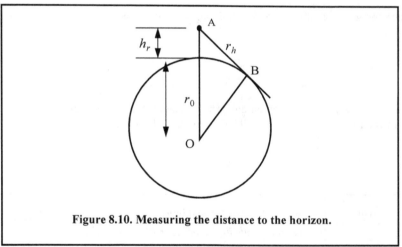

Figure 8.10. Measuring the distance to the horizon.

8.5. Ground Reflection

When radar waves are reflected from the earth's surface, they suffer a loss in amplitude and a change in phase. Three factors that contribute to these changes that are the overall ground reflection coefficient are the reflection coefficient for a flat surface, the divergence factor due to earth's curvature, and the surface roughness.

8.5.1. Smooth Surface Reflection Coefficient

The smooth surface reflection coefficient depends on the frequency, on the surface dielectric coefficient, and on the radar grazing angle. The vertical polarization and the horizontal polarization reflection coefficients are

$$\Gamma_v = \frac{\varepsilon \sin\psi_g - \sqrt{\varepsilon - (\cos\psi_g)^2}}{\varepsilon \sin\psi_g + \sqrt{\varepsilon - (\cos\psi_g)^2}}$$

Eq. (8.55)

$$\Gamma_h = \frac{\sin\psi_g - \sqrt{\varepsilon - (\cos\psi_g)^2}}{\sin\psi_g + \sqrt{\varepsilon - (\cos\psi_g)^2}}$$

Eq. (8.56)

where ψ_g is the grazing angle (incident angle) and ε is the complex dielectric constant of the surface, and are given by

$$\varepsilon = \varepsilon' - j\varepsilon'' = \varepsilon' - j60\lambda\sigma$$

Eq. (8.57)

where λ is the wavelength and σ the medium conductivity in mhos/meter. Typical values of ε' and ε'' can be found tabulated in the literature. Tables 8.3 through 8.5 show some typical values for the electromagnetic properties of soil, lake water, and seawater.

Note that when $\psi_g = 90°$ one gets

$$\Gamma_h = \frac{1 - \sqrt{\varepsilon}}{1 + \sqrt{\varepsilon}} = -\frac{\varepsilon - \sqrt{\varepsilon}}{\varepsilon + \sqrt{\varepsilon}} = -\Gamma_v$$

Eq. (8.58)

while when the grazing angle is very small ($\psi_g \approx 0$), one has

$$\Gamma_h = -1 = \Gamma_v$$

Eq. (8.59)

MATLAB Function "ref_coef.m"

The function *"ref_coef.m"* calculates the horizontal and vertical magnitude and phase response of the reflection coefficient. The syntax is as follows

[rh,rv] = ref_coef (psi, epsp, epspp)

where

Symbol	Description	Status
psi	grazing angle in degrees (can be a vector or a scalar)	input
epsp	ε'	input
epspp	ε''	input
rh	horizontal reflection coefficient complex vector	output
rv	vertical reflection coefficient complex vector	output

Fig. 8.11 shows the corresponding magnitude plots for Γ_h and Γ_v, while Fig. 8.12 shows the phase plots for seawater at $28°C$ where $\varepsilon' = 65$ and $\varepsilon'' = 30.7$ at the X-band. The plots shown in these figures show the general typical behavior of the reflection coefficient. Figures 8.13 and 8.14 show the magnitudes of the horizontal and vertical reflection coefficients as a function of grazing angle for four soils at $8GHz$.

Table 8.3. Electromagnetic properties of soil.

Frequency GHz	Moisture content by volume							
	0.3%		10%		20%		30%	
	ε'	ε''	ε'	ε''	ε'	ε''	ε'	ε''
0.3	2.9	0.071	6.0	0.45	10.5	0.75	16.7	1.2
3.0	2.9	0.027	6.0	0.40	10.5	1.1	16.7	2.0
8.0	2.8	0.032	5.8	0.87	10.3	2.5	15.3	4.1
14.0	2.8	0.350	5.6	1.14	9.4	3.7	12.6	6.3
24	2.6	0.030	4.9	1.15	7.7	4.8	9.6	8.5

Table 8.4. Electromagnetic properties of lake water.

Frequency GHz	Temperature					
	$T = 0°C$		$T = 10°C$		$T = 20°C$	
	ε'	ε''	ε'	ε''	ε'	ε''
0.1	85.9	68.4	83.0	91.8	79.1	115.2
1.0	84.9	15.66	82.5	15.12	78.8	15.84
2.0	82.1	20.7	81.1	16.2	78.1	14.4
3.0	77.9	26.4	78.9	20.6	76.9	16.2
4.0	72.6	31.5	75.9	24.8	75.3	19.4
6.0	61.1	39.0	68.7	33.0	71.0	24.9
8.0	50.3	40.5	60.7	36.0	65.9	29.3

Table 8.5. Electromagnetic properties of sea water.

Frequency GHz	Temperature					
	$T = 0°C$		$T = 10°C$		$T = 20°C$	
	ε'	ε''	ε'	ε''	ε'	ε''
0.1	77.8	522	75.6	684	72.5	864
1.0	77.0	59.4	75.2	73.8	72.3	90.0
2.0	74.0	41.4	74.0	45.0	71.6	50.4
3.0	71.0	38.4	72.1	38.4	70.5	40.2
4.0	66.5	39.6	69.5	36.9	69.1	36.0
6.0	56.5	42.0	63.2	39.0	65.4	36.0
8.0	47.0	42.8	56.2	40.5	60.8	36.0

Observation of Figs. 8.11 and 8.12 yields the following conclusions: (1) The magnitude of the reflection coefficient with horizontal polarization is equal to unity at very small grazing angles and it decreases monotonically as the angle is increased. (2) The magnitude of the vertical polarization has a well-defined minimum. The angle that corresponds to this condition is called Brewster's polarization angle. For this reason, airborne radars in the look-down mode utilize mainly vertical polarization to significantly reduce the terrain bounce reflections. (3) For horizontal polarization, the phase is almost π; however, for vertical polarization, the phase changes to zero around the Brewster's angle. (4) For very small angles (less than $2°$), both $|\Gamma_h|$ and $|\Gamma_v|$ are nearly one; $\angle\Gamma_h$ and $\angle\Gamma_v$ are nearly π. Thus, little difference in the propagation of horizontally or vertically polarized waves exists at low grazing angles.

Figures 8.11 and 8.12 can be reproduced using MATLAB program *"Fig8_11_12.m,"* listed in Appendix 8-A. Alternatively, Figs. 8.13 and 8.14 can be reproduced using MATLAB program *"Fig8_13_14.m,"* listed in Appendix 8-A.

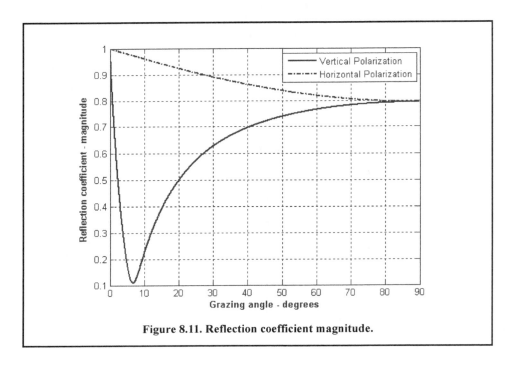

Figure 8.11. Reflection coefficient magnitude.

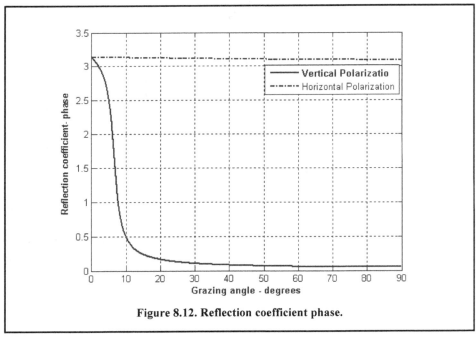

Figure 8.12. Reflection coefficient phase.

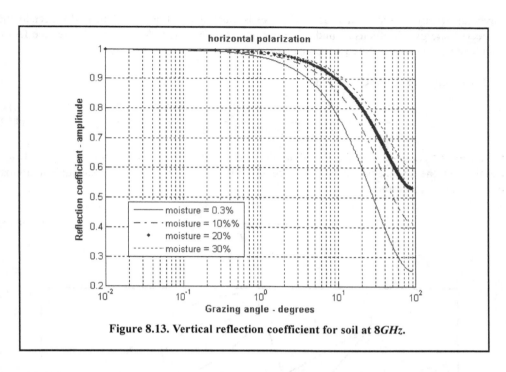

Figure 8.13. Vertical reflection coefficient for soil at 8*GHz*.

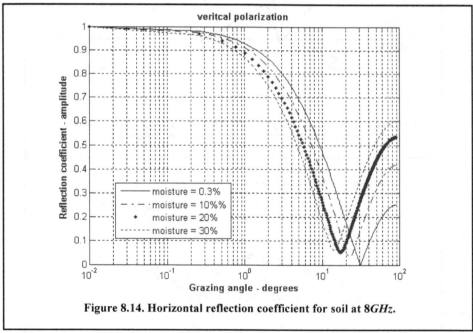

Figure 8.14. Horizontal reflection coefficient for soil at 8*GHz*.

8.5.2. Divergence

The overall reflection coefficient is also affected by the round earth divergence factor, D. When an electromagnetic wave is incident on a round earth surface, the reflected wave diverges because of the earth's curvature. This is illustrated in Fig. 8.15. Due to divergence, the reflected energy is defocused, and the radar power density is reduced. The divergence factor can be derived using geometrical considerations.

The divergence factor can be expressed as

$$D = \sqrt{\frac{r_e \, r \, \sin\psi_g}{[(2r_1r_2/\cos\psi_g) + r_e r\sin\psi_g](1 + h_r/r_e)(1 + h_t/r_e)}}$$ Eq. (8.60)

where all the parameters in Eq. (8.60) are defined in Fig. 8.16. Since the grazing ψ_g is always small when the divergence D is very large, the following approximation is adequate in almost most radar cases of interest,

$$D \approx \frac{1}{\sqrt{1 + \frac{4r_1r_2}{r_e r\sin 2\psi_g}}}$$ Eq. (8.61)

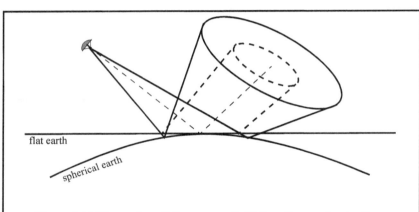

Figure 8.15.Illustration of divergence. Solid line: Ray perimeter for spherical earth. Dashed line: Ray perimeter for flat earth.

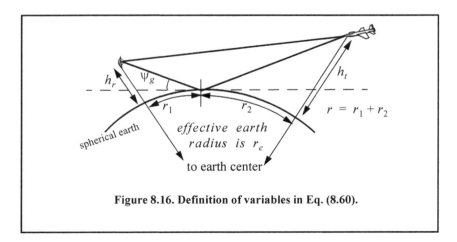

Figure 8.16. Definition of variables in Eq. (8.60).

MATLAB Function *"divergence.m"*

The MATLAB function *"divergence.m"* calculates the divergence using Eq. (8.60). The syntax is as follows:

$$D = divergence \ (r1, r2, hr, ht, psi)$$

where

Symbol	Description	Status
psi	*grazing angle in degrees (can be vector or scalar)*	*input*
r1	*ground range between radar and specular point in Km*	*input*
r2	*ground range between specular point and target in Km*	*input*
hr	*radar height in meters*	*input*
ht	*target height in meters*	*input*
D	*divergence*	*output*

8.5.3. Rough Surface Reflection

In addition to divergence, surface roughness also affects the reflection coefficient. Surface roughness is given by

$$S_r = e^{-2\left(\frac{2\pi h_{rms}\sin\psi_g}{\lambda}\right)^2}$$

Eq. (8.62)

where h_{rms} is the rms surface height irregularity. Another form for the rough surface reflection coefficient that is more consistent with experimental results is given by

$$S_r = e^{-z}I_0(z)$$

Eq. (8.63)

$$z = 2\left(\frac{2\pi h_{rms}\sin\psi_g}{\lambda}\right)^2$$

Eq. (8.64)

where I_0 is the modified Bessel function of order zero.

MATLAB Function *"surf_rough.m"*

The MATLAB function *"surf_rough.m"* calculates the surface roughness reflection coefficient as defined in Eq. (8.62). The syntax is as follows:

$$Sr = surf\_rough \ (hrms, freq, psi)$$

where

Symbol	Description	Status
hrms	*surface rms roughness value in meters*	*input*
freq	*frequency in Hz*	*input*
psi	*grazing angle in degrees*	*input*
Sr	*surface roughness coefficient*	*output*

Figure 8.17 shows a plot of the rough surface reflection coefficient versus $f_{MHz}h_{rms}\sin\psi_g$. The solid line uses Eq. (8.62) while the dashed line uses Eq. (8.63). This figure can be reproduced using MATLAB program *"Fig8_17.m,"* listed in Appendix 8-A.

8.5.4. Total Reflection Coefficient

In general, rays reflected from rough surfaces undergo changes in phase and amplitude, which results in the diffused (noncoherent) portion of the reflected signal. Combining the effects of smooth surface reflection coefficient, divergence, the rough surface reflection coefficient, one can express the total reflection coefficient Γ_t as

$$\Gamma_t = \Gamma_{(h, v)}DS_r,$$

Eq. (8.65)

where $\Gamma_{(h, v)}$ is the horizontal or vertical smooth surface reflection coefficient, D is divergence, and S_r is the rough surface reflection coefficient.

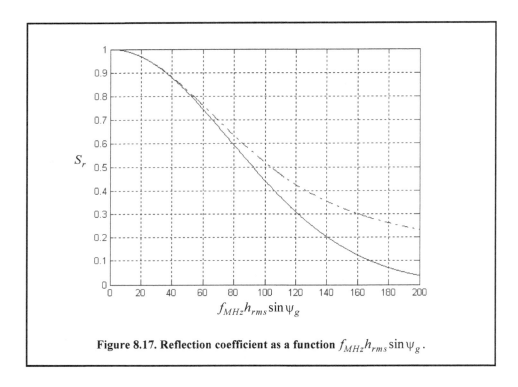

Figure 8.17. Reflection coefficient as a function $f_{MHz}h_{rms}\sin\psi_g$.

8.6. The Pattern Propagation Factor

In general, the pattern propagation factor is a term used to describe the wave propagation when free space conditions are not met. This factor is defined separately for the transmitting and receiving paths. The propagation factor also accounts for the radar antenna pattern effects. The basic definition of the propagation factor is

$$F_p = |E / E_0|,$$

Eq. (8.66)

where E is the electric field in the medium and E_0 is the free space electric field.

Near the surface of the earth, multipath propagation effects dominate the formation of the propagation factor. In this section, a general expression for the propagation factor due to multipath will be developed. In this sense, the propagation factor describes the constructive/destructive interference of the electromagnetic waves diffracted from the earth's surface (which can be either flat or curved). The subsequent sections derive the specific forms of the propagation factor due to flat and curved earth.

Consider the geometry shown in Fig. 8.18. The radar is located at height h_r. The target is at range R, and is located at a height h_t. The grazing angle is ψ_g. The radar energy emanating from its antenna will reach the target via two paths: the "direct path" AB and the "indirect path" ACB. The lengths of the paths AB and ACB are normally very close to one another, and thus the difference between the two paths is very small. Denote the direct path as R_d, the indirect path as R_i, and the difference as $\Delta R = R_i - R_d$. It follows that the phase difference between the two paths is given by

$$\Delta\Phi = \frac{2\pi}{\lambda}\Delta R$$

Eq. (8.67)

where λ is the radar wavelength.

The indirect signal amplitude arriving at the target is less than the signal amplitude arriving via the direct path. This is because the antenna gain in the direction of the indirect path is less than that along the direct path, and because the signal reflected from the earth's surface at point C is modified in amplitude and phase in accordance with the earth's reflection coefficient, Γ. The earth reflection coefficient is given by

$$\Gamma = \rho e^{j\varphi}$$

Eq. (8.68)

where ρ is less than unity and φ describes the phase shift induced on the indirect path signal due to surface roughness.

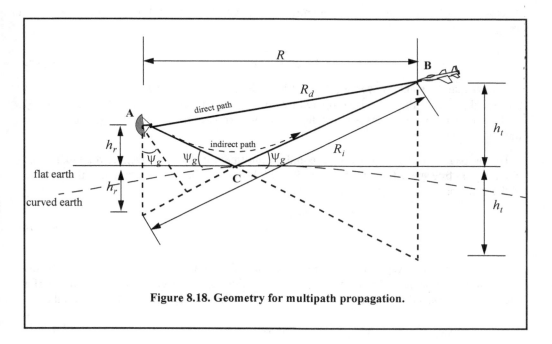

Figure 8.18. Geometry for multipath propagation.

The direct signal (in volts) arriving at the target via the direct path can be written as

$$E_d = e^{j\omega_0 t} e^{j\frac{2\pi}{\lambda}R_d}$$ Eq. (8.69)

where the time harmonic term $\exp(j\omega_0 t)$ represents the signal's time dependency, and the exponential term $\exp(j(2\pi/\lambda)R_d)$ represents the signal spatial phase. The indirect signal at the target is

$$E_i = \rho e^{j\varphi} e^{j\omega_0 t} e^{j\frac{2\pi}{\lambda}R_i}$$ Eq. (8.70)

where $\rho\exp(j\varphi)$ is the surface reflection coefficient. Therefore, the overall signal arriving at the target is

$$E = E_d + E_i = e^{j\omega_0 t} e^{j\frac{2\pi}{\lambda}R_d}\left(1 + \rho e^{j\left(\varphi + \frac{2\pi}{\lambda}(R_i - R_d)\right)}\right).$$ Eq. (8.71)

Due to reflections from the earth's surface, the overall signal strength is then modified at the target by the ratio of the signal strength in the presence of earth to the signal strength at the target in free space. By using Eqs. (8.69) and (8.71) into Eq. (8.66) the propagation factor is computed as

$$F_p = \left|\frac{E_d}{E_d + E_i}\right| = \left|1 + \rho e^{j\varphi} e^{j\Delta\Phi}\right|,$$ Eq. (8.72)

which can be rewritten as

$$F_p = \left|1 + \rho e^{j\alpha}\right|$$ Eq. (8.73)

where $\alpha = \Delta\Phi + \varphi$. Using Euler's identity ($e^{j\alpha} = \cos\alpha + j\sin\alpha$), Eq. (8.73) can be written as

$$F_p = \sqrt{1 + \rho^2 + 2\rho\cos\alpha}.$$ Eq. (8.74)

It follows that the signal power at the target is modified by the factor F_p^2. By using reciprocity, the signal power at the radar is computed by multiplying the radar equation by the factor F_p^4. In the following two sections we will develop exact expressions for the propagation factor for flat and curved earth.

The propagation factor for free space and no multipath is $F_p = 1$. Denote the radar detection range in free space (i.e., $F_p = 1$) as R_0. It follows that the detection range in the presence of the atmosphere and multipath interference is

$$R = \frac{R_0 F_p}{(L_a)^{1/4}}$$ Eq. (8.75)

where L_a is the two-way atmospheric loss at range R. Atmospheric attenuation will be discussed in a later section. Thus, for the purpose of illustrating the effect of multipath interference on the propagation factor, assume that $L_a = 1$. In this case, Eq. (8.75) is modified to

$$R = R_0 F_p.$$ Eq. (8.76)

Figure 8.19 shows the general effects of multipath interference on the propagation factor. Note that, due to the presence of surface reflections, the antenna elevation coverage is transformed into a lobed pattern structure. The lobe widths are directly proportional to λ, and inversely proportional to h_r. A target located at a maxima will be detected at twice its free space range. Alternatively, at other angles, the detection range will be less than that in free space.

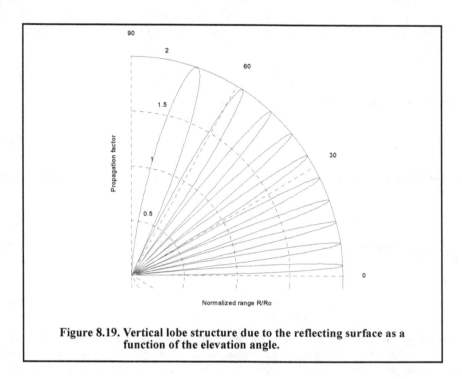

Figure 8.19. Vertical lobe structure due to the reflecting surface as a function of the elevation angle.

8.6.1. Flat Earth

Using the geometry of Fig. 8.18, the direct and indirect paths are computed as

$$R_d = \sqrt{R^2 + (h_t - h_r)^2}$$

Eq. (8.77)

$$R_i = \sqrt{R^2 + (h_t + h_r)^2}.$$

Eq. (8.78)

Eqs. (8.77) and (8.78) can be approximated using the truncated binomial series expansion as

$$R_d \approx R + \frac{(h_t - h_r)^2}{2R}$$

Eq. (8.79)

$$R_i \approx R + \frac{(h_t + h_r)^2}{2R}.$$

Eq. (8.80)

This approximation is valid for low grazing angles, where $R \gg h_t, h_r$. It follows that

$$\Delta R = R_i - R_d \approx \frac{2h_t h_r}{R}.$$

Eq. (8.81)

Substituting Eq. (8.81) into Eq. (8.67) yields the phase difference due to multipath propagation between the two signals (direct and indirect) arriving at the target. More precisely,

$$\Delta\Phi = \frac{2\pi}{\lambda}\Delta R \approx \frac{4\pi h_t h_r}{\lambda R}.$$

Eq. (8.82)

At this point, assume a smooth surface with reflection coefficient $\Gamma = -1$. This assumption means that waves reflected from the surface suffer no amplitude loss, and that the induced surface phase shift is equal to $180°$. Using Eq. (8.67) and Eq. (8.74) along with these assumptions yields

$$F_p^2 = 2 - 2\cos\Delta\Phi = 4(\sin(\Delta\Phi/2))^2.$$

Eq. (8.83)

Substituting Eq. (8.82) into Eq. (8.83) yields

$$F_p^2 = 4\left(\sin\frac{2\pi h_t h_r}{\lambda R}\right)^2.$$

Eq. (8.84)

By using reciprocity, the expression for the propagation factor at the radar is then given by

$$F_p^4 = 16\left(\sin\frac{2\pi h_t h_r}{\lambda R}\right)^4.$$

Eq. (8.85)

Finally, the signal power at the radar is computed by multiplying the radar equation by the factor F_p^4,

$$P_r = \frac{P_t G^2 \lambda^2 \sigma}{(4\pi)^3 R^4} 16\left(\sin\frac{2\pi h_t h_r}{\lambda R}\right)^4.$$

Eq. (8.86)

Since the sine function varies between 0 and 1, the signal power will then vary between 0 and 16. Therefore, the fourth power relation between signal power and the target range results in varying the target range from 0 to twice the actual range in free space. In addition to that, the field strength at the radar will now have holes that correspond to the nulls of the propagation factor.

The nulls of the propagation factor occur when the sine is equal to zero. More precisely,

$$\frac{2h_r h_t}{\lambda R} = n$$

Eq. (8.87)

where $n = \{0, 1, 2, ...\}$. The maxima occur at

$$\frac{4h_r h_t}{\lambda R} = n + 1.$$

Eq. (8.88)

The target heights that produce nulls in the propagation factor are $\{h_t = n(\lambda R/2h_r); n = 0, 1, 2, ...\}$, and the peaks are produced from target heights $\{h_t = n(\lambda R/4h_r); n = 1, 2, ...\}$. Therefore, due to the presence of surface reflections, the antenna elevation coverage is transformed into a lobed pattern structure as illustrated by Fig. 8.19. A target located at a maxima will be detected at twice its free space range. Alternatively, at other angles, the detection range will be less than that in free space. At angles defined by Eq. (8.87), there would be no measurable target returns.

For small angles, Eq. (8.86) can be approximated by

$$P_r \approx \frac{4\pi P_t G^2 \sigma}{\lambda^2 R^8} (h_t h_r)^4,$$

Eq. (8.89)

thus, the received signal power varies as the eighth power of the range instead of the fourth power. Also, the factor $G\lambda$ is now replaced by G/λ.

8.6.2. Spherical Earth

In order to model the effects of multipath propagation on radar performance more accurately, we need to remove the flat earth condition and account for the earth's curvature. When considering round earth, electromagnetic waves travel in curved paths because of the atmospheric refraction. And as mentioned earlier, the most commonly used approach to mitigating the effects of atmospheric refraction is to replace the actual earth by an imaginary earth such that electromagnetic waves travel in straight lines. The effective radius of the imaginary earth is

$$r_e = kr_0$$

Eq. (8.90)

where k is a constant and r_0 is the actual earth radius. Using the geometry in Fig. 8.20, the direct and indirect path difference is

$$\Delta R = R_1 + R_2 - R_d.$$

Eq. (8.91)

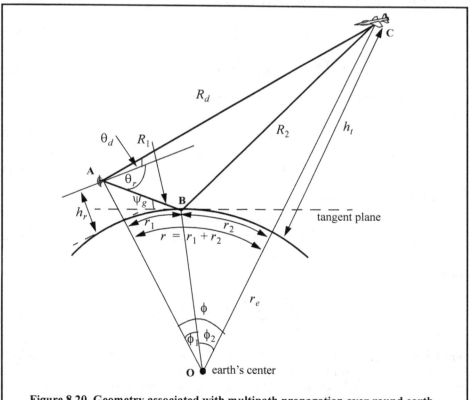

Figure 8.20. Geometry associated with multipath propagation over round earth.

The propagation factor is computed by using ΔR from Eq. (8.91) in Eq. (8.67) and substituting the result in Eq. (8.74). To compute (R_1, R_2, and R_d), the following cubic equation must first be solved for r_1:

$$2r_1^3 - 3rr_1^2 + (r^2 - 2r_e(h_r + h_t))r_1 + 2r_e h_r r = 0.$$

Eq. (8.92)

The solution is

$$r_1 = \frac{r}{2} - p\sin\frac{\xi}{3}$$

Eq. (8.93)

where

$$p = \frac{2}{\sqrt{3}}\sqrt{r_e(h_t + h_r) + \frac{r^2}{4}}$$

Eq. (8.94)

$$\xi = \text{asin}\left(\frac{2r_e r(h_t - h_r)}{p^3}\right).$$

Eq. (8.95)

Next, we solve for R_1, R_2, and R_d. From Fig. 8.20,

$$\phi_1 = r_1/r_e; \quad \phi_2 = r_2/r_e$$

Eq. (8.96)

$$\phi = r/r_e.$$

Eq. (8.97)

Using the law of cosines to the triangles ABO and BOC yields

$$R_1 = \sqrt{r_e^2 + (r_e + h_r)^2 - 2r_e(r_e + h_r)\cos\phi_1}$$

Eq. (8.98)

$$R_2 = \sqrt{r_e^2 + (r_e + h_t)^2 - 2r_e(r_e + h_t)\cos\phi_2}.$$

Eq. (8.99)

Eqs. (8.98) and (8.99) can be written in the following simpler forms:

$$R_1 = \sqrt{h_r^2 + 4r_e(r_e + h_r)(\sin(\phi_1/2))^2}$$

Eq. (8.100)

$$R_2 = \sqrt{h_t^2 + 4r_e(r_e + h_t)(\sin(\phi_2/2))^2}$$

Eq. (8.101)

Using the law of cosines on the triangle AOC yields

$$R_d = \sqrt{(h_r - h_t)^2 + 4(r_e + h_t)(r_e + h_r)\left(\sin\left(\frac{\phi_1 + \phi_2}{2}\right)\right)^2}.$$

Eq. (8.102)

Additionally

$$r = r_e\text{acos}\left(\sqrt{\frac{(r_e + h_r)^2 + (r_e + h_t)^2 - R_d^2}{2(r_e + h_r)(r_e + h_t)}}\right).$$

Eq. (8.103)

Substituting Eqs. (8.100) through (8.102) directly into Eq. (8.91) may not be conducive to numerical accuracy. A more suitable form for the computation of ΔR is then derived. The detailed derivation is in Blake (1986). The results are listed below. For better numerical accuracy, use the following expression to compute ΔR:

$$\Delta R = \frac{4R_1 R_2 (\sin \psi_g)^2}{R_1 + R_2 + R_d}$$

Eq. (8.104)

where

$$\psi_g = \text{asin}\left(\frac{2r_e h_r + h_r^2 - R_1^2}{2r_e R_1}\right) \approx \text{asin}\left(\frac{h_r}{R_1} - \frac{R_1}{2r_e}\right).$$

Eq. (8.105)

8.6.3. MATLAB Program "multipath.m"

The MATLAB program *"multipath.m"* calculates the two-way propagation factor using the 4/3 earth model for spherical earth. It assumes a known free space radar-to-target range. It can be easily modified to assume a known true spherical earth ground range between the radar and the target. Additionally, this program generates three types of plots. They are: (1) The propagation factor as a function of range, (2) the free space relative signal level versus range, and (3) the relative signal level with multipath effects included. This program uses the equations presented in the previous few sections.

This program includes the effects of divergence D and the total surface reflection coefficient Γ_t. Adding the effects of the radar antenna pattern on the signal level is left to the reader as an exercise. Finally, it can also be easily modified to plot the propagation factor versus target height at a fixed target range.

Using this program, Fig. 8.21 presents a plot for the propagation factor loss versus range using $f = 3GHz$; $h_r = 30.48m$; and $h_t = 60.96m$. In this case, the target reference range is at $R_o = 185.2Km$. Divergence effects are not included; neither is the reflection coefficient. More precisely, $D = \Gamma_t = 1$.

Figure 8.22 shows the relative signal level with and without multipath losses. Note that multipath losses affect the signal level by introducing numerous nulls in the signal level. These nulls will typically cause the radar to lose track of targets passing through such nulls. Figures 8.23 and 8.24 are similar to Figs. 8.21 and 8.22, except these new figures account for divergence. All plots assume vertical polarization.

8.7. Diffraction

Diffraction is a term used to describe the phenomenon of electromagnetic waves bending around obstacles. It is of major importance to radar systems operating at very low altitudes. Hills and ridges diffract radio energy and make it possible to perform detection in regions that are physically shadowed. In practice, experimental data measurements provide the dominant source of information available on this phenomenon. Some theoretical analyses of diffraction are also available. However, in these cases many assumptions are made, and perhaps the most important assumption is that obstacles are chosen to be perfect conductors.

The problem of propagation over a knife edge on a plane can be described with the help of Fig. 8.25. The target and radar heights are denoted, respectively, by h_t and h_r. The edge height is h_e. Denote the distance by which the radar rays clear (or do not clear) the tip of the edge by δ. As a matter of notation, δ is assumed to be positive when the direct rays clear the edge, and is negative otherwise. Because the ground reflection occurs on both sides of the edge, the propagation factor is composed of four distinct rays, as illustrated in Fig. 8.26.

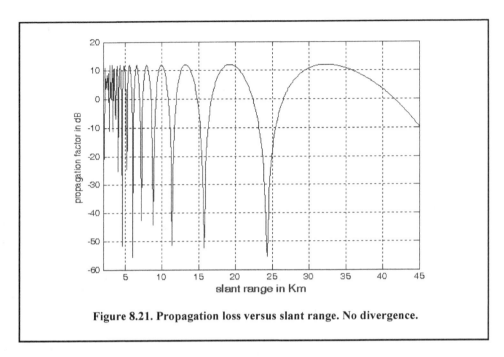

Figure 8.21. Propagation loss versus slant range. No divergence.

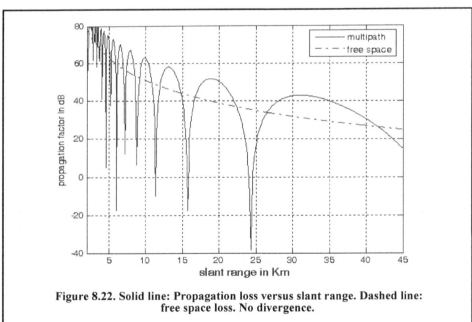

Figure 8.22. Solid line: Propagation loss versus slant range. Dashed line: free space loss. No divergence.

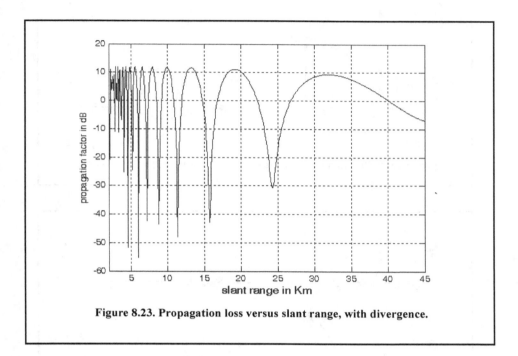

Figure 8.23. Propagation loss versus slant range, with divergence.

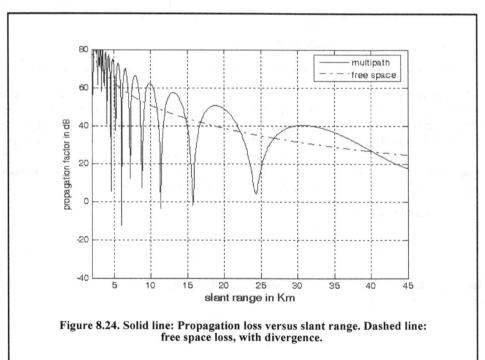

Figure 8.24. Solid line: Propagation loss versus slant range. Dashed line: free space loss, with divergence.

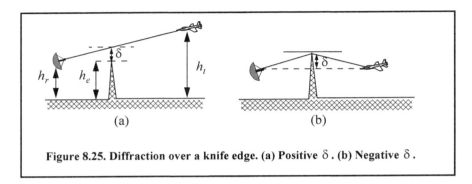

Figure 8.25. Diffraction over a knife edge. (a) Positive δ. (b) Negative δ.

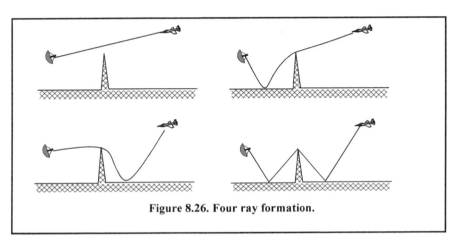

Figure 8.26. Four ray formation.

The analysis that led to creating the multipath model described in the previous section applies only to ground reflections from the intermediate region, as illustrated in Fig. 8.27. The effects of ground reflection below the radar horizon are governed by another physical phenomenon referred to as diffraction. The diffraction model requires calculations of the Airy function and its roots. For this purpose, the numerical approximation presented in Shatz and Polychronopoulos[1] is adopted. This numerical algorithm, described by Shatz and Polychronopoulos, is very accurate and its implementation using MATLAB is straightforward.

Define the following parameters,

$$x = \frac{R}{r_0} \quad , \quad y = \frac{h_r}{h_0} \quad , \quad t = \frac{h_t}{h_0}$$

Eq. (8.106)

where h_r is the radar altitude, h_t is target altitude, R is range to the target, h_0 and r_0 are normalizing factors given by

$$h_0 = \frac{1}{2}(r_e \lambda^2 / \pi^2)^{1/3}$$

Eq. (8.107)

---

1. Shatz, M. P., and Polychronopoulos, G. H., An Algorithm for Evaluation of Radar Propagation in the Spherical Earth Diffraction Region. *IEEE Transactions on Antenna and Propagation*, Vol. 38, August 1990, pp. 1249-1252.

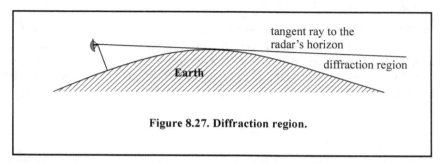

Figure 8.27. Diffraction region.

$$r_0 = \left(\frac{r_e^2 \lambda}{\pi}\right)^{1/3}.$$

Eq. (8.108)

λ is the wavelength and r_e is the effective earth radius. Let $A_i(u)$ denote the Airy function defined by

$$A_i(u) = \frac{1}{\pi} \int_0^\infty \cos\left(\frac{q^3}{3} + uq\right) dq.$$

Eq. (8.109)

The general expression for the propagation factor in the diffraction region is equal to

$$F = 2\sqrt{\pi x} \sum_{n=1}^\infty f_n(y) f_n(t) \exp\left[(e^{j\pi/6}) a_n x\right]$$

Eq. (8.110)

where (x, y, t) are defined in Eq. (8.106) and

$$f_n(u) = \frac{A_i(a_n + ue^{j\pi/3})}{e^{j\pi/3} A_i'(a_n)}$$

Eq. (8.111)

where a_n is the nth root of the Airy function and A_i' is the first derivative of the Airy function. Shatz and Polychronopoulos showed that Eq. (8.110) can be approximated by

$$F = 2\sqrt{\pi x} \sum_{n=1}^\infty \frac{\widehat{A_i}(a_n + ye^{j\pi/3})}{e^{j\pi/3} A_i'(a_n)} \frac{\widehat{A_i}(a_n + te^{j\pi/3})}{e^{j\pi/3} A_i'(a_n)}$$

Eq. (8.112)

$$\exp\left[\frac{1}{2}(\sqrt{3} + j) a_n x - \frac{2}{3}(a_n + ye^{j\pi/3})^{3/2} - \frac{2}{3}(a_n + te^{j\pi/3})^{3/2}\right]$$

where

$$\widehat{A_i}(u) = A_i(u) e^{j\frac{2}{3}u^{3/2}}.$$

Eq. (8.113)

Shatz and Polychronopoulos showed that the sum in Eq. (8.112) represents accurate computation of the propagation factor within the diffraction region.

MATLAB Function "diffraction.m"

The MATLAB function *"diffraction.m"* implements Eq. (8.112) where the sum is terminated at $n \leq 1500$ for accurate computation. It utilizes Shatz's model to calculate the propagation factor in the diffraction region. For this purpose, another MATLAB function called *"airyzol.m"* was used to compute the roots of the Airy function and the roots of its first derivative. The syntax for the function *"diffraction.m"* is as follows

$$F = diffraction(freq, hr, ht, R, nt);$$

where

Symbol	Description	Status
freq	*radar operating frequency*	*Hz*
hr	*radar height*	*meters*
ht	*target height*	*meters*
R	*range over which to calculate the propagation factor*	*Km*
nt	*number of data point is the series given in Eq. (1.186)*	*none*
F	*propagation factor in diffraction region*	*dB*

Figure 8.28 (after Shatz) shows a typical output generated by this program for $h_t = 1000m$, $h_r = 8000m$, and $frequency = 167MHz$. Figure 8.29 is similar to Fig. 8.28 except in this case the following parameters are used: $h_t = 3000m$, $h_r = 200m$, and $frequency = 428MHz$. Figure 8.30 shows a plot for the propagation factor using the same parameters in Fig. 8.29; however, in this figure, both intermediate and diffraction regions are shown. These figures can be reproduced using the MATLAB code listed in Appendix 8-A.

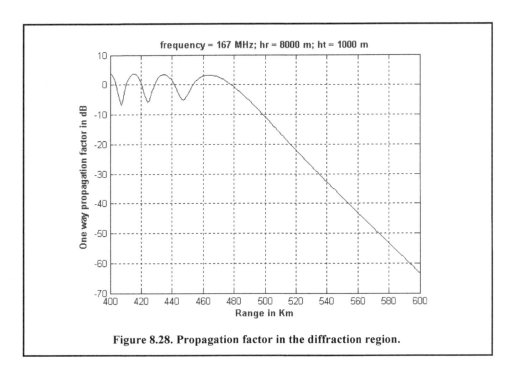

Figure 8.28. Propagation factor in the diffraction region.

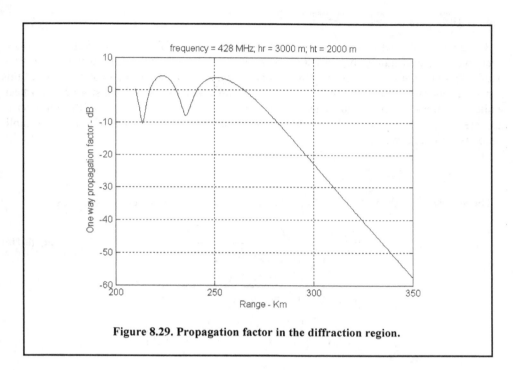

Figure 8.29. Propagation factor in the diffraction region.

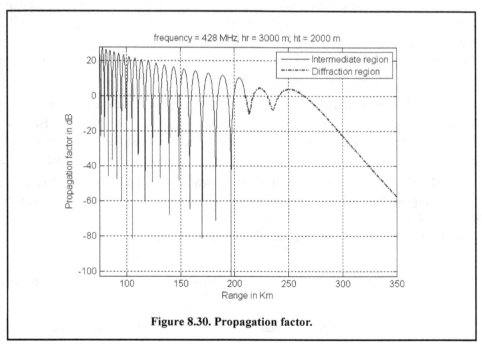

Figure 8.30. Propagation factor.

8.8. Atmospheric Attenuation

Radar electromagnetic waves travel in free space without suffering any energy loss. However, due to gases (mainly oxygen) and water vapor present along the radar wave propagation path, a loss in radar energy occurs. This loss is known as atmospheric attenuation. Most of this lost radar energy is normally absorbed by gases and water vapor and transformed into heat, while a small portion of this lost energy is used in molecular transformation of the atmosphere particles. This section will analyze atmospheric attenuation in the context of most radar application within the atmosphere.

8.8.1. Atmospheric Absorption

The atmospheric absorption due to oxygen is given by the Van Vleck[1] equation as

$$\gamma_O = 28.809 \frac{Pv^2}{T^2} \left\{ \frac{[(1.704 P v_1)/(\sqrt{T})]}{v^2 + [(1.704 \times 10^{-12} P v_1)/(\sqrt{T})]^2} \right.$$
$$+ \frac{[(1.704 P v_2)/(\sqrt{T})]}{(v_0 - v)^2 + [(1.704 \times 10^{-12} P v_2)/(\sqrt{T})]^2}$$
$$\left. + \frac{[(1.704 P v_2)/(\sqrt{T})]}{(v_0 + v)^2 + [(1.704 \times 10^{-12} P v_2)/(\sqrt{T})]^2} \right\} \qquad \text{Eq. (8.114)}$$

where γ_O is the total oxygen absorption in dB/Km; v is the wave number (reciprocal of the wavelength) in cm^{-1}, v_0 is the resonance wave number for oxygen and is equal to $2 cm^{-1}$, v_1 is a constant related to the non-resonance part of absorption in cm^{-1}, v_2 is a constant related to the resonance part of absorption in cm^{-1}, P is the atmospheric pressure in *millibars,* and T is the atmospheric temperature in degrees *Kelvin.*

Using data derived from his experiments, Van Vleck suggested using equal values for both v_1 and v_2; more specifically, he recommended using $v_1 = v_2 = 0.02 cm^{-1}$. However, a decade later after Van Vleck's work, Bean and Abbott[2] using more advanced experimentations determined more accurate values for both constants. They found that $v_1 = 0.018 cm^{-1}$ and $v_2 = 0.05 cm^{-1}$. Nonetheless, for most radar applications one can use Van Vleck's values without losing much accuracy. The relationship between v_1 and v_2 is rather complicated and has dependencies on pressure and temperature.

Equation (8.114) can be approximated by (see Problem 8.16)

$$\gamma_O = \left[0.4909 \frac{P^2}{T^{5/2}} v_1 \right] \left\{ \frac{1}{1 + 2.904 \times 10^{-4} \lambda^2 P^2 T^{-1} v_1^2} \right\} \left\{ 1 + \frac{0.5 v_2}{\lambda^2 v_1} \right\} \qquad \text{Eq. (8.115)}$$

where λ is the radar wavelength. Note that water vapor absorption is negligible below $3 GHz$.

1. Van Vleck, J. H., The Absorption of Microwaves by Oxygen, *Physical Review,* Vol. 71:413, 1947.
2. Bean, B. R., and Abbott, R., Oxygen and Water-Vapor Absorption of Radio Waves in the Atmospheric, *J. Appl. Phys.* 30:1417, 1959.

The Van Vleck[1] equation for water vapor absorption for frequencies over $3GHz$ is given by

$$\gamma_w = 1.012 \times 10^{-3}\frac{\rho_w Pv^2}{T}\left\{\frac{[1.689 \times 10^{-2}Pv_3/(\sqrt{T})]}{(v_w - v)^2 + [1.689 \times 10^{-2}Pv_3/(\sqrt{T})]^2} + \right.$$

$$\left.\frac{[(1.689 \times 10^{-2}Pv_3)/(\sqrt{T})]}{(v_w + v)^2 + [(1.689 \times 10^{-2}Pv_3)/(\sqrt{T})]^2}\right\} +$$

$$3.471 \times 10^{-3}\frac{\rho_w Pv^2}{T}[(1.689 \times 10^{-2}Pv_4)/(\sqrt{T})]$$

$$\text{Eq. (8.116)}$$

where all variables are as defined before in Eq. (8.114) except for: γ_w is the water vapor absorption in dB/Km, ρ_w is the water vapor density in m^{-3}, v_w is a constant equal to $0.742cm^{-1}$, v_3 is a constant related to water vapor resonance at $22.2GHz$, and v_4 is a constant related to water vapor resonance above $22.2GHz$. Van Vleck suggested using $v_3 = v_4 = 0.1cm^{-1}$, which was later updated by Bean and Abbott to the more accurate values of at $v_3 = 0.1cm^{-1}$ and $v_4 = 0.3cm^{-1}$. Equation (8.116) can be approximated by (see Problem 8.17)

$$\gamma_w = 1.852 \times 3.165 \times 10^{-6}\frac{\rho_w P^2}{T^{3/2}}\left\{\frac{1}{(1 - 0.742\lambda)^2 + 2.853 \times 10^{-6}\lambda^2 P^2 T^{-1}}\right. \qquad \text{Eq. (8.117)}$$

$$\left. + \frac{1}{(1 + 0.742\lambda)^2 + 2.853 \times 10^{-6}\lambda^2 P^2 T^{-1}} + \frac{3.43}{\lambda^2}\right\}$$

The atmospheric temperature for altitudes less than $12Km$ is given by

$$T = 288 - 6.5h \qquad \text{Eq. (8.118)}$$

where T is the temperature in degrees *Kelvin* and h is the altitude in *Km*. Assuming that air pressure at sea level is *1015 millibars*, then the air pressure in *millibars* at any altitude for up to $12Km$ is given by

$$P = 1015(1 - 0.02257h)^{5.2561} \qquad \text{Eq. (8.119)}$$

Using Eqs. (8.118) and (8.119), one can construct Table 8.6, which shows some representative data for air pressure, atmospheric pressure, and their corresponding water vapor density.

Table 8.6. Sample Atmospheric Data.

$h - Km$	P - millibars	T - degrees Kelvin	Water vapor density - g/m^3
0.0	1015.0	288.0	6.18
0.7620	925.86	282.89	4.93
1.5240	843.18	277.79	3.74

1. Van Vleck, J. H., The Absorption of Microwaves by Uncondensed Water Vapor, *Physical Review*, Vol. 71:425, 1947.

Table 8.6. Sample Atmospheric Data.

$h - Km$	P - millibars	T - degrees Kelvin	Water vapor density - g/m^3
3.0480	695.73	267.58	2.01
6.0960	463.10	247.16	0.34
9.1440	297.91	226.74	0.05
12.1920	184.04	206.31	<0.01

MATLAB Function "atmo_absorp.m"

The MATLAB function *"atmo_absorp.m"* implements Eqs. (8.115) and (8.117). It syntax is as follows:

[gammaO2, gammaH2O] = atmo_absorp (height, Wvd, freq)

where

Symbol	Description	Units	Status
height	altitude array	Km	input
Wvd	Water vapor density array	g/m^3	input
freq	radar frequency	Hz	input
gammaO2	oxygen absorption	dB	output
gammaH2O	water vapor absorption	dB	output

Figure 8.31 shows the total atmospheric absorption in *dB* and the attenuation due to oxygen alone versus range using the data in Table 8.6. This figure can be reproduced using the MATLAB program *"Fig8_31.m,"* listed in Appendix 8-A.

8.8.2. Atmospheric Attenuation Plots

To compute the total atmospheric attenuation experienced by a radar, one must first compute the two-way total absorption along the radar wave path, from the radar to the target and back. Then, the total atmospheric attenuation is computed from the integral of $\gamma_{atm} = \gamma_O + \gamma_w$ along the ray path. Clearly, γ_{atm} is not only a function of pressure, temperature, water vapor, and frequency, but it is also a function of the radar waves path and its initial elevation angle. More specifically, one would expect the radar wave ray to go through more atmosphere at lower elevation angles, and thus experience more atmospheric attenuation. The total two-way atmospheric attenuation at range R_i using the elevation angle β and the wavelength λ as parameters is then given by

$$\kappa_{atm}(R_i;\beta, \lambda) = 2 \int_0^{R_i} \gamma_{atm}(R_i;\beta, \lambda) \ dR \qquad \text{Eq. (8.120)}$$

where the factor 2 is used to account for the two-way loss or attenuation. The computation of Eq. (8.120) is complex. In this book, the computational power of MATLAB is utilized to generate plots of κ_{atm} versus range using the algorithm described in the next paragraph.

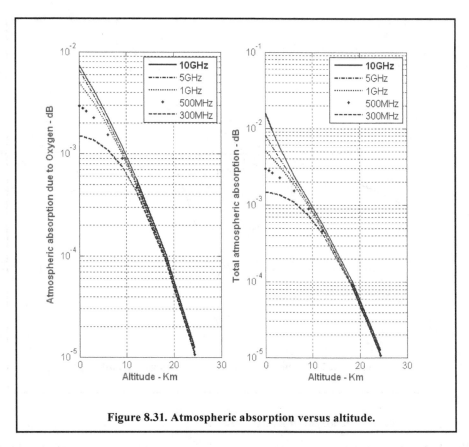

Figure 8.31. Atmospheric absorption versus altitude.

In the previous section, atmospheric absorption was computed and plotted versus target height. To calculate the same absorption versus range, consider the geometry shown in Fig. 8.32. Using the law of sines, one can compute the angle α, then using the law of cosines, one can compute the range R. The MATLAB function *"absorption_range.m"* is then used to generate data for plotting absorption versus range. Finally, the two-way atmospheric attenuation given in Eq. (8.120) is computed using numerical integration. Simply put, once the plot of absorption versus range is generated (see Fig. 8.33), the atmospheric attenuation is equal to the area under the curve.

Using the law of sines,

$$\alpha = \text{asin}\left(\frac{r_0}{r_0 + h}\cos\beta\right)$$ Eq. (8.121)

where the angle θ is

$$\theta = \frac{\pi}{2} - \beta - \alpha,$$ Eq. (8.122)

and from the law of cosines,

$$R = \sqrt{(r_0^2 + (r_0 + h)^2 - 2r_0(r_0 + h)\cos\theta)}.$$ Eq. (8.123)

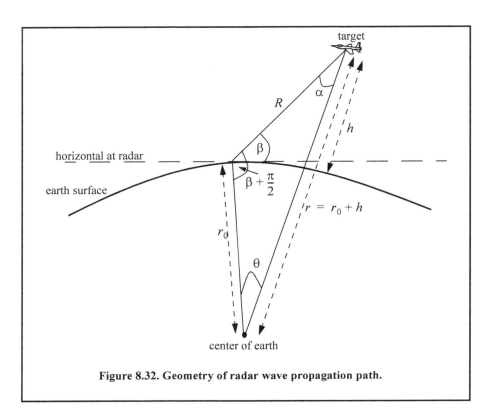

Figure 8.32. Geometry of radar wave propagation path.

MATLAB Function "absorption_range.m"

The MATLAB function *"absorption_range.m"* is a modified version of the function *"atmo_absorp.m."* In this case, the function will use Eqs. (81.21) to (8.123) to also return the total atmospheric absorption versus range. Its syntax is as follows:

[gammaO2, gammaH2O,range] = absorption_range (height, Wvd, freq,beta)

where

Symbol	Description	Units	Status
height	*altitude array*	*Km*	*input*
Wvd	*Water vapor density array*	*g/m^3*	*input*
freq	*radar frequency*	*Hz*	*input*
beta	*radar wave ray path elevation angle*	*degrees*	*input*
gammaO2	*oxygen absorption versus target height*	*dB*	*output*
gammaH2O	*water vapor absorption versus target height*	*dB*	*output*
range	*range array*	*Km*	*output*

Figure 8.33 shows plots of total atmospheric absorption versus range using the same atmospheric data used to generate Fig. 8.31. This figure can be reproduced using MATLAB program *"Fig8_33.m,"* listed in Appendix 8-A.

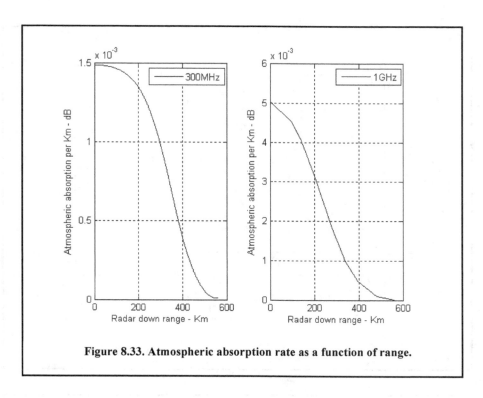

Figure 8.33. Atmospheric absorption rate as a function of range.

MATLAB Function "atmospheric_attn.m"

The MATLAB function *"atmospheric_attn.m"* uses, Riemann sums method to compute the area under the curves in Fig. 8.33. It also uses data generated using the function *"absorption_range.m"* to compute the two-way atmospheric attenuation along the radar wave ray path. Its syntax is as follows:

[Attn, rangei] = atmospheric_attn (gammaO2, gammaH2O, range)

where

Symbol	Description	Units	Status
gammaO2	oxygen absorption versus target height	dB	input
gammaH2O	water vapor absorption versus target height	dB	input
range	range array	Km	input
Attn	2-way atmospheric attenuation	dB	output
rangei	range array used in integration	Km	output

Figure 8.34 shows a typical two-way atmospheric attenuation plot versus range at $300 MHz$, with the elevation angle as a parameter. Figure. 8.35 is similar to Fig. 8.34, except it is for $3 GHz$. Both figures can be reproduced using MATLAB program *"Fig8_33_34.m,"* listed in Appendix 8-A.

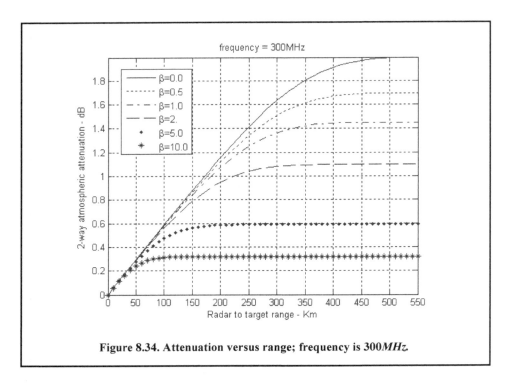

Figure 8.34. Attenuation versus range; frequency is 300*MHz*.

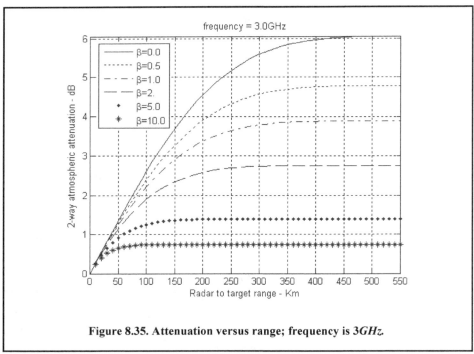

Figure 8.35. Attenuation versus range; frequency is 3*GHz*.

8.9. *Attenuation Due to Precipitation*

Radar waves propagating through rain precipitation suffer loss in signal power. This power loss is due to absorption by and scattering from the rain droplets. Clearly, heaver rain rate will result in more absorption and scattering, thus leading to more power loss. Attenuation due to rain is also a function of frequency or radar wavelength. For example, the one-way attenuation, measured in dB/Km, due to rain precipitation is given y

$$A_r = \begin{cases} 3.43 \times 10^{-4} r^{0.97} & \lambda = 10cm \\ 1.8 \times 10^{-3} r^{1.05} & \lambda = 5cm \\ 1.0 \times 10^{-2} r^{1.21} & \lambda = 3.2cm \end{cases}$$

Eq. (8.124)

where r is the rainfall rate in *mm/hr*. A more general formula for this attenuation is given by

$$A_r = K_A f^\alpha r$$

Eq. (8.125)

where f is the frequency in *GHz*, K_A and α are constants yet to be defined. Almost all open literature sources do not agree on specific values for these two constants, where α varies from about *2.39* to *3.84* while K_A varies from 1.21×10^{-5} to 8.33×10^{-6}. This author recommends using $K_A = 0.0002$ and $\alpha = 2.25$. It follows that

$$A_r = 0.0002 \ f^{2.25} r \ dB/Km.$$

Eq. (8.126)

Figure 8.36 illustrates the behavior of rain attenuation as a function of frequency. Clearly, and as one would expect, as the wavelength becomes smaller, the rain attenuation becomes more dominant. This figure can be reproduced using MATLAB program *"Fig8_36.m,"* listed in Appendix 8-A.

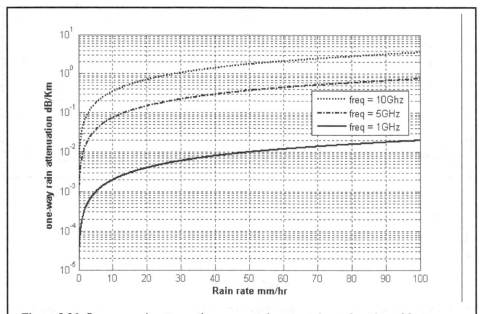

Figure 8.36. One-way rain attenuation versus rain rate and as a function of frequency.

The one-way attenuation in dB/Km due to snow precipitation has been reported in the literature as one of the following two formulas

$$A_s = \frac{0.035r^2}{\lambda^4} + \frac{0.0022r}{\lambda}$$ Eq. (8.127)

$$A_s = \frac{0.00349r^{1.6}}{\lambda^4} + \frac{0.00224r}{\lambda}$$ Eq. (8.128)

where r is the snow fall rate in *millimeters* of water content per hour and λ is the radar wavelength in *centimeters*. Both of Eqs. (8.127) and (8.128) give fairly accurate results with Eq. (8.127) having the edge.

Figure 8.37 illustrates the behavior of snow attenuation as a function of frequency. Clearly, and as one would expect, as the wavelength becomes smaller, the snow attenuation becomes more dominant. This figure can be reproduced using MATLAB program *"Fig8_37.m,"* listed in Appendix 8-A.

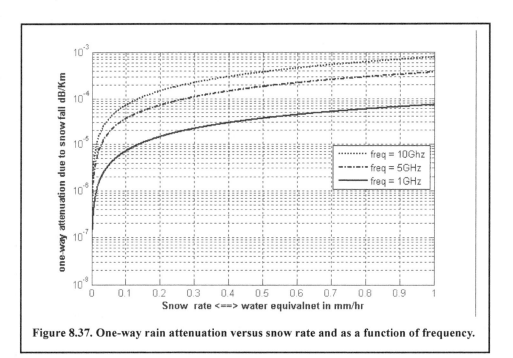

Figure 8.37. One-way rain attenuation versus snow rate and as a function of frequency.

Problems

8.1. Using Eq. (8.50), determine h when $h_r = 15m$ and $R = 35Km$.

8.2. An exponential expression for the index of refraction is given by

$$n = 1 + 315 \times 10^{-6}\exp(-0.136h)$$

where the altitude h is in Km. Calculate the index of refraction for a well-mixed atmosphere at 10% and 50% of the troposphere.

8.3. Validate Eq. (8.20) and Eq. (8.25).

8.4. Derive Eq. (8.24).

8.5. Using Snell's law (i.e., $n_o r_o \cos\theta_0 = n_1 r_1 \cos\theta_1$), show that

$$\left(\sin\left(\frac{\theta_1}{2}\right)\right)^2 = \frac{r_o}{2r_1}\left[2\left(\sin\left(\frac{\theta_0}{2}\right)\right)^2 + \frac{r_1 - r_o}{r_o} - \frac{(N_o - N_1)\cdot 10^{-6}}{n_1}\cos\theta_0\right].$$

8.6. Reproduce Figs. 8.11 and 8.12 by using $f = 8GHz$ and (a) $\varepsilon' = 2.8$ and $\varepsilon'' = 0.032$ (dry soil); (b) $\varepsilon' = 47$ and $\varepsilon'' = 19$ (seawater at $0°C$); (c) $\varepsilon' = 50.3$ and $\varepsilon'' = 18$ (lake water at $0°C$).

8.7. Derive an asymptotic form for Γ_h and Γ_v when the grazing angle is very small.

8.8. Starting with Eq. (8.60), derive Eq. (8.61).

8.9. Calculate the range to the horizon corresponding to a radar at $5Km$ and $10Km$ of altitude. Assume 4/3 earth.

8.10. In reference to Fig. 8.18, assume a radar height of $h_r = 100m$ and a target height of $h_t = 500m$. The range is $R = 20Km$. (a) Calculate the lengths of the direct and indirect paths. (b) Calculate how long it will take a pulse to reach the target via the direct and indirect paths.

8.11. In the previous problem, assuming that you may be able to use the small grazing angle approximation. (a) Calculate the ratio of the direct to the indirect signal strengths at the target. (b) If the target is closing on the radar with velocity $v = 300m/s$, calculate the Doppler shift along the direct and indirect paths. Assume $\lambda = 3cm$.

8.12. Assume a radar at altitude $h_r = 10m$ and a target at altitude $h_t = 300m$, and assuming a spherical earth, calculate r_1, r_2, and ψ_g.

8.13. Derive Eq. (8.103).

8.14. Modify the MATLAB program *"multipath.m"* so that it uses the true spherical ground range between the radar and the target.

8.15. Modify the MATLAB program *"multipath.m"* so that it accounts for the radar antenna.

8.16. Starting with Eq. (8.114), derive Eq. (8.115), assume $v_o = 2cm^{-1}$. In your analysis you may assume that $\left(4 \pm \frac{1}{\lambda^2}\right) \approx 4$.

8.17. Derive Eq. (8.117) from Eq. (8.116).

Appendix 8-A: Chapter 8 MATLAB Code Listings

The MATLAB code provided in this chapter was designed as an academic standalone tool and is not adequate for other purposes. The code was written in a way to assist the reader in gaining a better understanding of the theory. The code was not developed, nor is it intended to be used as part of an open-loop or a closed-loop simulation of any kind. The MATLAB code found in this textbook can be downloaded from this book's web page on the CRC Press web-site. Simply use your favorite web browser, go to *www.crcpress.com*, and search for keyword *"Mahafza"* to locate this book's web page.

MATLAB Function "refraction.m" Listing

```
function [deltaR, Rm, Rt] = refraction(Rmax, el,H, No, Ce, pmax, hm, f)
% Compute the apparent range, range error, and the time delay due to
% refraction; Implements a stratified atmospheric refraction model.
%% Inputs:
    % Rmax     == true range maximum (km)
    % el       == true initial elevation angle (deg)
    % H        == height scale factor in km
    % No       == refractivity at earth surface
    % Ce       == constant im km^-1
    % pmax     == maximum electron density at hm
    % hm       == height for maximum electron contents in Kmkm
    % f        == hz, center frequency
%% Outputs:
    % deltaR   == range error (m)
    % Rm       == apparent range (m)
    % Rt       == time delay (sec)
% initizlize some variables
c = 299792.458;    % km/s, speed of light
Re = 6375;         % km, Earth equatorial radius
% compute object altitude using the law of cosines
hmax = sqrt(Re^2 + Rmax^2 - 2*Re*Rmax*cosd(90 + el)) - Re;
% compute the distance from Earth's center to top of each stratified layer
alt = linspace(0, hmax, ceil(hmax));
r = Re + alt;
% get the altitude indices for both the troposphere and ionosphere
Tindx = find(alt <= 50);
Iindx = find(alt > 50);
% compute the index of refraction for each layer
Ntropo = No * exp(-Ce * alt(Tindx));    % eqn 8.7
z = (alt(Iindx) - hm)/H;                % eqn 8.11
pe = pmax * exp((1-z-exp(-z))/2);       % eqn 8.10
Niono = -40.3 * pe * 1e6 / f^2;         % eqn 8.16
n = 1 + 1e-6*[Ntropo, Niono];           % eqn 8.4
% compute Bm from eqn 8.31 in degrees
Bm = el;
for k = 2:length(alt)-1
    j = k - 1;
    Bm(k) = acosd(cosd(Bm(j)).*n(j).*r(j)./n(k)./r(k));
end
% compute Am from eqn 8.32 in degrees
rm = r(1:end-1);
```

```
rmp1 = r(2:end);
Am = asind(cosd(Bm).*rm./rmp1);
% compute Theta from eqn 8.35 in degrees
theta = 90 - Bm - Am;
% compute Rom from eqn 8.34 in km
Rom = sqrt(Re^2 + rmp1.^2 - 2*Re*rmp1.*cosd(cumsum(theta)));
% compute Rm from from eqn 8.44 in km
Rm = sqrt(rm.^2 + rmp1.^2 - 2*rm.*rmp1.*cosd(theta));
% compute deltaR from eqns 8.43 & 8.47 in km
nR = [n(Tindx).*Rm(Tindx), Rm(Iindx(1:end-1))./n(Iindx(1:end-1))];
deltaR = cumsum(nR) - Rom;
% compute the time delay in seconds
tT = sum(nR) / c;        % eqns 8.41, 8.46
toM = Rom / c;           % eqn 8.42
Rt = tT - toM(end);
return
```

MATLAB Program "Fig8_7.m" Listing

```
% this program reproduces Fig. 8.7 of text
clc
close all
clear all
Rmax = 1600;          % Km
el = [ 1 2 5 10];     % elevation angle in deg
H = 78.11;            % km
No = 313;             % refractivity at earth surface
Ce = 0.1439;         % km^-1
pmax = 1.25e5;       % maximum electron density at hm
hm = 300.73;         % % height for maximum electron contents in Kmkm
f = 9.5e9;           % hz, center frequency
[deltaR, Rm, Rt] = refraction(Rmax, el(1),H, No, Ce, pmax, hm, f);
figure
plot(cumsum(Rm), deltaR .*1000, 'k--')
hold on
[deltaR, Rm, Rt] = refraction(Rmax, el(2),H, No, Ce, pmax, hm, f);
plot(cumsum(Rm), deltaR .*1000, '.-k')
hold on
[deltaR, Rm, Rt] = refraction(Rmax, el(3),H, No, Ce, pmax, hm, f);
plot(cumsum(Rm), deltaR .*1000, 'k:','linewidth',1.5)
hold on
[deltaR, Rm, Rt] = refraction(Rmax, el(4),H, No, Ce, pmax, hm, f);
plot(cumsum(Rm), deltaR .*1000, 'k')
hold off
grid on
xlabel('\bfRange - Km ')
ylabel('\bfRange Error - meters')
legend('\beta=1deg','\beta=2deg','\beta=5deg','\beta=10deg')
title('frequency = 9.5GHz')
```

MATLAB Function "ref_coef.m" Listing

```
function [rh,rv] = ref_coef (psi, epsp, epspp)
% This function calculates the horizontal and vertical magnitude and phase
```

```
% response of the reflection coefficient.
%%% Inputs
   % psi        == grazing angle in degrees (a vector or a scalar)
   % epsp       == epsilon prime
   % epspp      == epsilon double prime
%%% Output
   % rh         == horizontal reflection coefficient complex vector
   % rv         == vertical reflection coefficient complex vector
eps = epsp - i .* epspp;
psirad = psi.*(pi./180.);
arg1 = eps - (cos(psirad).^2);
arg2 = sqrt(arg1);
arg3 = sin(psirad);
arg4 = eps.*arg3;
rv = (arg4-arg2)./(arg4+arg2);
rh = (arg3-arg2)./(arg3+arg2);
return
```

MATLAB Program *"Fig8_11_12.m"* Listing

```
% this program generates Figs. 8.11 and 8.12 of text
close all
clear all
psi = 0.01:0.05:90;
[rh,rv] = ref_coef(psi, 65,30.7);
gamamodv = abs(rv);
gamamodh = abs(rh);
figure
plot(psi,gamamodv,'k',psi,gamamodh,'k -.','linewidth',1.5);
grid
legend ('Vertical Polarization','Horizontal Polarization')
xlabel('\bfGrazing angle - degrees');
ylabel('\bfReflection coefficient - magnitude')
pv = -angle(rv);
ph = angle(rh);
figure
plot(psi,pv,'k',psi,ph,'k -.','linewidth',1.5);
grid
legend ('\bfVertical Polarizatio','Horizontal Polarization')
xlabel('\bfGrazing angle - degrees');
ylabel('\bfReflection coefficient- phase')
```

MATLAB Program *"Fig8_13_14.m"* Listing

```
% this program generates Fig. 8.13 and 8.14 of text
close all
clear all
psi = 0.01:0.25:90;
epsp = [2.8];
epspp = [0.032];% 0.87 2.5 4.1];
[rh1,rv1] = ref_coef(psi, epsp,epspp);
gamamodv1 = abs(rv1);
gamamodh1 = abs(rh1);
epsp = [5.8] ;
```

```
epspp = [0.87];
[rh2,rv2] = ref_coef(psi, epsp,epspp);
gamamodv2 = abs(rv2);
gamamodh2 = abs(rh2);
epsp = [10.3];
epspp = [2.5];
[rh3,rv3] = ref_coef(psi, epsp,epspp);
gamamodv3 = abs(rv3);
gamamodh3 = abs(rh3);
epsp = [15.3]; epspp = [4.1];
[rh4,rv4] = ref_coef(psi, epsp,epspp);
gamamodv4 = abs(rv4);
gamamodh4 = abs(rh4);
figure(1)
semilogx(psi,gamamodh1,'k',psi,gamamodh2,'k-.',psi,gamamodh3,'k.',psi,gamamodh4,'k:','linewidth',1);
grid
xlabel('\bfGrazing angle - degrees');
ylabel('\bfReflection coefficient - amplitude')
legend('moisture = 0.3%','moisture = 10%%','moisture = 20%','moisture = 30%')
title('\bfhorizontal polarization')
figure(2)
semilogx(psi,gamamodv1,'k',psi,gamamodv2,'k-.',psi,gamamodv3,'k.',psi,gamamodv4,'k:','linewidth',1);
grid
xlabel('\bfGrazing angle - degrees');
ylabel('\bfReflection coefficient - amplitude')
legend('moisture = 0.3%','moisture = 10%%','moisture = 20%','moisture = 30%')
title('\bfveritcal polarization')
```

MATLAB Function "divergence.m" Listing

```
function [D] = divergence(r1, r2, ht, hr, psi)
% calculates divergence
% Inputs
    % r1        == ground range between radar and specular point in Km
    % r2        == ground range between specular point and target in Km
    % hr        == radar height in meters
    % ht        == target height in meters
    % psi       == grazing angle in degrees
% Output
    % D         == divergence
psi = psi .* pi ./180; % psi in radians
re = (4/3) * 6375e3;
r = r1 + r2;
arg1 = re.* r .* sin(psi) .*cos(psi);
arg2 = ((2.*r1.*r2./cos(psi)) + re.*r.*sin(psi)) .* ...
    (1+hr./re) .* (1+hr./re);
D = sqrt(arg1 ./ arg2);
return
```

MATLAB Function "surf_rough.m" Listing

```
function Sr = surf_rough(hrms, freq, psi)
clight = 3e8;
psi = psi .* pi ./ 180; % angle in radians
```

```
lambda = clight / freq; % wavelength
g = (2.* pi .* hrms .* sin(psi) ./ lambda).^2;
Sr = exp(-2 .* g);
return
```

MATLAB Program "Fig8_17.m" Listing

```
% this program generates Fig. 8.17 of text
clear all
close all
clight = 3.0e8;
gg = linspace(0,200,500);
zz = 2.* (2*pi.* gg .* .3048/300).^2;
val1 = besseli(0,zz);
% index= find(val1 >1e20);
% val1(index) = 1e-12;
Sr = exp(-zz) ;
Srr = exp(-zz);
Srr1 = val1 .* Sr;
figure(1)
plot(gg,Sr,'k',gg,Srr1,'k-.','linewidth',1)
grid
```

MATLAB Program "multipath.m" Listing

```
% This program calculates and plots the propagation factor versus
% target range with a fixed target hieght.
% The free space radar-to-target range is assumed to be known.
fprintf('****** WARNING ****** \n')
fprintf('Diffraction is not accounted for in this routine')
clear all ; close all
eps = 0.0015;
%%%%%%%%%%%%%%% input %%%%%%%%%%%%%%%%%%%%%%
ro = 6375e3; % earth radius
re = ro * 4 /3; % 4/3 earth radius
freq = 3000e6; % frequnecy
lambda = 3.0e8 / freq; % wavelength
hr = 100*.3048; % radar height in meters
ht = 200*.3048; % target height in meters
Rd1 = linspace(2e3, 45e3, 500); % slant range 3 to 55 Km 500 points
%%%%%%%%%%%%%%%%%%%%%%%%%%%%%%%%%%%%%%%%%%%%%%%%
%determine whether the target is beyond the radar's line of sight
range_to_horizon = sqrt(2*re) * (sqrt(ht) + sqrt(hr)); % range to horizon
index = find(Rd1 > range_to_horizon);
if isempty(index);
   Rd = Rd1;
else
   Rd = Rd1(1:index(1)-1);
   fprintf('****** WARNING ****** \n')
   fprintf('Maximum range is beyond radar line of sight. \n')
   fprintf('****** WARNING ****** \n')
end
val1 = (re + hr).^2 + (re + ht).^2 - Rd.^2;
val2 = 2 .* (re +hr) .* (re + ht);
```

```
phi = acos(val1./val2); % Eq. (8.77)
r = re .* phi; % Eq. (8.71)
p = sqrt(re .* (ht + hr) + (r.^2 ./4)) .* 2 ./ sqrt(3); %Eq.(8.68)
exci = asin((2 .* re .* r .* (ht - hr) ./ p.^3)); % Eq. (8.69)
r1 = (r ./ 2) - p .* sin(exci ./3);
phi1 = r1 ./ re; % Eq. (8.70)
r2 = r - r1;
phi2 = r2 ./ re; % Eq. (8.70)
R1 = sqrt(hr.^2 + 4 .* re .* (re + hr) .* (sin(phi1./2)).^2); % Eq. (8.74)
R2 = sqrt(ht.^2 + 4 .* re .* (re + ht) .* (sin(phi2./2)).^2); % Eq. (8.75)
psi = asin((2 .* re .* hr + hr.^2 - R1.^2) ./ (2 .* re .* R1));
deltaR = (4 .* R1 .* R2 .* (sin(psi)).^2) ./ (R1 + R2 + Rd); % Eq. (8.65)
%%%%%%%%%%%%%%%% input surface roughness %%%%%%%%%%%%%%%%%%%
hrms = 1; %
psi = psi .* 180 ./ pi;
[Sr] = surf_rough(hrms, freq, psi);
%%%%%%%%%%%%%%%% input divergence %%%%%%%%%%%%%%%%%%%%%%
[D] = divergence(r1, r2, ht, hr, psi);
%%%%%%%%%%%%%%%% input smooth earth ref. coefficient %%%%%%%%%%%%%
epsp = 13.7;
epspp = .01;
[rh,rv] = ref_coef (psi, epsp, epspp);
%D = 1;
Sr =1;
gamav = abs(rv);
phv = angle(rv);
gamah = abs(rh);
phh = angle (rh);
 gamav =1;
 phv = -pi;
Gamma_mod = abs(gamav .* D .* Sr); % Eq. (8.39)
Gamma_phase = phv; %
rho = Gamma_mod;
delta_phi = 2 .* pi .* deltaR ./ lambda; % Eq. (8.56)
alpha = delta_phi + phv;
F = ( 1 + rho.^2 + 2 .* rho .* cos( alpha)); % Eq. (8.48)
Ro = 185.2e3; % refrence range in Km
F_free = 40 .* log10(Ro ./ Rd);
F_dbr = 20 .* log10( F ) + F_free;
F_db = 20 .* log10( eps + F );
figure(1)
plot(Rd./1000, F_db,'k','linewidth',1)
grid
xlabel('slant range in Km')
ylabel('propagation factor in dB')
axis tight
axis([2 Rd(end)/1000 -60 20])
figure(2)
plot(Rd./1000, F_dbr,'k',Rd./1000, F_free,'k-.','linewidth',1)
grid
xlabel('slant range in Km')
ylabel('propagation factor in dB')
axis tight
axis([2 Rd(end)/1000 -40 80])
```

legend('multipath','free space')

MATLAB Program "diffraction.m" Listing

```
function F = diffraction(freq, hr, ht,R,nt);
%   Generalized spherical earth propagation factor calculations
%   After Shatz: Michael P. Shatz, and George H. Polychronopoulos, An
%   Algorithm for Elevation of Radar Propagation in the Spherical Earth
%   Diffraction Region. IEEE Transactions on Antenna and Propagation,
%   VOL. 38, NO.8, August 1990.
format long
re = 6373e3 * (4/3); % 4/3 earth radius in Km
[an] = airyzo1(nt);% calculate the roots of the Airy function
c = 3.0e8; % speed of light
lambda = c/freq; % wavelength
r0 = (re*re*lambda / pi)^(1/3);
h0 = 0.5 * (re*lambda*lambda/pi/pi)^(1/3);
y = hr / h0;
z = ht / h0;
%%%%%%%%%%%%%
par1 = exp(sqrt(-1)*pi/3);
pary1 = ((2/3).*(an + y .* par1).^(1.5));
   pary = exp(pary1);
   parz1 = ((2/3).*(an + z .* par1).^(1.5));
   parz = exp(parz1);
   f1n = airy(an + y * par1) .* airy(an + z * par1) .* pary .*parz ;
   f1d = par1 .* par1 .* airy(1,an) .* airy(1,an);
   f1 = f1n ./ f1d;
   index = find(f1<1e6);
%%%%%%%%%%%%%
F = zeros(1,size(R,2));
for range = 1:size(R,2)
   x(range) = R(range)/r0;
   f2 = exp(0.5 .* (sqrt(3) +sqrt(-1)) .*an.*x(range) - pary1 -parz1);
   victor = f1(index) .* f2(index);
   fsum = sum(victor);
   F(range) = 2 .*sqrt(pi.*x(range)) .* fsum;
end
```

MATLAB Program "airyzo1.m" Listing

```
function [an] = airyzo1(nt)
%   This program is a modified version of a function obtained from
%   free internet source www.mathworks.com/matlabcentral/fileexchange/
%   modified by B. Mahafza (bmahafza@dbresearch.net) in 2005
%   ================================
%   Purpose: This program computes the first nt zeros of Airy
%   functions Ai(x)
%   Input :  nt  --- Total number of zeros
%   Output: an ---  first nt roots for Ai(x)
format long
an = zeros(1,nt);
xb = zeros(1,nt);
ii = linspace(1,nt,nt);
```

```
u = 3.0.*pi.*(4.0.*ii-1)./8.0;
u1 = 1./(u.*u);
rt0 = -(u.*u).^(1.0./3.0).*(((((-15.5902.*u1+.929844).* ...
u1-.138889).*u1+.10416667).*u1+1.0);
rt = 1.0e100;
while(abs((rt-rt0)./rt)> 1.e-12);
x = rt0;
ai = airy(0,x);
ad = airy(1,x);
rt=rt0-ai./ad;
if(abs((rt-rt0)./rt)> 1.e-12);
rt0 = rt;
end;
end;
an(ii)= rt;
end
```

MATLAB Program "Fig8_29.m" Listing

```
% Figure 8.28 or Figure 8.29
clc
clear all
close all
freq =167e6;
hr = 8000;
ht = 1000;
R = linspace(400e3,600e3,200); % range in Km
nt =1500; % number of point used in calculating infinite series
F = diffraction(freq, hr, ht, R, nt);
figure(1)
plot(R/1000,10*log10(abs(F).^2),'k','linewidth',1)
grid
xlabel('Range in Km')
ylabel('One way propagation factor in dB')
title('frequency = 167MHz; hr = 8000 m; ht = 1000m')
```

MATLAB Program "Fig8_30.m" Listing

```
% generates Fig. 8.30 of text
clc; clear all; close all
freq =428e6;
hr = 3000;
ht = 200;
%%%%%%%%%%%%%%%% input %%%%%%%%%%%%%%%%%%%%%
ro = 6375e3; % earth radius
re = ro * 4 /3; % 4/3 earth radius
lambda = 3.0e8 / freq; % wavelength
Rd1 = linspace(75e3, 210.1e3, 800); % slant range 3 to 55 Km 500 points
%%%%%%%%%%%%%%%%%%%%%%%%%%%%%%%%%%%%%%%%%%%%
% determine whether the traget is beyond the radar's line of sight
range_to_horizon = sqrt(2*re) * (sqrt(ht) + sqrt(hr)); % range to horizon
index =find(Rd1 > range_to_horizon);
if isempty(index);
   Rd = Rd1;
```

```
else
   Rd = Rd1(1:index(1)-1);
   fprintf('****** WARNING ****** \n')
   fprintf('Maximum range is beyond radar line of sight. \n')
   fprintf('Traget is in diffraction region \n')
   fprintf('****** WARNING ****** \n')
end
%%%%%%%%%%%%%%%%%%%%%%%%%%%%%%%%%%%%%%%%%%%%%%%%
val1 = Rd.^2 - (ht -hr).^2;
val2 = 4 .* (re + hr) .* (re + ht);
r = 2 .* re .* asin(sqrt(val1 ./ val2));
phi = r ./ re;
p = sqrt(re .* (ht + hr) + (r.^2 ./4)) .* 2 ./ sqrt(3);
exci = asin((2 .* re .* r .* (ht - hr) ./ p.^3));
r1 = (r ./ 2) - p .* sin(exci ./3);
phi1 = r1 ./ re;
r2 = r - r1;
phi2 = r2 ./ re;
R1 = sqrt( re.^2 + (re + hr).^2 - 2 .* re .* (re + hr) .* cos(phi1));
R2 = sqrt( re.^2 + (re + ht).^2 - 2 .* re .* (re + ht) .* cos(phi2));
psi = asin((2 .* re .* hr + hr^2 - R1.^2) ./ (2 .* re .* R1));
deltaR = R1 + R2 - Rd;
%%%%%%%%%%%%%% input surface roughness %%%%%%%%%%%%%%%%%
hrms = 1; %
psi = psi .* 180 ./ pi;
[Sr] = surf_rough(hrms, freq, psi);
%%%%%%%%%%%%%%%% input divergence %%%%%%%%%%%%%%%%%%%%%
[D] = divergence(r1, r2, ht, hr, psi);
%%%%%%%%%%%%%%%% input smooth earth ref. coefficient %%%%%%%%%%%%%
epsp = 50;
epspp = 15;
[rh,rv] = ref_coef (psi, epsp, epspp);
D = 1;
 Sr =1;
gamav = abs(rv);
phv = angle(rv);
gamah = abs(rh);
phh = angle (rh);
 gamav =1;
 phv = pi;
Gamma_mod = gamav .* D .* Sr;
Gamma_phase = phv; %
rho = Gamma_mod;
delta_phi = 2 .* pi .* deltaR ./ lambda;
alpha = delta_phi + phv;
F = sqrt( 1 + rho.^2 + 2 .* rho .* cos( alpha));
Ro = 185.2e3; % refrence range in Km
F_free = 40 .* log10(Ro ./ Rd);
F_dbr = 40 .* log10( F .* Ro ./ Rd);
F_db = 40 .* log10( eps + F );
 figure(2)
plot(Rd./1000, F_dbr,'r','linewidth',1)
grid
xlabel('\bfslant range in Km')
```

```
ylabel('\bfPropagation factor in dB')
axis tight
title('\bffrequency = 428 MHz; ht = 3000 m; hr = 200 m')
 R = linspace(210.1e3,350e3,200); % range in Km
nt =1500; % number of point used in calculting infinite series
F = diffraction(freq, hr, ht,R,nt);
figure(3)
plot(R/1000,10*log10(abs(F).^2),'k','linewidth',1)
grid
xlabel('\bfRange - Km')
ylabel('\bfOne way propagation factor - dB')
title('\bffrequency = 428 MHz; hr = 3000 m; ht = 2000 m')
figure(4)
plot(Rd./1000, F_dbr,'k','linewidth',1.)
hold on
plot(R/1000,10*log10(abs(F).^2),'k-.','linewidth',1.5)
grid on
hold off
axis tight
title('\bffrequency = 428 MHz; hr = 3000 m; ht = 2000 m')
legend('Intermediate region', 'Diffraction region')
ylabel('\bfPropagation factor in dB')
xlabel('\bfRange in Km')
```

MATLAB Function "atmo_absorp.m" Listing

```
function [gammaO2, gammaH2O] = atmo_absorp(height,Wvd, freq)
% This function computes the atmospheric attenuation as a function of
% target height for up to 12 Km
%% Inputs
    % height      == target height array in Km
    % Wvd         == water vapor density array in g/m^3
    % freq        == radar operating frequency in Hz
%% Outputs
    % gammaO2   == atmospheric attention due to oxygen in dB
    % gammaH2O  == atmospheric attention due to water vapor in dB
%format long
format short
ro = 6375;
v1 = 0.018; v2 = .05;
v3 = 0.1; v4 = 0.3;
lambda = 3e10/freq; % wavelength in cm
height = height ./1000;
T = 288 -6.7 .* height; % compute temperature array at different heights
pressure = 1015 .* (1-0.02275.*height).^5.2561;% compute air pressure array at different
heights
% implement Eq. (8.115)
P = (v1 * 0.4909 .* pressure.^2) ./ (T.^(5/2));
Q = v1^2 * 2.904e-4 .* pressure.^2 ./ T;
gammaO2 = P .* (1./(1+Q.*lambda^2)) .* (1+ (1.39/lambda^2));
```

% implement Eq. (8.117)
P = 1.852 * 3.165e-6 .* Wvd .*pressure.^2 ./ (T.^(3/2));
Q1 = (1 - 0.742 * lambda)^2;
Q2 = (1 + 0.742 * lambda)^2;
Q = 2.853e-6 .* pressure.^2 ./T;
gammaH2O = P .*((1./(Q1 + Q .*lambda^2)) + (1./(Q2 + Q .*lambda^2)) + 3.43/lambda^2);
end

MATLAB Program "Fig8_31.m" Listing

% this program reproduces Fig 8.31 of text book
clc
clear all
close all
format long
h_ft = [0 2500 5000 10000 20000 30000 40000 60000 80000];
height = 0.3048 .* h_ft ;
Wvd = [6.18 4.93 3.74 2.01 0.34 0.05 .009 eps eps];
freq = 300e6;
[gammaO21, gammaH2O1] = atmo_absorp(height,Wvd, freq);
gamma1 = gammaO21 + gammaH2O1;
freq = 500e6;
[gammaO22, gammaH2O2] = atmo_absorp(height,Wvd, freq);
gamma2 = gammaO22 + gammaH2O2;
freq = 1e9;
[gammaO23, gammaH2O3] = atmo_absorp(height,Wvd, freq);
gamma3 = gammaO23 + gammaH2O3;
freq = 5e9;
[gammaO24, gammaH2O4] = atmo_absorp(height,Wvd, freq);
gamma4 = gammaO24 + gammaH2O4;
freq = 10e9;
[gammaO25, gammaH2O5] = atmo_absorp(height,Wvd, freq);
gamma5 = gammaO25 + gammaH2O5;
figure
height = height ./1000;
subplot(1,2,1)
semilogy (height, gammaO25,'k',height, gammaO24,'k-.',height, gammaO23,'k:',height,...
*   gammaO22,'k.',height, gammaO21,'k--','linewidth', 1.5)*
grid
legend('\bf10GHz','5GHz','1GHz','500MHz','300MHz')
ylabel('\bfAtmospheric absorption due to Oxygen - dB')
xlabel('\bfAltitude - Km')
subplot(1,2,2)
semilogy (height, gamma5,'k',height, gamma4,'k-.',height, gamma3,'k:',height,...
*   gamma2,'k.',height, gamma1,'k--','linewidth', 1.5)*
grid
legend('\bf10GHz','5GHz','1GHz','500MHz','300MHz')
ylabel('\bfTotal atmospheric absorption - dB')
xlabel('\bfAltitude - Km')

MATLAB Function "absorption_range.m" Listing

function [gammaO2, gammaH2O,range] = absorption_range(height,Wvd, freq,beta)

```
% This function computes the atmospheric absorption as a function of
% target height and range
% % Inputs
  % height         == target height array in Km
  % Wvd            == water vapor density array in g/m^3
  % freq           == radar operating frequency in Hz
  % beta           == intial elevation angle in degrees
%% Outputs
  % gammaO2        == atmospheric absorption due to oxygen in dB
  % gammaH2O       == atmospheric absorption due to water vapor in dB
  % A_km           == atmospheric absorption versu range
%
format long
ro = 6375;
v1 = 0.018;
v2 = .05;
v3 = 0.1;
v4 = 0.3;
lambda = 3e10/freq; % wavelength in cm
height = height ./1000;
T = 288 -6.7 .* height; % compute temperature array at different heights
pressure = 1015 .* (1-0.02275.*height).^5.2561;% compute air pressure array at different heights
% implement Eq. (8.115)
P = (v1 * 0.4909 .* pressure.^2) ./ (T.^(5/2));
Q = v1^2 * 2.904e-4 .* pressure.^2 ./ T;
gammaO2 = P .* (1./(1+Q.*lambda^2)) .* (1+ (1.39/lambda^2));
% implement Eq. (8.117)
P = 1.852 * 3.165e-6 .* Wvd .*pressure.^2 ./ (T.^(3/2));
Q1 = (1 - 0.742 * lambda)^2;
Q2 = (1 + 0.742 * lambda)^2;
Q = 2.853e-6 .* pressure.^2 ./T;
gammaH2O = P .*((1./(Q1 + Q .*lambda^2)) + (1./(Q2 + Q .*lambda^2)) + 3.43/lambda^2);
% convert beta into radian
beta = beta * pi /180.;
% calcualte array of r0 plus target height
r = ro + height;
alpha =asin(cos(beta) * ro ./r);
theta = (pi/2) - beta - alpha;
% range = sqrt(ro^2 + r.^2 - 2 * cos(theta) * ro .* r);
range = r .* sin(theta) / cos(beta);
end
```

MATLAB Program "Fig8_33.m" Listing

```
% this program reproduces Figs 8.33
clc
clear all
close all
format long
h_ft = [0 2500 5000 10000 20000 30000 40000 60000 80000];
height = 0.3048 .* h_ft ;
```

```
Wvd = [6.18 4.93 3.74 2.01 0.34 0.05 .009 eps eps];
freq = 300e6;
beta = .0;
[gammaO2, gammaH2O,range] = absorption_range(height,Wvd, freq,beta);
Akm1 = gammaO2 + gammaH2O;
xx = 0:.1:range(end);
yy1 = spline(range,Akm1,xx);
freq = 1e9;
[gammaO2, gammaH2O,range] = absorption_range(height,Wvd, freq,beta);
Akm2 = gammaO2 + gammaH2O;
yy2 = spline(range,Akm2,xx);
figure
height = height ./1000;
subplot(1,2,1)
plot(xx,yy1,'k','linewidth',1)
grid
legend('300MHz')
ylabel('Atmospheric absorption per Km - dB')
xlabel('Radar down range - Km')
subplot(1,2,2)
plot(range,Akm2,'k','linewidth',1)
grid
legend('1GHz')
ylabel('Atmospheric absorption per Km - dB')
xlabel('Radar down range - Km')
```

MATLAB Function "atmospheric_attn.m" Listing

```
function [Attn,rangei] = atmospheric_attn(gammaO2,gammaH2O,range)
% this function usse Rieman sums to calculate area under the
% total abosrption curve veruses range
sum = gammaO2 + gammaH2O;
delr = 10;
rangei = 0:delr:range(end);
Attn = zeros(1,size(rangei,2));
yy1 = spline(range,sum,rangei);
yint(1) = 0;
n = 2;
N = size(rangei,2);
while n<=N
    yint(n) = yint(n-1) + delr * (yy1(n-1) + yy1(n));
    n = n+1;
end
% use 1.75 instead of 2 for the 2-way because of inaccuracies of Riemann
% sums method
Attn = 1.75 .* yint;
end
```

MATLAB Program "Fig_34_35.m" Listing

```
% this program reproduces Figs 8.34 and 8.35 of text book
clc
clear all
close all
```

```
format long
h_ft = [0 2500 5000 10000 20000 30000 40000 60000 80000];
height = 0.3048 .* h_ft ;
Wvd = [6.18 4.93 3.74 2.01 0.34 0.05 .009 eps eps];
figure(1)
freq = 500e6;
beta = .0;
[gammaO2, gammaH2O,range] = absorption_range(height,Wvd, freq,beta);
[Attn rangei1] = atmospheric_attn(gammaO2,gammaH2O,range);
M = size(Attn,2);
plot(rangei1,Attn,'K', 'linewidth',1.5)
hold on
beta = 0.5;
[gammaO2, gammaH2O,range] = absorption_range(height,Wvd, freq,beta);
[Attn rangei] = atmospheric_attn(gammaO2,gammaH2O,range);
Attn (end:M) = Attn(end);
plot(rangei1,Attn,'k:','linewidth',1.5)
hold on
beta = 1.0;
[gammaO2, gammaH2O,range] = absorption_range(height,Wvd, freq,beta);
[Attn rangei] = atmospheric_attn(gammaO2,gammaH2O,range);
Attn (end:M) = Attn(end);
plot(rangei1,Attn,'k-.', 'linewidth',1.5)
hold on
beta = 2;
[gammaO2, gammaH2O,range] = absorption_range(height,Wvd, freq,beta);
[Attn rangei] = atmospheric_attn(gammaO2,gammaH2O,range);
Attn (end:M) = Attn(end);
plot(rangei1,Attn,'k--', 'linewidth',1.5)
hold on
beta = 5;
[gammaO2, gammaH2O,range] = absorption_range(height,Wvd, freq,beta);
[Attn rangei] = atmospheric_attn(gammaO2,gammaH2O,range);
Attn (end:M) = Attn(end);
plot(rangei1,Attn,'k.', 'linewidth',1.5)
hold on
beta = 10;
[gammaO2, gammaH2O,range] = absorption_range(height,Wvd, freq,beta);
[Attn rangei] = atmospheric_attn(gammaO2,gammaH2O,range);
Attn (end:M) = Attn(end);
plot(rangei1,Attn,'k*')
hold off
legend('\beta=0.0','\beta=0.5','\beta=1.0','\beta=2.','\beta=5.0','\beta=10.0')
xlabel('Radar to target range - Km')
ylabel('2-way atmospheric attenuation - dB')
title('frequency = 3.0GHz')
axis tight
grid on
```

MATLAB Program "Fig8_36.m" Listing

```
% geerates Fig 8.36 of text
clc
clear all
```

```
close all
format long
alpha = 0.0002;
beta = 2.25;
freq = [1 10 20];
f = freq.^beta;
r = linspace(0,100,1000); % rai fall rate i mm/hr
Att1 = (alpha * f(1) .* r);
Att2 = (alpha * f(2) .* r);
Att3 = (alpha * f(3) .* r);
figure(1)
semilogy(r,Att3, 'k:',r,Att2,'k-.',r,Att1,'k','linewidth',1.5)
xlabel('\bf Rain rate mm/hr')
ylabel('\bfone-way rain attenuation dB/Km')
grid on
legend('freq = 10Ghz','freq = 5GHz','freq = 1GHz')
```

MATLAB Program "Fig8_37.m" Listing

```
% generates Fig 8.37 of text
clc
clear all
close all
format long
alpha = 0.00349;
beta = 0.00224;
freq = [1e9 5e9 10e9];
lambda = 3e10 ./ freq; % wavelength in cm;
r = linspace(0,1,1000); % rai fall rate i mm/hr
Att1 = (0.0035 .* r.^2 ./lambda(1)^4) + 0.0022 .* r ./ lambda(1);
Att2 = (0.0035 .* r.^2 ./lambda(2)^4) + 0.0022 .* r ./ lambda(2);
Att3 = (0.0035 .* r.^2 ./lambda(3)^4) + 0.0022 .* r ./ lambda(3);
figure(1)
semilogy(r,Att3, 'k:',r,Att2,'k-.',r,Att1,'k','linewidth',1.5)
xlabel('\bf Snow  rate <==> water equivalnet in mm/hr')
ylabel('\bfone-way attenuation due to snow fall dB/Km')
grid on
legend('freq = 10Ghz','freq = 5GHz','freq = 1GHz')
```

Chapter 9

Radar Clutter

9.1. Clutter Definition

Clutter is a term used to describe any object that may generate unwanted radar returns that may interfere with normal radar operations. Parasitic returns that enter the radar through the antenna's mainlobe are called mainlobe clutter; otherwise they are called sidelobe clutter. Clutter can be classified into two main categories: surface clutter and airborne or volume clutter. Surface clutter includes trees, vegetation, ground terrain, man-made structures, and sea surface (sea clutter). Volume clutter normally has a large extent (size) and includes chaff, rain, birds, and insects. Surface clutter changes from one area to another, while volume clutter may be more predictable.

Clutter echoes are random and have thermal noise-like characteristics because the individual clutter components (scatterers) have random phases and amplitudes. In many cases, the clutter signal level is much higher than the receiver noise level. Thus, the radar's ability to detect targets embedded in high clutter background depends on the Signal-to-Clutter Ratio (SCR) rather than the SNR. White noise normally introduces the same amount of noise power across all radar range bins, while clutter power may vary within a single range bin. Since clutter returns are target-like echoes, the only way a radar can distinguish target returns from clutter echoes is based on the target RCS σ_t, and the anticipated clutter RCS σ_c (via clutter map). Clutter RCS can be defined as the equivalent radar cross section attributed to reflections from a clutter area, A_c. The average clutter RCS is given by

$$\sigma_c = \sigma^0 A_c \qquad \text{Eq. (9.1)}$$

where $\sigma^0 (m^2/m^2)$ is the clutter scattering coefficient, a dimensionless quantity that is often expressed in *dB*. Some radar engineers express σ^0 in terms of squared centimeters per squared meter. In these cases, σ^0 is $40 dB$ higher than normal.

9.2. Surface Clutter

Surface clutter includes both land and sea clutter, and is often called area clutter. Area clutter manifests itself in airborne radars in the look-down mode. It is also a major concern for ground-based radars when searching for targets at low grazing angles. The grazing angle ψ_g is the angle from the surface of the earth to the main axis of the illuminating beam, as illustrated in Fig. 9.1.

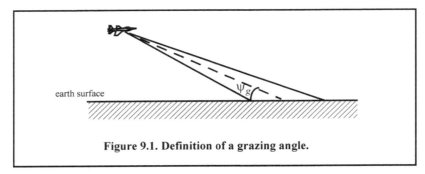

Figure 9.1. Definition of a grazing angle.

Three factors affect the amount of clutter in the radar beam. They are the grazing angle, surface roughness, and the radar wavelength. Typically, the clutter scattering coefficient σ^0 is larger for smaller wavelengths. Fig. 9.2 shows a sketch describing the dependency of σ^0 on the grazing angle. Three regions are identified; they are the low grazing angle region, flat or plateau region, and the high grazing angle region.

The low grazing angle region extends from zero to about the critical angle. The critical angle is defined by Rayleigh as the angle below which a surface is considered to be smooth, and above which a surface is considered to be rough; Denote the root mean square (rms) of a surface height irregularity as h_{rms}, then according to the Rayleigh criteria, the surface is considered to be smooth if

$$\frac{\{(4\pi h_{rms})\sin\psi_g\}}{\lambda} < \frac{\pi}{2} .$$ Eq. (9.2)

Consider a wave incident on a rough surface, as shown in Fig. 9.3. Due to surface height irregularity (surface roughness), the "rough path" is longer than the "smooth path" by a distance $2h_{rms}\sin\psi_g$. This path difference translates into a phase differential $\Delta\psi$:

$$\Delta\psi = \{(2\pi)\ 2h_{rms}\sin\psi_g\}/\lambda .$$ Eq. (9.3)

The critical angle ψ_{gc} is then computed when $\Delta\psi = \pi$ (first null), thus

$$\frac{(4\pi h_{rms})\ \sin\psi_{gc}}{\lambda} = \pi$$ Eq. (9.4)

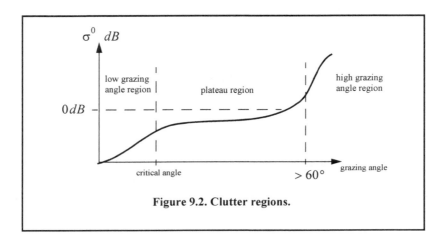

Figure 9.2. Clutter regions.

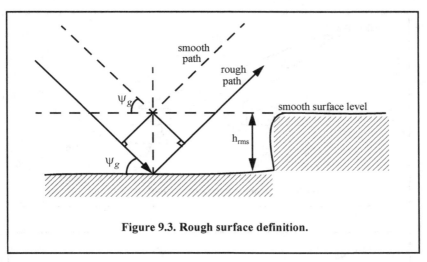

Figure 9.3. Rough surface definition.

or equivalently,

$$\psi_{gc} = \operatorname{asin}\frac{\lambda}{4h_{rms}}.$$

Eq. (9.5)

In the case of sea clutter, for example, the rms surface height irregularity is

$$h_{rms} \approx 0.025 + 0.046 \ S_{state}^{1.72}$$

Eq. (9.6)

where S_{state} is the sea state, which is tabulated in several cited references. The sea state is characterized by the wave height, period, length, particle velocity, and wind velocity. For example, $S_{state} = 3$ refers to a moderate sea state, where in this case the wave height is approximately between 0.9144 *to* 1.2192 *m*, the wave period 6.5 *to* 4.5 seconds, wave length 1.9812 *to* 33.528 *m*, wave velocity 20.372 *to* 25.928 *Km/hr*, and wind velocity 22.224 *to* 29.632 *Km/hr*.

Clutter at low grazing angles is often referred to as diffuse clutter, where there are a large number of clutter returns in the radar beam (noncoherent reflections). In the flat region the dependency of σ^0 on the grazing angle is minimal. Clutter in the high grazing angle region is more specular (coherent reflections) and the diffuse clutter components disappear. In this region the smooth surfaces have larger σ^0 than rough surfaces, the opposite of the low grazing angle region.

9.2.1. Radar Equation for Area Clutter - Airborne Radar

Consider an airborne radar in the look-down mode shown in Fig. 9.4. The intersection of the antenna beam with the ground defines an elliptically shaped footprint. The size of the footprint is a function of the grazing angle and the antenna $3dB$ beamwidth θ_{3dB}, as illustrated in Fig. 9.5. The footprint is divided into many ground range bins each of size $(c\tau/2)\sec\psi_g$, where τ is the pulse width.

From Fig. 9.5, the clutter area A_c is

$$A_c \approx R\theta_{3dB}\frac{c\tau}{2}\sec\psi_g.$$

Eq. (9.7)

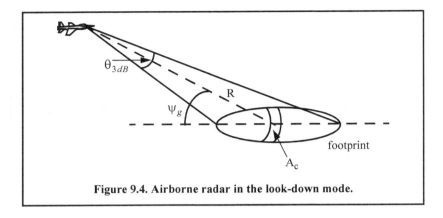

Figure 9.4. Airborne radar in the look-down mode.

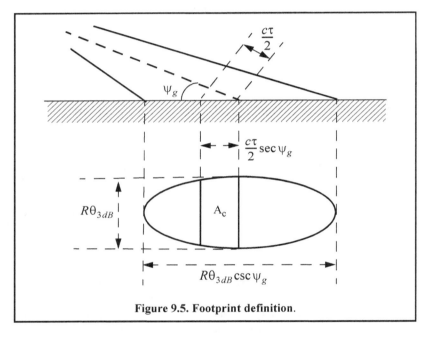

Figure 9.5. Footprint definition.

The power received by the radar from a scatterer within A_c is given by the radar equation as

$$S_t = \frac{P_t G^2 \lambda^2 \sigma_t}{(4\pi)^3 R^4}$$ Eq. (9.8)

where, as usual, P_t is the peak transmitted power, G is the antenna gain, λ is the wavelength, and σ_t is the target RCS. Similarly, the received power from clutter is

$$S_C = \frac{P_t G^2 \lambda^2 \sigma_c}{(4\pi)^3 R^4}$$ Eq. (9.9)

where the subscript C is used for area clutter. Substituting Eq. (9.1) for σ_c and Eq. (9.7) for A_c into Eq. (9.9), one can then obtain the SCR for area clutter by dividing Eq. (9.8) by Eq. (9.9). More precisely,

$$(SCR)_C = \frac{2\sigma_t \cos \psi_g}{\sigma^0 \theta_{3dB} Rc\tau} \, .$$ Eq. (9.10)

Example:

Consider an airborne radar shown in Fig. 9.4. Let the antenna $3dB$ beamwidth be $\theta_{3dB} = 0.02 rad$, the pulse width $\tau = 2\mu s$, range $R = 20 Km$, and grazing angle $\psi_g = 20°$. The target RCS is $\sigma_t = 1 m^2$. Assume that the clutter reflection coefficient is $\sigma^0 = 0.0136$. Compute the SCR.

Solution:

The SCR is given by Eq. (9.10) as

$$(SCR)_C = \frac{2\sigma_t \cos \psi_g}{\sigma^0 \theta_{3dB} Rc\tau} \Rightarrow$$

$$(SCR)_C = \frac{(2)(1)(\cos 20°)}{(0.0136)(0.02)(20000)(3 \times 10^8)(2 \times 10^{-6})} = 5.76 \times 10^{-4} \, .$$

It follows that

$$(SCR)_C = -32.4 dB \, .$$

Thus, for reliable detection, the radar must somehow increase its SCR by at least $(32 + X)dB$, where X is on the order of 13 to $15 dB$ or better.

9.3. Volume Clutter

Volume clutter has large extents and includes rain (weather), chaff, birds, and insects. The volume clutter coefficient is normally expressed in square meters (RCS per resolution volume). Birds, insects, and other flying particles are often referred to as angle clutter or biological clutter.

Weather or rain clutter can be suppressed by treating the rain droplets as perfect small spheres. We can use the Rayleigh approximation of a perfect sphere to estimate the rain droplets' RCS. The Rayleigh approximation, without regard to the propagation medium index of refraction is

$$\sigma = 9\pi r^2 (kr)^4 \qquad r \ll \lambda$$ Eq. (9.11)

where $k = 2\pi / \lambda$, and r is radius of a rain droplet.

Electromagnetic waves, when reflected from a perfect sphere, become strongly co-polarized (have the same polarization as the incident waves). Consequently, if the radar transmits, for example, a right-hand-circular (RHC) polarized wave, then the received waves are left-hand-circular (LHC) polarized because they are propagating in the opposite direction. Therefore, the back-scattered energy from rain droplets retains the same wave rotation (polarization) as the incident wave, but has a reversed direction of propagation. It follows that radars can suppress rain clutter by co-polarizing the radar transmit and receive antennas.

Denote σ_w as RCS per unit resolution volume V_w. It is computed as the sum of all individual scatterers RCS within the volume

$$\sigma_w = \sum_{i=1}^{N} \sigma_i \qquad \text{Eq. (9.12)}$$

where N is the total number of scatterers within the resolution volume. Thus, the total RCS of a single resolution volume is

$$\sigma_w = \sum_{i=1}^{N} \sigma_i V_w . \qquad \text{Eq. (9.13)}$$

A resolution volume is shown in Fig. 9.6 and is approximated by

$$V_w \approx \frac{\pi}{8} \theta_a \theta_e R^2 c\tau \qquad \text{Eq. (9.14)}$$

where θ_a and θ_e are, respectively, the antenna azimuth and elevation beamwidths in radians, τ is the pulse width in seconds, c is the speed of light, and R is range.

Consider a propagation medium with an index of refraction m. The *ith* rain droplet RCS approximation in this medium is

$$\sigma_i \approx \frac{\pi^5}{\lambda^4} K^2 D_i^6 \qquad \text{Eq. (9.15)}$$

where

$$K^2 = \left| \frac{m^2 - 1}{m^2 + 2} \right|^2 \qquad \text{Eq. (9.16)}$$

and D_i is the *ith* droplet diameter. For example, temperatures between $32°F$ and $68°F$ yield

$$\sigma_i \approx 0.93 \frac{\pi^5}{\lambda^4} D_i^6 . \qquad \text{Eq. (9.17)}$$

and for ice, Eq. (9.17) can be approximated by

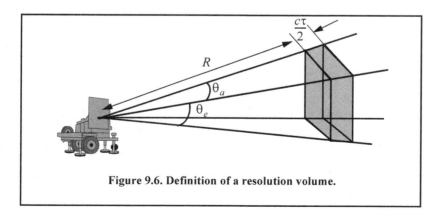

Figure 9.6. Definition of a resolution volume.

$$\sigma_i \approx 0.2\frac{\pi^5}{\lambda^4}D_i^6.$$ **Eq. (9.18)**

Substituting Eq. (9.18) into Eq. (9.13) yields

$$\sigma_w = \frac{\pi^5}{\lambda^4}K^2 Z$$ **Eq. (9.19)**

where the weather clutter backscatter coefficient Z is defined as

$$Z = \sum_{i=1}^{N} D_i^6.$$ **Eq. (9.20)**

In general, a rain droplet diameter is given in millimeters and the radar resolution volume is expressed in cubic meters; thus the units of Z are often expressed in $millimeter^6/m^3$.

9.3.1. Radar Equation for Volume Clutter

The radar equation gives the total power received by the radar from a σ_t target at range R as

$$S_t = \frac{P_t G^2 \lambda^2 \sigma_t}{(4\pi)^3 R^4}$$ **Eq. (9.21)**

where all parameters in Eq. (9.21) have been defined earlier. The weather clutter power received by the radar is

$$S_w = \frac{P_t G^2 \lambda^2 \sigma_w}{(4\pi)^3 R^4}.$$ **Eq. (9.22)**

It follows that

$$S_w = \frac{P_t G^2 \lambda^2}{(4\pi)^3 R^4} \frac{\pi}{8}R^2 \theta_a \theta_e c\tau \sum_{i=1}^{N} \sigma_i.$$ **Eq. (9.23)**

The SCR for weather clutter is then computed by dividing Eq. (9.21) by Eq. (9.23). More precisely,

$$(SCR)_V = \frac{S_t}{S_w} = \frac{(8\sigma_t)}{\left(\pi\theta_a \theta_e c\tau R^2 \sum_{i=1}^{N} \sigma_i\right)}$$ **Eq. (9.24)**

where the subscript V is used to denote volume clutter.

Example:

A certain radar has target RCS $\sigma_t = 0.1m^2$, pulse width $\tau = 0.2\mu s$, antenna beamwidth $\theta_a = \theta_e = 0.02 radians$. Assume the detection range to be $R = 50Km$, and compute the SCR if $\sum \sigma_i = 1.6 \times 10^{-8}(m^2/m^3)$.

Solution:

From Eq. (9.24) we have

$$(SCR)_V = \frac{8\sigma_t}{\pi\theta_a\theta_e c\tau R^2 \sum\limits_{N} \sigma_i}.$$

Substituting the proper values we get

$$(SCR)_V = \frac{(8)(0.1)}{\pi(0.02)^2(3\times10^8)(0.2\times10^{-6})(50\times10^3)^2(1.6\times10^{-8})} = 0.265$$

$$(SCR)_V = -5.76dB.$$

9.4. Surface Clutter RCS

9.4.1. Single Pulse - Low PRF Case

In this case, the received power from clutter is calculated using Eq. (9.9). However, the clutter RCS σ_c is now computed differently. It is

$$\sigma_c = \sigma_{MBc} + \sigma_{SLc} \qquad \qquad \textbf{Eq. (9.25)}$$

where σ_{MBc} is the main-beam clutter RCS and σ_{SLc} is the sidelobe clutter RCS, as illustrated in Fig. 9.7.

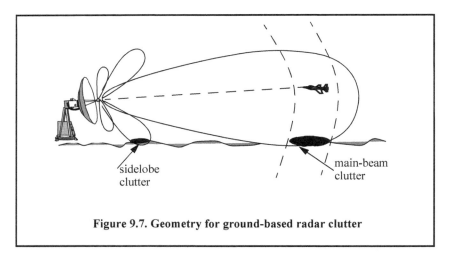

Figure 9.7. Geometry for ground-based radar clutter

In order to calculate the total clutter RCS given in Eq. (9.125), one must first compute the corresponding clutter areas for both the main beam and the sidelobes. For this purpose, consider the geometry shown in Fig. 9.8. The angles θ_A *and* θ_E represent the antenna 3*dB* azimuth and elevation beamwidths, respectively. The radar height (from the ground to the phase center of the antenna) is denoted by h_r, while the target height is denoted by h_t. The radar slant range is R, and its ground projection is R_g. The range resolution is ΔR and its ground projection is ΔR_g. The main beam clutter area is denoted by A_{MBc} and the sidelobe clutter area is denoted by A_{SLc}.

From Fig. 9.8, the following relations can be derived

$$\theta_r = \mathrm{asin}(h_r/R)$$

Eq. (9.26)

$$\theta_e = \mathrm{asin}((h_t - h_r)/R)$$

Eq. (9.27)

$$\Delta R_g = \Delta R \cos\theta_r$$

Eq. (9.28)

where ΔR is the radar range resolution. The slant range ground projection is

$$R_g = R\cos\theta_r.$$

Eq. (9.29)

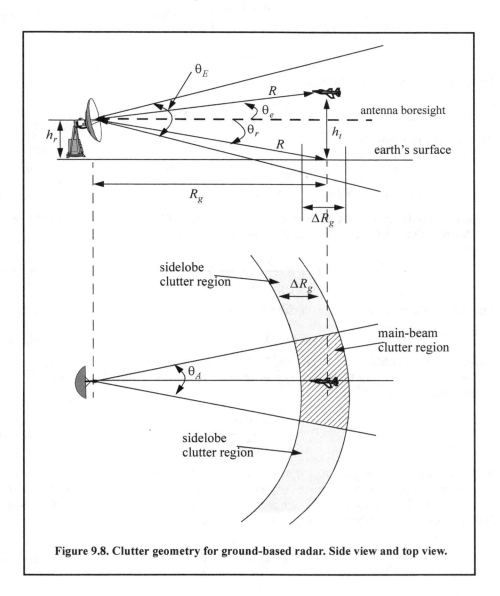

Figure 9.8. Clutter geometry for ground-based radar. Side view and top view.

It follows that the main beam and the sidelobe clutter areas are

$$A_{MBc} = \Delta R_g \ R_g \ \theta_A \qquad\qquad \text{Eq. (9.30)}$$

$$A_{SLc} = \Delta R_g \ \pi R_g. \qquad\qquad \text{Eq. (9.31)}$$

Assume a radar antenna beam $G(\theta)$ of the form

$$G(\theta) = \exp\left(-\frac{2.776\theta^2}{\theta_E^2}\right) \Rightarrow Gaussian \qquad\qquad \text{Eq. (9.32)}$$

$$G(\theta) = \left\{ \begin{array}{ll} \left(\dfrac{\sin\left(\dfrac{\theta}{\theta_E}\right)}{\left(\dfrac{\theta}{\theta_E}\right)}\right)^2 & ;|\theta| \le \dfrac{\pi\theta_E}{2.78} \\ \\ 0 & ;elsewhere \end{array} \right\} \Rightarrow \left(\frac{\sin(x)}{x}\right)^2. \qquad\qquad \text{Eq. (9.33)}$$

Then the main-beam clutter RCS is

$$\sigma_{MBc} = \sigma^0 A_{MBc} G^2(\theta_e + \theta_r) = \sigma^0 \Delta R_g \ R_g \ \theta_A G^2(\theta_e + \theta_r) \qquad\qquad \text{Eq. (9.34)}$$

and the sidelobe clutter RCS is

$$\sigma_{SLc} = \sigma^0 A_{SLc}(SL_{rms})^2 = \sigma^0 \Delta R_g \ \pi R_g (SL_{rms})^2 \qquad\qquad \text{Eq. (9.35)}$$

where the quantity SL_{rms} is the rms for the antenna sidelobe level.

Finally, in order to account for the variation of the clutter RCS versus range, one can calculate the total clutter RCS as a function of range. It is given by

$$\sigma_c(R) = \frac{\sigma_{MBc} + \sigma_{SLc}}{(1 + (R/R_h)^4)} \qquad\qquad \text{Eq. (9.36)}$$

where R_h is the radar range to the horizon calculated as

$$R_h = \sqrt{8 h_r r_o / 3} \qquad\qquad \text{Eq. (9.37)}$$

where r_o is the Earth's radius equal to $6375 Km$. The denominator in Eq. (9.36) is put in that format in order to account for refraction and for round (spherical) Earth effects. The radar SNR due to a target at range R is

$$SNR = \frac{P_t G^2 \lambda^2 \sigma_t}{(4\pi)^3 R^4 k T_o B F L} \qquad\qquad \text{Eq. (9.38)}$$

where, as usual, P_t is the peak transmitted power, G is the antenna gain, λ is the wavelength, σ_t is the target RCS, k is Boltzmann's constant, T_0 is the effective noise temperature, B is the radar operating bandwidth, F is the receiver noise figure, and L is the total radar losses. Similarly, the Clutter-to-Noise Ratio (CNR) at the radar is

$$CNR = \frac{P_t G^2 \lambda^2 \sigma_c}{(4\pi)^3 R^4 k T_o B F L} \qquad\qquad \text{Eq. (9.39)}$$

where the σ_c is calculated using Eq. (9.36).

When the clutter statistic is Gaussian, the clutter signal return and the noise return can be combined, and a new value for determining the radar measurement accuracy is derived from the Signal-to-Clutter+Noise Ratio, denoted by SIR. It is given by

$$SIR = \frac{SNR}{1 + CNR}.$$

Eq. (9.40)

Note that the *CNR* is computed from Eq. (9.439).

MATLAB Function "clutter_rcs.m"

The function *"clutter_rcs.m"* implements Eq. (9.36). It generates plots of the clutter RCS and the CNR versus the radar range. Its outputs include the clutter RCS in *dBsm* and the CNR in *dB*. The function *"clutter_rcs.m"* is listed in Appendix 9-A, and its syntax is as follows:

sigmac = clutter_rcs(sigma0,thetaE,thetaA,SL,range,hr,ht,b,ant_id)

where

Symbol	Description	Units	Status
sigma0	clutter back scatterer coefficient	dB	input
thetaE	antenna 3dB elevation beamwidth	degrees	input
thetaA	antenna 3dB azimuth beamwidth	degrees	input
SL	antenna sidelobe level	dB	input
range	range; can be a vector or a single value	Km	input
hr	radar height	meters	input
ht	target height	meters	input
b	bandwidth	Hz	input
ant_id	1 for (sin(x)/x)^2 pattern; 2 for Gaussian pattern	none	input
sigmac	clutter RCS; vector or single value depending on "range"	dB	output

As an example consider the case with the following parameters

clutter back scatterer coefficient	-20 dB
antenna 3dB elevation beamwidth	1.5 degrees
antenna 3dB azimuth beamwidth	2 degrees
antenna sidelobe level	-25 dB
radar height	3 meters
target height	150 meters
pulse width	1 micro sec
range	2 - 45Km
target RCS	-10 dBsm
radar center frequency	5 GHz

Figure 9.9a shows the clutter RCS versus range when a *sin(x)/x* antenna pattern is used, and Fig. 9.9b shows the resulting SNR, CNR, and SCR. Figure 9.10 is similar to Fig. 9.9, except in this case the antenna has a Gaussian shape. These plots can be reproduced using the MATLAB program *"Fig9_9_10.m,"* listed in Appendix 9-A.

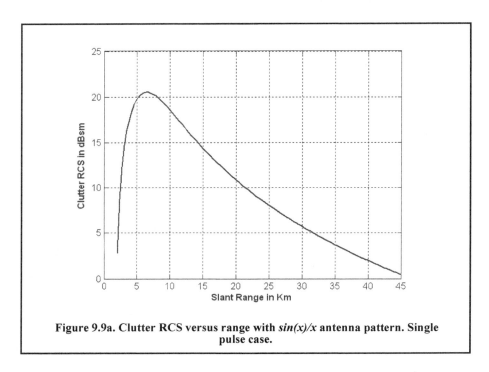

Figure 9.9a. Clutter RCS versus range with *sin(x)/x* antenna pattern. Single pulse case.

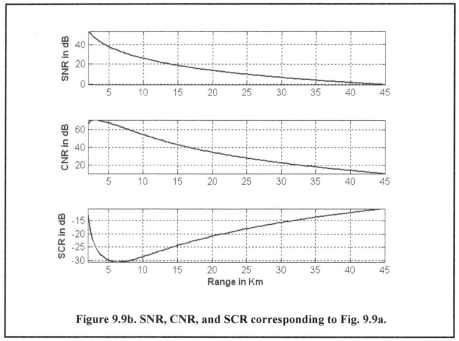

Figure 9.9b. SNR, CNR, and SCR corresponding to Fig. 9.9a.

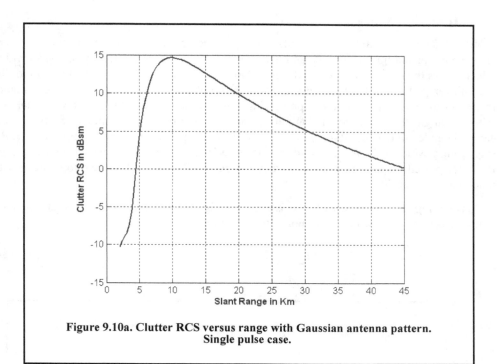

Figure 9.10a. Clutter RCS versus range with Gaussian antenna pattern. Single pulse case.

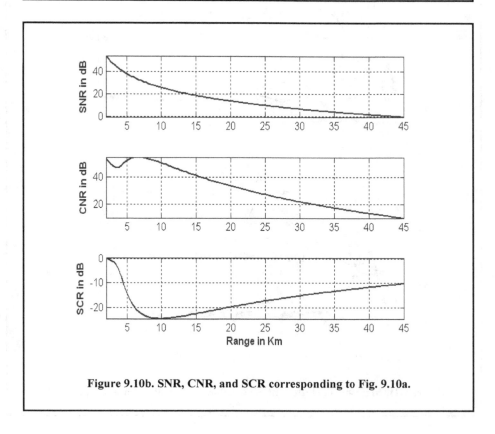

Figure 9.10b. SNR, CNR, and SCR corresponding to Fig. 9.10a.

9.4.2. High PRF Case

High PRFs are typically used by pulsed Doppler radars. Pulsed Doppler radars use a very short unmodulated train of pulses, and hence, range resolution is limited by the pulse width, which forces the radar to use extremely short duration pulses. High PRF radars make up for the loss of average transmitted power due to using short pulses by coherently processing a train of these pulses within one coherent processing interval (integration time or dwell interval). Although high PRF radars are ambiguous in range, they provide excellent capability to measuring Doppler frequency. Range ambiguity can be dealt with by using multiple PRF (PRF staggering), which will be addressed in a later section. One major drawback of using high PRFs (or pulsed Doppler radars) is the fact that pulsed Doppler radars have to contend with much more clutter than do low PRF radars.

Consider Fig. 9.11; the low PRF case is shown in Fig. 9.11a. In this case, the target is at maximum detection range, which corresponds to an unambiguous range

$$R_u = \frac{cT}{2} = \frac{c}{2f_r} \qquad\qquad \textbf{Eq. (9.41)}$$

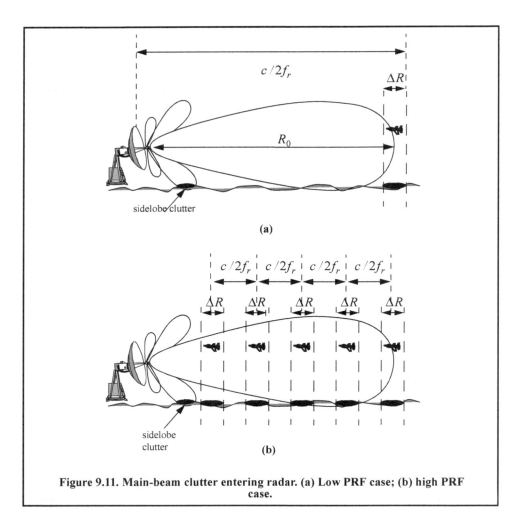

Figure 9.11. Main-beam clutter entering radar. (a) Low PRF case; (b) high PRF case.

where T is the pulse repetition interval and f_r is the radar PRF. The amount of clutter entering the radar through its main-beam corresponds only to the clutter patch located at the target's range. Alternatively, in Fig. 9.11b the high PRF case is depicted. In this case, the radar is range ambiguous and the amount of main-beam clutter entering the radar corresponds to many more clutter patches, as shown in Fig. 9.11b. Consequently, the amount of clutter competing with target detection is an order of magnitude larger than the case of low PRF. This is typically referred to as clutter folding.

Denote the clutter power entering the radar due to a single pulse for the target at range R_0 as P_{C_1}, then because of the high PRF operation, the total clutter power entering the radar is

$$P_{C_{folded}} = \sum_{n=0}^{N-1} P_{C_1} Rect\left(\frac{t-nT}{\tau_0}\right) \qquad \text{Eq. (9.42)}$$

where N is the number of pulses in one coherent processing interval (dwell), T is the PRI, and τ_0 is the pulse width. Note that since the radar receiver is shut off during transmission of a given pulse, Eq. (9.42) is computed only at delays (range) that correspond to

$$\{(nT + 2\tau_0) < t < (n+1)T - \tau_0; \ 0 \le n \le N-1\} \qquad \text{Eq. (9.43)}$$

where in this case, the transmitter is assumed to be shut off not only during the transmission of each pulse, but also for one pulse width before and after each transmission. Thus, one would expect the folded clutter RCS to not be continuous versus the range, but rather to exist over intervals of length T seconds with gaps that correspond to three times the pulse width. This is illustrated in the following few examples for both low and high PRF cases.

Figure 9.12 shows the SIR, SCR, CNR, and SNR for the high PRF using the same data used in Figs. 9.9 and 9.10. In this figure the antenna pattern has a *sin(x)/x* shape. Figure 9.13 is similar to Fig. 9.12, except in this case the antenna pattern is Gaussian. These plots can be reproduced using MATLAB program *"Fig9_12_13.m,"* listed in Appendix 9-A.

9.5. *Clutter Components*

It was established earlier that the complex envelope of the signal received by the radar comprise the target returns and additive band-limited white noise. In the presence of clutter, the complex envelope is now composed of target, noise, and clutter returns. That is,

$$\tilde{x}(t) = \tilde{s}(t) + \tilde{n}(t) + \tilde{w}(t) \qquad \text{Eq. (9.44)}$$

where $\tilde{s}(t)$, $\tilde{n}(t)$, and $\tilde{w}(t)$ are, respectively, the target, noise, and clutter complex envelope echoes. Noise is typically modeled (as discussed in earlier chapters) as a bandlimited white Gaussian random process. Furthermore, noise samples are considered statistically independent of each other and of clutter measurements.

Clutter arises from reflections of unwanted objects within the radar beam. Since many objects comprise the clutter returns, clutter may also be molded as a Gaussian random process. In other words, clutter samples from one radar measurement to another constitute a joint set of Gaussian random variables. However, because of the clutter fluctuation and due to antenna mechanical scanning, wind speed, and radar platform motion (if applicable), these random variables are not statistically independent.

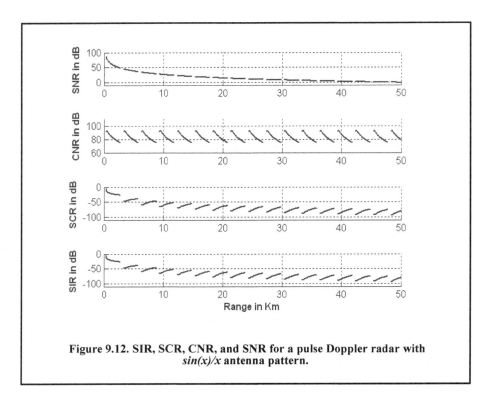

Figure 9.12. SIR, SCR, CNR, and SNR for a pulse Doppler radar with
sin(x)/x **antenna pattern.**

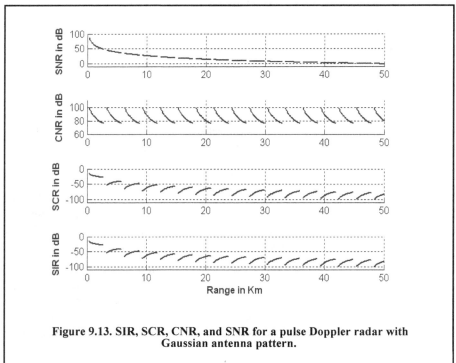

Figure 9.13. SIR, SCR, CNR, and SNR for a pulse Doppler radar with
Gaussian antenna pattern.

More precisely, because of the antenna mechanical scanning, clutter returns in the radar main-beam do not have the same amplitude from pulse to pulse. This will effectively add amplitude modulation to the clutter returns. This additional modulation is governed by the shape of the antenna pattern, the rate of mechanical scanning, and the radar PRF. Denote the antenna two-way azimuth $3\,dB$ beamwidth as θ_a and the antenna scan rate as $\dot{\theta}_{scan}$. It follows that the contribution of antenna scanning to the standard deviation of the clutter fluctuation is

$$\sigma_s = 0.399 \frac{\dot{\theta}_{scan}}{\theta_a}. \qquad \text{Eq. (9.45)}$$

Another contributor to the clutter spectral spreading is caused by motion of the clutter itself, due to wind. Trees, vegetation, and sea waves are the main contributors to this effect. This relative motion, although relatively small, introduces additional Doppler shift in the clutter returns. Earlier, it was established that Doppler frequency due to a relative velocity v is given by

$$f_d = 2v/\lambda \qquad \text{Eq. (9.46)}$$

where λ is the radar operating wavelength. It follows that if the apparent rms velocity due to wind is v_{rms}, then the standard deviation is

$$\sigma_w = 2v_{rms}/\lambda. \qquad \text{Eq. (9.47)}$$

Finally, if the radar platform is in motion, then the relative motion between the platform and the stationary clutter will cause a Doppler shift given by

$$f_c = (2v_{radar}\cos\theta)/\lambda \qquad \text{Eq. (9.48)}$$

where $v_{radar}\cos\theta$ is the radial velocity component of the platform in the direction of clutter. Since the radar beam has a finite width, not all clutter components have the same radial velocity at all times. More specifically, if the angles θ_1 and θ_2 represent the edges of the radar beam, then Eq. (9.48) can be written as

$$f_c = \frac{2v_{radar}}{\lambda}(\cos\theta_2 - \cos\theta_1) \approx \frac{2v_{radar}}{\lambda}\theta_a\sin\theta \qquad \text{Eq. (9.49)}$$

and the standard deviation due to platform motion is given by

$$\sigma_v = \frac{v_{radar}}{\lambda}\sin\theta. \qquad \text{Eq. (9.50)}$$

Finally, the overall clutter spreading is denoted by σ_f, where

$$\sigma_f^2 = \sigma_v^2 + \sigma_s^2 + \sigma_w^2. \qquad \text{Eq. (9.51)}$$

9.6. Clutter Backscatter Coefficient Statistical Models

Assessing radar performance in the presence of clutter depends heavily on one's ability to accurately estimate or measure the backscatter coefficient σ°. Since clutter within a resolution or volume cell is composed of a large number of scatterers with random phases and amplitudes, the backscatter coefficient is typically described statistically by a probability distribution function. The type of distribution depends on the nature of clutter itself (sea, land, volume), the radar operating frequency, and the grazing angle.

9.6.1. Surface Clutter Case

The most common statistical model used to describe $\sigma°$ for surface clutter is the log-normal and exponential (i.e., Rayleigh amplitude) probability density functions. Although the log-normal distribution will provide an accurate measure of $\sigma°$ at large grazing angles, it is not as accurate at low grazing angles less than 5-7 degrees. In this case, the Rayleigh distribution (which is a special case of the Weibull distribution) provides more accurate statistical estimates of $\sigma°$. Another probability density function widely used to estimate $\sigma°$ is the Weibull distribution.

The Weibull probability density function can be written as

$$f(\sigma°) = \frac{b(\sigma°)^{b-1}}{\alpha} \exp\left(-\frac{(\sigma°)^b}{\alpha}\right); \quad \sigma° \geq 0 \qquad \text{Eq. (9.52)}$$

where b, α are the Weibull distribution parameters. Define the Weibull distribution slope a as $1/b$, and the parameter α as

$$\alpha = \frac{(\sigma_m^o)^b}{\ln 2} \qquad \text{Eq. (9.53)}$$

where σ_m^o is the median value for σ^o. The proof of Eq. (9.53) is left as an exercise (see Problem 9.6) It follows that Eq. (9.52) can be written as

$$f(\sigma°) = \frac{\ln 2 \ (\sigma°)^{\frac{1}{a}-1}}{a(\sigma_m^o)^{1/a}} \exp\left(-\ln 2\left(\frac{\sigma°}{\sigma_m^o}\right)^{1/a}\right); \quad \sigma° \geq 0. \qquad \text{Eq. (9.54)}$$

Note that when $b=1$, then Eq. (9.52) becomes the exponential (or Rayleigh amplitude) probability density function,

$$f(\sigma°) = \frac{1}{\overline{\sigma°}} \exp\left(-\frac{\sigma°}{\overline{\sigma°}}\right); \quad \sigma° \geq 0 \qquad \text{Eq. (9.55)}$$

where $\alpha = \overline{\sigma°}$ is the average value for σ^o.

The mean value for σ^o can be determined from the integral

$$\overline{\sigma°} = \int_0^\infty \sigma° f(\sigma°) d\sigma° \qquad \text{Eq. (9.56)}$$

by making the change of variable $q = \sigma°^b/\alpha$, and by using $a = 1/b$ yields

$$\overline{\sigma°} = \alpha^a \int_0^\infty q^a e^{-q} dq, \qquad \text{Eq. (9.57)}$$

which is the incomplete Gamma integral. More precisely,

$$\overline{\sigma^\circ} = \alpha^a \Gamma(1 + a).$$ **Eq. (9.58)**

The probability that an actual clutter radar cross section per unit area will not exceed the value σ^o is

$$Pr(\sigma_c^o \le \sigma^o) = \int_0^{\sigma^o} f(\sigma^\circ) d\sigma^\circ.$$ **Eq. (9.59)**

Substituting Eq. (9.52) into Eq. (9.59) and performing the integration yields:

$$Pr(\sigma_c^o \le \sigma^o) = 1 - \exp\left(\frac{\sigma^{o^b}}{\alpha}\right)$$ **Eq. (9.60)**

Eq. (9.60) can now be used to solve for σ^o, that is

$$\sigma^o = \left[\alpha \ln\left(\frac{1}{1 - Pr(\sigma_c^o \le \sigma^o)}\right)\right]^{1/b} = \left[\alpha \ln\left(\frac{1}{1 - Pr(\sigma_c^o \le \sigma^o)}\right)\right]^a.$$ **Eq. (9.61)**

The median value for σ^o is compute be setting $Pr(\sigma_c^o \le \sigma^o) = 0.5$ in Eq. (9.61). In this case,

$$\sigma_m^o = [\alpha \ln 2]^a.$$ **Eq. (9.62)**

Using Eqs. (9.53) and (9.62) into Eq. (9.60) yields

$$Pr(\sigma_c^o \le \sigma^o) = 1 - \exp\left(\ln 2 \left(\frac{\sigma^o}{\sigma_m^o}\right)^b\right).$$ **Eq. (9.63)**

To obtain a simpler formula for σ^o in decibels, substitute Eq. (9.53) into Eq. (9.61) to get

$$\sigma^o\big|_{dB} = 10 \log \sigma_m^o - 10 a \log(\ln 2) + 10 a \log\left(\ln\left(\frac{1}{1 - Pr(\sigma_c^o \le \sigma^o)}\right)\right),$$ **Eq. (9.64)**

which can be rewritten as

$$\sigma^o\big|_{dB} = \sigma_m^o\big|_{dB} - 10 \log(\Gamma(1 + a)) + 10 a \log\{-\ln[1 - Pr(\sigma_c^o \le \sigma^o)]\}.$$ **Eq. (9.65)**

Figure 9.14 shows some typical plots for σ^o against the probability defined in Eq. (9.60). Note that only values where $Pr > 0.2$ *and* $Pr < 0.9$ are used because values of σ^o corresponding to very low probabilities are typically below the radar's noise level. Alternatively, values for σ^o corresponding to high probabilities are typically too high for an MTI radar to suppress (the next chapter addresses MTI radars in details). Figure 9.14 can be reproduced using MATLAB program *"Fig9_14.m,"* listed in Appendix 9-A.

9.6.2. Volume Clutter Case

The backscatter coefficient, Z, defined in Eq. (9.20) of Section 9.3, is often used by meteorologists and less often by radar engineers. In radar applications, it is more meaningful to use a precipitation backscatter coefficient that is measured in squared meters per cubic meter instead of $millimeter^6/m^3$. For this purpose, define a new precipitation backscatter coefficient η as

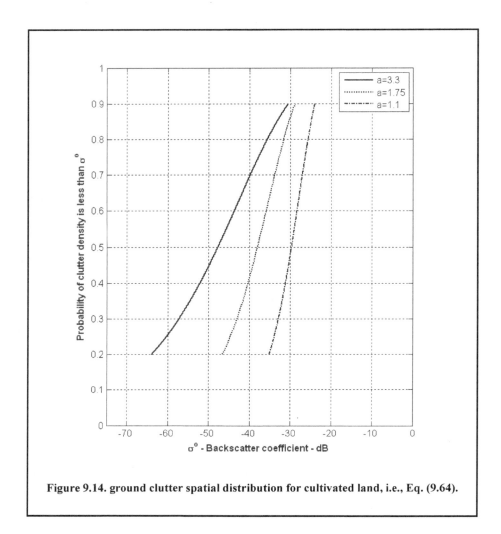

Figure 9.14. ground clutter spatial distribution for cultivated land, i.e., Eq. (9.64).

$$\eta = \frac{5.63 \times 10^{-14} \ r^{1.6}}{\lambda} \quad m^2/m^3 \qquad \text{Eq. (9.66)}$$

where r is the rate of precipitation in *millimeter/hr* and λ is the radar operating wavelength in *meters*.

The value of the exponent in Eq. (9.66) varies from 0.95 at tropical latitudes and frequencies above $10GHz$ to about 1.6, which is more applicable to temperate latitudes. Additionally, radar waves using circular polarization and wavelengths comparable to the rain droplets' average diameter will result in less backscattering than is the case for linearly polarized waves. To explain this observation further, consider a right circular polarized radar whose wavelength is comparable to the average rain droplet diameter. The reflected waves from the rain droplets will also be right circularly polarized waves but traveling in the opposite direction (i.e., from the point view of the radar they will be left circularly polarized). Therefore, most of the reflected energy will be denied entry into the radar receiver by its antenna, resulting in less backscatter energy in the radar signal and data processors. The average ratio of a circularly polarized to a linearly polarized backscatter coefficient is

$$\frac{\eta_{cp}}{\eta_{lp}}\bigg|_{dB} \approx \begin{cases} -15 & for \quad rain \\ -10 & for \quad bright \quad land \end{cases}$$

Eq. (9.67)

where bright land is defined as the transitional region between ice or snow and water resulting from melting.

Problems

9.1. Compute the signal-to-clutter ratio (SCR) for the radar described in Section 9.2.1. In this case, assume antenna $3dB$ beam width $\theta_{3dB} = 0.03rad$, pulse width $\tau = 10\mu s$, range $R = 50Km$, grazing angle $\psi_g = 15°$, target RCS $\sigma_t = 0.1m^2$, and clutter reflection coefficient $\sigma^0 = 0.02(m^2/m^2)$.

9.2. Repeat the example of Section 9.3 for target RCS $\sigma_t = 0.15m^2$, pulse width $\tau = 0.1\mu s$, antenna beam width $\theta_a = \theta_e = 0.03radians$; the detection range is $R = 100Km$, and $\sum \sigma_i = 1.6 \times 10^{-9}(m^2/m^3)$.

9.3. The quadrature components of the clutter power spectrum are, respectively, given by

$$\bar{S}_I(f) = \delta(f) + \frac{C}{\sqrt{2\pi}\sigma_c}\exp(-f^2/2\sigma_c^2)$$

$$\bar{S}_Q(f) = \frac{C}{\sqrt{2\pi}\sigma_c}\exp(-f^2/2\sigma_c^2).$$

Compute the D.C. and A.C. power of the clutter. Let $\sigma_c = 10Hz$.

9.4. A certain radar has the following specifications: pulse width $\tau' = 1\mu s$, antenna beam width $\Omega = 1.5°$, and wavelength $\lambda = 3cm$. The radar antenna is $7.5m$ high. A certain target is simulated by two point targets (scatterers). The first scatterer is $4m$ high and has RCS $\sigma_1 = 20m^2$. The second scatterer is $12m$ high and has RCS $\sigma_2 = 1m^2$. If the target is detected at $10Km$, compute (a) the SCR when both scatterers are observed by the radar; (b) the SCR when only the first scatterer is observed by the radar. Assume a reflection coefficient of -1, and $\sigma^0 = -30dB$.

9.5. A certain radar has range resolution of $300m$ and is observing a target somewhere in a line of high towers each having RCS $\sigma_{tower} = 10^6 m^2$. If the target has RCS $\sigma_t = 1m^2$, (a) how much signal-to-clutter ratio should the radar have? (b) Repeat part (a) for range resolution of $30m$.

9.6. Prove that the Weibull distribution α is given by $\alpha = \frac{(\sigma_m^0)^b}{\ln 2}$ where σ_m^0 is the median value for σ^0.

Appendix 9-A: Chapter 9 MATLAB Code Listings

The MATLAB code provided in this chapter was designed as an academic standalone tool and is not adequate for other purposes. The code was written in a way to assist the reader in gaining a better understanding of the theory. The code was not developed, nor is it intended to be used as part of an open-loop or a closed-loop simulation of any kind. The MATLAB code found in this textbook can be downloaded from this book's web page on the CRC Press website. Simply use your favorite web browser, go to *www.crcpress.com*, and search for keyword *"Mahafza"* to locate this book's web page.

MATLAB Function "clutter_rcs.m" Listing

```
function [sigmaC] = clutter_rcs(sigma0, thetaE, thetaA, SL, range, hr, ht, b,ant_id)
% This unction calculates the clutter RCS and the CNR for a ground based radar.
%% Inputs
  % sigma0     == clutter back scatterer coefficient   dB
  % thetaE     == antenna 3dB elevation beamwidth  degrees
  % thetaA     == antenna 3dB azimuth beamwidth    degrees
  % SL         == antenna sidelobe level  dB
  % range      == range; can be a vector or a single value Km
  % hr         == radar height meters
  % ht         == target height    meters
  % b          == bandwidth   Hz
  % ant_id     == 1 for (sin(x)/x)^2 pattern; 2 for Gaussian pattern
%% Outputs
  % sigmac     == clutter RCS; vector or single value depending on "range" dB
%
thetaA = thetaA * pi /180; % antenna azimuth beamwidth in radians
thetaE = thetaE * pi /180.; % antenna elevation beamwidth in radians
re = 6371000; % earth radius in meter
rh = sqrt(8.0*hr*re/3.); % range to horizon in meters
SLv = 10.0^(SL/10); % radar rms sidelobes in volts
sigma0v = 10.0^(sigma0/10); % clutter backscatter coefficient
deltar = 3e8 / 2 / b; % range resolution for unmodulated pulse
range_m = 1000 .* range;  % range in meters
thetar = asin(hr ./ range_m);
thetae = asin((ht-hr) ./ range_m);
propag_atten = 1. + ((range_m ./ rh).^4); % propagation attenuation due to round earth
Rg = range_m .* cos(thetar);
deltaRg = deltar .* cos(thetar);
theta_sum = thetae + thetar;
% use sinc^2 antenna pattern when ant_id=1
% use Gaussian antenna pattern when ant_id=2
if(ant_id ==1) % use sinc^2 antenna pattern
   ant_arg = (theta_sum ) ./ (pi*thetaE);
   gain = (sinc(ant_arg)).^2;
else
   gain = exp(-2.776 .*(theta_sum./thetaE).^2);
end
% compute sigmac
sigmac = (sigma0v .* Rg .* deltaRg) .* (pi * SLv * SLv + thetaA .* gain.^2) ./ propag_atten;
sigmaC = 10*log10(sigmac);
figure
```

```
plot(range, sigmaC,'linewidth',1.5)
grid
xlabel('\bfSlant Range in Km')
ylabel('\bfClutter RCS in dBsm')
```

MATLAB Program "Fig9_9_10.m" Listing

```
% Use this code to generate Fig. 9.9 or Fig 9.10 of text
clc
clear all
close all
k = 1.38e-23; % Boltzman's constant
pt = 45e3;
theta_AZ = 1.5;
theta_EL = 2;
F = 6;
L = 10;
tau = 1e-6;
B = 1/tau;
sigmmat = -10;
sigmma0 = -20;
SL = -25;
hr = 3;
ht = 150;
f0 = 5e9;
lambda = 3e8/f0;
range = linspace(2,45, 120);
[sigmmaC] = clutter_rcs(sigmma0, theta_EL, theta_AZ, SL, range, hr, ht, B,1);
sigmmaC = 10.^(sigmmaC./10);
range_m = 1000 .* range;
F = 10.^(F/10); % noise figure is 6 dB
T0 = 290; % noise temperature 290K
g = 26000 /theta_AZ /theta_EL; % antenna gain
Lt = 10.^(L/10); % total radar losses 13 dB
sigmmat = 10^(sigmmat/10)
CNR = pt*g*g*lambda^2 .* sigmmaC ./ ((4*pi)^3 .* (range_m).^4 .* k*T0*F*Lt*B); % CNR
SNR = pt*g*g*lambda^2 .* sigmmat ./ ((4*pi)^3 .* (range_m).^4 .* k*T0*F*L*B); % SNR
SCR = SNR ./ CNR; % Signal to clutter ratio
SIR = SNR ./ (1+CNR); % Signal to interference ratio
figure(2)
subplot(3,1,1)
plot(range,10*log10(SNR),'linewidth',1.5);
ylabel('\bfSNR in dB');
grid on;
axis tight
subplot(3,1,2)
plot(range,10*log10(CNR),'linewidth',1.5);
ylabel('\bfCNR in dB');
grid on;
axis tight
subplot(3,1,3)
plot(range,10*log10(SCR),'linewidth',1.5);
ylabel('\bfSCR in dB') ;
grid on;
```

axis tight
xlabel('\bfRange in Km')

MATLAB Program "Fig9_12_13.m" Listing

```
% Use this code to generate Fig. 9.12 or 9. 13 of text
clear all
close all
k = 1.38e-23; % Boltzmann's constant
T0 = 290; % degrees Kelvin
ant_id = 2; % use 1 for sin(x)/x antenna pattern and use 2 for Gaussian pattern
theta_ref = 0.75; % reference angle of radar antenna in degrees
re = 6371000 * 4 /3; %4 3rd earth radius in Km
c = 3e8; % speed of light
theta_EL = 1.5; % Antenna elevation beamwidth in degrees
theta_AZ = 2; % Antenna azimuth beamwidth in degrees
SL_dB = -25; % Antenna RMS sidelobe level
hr = 3; % Radar antenna hieght in meters
ht = 150; % Target hieght in meters
Sigmmat = -10; % Target RCS in dB
Sigmma0 = -20; % Clutter backscatter coefficient
P = 45e3; % Radar peak power in Watts
tau = 1e-6; % Pulse width (unmodulated)
fr = 50e3; % PRF in Hz
f0 = 5e9; % Radar center frequency
F = 6; % Noise figure in dB
L = 10; % Radar losses in dB
lambda = c /f0;
SL = 10^(SL_dB/10);
sigmma0 = 10^(Sigmma0/10);
F = 10^(F/10);
L = L^(L/10);
sigmmat = 10^(Sigmmat/10);
T = 1/fr; % PRI
B = 1/tau; % Bandwidth
delr = c * tau /2; % Range resolution;
Rh = sqrt(2*re*hr); % Range to Horizon
R1 = [2*delr:delr:c/2*(T-tau)];
Rclut = sqrt(R1.^2 + hr^2); % Range to clutter patches
G = 26000 /theta_EL /theta_AZ; %Antenna gain
for j = 0:40
    Rtgt = [c/2*(j*T+2*tau):delr:c/2*((j+1)*T-tau)];
    thetaR = asin(hr./Rclut); % Elevation angle from radar to clutter patch where traget is present
    thetae = theta_ref *pi/180;
    d = Rclut .* cos(thetaR); % Ground range to center of clutter at range Rclut
    del_d = delr .* cos(thetaR);
    % claculte clutter RCS
    theta_sum = thetaR+thetae;
    if(ant_id ==1) % use sinc^2 antenna pattern
        ant_arg = ( theta_sum ) ./ (pi*theta_EL/180);
        gain = (sinc(ant_arg)).^2;
    else
        gain = exp(-2.776 .*(theta_sum./(pi*theta_EL/180)).^2);
    end
```

```
% clutter RCS
sigmmac = (pi*SL^2+(theta_AZ*pi/180).*gain.*sigmma0.*d.*del_d) ./ (1+(Rclut/Rh).^4);
CNR = P*G*G*lambda^2 .* sigmmac ./ ((4*pi)^3 .* Rclut.^4 .* k*T0*F*L*B); % CNR
SNR = P*G*G*lambda^2 .* sigmmat ./ ((4*pi)^3 .* Rtgt.^4 .* k*T0*F*L*B); % SNR
SCR = SNR ./ CNR; % Signal to clutter ratio
SIR = SNR ./ (1+CNR); % Signal to interfernce ratio
figure(2)
  subplot(4,1,1), hold on
  plot(Rtgt/1000,10*log10(SNR),'linewidth',1.5);
  ylabel('\bfSNR in dB');
grid on
  subplot(4,1,2), hold on
  plot(Rtgt/1000,10*log10(CNR),'linewidth',1.5);
  ylabel('\bfCNR in dB');
grid on
  subplot(4,1,3), hold on
  plot(Rtgt/1000,10*log10(SCR),'linewidth',1.5);
  ylabel('\bfSCR in dB') ;
grid on
  subplot(4,1,4), hold on
  plot(Rtgt/1000,10*log10(SIR),'linewidth',1.5);
  xlabel('\bfRange in Km')
  ylabel('\bfSIR in dB');
grid on
 end
subplot(4,1,1)
axis([0 50 -10 100])
subplot(4,1,2)
axis([0 50 60 110]);
subplot(4,1,3)
axis([0 50 -110 0])
subplot(4,1,4)
axis([0 50 -110 0])
```

MATLAB Program "Fig9_14.m" Listing

```
% reproduce Fig 9.14 of text
clc
clear all
close all
P = linspace(.001,.999,10000);
sigmam = -47.75;
a =3.3
sigmao = sigmam + 1.5917 * a + 10 * a .* log10(-log((1-P)));
figure
index = find (P >=.2 & P <=.9);
plot(sigmao(index),P(index),'k','linewidth',1.5)
hold on
sigmam = -38.;
a =1.75
sigmao = sigmam + 1.5917 * a + 10 * a .* log10(-log((1-P)));
index = find (P >=.20 & P <=.9);
plot(sigmao(index),P(index),'k:','linewidth',1.5)
sigmam = -29.8;
```

```
a =1.1
sigmao = sigmam + 1.6917 * a + 10 * a .* log10(-log((1-P)));
index = find (P >=.2 & P <=.9);
plot(sigmao(index),P(index),'k-.','linewidth',1.5);
hold off
axis([-75 0 0 1])
legend('a=3.3','a=1.75','a=1.1')
xlabel('\bf\sigma^o - Backscatter coefficient - dB')
ylabel('\bf Probability of clutter density is less than \sigma^o')
grid on
```

Chapter 10

Moving Target Indicator (MTI) and Pulsed Doppler Radars

10.1. Clutter Power Spectrum Density

Clutter primarily comprises unwanted stationary ground reflections with limited relative motion with respect to the radar. Therefore, its power spectrum density will be concentrated around $f = 0$. However, because the overall clutter spreading σ_f (derived in Chapter 9 and repeated here as Eq. (10.1) for convenience) is not always zero, clutter actually exhibits some Doppler frequency spread. The overall clutter spreading is denoted by σ_f, and is given by

$$\sigma_f^2 = \sigma_v^2 + \sigma_s^2 + \sigma_w^2 . \qquad \text{Eq. (10.1)}$$

σ_v accounts for clutter spread due to platform motion, σ_s accounts for the antenna scan rate, and σ_w accounts for the clutter spread due to wind.

The clutter power spectrum can be written as the sum of fixed (stationary) and random (due to frequency spreading) components, as

$$S_c(f) = \frac{P_c}{T\sigma_f\sqrt{2\pi}} \sum_{k=-\infty}^{\infty} \exp\left(-\frac{(f-k/T)^2}{2\sigma_f^2}\right) \qquad \text{Eq. (10.2)}$$

where T is the PRI (i.e., $1/f_r$, f_r is the PRF), P_c is the clutter power or clutter mean square value, and σ_f is the clutter spectral spreading parameter as defined in Eq. (10.1). As clearly indicated by Eq. (10.2), the clutter PSD is periodic with period equal to f_r. Furthermore, the clutter PSD extends about each multiple integer of the PRF. It must be noted that this spread is relatively small and thus the relation $\sigma_f \ll f_r$ is always true. This is illustrated in Fig. 10.1. The mean square value can be calculated from

$$P_c = T \int_{-f_r/2}^{f_r/2} S_c(f)df . \qquad \text{Eq. (10.3)}$$

Let $S_{c0}(f)$ denote the central portion of Eq. (10.2) (i.e., $k = 0$); then P_c is be expressed by

$$P_c = T \int_{-\infty}^{\infty} S_{c0}(f)df \qquad \text{Eq. (10.4)}$$

where $S_{c0}(f)$ is a Gaussian shape function given by

$$S_{c0}(f) = \frac{k}{\sigma_f \sqrt{2\pi}} \exp\left(-\frac{f^2}{2\sigma_f^2}\right)$$ **Eq. (10.5)**

and $k = P_c/T$.

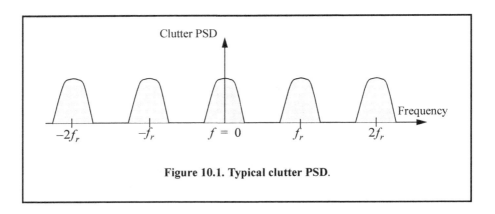

Figure 10.1. Typical clutter PSD.

10.2. Concept of a Moving Target Indicator (MTI)

The clutter spectrum is concentrated around DC ($f = 0$) and multiple integers of the radar PRF f_r, as was illustrated in Fig. 10.1. In CW radars, clutter is avoided or suppressed by ignoring the receiver output around DC, since most of the clutter power is concentrated about the zero frequency band. Pulsed radar systems may utilize special filters that can distinguish between slow-moving or stationary targets and fast-moving ones. This class of filter is known as the Moving Target Indicator (MTI). In simple words, the purpose of an MTI filter is to suppress target-like returns produced by clutter and allow returns from moving targets to pass through with little or no degradation. In order to effectively suppress clutter returns, an MTI filter needs to have a deep stopband at DC and at integer multiples of the PRF. Figure 10.2b shows a typical sketch of an MTI filter response, while Fig. 10.2c shows its output when the PSD shown in Fig. 10.2a is the input.

MTI filters can be implemented using delay line cancelers. As we will show later in this chapter, the frequency response of this class of MTI filter is periodic, with nulls at integer multiples of the PRF. Thus, targets with Doppler frequencies equal to nf_r are severely attenuated. Since Doppler is proportional to target velocity ($f_d = 2v/\lambda$), target speeds that produce Doppler frequencies equal to integer multiples of f_r are known as blind speeds. More precisely,

$$v_{blind} = (n\lambda f_r)/2; \quad n \geq 0.$$ **Eq. (10.6)**

Radar systems can minimize the occurrence of blind speeds either by employing multiple PRF schemes (PRF staggering) or by using high PRFs in which the radar may become range ambiguous. The main difference between PRF staggering and PRF agility is that the pulse repetition interval (within an integration interval) can be changed between consecutive pulses for the case of PRF staggering.

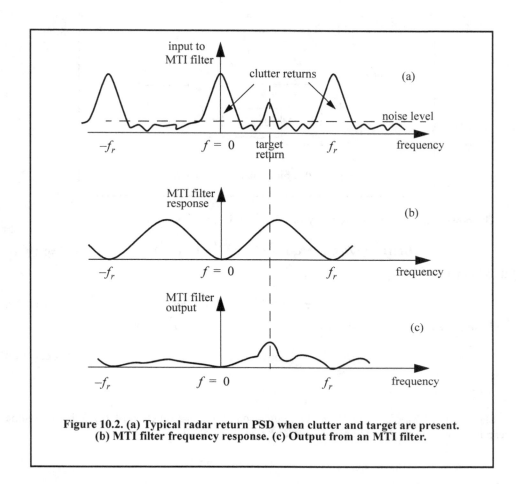

Figure 10.2. (a) Typical radar return PSD when clutter and target are present. (b) MTI filter frequency response. (c) Output from an MTI filter.

10.2.1. Single Delay Line Canceler

A single delay line canceler can be implemented as shown in Fig. 10.3. The canceler's impulse response is denoted as $h(t)$. The output $y(t)$ is equal to the convolution between the impulse response $h(t)$ and the input $x(t)$. The single delay canceler is often called a two-pulse canceler since it requires two distinct input pulses before an output can be read.

The delay T is equal to the radar PRI ($1 / f_r$). The output signal $y(t)$ is

$$y(t) = x(t) - x(t - T).$$

Eq. (10.7)

The impulse response of the canceler is given by

$$h(t) = \delta(t) - \delta(t - T)$$

Eq. (10.8)

where $\delta(\ )$ is the delta function. It follows that the Fourier transform (FT) of $h(t)$ is

$$H(\omega) = 1 - e^{-j\omega T}$$

Eq. (10.9)

where $\omega = 2\pi f$. In the z-domain, the single delay line canceler response is

$$H(z) = 1 - z^{-1}.$$

Eq. (10.10)

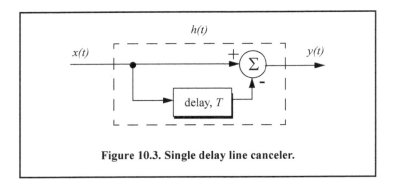

Figure 10.3. Single delay line canceler.

The power gain for the single delay line canceler is given by

$$|H(\omega)|^2 = H(\omega)H^*(\omega) = (1 - e^{-j\omega T})(1 - e^{j\omega T}).$$
    Eq. (10.11)

It follows that

$$|H(\omega)|^2 = 1 + 1 - (e^{j\omega T} + e^{-j\omega T}) = 2(1 - \cos\omega T)$$
    Eq. (10.12)

and using the trigonometric identity $(2 - 2\cos 2\vartheta) = 4(\sin\vartheta)^2$ yields

$$|H(\omega)|^2 = 4(\sin(\omega T/2))^2.$$
    Eq. (10.13)

MATLAB Function "single_canceler.m"

The function *"single_canceler.m"* computes and plots (as a function of f/f_r) the amplitude response for a single delay line canceler. The syntax is as follows:

[resp] = single_canceler (fofr)

where *"fofr"* is the number of periods desired. It is Listed in Appendix 10-A.

The amplitude frequency response for a single delay line canceller is shown in Fig. 10.4. Clearly, the frequency response of a single canceler is periodic with a period equal to f_r. The peaks occur at $f = (2n + 1)/(2f_r)$, and the nulls are at $f = nf_r$, where $n \geq 0$. In most radar applications the response of a single canceler is not acceptable since it does not have a wide notch in the stopband. A double delay line canceler has better response in both the stop- and passbands, and thus it is more frequently used than a single canceler. In this book, we will use the names *single delay line canceler* and *single canceler* interchangeably.

10.2.2. Double Delay Line Canceler

Two basic configurations of a double delay line canceler are shown in Fig. 10.5. Double cancelers are often called three-pulse cancelers since they require three distinct input pulses before an output can be read. The double line canceler impulse response is given by

$$h(t) = \delta(t) - 2\delta(t - T) + \delta(t - 2T).$$
    Eq. (10.14)

Again, the names *double delay line canceler* and *double canceler* will be used interchangeably. The power gain for the double delay line canceler is

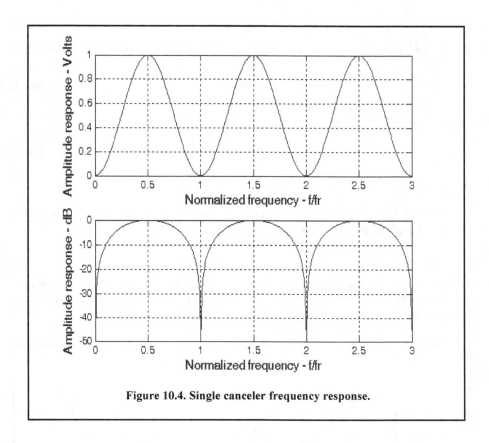

Figure 10.4. Single canceler frequency response.

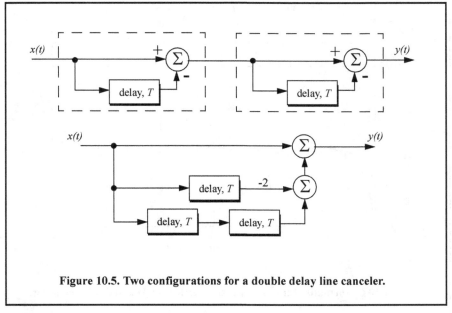

Figure 10.5. Two configurations for a double delay line canceler.

$$|H(\omega)|^2 = |H_1(\omega)|^2 |H_1(\omega)|^2 \qquad \textbf{Eq. (10.15)}$$

where $|H_1(\omega)|^2$ is the single line canceler power gain given in Eq. (10.13). It follows that

$$|H(\omega)|^2 = 16\left(\sin\left(\omega\frac{T}{2}\right)\right)^4. \qquad \textbf{Eq. (10.16)}$$

And in the z-domain, we have

$$H(z) = (1 - z^{-1})^2 = 1 - 2z^{-1} + z^{-2}. \qquad \textbf{Eq. (10.17)}$$

MATLAB Function "double_canceler.m"

The function *"double_canceler.m"* computes and plots (as a function of f/f_r) the amplitude response for a double delay line canceler. The syntax is as follows:

[resp] = double_canceler (fofr)

where *"fofr"* is the number of periods desired. Figure 10.6 shows typical output from this function. Note that the double canceler has a better response than the single canceler (deeper notch and flatter passband response). This function is listed in Appendix 10-A.

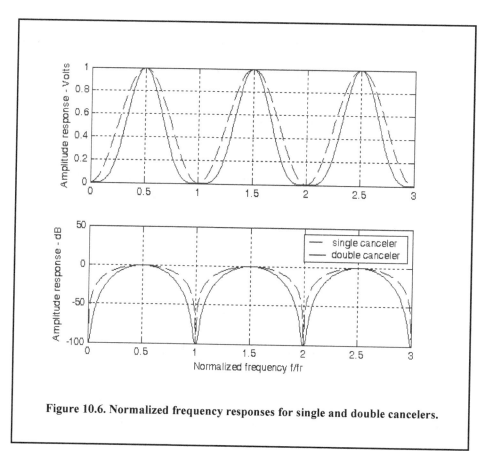

Figure 10.6. Normalized frequency responses for single and double cancelers.

10.2.3. Delay Lines with Feedback (Recursive Filters)

Delay line cancelers with feedback loops are known as recursive filters. The advantage of a recursive filter is that through a feedback loop, we will be able to shape the frequency response of the filter. As an example, consider the single canceler shown in Fig. 10.7. From the figure we can write

$$y(t) = x(t) - (1 - K)w(t) \qquad \text{Eq. (10.18)}$$

$$v(t) = y(t) + w(t) \qquad \text{Eq. (10.19)}$$

$$w(t) = v(t - T). \qquad \text{Eq. (10.20)}$$

Applying the z-transform to the above three equations yields

$$Y(z) = X(z) - (1 - K)W(z) \qquad \text{Eq. (10.21)}$$

$$V(z) = Y(z) + W(z) \qquad \text{Eq. (10.22)}$$

$$W(z) = z^{-1}V(z). \qquad \text{Eq. (10.23)}$$

Solving for the transfer function $H(z) = Y(z)/X(z)$ yields

$$H(z) = \frac{1 - z^{-1}}{1 - Kz^{-1}}. \qquad \text{Eq. (10.24)}$$

The modulus square of $H(z)$ is then equal to

$$|H(z)|^2 = \frac{(1 - z^{-1})(1 - z)}{(1 - Kz^{-1})(1 - Kz)} = \frac{2 - (z + z^{-1})}{(1 + K^2) - K(z + z^{-1})}. \qquad \text{Eq. (10.25)}$$

Using the transformation $z = e^{j\omega T}$ yields

$$z + z^{-1} = 2\cos \omega T. \qquad \text{Eq. (10.26)}$$

Thus, Eq. (10.24) can now be rewritten as

$$|H(e^{j\omega T})|^2 = \frac{2(1 - \cos \omega T)}{(1 + K^2) - 2K\cos(\omega T)}. \qquad \text{Eq. (10.27)}$$

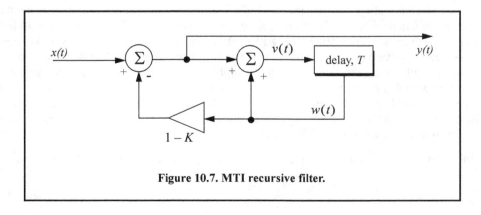

Figure 10.7. MTI recursive filter.

Note that when $K = 0$, Eq. (10.27) collapses to Eq. (10.11) (single line canceler). Figure 10.8 shows a plot of Eq. (10.27) for $K = 0.25, 0.7, 0.9$. Clearly, by changing the gain factor K, one can control the filter response. This plot can be reproduced using the MATLAB program *"Fig10_8.m,"* listed in Appendix 10-A.

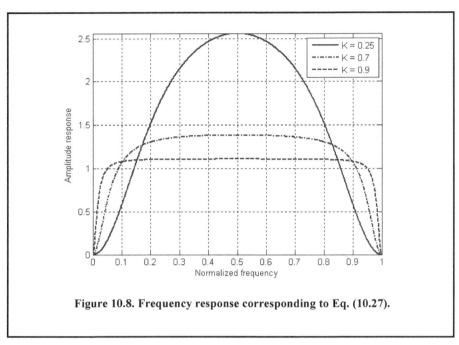

Figure 10.8. Frequency response corresponding to Eq. (10.27).

In order to avoid oscillation due to the positive feedback, the value of K should be less than unity. The value $(1 - K)^{-1}$ is normally equal to the number of pulses received from the target. For example, $K = 0.9$ corresponds to ten pulses, while $K = 0.98$ corresponds to about fifty pulses.

10.3. PRF Staggering

Target velocities that correspond to multiple integers of the PRF are referred to as blind speeds. This terminology is used since an MTI filter response is equal to zero at these values. Blind speeds can pose serious limitations on the performance of MTI radars and their ability to perform adequate target detection. Using PRF agility by changing the pulse repetition interval between consecutive pulses can extend the first blind speed to more tolerable values. In order to show how PRF staggering can alleviate the problem of blind speeds, let us first assume that two radars with distinct PRFs are utilized for detection. Since blind speeds are proportional to the PRF, the blind speeds of the two radars would be different. However, using two radars to alleviate the problem of blind speeds is a very costly option. A more practical solution is to use a single radar with two or more different PRFs.

For example, consider a radar system with two interpulse periods T_1 and T_2, such that

$$\frac{T_1}{T_2} = \frac{n_1}{n_2} \qquad \qquad \text{Eq. (10.28)}$$

where n_1 and n_2 are integers. The first true blind speed occurs when

$$\frac{n_1}{T_1} = \frac{n_2}{T_2}.$$

Eq. (10.29)

This is illustrated in Fig. 10.9 for $n_1 = 4$ and $n_2 = 5$. This figure can be reproduced using the MATLAB program *"Fig10_9.m,"* listed in Appendix 10-A.

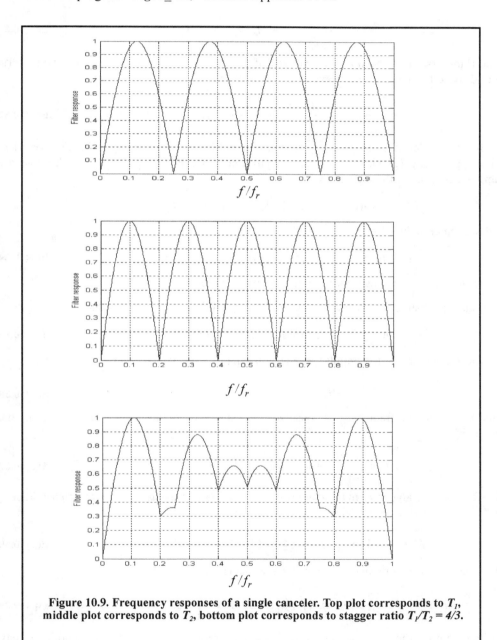

Figure 10.9. Frequency responses of a single canceler. Top plot corresponds to T_1, middle plot corresponds to T_2, bottom plot corresponds to stagger ratio $T_1/T_2 = 4/3$.

The ratio

$$k_s = n_1 / n_2 \qquad\qquad \textbf{Eq. (10.30)}$$

is known as the stagger ratio. Using staggering ratios closer to unity pushes the first true blind speed farther out. However, the dip in the vicinity of $1/T_1$ becomes deeper. In general, if there are N PRFs related by

$$\frac{n_1}{T_1} = \frac{n_2}{T_2} = \dots = \frac{n_N}{T_N}. \qquad\qquad \textbf{Eq. (10.31)}$$

and if the first blind speed to occur for any of the individual PRFs is v_{blind1}, then the first true blind speed for the staggered waveform is

$$v_{blind} = \frac{n_1 + n_2 + \dots + n_N}{N} v_{blind1}. \qquad\qquad \textbf{Eq. (10.32)}$$

To better determine the frequency response of an MTI filter with staggered PRFs, consider a three-pulse canceler with two PRFs, or equivalently two PRIs, T_1 and T_2. In this case, the impulse response will be given by

$$h(t) = [\delta(t) - \delta(t - T_1)] - [\delta(t - T_1) - \delta(t - T_1 - T_2)] \qquad\qquad \textbf{Eq. (10.33)}$$

which can be written as

$$h(t) = \delta(t) - 2\delta(t - T_1) + \delta(t - T_1 - T_2). \qquad\qquad \textbf{Eq. (10.34)}$$

Note that PRF staggering requires a minimum of two PRFs.

Make the change of variables $u = t - T_1$ in Eq. (10.34), and it follows that

$$h(u + T_1) = \delta(u + T_1) - 2\delta(u) + \delta(u - T_2). \qquad\qquad \textbf{Eq. (10.35)}$$

The Z-transform of the impulse response in Eq. (10.35) is then given by

$$H(z)z^{-T_1} = z^{T_1} - 2 + z^{-T_2} \qquad\qquad \textbf{Eq. (10.36)}$$

and the amplitude frequency response for the staggered double delay line canceller is then given by

$$|H(z)|^2\big|_{z = e^{j\omega T}} = (z^{T_1} - 2 + z^{-T_2})(z^{-T_1} - 2 + z^{T_2}). \qquad\qquad \textbf{Eq. (10.37)}$$

Performing the algebraic manipulation in Eq. (10.37) and using the t trigonometric identity $(e^{j\omega T} + e^{-j\omega T}) = 2\cos\omega T$ yields

$$|H(\omega)|^2 = 6 - 4\cos(2\pi f T_1) - 4\cos(2\pi f T_2) + 2\cos(2\pi f(T_1 + T_2)). \qquad\qquad \textbf{Eq. (10.38)}$$

It is customary to normalize the amplitude frequency response, thus

$$|H(\omega)|^2 = 1 - \frac{2}{3}\cos(2\pi f T_1) - \frac{2}{3}\cos(2\pi f T_2) + \frac{1}{3}\cos(2\pi f(T_1 + T_2)). \qquad\qquad \textbf{Eq. (10.39)}$$

To determine the characteristics of higher stagger ratio MTI filters, adopt the notion of having several MTI filters, one for each combination of two staggered PRFs. Then the overall filter response is computed as the average of all individual filters. For example, consider the case where a PRF stagger is required with PRIs T_1, T_2, T_3, and T_4. First, compute the filter response using T_1 T_2 and denote by H_1. Then compute H_2 using T_2 and T_3, the filter H_3 is computed using T_3 T_4 and the filter H_4 is computed using T_4 and T_1. Finally compute the overall response as

$$H(f) = \frac{1}{4}[H_1(f) + H_2(f) + H_3(f) + H_4(f)].$$ **Eq. (10.40)**

Figure 10.10 shows the MTI filter response for a 4-stagger-ratio defined. The overall response is computed as the average of 4 individual filters, each corresponding to one combination of the stagger ratio. In the top portion of the figure the individual filters used were 2-pulse MTIs, while the bottom portion used 4-pulse individual MTI filters. This plot can be reproduced using the MATLAB program *"Fig10_10.m,"* listed in Appendix 10-A.

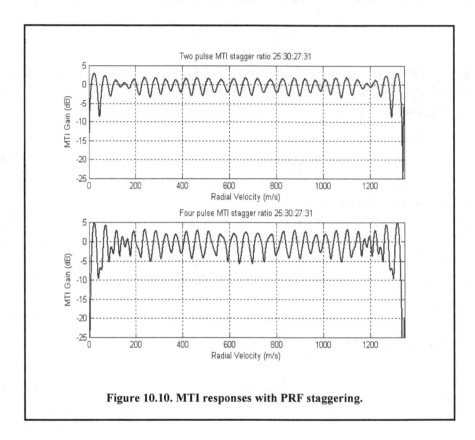

Figure 10.10. MTI responses with PRF staggering.

10.4. MTI Improvement Factor

In this section, two quantities that are normally used to define the performance of MTI systems are introduced. They are Clutter Attenuation (CA) and the Improvement Factor. The MTI CA is defined as the ratio between the MTI filter input clutter power C_i to the output clutter power C_o,

$$CA = C_i / C_o.$$ **Eq. (10.41)**

The MTI improvement factor is defined as the ratio of the SCR at the output to the SCR at the input,

$$I = \left(\frac{S_o}{C_o}\right) / \left(\frac{S_i}{C_i}\right),$$

Eq. (10.42)

which can be rewritten as

$$I = \frac{S_o}{S_i} CA.$$

Eq. (10.43)

The ratio S_o / S_i is the average power gain of the MTI filter, and it is equal to $|H(\omega)|^2$. In this section, a closed form expression for the improvement factor using a Gaussian-shaped power spectrum is developed. A Gaussian-shaped clutter power spectrum is given by

$$S(f) = \frac{P_c}{\sqrt{2\pi}\,\sigma_f} \exp(-f^2/2\sigma_f^2)$$

Eq. (10.44)

where P_c is the clutter power (constant), and σ_f is the clutter rms frequency (which describes the clutter spectrum spread in the frequency domain, see Eq. (10.1)).

The clutter power at the input of an MTI filter is

$$C_i = \int_{-\infty}^{\infty} \frac{P_c}{\sqrt{2\pi}\,\sigma_f} \exp\left(-\frac{f^2}{2\sigma_f^2}\right) df.$$

Eq. (10.45)

Factoring out the constant P_c yields

$$C_i = P_c \int_{-\infty}^{\infty} \frac{1}{\sqrt{2\pi}\sigma_f} \exp\left(-\frac{f^2}{2\sigma_f^2}\right) df.$$

Eq. (10.46)

It follows that

$$C_i = P_c.$$

Eq. (10.47)

The clutter power at the output of an MTI is

$$C_o = \int_{-\infty}^{\infty} S(f)|H(f)|^2 \, df.$$

Eq. (10.48)

10.4.1. Two-Pulse MTI Case

In this section we will continue the analysis using a single delay line canceler. The frequency response for a single delay line canceler is

$$|H(f)|^2 = 4\left(\sin\left(\frac{\pi f}{f_r}\right)\right)^2.$$

Eq. (10.49)

It follows that

$$C_o = \int_{-\infty}^{\infty} \frac{P_c}{\sqrt{2\pi}\ \sigma_f} \exp\left(-\frac{f^2}{2\sigma_f^2}\right) 4\left(\sin\left(\frac{\pi f}{f_r}\right)\right)^2 df.$$ Eq. (10.50)

Since clutter power will only be significant for small f, the ratio f/f_r is very small. Consequently, by using the small angle approximation, Eq. (10.50) is approximated by

$$C_o \approx \int_{-\infty}^{\infty} \frac{P_c}{\sqrt{2\pi}\ \sigma_f} \exp\left(-\frac{f^2}{2\sigma_f^2}\right) 4\left(\frac{\pi f}{f_r}\right)^2 df,$$ Eq. (10.51)

which can be rewritten as

$$C_o = \frac{4P_c\pi^2}{f_r^2} \int_{-\infty}^{\infty} \frac{1}{\sqrt{2\pi\sigma_f^2}} \exp\left(-\frac{f^2}{2\sigma_f^2}\right) f^2\ df.$$ Eq. (10.52)

The integral part in Eq. (10.52) is the second moment of a zero-mean Gaussian distribution with variance σ_f^2. Replacing the integral in Eq. (10.52) by σ_f^2 yields

$$C_o = \frac{4P_c\pi^2}{f_r^2} \sigma_f^2.$$ Eq. (10.53)

Substituting Eq. (10.53) and Eq. (10.47) into Eq. (10.41) produces

$$CA = \frac{C_i}{C_o} = \left(\frac{f_r}{2\pi\sigma_f}\right)^2.$$ Eq. (10.54)

It follows that the improvement factor for a single canceler is

$$I = \left(\frac{f_r}{2\pi\sigma_f}\right)^2 \frac{S_o}{S_i}.$$ Eq. (10.55)

The power gain ratio for a single canceler is (remember that $|H(f)|$ is periodic with period f_r)

$$\frac{S_o}{S_i} = |H(f)|^2 = \frac{1}{f_r} \int_{-f_r/2}^{f_r/2} 4\left(\sin\frac{\pi f}{f_r}\right)^2 df.$$ Eq. (10.56)

Using the trigonometric identity $(2 - 2\cos 2\vartheta) = 4(\sin\vartheta)^2$ yields

$$|H(f)|^2 = \frac{1}{f_r} \int_{-f_r/2}^{f_r/2} \left(2 - 2\cos\frac{2\pi f}{f_r}\right) df = 2.$$ Eq. (10.57)

It follows that

$$I = 2(f_r/2\pi\sigma_f)^2.$$ Eq. (10.58)

The expression given in Eq. (10.58) is an approximation valid only for $\sigma_f \ll f_r$. When the condition $\sigma_f \ll f_r$ is not true, then the autocorrelation function needs to be used in order to develop an exact expression for the improvement factor. Furthermore, when taking into account Eq. (10.1) (i.e., account for antenna scan rate, wind, and platform motion) the improvement factor is reduced since σ_f becomes larger.

Example:

A certain radar has $f_r = 800\,Hz$. If the clutter rms is $\sigma_f = 6.4\,Hz$, find the improvement factor when a single delay line canceler is used.

Solution:

The clutter attenuation CA is

$$CA = \left(\frac{f_r}{2\pi\sigma_f}\right)^2 = \left(\frac{800}{(2\pi)(6.4)}\right)^2 = 395.771 = 25.974\,dB$$

and since $S_o/S_i = 2 = 3\,dB$ one gets

$$I_{dB} = (CA + S_o/S_i)_{dB} = 3 + 25.97 = 28.974\,dB.$$

10.4.2. The General Case

A general expression for the improvement factor for the n-pulse MTI (shown for a 2-pulse MTI in Eq. (10.58)) is given by

$$I = \frac{1}{Q^2(2(n-1)-1)!!}\left(\frac{f_r}{2\pi\sigma_f}\right)^{2(n-1)}$$

Eq. (10.59)

where the double factorial notation is defined by

$$(2n-1)!! = 1 \times 3 \times 5 \times \ldots \times (2n-1)$$

Eq. (10.60)

$$(2n)!! = 2 \times 4 \times \ldots \times 2n$$

Eq. (10.61)

Of course $0!! = 1$; Q is defined by

$$Q^2 = 1/\left(\sum_{i=1}^{n} A_i^2\right)$$

Eq. (10.62)

where A_i are the binomial coefficients for the MTI filter. It follows that Q^2 for a 2-pulse, 3-pulse, and 4-pulse MTI are, respectively,

$$\left\{\frac{1}{2}, \frac{1}{20}, \frac{1}{70}\right\}.$$

Eq. (10.63)

Using this notation, the improvement factor for a 3-pulse and 4-pulse MTI are, respectively, given by

$$I_{3-pulse} = 2\left(\frac{f_r}{2\pi\sigma_f}\right)^4$$

Eq. (10.64)

$$I_{4-pulse} = \frac{4}{3}\left(\frac{f_r}{2\pi\sigma_f}\right)^6.$$

Eq. (10.65)

10.5. Subclutter Visibility (SCV)

Subclutter Visibility (SCV) describes the radar's ability to detect nonstationary targets embedded in a strong clutter background, for some probabilities of detection and false alarm. It is often used as a measure of MTI performance. For example, a radar with $10dB$ SCV will be able to detect moving targets whose returns are ten times smaller than those of clutter. A sketch illustrating the concept of SCV is shown in Fig. 10.11.

If a radar system can resolve the areas of strong and weak clutter within its field of view, then Interclutter Visibility (ICV) describes the radar's ability to detect nonstationary targets between strong clutter points. The subclutter visibility is expressed as the ratio of the improvement factor to the minimum MTI output SCR required for proper detection for a given probability of detection. More precisely,

$$SCV = I/(SCR)_o .$$

Eq. (10.66)

When comparing the performance of different radar systems on the basis of SCV, one should use caution since the amount of clutter power is dependent on the radar resolution cell (or volume), which may be different from one radar to another. Thus, only if the different radars have the same beamwidths and the same pulse widths, can SCV be used as a basis of performance comparison.

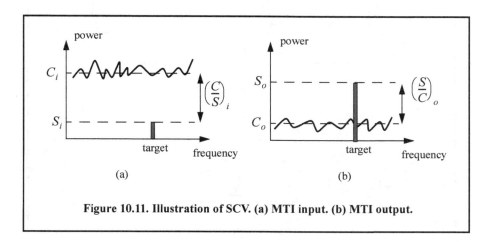

Figure 10.11. Illustration of SCV. (a) MTI input. (b) MTI output.

10.6. Delay Line Cancelers with Optimal Weights

The delay line cancelers discussed in this chapter belong to a family of transversal Finite Impulse Response (FIR) filters widely known as the "tapped delay line" filters. Figure 10.12 shows an N-stage tapped delay line implementation. When the weights are chosen such that they are the binomial coefficients (i.e., the coefficients of the expansion $(1-x)^N$) with alternating signs, then the resultant MTI filter is equivalent to N-stage cascaded single line cancelers. This is illustrated in Fig. 10.13 for $N = 4$. In general, the binomial coefficients are given by

$$w_i = (-1)^{i-1} \frac{N!}{(N-i+1)!(i-1)!} \; ; \; i = 1, ..., N+1 .$$

Eq. (10.67)

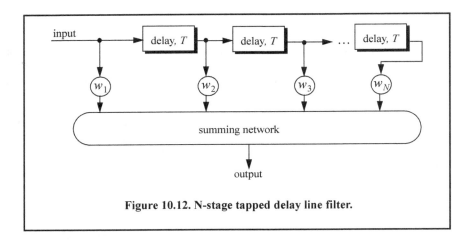

Figure 10.12. N-stage tapped delay line filter.

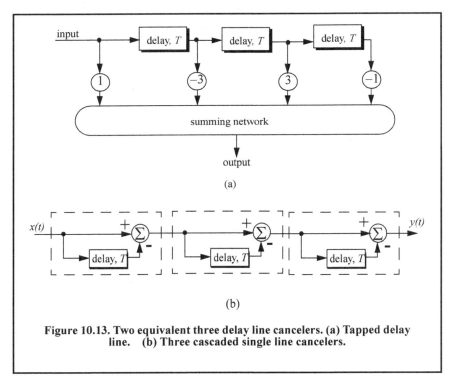

(a)

(b)

Figure 10.13. Two equivalent three delay line cancelers. (a) Tapped delay line. (b) Three cascaded single line cancelers.

Using the binomial coefficients with alternating signs produces an MTI filter that closely approximates the optimal filter in the sense that it maximizes the improvement factor, as well as the probability of detection. In fact, the difference between an optimal filter and one with binomial coefficients is so small that the latter one is considered to be optimal by most radar designers. However, being optimal in the sense of the improvement factor does not guarantee a deep notch or a flat passband in the MTI filter response. Consequently, many researchers have been investigating other weights that can produce a deeper notch around DC, as well as a better passband response.

In general, the average power gain for an N-stage delay line canceler is

$$\frac{S_o}{S_i} = \prod_{i=1}^{N} |H_1(f)|^2 = \prod_{i=1}^{N} 4\left(\sin\left(\frac{\pi f}{f_r}\right)\right)^2.$$

Eq. (10.68)

For example, $N = 2$ (double delay line canceler) gives

$$\frac{S_o}{S_i} = 16\left(\sin\left(\frac{\pi f}{f_r}\right)\right)^4.$$

Eq. (10.69)

Equation (10.69) can be rewritten as

$$\frac{S_o}{S_i} = |H_1(f)|^{2N} = 2^{2N}\left(\sin\left(\frac{\pi f}{f_r}\right)\right)^{2N}.$$

Eq. (10.70)

As indicated by Eq. (10.70), blind speeds for an N-stage delay canceler are identical to those of a single canceler. It follows that blind speeds are independent from the number of cancelers used. It is possible to show that Eq. (10.70) can be written as

$$\frac{S_o}{S_i} = 1 + N^2 + \left(\frac{N(N-1)}{2!}\right)^2 + \left(\frac{N(N-1)(N-2)}{3!}\right)^2 + \dots$$

Eq. (10.71)

A general expression for the improvement factor of an N-stage tapped delay line canceler is

$$I = \frac{(S_o/S_i)}{\displaystyle\sum_{k=1}^{N}\sum_{j=1}^{N} w_k w_j^* \rho\left(\frac{(k-j)}{f_r}\right)}$$

Eq. (10.72)

where the weights w_k and w_j are those of a tapped delay line canceler, and $\rho((k-j)/f_r)$ is the correlation coefficient between the kth and jth samples. For example, $N = 2$ produces

$$I = \frac{1}{1 - \frac{4}{3}\rho T + \frac{1}{3}\rho 2 T}.$$

Eq. (10.73)

10.7. Pulsed Doppler Radars

Pulsed radars transmit and receive a train of modulated pulses. Range is extracted from the two-way time delay between a transmitted and received pulse. Doppler measurements can be made in two ways. If accurate range measurements are available between consecutive pulses, then Doppler frequency can be extracted from the range rate $\dot{R} = \Delta R / \Delta t$. This approach works fine as long as the range is not changing drastically over the interval Δt. Otherwise, pulsed radars utilize a Doppler filter bank.

Pulsed radar waveforms can be completely defined by the following: (1) carrier frequency, which may vary depending on the design requirements and radar mission; (2) pulse width, which is closely related to the bandwidth and defines the range resolution; (3) modulation; and finally (4) the pulse repetition frequency. Different modulation techniques are usually utilized to enhance the radar performance, or to add more capabilities to the radar that otherwise would not have been possible. The PRF must be chosen to avoid Doppler and range ambiguities as well as maximize the average transmitted power.

Radar systems employ low, medium, and high PRF schemes. Low PRF waveforms can provide accurate, long, unambiguous range measurements, but exert severe Doppler ambiguities. Medium PRF waveforms must resolve both range and Doppler ambiguities; however, they provide adequate average transmitted power as compared to low PRFs. High PRF waveforms can provide superior average transmitted power and excellent clutter rejection capabilities. Alternatively, high PRF waveforms are extremely ambiguous in range. Radar systems utilizing high PRFs are often called Pulsed Doppler Radars (PDR). Range and Doppler ambiguities for different PRFs are summarized in Table 10.1.

Distinction of a certain PRF as low, medium, or high PRF is almost arbitrary and depends on the radar mode of operations. For example, a $3 KHz$ PRF is considered low if the maximum detection range is less than $30 Km$. However, the same PRF would be considered medium if the maximum detection range is well beyond $30 Km$.

Radars can utilize constant and varying (agile) PRFs. For example, Moving Target Indicator (MTI) radars use PRF agility to avoid blind speeds, as discussed in Chapter 9. This kind of agility is known as PRF staggering. PRF agility is also used to avoid range and Doppler ambiguities, as will be explained in the next three sections. Additionally, PRF agility is used to prevent jammers from locking onto the radar's PRF. These two last forms of PRF agility are sometimes referred to as PRF jitter.

Figure 10.14 shows a simplified pulsed radar block diagram. The range gates can be implemented as filters that open and close at time intervals that correspond to the detection range. The width of such an interval corresponds to the desired range resolution. The radar receiver is often implemented as a series of contiguous (in time) range gates, where the width of each gate is achieved through pulse compression. The clutter rejection can be implemented using MTI or other forms of clutter rejection techniques. The NBF bank is normally implemented using an FFT, where bandwidth of the individual filters corresponds to the FFT frequency resolution.

In ground-based radars, the amount of clutter in the radar receiver depends heavily on the radar-to-target geometry. The amount of clutter is considerably higher when the radar beam has to face toward the ground. Radars employing high PRFs have to deal with an increased amount of clutter due to folding in range. Clutter introduces additional difficulties for airborne radars when detecting ground targets and other targets flying at low altitudes. This is illustrated in Fig. 10.15. Returns from ground clutter emanate from ranges equal to the radar altitude to those which exceed the slant range along the mainbeam, with considerable clutter returns in the sidelobes and mainbeam. The presence of such large amounts of clutter interferes with radar detection capabilities and makes it extremely difficult to detect targets in the look-down mode. This difficulty in detecting ground or low-altitude targets has led to the development of pulse Doppler radars where other targets, kinematics such as Doppler effects, are exploited to enhance detection.

TABLE 10.1. PRF Ambiguities.

PRF	Range Ambiguous	Doppler Ambiguous
Low PRF	No	Yes
Medium PRF	Yes	Yes
High PRF	Yes	No

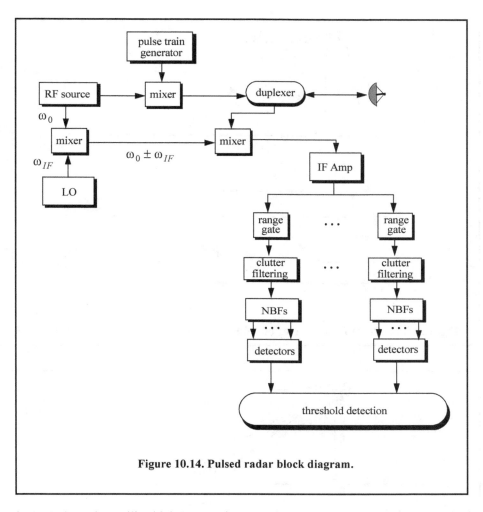

Figure 10.14. Pulsed radar block diagram.

Pulse Doppler radars utilize high PRFs to increase the average transmitted power and rely on the target's Doppler frequency for detection. The increase in the average transmitted power leads to an improved SNR, which helps the detection process. However, using high PRFs compromises the radar's ability to detect long-range targets because of range ambiguities associated with high PRF applications.

Pulse Doppler radars (or high PRF radars) have to deal with the additional increase in clutter power due to clutter folding. This has led to the development of a special class of airborne MTI filters, often referred to as AMTI. Techniques such as using specialized Doppler filters to reject clutter are very effective and are often employed by pulse Doppler radars. Pulse Doppler radars can measure target Doppler frequency (or its range rate) fairly accurately and use the fact that ground clutter typically possesses limited Doppler shift when compared with moving targets to separate the two returns. This is illustrated in Fig. 10.16. Clutter filtering (i.e., AMTI) is used to remove both main-beam and altitude clutter returns, and fast-moving target detection is done effectively by exploiting its Doppler frequency. In many modern pulse Doppler radars, the limiting factor in detecting slow-moving targets is not clutter but rather another source of noise, referred to as phase noise, generated from the receiver local oscillator instabilities.

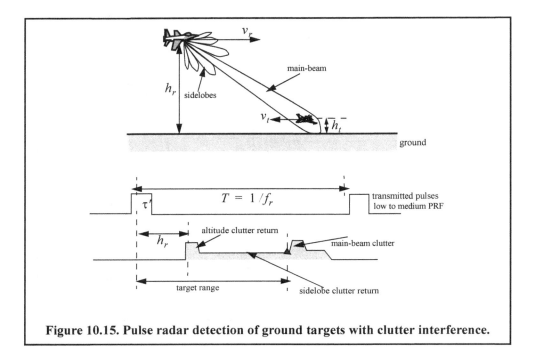

Figure 10.15. Pulse radar detection of ground targets with clutter interference.

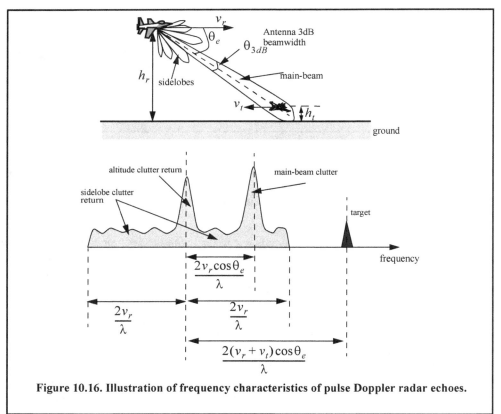

Figure 10.16. Illustration of frequency characteristics of pulse Doppler radar echoes.

10.7.1. Pulse Doppler Radar Signal Processing

The main idea behind pulse Doppler radar signal processing is to divide the footprint (the intersection of the antenna $3dB$ beamwidth with the ground) into resolution cells that constitute a range Doppler map, MAP. The sides of this map are range and Doppler, as illustrated in Fig. 10.17. Fine range resolution, ΔR, is accomplished in real time by utilizing range gating and pulse compression. Frequency (Doppler) resolution is obtained from the coherent processing interval.

To further illustrate this concept, consider the case where N_a is the number of azimuth (Doppler) cells, and N_r is the number of range bins. Hence, the MAP is of size $N_a \times N_r$, where the columns refer to range bins and the rows refer to azimuth cells. For each transmitted pulse within the dwell, the echoes from consecutive range bins are recorded sequentially in the first row of MAP. Once the first row is completely filled (i.e., returns from all range bins have been received), all data (in all rows) are shifted downward one row before the next pulse is transmitted. Thus, one row of MAP is generated for every transmitted pulse. Consequently, for the current observation interval, returns from the first transmitted pulse will be located in the bottom row of MAP, and returns from the last transmitted pulse will be in the top row of MAP.

Fine range resolution is achieved using the matched filter. Clutter rejection (filtering) is performed on each range bin (i.e., rows in the MAP). Then all samples from one dwell within each range bin are processed using an FFT to resolve targets in Doppler. It follows that a peak in a given resolution cell corresponds to a specific target detection at that range and Doppler frequency. Selection of the proper size FFT and its associated parameters were discussed in Chapter 3.

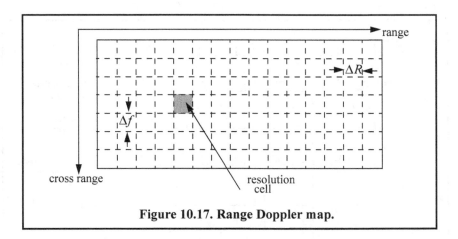

Figure 10.17. Range Doppler map.

10.7.2. Resolving Range Ambiguities

Pulse Doppler radars exhibit serve range ambiguities because they use high PRF pulse streams. In order to resolve these ambiguities, pulse Doppler radars utilize multiple high PRFs (PRF staggering) within each processing interval (dwell). For this purpose, consider a pulse Doppler radar that uses two PRFs, f_{r1} and f_{r2}, on transmit to resolve range ambiguity, as shown in Fig. 10.18. Denote R_{u1} and R_{u2} as the unambiguous ranges for the two PRFs, respectively. Normally, these unambiguous ranges are relatively small and are short of the desired radar unambiguous range R_u (where $R_u \gg R_{u1}, R_{u2}$). Denote the radar desired PRF that corresponds to R_u as f_{rd}.

The choice of f_{r1} and f_{r2} is such that they are relatively prime with respect to one another. One choice is to select $f_{r1} = N f_{rd}$ and $f_{r2} = (N+1) f_{rd}$ for some integer N. Within one period of the desired PRI ($T_d = 1/f_{rd}$), the two PRFs f_{r1} and f_{r2} coincide only at one location, which is the true unambiguous target position. The time delay T_d establishes the desired unambiguous range. The time delays t_1 and t_2 correspond to the time between the transmit of a pulse on each PRF and receipt of a target return due to the same pulse.

Let M_1 be the number of PRF1 intervals between transmit of a pulse and receipt of the true target return. The quantity M_2 is similar to M_1 except it is for PRF2. It follows that over the interval 0 to T_d, the only possible results are $M_1 = M_2 = M$ or $M_1 + 1 = M_2$. The radar needs only to measure t_1 and t_2. First, consider the case when $t_1 < t_2$. In this case,

$$t_1 + \frac{M}{f_{r1}} = t_2 + \frac{M}{f_{r2}}$$

Eq. (10.74)

for which we get

$$M = \frac{t_2 - t_1}{T_1 - T_2}$$

Eq. 10.75)

where $T_1 = 1/f_{r1}$ and $T_2 = 1/f_{r2}$. It follows that the round-trip time to the true target location is

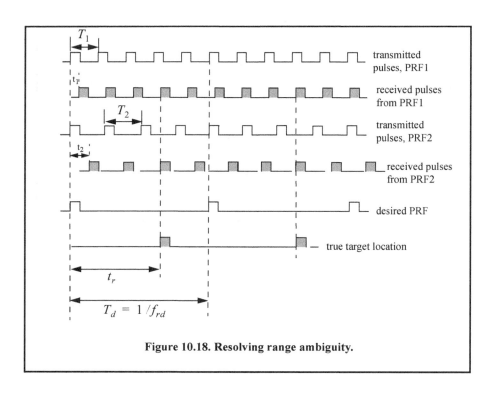

Figure 10.18. Resolving range ambiguity.

$$t_r = MT_1 + t_1$$
$$t_r = MT_2 + t_2$$

Eq. (10.76)

and the true target range is

$$R = ct_r/2.$$

Eq. (10.77)

Now, if $t_1 > t_2$, then

$$t_1 + \frac{M}{f_{r1}} = t_2 + \frac{M+1}{f_{r2}}.$$

Eq. (10.78)

Solving for M we get

$$M = \frac{(t_2 - t_1) + T_2}{T_1 - T_2}$$

Eq. (10.79)

and the round-trip time to the true target location is

$$t_{r1} = MT_1 + t_1,$$

Eq. (10.80)

and in this case, the true target range is

$$R = \frac{ct_{r1}}{2}.$$

Eq. (10.81)

Finally, if $t_1 = t_2$, then the target is in the first ambiguity. It follows that

$$t_{r2} = t_1 = t_2$$

Eq. (10.82)

and

$$R = ct_{r2}/2$$

Eq. (10.83)

Since a pulse cannot be received while the following pulse is being transmitted, these times correspond to blind ranges. This problem can be resolved by using a third PRF. In this case, once an integer N is selected, then in order to guarantee that the three PRFs are relatively prime with respect to one another, we may choose $f_{r1} = N(N+1)f_{rd}$, $f_{r2} = N(N+2)f_{rd}$, and $f_{r3} = (N+1)(N+2)f_{rd}$.

10.7.3. Resolving Doppler Ambiguity

In the case where the pulse Doppler radar is utilizing medium PRFs, it will be ambiguous in both range and Doppler. Resolving range ambiguities was discussed in the previous section. In this section, Doppler ambiguity is addressed. Remember that the line spectrum of a train of pulses has $\sin x / x$ envelope, and the line spectra are separated by the PRF, f_r, as illustrated in Fig. 10.19. The Doppler filter bank is capable of resolving target Doppler as long as the anticipated Doppler shift is less than one half the bandwidth of the individual filters (i.e., one half the width of an FFT bin). Thus, pulsed radars are designed such that

$$f_r = 2f_{dmax} = (2v_{rmax})/\lambda$$

Eq. (10.84)

where f_{dmax} is the maximum anticipated target Doppler frequency, v_{rmax} is the maximum anticipated target radial velocity, and λ is the radar wavelength.

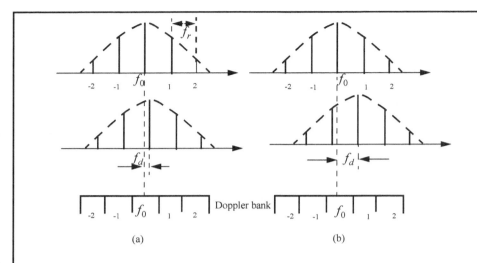

Figure 10.19. Spectra of transmitted and received waveforms, and Doppler bank. (a) Doppler is resolved. (b) Spectral·lines have moved into the next Doppler filter. This results in an ambiguous Doppler measurement.

If the Doppler frequency of the target is high enough to make an adjacent spectral line move inside the Doppler band of interest, the radar can be Doppler ambiguous. Therefore, in order to avoid Doppler ambiguities, radar systems require high PRF rates when detecting high-speed targets. When a long-range radar is required to detect a high-speed target, it may not be possible to be both range and Doppler unambiguous. This problem can be resolved by using multiple PRFs. Multiple PRF schemes can be incorporated sequentially within each dwell interval (scan or integration frame) or the radar can use a single PRF in one scan and resolve ambiguity in the next. The latter technique, however, may have problems due to changing target dynamics from one scan to the next.

The Doppler ambiguity problem is analogous to that of range ambiguity. Therefore, the same methodology can be used to resolve Doppler ambiguity. In this case, we measure the Doppler frequencies f_{d1} and f_{d2} instead of t_1 and t_2. If $f_{d1} > f_{d2}$, then we have

$$M = \frac{(f_{d2} - f_{d1}) + f_{r2}}{f_{r1} - f_{r2}}.$$

Eq. (10.85)

And if $f_{d1} < f_{d2}$,

$$M = \frac{f_{d2} - f_{d1}}{f_{r1} - f_{r2}}$$

Eq. (10.86)

and the true Doppler is

$$f_d = M f_{r1} + f_{d1} \qquad ; f_d = M f_{r2} + f_{d2}.$$

Eq. (10.87)

Finally, if $f_{d1} = f_{d2}$, then

$$f_d = f_{d1} = f_{d2}.$$

Eq. (10.88)

Again, blind Dopplers can occur, which can be resolved using a third PRF.

Example:

A certain radar uses two PRFs to resolve range ambiguities. The desired unambiguous range is $R_u = 100 Km$. Choose $N = 59$. Compute f_{r1}, f_{r2}, R_{u1}, and R_{u2}.

Solution:

First let us compute the desired PRF, f_{rd}

$$f_{rd} = \frac{c}{2R_u} = \frac{3 \times 10^8}{200 \times 10^3} = 1.5 KHz.$$

It follows that

$$f_{r1} = Nf_{rd} = (59)(1500) = 88.5 KHz$$

$$f_{r2} = (N+1)f_{rd} = (59+1)(1500) = 90 KHz$$

$$R_{u1} = \frac{c}{2f_{r1}} = \frac{3 \times 10^8}{2 \times 88.5 \times 10^3} = 1.695 Km$$

$$R_{u2} = \frac{c}{2f_{r2}} = \frac{3 \times 10^8}{2 \times 90 \times 10^3} = 1.667 Km.$$

Example:

Consider a radar with three PRFs; $f_{r1} = 15 KHz$, $f_{r2} = 18 KHz$, and $f_{r3} = 21 KHz$. Assume $f_0 = 9 GHz$. Calculate the frequency position of each PRF for a target whose velocity is $550 m/s$. Calculate f_d (Doppler frequency) for another target appearing at $8 KHz$, $2 KHz$, and $17 KHz$ for each PRF.

Solution:

The Doppler frequency is

$$f_d = 2\frac{vf_0}{c} = \frac{2 \times 550 \times 9 \times 10^9}{3 \times 10^8} = 33 KHz.$$

Then by using Eq. (10.87) $n_i f_{ri} + f_{di} = f_d$ where $i = 1, 2, 3$, we can write

$$n_1 f_{r1} + f_{d1} = 15n_1 + f_{d1} = 33$$

$$n_2 f_{r2} + f_{d2} = 18n_2 + f_{d2} = 33$$

$$n_3 f_{r3} + f_{d3} = 21n_3 + f_{d3} = 33.$$

We will show here how to compute n_1, and leave the computations of n_2 and n_3 to the reader. First, if we choose $n_1 = 0$, that means $f_{d1} = 33 KHz$, which cannot be true since f_{d1} cannot be greater than f_{r1}. Choosing $n_1 = 1$ is also invalid since $f_{d1} = 18 KHz$ cannot be true either. Finally, if we choose $n_1 = 2$, we get $f_{d1} = 3 KHz$, which is an acceptable value. It follows that the minimum n_1, n_2, n_3 that may satisfy the above three relations are $n_1 = 2$, $n_2 = 1$, and $n_3 = 1$. Thus, the apparent Doppler frequencies are $f_{d1} = 3 KHz$, $f_{d2} = 15 KHz$, and $f_{d3} = 12 KHz$, as seen below.

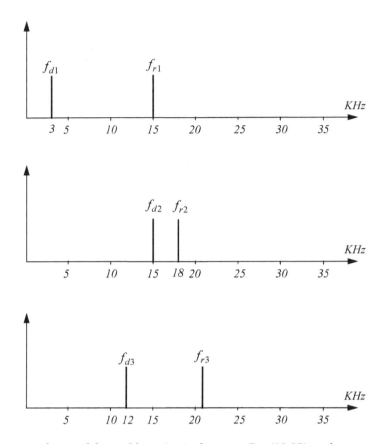

Now for the second part of the problem. Again, by using Eq. (10.87) we have

$$n_1 f_{r1} + f_{d1} = f_d = 15n_1 + 8$$

$$n_2 f_{r2} + f_{d2} = f_d = 18n_2 + 2$$

$$n_3 f_{r3} + f_{d3} = f_d = 21n_3 + 17.$$

We can now solve for the smallest integers n_1, n_2, n_3 that satisfy the above three relations. See the table below. Thus, $n_1 = 2 = n_2$, and $n_3 = 1$, and the true target Doppler is $f_d = 38KHz$. It follows that

$$v_r = 38000 \times \frac{0.0333}{2} = 632.7 \frac{m}{\sec}.$$

n	0	1	2	3	4
f_d from f_{r1}	8	23	<u>38</u>	53	68
f_d from f_{r2}	2	20	<u>38</u>	56	
f_d from f_{r3}	17	<u>38</u>	39		

10.8. Phase Noise

It was determined in earlier chapters that the radar performance is improved as the SNR becomes larger, and in Chapter 2, an expression of the SNR against thermal noise was derived and analyzed. It was also established that in the presence of clutter, the radar performance will degrade beyond the impact of thermal noise alone, and in this case, the SCR becomes more critical than the SNR. Another source of noise that greatly limits MTI and pulsed Doppler radar performance is known as phase noise. Phase noise, sometimes called flicker noise, is random in nature and is caused by instabilities within the radar's local oscillator.

Phase noise may limit, depending on its actual value, pulsed Doppler radars' ability to detect very slow moving targets whose RCS is relatively small. This is illustrated in Fig. 10.20. In this case, when the slow-moving target return signal is close to zero (or multiple integers of the PRF) it will likely be masked by the phase noise power caused by a noisy radar local oscillator. In addition to the masking problem illustrated in Fig. 10.20, the MTI improvement factor will also be reduced by some appreciable values corresponding to the amount of phase noise present in the radar receiver, as will be explained later in this section.

Simply put, phase noise is a term used to describe the random frequency perturbation (relatively small in nature) occurring around the signal carrier frequency, thus causing a new instantaneous signal frequency that is slightly different form the original value. For example, consider the signal defined by the simple sinusoid

$$x(t) = r(t)\sin(2\pi f_c t) \qquad \text{Eq. (10.89)}$$

where $r(t)$ is the amplitude modulation and f_c is the carrier frequency. The perturbed signal due to amplitude and phase instabilities of the local oscillator will take on the form

$$x(t) = r(t + a(t))\sin(2\pi f_c t + \phi(t)) \qquad \text{Eq. (10.90)}$$

where $a(t)$ is the amplitude perturbation and $\phi(t)$ is the phase perturbation or fluctuation. Recalling that the instantaneous frequency is the derivative of phase with respect to time divided by 2π, then the phase perturbation will change the center frequency from f_c to $f_c + \delta f$ where δf is a fractional frequency deviation away from the carrier. More specifically,

$$\delta f = \frac{1}{2\pi}\frac{d}{dt}\phi(t). \qquad \text{Eq. (10.91)}$$

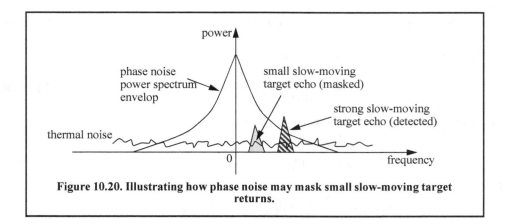

Figure 10.20. Illustrating how phase noise may mask small slow-moving target returns.

The notation commonly used in the literature to describe the phase noise power spectrum density is

$$S_\phi(f) = 2\mathcal{L}(f).$$

Eq. (10.92)

The factor of 2 accounts for both sidebands of the spectrum (lower and upper sidebands around f_c) and $\mathcal{L}(f)$ is the ratio of noise power in a $1Hz$ bandwidth at an offset from the carrier signal power measured in dBc/Hz. In this notation, $\mathcal{L}(f)$ is often used to denote phase noise. For example, consider a frequency-modulated signal as described in Eq. (2.117b) in Chapter 2, which is repeated here for convenience as Eq. (10.93),

$$x(t) = A\left\{ J_0(\beta)\cos 2\pi f_c t + \right.$$

Eq. (10.93)

$$\left[\sum_{n=2(even)}^{\infty} J_n(\beta)[\cos(2\pi f_c + 2n\pi f_m)t + \cos(2\pi f_c - 2n\pi f_m)t] \right] +$$

$$\left. \left[\sum_{q=1(odd)}^{\infty} J_q(\beta)[\cos(2\pi f_c + 2q\pi f_m)t - \cos(2\pi f_c - 2q\pi f_m)t] \right] \right\}$$

where β is the modulation index and A is the amplitude. In this case, phase noise is defined as the ratio of the sideband power to the carrier power at a certain modulation frequency offset from the carrier. More specifically,

$$\mathcal{L}(f)\big|_{dBc/Hz} = 10 \times \log\left\{ \frac{|J_1(\beta)|^2}{|J_0(\beta)|^2} \right\}.$$

Eq. (10.94)

In general, the phase noise $\mathcal{L}(f)$ decreases with frequency as a function of $(1/f^3)$, $(1/f^2)$, and $(1/f)$. Figure 10.21 shows an illustration plot for $\mathcal{L}(f)$ versus the *log* of the frequency. Typically, the manufacturer of a given oscillator will measure and publish the phase noise characteristics (plots similar to Fig. 10.21) as part of their product documentation. Observation of Fig. 10.21 shows that this plot is a piece-wise linear function. It follows that the formula for a given segment of this plot is given by

$$\mathcal{L}(f; f_{i+1} - f_i)\big|_{dBc/Hz} = m_i \log(f) - m_i \log(f_i) + \mathcal{L}(f_i)$$

Eq. (10.95)

where m_i is the slope of the *ith* segment defined by

$$m_i = \frac{\mathcal{L}(f_{i+1}) - \mathcal{L}(f_i)}{\log(f_{i+1}) - \log(f_i)}.$$

Eq. (10.96)

Eq. (10.95) can be written in a more compact form as

$$\mathcal{L}(f) = \mathcal{L}(f_i) \times 10^{\left[\frac{m_i}{10} \log\left(\frac{f}{f_i} \right) \right]}.$$

Eq. (10.97)

Figure 10.22 shows an actual plot for $\mathcal{L}(f)$ using the following values:

$$\{\mathcal{L}(f_1), \mathcal{L}(f_2), \mathcal{L}(f_3), \mathcal{L}(f_4), \mathcal{L}(f_5)\} = \{-55, -85, -105, -115, -115\}.$$

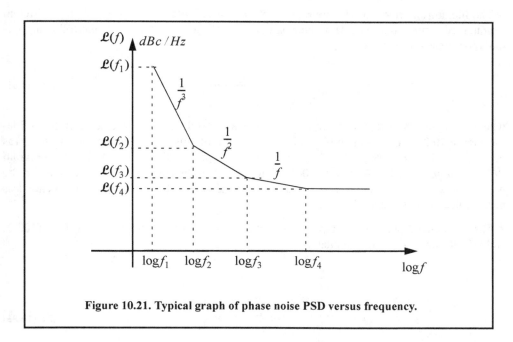

Figure 10.21. Typical graph of phase noise PSD versus frequency.

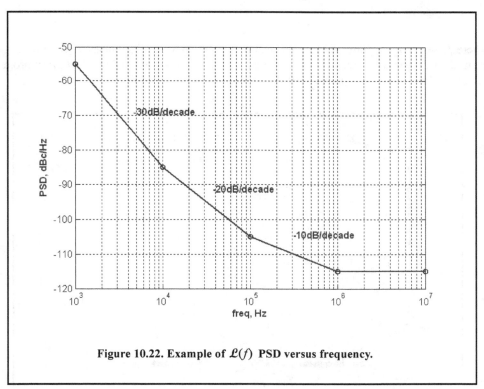

Figure 10.22. Example of $\mathcal{L}(f)$ PSD versus frequency.

The literature is flooded with sources on phase noise. Different users use slightly different formulas to express phase noise in their particular application. In this book, the following formula for phase noise is adopted,

$$\mathcal{L}(f) = \frac{1}{\pi} \frac{c_p \pi f_c^2}{(f - c_p \pi f_c)^2 + (c_p \pi f_c^2)^2}$$

Eq. (10.98)

where f_c is the carrier frequency and c_p is a constant that describes phase noise of the oscillator. Figure 10.23 shows some typical plots of the ratio of noise power to the carrier power (as defined in Eq. (10.98)). This figure can be reproduced using MATLAB Program *"Fig10_23.m,"* listed in Appendix 10-A. Note that when the constant c_p becomes larger, the noise ratio spectrum becomes wider with lower amplitude of the main beam, and hence less phase noise in the system.

The normalized phase noise power spectrum density can be computed using Eqs. (10.92) and (10.97), and can be approximated as

$$\mathcal{L}(f) = \begin{cases} \dfrac{1}{\pi f_b} & ;f \text{ approaches } f_c \\[2ex] \dfrac{1}{2\pi f_b} & ;f = f_c + f_b \\[2ex] \dfrac{f_b}{\pi f^2} & ;f \gg f_c + f_b \end{cases}$$

Eq. (10.99)

where $f_b = c_p \pi f_c^2$. Figure 10.24 shows the corresponding normalized phase noise power spectrum density versus frequency. This figure can be reproduced using MATLAB program *"Fig10_24.m,"* listed in Appendix 10-A.

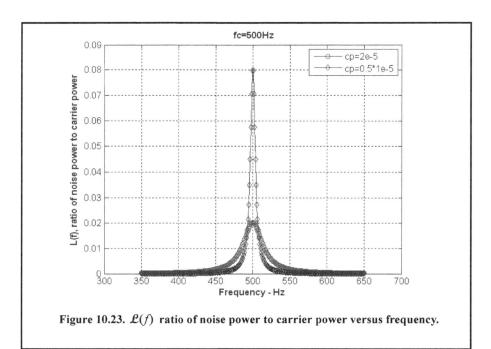

Figure 10.23. $\mathcal{L}(f)$ ratio of noise power to carrier power versus frequency.

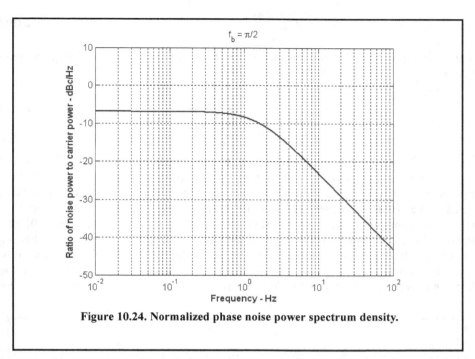

Figure 10.24. Normalized phase noise power spectrum density.

As indicated by Fig. 10.24, quiet oscillators will almost always have phase noise of less than $0 dBc / Hz$ at frequency offsets of more than $1Hz$ away from the carrier. However, at some small frequency bandwidth of less than $1Hz$, phase noise maybe greater than $0 dBc / Hz$.

The power spectral density function of phase noise at the output of the radar's matched filter can be expressed as

$$S_\phi(f) = \mathcal{L}_0 P_c \left(\frac{\sin(\pi f \tau)}{\pi f \tau} \right)^2 \qquad \text{Eq. (10.100)}$$

where $\mathcal{L}_0$ is the phase noise ratio relative to the carrier, P_c was defined in Eq. (10.2) as the clutter power, and τ is the radar pulsewidth. $\mathcal{L}_0$ can be computed from the analysis presented earlier; however, an acceptable range for $\mathcal{L}_0$ varies between 10^{-9} to 10^{-15} dBc/Hz. The clutter attenuation was defined in Eq. (10.41) as

$$CA = C_i / C_o \qquad \text{Eq. (10.101)}$$

where C_i is equal to P_c (see the derivation of Eq. (10.47)).

Ignoring the phase noise, the clutter power spectrum was given in Eq. (10.44), but when phase noise is taken into consideration, Eq. (10.44) is modified to

$$S_t(f) = S(f) + S_\phi(f) = \frac{P_c}{\sqrt{2\pi}\ \sigma_f} \exp(-f^2 / 2\sigma_f^2) + \mathcal{L}_0 P_c \left(\frac{\sin(\pi f \tau)}{\pi f \tau} \right)^2 \qquad \text{Eq. (10.102)}$$

and the clutter output of the MTI filter is computed in a manner similar to that described in Section 10.4, except in this case $S(f)$ is replaced by $S_t(f)$ in Eq. (10.102). Performing the integration and collecting terms (assuming a 2-pulse MTI filter), yields

$$C_o = P_c \frac{1}{2} \left(\frac{2\pi\sigma_f}{f_r} \right)^2 + P_c \frac{\mathcal{L}_0}{\tau}$$ Eq. (10.103)

where f_r is the PRF. It follows that the clutter attenuation in the presence of phase noise is given by

$$CA = \frac{C_i}{C_o} = \frac{1}{\frac{1}{2}\left(\frac{2\pi\sigma_f}{f_r} \right)^2 + \frac{\mathcal{L}_0}{\tau}} .$$ Eq. (10.104)

Figure 10.25 shows the clutter attenuation for two values of σ_f, with $f_r = 2.5KHz$ and $\tau = 1\mu s$ using Eq. (10.104). Observation of Fig. 10.26 leads to the following conclusions: Larger values of σ_f will result in less clutter attenuation, so if more clutter attenuation is desired then a 3-pulse or higher order MTI filter ought to be used. Next, phase noise does not start to affect the performance of the MTI filter until it becomes higher than $-100dBc/Hz$. Clearly, using higher-order MTI filters will increase the amount of clutter attenuation; however, the question that remain, is how does phase noise affect the MTI performance when higher-order filters are used? This analysis is left as an exercise (see Problem 10.20). Figure 10.25 can be reproduced using MATLAB program *"Fig10_25.m,"* listed in Appendix 10-A.

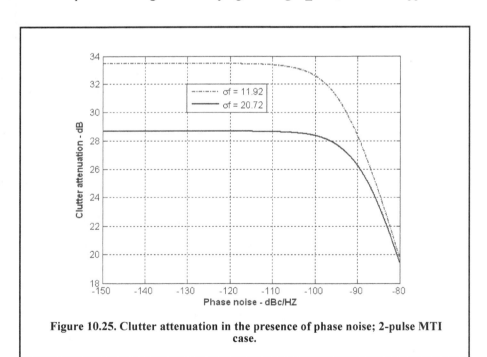

Figure 10.25. Clutter attenuation in the presence of phase noise; 2-pulse MTI case.

Problems

10.1. (a) Derive an expression for the impulse response of a single delay line canceler. (b) Repeat for a double delay line canceler.

10.2. One implementation of a single delay line canceler with feedback is shown below.

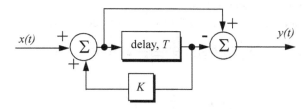

(a) What is the transfer function, $H(z)$? (b) If the clutter power spectrum is $W(f) = w_0 \exp(-f^2/2\sigma_c^2)$, find an exact expression for the filter power gain. (c) Repeat part (b) for small values of frequency, f. (d) Compute the clutter attenuation and the improvement factor in terms of K and σ_c.

10.3. Plot the frequency response for the filter described in the previous problem for $K = -0.5, 0, and\ 0.5$.

10.4. An implementation of a double delay line canceler with feedback is shown below.

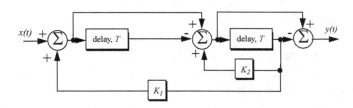

(a) What is the transfer function, $H(z)$? (b) Plot the frequency response for $K_1 = 0 = K_2$, and $K_1 = 0.2, K_2 = 0.5$.

10.5. Consider a single delay line canceler. Calculate the clutter attenuation and the improvement factor. Assume that $\sigma_c = 4Hz$ and PRF $f_r = 450Hz$.

10.6. Develop an expression for the improvement factor of a double delay line canceler.

10.7. Repeat Problem 9.10 for a double delay line canceler.

10.8. An experimental expression for the clutter power spectrum density is

$$W(f) = w_0 \exp(-f^2/2\sigma_c^2)$$

where w_0 is a constant. Show that using this expression leads to the same result obtained for the improvement factor as developed in Section 10.4.

10.9. A certain radar uses two PRFs with a stagger ratio 63/64. If the first PRF is $f_{r1} = 500Hz$. Compute the blind speeds for both PRFs and for the resultant composite PRF. Assume $\lambda = 3cm$.

10.10. Using PRI ratios 25:30:27:31, generate the MTI response for a 3-pulse MTI.

10.11. A certain filter used for clutter rejection has an impulse response $h(n) = \delta(n) - 3\delta(n-1) + 3\delta(n-2) - \delta(n-3)$. (a) Show an implementation of this filter using delay lines and adders. (b) What is the transfer function? (c) Plot the frequency response of this filter. (d) Calculate the output when the input is the unit step sequence.

10.12. The quadrature components of the clutter power spectrum are given in Problem 9.3. Let $\sigma_c = 10Hz$ and $f_r = 500Hz$. Compute the improvement of the signal-to-clutter ratio when a double delay line canceler is utilized.

10.13. The quadrature components of the clutter power spectrum are

$$\bar{S}_I(f) = \delta(f) + \frac{C}{\sqrt{2\pi}\sigma_c} \exp(-f^2/2\sigma_c^2)$$

$$\bar{S}_Q(f) = \frac{C}{\sqrt{2\pi}\sigma_c} \exp(-f^2/2\sigma_c^2).$$

Let $\sigma_c = 10Hz$ and $f_r = 500Hz$. Compute the improvement of the signal-to-clutter ratio when a double delay line canceler is utilized.

10.14. Develop an expression for the clutter improvement factor for single and double line cancelers using the clutter autocorrelation function.

10.15. Consider a medium PRF radar on board an aircraft moving at a speed of 350 m/s with PRFs $f_{r1} = 10KHz$, $f_{r2} = 15KHz$, and $f_{r3} = 20KHz$; the radar operating frequency is $9.5GHz$. Calculate the frequency position of a nose-on target with a speed of 300 m/s. Also calculate the closing rate of a target appearing at 6, 5, and $18KHz$ away from the center line of PRF 10, 15, and $20KHz$, respectively.

10.16. Repeat Problem 10.13 when the target is $15°$ off the radar line of sight.

10.17. A certain radar operates at two PRFs, f_{r1} and f_{r2}, where $T_{r1} = (1/f_{r1}) = T/5$ and $T_{r2} = (1/f_{r2}) = T/6$. Show that this multiple PRF scheme will give the same range ambiguity as that of a single PRF with PRI T.

10.18. Consider an X-band radar with wavelength $\lambda = 3cm$ and bandwidth $B = 10MHz$. The radar uses two PRFs, $f_{r1} = 50KHz$ and $f_{r2} = 55.55KHz$. A target is detected at range bin 46 for f_{r1} and at bin 12 for f_{r2}. Determine the actual target range.

10.19. A certain radar uses two PRFs to resolve range ambiguities. The desired unambiguous range is $R_u = 150Km$. Select a reasonable value for N. Compute the corresponding f_{r1}, f_{r2}, R_{u1}, and R_{u2}.

10.20. A certain radar uses three PRFs to resolve range ambiguities. The desired unambiguous range is $R_u = 250Km$. Select $N = 43$. Compute the corresponding f_{r1}, f_{r2}, f_{r3}, R_{u1}, R_{u2}, and R_{u3}.

10.21. Starting with Eq. (10.94), derive a closed form expression for the phase noise of an FM modulated waveform.

10.22. Reproduce Fig. 10.26 for a 3-pulse and a 4-pulse MTI system, and briefly discuss your output in terms how much phase noise affects the performance of each system,

Appendix 10-A: Chapter 10 MATLAB Code Listings

The MATLAB code provided in this chapter was designed as an academic standalone tool and is not adequate for other purposes. The code was written in a way to assist the reader in gaining a better understanding of the theory. The code was not developed, nor is it intended to be used as part of an open-loop or a closed-loop simulation of any kind. The MATLAB code found in this textbook can be downloaded from this book's web page on the CRC Press website. Simply use your favorite web browser, go to *www.crcpress.com*, and search for keyword *"Mahafza"* to locate this book's web page.

MATLAB Function *"single_canceler.m"* Listing

```
function [resp] = single_canceler (fofr1)
% single delay canceller
eps = 0.00001;
fofr = 0:0.01:fofr1;
arg1 = pi .* fofr;
resp = 4.0 .*((sin(arg1)).^2);
max1 = max(resp);
resp = resp ./ max1;
subplot(2,1,1)
plot(fofr,resp,'k')
xlabel ('Normalized frequency in f/fr')
ylabel( 'Amplitude response in Volts')
grid
subplot(2,1,2)
resp=10.*log10(resp+eps);
plot(fofr,resp,'k');
axis tight
grid
xlabel ('Normalized frequency in f/fr')
ylabel( 'Amplitude response in dB')
end
```

MATLAB Function *"double_canceler.m"* Listing

```
function [resp] = double_canceler(fofr1)
eps = 0.00001;
fofr = 0:0.01:fofr1;
arg1 = pi .* fofr;
resp = 4.0 .* ((sin(arg1)).^2);
max1 = max(resp);
resp = resp ./ max1;
resp2 = resp .* resp;
subplot(2,1,1);
plot(fofr,resp,'k--',fofr, resp2,'k');
ylabel ('Amplitude response - Volts')
resp2 = 20. .* log10(resp2+eps);
resp1 = 20. .* log10(resp+eps);
subplot(2,1,2)
plot(fofr,resp1,'k--',fofr,resp2,'k');
legend ('single canceler','double canceler')
xlabel ('Normalized frequency f/fr')
ylabel ('Amplitude response in dB')
```

end

MATLAB Program "Fig10_8.m" Listing

```
% generates Fig. 10.8 of text
clear all;
fofr = 0:0.001:1;
arg = 2.*pi.*fofr;
nume = 2.*(1.-cos(arg));
den11 = (1. + 0.25 * 0.25);
den12 = (2. * 0.25) .* cos(arg);
den1 = den11 - den12;
den21 = 1.0 + 0.7 * 0.7;
den22 = (2. * 0.7) .* cos(arg);
den2 = den21 - den22;
den31 = (1.0 + 0.9 * 0.9);
den32 = ((2. * 0.9) .* cos(arg));
den3 = den31 - den32;
resp1 = nume ./ den1;
resp2 = nume ./ den2;
resp3 = nume ./ den3;
plot(fofr,resp1,'k',fofr,resp2,'k-.',fofr,resp3,'k--');
xlabel('Normalized frequency')
ylabel('Amplitude response')
legend('K=0.25','K=0.7','K=0.9')
grid
axis tight
```

MATLAB Program "Fig10_9.m" Listing

```
% generates Fig 9.10 of text
clc
close all
clear all
fofr=0:0.001:1;
f1=4.*fofr;
f2=5.*fofr;
arg1=pi.*f1;
arg2=pi.*f2;
resp1=abs(sin(arg1));
resp2=abs(sin(arg2));
resp=resp1+resp2;
max1=max(resp);
resp=resp./max1;
subplot(3,1,1)
plot(fofr,resp1);
ylabel('\bfFilter response')
grid on
subplot(3,1,2)
plot(fofr,resp2);
ylabel('\bfFilter response')
grid on
subplot(3,1,3)
plot(fofr,resp);
```

```
ylabel('\bfFilter response')
xlabel('\bff/fr')
grid on
```

MATLAB Program "Fig10_10.m" Listing

```
% generates Fig 10.10 of text
k = .00035/25; a = 25*k; b = 30*k; c = 27*k; d = 31*k;
v2 = linspace(0,1345,10000);
f2 = (2.*v2)/.0375;
% H1(f)
T1 = exp(-j*2*pi.*f2*a); X1 = 1/2.*(1 - T1).*conj(1 - T1); H1 = 10*log10(abs(X1));
% H2(f)
T2 = exp(-j*2*pi.*f2*b); X2 = 1/2.*(1 - T2).*conj(1 - T2); H2 = 10*log10(abs(X2));
% H3(f)
T3 = exp(-j*2*pi.*f2*c); X3 = 1/2.*(1 - T3).*conj(1 - T3); H3 = 10*log10(abs(X3));
% H4(f)
T4 = exp(-j*2*pi.*f2*d); X4 = 1/2.*(1 - T4).*conj(1 - T4); H4 = 10*log10(abs(X4));
% Plot of the four components of H(f)
figure(1)
subplot(2,1,1)
% H(f) Average
ave2 = abs((X1 + X2 + X3 + X4)./4);
Have2 = 10*log10(abs((X1 + X2 + X3 + X4)./4));
plot(v2,Have2);
axis([0 1345 -25 5]);
 title('Two pulse MTI stagger ratio 25:30:27:31');
xlabel('Radial Velocity (m/s)');
 ylabel('MTI Gain (dB)'); grid on
% %Mean value of H(f)
v4 = v2; f4 = (2.*v4)/.0375;
% H1(f)
T1 = exp(-j*2*pi.*f4*a);
 T2 = exp(-j*2*pi.*f4*(a + b));
T3 = exp(-j*2*pi.*f4*(a + b + c));
X1 = 1/20.*(1 - 3.*T1 + 3.*T2 - T3).*conj(1 - 3.*T1 + 3.*T2 - T3);
H1 = 10*log10(abs(X1));
% H2(f)
T3 = exp(-j*2*pi.*f4*b);
T4 = exp(-j*2*pi.*f4*(b + c));
T5 = exp(-j*2*pi.*f4*(b + c + d));
X2 = 1/20.*(1 - 3.*T3 + 3.*T4 - T5).*conj(1 - 3.*T3 + 3.*T4 - T5);
H2 = 10*log10(abs(X2));
% H3(f)
T6 = exp(-j*2*pi.*f4*c);
T7 = exp(-j*2*pi.*f4*(c + d));
T8 = exp(-j*2*pi.*f4*(c + d + a));
X3 = 1/20.*(1 - 3.*T6 + 3.*T7 - T8).*conj(1 - 3.*T6 + 3.*T7 - T8);
H3 = 10*log10(abs(X3));
% H4(f)
T9 = exp(-j*2*pi.*f4*d); T10 = exp(-j*2*pi.*f4*(d + a));
T11 = exp(-j*2*pi.*f4*(d + a + b));
X4 = 1/20.*(1 - 3.*T9 + 3.*T10 - T11).*conj(1 - 3.*T9 + 3.*T10 - T11);
H4 = 10*log10(abs(X4));
```

```
% H(f) Average
ave4 = abs((X1 + X2 + X3 + X4)./4);
Have4 = 10*log10(abs((X1 + X2 + X3 + X4)./4));
% Plot of H(f) Average
subplot(2,1,2)
plot(v4,Have4);
axis([0 1345 -25 5]);
title('Four pulse MTI stagger ratio 25:30:27:31');
xlabel('Radial Velocity (m/s)');
ylabel('MTI Gain (dB)');
grid on
```

MATLAB Program "Fig10_23.m" Listing

```
% generates Fig. 10.23 of text
clc
close all;
clear all;
fc = 500;
f = linspace(350,650, 300);
c1 = 2*1e-5;
fc1 = c1*pi*fc^2;
L1f = 1/pi*fc1./(fc1^2 + (f-fc).^2);
c2 = 0.5*1e-5;
fc2 = c2*pi*fc^2;
L2f = 1/pi*fc2./(fc2^2 + (f-fc).^2);
plot(f,L1f,'ro-','linewidth',1.);
hold on;
plot(f,L2f,'bd-','linewidth',1.);
xlabel('\bfFrequency - Hz');
ylabel('\bfL(f), ratio of noise power to carrier power');
axis([300 700 0 0.09]);
title('\bf fc=500Hz')
grid on;
legend('cp=2e-5','cp=0.5*1e-5');
```

MATLAB Program "Fig10_24.m" Listing

```
% generates Fig. 10.24 of text
clc; close all
clear all;
fc = 500;
f = [0.01:.01:100];
fb = pi/2;
Lf = 1/pi*fb./(fb^2 + (f-fc+fc).^2);
semilogx((f),10*log10(Lf),'b','linewidth',1.5);
xlabel('\bfFrequency - Hz');
ylabel('\bfRatio of noise power to carrier power - dBc/Hz');
title('\bf fb= \pi /2')
axis([0.01 100 -50 10]); grid on;
```

MATLAB Program "Fig10_25.m" Listing

```
% generates Fig. 10.25 of text
```

```
clc
clear all
close all
format long
PRF = 1/400e-6;
taup = 1e-6;
sigma1 = 11.93;
sigma2 = 20.72;
phi0L = 10^-15.00;
phi0U = 10^-8;
phi0 = linspace(phi0L,phi0U,150000);
phi_ratio = phi0 ./ taup;
 % two-pulse MTI
%%%% sigma1 %%%%
gns = 1/2 * (2*pi*sigma1/PRF)^2
den = gns + phi_ratio;
CA_NS = 10*log10(1.0 ./ den);
%%%% sigma2 %%%%%
gs = 1/2 * (2*pi*sigma2/PRF)^2;
den = gs + phi_ratio;
CA_S = 10*log10(1.0 ./ den);
x_axis = 10*log10(phi0);
figure(1)
plot(x_axis,CA_NS,'r-.',x_axis,CA_S,'b','linewidth',1.5)
grid
% axis([-135 -90 20 35])
xlabel('\bfPhase noise - dBc/HZ')
ylabel('\bfClutter attenuation - dB')
legend('\sigmaf = 11.92', '\sigmaf = 20.72')
```

Part IV

Radar Detection

Chapter 11

Random Variables and Random Processes

The material in Part IV of this book, requires a strong background in random variables and random processes. Users of this book are advised to use this chapter as means for a quick top-level review of random variables and random processes. Instructors using this book as a text may assign Chapter 11 as a reading assignment to their students. This chapter is written in such a way that it only highlights the major points of the subject.

11.1. Random Variables

Consider an experiment with outcomes defined by a certain sample space. The rule or functional relationship that maps each point in this sample space into a real number is called a random variable. Random variables are designated by capital letters (e.g., $X, Y, \ldots$), and a particular value of a random variable is denoted by a lowercase letter (e.g., $x, y, \ldots$).

The Cumulative Distribution Function (*cdf*) associated with the random variable X is denoted as $F_X(x)$ and is interpreted as the total probability that the random variable X is less than or equal to the value x. More precisely,

$$F_X(x) = Pr\{X \le x\}.$$

Eq. (11.1)

The probability that the random variable X is in the interval (x_1, x_2) is then given by

$$F_X(x_2) - F_X(x_1) = Pr\{x_1 \le X \le x_2\}.$$

Eq. (11.2)

It is often practical to describe a random variable by the derivative of its *cdf*, which is called the Probability Density Function *(pdf)*. The *pdf* of the random variable X is

$$f_X(x) = \frac{\mathrm{d}}{\mathrm{d}x} F_X(x)$$

Eq. (11.3)

or, equivalently,

$$F_X(x) = Pr\{X \le x\} = \int_{-\infty}^{x} f_X(\lambda) d\lambda.$$

Eq. (11.4)

It follows that Eq. (11.2) can be written in the following equivalent form

$$F_X(x_2) - F_X(x_1) = Pr\{x_1 \leq X \leq x_2\} = \int_{x_1}^{x_2} f_X(x)dx.$$
Eq. (11.5)

The *cdf* has the following properties:

$$0 \leq F_X(x) \leq 1$$
$$F_X(-\infty) = 0$$
$$F_X(\infty) = 1$$
$$F_X(x_1) \leq F_X(x_2) \Leftrightarrow x_1 \leq x_2$$
Eq. (11.6)

Define the *nth* moment for the random variable X as

$$E[X^n] = \overline{X^n} = \int_{-\infty}^{\infty} x^n f_X(x)dx.$$
Eq. (11.7)

The first moment, $E[X]$, is called the mean value, while the second moment, $E[X^2]$, is called the mean squared value. When the random variable X represents an electrical signal across a 1Ω resistor, then $E[X]$ is the DC component, and $E[X^2]$ is the total average power.

The *nth* central moment is defined as

$$E[(X-\overline{X})^n] = \overline{(X-\overline{X})^n} = \int_{-\infty}^{\infty} (x-\bar{x})^n f_X(x)dx.$$
Eq. (11.8)

and thus the first central moment is zero. The second central moment is called the variance and is denoted by the symbol σ_X^2,

$$\sigma_X^2 = E[(X-\overline{X})^2] = \overline{(X-\overline{X})^2}.$$
Eq. (11.9)

In practice, the random nature of an electrical signal may need to be described by more than one random variable. In this case, the joint *cdf* and *pdf* functions need to be considered. The joint *cdf* and *pdf* for the two random variables X and Y are, respectively, defined by

$$F_{XY}(x,y) = Pr\{X \leq x; Y \leq y\}$$
Eq. (11.10)

$$f_{XY}(x,y) = \frac{\partial^2}{\partial x \partial y} F_{XY}(x,y).$$
Eq. (11.11)

The marginal *cdf*s are obtained as follows:

$$F_X(x) = \int_{-\infty}^{\infty} \int_{-\infty}^{x} f_{UV}(u,v)dudv = F_{XY}(x,\infty)$$
$$F_Y(y) = \int_{-\infty}^{\infty} \int_{-\infty}^{y} f_{UV}(u,v)dvdu = F_{XY}(\infty,y)$$
Eq. (11.12)

If the two random variables are statistically independent, then the joint *cdf*s and *pdf*s are, respectively, given by

$$F_{XY}(x, y) = F_X(x)F_Y(y) \qquad \text{Eq. (11.13)}$$

$$f_{XY}(x, y) = f_X(x)f_Y(y). \qquad \text{Eq. (11.14)}$$

Consider a case when the two random variables X and Y are mapped into two new variables U and V through some transformations T_1 and T_2 defined by

$$U = T_1(X, Y) \qquad ; \quad V = T_2(X, Y). \qquad \text{Eq. (11.15)}$$

The joint *pdf*, $f_{UV}(u, v)$, may be computed based on the invariance of probability under the transformation. For this purpose, one must first compute the matrix of derivatives; then the new joint *pdf* is computed as

$$f_{UV}(u, v) = f_{XY}(x, y)|\mathbf{J}| \qquad \text{Eq. (11.16)}$$

$$|\mathbf{J}| = \begin{vmatrix} \dfrac{\partial x}{\partial u} & \dfrac{\partial x}{\partial v} \\ \dfrac{\partial y}{\partial u} & \dfrac{\partial y}{\partial v} \end{vmatrix} \qquad \text{Eq. (11.17)}$$

where the determinant of the matrix of derivatives $|\mathbf{J}|$ is called the Jacobian. The characteristic function for the random variable X is defined as

$$C_X(\omega) = E[e^{j\omega X}] = \int_{-\infty}^{\infty} f_X(x)e^{j\omega x}\, dx. \qquad \text{Eq. (11.18)}$$

The characteristic function can be used to compute the *pdf* for a sum of independent random variables. More precisely, let the random variable Y be

$$Y = X_1 + X_2 + \dots + X_M \qquad \text{Eq. (11.19)}$$

where $\{X_m \; ; \; i = 1, \dots, M\}$ is a set of independent random variables. It can be shown that

$$C_Y(\omega) = C_{X_1}(\omega)C_{X_2}(\omega)\dots C_{X_M}(\omega) \qquad \text{Eq. (11.20)}$$

and the *pdf* $f_Y(y)$ is computed as the inverse Fourier transform of $C_Y(\omega)$ is

$$f_Y(y) = \frac{1}{2\pi} \int_{-\infty}^{\infty} C_Y(\omega)e^{-j\omega y}\, d\omega. \qquad \text{Eq. (11.21)}$$

The characteristic function may also be used to compute the *nth* moment for the random variable X as

$$E[X^n] = (-j)^n \frac{d^n}{d\omega^n} C_X(\omega) \Big|_{\omega = 0}. \qquad \text{Eq. (11.22)}$$

11.2. Multivariate Gaussian Random Vector

Consider a joint probability for M random variables, $X_1, X_2, ..., X_M$. These variables can be represented as components of an $M \times 1$ random column vector, $\mathbf{x}$. More precisely,

$$\mathbf{x} = \begin{bmatrix} X_1 \\ X_2 \\ \vdots \\ X_M \end{bmatrix}.$$

Eq. (11.23)

The joint *pdf* for the vector $\mathbf{x}$ is

$$f_X(\mathbf{x}) = f_{X_1, X_2, ..., X_M}(x_1, x_2, ..., x_M).$$

Eq. (11.24)

The mean vector is defined as

$$\mu_{\mathbf{x}} = \begin{bmatrix} E[X_1] \\ E[X_2] \\ \vdots \\ E[X_M] \end{bmatrix}$$

Eq. (11.25)

and the covariance is an $M \times M$ matrix given by

$$\mathbf{C}_X = E[\mathbf{x}\ \mathbf{x}^t] - \mu_{\mathbf{x}}\ \mu_{\mathbf{x}}^t$$

Eq. (11.26)

where the superscript t indicates the transpose operation. Note that if the elements of the vector $\mathbf{x}$ are independent, then the covariance matrix is a diagonal matrix.

A random vector $\mathbf{x}$ is multivariate Gaussian if its *pdf* is of the form

$$f_X(\mathbf{x}) = \frac{1}{\sqrt{(2\pi)^M |\mathbf{C}_X|}}\ \exp\left(-\frac{1}{2}(\mathbf{x} - \mu_{\mathbf{x}})^t \mathbf{C}_X^{-1}(\mathbf{x} - \mu_{\mathbf{x}})\right)$$

Eq. (11.27)

where $\mu_{\mathbf{x}}$ is the mean vector, $\mathbf{C}_X$ is the covariance matrix, $\mathbf{C}_X^{-1}$ is inverse of the covariance matrix, $|\mathbf{C}_X|$ is its determinant, and $\mathbf{x}$ is of dimension $M \times 1$. If $\mathbf{A}$ is a $K \times M$ matrix of rank K, then the random vector $\mathbf{y} = \mathbf{A}\mathbf{x}$ is a K-variate Gaussian vector with

$$\mu_{\mathbf{y}} = \mathbf{A}\mu_{\mathbf{x}}$$

Eq. (11.28)

$$\mathbf{C}_Y = \mathbf{A}\ \mathbf{C}_X\ \mathbf{A}^t.$$

Eq. (11.29)

The characteristic function for a multivariate Gaussian *pdf* is defined by

$$C_X = E[\exp\{j(\omega_1 X_1 + \omega_2 X_2 + ... + \omega_M X_M)\}] = \exp\left\{j\mu_{\mathbf{x}}^t\omega - \frac{1}{2}\omega^t\mathbf{C}_X\omega\right\}.$$

Eq. (11.30)

Then the moments for the joint distribution can be obtained by partial differentiation. For example,

$$E[X_1 X_2 X_3] = \frac{\partial^3}{\partial \omega_1 \partial \omega_2 \partial \omega_3} C_X(\omega_1, \omega_2, \omega_3) \quad at \quad \omega = \mathbf{0}. \qquad \text{Eq. (11.31)}$$

A special case of Eq. (11.29) is when the matrix **A** is given by

$$\mathbf{A} = \begin{bmatrix} a_1 a_2 \ \dots \ a_M \end{bmatrix}. \qquad \text{Eq. (11.32)}$$

It follows that $\mathbf{y} = \mathbf{A}\mathbf{x}$ is a sum of random variables X_m, that is

$$Y = \sum_{m=1}^{M} a_m \ X_m. \qquad \text{Eq. (11.33)}$$

The finding in Eq. (11.33) leads to the conclusion that the linear sum of Gaussian variables is also a Gaussian variable with mean and variance given by

$$\bar{Y} = a_1 \bar{X}_1 + a_2 \bar{X}_2 + \dots + a_M \bar{X}_M \qquad \text{Eq. (11.34)}$$

$$\sigma_Y^2 = E[(X - \bar{X})^2] = E[a_1^2 (X_1 - \bar{X}_1)^2] + \qquad , \qquad \text{Eq. (11.35)}$$
$$E[a_2^2 (X_2 - \bar{X}_2)^2] + \dots + E[a_M^2 (X_M - \bar{X}_M)^2]$$

and if the variables X_i are independent then Eq. (11.35) reduces to

$$\sigma_Y^2 = a_1^2 \sigma_{X_1}^2 + a_2^2 \sigma_{X_2}^2 + \dots + a_M^2 \sigma_{X_M}^2. \qquad \text{Eq. (11.36)}$$

Finally, in this case, the probability density function $f_Y(y)$ is given by (which can also be derived from Eq. (11.20))

$$f_Y(y) = f_{X_1}(x_1) \otimes f_{X_2}(x_2) \otimes \dots \otimes f_{X_M}(x_M) \qquad \text{Eq. (11.37)}$$

where $\otimes$ indicates convolution.

Example:

Define

$$\mathbf{x}_1 = \begin{bmatrix} X_1 \\ X_2 \end{bmatrix} \qquad \mathbf{x}_2 = \begin{bmatrix} X_3 \\ X_4 \end{bmatrix}$$

and the vector **x** *as a 4-variate Gaussian with*

$$\mu_\mathbf{x} = \begin{bmatrix} 2 \\ 1 \\ 1 \\ 0 \end{bmatrix} \ and \ \mathbf{C}_X = \begin{bmatrix} 6 & 3 & 2 & 1 \\ 3 & 4 & 3 & 2 \\ 2 & 3 & 4 & 3 \\ 1 & 2 & 3 & 3 \end{bmatrix}.$$

Find the distribution of **x**$_1$ *and the distribution of*

$$\mathbf{y} = \begin{bmatrix} 2\mathbf{x}_1 \\ \mathbf{x}_1 + 2\mathbf{x}_2 \\ \mathbf{x}_3 + \mathbf{x}_4 \end{bmatrix}.$$

Solution:

$\mathbf{x}_1$ *has a bivariate Gaussian distribution with*

$$\mu_{\mathbf{x}_1} = \begin{bmatrix} 2 \\ 1 \end{bmatrix} \qquad \mathbf{C}_{X_1} = \begin{bmatrix} 6 & 3 \\ 3 & 4 \end{bmatrix}.$$

The vector $\mathbf{y}$ *can be expressed as*

$$\mathbf{y} = \begin{bmatrix} 2 & 0 & 0 & 0 \\ 1 & 2 & 0 & 0 \\ 0 & 0 & 1 & 1 \end{bmatrix} \begin{bmatrix} \mathbf{x}_1 \\ \mathbf{x}_2 \\ \mathbf{x}_3 \\ \mathbf{x}_4 \end{bmatrix} = \mathbf{Ax}.$$

It follows that

$$\mu_{\mathbf{y}} = \mathbf{A}\mu_{\mathbf{x}} = \begin{bmatrix} 2 & 0 & 0 & 0 \\ 1 & 2 & 0 & 0 \\ 0 & 0 & 1 & 1 \end{bmatrix} \begin{bmatrix} 2 \\ 1 \\ 1 \\ 0 \end{bmatrix} = \begin{bmatrix} 4 \\ 4 \\ 1 \end{bmatrix}$$

$$\mathbf{C}_Y = \mathbf{A}\mathbf{C}_X\mathbf{A}^t = \begin{bmatrix} 2 & 0 & 0 & 0 \\ 1 & 2 & 0 & 0 \\ 0 & 0 & 1 & 1 \end{bmatrix} \begin{bmatrix} 6 & 3 & 2 & 1 \\ 3 & 4 & 3 & 2 \\ 2 & 3 & 4 & 3 \\ 1 & 2 & 3 & 3 \end{bmatrix} \begin{bmatrix} 2 & 1 & 0 \\ 0 & 2 & 0 \\ 0 & 0 & 1 \\ 0 & 0 & 1 \end{bmatrix} = \begin{bmatrix} 24 & 24 & 6 \\ 24 & 34 & 13 \\ 6 & 13 & 13 \end{bmatrix}.$$

11.2.1. Complex Multivariate Gaussian Random Vector

Consider the complex envelope for the $M \times 1$ vector random variable $\tilde{X}$ is,

$$\tilde{\mathbf{x}} = \mathbf{x}_I + j\mathbf{x}_Q \qquad\qquad \text{Eq. (11.38)}$$

where $\mathbf{x}_I$ and $\mathbf{x}_Q$ are real random multivariate Gaussian random vectors. The joint *pdf* for the complex random vector $\tilde{\mathbf{x}}$ is computed from the joint *pdf* of the two real vectors. The mean for the vector $\tilde{\mathbf{x}}$ is

$$E[\tilde{\mathbf{x}}] = E[\mathbf{x}_I] + jE[\mathbf{x}_Q]. \qquad\qquad \text{Eq. (11.39)}$$

The covariance matrix is also defined by

$$\mathbf{C}_{\tilde{X}} = E[(\tilde{\mathbf{x}} - E[\tilde{\mathbf{x}}])(\tilde{\mathbf{x}} - E[\tilde{\mathbf{x}}])^\dagger] \qquad\qquad \text{Eq. (11.40)}$$

where the operator $\dagger$ indicates complex conjugate transpose.

The *pdf* for the vector $\tilde{\mathbf{x}}$ is

$$f_{\tilde{X}}(\tilde{\mathbf{x}}) = \frac{\exp[-(\tilde{\mathbf{x}} - E[\tilde{\mathbf{x}}])^{\dagger}\mathbf{C}_{\tilde{X}}^{-1}(\tilde{\mathbf{x}} - E[\tilde{\mathbf{x}}])]}{\pi^{M}|\mathbf{C}_{\tilde{X}}|}$$

Eq. (11.41)

with the following three conditions holding true

$$E[(\mathbf{x}_{I_i} - E[\mathbf{x}_{I_i}])(\mathbf{x}_{Q_i} - E[\mathbf{x}_{Q_i}])^{\dagger}] = \mathbf{0}$$

Eq. (11.42)

$$E[(\mathbf{x}_{I_i} - E[\mathbf{x}_{I_i}])(\mathbf{x}_{I_k} - E[\mathbf{x}_{I_k}])^{\dagger}] = E[(\mathbf{x}_{Q_i} - E[\mathbf{x}_{Q_i}])(\mathbf{x}_{Q_k} - E[\mathbf{x}_{Q_k}])^{\dagger}] \; ;all \; i, k$$

Eq. (11.43)

$$E[(\mathbf{x}_{I_i} - E[\mathbf{x}_{I_i}])(\mathbf{x}_{Q_k} - E[\mathbf{x}_{Q_k}])^{\dagger}] = -E[(\mathbf{x}_{Q_i} - E[\mathbf{x}_{Q_i}])(\mathbf{x}_{I_k} - E[\mathbf{x}_{I_k}])^{\dagger}] \; ; \; i \neq k.$$

Eq. (11.44)

11.3. Rayleigh Random Variables

Let X_I and X_Q be zero mean independent Gaussian random variables with zero mean and variance σ^2. Define two new random variables R and Φ as

$$\begin{aligned} X_I &= R\cos\Phi \\ X_Q &= R\sin\Phi \end{aligned}.$$

Eq. (11.45)

The joint *pdf* of the two random variables $X_I;X_Q$ is

$$f_{X_I X_Q}(x_I, x_Q) = \frac{1}{2\pi\sigma^2}\exp\left(-\frac{x_I^2 + x_Q^2}{2\sigma^2}\right) = \frac{1}{2\pi\sigma^2}\exp\left(-\frac{(r\cos\varphi)^2 + (r\sin\varphi)^2}{2\sigma^2}\right).$$

Eq. (11.46)

The joint *pdf* for the two random variables $R;\Phi$ is given by

$$f_{R\Phi}(r, \varphi) = f_{X_I X_Q}(x_I, x_Q)|\mathbf{J}|$$

Eq. (11.47)

where $[\mathbf{J}]$ is a matrix of derivatives defined by

$$[\mathbf{J}] = \begin{bmatrix} \dfrac{\partial x_I}{\partial r} & \dfrac{\partial x_I}{\partial \varphi} \\ \dfrac{\partial x_Q}{\partial r} & \dfrac{\partial x_Q}{\partial \varphi} \end{bmatrix} = \begin{bmatrix} \cos\varphi & -r\sin\varphi \\ \sin\varphi & r\cos\varphi \end{bmatrix}.$$

Eq. (11.48)

The determinant of the matrix of derivatives is called the Jacobian, and in this case it is equal to

$$|\mathbf{J}| = r$$

Eq. (11.49)

Substituting Eqs. (11.46) and (11.49) into Eq. (11.47) and collecting terms yields

$$f_{R\Phi}(r, \varphi) = \frac{r}{2\pi\sigma^2}\exp\left(-\frac{(r\cos\varphi)^2 + (r\sin\varphi)^2}{2\sigma^2}\right) = \frac{r}{2\pi\sigma^2}\exp\left(-\frac{r^2}{2\sigma^2}\right).$$

Eq. (11.50)

The *pdf* for R alone is obtained by integrating Eq. (11.50) over φ

$$f_R(r) = \int_0^{2\pi} f_{R\Phi}(r, \varphi)d\varphi = \frac{r}{\sigma^2}\exp\left(-\frac{r^2}{2\sigma^2}\right)\frac{1}{2\pi}\int_0^{2\pi} d\varphi \qquad \text{Eq. (11.51)}$$

where the integral inside Eq. (11.51) is equal to 2π ; thus,

$$f_R(r) = \frac{r}{\sigma^2}\exp\left(-\frac{r^2}{2\sigma^2}\right) \quad ;r \geq 0 \ . \qquad \text{Eq. (11.52)}$$

The *pdf* described in Eq. (11.52) is referred to as a Rayleigh probability density function. The density function for the random variable Φ is obtained from

$$f_\Phi(\varphi) = \int_0^r f(r, \varphi) \ dr \ . \qquad \text{Eq. (11.53)}$$

substituting Eq. (11.50) into Eq. (11.53) and performing integration by parts yields

$$f_\Phi(\varphi) = \frac{1}{2\pi} \quad ; \ 0 < \varphi < 2\pi \ , \qquad \text{Eq. (11.54)}$$

which is a uniform probability density function.

11.4. The Chi-Square Random Variables

11.4.1. Central Chi-Square Random Variable with N Degrees of Freedom

Let the random variables $\{X_1, X_2, ..., X_M\}$ be zero mean, statistically independent Gaussian random variable with unity variance. The variable

$$\chi_N^2 = \sum_{m=1}^M X_m^2 \qquad \text{Eq. (11.55)}$$

is the central chi-square random variable with M degrees of freedom. The chi-square *pdf* is

$$f_{\chi_M^2}(x) = \begin{cases} \dfrac{x^{(M-2)/2}\ e^{(-x/2)}}{2^{M/2}\ \Gamma(M/2)} & x \geq 0 \\ 0 & x < 0 \end{cases} \qquad \text{Eq. (11.56)}$$

where the Gamma function is defined as

$$\Gamma(m) = \int_0^\infty \lambda^{m-1} e^{-\lambda} \ d\lambda \ ; \ m > 0 \qquad \text{Eq. (11.57)}$$

with the following recursion

$$\Gamma(m+1) = m\Gamma(m) \qquad \text{Eq. (11.58)}$$

and

$$\Gamma(m+1) = m! \qquad ; \; m = 0, 1, 2, ..., \; and \; 0! = 1.$$

Eq. (11.59)

The mean and variance for the central chi-square are, respectively, given by

$$E[\chi_M^2] = M$$

Eq. (11.60)

$$\sigma_{\chi_M^2}^2 = 2M.$$

Eq. (11.61)

Hence, the degrees of freedom M is the ratio of twice the squared mean to the variance

$$M = (2E^2[\chi_M^2]) / \sigma_{\chi_M^2}^2.$$

Eq. (11.62)

11.4.2. Non-Central Chi-Square Random Variable with N Degrees of Freedom

In the general, the chi-square random variable requires that the Gaussian random variables $\{X_1, X_2, ..., X_M\}$ do not have zero means. Define a multivariate random variable $\mathbf{y}$ such that

$$Y_m = X_m + \mu_{X_m} \; ; m = 1, 2, ..., M$$

Eq. (11.63a)

$$\mathbf{y} = \mathbf{x} + \mu_\mathbf{x}.$$

Eq. (11.63b)

Consider the random variable

$$\chi'^2_M = \sum_{m=1}^{M} Y_m^2 = \sum_{m=1}^{M} (X_m + \mu_{X_m})^2.$$

Eq. (11.64)

the variable χ'^2_M is called the non-central chi-square random variable with M degrees of freedom and with a non-central parameter λ, where

$$\lambda = \sum_{m=1}^{M} \mu_{X_m}^2 = \sum_{m=1}^{M} E^2[Y_m].$$

Eq. (11.65)

The non-central chi-square *pdf* is

$$f_{\chi'^2_M}(x) = \begin{cases} \left(\dfrac{1}{2}\right)\left(\dfrac{x}{\lambda}\right)^{(M-2)/4} e^{[-(x+\lambda)/2]} I_{(M-2)/2}(\sqrt{\lambda x}) & x \geq 0 \\ \\ 0 & x < 0 \end{cases}$$

Eq. (11.66)

where I is the modified Bessel function (or occasionally called the hyperbolic Bessel function) of the first kind; and the subscripts are referred to as its order.

11.5. Random Processes

A random variable X is by definition a mapping of all possible outcomes of a random experiment to numbers. When the random variable becomes a function of both the outcome of the experiment and time, it is called a random process and is denoted by $X(t)$. Thus, one can view a random process as an ensemble of time-domain functions that are the outcome of a certain random experiment, as compared with single real numbers in the case of a random variable.

Since the *cdf* and *pdf* of a random process are time dependent, we will denote them as $F_X(x;t)$ and $f_X(x;t)$, respectively. The *nth* moment for the random process $X(t)$ is

$$E[X^n(t)] = \int_{-\infty}^{\infty} x^n f_X(x;t) dx.$$

Eq. (11.67)

A random process $X(t)$ is referred to as stationary to order one if all its statistical properties do not change with time. Consequently, $E[X(t)] = \bar{X}$, where $\bar{X}$ is a constant. A random process $X(t)$ is called stationary to order two (or wide-sense stationary) if

$$f_X(x_1, x_2; t_1, t_2) = f_X(x_1, x_2; t_1 + \Delta t, t_2 + \Delta t)$$

Eq. (11.68)

for all t_1, t_2 and Δt.

Define the statistical autocorrelation function for the random process $X(t)$ as

$$\Re_X(t_1, t_2) = E[X(t_1)X(t_2)].$$

Eq. (11.69)

The correlation $E[X(t_1)X(t_2)]$ is, in general, a function of (t_1, t_2). As a consequence of the wide-sense stationary definition, the autocorrelation function depends on the time difference $\tau = t_2 - t_1$, rather than on absolute time; and thus, for a wide-sense stationary process we have

$$\begin{aligned} E[X(t)] &= \bar{X} \\ \Re_X(\tau) &= E[X(t)X(t+\tau)] \end{aligned}.$$

Eq. (11.70)

If the time average and time correlation functions are equal to the statistical average and statistical correlation functions, the random process is referred to as an ergodic random process. The following is true for an ergodic process:

$$\lim_{T \to \infty} \frac{1}{T} \int_{-T/2}^{T/2} x(t) dt = E[X(t)] = \bar{X}$$

Eq. (11.71)

$$\lim_{T \to \infty} \frac{1}{T} \int_{-T/2}^{T/2} x^*(t)x(t+\tau) dt = \Re_X(\tau).$$

Eq. (11.72)

The covariance of two random processes $X(t)$ and $Y(t)$ is defined by

$$C_{XY}(t, t+\tau) = E[\{X(t) - E[X(t)]\}\{Y(t+\tau) - E[Y(t+\tau)]\}],$$

Eq. (11.73)

which can also be written as

$$C_{XY}(t, t+\tau) = \Re_{XY}(\tau) - \bar{X}\bar{Y}.$$

Eq. (11.74)

11.6. The Gaussian Random Process

Let $X(t)$ be a random process defined over the interval $\{0, T\}$, then $X(t)$ is said to be a Gaussian random process if every possible outcome of this process over this interval is a Gaussian process, provided that the mean square value of this process is finite. More precisely, $Y(t)$ will be a random process over the same interval $\{0, T\}$ defined by

$$Y(t) = \int_0^T x(t)z(t) \; dt \qquad \qquad \textbf{Eq. (11.75)}$$

where $z(t)$ is any function that yields $E[|Y|^2] < \infty$.

Gaussian random processes have a few unique properties that distinguish them from other types of random processes. (1) If the input to an LTI system is said to be a Gaussian random process, then its output is also a Gaussian random process. (2) If $X(t)$ is a Gaussian random process for any set of time occurrences $\{t_1, t_2, ..., t_M\}$, then the random variables $\{X(t_1), X(t_2), ..., X(t_M)\}$ are jointly Gaussian random variables. Finally, (3) any linear combination of a Gaussian process yields another jointly Gaussian random variable.

11.6.1. Lowpass Gaussian Random Processes

Let $X(t)$ be a real-valued Gaussian random process. If this process is an essentially band-limited process (recall the definition of an essentially band-limited signals in Chapter3) over the frequency interval $\{-B, B\}$, then the minimal number of samples required to represent this process is $M = 2TB$ real samples. Therefore, over the interval $\{0, T\}$, there are M random variables represented by the vector $\mathbf{x}$ made of M random variables, that is

$$\mathbf{x} = \begin{bmatrix} X(1/2B) \\ X(1/B) \\ \vdots \\ X(M/2B) \end{bmatrix} = \begin{bmatrix} X_1 \\ X_2 \\ \vdots \\ X_M \end{bmatrix}. \qquad \qquad \textbf{Eq. (11.76)}$$

If the random process $X(t)$ is a complex lowpass Gaussian process, represented by its complex envelop $\tilde{X}(t)$, then in this case, the minimal number of samples required to represent this process is $M = TB$ complex samples. The resulting jointly Gaussian complex random vector comprising $M = TB$ complex random variables is

$$\tilde{\mathbf{x}} = \begin{bmatrix} \tilde{X}(1/B) \\ \tilde{X}(2/B) \\ \vdots \\ \tilde{X}(M/B) \end{bmatrix} = \begin{bmatrix} \tilde{X}_1 \\ \tilde{X}_2 \\ \vdots \\ \tilde{X}_M \end{bmatrix}. \qquad \qquad \textbf{Eq. (11.77)}$$

If the power spectral destiny of a real Gaussian random process $X(t)$ is defined by

$$S_X(f) = \begin{pmatrix} S_o & ; |f| < B \\ 0 & ; otherwise \end{pmatrix}, \qquad \qquad \textbf{Eq. (11.78)}$$

then the probability density function of the vector **x** is given by

$$f_X(x(t)) = \frac{1}{(4\pi S_o B)^{M/2}} \exp\left[\left(-\frac{1}{4S_o B}\right)\sum_{m=1}^{M} X_m\right] = \frac{1}{(4\pi S_o B)^{M/2}}\exp\left(-\frac{\mathbf{x}'\mathbf{x}}{4S_o B}\right).$$ Eq. (11.79)

The mean of the random process defined in Eq. (11.76) is

$$\mu_{\mathbf{x}} = \begin{bmatrix} E[X_1] \\ E[X_2] \\ \vdots \\ E[X_M] \end{bmatrix} = \begin{bmatrix} \mu_1 \\ \mu_2 \\ \vdots \\ \mu_M \end{bmatrix}.$$ Eq. (11.80)

When the power spectral density of the process is non-white over the bandwidth, then in this case the random variables defined in Eq. (11.76) are no longer independent. Therefore, the *pdf* given in Eq. (11.79) is modified to

$$f_X(x(t)) = \frac{1}{\sqrt{(2\pi)^M \, \mathbf{C}_X}} \exp\left[\left(-\frac{1}{2}\right)(\mathbf{x}-\mu_{\mathbf{x}})'\mathbf{C}_X^{-1}(\mathbf{x}-\mu_{\mathbf{x}})\right]$$ Eq. (11.81)

where the covariance matrix is

$$\mathbf{C}_X = E[(\mathbf{x}-\mu_{\mathbf{x}})(\mathbf{x}-\mu_{\mathbf{x}})'].$$ Eq. (11.82)

11.6.2. Bandpass Gaussian Random Processes

It is customary to define the bandpass Gaussian random process through its complex envelope as

$$\tilde{X}(t) = X_I(t) + jX_Q(t)$$ Eq. (11.83)

where both $X_I(t)$ and $X_Q(t)$ are jointly lowpass statistically independent and stationary Gaussian random processes with zero mean and equal variance σ^2. The *pdf* for a sample $\tilde{X}(t_0)$ of the complex envelope is the joint *pdf* for $X_I(t)$ and $X_Q(t)$. That is,

$$f_X(\tilde{x}(t_0)) = \frac{1}{2\pi\sigma^2}\exp\left[-\frac{x_I^2(t_0) + x_Q^2(t_0)}{2\sigma^2}\right] = \frac{1}{2\pi\sigma^2}\exp\left[-\frac{|\tilde{x}(t_0)|^2}{2\sigma^2}\right].$$ Eq. (11.84)

Now, if both lowpass processes do not have zero mean and instead have a mean defined by

$$\mu(t) = \mu_I(t)\cos(2\pi f_0 t) + j\mu_Q(t)\sin(2\pi f_0 t),$$ Eq. (11.85)

the mean complex envelope is

$$\tilde{\mu}(t) = \mu_I(t) + j\mu_Q(t).$$ Eq. (11.86)

It follows that Eq. (11.84) can be rewritten as

$$f_X(\tilde{x}(t_0)) = \frac{1}{2\pi\sigma^2}\exp\left[-\frac{[x_I(t_0) - \mu_I(t_0)]^2 + [x_Q(t_0) - \mu_Q(t_0)]^2}{2\sigma^2}\right] = \qquad \text{Eq. (11.87)}$$

$$\frac{1}{2\pi\sigma^2}\exp\left[\frac{|\tilde{x}(t_0) - \tilde{\mu}(t_0)|^2}{2\sigma^2}\right]$$

Consider a duration of the process that spans the interval $\{0, T\}$. Then this segment of the complex envelope of the random process can be represented using a complex random variable vector of at least $M = BT$ elements where B is the bandwidth of the process. Define

$$\tilde{X}_i = \tilde{X}\left(\frac{m}{B}\right) \quad ;m = 1, 2, \ldots, BT = M \qquad \text{Eq. (11.88)}$$

$$\tilde{\mathbf{x}} = \begin{bmatrix} \tilde{X}_1 \\ \tilde{X}_2 \\ \vdots \\ \tilde{X}_M \end{bmatrix}. \qquad \text{Eq. (11.89)}$$

By definition, the covariance matrix $\mathbf{C}$ is

$$\tilde{\mathbf{C}}_X = E[(\tilde{\mathbf{x}} - \tilde{\mu}_\mathbf{x})(\tilde{\mathbf{x}} - \tilde{\mu}_\mathbf{x})^\dagger] = 2(\tilde{\mathbf{C}}_{X_I} + j\tilde{\mathbf{C}}_{X_{IQ}}) \qquad \text{Eq. (11.90)}$$

where

$$\tilde{\mathbf{C}}_{X_I} = E[(\tilde{\mathbf{x}}_I - \tilde{\mu}_{\mathbf{x}_I})(\tilde{\mathbf{x}}_I - \tilde{\mu}_{\mathbf{x}_I})^\dagger] \qquad \text{Eq. (11.91)}$$

$$\tilde{\mathbf{C}}_{X_{IQ}} = E[(\tilde{\mathbf{x}}_I - \tilde{\mu}_{\mathbf{x}_I})(\tilde{\mathbf{x}}_Q - \tilde{\mu}_{\mathbf{x}_Q})^\dagger]. \qquad \text{Eq. (11.92)}$$

Therefore, the *pdf* for the segment $\{\tilde{X}(t) \ ;0 < t < T\}$ is

$$f_X(\tilde{\mathbf{x}}) = \frac{\exp[-(\tilde{\mathbf{x}} - \tilde{\mu}_\mathbf{x})^\dagger \tilde{\mathbf{C}}_X^{-1}(\tilde{\mathbf{x}} - \tilde{\mu}_\mathbf{x})]}{\pi^N|\tilde{\mathbf{C}}_X|}. \qquad \text{Eq. (11.93)}$$

11.6.3. The Envelope of a Bandpass Gaussian Process

Consider the *pdf* of a segment of the envelope of a bandpass Gaussian random process. This process can be expressed as

$$X(t) = X_I(t)\cos(2\pi f_0 t) - X_Q(t)\sin(2\pi f_0 t) \qquad \text{Eq. (11.94)}$$

where $X_I(t)$ and $X_Q(t)$ are zero mean independent lowpass Gaussian processes. The envelope and phase are respectively denoted by $R(t)$ and $\Phi(t)$, where

$$R(t) = \sqrt{X_I(t)^2 + X_Q(t)^2} \qquad \text{Eq. (11.95)}$$

and

$$\Phi(t) = \left[\tan\left(\frac{X_Q(t)}{X_I(t)}\right) \right]^{-1}$$ **Eq. (11.96)**

where

$$X_I(t) = R(t)\cos(\Phi(t))$$
$$X_Q(t) = R(t)\sin(\Phi(t))$$ **Eq. (11.97)**

The two processes $R(t)$ and $\Phi(t)$ are also independent, and their respective *pdfs* were derived in Section 11.3 and were given in Eqs. (11.52) and (11.54), respectively.

Problems

11.1. Suppose you want to determine an unknown DC voltage v_{dc} in the presence of additive white Gaussian noise $n(t)$ of zero mean and variance σ_n^2. The measured signal is $x(t) = v_{dc} + n(t)$. An estimate of v_{dc} is computed by making three independent measurements of $x(t)$ and computing the arithmetic mean, $v_{dc} \approx (x_1 + x_2 + x_3)/3$. (a) Find the mean and variance of the random variable v_{dc}. (b) Does the estimate of v_{dc} get better by using ten measurements instead of three? Why?

11.2. Assume the X and Y miss distances of darts thrown at a bulls-eye dart board are Gaussian with zero mean and variance σ^2. (a) Determine the probability that a dart will fall between 0.8σ and 1.2σ. (b) Determine the radius of a circle about the bull's-eye that contains 80% of the darts thrown. (c) Consider a square with side s in the first quadrant of the board. Determine s so that the probability that a dart will fall within the square is 0.07.

11.3. (a) A random voltage $v(t)$ has an exponential distribution function $f_V(v) = a\exp(-av)$, where $(a > 0); (0 \le v < \infty)$. The expected value $E[V] = 0.5$. Determine $Pr\{V > 0.5\}$.

11.4. Consider the network shown in the figure below, where $x(t)$ is a random voltage with zero mean and autocorrelation function $\Re_x(\tau) = 1 + \exp(-a|t|)$. Find the power spectrum $S_x(\omega)$. What is the transfer function? Find the power spectrum $S_v(\omega)$.

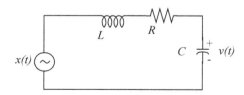

11.5. Let $\bar{S}_X(\omega)$ be the PSD function for the stationary random process $X(t)$. Compute an expression for the PSD function of

$$Y(t) = X(t) - 2X(t - T).$$

11.6. Let X be a random variable with

$$f_X(x) = \begin{cases} \dfrac{1}{\sigma} t^3 e^{-t} & t \geq 0 \\ 0 & elsewhere \end{cases}.$$

(a) Determine the characteristic function $C_X(\omega)$. (b) Using $C_X(\omega)$, validate that $f_X(x)$ is a proper *pdf*. (c) Use $C_X(\omega)$ to determine the first two moments of X. (d) Calculate the variance of X.

11.7. Let the random variable Z be written in terms of two other random variables X and Y as follows: $Z = X + 3Y$. Find the mean and variance for the new random variable in terms of the other two.

11.8. Suppose you have the following sequences of statistically independent Gaussian random variables with zero means and variances σ^2. if

$$X_1, X_2, ..., X_N \; ; \; X_i = A_i \cos\Theta_i \text{ and } Y_1, Y_2, ..., Y_N \; ; \; Y_i = A_i \sin\Theta_i.$$

Define $Z = \displaystyle\sum_{i=1}^{N} A_i^2$. Find an expression where Z exceeds a threshold value v_T.

11.9. Repeat the previous problem when two single delay line cancelers are cascaded to produce a double delay line canceler. Let $X(t)$ be a stationary random process, $E[X(t)] = 1$ and the autocorrelation $\Re_x(\tau) = 3 + \exp(-|\tau|)$. Define a new random variola Y as

$$Y = \int_0^2 x(t)dt.$$ **Eq. (11.98)**

Compute $E[Y(t)]$ and σ_Y^2.

11.10. Consider the single delay line canceler in the figure below. The input $x(t)$ is a wide-sense stationary random process with variance σ_x^2 and mean μ_x and a covariance matrix Λ. Find the mean and variance and the autocorrelation function of the output $y(t)$.

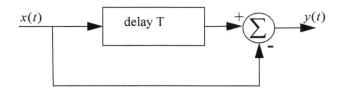

Chapter 12

Single Pulse Detection

12.1. Single Pulse with Known Parameters

In its simplest form, a radar signal can be represented by a single pulse comprising a sinusoid of known amplitude and phase. Consequently, a retuned signal will also comprise a sinusoid. Under the assumption of completely known signal parameters, a returned pulse from a target has known amplitude and known phase with no random components; and the radar signal processor will attempt to maximize the probability of detection for a given probability of false alarm. In this case, detection is referred to as coherent detection or coherent demodulation. A radar system will declare detection with a certain probability of detection if the received voltage signal envelope exceeds a pre-set threshold value. For this purpose, the radar receiver is said to employ an envelope detector.

Figure 12.1 shows a simplified block diagram of a radar matched filter receiver followed by a threshold decision logic. The signal at the input of the matched filter $s(t)$ is composed of the target echo signal $x(t)$ and additive zero mean Gaussian noise (white noise is assumed in the analysis presented in this chapter) random process $n(t)$, with variance σ^2. The input noise is assumed to be spatially incoherent and uncorrelated with the signal. The matched filter impulse response is $h(t)$, and its output is denoted by the signal $v(t)$, which was derived in Chapter 4; it is given by

$$v(t) = \int_{-\infty}^{\infty} s(t)h(t-u) \ du. \qquad \text{Eq. (12.1)}$$

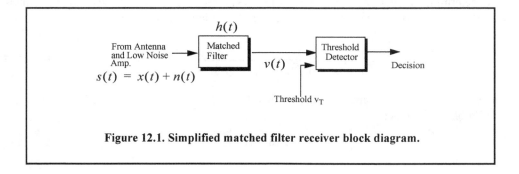

Figure 12.1. Simplified matched filter receiver block diagram.

Recall that if $n(t)$ is a Gaussian random process, then so is $s(t)$, since $x(t)$ is a deterministic signal; its only effect is a shift of the mean of the random process. Following the definition of the Gaussian process developed in Section 11.6 of Chapter 11, one concludes that the signal $v(t)$ is also a Gaussian random process, and over a coherent processing interval $\{0, T\}$, two hypotheses are considered, they are:

H_0 when the signal $s(t)$ is made of noise only, and

H_1 when the signal $s(t)$ is made of signal plus noise.

More specifically, by following the analysis in Section 11.6.1 of Chapter 11, one gets

$$H_0 \Leftrightarrow \mathbf{s} = \mathbf{n} \qquad ; 0 < t < T \qquad\qquad \text{Eq. (12.2)}$$

$$H_1 \Leftrightarrow \mathbf{s} = \mathbf{n} + \mathbf{x} \qquad ; 0 < t < T \qquad\qquad \text{Eq. (12.3)}$$

where all vectors are of size $M \times 1$ ($M = 2TB$), and B is the operating bandwidth of the receiver. It follows that,

$$E[\mathbf{n}] = \mathbf{0}$$
$$E[\mathbf{n}\mathbf{n}^\dagger] = \mathbf{C}_n \qquad\qquad \text{Eq. (12.4)}$$
$$E[\mathbf{s}/H_1] = \mathbf{x}$$

where $\mathbf{C}_n$ is the noise covariance matrix.

When the noise $n(t)$ is white, or it is band-limited white over the frequency band $\{-B, B\}$, then its power spectrum density is given by

$$\bar{S}_n(f) = \frac{\eta_o}{2} \qquad ; -B < f < B \qquad\qquad \text{Eq. (12.5)}$$

where η_o is a constant. Analysis of the non-white noise case is left as an exercise. The conditional probability for the H_0 case was derived in Eq. (11.79) of Chapter 11; it is

$$f(\mathbf{s}/H_0) = \frac{1}{(2\pi\eta_o B)^{M/2}} \exp\left(-\frac{\mathbf{s}^\dagger\mathbf{s}}{2\eta_o B}\right). \qquad\qquad \text{Eq. (12.6)}$$

Alternatively, the conditional probability for H_1 is identical to Eq. (12.6) except in this case, one must replace $\mathbf{s}$ by $\mathbf{s} - \mathbf{x}$. It follows that

$$f(\mathbf{s}/H_1) = \frac{1}{(2\pi\eta_o B)^{M/2}} \exp\left(-\frac{(\mathbf{s}-\mathbf{x})^\dagger(\mathbf{s}-\mathbf{x})}{2\eta_o B}\right). \qquad\qquad \text{Eq. (12.7)}$$

As determined earlier, the statistics associated with the random process $v(t)$ over the interval $\{0, T\}$ is Gaussian. In general, a Gaussian *pdf* function is given by

$$f_V(v) = \frac{1}{\sigma\sqrt{2\pi}} \exp\left(-\frac{(v-\bar{V})^2}{2\sigma^2}\right) \qquad\qquad \text{Eq. (12.8)}$$

where σ^2 is the variance and $\bar{V}$ is the mean value. It follows (the proof is left as an exercise (see Problem 12.1) that

$$E[V/H_0] = 0 \qquad\qquad \text{Eq. (12.9)}$$

$$Var[V/H_0] = \frac{E_x \eta_o}{2} = \sigma^2 \qquad \text{Eq. (12.10)}$$

$$E[V/H_1] = \int_0^T x^*(t)x(t) \ dt = E_x \qquad \text{Eq. (12.11)}$$

$$Var[V/H_1] = \frac{E_x \eta_o}{2} = \sigma^2 \qquad \text{Eq. (12.12)}$$

where E_x is the signal's energy.

Assuming the H_0 hypothesis, then the probability of a false P_{fa} alarm is computed from Eq. (12.8) when the signal $v(t)$ exceeds a set threshold value V_T. More specifically,

$$P_{fa} = Pr\{v(t) > V_T/H_0\} = \int_{V_T}^{\infty} \frac{1}{\sigma\sqrt{2\pi}} \exp\left(-\frac{v^2}{2\sigma^2}\right) dv. \qquad \text{Eq. (12.13)}$$

Substituting the variance as computed in Eq. (12.10) into Eq. (12.13) yields,

$$P_{fa} = \int_{V_T}^{\infty} \frac{1}{\sqrt{\pi E_x \eta_o}} \exp\left(-\frac{v^2}{E_x \eta_o}\right) dv. \qquad \text{Eq. (12.14)}$$

Making the change of variable $\zeta = v/(\sqrt{E_x \eta_o})$ yields

$$P_{fa} = \int_{\frac{V_T}{\sqrt{E_x \eta_o}}}^{\infty} \frac{1}{\sqrt{\pi}} e^{-\zeta^2} d\zeta. \qquad \text{Eq. (12.15)}$$

Multiplying and dividing Eq. (12.15) by 2 yields

$$P_{fa} = \frac{2}{2}\frac{1}{\sqrt{\pi}} \int_{\frac{V_T}{\sqrt{E_x \eta_o}}}^{\infty} e^{-\zeta^2} d\zeta = \frac{1}{2} erfc\left(\frac{V_T}{\sqrt{E_x \eta_o}}\right) \qquad \text{Eq. (12.16)}$$

where $erfc$ is the complimentary error function defined by

$$erfc(z) = \frac{2}{\sqrt{\pi}} \int_z^{\infty} e^{-\zeta^2} d\zeta. \qquad \text{Eq. (12.17)}$$

The error function erf is related to the complimentary error function using the relation

$$erfc(z) = 1 - erf(z) = 1 - \frac{2}{\sqrt{\pi}} \int_0^z e^{-\zeta^2} d\zeta. \qquad \text{Eq. (12.18)}$$

Both *erf* and *erfc* are intrinsic MATLAB functions found within the Signal Processing Toolbox. Using similar analysis, one can derive the probability of detection as

$$P_D = Pr\{v(t) > V_T / H_1\} = \frac{1}{2}erfc\left(\frac{V_T - Ex}{\sqrt{E_x \eta_o}}\right).$$

Eq. (12.19)

The derivation of Eq. (12.19) is left as an exercise (see Problem 12.2).

Table 12.1 gives samples of the single pulse SNR corresponding to few values of P_D and P_{fa}, using Eq. (12.19). For example, if $P_D = 0.99$ and $P_{fa} = 10^{-10}$, then the minimum single pulse SNR required to accomplish this combination of P_D and P_{fa} is $SNR = 16.12 dB$.

TABLE 12.1. Single Pulse SNR (dB)

P_D	10^{-3}	10^{-4}	10^{-5}	10^{-6}	10^{-7}	10^{-8}	10^{-9}	10^{-10}	10^{-11}	10^{-12}
.1	4.00	6.19	7.85	8.95	9.94	10.44	11.12	11.62	12.16	12.65
.2	5.57	7.35	8.75	9.81	10.50	11.19	11.87	12.31	12.85	13.25
.3	6.75	8.25	9.50	10.44	11.10	11.75	12.37	12.81	13.25	13.65
.4	7.87	8.85	10.18	10.87	11.56	12.18	12.75	13.25	13.65	14.00
.5	8.44	9.45	10.62	11.25	11.95	12.60	13.11	13.52	14.00	14.35
.6	8.75	9.95	11.00	11.75	12.37	12.88	13.50	13.87	14.25	14.62
.7	9.56	10.50	11.50	12.31	12.75	13.31	13.87	14.20	14.59	14.95
.8	10.18	11.12	12.05	12.62	13.25	13.75	14.25	14.55	14.87	15.25
.9	10.95	11.85	12.65	13.31	13.85	14.25	14.62	15.00	15.45	15.75
.95	11.50	12.40	13.12	13.65	14.25	14.64	15.10	15.45	15.75	16.12
.98	12.18	13.00	13.62	14.25	14.62	15.12	15.47	15.85	16.25	16.50
.99	12.62	13.37	14.05	14.50	15.00	15.38	15.75	16.12	16.47	16.75
.995	12.85	13.65	14.31	14.75	15.25	15.71	16.06	16.37	16.65	17.00
.998	13.31	14.05	14.62	15.06	15.53	16.05	16.37	16.7	16.89	17.25
.999	13.62	14.25	14.88	15.25	15.85	16.13	16.50	16.85	17.12	17.44
.9995	13.84	14.50	15.06	15.55	15.99	16.35	16.70	16.98	17.35	17.55
.9999	14.38	14.94	15.44	16.12	16.50	16.87	17.12	17.35	17.62	17.87

12.2. Single Pulse with Known Amplitude and Unknown Phase

In this case, the retuned radar signal comprises a sinusoid of a deterministic amplitude and random phase whose *pdf* is uniform over the interval $\{0, 2\pi\}$. The output of the matched filter receiver that employs an envelope detector is denoted by $v(t)$ (see Fig. 12.2), and it can be written as a bandpass random process as

$$v(t) = v_I(t)\cos\omega_0 t + v_Q(t)\sin\omega_0 t = r(t)\cos(\omega_0 t - \Phi(t))$$

$$v_I(t) = r(t)\cos\Phi(t)$$

Eq. (12.20)

$$v_Q(t) = r(t)\sin\Phi(t)$$

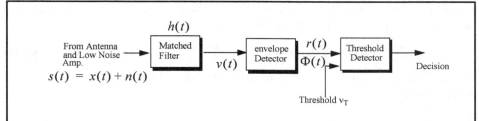

Figure 12.2. Simplified matched filter receiver employing an envelope detector.

$$r(t) = \sqrt{[v_I(t)]^2 + [v_Q(t)]^2}$$

$$\Phi(t) = \left[\tan\left(\frac{v_Q(t)}{v_I(t)}\right)\right]^{-1}$$

Eq. (12.21)

where $\omega_0 = 2\pi f_0$ is the radar operating frequency, $r(t)$ is the envelope of $v(t)$, the phase is $\Phi(t) = \operatorname{atan}(v_Q / v_I)$, and the subscripts I, and Q, respectively, refer to the in-phase and quadrature components. A target is detected when $r(t)$ exceeds the threshold value v_T, where the decision hypotheses are

$$H_1 \Leftrightarrow s(t) = x(t) + n(t) \qquad and \qquad r(t) > v_T \Rightarrow Detection$$
$$H_0 \Leftrightarrow s(t) = n(t) \qquad and \qquad r(t) > v_T \Rightarrow False\ alarm$$

Eq. (12.22)

The case when the noise subtracts from the signal (while a target is present) to make $r(t)$ smaller than the threshold is called a miss. The matched filter output is a complex random variable that comprises either noise alone or noise plus target returns (i.e., sine wave of amplitude A and random phase). The quadrature components corresponding to the case of noise alone are

$$v_I(t) = n_I(t)$$
$$v_Q(t) = n_Q(t)$$

Eq. (12.23)

where the noise quadrature components $n_I(t)$ and $n_Q(t)$ are uncorrelated zero mean lowpass Gaussian noise with equal variances, σ^2. In the second case the quadrature components are

$$v_I(t) = A + n_I(t) = r(t)\cos\Phi(t) \Rightarrow n_I(t) = r(t)\cos\Phi(t) - A$$
$$v_Q(t) = n_Q(t) = r(t)\sin\Phi(t)$$

Eq. (12.24)

The joint probability density function (*pdf*) of the two random variables $n_I; n_Q$ is

$$f_{n_I n_Q}(n_I, n_Q) = \frac{1}{2\pi\sigma^2}\exp\left(-\frac{n_I^2 + n_Q^2}{2\sigma^2}\right) = \frac{1}{2\pi\sigma^2}\exp\left(-\frac{(r\cos\varphi - A)^2 + (r\sin\varphi)^2}{2\sigma^2}\right),$$

Eq. (12.25)

which can be written as

$$f_{n_I n_Q}(n_I, n_Q) = \frac{1}{2\pi\sigma^2}\exp\left(-\frac{(r\cos\varphi - A)^2 + (r\sin\varphi)^2}{2\sigma^2}\right).$$

Eq. (12.26)

The *pdfs* of the random variables $r(t)$ and $\Phi(t)$, respectively, represent the modulus and phase of $v(t)$. The joint *pdf* for the two random variables $r(t);\Phi(t)$ are derived using a similar approach to that developed in Chapter 11. More precisely,

$$f_{R\Phi}(r, \varphi) = f_{n_I n_Q}(n_I, n_Q)|\mathbf{J}| \qquad \text{Eq. (12.27)}$$

where $|\mathbf{J}|$ is determinant of the matrix of derivatives $\mathbf{J}$ and referred to as the Jacobian. The matrix of derivatives is given by

$$\mathbf{J} = \begin{bmatrix} \dfrac{\partial n_I}{\partial r} & \dfrac{\partial n_I}{\partial \varphi} \\ \dfrac{\partial n_Q}{\partial r} & \dfrac{\partial n_Q}{\partial \varphi} \end{bmatrix} = \begin{bmatrix} \cos\varphi & -r\sin\varphi \\ \sin\varphi & r\cos\varphi \end{bmatrix}. \qquad \text{Eq. (12.28)}$$

It follows that the Jacobian is

$$|\mathbf{J}| = r(t). \qquad \text{Eq. (12.29)}$$

Substituting Eq. (12.25) and Eq. (12.29) into Eq. (12.27) and collecting terms yields

$$f_{R\Phi}(r, \varphi) = \frac{r}{2\pi\sigma^2} \exp\left(-\frac{r^2 + A^2}{2\sigma^2}\right) \exp\left(\frac{rA\cos\varphi}{\sigma^2}\right). \qquad \text{Eq. (12.30)}$$

The *pdf* for $r(t)$ alone is obtained by integrating Eq. (12.30) over φ. That is,

$$f_R(r) = \int_0^{2\pi} f_{R\Phi}(r, \varphi)d\varphi = \frac{r}{\sigma^2}\exp\left(-\frac{r^2 + A^2}{2\sigma^2}\right) \frac{1}{2\pi}\int_0^{2\pi}\exp\left(\frac{rA\cos\varphi}{\sigma^2}\right)d\varphi \qquad \text{Eq. (12.31)}$$

where the integral inside Eq. (12.31) is known as the modified Bessel function of zero order,

$$I_0(\beta) = \frac{1}{2\pi}\int_0^{2\pi} e^{\beta\cos\theta} \, d\theta. \qquad \text{Eq. (12.32)}$$

Thus,

$$f_R(r) = \frac{r}{\sigma^2} I_0\left(\frac{rA}{\sigma^2}\right) \exp\left(-\frac{r^2 + A^2}{2\sigma^2}\right), \qquad \text{Eq. (12.33)}$$

which is the Rician probability density function. The case when $A/\sigma^2 = 0$ (noise alone) was analyzed in Chapter 11 and the resulting *pdf* is a Rayleigh probability density function

$$f_R(r) = \frac{r}{\sigma^2}\exp\left(-\frac{r^2}{2\sigma^2}\right). \qquad \text{Eq. (12.34)}$$

When (A/σ^2) is very large, Eq. (12.33) becomes a Gaussian probability density function of mean A and variance σ^2:

$$f_R(r) \approx \frac{1}{\sqrt{2\pi\sigma^2}}\exp\left(-\frac{(r-A)^2}{2\sigma^2}\right). \qquad \text{Eq. (12.35)}$$

Figure 12.3 shows plots for the Rayleigh and Gaussian densities. The density function for the random variable Φ is obtained from

$$f_\Phi(\varphi) = \int_0^r f_{R\Phi}(r, \varphi) \, dr.$$

Eq. (12.36)

While the detailed derivation is left as an exercise, the result is

$$f_\Phi(\varphi) = \frac{1}{2\pi} \exp\left(\frac{-A^2}{2\sigma^2}\right) + \frac{A\cos\varphi}{\sqrt{2\pi\sigma^2}} \exp\left(\frac{-(A\sin\varphi)^2}{2\sigma^2}\right) F\left(\frac{A\cos\varphi}{\sigma}\right)$$

Eq. (12.37)

where

$$F(x) = \int_{-\infty}^x \frac{1}{\sqrt{2\pi}} e^{-\zeta^2/2} \, d\zeta.$$

Eq. (12.38)

The function $F(x)$ can be found tabulated in most mathematical formula reference books. Note that for the case of noise alone ($A = 0$), Eq. (12.37) collapses to a uniform *pdf* over the interval $\{0, 2\pi\}$. One excellent approximation for the function $F(x)$ is

$$F(x) = 1 - \left(\frac{1}{0.661x + 0.339\sqrt{x^2 + 5.51}}\right) \frac{1}{\sqrt{2\pi}} e^{-x^2/2} \qquad x \ge 0.$$

Eq. (12.39)

and for negative values of x

$$F(-x) = 1 - F(x).$$

Eq. (12.40)

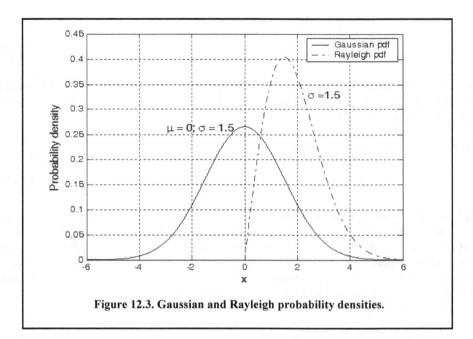

Figure 12.3. Gaussian and Rayleigh probability densities.

MATALAB Function "que_func.m"

The MATLAB function *"que_function.m"* calculates Eq.(12.38) using the approximation in Eqs. (12.39) and (12.40). Its syntax is as follows:

```
fofx = que_func(x).
```

12.2.1. Probability of False Alarm

The probability of false alarm P_{fa} is defined as the probability that a sample r of the signal $r(t)$ will exceed the threshold voltage v_T when noise alone is present in the radar:

$$P_{fa} = \int_{v_T}^{\infty} \frac{r}{\sigma^2} \exp\left(-\frac{r^2}{2\sigma^2}\right) \, dr = \exp\left(\frac{-v_T^2}{2\sigma^2}\right) \qquad \text{Eq. (12.41)}$$

$$v_T = \sqrt{2\sigma^2 \ln\left(\frac{1}{P_{fa}}\right)}. \qquad \text{Eq. (12.42)}$$

Figure 12.4 shows a plot of the normalized threshold versus the probability of false alarm. It is evident from this figure that P_{fa} is very sensitive to small changes in the threshold value. The false alarm time T_{fa} is related to the probability of false alarm by

$$T_{fa} = \frac{t_{int}}{P_{fa}} \qquad \text{Eq. (12.43)}$$

where t_{int} represents the radar integration time, or the average time that the output of the envelope detector will pass the threshold voltage. Since the radar operating bandwidth B is the inverse of t_{int}, by using the right-hand side of Eq. (12.41) and Eq. (12.42), one can rewrite T_{fa} as

$$T_{fa} = \frac{1}{B} \exp\left(\frac{v_T^2}{2\sigma^2}\right) \qquad \text{Eq. (12.44)}$$

Minimizing T_{fa} means increasing the threshold value, and as a result, the radar maximum detection range is decreased. The choice of an acceptable value for T_{fa} becomes a compromise depending on the radar mode of operation.

The false alarm number is defined as

$$n_{fa} = \frac{-\ln(2)}{\ln(1 - P_{fa})} \approx \frac{\ln(2)}{P_{fa}}. \qquad \text{Eq. (12.45)}$$

Other slightly different definitions for the false alarm number exist in the literature, causing a source of confusion for many non-expert readers. Other than the definition in Eq. (12.45), the most commonly used definition for the false alarm number is the one introduced by Marcum (1960). Marcum defines the false alarm number as the reciprocal of P_{fa}. In this text, the definition given in Eq. (12.45) is always assumed. Hence, a clear distinction is made between Marcum's definition of the false alarm number and the definition in Eq. (12.45).

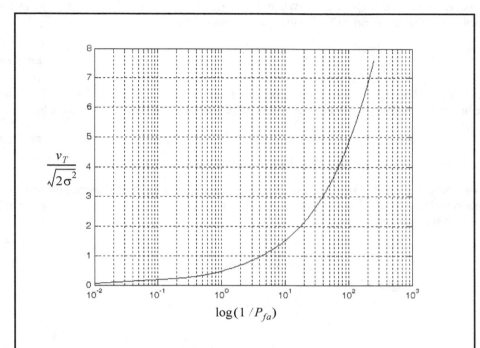

Figure 12.4. Normalized detection threshold versus probability of false alarm.

12.2.2. Probability of Detection

The probability of detection P_D is the probability that a sample r of $r(t)$ will exceed the threshold voltage in the case of noise plus signal,

$$P_D = \int_{v_T}^{\infty} \frac{r}{\sigma^2} I_0\left(\frac{rA}{\sigma^2}\right) \exp\left(-\frac{r^2 + A^2}{2\sigma^2}\right) dr .$$

Eq. (12.46)

Assuming that the radar signal is a sinusoid of amplitude A (completely known), then its power is $A^2/2$. Now, by using $SNR = A^2/2\sigma^2$ (single-pulse SNR) and $(v_T^2/2\sigma^2) = \ln(1/P_{fa})$, then Eq. (12.46) can be rewritten as

$$P_D = \int_{\sqrt{2\sigma^2 \ln(1/P_{fa})}}^{\infty} \frac{r}{\sigma^2} I_0\left(\frac{rA}{\sigma^2}\right) \exp\left(-\frac{r^2 + A^2}{2\sigma^2}\right) dr = Q\left[\sqrt{\frac{A^2}{\sigma^2}}, \sqrt{2\ln\left(\frac{1}{P_{fa}}\right)}\right]$$

Eq. (12.47)

where

$$Q[a, b] = \int_{b}^{\infty} \zeta I_0(a\zeta) e^{-(\zeta^2 + a^2)/2} \, d\zeta .$$

Eq. (12.48)

Q is called Marcum's Q-function. When P_{fa} is small and P_D is relatively large so that the threshold is also large, Eq. (12.47) can be approximated by

$$P_D \approx F\left(\frac{A}{\sigma} - \sqrt{2\ln\left(\frac{1}{P_{fa}}\right)}\right).$$

Eq. (12.49)

$F(x)$ is given by Eq. (12.38). Many approximations for Eq. (12.47) can be found throughout the literature. One very accurate approximation presented by North (1963) is given by

$$P_D \approx 0.5 \times erfc(\sqrt{-\ln P_{fa}} - \sqrt{SNR + 0.5})$$

Eq. (12.50)

where the complementary error function was defined in Eq. (12.17).

The integral given in Eq. (12.47) is complicated and can be computed using numerical integration techniques. Parl[1] developed an excellent algorithm to numerically compute this integral. It is summarized as follows:

$$Q[a, b] = \begin{cases} \dfrac{\alpha_n}{2\beta_n}\exp\left(\dfrac{(a-b)^2}{2}\right) & a < b \\[3mm] 1 - \left(\dfrac{\alpha_n}{2\beta_n}\exp\left(\dfrac{(a-b)^2}{2}\right)\right) & a \geq b \end{cases}$$

Eq. (12.51)

$$\alpha_n = d_n + \frac{2n}{ab}\alpha_{n-1} + \alpha_{n-2}$$

Eq. (12.52)

$$\beta_n = 1 + \frac{2n}{ab}\beta_{n-1} + \beta_{n-2}$$

Eq. (12.53)

$$d_{n+1} = d_n d_1$$

Eq. (12.54)

$$\alpha_0 = \begin{cases} 1 & a < b \\ 0 & a \geq b \end{cases}$$

Eq. (12.55)

$$d_1 = \begin{cases} a/b & a < b \\ b/a & a \geq b \end{cases}.$$

Eq. (12.56)

$\alpha_{-1} = 0.0$, $\beta_0 = 0.5$, and $\beta_{-1} = 0$. The recursive Eq. (12.51) through Eq. (12.56) are computed continuously until $\beta_n > 10^p$ for values of $p \geq 3$. The accuracy of the algorithm is enhanced as the value of p is increased.

MATLAB Function "marcumsq.m"

The MATLAB function *"marcumsq.m"* implements Parl's algorithm to calculate the probability of detection defined in Eq. (12.47). The syntax is as follows:

$$Pd = marcumsq\ (a, b)$$

---

1. Parl, S., A New Method of Calculating the Generalized Q Function, *IEEE Trans. Information Theory*, Vol. IT-26, January 1980, pp. 121-124.

where

Symbol	Description	Units	Status
a	A/σ	*dB*	*input*
b	$\sqrt{-2\ln(P_{fa})}$	*none*	*input*
Pd	*signal pulse probability of detection*	*none*	*output*

Figure 12.5 shows plots of the probability of detection, P_D, versus the single pulse SNR, with the P_{fa} as a parameter using this MATLAB function. This figure can be reproduced using MATLAB program *"Fig12_5.m,"* listed in Appendix 12-A.

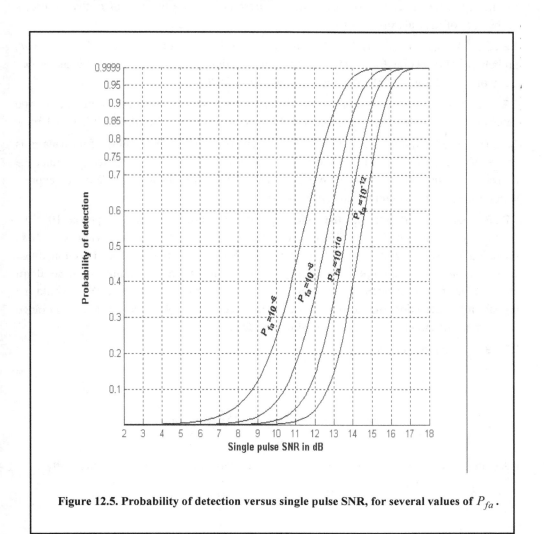

Figure 12.5. Probability of detection versus single pulse SNR, for several values of P_{fa}.

Problems

12.1. Prove the results given in Eqs. (12.9) through (12.12).

12.2. Derive Eq. (12.19).

12.3. Consider the matched filter receiver shown in Fig. 12.1. Develop expressions for the single pulse of known parameters probability of detection P_D and probability of false alarm P_{fa}

12.4. In the case of noise alone, the quadrature components of a radar return are independent Gaussian random variables with zero mean and variance σ^2. Assume that the radar processing consists of envelope detection followed by threshold decision. (a) Write an expression for the *pdf* of the envelope; (b) determine the threshold V_T as a function of σ that ensures a probability of false alarm $P_{fa} \leq 10^{-8}$.

12.5. A pulsed radar has the following specifications: time of false alarm $T_{fa} = 10 min$, probability of detection $P_D = 0.95$, operating bandwidth $B = 1 MHz$. (a) What is the probability of false alarm P_{fa}? (b) What is the single pulse SNR?

12.6. Show that when computing the probability of detection at the output of an envelope detector, it is possible to use Gaussian probability approximation when the SNR is very large.

12.7. A radar system uses a threshold detection criterion. The probability of false alarm is $P_{fa} = 10^{-10}$. (a) What must be the average SNR at the input of a linear detector so that the probability of miss is $P_m = 0.15$? Assume a large SNR approximation. (b) Write an expression for the *pdf* at the output of the envelope detector.

12.8. An X-band radar has the following specifications: received peak power $10^{-10}W$, probability of detection $P_D = 0.95$, time of false alarm $T_{fa} = 8 min$, pulse width $\tau = 2 \mu s$, operating bandwidth $B = 2 MHz$, operating frequency $f_0 = 10 GHz$, and detection range $R = 100 Km$. Assume single pulse processing. (a) Compute the probability of false alarm P_{fa}. (b) Determine the SNR at the output of the matched filter. (c) At what SNR would the probability of detection drop to 0.9 (P_{fa} does not change)? (d) What is the increase in range that corresponds to this drop in the probability of detection?

12.9. Using the equation

$$P_D = 1 - e^{-SNR} \int_{P_{fa}}^{1} I_0(\sqrt{-4SNR\ln u})\,du \,.$$

calculate P_D when $SNR = 10 dB$ and $P_{fa} = 0.01$. Perform the integration numerically.

Appendix 12-A: Chapter 12 MATLAB Code Listings

The MATLAB code provided in this chapter was designed as an academic standalone tool and is not adequate for other purposes. The code was written in a way to assist the reader in gaining a better understanding of the theory. The code was not developed, nor is it intended to be used as part of an open-loop or a closed-loop simulation of any kind. The MATLAB code found in this textbook can be downloaded from this book's web page on the CRC Press website. Simply use your favorite web browser, go to *www.crcpress.com*, and search for keyword *"Mahafza"* to locate this book's web page.

MATLAB Function "que_func.m" Listing

```
function fofx = que_func(x)
% This function computes the value of the Q-function
% It uses the approximation in Eqs. (12.39) and (12.40)
if (x >= 0)
    denom = 0.661 * x + 0.339 * sqrt(x^2 + 5.51);
    expo = exp(-x^2 /2.0);
    fofx = 1.0 - (1.0 / sqrt(2.0 * pi)) * (1.0 / denom) * expo;
else
    denom = 0.661 * x + 0.339 * sqrt(x^2 + 5.51);
    expo = exp(-x^2 /2.0);
    value = 1.0 - (1.0 / sqrt(2.0 * pi)) * (1.0 / denom) * expo;
    fofx = 1.0 - value;
end
```

MATLAB Function "marcumsq.m" Listing

```
function PD = marcumsq (a,b)
% This function uses Parl's method to compute PD
% Inputs
%   a   == sqrt(2.0 * 10^(.1*snr))
%   b   == sqrt(-2.0 * log(10^(-nfa)));
%%%Output
%   PD  == single pulse probability of detection
if (a < b)
    alphan0 = 1.0;
    dn = a ./ b;
else
    alphan0 = 0.;
    dn = b ./ a;
end
alphan_1 = 0.;
betan0 = 0.5;
betan_1 = 0.;
D1 = dn;
n = 0;
ratio = 2.0 ./ (a .* b);
r1 = 0.0;
betan = 0.0;
alphan = 0.0;
```

```
while betan < 1000.,
  n = n + 1;
  alphan = dn + ratio .* n .* alphan0 + alphan_1;
  betan = 1.0 + ratio .* n .* betan0 + betan_1;
  alphan_1 = alphan0;
  alphan0 = alphan;
  betan_1 = betan0;
  betan0 = betan;
  dn = dn .* D1;
end
PD = (alphan0 / (2.0 * betan0)) * exp( -(a-b).^2 / 2.0);
if ( a >= b)
  PD = 1.0 - PD;
end
return
```

MATLAB Program "Fig12_5.m" Listing

```
% This program is used to produce Fig. 12.5
close all
clear all
for nfa = 6:2:12
  b = sqrt(-2.0 * log(10^(-nfa)));
  index = 0;
  hold on
  for snr = 2:.1:18
    index = index +1;
    a = sqrt(2.0 * 10^(.1*snr));
    pro(index) = marcumsq(a,b);
  end
  x = 2:.1:18;
  set(gca,'ytick',[.1 .2 .3 .4 .5 .6 .7 .75 .8 .85 .9 .95 .9999])
  set(gca,'xtick',[2 3 4 5 6 7 8 9 10 11 12 13 14 15 16 17 18])
  plot(x, pro,'k');
end
hold off
xlabel ('\bfSingle pulse SNR in dB'); ylabel ('\bfProbability of detection')
grid on
gtext('\bfP_f_a=10^-^6','rotation',65)
gtext('\bfP_f_a=10^-^8','rotation', 68)
gtext('\bfP_f_a=10^-^1^0','rotation', 70)
gtext('\bfP_f_a=10^-^1^2','rotation', 72)
```

Chapter 13

Detection of Fluctuating Targets

13.1. Introduction

In the previous chapter target detection was introduced in the context of single pulse detection with completely known (i.e., deterministic) amplitude and phase in one case, and known amplitude with random phase in another. The underlying assumption was that radar targets were made of non-varying (non-fluctuating) scatterers. However, in practice that it is rarely the case. First, one would expect the radar to receive multiple returns (pulses) from any given target in its field of view. Furthermore, real-world targets will fluctuate over the duration of a single pulse or from pulse to pulse. This chapter extends the analysis of Chapter 12 to account for target fluctuation as well as for target detection where multiple returned pulses are taken into consideration. Multiple returned pulses can be integrated (combined) coherently or non-coherently. The process of combining radar returns from many pulses is called radar pulse integration. Pulse integration can be performed on the quadrature components prior to the envelope detector. This is called coherent integration or predetection integration. Coherent integration preserves the phase relationship between the received pulses. Thus a buildup in the signal amplitude is expected. Alternatively, pulse integration performed after the envelope detector (where the phase relation is lost) is called noncoherent or post-detection integration, and a buildup in the signal amplitude is guaranteed.

13.2. Pulse Integration

Combining the returns from all pulses returned by a given target during a single scan is very likely to increase the radar sensitivity (i.e., SNR). The number of returned pulses from a given target depends on the antenna scan rate, the antenna beamwidth, and the radar PRF. More precisely, the number of pulses returned from a given target is given by

$$n_P = \frac{\theta_a T_{sc} f_r}{2\pi}$$

Eq. (13.1)

where θ_a is the azimuth antenna beamwidth, T_{sc} is the scan time, and f_r is the radar PRF. The number of reflected pulses may also be expressed as

$$n_P = \frac{\theta_a f_r}{\dot{\theta}_{scan}}$$

Eq. (13.2)

where $\dot{\theta}_{scan}$ is the antenna scan rate in degrees per second. Note that when using Eq. (13.1), θ_a is expressed in radians, while when using Eq. (13.2), it is expressed in degrees. As an example, consider a radar with an azimuth antenna beamwidth $\theta_a = 3°$, antenna scan rate $\dot{\theta}_{scan} = 45°/\sec$ (antenna scan time, $T_{sc} = 8\sec$), and a PRF $f_r = 300Hz$. Using either Eq. (13.1) or Eq. (13.2) yields $n_P = 20$ pulses.

As it was described in Chapter 2, pulse integration will very likely improve the receiver SNR. Nonetheless, caution should be exercised when attempting to account for how much SNR is attained through pulse integration. This is true for the following reasons: First, during a antenna scan, a given target will not always be located at the center of the radar beam (i.e., have maximum gain). In fact, during a scan, a target will first enter the antenna beam at the $3dB$ point, reach maximum gain, and finally leave the beam at the $3dB$ point again. Thus, the returns do not have the same amplitude even though the target RCS may be constant and all other factors that may introduce signal loss remain the same.

Other factors that may introduce further variation to the amplitude of the returned pulses include target RCS and propagation path fluctuations. Additionally, when the radar employs a very fast scan rate, an additional loss term is introduced due to the motion of the beam between transmission and reception. This is referred to as scan loss. A distinction should be made between scan loss due to a rotating antenna (which is described here) and the term scan loss that is normally associated with phased array antennas (which takes on a different meaning in that context).

Finally, since coherent integration utilizes the phase information from all integrated pulses, it is critical that any phase variation between all integrated pulses be known with a great level of confidence. Consequently, target dynamics (such as target range, range rate, tumble rate, RCS fluctuation) must be estimated or computed accurately so that coherent integration can be meaningful. In fact, if a radar coherently integrates pulses from targets without proper knowledge of the target dynamics, it suffers a loss in SNR rather than the expected SNR buildup. Knowledge of target dynamics is not as critical when employing noncoherent integration; nonetheless, target range rate must be estimated so that only the returns from a given target within a specific range bin are integrated. In other words, one must avoid range walk (i.e., having a target cross between adjacent range bins during a single scan).

A comprehensive analysis of pulse integration should also take into account issues such as the probability of detection P_D, probability of false alarm P_{fa}, the target statistical fluctuation model, and the noise or interference of statistical models. This is the subject of the rest of this chapter.

13.2.1. Coherent Integration

In coherent integration, and when a perfect integrator is used (100% efficiency) to integrate n_P pulses, the SNR is improved by the same factor. Otherwise, integration loss occurs, which is always the case for noncoherent integration. Coherent integration loss occurs when the integration process is not optimum. This could be due to target fluctuation, instability in the radar local oscillator, or propagation path changes.

Denote the single pulse SNR required to produce a given probability of detection as $(SNR)_1$. The SNR resulting from coherently integrating n_P pulses is then given by

$$(SNR)_{CI} = n_P(SNR)_1 . \qquad \text{Eq. (13.3)}$$

Coherent integration cannot be applied over a long interval of time, particularly if the target RCS is varying rapidly. If the target radial velocity is known and no acceleration is assumed, the maximum coherent integration time is limited to

$$t_{CI} = \sqrt{\frac{\lambda}{2a_r}}$$

Eq. (13.4)

where λ is the radar wavelength and a_r is the target radial acceleration. Coherent integration time can be extended if the target radial acceleration can be compensated for by the radar.

In order to demonstrate the improvement in the SNR using coherent integration, consider the case where the radar return signal contains both signal plus additive noise. The *mth* pulse is

$$y_m(t) = s(t) + n_m(t)$$

Eq. (13.5)

where $s(t)$ is the radar signal return of interest and $n_m(t)$ is white uncorrelated additive noise signal with variance σ^2. Coherent integration of n_P pulses yields

$$z(t) = \frac{1}{n_P} \sum_{m=1}^{n_P} y_m(t) = \sum_{m=1}^{n_P} \frac{1}{n_P}[s(t) + n_m(t)] = \\ s(t) + \sum_{m=1}^{n_P} \frac{1}{n_P} n_m(t)$$

Eq. (13.6)

The total noise power in $z(t)$ is equal to the variance. More precisely,

$$\sigma_{n_P}^2 = E\left[\left(\sum_{m=1}^{n_P} \frac{1}{n_P} n_m(t)\right)\left(\sum_{l=1}^{n_P} \frac{1}{n_P} n_l(t)\right)^*\right]$$

Eq. (13.7)

where E is the expected value operator. It follows that

$$\sigma_{n_P}^2 = \frac{1}{n_P^2} \sum_{m,l=1}^{n_P} E[n_m(t)n_l^*(t)] = \frac{1}{n_P^2} \sum_{m,l=1}^{n_P} \sigma_{ny}^2 \delta_{ml} = \frac{1}{n_P} \sigma_{ny}^2$$

Eq. (13.8)

where σ_{ny}^2 is the single pulse noise power and δ_{ml} is equal to zero for $m \neq l$ and unity for $m = l$. Observation of Eqs. (13.6) and (13.8) indicates that the desired signal power after coherent integration is unchanged, while the noise power is reduced by the factor $1/n_P$. Thus, the SNR after coherent integration is improved by n_P.

13.2.2. Noncoherent Integration

When the phase of the integrated pulses is not known, so that coherent integration is no longer possible, another form of pulse integration is done. In this case, pulse integration is performed by adding (integrating) the individual pulses' envelopes or the square of their envelopes. Thus, the term noncoherent integration is adopted. A block diagram of a radar receiver utilizing noncoherent integration is illustrated in Fig. 13.1.

The performance difference (measured in SNR) between the linear envelope detector and the quadratic (square law) detector is practically negligible. Robertson (1967) showed that this difference is typically less than $0.2\,dB$; he showed that the performance difference is higher than $0.2\,dB$ only for cases where $n_P > 100$ and $P_D < 0.01$. Both of these conditions are of no practical significance in radar applications. It is much easier to analyze and implement the square law detector in real hardware than is the case for the envelope detector. Therefore, most authors make no distinction between the type of detector used when referring to noncoherent integration, and the square law detector is almost always assumed. The analysis presented in this book will always assume, unless indicated otherwise, noncoherent integration using the square law detector.

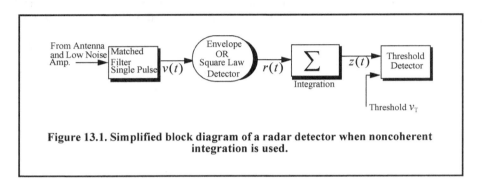

Figure 13.1. Simplified block diagram of a radar detector when noncoherent integration is used.

13.2.3. Improvement Factor and Integration Loss

Noncoherent integration is less efficient than coherent integration. Actually, the noncoherent integration gain is always smaller than the number of noncoherently integrated pulses. This loss in integration is referred to as post-detection or square-law detector loss.

Define $(SNR)_{NCI}$ as the SNR required to achieve a specific P_D given a particular P_{fa} when n_P pulses are integrated noncoherently. Also denote the single pulse SNR as $(SNR)_1$. It follows that

$$(SNR)_{NCI} = (SNR)_1 \times I(n_P)$$ **Eq. (13.9)**

where $I(n_P)$ is called the integration improvement factor. An empirically derived expression for the improvement factor that is accurate within $0.8\,dB$ is reported in Peebles (1998) as

$$[I(n_P)]_{dB} = 6.79(1 + 0.253P_D)\left(1 + \frac{\log(1/P_{fa})}{46.6}\right)\log(n_P) \quad .$$ **Eq. (13.10)**

$$(1 - 0.140\log(n_P) + 0.018310(\log n_P)^2)$$

The integration loss in dB is defined as

$$[L_{NCI}]_{dB} = 10\log n_P - [I(n_P)]_{dB}.$$ **Eq. (13.11)**

MATLAB Function "impmrov_fact.m"

The function *"improv_fac.m"* calculates the improvement factor using Eq. (13.10). The syntax is as follows:

[impr_of_np] = improv_fac (np, pfa, pd)

where

Symbol	Description	Units	Status
np	*number of integrated pulses*	*none*	*input*
pfa	*probability of false alarms*	*none*	*input*
pd	*probability of detection*	*none*	*input*
impr_of_np	*improvement factor*	*output*	*dB*

Figure 13.2 shows plots of the improvement factor versus the number of integrated pulses using different combinations of P_D and P_{fa}. The top part of Fig. 13.2 shows plots of the integration improvement factor as a function of the number of integrated pulses with P_D and P_{fa} as parameters using Eq. (13.10). While, the lower part of Fig. 13.2 shows plots of the corresponding integration loss versus n_P with P_D and P_{fa} as parameters. This figure can be reproduced using the MATLAB program *"Fig13_2.m,"* listed in Appendix 13-B.

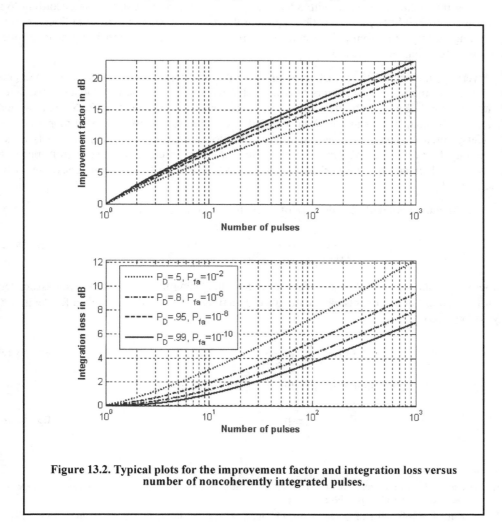

Figure 13.2. Typical plots for the improvement factor and integration loss versus number of noncoherently integrated pulses.

13.3. Target Fluctuation: The Chi-Square Family of Targets

Target detection utilizing the square law detector was first analyzed by Marcum[1], where he assumed a constant RCS (nonfluctuating target). This work was extended by Swerling[2] to four distinct cases of target RCS fluctuation. These cases have come to be known as Swerling models. They are Swerling I, Swerling II, Swerling III, and Swerling IV. The constant RCS case analyzed by Marcum is widely known as Swerling 0 or equivalently as Swerling V. Target fluctuation introduces an additional loss factor in the SNR as compared to the case where fluctuation is not present, given the same P_D and P_{fa}.

Swerling V targets have constant amplitude over one antenna scan or observation interval; however, a Swerling I target amplitude varies independently from scan to scan according to a chi-square probability density function with two degrees of freedom. The amplitude of Swerling II targets fluctuates independently from pulse to pulse according to a chi-square probability density function with two degrees of freedom.

Target fluctuation associated with a Swerling III model is from scan to scan according to a chi-square probability density function with four degrees of freedom. Finally, the fluctuation of Swerling IV targets is from pulse to pulse according to a chi-square probability density function with four degrees of freedom.

Swerling showed that the statistics associated with Swerling I and II models apply to targets consisting of many small scatterers of comparable RCS values, while the statistics associated with Swerling III and IV models apply to targets consisting of one large RCS scatterer and many small equal RCS scatterers. Noncoherent integration can be applied to all four Swerling models; however, coherent integration cannot be used when the target fluctuation is either Swerling II or Swerling IV. This is because the target amplitude decorrelates from pulse to pulse (fast fluctuation) for Swerling II and IV models, and thus phase coherency cannot be maintained.

The chi-square *pdf* with $2N$ degrees of freedom can be written as

$$f_X(x) = \frac{N}{(N-1)! \sqrt{\sigma_x^2}} \left(\frac{Nx}{\sigma_x}\right)^{N-1} \exp\left(-\frac{Nx}{\sigma_x}\right) \qquad \textbf{Eq. (13.12)}$$

where σ_x is the standard deviation for the RCS value. Using this equation, the *pdf* associated with Swerling I and II targets can be obtained by letting $N = 1$, which yields a Rayleigh *pdf*. More precisely,

$$f_X(x) = \frac{1}{\sigma_x}\exp\left(-\frac{x}{\sigma_x}\right) \qquad x \geq 0 . \qquad \textbf{Eq. (13.13)}$$

Letting $N = 2$ yields the *pdf* for Swerling III and IV type targets,

$$f_X(x) = \frac{4x}{\sigma_x^2}\exp\left(-\frac{2x}{\sigma_x}\right) \qquad x \geq 0 . \qquad \textbf{Eq. (13.14)}$$

1. Marcum, J. I., A Statistical Theory of Target Detection by Pulsed Radar, *IRE Transactions on Information Theory*, Vol IT-6, pp. 59-267, April 1960.

2. Swerling, P., Probability of Detection for Fluctuating Targets, *IRE Transactions on Information Theory*, Vol IT-6, pp. 269-308, April 1960.

13.4. Probability of False Alarm Formulation for a Square Law Detector

Computation of the general formula for the probability of false alarm P_{fa} and subsequently the rest of square law detection theory requires knowledge and a good understating of the incomplete Gamma function. Hence, those readers who are not familiar with this function are advised to read Appendix 13-A before proceeding with the rest of this chapter.

DiFranco and Rubin[1] derived a general form relating the threshold and P_{fa} for any number of pulses when noncoherent integration is used. The square law detector under consideration is shown in Fig. 13.3. There are $n_P \geq 2$ pulses integrated noncoherently and the noise power (variance) is σ^2.

The complex envelope in terms of the quadrature components is given by

$$\tilde{r}(t) = r_I(t) + jr_Q(t),$$

Eq. (13.15)

thus the square of the complex envelope is

$$|\tilde{r}(t)|^2 = r_I^2(t) + r_Q^2(t).$$

Eq. (13.16)

The samples $|\tilde{r}_k|^2$ are computed from the samples of $\tilde{r}(t)$ evaluated at $t = t_k;\ k = 1, 2, \ldots, n_P$. It follows that

$$Z = \frac{1}{2\sigma^2} \sum_{k=1}^{n_P} [r_I^2(t_k) + r_Q^2(t_k)].$$

Eq. (13.17)

The random variable Z is the sum of $2n_P$ squares of random variables, each of which is a Gaussian random variable with variance σ^2. Thus, using the analysis developed in Chapter 3, the *pdf* for the random variable Z is given by

$$f_Z(z) = \begin{cases} \dfrac{z^{n_P-1}e^{-z}}{\Gamma(n_P)} & z \geq 0 \\ 0 & z < 0 \end{cases}.$$

Eq. (13.18)

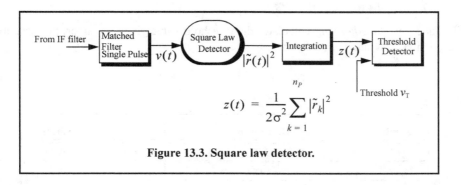

Figure 13.3. Square law detector.

---

1. DiFranco, J. V. and Rubin, W. L., *Radar Detection*, Artech House, Norwood, MA 1980.

Consequently, the probability of false alarm given a threshold value v_T is

$$P_{fa} = Prob\{Z \geq v_T\} = \int_{v_T}^{\infty} \frac{z^{n_P-1}e^{-z}}{\Gamma(n_P)} \, dz \, .$$ **Eq. (13.19)**

and using analysis provided in Appendix 13-A yields

$$P_{fa} = 1 - \Gamma_I\left(\frac{v_T}{\sqrt{n_P}}, n_P - 1\right) \, .$$ **Eq. (13.20)**

Using the algebraic expression for the incomplete Gamma function, Eq. (13.20) can be written as

$$P_{fa} = e^{-v_T} \sum_{k=0}^{n_P-1} \frac{v_T^k}{k!} = 1 - e^{-v_T} \sum_{k=n_P}^{\infty} \frac{v_T^k}{k!} \, .$$ **Eq. (13.21)**

The threshold value v_T can then be approximated by the recursive formula used in the Newton-Raphson method. More precisely,

$$v_{T,m} = v_{T,m-1} - \frac{G(v_{T,m-1})}{G'(v_{T,m-1})} \qquad ; \quad m = 1, 2, 3, \ldots$$ **Eq. (13.22)**

The iteration is terminated when $|v_{T,m} - v_{T,m-1}| < v_{T,m-1}/10000.0$. The functions G and G' are

$$G(v_{T,m}) = (0.5)^{n_P/n_{fa}} - \Gamma_I(v_T, n_P)$$ **Eq. (13.23)**

$$G'(v_{T,m}) = -\frac{e^{-v_T} v_T^{n_P-1}}{(n_P-1)!} \, .$$ **Eq. (13.24)**

The initial value for the recursion is

$$v_{T,0} = n_P - \sqrt{n_P} + 2.3 \sqrt{-\log P_{fa}} \left(\sqrt{-\log P_{fa}} + \sqrt{n_P} - 1\right) \, .$$ **Eq. (13.25)**

MATLAB Function "threshold.m"

The function *"threshold.m"* calculates the threshold value given the algorithm described in this section. The syntax is as follows:

$$[pfa, vt] = threshold \, (nfa, np)$$

where

Symbol	Description	Units	Status
nfa	number of false alarm	none	input
np	number of pulses	none	input
pfa	probability of alarm	none	output
vt	threshold value	none	output

Figure 13.4 shows plots of the threshold value versus the number of integrated pulses for several values of n_{fa}; remember that $P_{fa} \approx \ln(2)/n_{fa}$. This figure can be reproduced using the following MATLAB program, *"Fig13_4.m,"* listed in Appendix 13-B.

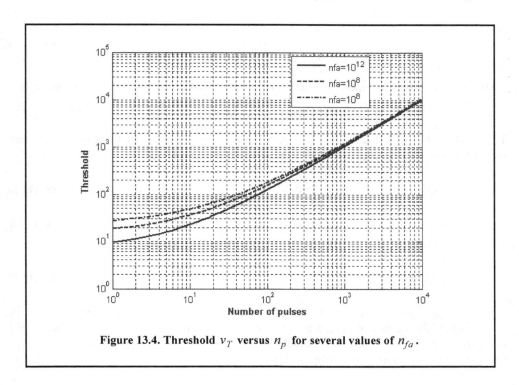

Figure 13.4. Threshold v_T versus n_p for several values of n_{fa}.

13.4.1. Square Law Detection

The *pdf* for the linear envelope $r(t)$ was derived in Chapter 12. Define a new dimensionless variable y as

$$y_n = r_n/\sigma \qquad \text{Eq. (13.26)}$$

where the subscript n denotes the *nth* pulse. Also define

$$\Re_p = A^2/\sigma^2 = 2SNR. \qquad \text{Eq. (13.27)}$$

σ^2 is the noise variance. It follows that the *pdf* for the new variable is

$$f_{Y_n}(y_n) = f_{R_n}(r_n)\left|\frac{dr_n}{dy_n}\right| = y_n\ I_0(y_n\sqrt{\Re_p})\ \exp\left(\frac{-(y_n^2 + \Re_p)}{2}\right). \qquad \text{Eq. (13.28)}$$

The output of a square law detector for the *nth* pulse is proportional to the square of its input. Thus, it is convenient to define a new change variable,

$$z_n = \frac{1}{2}y_n^2. \qquad \text{Eq. (13.29)}$$

The *pdf* for the variable at the output of the square law detector is given by

$$f_{Z_n}(x_n) = f(y_n)\left|\frac{dy_n}{dz_n}\right| = \exp\left(-\left(z_n + \frac{\mathfrak{R}_p}{2}\right)\right)I_0(\sqrt{2z_n\mathfrak{R}_p}) \,.$$

Eq. (13.30)

Noncoherent integration of n_p pulses is implemented as

$$z = \sum_{n=1}^{n_P} \frac{1}{2}y_n^2 \,.$$

Eq. (13.31)

Again, $n_P \geq 2$. Since the random variables y_n are independent, the *pdf* for the variable z is

$$f(z) = f((y_1) \otimes f(y_2) \otimes \dots \otimes f(y_{n_p})) \,.$$

Eq. (13.32)

The operator $\otimes$ symbolically indicates convolution. The characteristic functions for the individual *pdf*s can then be used to compute the joint *pdf* for Eq. (13.32). The result is

$$f_Z(z) = \left(\frac{2z}{n_P\mathfrak{R}_p}\right)^{(n_P-1)/2} \exp\left(-z - \frac{1}{2}n_P\mathfrak{R}_p\right)I_{n_p-1}(\sqrt{2n_Pz\mathfrak{R}_p}) \,.$$

Eq. (13.33)

I_{n_p-1} is the modified Bessel function of order $n_P - 1$. Substituting Eq. (13.27) into (13.33) yields

$$f_Z(z) = \left(\frac{z}{n_PSNR}\right)^{(n_P-1)/2} e^{(-z-n_PSNR)}I_{n_p-1}(2\sqrt{n_PzSNR}) \,.$$

Eq. (13.34)

When target fluctuation is not present (referred to as Swerling 0 or Swerling V target), the probability of detection is obtained by integrating $f_Z(z)$ from the threshold value to infinity. The probability of false alarm is obtained by letting $\mathfrak{R}_p$ be zero and integrating the *pdf* from the threshold value to infinity. More specifically,

$$\left. P_D \right|_{SNR} = \int_{v_T}^{\infty} \left(\frac{z}{n_PSNR}\right)^{(n_P-1)/2} e^{(-z-n_PSNR)}I_{n_p-1}(2\sqrt{n_PzSNR})dz \,,$$

Eq. (13.35)

which can be rewritten as

$$\left. P_D \right|_{SNR} = e^{-n_PSNR}\left(\sum_{k=0}^{\infty} \frac{(n_PSNR)^k}{k!}\right)\left(\sum_{j=0}^{n_P-1+k} \frac{e^{-v_T}v_T^j}{j!}\right) \,.$$

Eq. (13.36)

Alternatively, when target fluctuation is present, the *pdf* is calculated using the conditional probability density function of Eq. (13.35) with respect to the SNR value of the target fluctuation type. In general, given a fluctuating target with SNR^F, where the superscript indicates fluctuation, the expression for the probability of detection is

$$\left. P_D \right|_{SNR^F} = \int_0^{\infty} \left. P_D \right|_{SNR} f_Z(z^F/SNR^F)dz =$$

Eq. (13.37)

$$\int_0^{\infty} \left. P_D \right|_{SNR}\left(\frac{z^F}{n_PSNR^F}\right)^{(n_P-1)/2} e^{(-z^F-n_PSNR^F)}I_{n_p-1}(2\sqrt{n_Pz^FSNR^F})dz$$

Remember that target fluctuation introduces an additional loss term in the SNR. It follows that for the same P_D given the same P_{fa} and the same n_P, $SNR^F > SNR$. One way to calculate this additional SNR is to first compute the required SNR given no fluctuation, then add to it the amount of target fluctuation loss to get the required value for SNR^F. How to calculate this fluctuation loss will be addressed later on in this chapter. Meanwhile, hereinafter, the superscript $\{^F\}$ will be dropped and it will always be assumed.

13.5. Probability of Detection Calculation

Marcum defined the probability of false alarm for the case when $n_P > 1$ as

$$P_{fa} \approx \ln(2)\left(\frac{n_P}{n_{fa}}\right). \qquad \text{Eq. (13.38)}$$

The single pulse probability of detection for nonfluctuating targets was derived in Chapter 12. When $n_P > 1$, the probability of detection is computed using the Gram-Charlier series. In this case, the probability of detection is

$$P_D \cong \frac{erfc(V/\sqrt{2})}{2} - \frac{e^{-V^2/2}}{\sqrt{2\pi}}[C_3(V^2-1) + C_4 V(3-V^2) - C_6 V(V^4 - 10V^2 + 15)] \quad \text{Eq. (13.39)}$$

where the constants C_3, C_4, and C_6 are the Gram-Charlier series coefficients, and the variable V is

$$V = \frac{v_T - n_P(1 + SNR)}{\varpi}. \qquad \text{Eq. (13.40)}$$

In general, values for C_3, C_4, C_6, and ϖ vary depending on the target fluctuation type.

13.5.1. Detection of Swerling 0 (Swerling V) Targets

For Swerling 0 (Swerling V) target fluctuations, the probability of detection is calculated using Eq. (13.39). In this case, the Gram-Charlier series coefficients are

$$C_3 = -\frac{SNR + 1/3}{\sqrt{n_p}(2SNR+1)^{1.5}} \qquad \text{Eq. (13.41)}$$

$$C_4 = \frac{SNR + 1/4}{n_p(2SNR+1)^2} \qquad \text{Eq. (13.42)}$$

$$C_6 = C_3^2/2 \qquad \text{Eq. (13.43)}$$

$$\varpi = \sqrt{n_p(2SNR+1)}. \qquad \text{Eq. (13.44)}$$

MATLAB Function "pd_swerling5.m"

The function *"pd_swerling5.m"* calculates the probability of detection for Swerling 0 targets. The syntax is as follows:

[pd] = pd_swerling5 (input1, indicator, np, snr)

where

Symbol	Description	Units	Status
input1	P_{fa} *or* n_{fa}	*none*	*input*
indicator	*1 when input1* $= P_{fa}$ *2 when input1* $= n_{fa}$	*none*	*input*
np	*number of integrated pulses*	*none*	*input*
snr	*SNR*	*dB*	*input*
pd	*probability of detection*	*none*	*output*

Figure 13.5 shows a plot for the probability of detection versus SNR for cases $n_p = 1, 10$. Note that it requires less SNR, with ten pulses integrated noncoherently, to achieve the same probability of detection as in the case of a single pulse. Hence, for any given P_D, the SNR improvement can be read from the plot. Equivalently, using the function *"improv_fac.m"* leads to about the same result.

For example, when $P_D = 0.8$, the function *"improv_fac.m"* gives an SNR improvement factor of $I(10) \approx 8.55 dB$. Figure 13.5 shows that the ten pulse SNR is about $6.03 dB$. Therefore, the single pulse SNR is about $14.5 dB$, which can be read from the figure. This figure can be reproduced using MATLAB program *"Fig13_5.m,"* listed in Appendix 13-B.

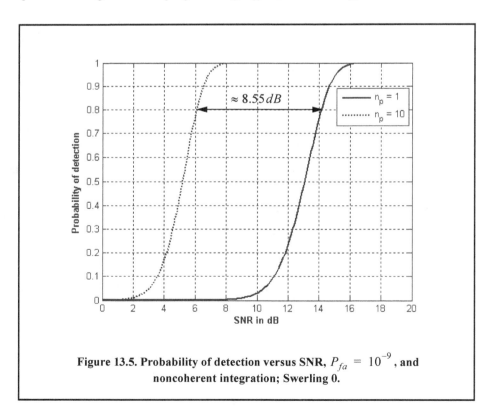

Figure 13.5. Probability of detection versus SNR, $P_{fa} = 10^{-9}$, and noncoherent integration; Swerling 0.

13.5.2. Detection of Swerling I Targets

The exact formula for the probability of detection for Swerling I type targets was derived by Swerling. It is

$$P_D = e^{-(v_T)/(1 + SNR)} \qquad ; \ n_P = 1 \qquad \text{Eq. (13.45)}$$

$$P_D = 1 - \Gamma_I(v_T, n_P - 1) + \left(1 + \frac{1}{n_P SNR}\right)^{n_P - 1} \Gamma_I\left(\frac{v_T}{1 + \frac{1}{n_P SNR}}, n_P - 1\right) \qquad \text{Eq. (13.46)}$$

$$\times \ e^{-v_T/(1 + n_P SNR)} \qquad ; \ n_P > 1.$$

MATLAB Function "pd_swerling1.m"

The function *"pd_swerling1.m"* calculates the probability of detection for Swerling I type targets. The syntax is as follows:

$$[pd] = pd\_swerling1 \ (nfa, np, snr)$$

where

Symbol	Description	Units	Status
nfa	*Marcum's false alarm number*	*none*	*input*
np	*number of integrated pulses*	*none*	*input*
snr	SNR	*dB*	*input*
pd	*probability of detection*	*none*	*output*

Figure 13.6 shows a plot of the probability of detection as a function of SNR for $n_p = 1$ and $P_{fa} = 10^{-9}$ for both Swerling I and V (Swerling 0) type fluctuations. Note that it requires more SNR, with fluctuation, to achieve the same P_D as in the case with no fluctuation. This figure can be reproduced using the MATLAB program *"Fig13_6.m,"* listed in Appendix 13-B. Figure 13.7 is similar to Fig. 13.6, except in this case $P_{fa} = 10^{-6}$ and $n_P = 5$. This figure can be reproduced using the following MATLAB program, *"Fig13_7.m,"* listed in Appendix 13-B.

13.5.3. Detection of Swerling II Targets

In the case of Swerling II targets, the probability of detection is given by

$$P_D = 1 - \Gamma_I\left(\frac{v_T}{(1 + SNR)}, n_P\right) \qquad ; \ n_P \leq 50. \qquad \text{Eq. (13.47)}$$

For the case when $n_P > 50$ the probability of detection is computed using the Gram-Charlier series. In this case,

$$C_3 = -\frac{1}{3\sqrt{n_p}} \qquad , \ C_6 = \frac{C_3^2}{2} \qquad \text{Eq. (13.48)}$$

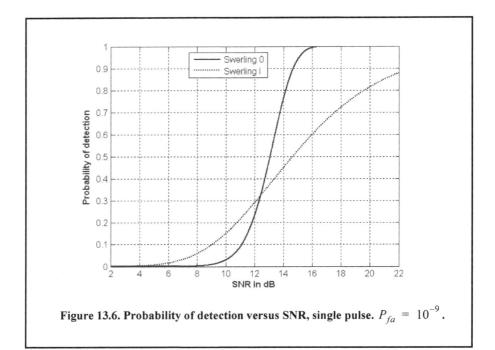

Figure 13.6. Probability of detection versus SNR, single pulse. $P_{fa} = 10^{-9}$.

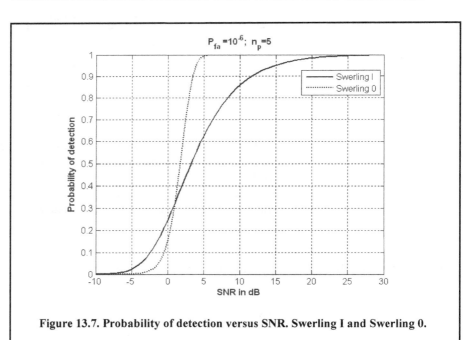

Figure 13.7. Probability of detection versus SNR. Swerling I and Swerling 0.

$$C_4 = \frac{1}{4n_P} \qquad \text{Eq. (13.49)}$$

$$\varpi = \sqrt{n_P} \ (1 + SNR). \qquad \text{Eq. (13.50)}$$

MATLAB Function "pd_swerling2.m"

The function *"pd_swerling2.m"* calculates P_D for Swerling II type targets. The syntax is as follows:

$$[pd] = pd\_swerling2 \ (nfa, \ np, \ snr)$$

where

Symbol	Description	Units	Status
nfa	*Marcum's false alarm number*	*none*	*input*
np	*number of integrated pulses*	*none*	*input*
snr	*SNR*	*dB*	*input*
pd	*probability of detection*	*none*	*output*

Figure 13.8 shows a plot of the probability of detection for Swerling 0, Swerling I, and Swerling II with $n_P = 5$, where $P_{fa} = 10^{-7}$. Figure 13.9 is similar to Fig. 13.8 except in this case $n_P = 2$. Both figures can be reproduced, respectively, using the MATLAB programs *"Fig13_8.m"* and *"Fig13_9.m,"* listed in Appendix 13-B.

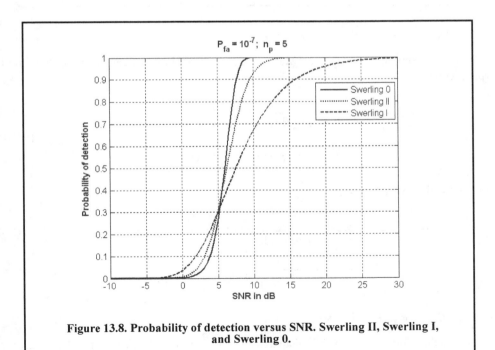

Figure 13.8. Probability of detection versus SNR. Swerling II, Swerling I, and Swerling 0.

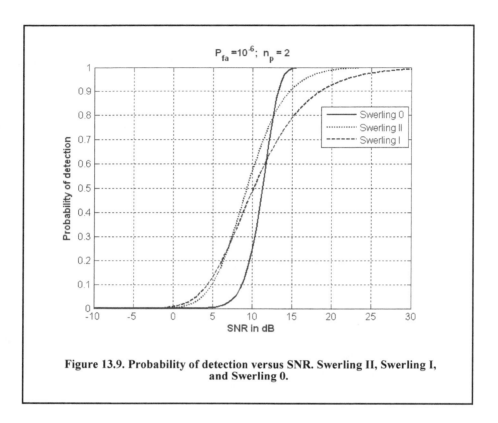

Figure 13.9. Probability of detection versus SNR. Swerling II, Swerling I, and Swerling 0.

13.5.4. Detection of Swerling III Targets

The exact formulas, developed by Marcum, for the probability of detection for Swerling III type targets when $n_P = 1, 2$

$$P_D = \exp\left(\frac{-v_T}{1 + n_P SNR / 2}\right)\left(1 + \frac{2}{n_P SNR}\right)^{n_P - 2} \times K_0$$

$$K_0 = 1 + \frac{v_T}{1 + n_P SNR / 2} - \frac{2}{n_P SNR}(n_P - 2)$$

Eq. (13.51)

For $n_P > 2$ the expression is

$$P_D = \frac{v_T^{n_P - 1} e^{-v_T}}{(1 + n_P SNR / 2)(n_P - 2)!} + 1 - \Gamma_I(v_T, n_P - 1) + K_0 \quad .$$

$$\times \ \Gamma_I\left(\frac{v_T}{1 + 2 / n_P SNR}, n_P - 1\right)$$

Eq. (13.52)

MATLAB Function "pd_swerling3.m"

The function *"pd_swerling3.m"* calculates P_D for Swerling III type targets. The syntax is as follows:

$$[pd] = pd\_swerling3 \ (nfa, \ np, \ snr)$$

where

Symbol	Description	Units	Status
nfa	*Marcum's false alarm number*	*none*	*input*
np	*number of integrated pulses*	*none*	*input*
snr	*SNR*	*dB*	*input*
pd	*probability of detection*	*none*	*output*

Figure 13.10 shows a plot of the probability of detection as a function of SNR for $n_P = 1, 10, 50, 100$, where $P_{fa} = 10^{-9}$. Figure 13.11 shows a plot of the probability of detection for Swerling 0, Swerling I, Swerling II, and Swerling III with $n_P = 5$ and $P_{fa} = 10^{-7}$.

Notice that (see Fig. 13.11) as the target fluctuation becomes more rapid, as in the case of Swerling I type targets, it requires more SNR to achieve the same probability of detection when considering lesser fluctuating targets as in the case of Swerling 0, for example. Figures 13.10 and 13.11 can be reproduced, respectively, using the MATLAB programs *"Fig13_10.m"* and *"Fig13_11.m,"* listed in Appendix 13-B.

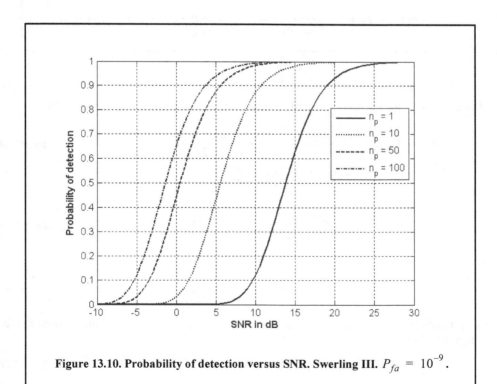

Figure 13.10. Probability of detection versus SNR. Swerling III. $P_{fa} = 10^{-9}$.

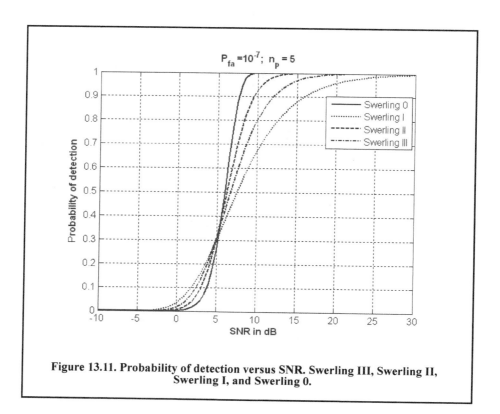

Figure 13.11. Probability of detection versus SNR. Swerling III, Swerling II, Swerling I, and Swerling 0.

13.5.5. Detection of Swerling IV Targets

The expression for the probability of detection for Swerling IV targets for $n_P < 50$ is

$$P_D = 1 - \left[\gamma_0 + \left(\frac{SNR}{2}\right) n_P \gamma_1 + \left(\frac{SNR}{2}\right)^2 \frac{n_P(n_P - 1)}{2!} \gamma_2 + \ldots + \right.$$
$$\left. \left(\frac{SNR}{2}\right)^{n_P} \gamma_{n_P} \right] / \left(1 + \frac{SNR}{2}\right)^{-n_P}$$

$$\text{Eq. (13.53)}$$

$$\gamma_i = \Gamma_I\left(\frac{v_T}{1 + (SNR)/2}, n_P + i\right). \qquad \text{Eq. (13.54)}$$

By using the recursive formula

$$\Gamma_I(x, i+1) = \Gamma_I(x, i) - \frac{x^i}{i! \exp(x)}, \qquad \text{Eq. (13.55)}$$

only γ_0 needs to be calculated using Eq. (13.54), and the rest of γ_i are calculated from the following recursion:

$$\gamma_i = \gamma_{i-1} - A_i \qquad ; \; i > 0 \qquad \text{Eq. (13.56)}$$

$$A_i = \frac{v_T/(1 + (SNR)/2)}{n_P + i - 1} A_{i-1} \qquad ; \; i > 1 \qquad \text{Eq. (13.57)}$$

$$A_1 = \frac{(v_T / (1 + (SNR)/2))^{n_P}}{n_P! \exp(v_T / (1 + (SNR)/2))}$$ 　　Eq. (13.58)

$$\gamma_0 = \Gamma_I\left(\frac{v_T}{(1 + (SNR)/2)}, n_P\right).$$ 　　Eq. (13.59)

For the case when $n_P \geq 50$, the Gram-Charlier series can be used to calculate the probability of detection. In this case,

$$C_3 = \frac{1}{3\sqrt{n_P}} \frac{2\beta^3 - 1}{(2\beta^2 - 1)^{1.5}} \quad ; \quad C_6 = \frac{C_3^2}{2}$$ 　　Eq. (13.60)

$$C_4 = \frac{1}{4n_P} \frac{2\beta^4 - 1}{(2\beta^2 - 1)^2}$$ 　　Eq. (13.61)

$$\varpi = \sqrt{n_P(2\beta^2 - 1)}$$ 　　Eq. (13.62)

$$\beta = 1 + (SNR)/2.$$ 　　Eq. (13.63)

MATLAB Function "pd_swerling4.m"

The function *"pd_swerling4.m"* calculates P_D for Swerling IV type targets. The syntax is as follows:

[pd] = pd_swerling4 (nfa, np, snr)

where

Symbol	Description	Units	Status
nfa	*Marcum's false alarm number*	*none*	*input*
np	*number of integrated pulses*	*none*	*input*
snr	*SNR*	*dB*	*input*
pd	*probability of detection*	*none*	*output*

Figure 13.12 shows plots of the probability of detection as a function of SNR for $n_P = 1, 10, 25, 75$, where $P_{fa} = 10^{-6}$. This figure can be reproduced using the MATLAB program *"Fig13_12.m,"* listed in Appendix 13-B.

13.6. Computation of the Fluctuation Loss

The fluctuation loss, L_f, can be viewed as the amount of additional SNR required to compensate for the SNR loss due to target fluctuation, given a specific P_D value. Kanter[1] developed an exact analysis for calculating the fluctuation loss. In this text, this author will take advantage of the computational power of MATLAB and the MATLAB functions developed in this text to numerically calculate the amount of fluctuation loss.

---

1. Kanter, I., Exact Detection Probability for Partially Correlated Rayleigh Targets, *IEEE Trans*, AES-22, pp. 184-196, March 1986.

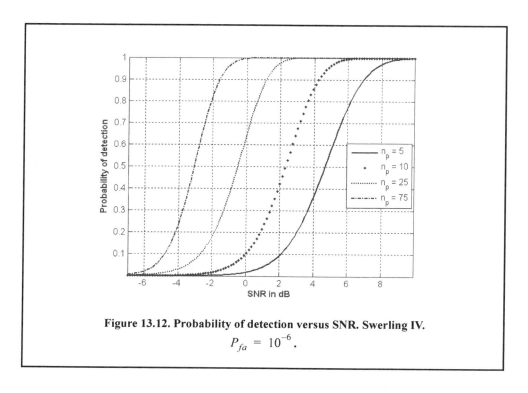

Figure 13.12. Probability of detection versus SNR. Swerling IV.
$$P_{fa} = 10^{-6}.$$

MATLAB Function "fluct.m"

To calculate the amount of fluctuation loss, the MATALB function "*fluct.m*" was developed. Its syntax is as follows:

$$[SNR] = fluct(pd, pfa, np, sw\_case)$$

where

Symbol	Description	Units	Status
pd	*desired probability of detection*	*none*	*input*
nfa	*desired number of false alarms*	*none*	*input*
np	*number of pulses*	*none*	*input*
sw_case	*0, 1, 2, 3, or 4 depending on the desired Swerling case*	*none*	*input*
SNR	*Resulting SNR*	*dB*	*output*

For example, using the syntax

$$[SNR0] = fluct(0.8, 1e6, 5, 0)$$

will calculate the *SNR0* corresponding to a Swerling 0. If one would use this *SNR* in the function "*pd_swerling5.m*" with following syntax

$$[pd] = pd\_swerling5\ (1e6, 2, 5, SNR0),$$

the resulting P_D will be equal to 0.8. Similarly, if the following syntax is used

$$[SNR1] = fluct(0.8, 1e-6, 5, 1),$$

then the value $SNR1$ will be that of Swerling 1. Of course, if one would use this $SNR1$ value in the function *"pd_swerling1.m"* with following syntax

$$[pd] = pd\_swerling1(1e6, 5, 0.8, SNR1),$$

the same P_D of 0.8 will be calculated. Therefore, the fluctuation loss for this case is equal to $SNR0 - SNR1$. Figure 13.13 shows a plot for the additional SNR (or fluctuation loss) required to achieve a certain probability of detection. This figure can be reproduced using MATLAB program *"Fig13_13.m,"* listed in Appendix 13-B.

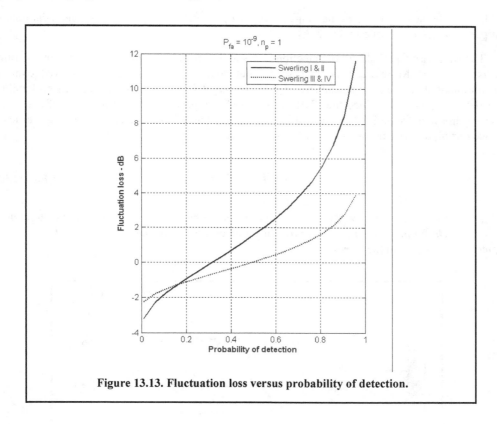

Figure 13.13. Fluctuation loss versus probability of detection.

13.7. Cumulative Probability of Detection

Denote the range at which the single pulse SNR is unity (0 dB) as R_0, and refer to it as the reference range. Then, for a specific radar, the single pulse SNR at R_0 is defined by the radar equation and is given by

$$(SNR)_{R_0} = \frac{P_t G^2 \lambda^2 \sigma}{(4\pi)^3 k T_0 BFLR_0^4} = 1 .$$

Eq. (13.64)

The single pulse SNR at any range R is

$$SNR = \frac{P_t G^2 \lambda^2 \sigma}{(4\pi)^3 k T_0 BFLR^4}.$$ **Eq. (13.65)**

Dividing Eq. (13.165) by Eq. (13.64) yields

$$\frac{SNR}{(SNR)_{R_0}} = \left(\frac{R_0}{R}\right)^4.$$ **Eq. (13.66)**

Therefore, if the range R_0 is known, then the SNR at any other range R is

$$(SNR)_{dB} = 40\log\left(\frac{R_0}{R}\right).$$ **Eq. (13.67)**

Also, define the range R_{50} as the range at which $P_D = 0.5 = P_{50}$. Normally, the radar unambiguous range R_u is set equal to $2R_{50}$.

The cumulative probability of detection refers to detecting the target at least once by the time it is at range R. More precisely, consider a target closing on a scanning radar, where the target is illuminated only during a scan (frame). As the target gets closer to the radar, its probability of detection increases since the SNR is increased. Suppose that the probability of detection during the nth frame is P_{D_n}; then, the cumulative probability of detecting the target at least once during the nth frame (see Fig. 13.14) is given by

$$P_{C_n} = 1 - \prod_{i=1}^{n}(1 - P_{D_i}).$$ **Eq. (13.68)**

P_{D_1} is usually selected to be very small. Clearly, the probability of not detecting the target during the nth frame is $1 - P_{C_n}$. The probability of detection for the ith frame, P_{D_i}, is computed as discussed in the previous section.

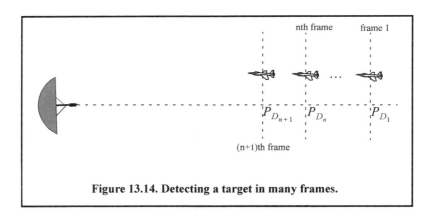

Figure 13.14. Detecting a target in many frames.

Example:

A radar detects a closing target at $R = 10 Km$, with probability of detection P_D equal to 0.5. Assume $P_{fa} = 10^{-7}$. Compute and sketch the single look probability of detection as a function of normalized range (with respect to $R = 10 Km$), over the interval $(2 - 20)Km$. If the range between two successive frames is $1 Km$, what is the cumulative probability of detection at $R = 8 Km$?

Solution:

From the function *"marcumsq.m,"* the SNR corresponding to $P_D = 0.5$ and $P_{fa} = 10^{-7}$ is approximately $12dB$. By using a similar analysis to that which led to Eq. (13.67), we can express the SNR at any range R as

$$(SNR)_R = (SNR)_{10} + 40 \ \log\frac{10}{R} = 52 - 40 \ \log R.$$

By using the function *"marcumsq.m,"* we can construct the following table:

R *Km*	(SNR) dB	P_D
2	39.09	0.999
4	27.9	0.999
6	20.9	0.999
8	15.9	0.999
9	13.8	0.9
10	12.0	0.5
11	10.3	0.25
12	8.8	0.07
14	6.1	0.01
16	3.8	ε
20	0.01	ε

where ε is very small. A sketch of P_D versus normalized range is shown in the figure below.

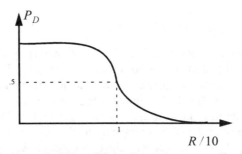

Cumulative probability of detection versus normalized range.

The cumulative probability of detection is given in Eq. (13.68), where the probability of detection of the first frame is selected to be very small. Thus, we can arbitrarily choose frame 1 to be at $R = 16Km$. Note that selecting a different starting point for frame 1 would have a negligible effect on the cumulative probability (we only need P_{D_1} to be very small). Below is a range listing for frames 1 through 9, where frame 9 corresponds to $R = 8Km$.

frame	1	2	3	4	5	6	7	8	9
range in Km	16	15	14	13	12	11	10	9	8

The cumulative probability of detection at 8Km is then

$$P_{C_9} = 1 - (1 - 0.999)(1 - 0.9)(1 - 0.5)(1 - 0.25)(1 - 0.07) \quad .$$
$$(1 - 0.01)(1 - \varepsilon)^2 \approx 0.9998$$

13.8. Constant False Alarm Rate (CFAR)

The detection threshold is computed so that the radar receiver maintains a constant predetermined probability of false alarm. The relationship between the threshold value V_T and the probability of false alarm P_{fa} was derived in Chapter 12, and for convenience is repeated here as Eq. (13.69):

$$v_T = \sqrt{2\sigma^2 \ln\left(\frac{1}{P_{fa}}\right)} . \qquad \text{Eq. (13.69)}$$

If the noise power σ^2 is constant, then a fixed threshold can satisfy Eq. (13.69). However, due to many reasons, this condition is rarely true. Thus, in order to maintain a constant probability of false alarm, the threshold value must be continuously updated based on the estimates of the noise variance. The process of continuously changing the threshold value to maintain a constant probability of false alarm is known as the Constant False Alarm Rate (CFAR).

Three different types of CFAR processors are primarily used. They are adaptive threshold CFAR, nonparametric CFAR, and nonlinear receiver techniques. Adaptive CFAR assumes that the interference distribution is known and approximates the unknown parameters associated with these distributions. Nonparametric CFAR processors tend to accommodate unknown interference distributions. Nonlinear receiver techniques attempt to normalize the root-mean-square amplitude of the interference. In this book, only the analog Cell-Averaging CFAR (CA-CFAR) technique is examined. The analysis presented in this section closely follows Urkowitz[1].

13.8.1. Cell-Averaging CFAR (Single Pulse)

The CA-CFAR processor is shown in Fig. 13.15. Cell averaging is performed on a series of range and/or Doppler bins (cells). The echo return for each pulse is detected by a square-law detector. In analog implementation, these cells are obtained from a tapped delay line. The Cell Under Test (CUT) is the central cell. The immediate neighbors of the CUT are excluded from the averaging process due to a possible spillover from the CUT. The output of M reference cells ($M/2$ on each side of the CUT) is averaged. The threshold value is obtained by multiplying the averaged estimate from all reference cells by a constant K_0 (used for scaling). A detection is declared in the CUT if

$$Y_1 \geq K_0 Z . \qquad \text{Eq. (13.70)}$$

CA-CFAR assumes that the target of interest is in the CUT and all reference cells contain zero-mean independent Gaussian noise of variance σ^2. Therefore, the output of the reference cells, Z, represents a random variable with gamma probability density function (special case of the chi-square) with $2M$ degrees of freedom. In this case, the gamma *pdf* is

---

1. Urkowitz, H., Decision and Detection Theory, unpublished lecture notes. Lockheed Martin Co., Moorestown, NJ.

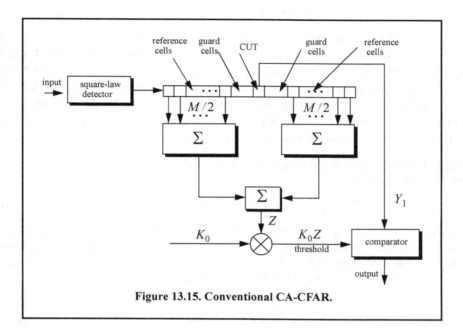

Figure 13.15. Conventional CA-CFAR.

$$f(z) = \frac{z^{(M/2)-1}e^{(-z/2\sigma^2)}}{2^{M/2}\,\sigma^M\Gamma(M/2)} \quad ; \; z > 0. \qquad \text{Eq. (13.71)}$$

The probability of false alarm corresponding to a fixed threshold was derived earlier. When CA-CFAR is implemented, then the probability of false alarm can be derived from the conditional false alarm probability, which is averaged over all possible values of the threshold in order to achieve an unconditional false alarm probability. The conditional probability of false alarm when $y = V_T$ can be written as

$$P_{fa}(v_T = y) = e^{-y/2\sigma^2}. \qquad \text{Eq. (13.72)}$$

It follows that the unconditional probability of false alarm is

$$P_{fa} = \int_0^\infty P_{fa}(v_T = y)f(y)dy \qquad \text{Eq. (13.73)}$$

where $f(y)$ is the *pdf* of the threshold, which except for the constant K_0, is the same as that defined in Eq. (13.71). Therefore,

$$f(y) = \frac{y^{M-1}e^{(-y/2K_0\sigma^2)}}{(2K_0\sigma^2)^M\Gamma(M)} \quad ; \; y \geq 0. \qquad \text{Eq. (13.74)}$$

Performing the integration in Eq. (13.73) yields

$$P_{fa} = 1/(1 + K_0)^M. \qquad \text{Eq. (13.75)}$$

Observation of Eq. (13.75) shows that the probability of false alarm is now independent of the noise power, which is the objective of CFAR processing.

13.8.2. Cell-Averaging CFAR with Noncoherent Integration

In practice, CFAR averaging is often implemented after noncoherent integration, as illustrated in Fig. 13.16. Now, the output of each reference cell is the sum of n_p squared envelopes. It follows that the total number of summed reference samples is Mn_p. The output Y_1 is also the sum of n_p squared envelopes. When noise alone is present in the CUT, Y_1 is a random variable whose *pdf* is a gamma distribution with $2n_p$ degrees of freedom. Additionally, the summed output of the reference cells is the sum of Mn_p squared envelopes. Thus, Z is also a random variable which has a gamma *pdf* with $2Mn_p$ degrees of freedom.

The probability of false alarm is then equal to the probability that the ratio Y_1/Z exceeds the threshold. More precisely,

$$P_{fa} = Prob\{Y_1/Z > K_1\}.$$ **Eq. (13.76)**

Equation (12.76) implies that one must first find the joint *pdf* for the ratio Y_1/Z. However, this can be avoided if P_{fa} is first computed for a fixed threshold value V_T, then averaged over all possible values of the threshold. Therefore, let the conditional probability of false alarm when $y = v_T$ be $P_{fa}(v_T = y)$. It follows that the unconditional false alarm probability is

$$P_{fa} = \int_0^\infty P_{fa}(v_T = y)f(y)\,dy$$ **Eq. (13.77)**

where $f(y)$ is the *pdf* of the threshold. In view of this, the probability density function describing the random variable $K_1 Z$ is given by

$$f(y) = \frac{(y/K_1)^{Mn_p-1}\,e^{(-y/2K_1\sigma^2)}}{(2\sigma^2)^{Mn_p} K_1\,\Gamma(Mn_p)} \quad ; \; y \geq 0.$$ **Eq. (13.78)**

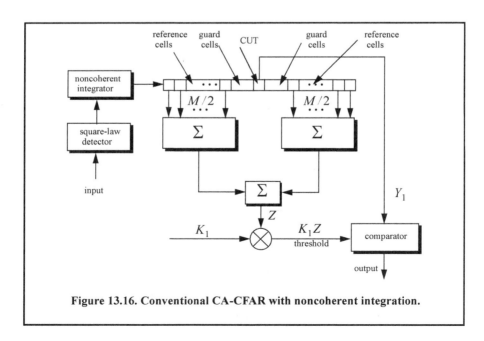

Figure 13.16. Conventional CA-CFAR with noncoherent integration.

It can be shown that in this case the probability of false alarm is independent of the noise power and is given by

$$P_{fa} = \frac{1}{(1+K_1)^{Mn_P}} \sum_{k=0}^{n_P-1} \frac{1}{k!} \frac{\Gamma(Mn_P+k)}{\Gamma(Mn_P)} \left(\frac{K_1}{1+K_1}\right)^k,$$

Eq. (13.79)

which is identical to Eq. (13.75) when $K_1 = K_0$ and $n_P = 1$.

13.9. M-out-of-N Detection

A few sources in the literature refer to the *M-out-of-N* detection as *binary integration* and / or as *double threshold* detection; nonetheless, *M-out-of-N* is the most commonly used name. The basic idea behind the *M-out-of-N* detection technique is as follows: In any given resolution cell (range, Doppler, or angle) the detection process is repeated N times, where the outcome of each decision cycle is either a "detection" or "no detection," hence the term *binary* is used in the literature. For each decision cycle, the probability of detection and the probability of false alarm are computed. The final decision criterion declares a target detection if M out of N decision cycles have resulted in a detection. Clearly, the decision criterion associated with this technique follows a binomial distribution.

To elaborate further on this concept of detection, assume a non-fluctuating target whose single trial probability of detection is P_D and its probability of false alarm is P_{fa}. Denote the total probability of detection resulting from the *M-out-of-N* detection technique as P_{Dmn}. It follows that after N independent trials of detection one gets

$$P_{Dmn} = 1 - (1 - P_D)^N.$$

Eq. (13.80)

Similarly, the probability of false alarm after the same number of trials is

$$P_{FA} = 1 - (1 - P_{fa})^N.$$

Eq. (13.81)

For example, if the desired P_{Dmn} is *0.99*, then by using Eq. (13.80), one finds that a $P_D = 0.9$ will accomplish the desired P_{Dmn} after 2 trials (i.e., *N=2*); alternatively, when using a $P_D = 0.2$, it will take 20 trials to reach the desired P_{Dmn}. Furthermore, Eq. (13.80) implicitly indicates that as the number of trials increases so does P_{Dmn}, but this buildup in detection probability is somewhat costly. That is true because as the number of trials is increased, the overall probability of false alarm P_{FA} is also increased. Obviously, a very undesirable result (the proof is left as an exercise, see Problem 13.20).

A slight modification to the *M-out-of-N* detection process that guarantees an increase or buildup in P_{Dmn} while simultaneously keeping P_{FA} in check is as follows:

1. A specific P_{fa} value is chosen; typically it is a design constraint.
2. For each value M, compute the corresponding P_{FA} from Eq. (13.83).
3. Using any of the techniques developed in this book to calculate the threshold value V_T so that P_{fa} is maintained, compute its corresponding SNR.
4. Calculate P_D that corresponds to the SNR computed in step 3.
5. Use Eq. (13.82) to compute the probability of detection P_{Dmn}, and from any of the techniques developed in this book, compute the corresponding SNR so that the threshold value computed in step 3 is maintained, therefore, P_{Dmn} is also maintained.

6. Repeat for each M to establish the specific combination of M (i.e., yielding P_{FA}) so that the SNR is minimized for a given P_{Dmn}.

Following this modified approach, P_{Dmn} and P_{FA} are given by

$$P_{Dmn} = \sum_{k=M}^{N} C_k^N \ P_D^k \ (1 - P_D)^{N-k} \qquad\qquad \text{Eq. (13.82)}$$

$$P_{FA} = \sum_{k=M}^{N} C_k^N \ P_{fa}^k \ (1 - P_{fa})^{N-k} \qquad\qquad \text{Eq. (13.83)}$$

where

$$C_k^N = \frac{N!}{k!(N-k)!} \ . \qquad\qquad \text{Eq. (13.84)}$$

For small values of P_D, Eq. (13.82) keeps the overall detection probability P_{Dmn} to less than or equal to P_D. Alternatively, for larger values of P_D, a quick buildup in the value of P_{Dmn} occurs.

Selecting the specific combination of N and M that yields a desired P_{Dmn} is typically a design constraint. In any case, once the choice is made, one must take target fluctuating into account. In this case, the optimal value for M is

$$M_{opt} = 10^\alpha N^\beta \qquad\qquad \text{Eq. (13.85)}$$

where α and β are constants that vary depending on the target fluctuation type, Table 13.1 shows their values corresponding to different Swerling targets.

Table 13.1. Parameters of Eq. (13.85)

Fluctuation Type	α	β	Range of N
Swerling 0	0.8	-0.02	5-700
Swerling I	0.8	-0.02	6-700
Swerling II	0.91	-0.38	9-700
Swerling III	0.8	-0.02	6-700
Swerling IV	0.873	-0.27	10-700

13.10. The Radar Equation Revisited

The radar equation developed in Chapter 2 assumed a constant target RCS and did not account for integration loss. In this section, a more comprehensive form of the radar equation is introduced. In this case, the radar equation is given by

$$R^4 = \frac{P_{av} G_t G_r \lambda^2 \sigma I(n_P)}{(4\pi)^3 k T_o F B \tau f_r L_t L_f \ (SNR)_1} \qquad\qquad \textbf{(4.86)}$$

where $P_{av} = P_t \tau f_r$ is the average transmitted power, P_t is the peak transmitted power, τ is the pulse width, f_r is PRF, G_t is the transmitting antenna gain, G_r is the receiving antenna gain, λ is the wavelength, σ is the target cross section, $I(n_p)$ is the improvement factor, n_p is the number of integrated pulses, k is Boltzman's constant, T_o is 290 Kelvin, F is the system noise figure, B is the receiver bandwidth, L_t is the total system losses including integration loss, L_f is the loss due to target fluctuation, and $(SNR)_1$ is the minimum single pulse SNR required for detection.

Assuming that the radar parameters such as power, antenna gain, wavelength, losses, band-width, effective temperature, and noise figure are known, the steps one should follow to solve for range are shown in Fig. 13.17. Note that both sides of the bottom half of Fig. 13.17 are identical. Nevertheless, two paths are purposely shown so that a distinction between scintillating and nonfluctuating targets is made.

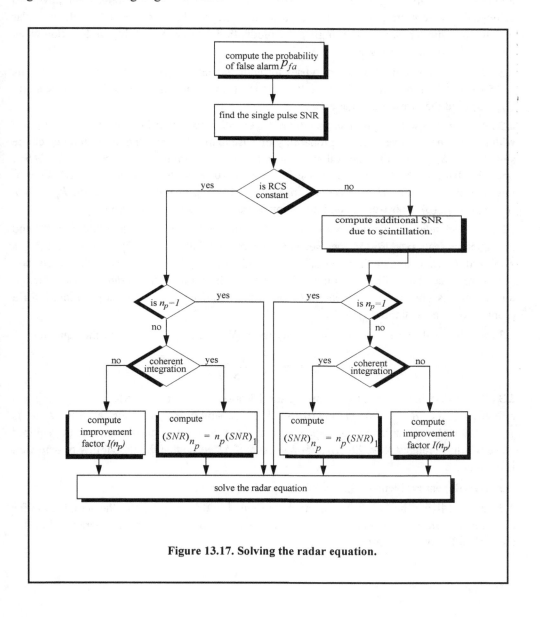

Figure 13.17. Solving the radar equation.

Problems

13.1. A pulsed radar has the following specifications: time of false alarm $T_{fa} = 10$ *min*, probability of detection $P_D = 0.95$, operating bandwidth $B = 1MHz$. (a) What is the probability of false alarm P_{fa}? (b) What is the single pulse SNR? (c) Assuming noncoherent integration of 100 pulses, what is the SNR reduction so that P_D and P_{fa} remain unchanged?

13.2. An L-band radar has the following specifications: operating frequency $f_0 = 1.5GHz$, operating bandwidth $B = 2MHz$, noise figure $F = 8dB$, system losses $L = 4dB$, time of false alarm $T_{fa} = 12$ *minutes*, detection range $R = 12Km$, probability of detection $P_D = 0.5$, antenna gain $G = 5000$, and target RCS $\sigma = 1m^2$. (a) Determine the PRF f_r, the pulse width τ, the peak power P_t, the probability of false alarm P_{fa}, and the minimum detectable signal level S_{min}. (b) How can you reduce the transmitter power to achieve the same performance when 10 pulses are integrated noncoherently? (c) If the radar operates at a shorter range in the single pulse mode, find the new probability of detection when the range decreases to $9Km$.

13.3. A certain radar utilizes 10 pulses for noncoherent integration. The single pulse SNR is $15dB$ and the probability of miss is $P_m = 0.15$. (a) Compute the probability of false alarm P_{fa}. (b) Find the threshold voltage V_T.

13.4. (a) Show how you can use the radar equation to determine the PRF f_r, the pulse width τ, the peak power P_t, the probability of false alarm P_{fa}, and the minimum detectable signal level S_{min}. Assume the following specifications: operating frequency $f_0 = 1.5MHz$, operating bandwidth $B = 1MHz$, noise figure $F = 10dB$, system losses $L = 5dB$, time of false alarm $T_{fa} = 20$ *min*, detection range $R = 12Km$, probability of detection $P_D = 0.5$ (three pulses). (b) If post-detection integration is assumed, determine the SNR.

13.5. Consider a scanning low PRF radar. The antenna half-power beam width is $1.5°$, and the antenna scan rate is $35°$ per second. The pulse width is $\tau = 2\mu s$, and the PRF is $f_r = 400Hz$. (a) Compute the radar operating bandwidth. (b) Calculate the number of returned pulses from each target illumination. (c) Compute the SNR improvement due to post-detection integration (assume 100% efficiency). (d) Find the number of false alarms per minute for a probability of false alarm $P_{fa} = 10^{-6}$.

13.6. Show that the detection probability for a SW 1&2 target is given by the equation

$$P_D = \exp\left\{\frac{\ln(P_{fa})}{1 + SNR}\right\}.$$

13.7. A certain radar has the following specifications: single pulse SNR corresponding to a reference range $R_0 = 200Km$ is $10dB$. The probability of detection at this range is $P_D = 0.95$. Assume a Swerling I type target. Use the radar equation to compute the required pulse widths at ranges $R = 220Km, 250Km$, and $175Km$, so that the probability of detection is maintained.

13.8. Repeat Problem 13.8 for a Swerling IV type target.

13.9. Utilizing the MATLAB functions presented in this chapter, plot the actual value for the improvement factor versus the number of integrated pulses. Pick three different values for the probability of false alarm.

13.10. A circularly scanning, fan beam radar has a rotation rate of 2 seconds per revolution. The azimuth beamwidth is 1.5 degrees and the radar uses a PRF of 12.5KHz. The radar uses an unmodulated pulse with a width of 1.2μs and searches a range window that extends from 15Km to 100Km. The range cells used during search are separated by one pulse width. It is desired that the false alarm probability be set so that the radar experiences only one false alarm every 2min. What is the required P_{fa} for each range cell? What is the threshold-to-noise ratio, in dB, needed to maintain that P_{fa}?

13.11. The probability of recording a detection in a particular range-angle cell of the scanning radar of Problem 13.11 is 0.7. What is the cumulative detection probability if the cell is checked on three successive scans? If the false alarm probability for a certain range-angle cell of the same radar is 10^{-6} what is the cumulative false alarm probability for that cell over three scans?

13.12. A radar with a phased array antenna conducts a search using a 1500-beam search raster. That is, it steps through 1500 beam positions that span a certain angular area. It transmits one pulse per beam. The radar uses range gates separated by 10m. The output of each range gate is sent to a bank of Doppler filters with a width of 1000Hz each. Thus, the signal processor consists of a set of range gates with a bank of Doppler filters connected to each range gate output. The output of the signal processor consists of a range-Doppler array of signals that consists of MN elements where M is the number of range gates and N is the number of Doppler filter outputs. During the particular search of interest, the detection processor covers a range extent of 10Km and a Doppler extent of 25Km. The design specifications state that, in this mode, the radar must have less than one false alarm every 10 scans through the search raster. What is the required P_{fa} in each range-Doppler-beam cell needed to support this requirement?

13.13. A certain radar employs a noncoherent integrator that integrates 25 pulses. What are the integrator gains, in dB, for a SW0, a SW1, a SW2, a SW3, and a SW4 target? Briefly discuss how you arrived at each of your answers. If needed, assume that the radar is to operate with a desired detection probability of 0.9.

13.14. A certain radar has the following parameters: Peak power P_t = 500KW, total losses L = 12dB, operating frequency f_o = 5.6GHZ, PRF f_r = 2KHz, pulse width τ = 0.5μs, antenna beamwidth θ_{az} = 2° and θ_{el} = 7°, noise figure F = 6dB, and scan time T_{sc} = 2s. The radar can experience one false alarm per scan. (a) What is the probability of false alarm? Assume that the radar searches a minimum range of 10Km to its maximum unambiguous range. (b) Plot the detection range versus RCS in dBsm. The detection range is defined as the range at which the single scan probability of detection is equal to 0.94. Generate curves for Swerling I, II, III, and IV type targets. (c) Repeat part (b) above when noncoherent integration is used.

13.15. A certain circularly scanning radar with a fan beam has a rotation rate of 3 seconds per revolution. The azimuth beamwidth is 3 degrees, and the radar uses a PRI of 600 microseconds. The radar pulse width is 2 microseconds and the radar searches a range window that extends from 15Km to 100Km. It is desired that the false alarm rate not be higher than two false alarms per revolution. What is the required probability of false alarm? What is the minimum SNR so that the minimum probability of false alarm can be maintained?

13.16. Write a MATLAB program to compute the CA-CFAR threshold value. Use similar approach to that used in the case of a fixed threshold.

13.17. Develop a MATLAB program to calculate the cumulative probability of detection.

13.18. Derive Eq. (13.79).

13.19. The sum inside Eq. (13.79) presents a very formidable challenge. It can be, however, computed recursively with relative ease. Develop a recursive algorithm to calculate this sum.

13.20. Starting with Eq. (13.81), show that as N is increased so is the over all probability of false alarm. More specifically, prove that $P_{FA} \approx N P_{fa}$.

Appendix 13-A The Incomplete Gamma Function

The Gamma Function

Define the Gamma function (not the incomplete Gamma function) of the variable z (generally complex) as

$$\Gamma(z) = \int_0^\infty x^{z-1} e^{-x} \, dx \qquad \text{Eq. (13.87)}$$

and when z is a positive integer, then

$$\Gamma(z) = (z-1)! \, . \qquad \text{Eq. (13.88)}$$

One very useful and frequently used property is

$$\Gamma(z+1) = z\Gamma(z) \qquad \text{Eq. (13.89)}$$

The Incomplete Gamma Function

The incomplete gamma function $\Gamma_I(u, q)$ used in this text is given by

$$\Gamma_I(u, q) = \int_0^{u\sqrt{q+1}} \frac{e^{-x} x^q}{q!} \, dx \, . \qquad \text{Eq. (13.90)}$$

Another definition, which is often used in the literature, for the incomplete Gamma function is

$$\Gamma_I[z, q] = \int_q^\infty x^{z-1} e^{-x} \, dx \, . \qquad \text{Eq. (13.91)}$$

It follows that

$$\Gamma(z) = \Gamma_I[z, 0] = \int_0^\infty x^{z-1} e^{-x} dx \, , \qquad \text{Eq. (13.92)}$$

which is the same as Eq. (13.80). Furthermore, for a positive integer n, the incomplete Gamma function can be represented by

$$\Gamma_I[n, z] = (n-1)! e^{-z} \sum_{k=0}^{n-1} \frac{z^k}{k!} \, . \qquad \text{Eq. (13.93)}$$

In order to relate $\Gamma_I[n, z]$ and $\Gamma_I(u, q)$, compute the following relation

$$\Gamma_I[a, 0] - \Gamma_I[a, z] = \int_0^\infty x^{a-1} e^{-x} dx - \int_z^\infty x^{a-1} e^{-x} dx = \int_0^z x^{a-1} e^{-x} dx \, . \qquad \text{Eq. (13.94)}$$

Applying the change of variables $a = q+1$ and $z = u\sqrt{q+1}$ yields

$$\Gamma_I[q+1, 0] - \Gamma_I[q+1, u\sqrt{q+1}] = \int_0^{u\sqrt{q+1}} x^q \, e^{-x} dx,$$

Eq. (13.95)

and if q is a positive integer then

$$\frac{\Gamma_I[q+1, 0] - \Gamma_I[q+1, u\sqrt{q+1}]}{q!} = \int_0^{u\sqrt{q+1}} \frac{x^q \, e^{-x}}{q!} dx = \Gamma_I(u, q).$$

Eq. (13.96)

Using Eqs. (13.81) and (7.86) in Eq. (13.89) yields

$$\Gamma_I(u, q) = 1 - \frac{(q+1-1)! \, e^{-u\sqrt{q+1}}}{q!} \sum_{k=0}^{q} \frac{(u\sqrt{q+1})^k}{k!}.$$

Eq. (13.97)

Finally, the incomplete Gamma function can be written as

$$\Gamma_I(u, q) = 1 - e^{-u\sqrt{q+1}} \sum_{k=0}^{q} \frac{(u\sqrt{q+1})^k}{k!}.$$

Eq. (13.98)

The two limiting values for Eq. (13.91) are

$$\Gamma_I(0, q) = 0 \qquad \Gamma_I(\infty, q) = 1.$$

Eq. (13.99)

Figure 13A.1 shows the incomplete gamma function for $q = 1, 3, 5, 8$. This figure can be reproduced using the MATLAB program *"Fig13A_1.m"* listed in Appendix 13-B, which utilizes the built-in MATLAB function *"gammainc.m."*

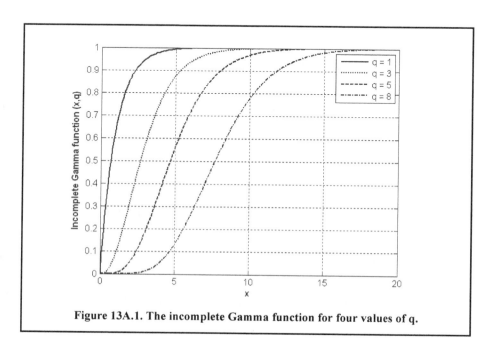

Figure 13A.1. The incomplete Gamma function for four values of q.

Appendix 13-B: Chapter 13 MATLAB Code Listings

The MATLAB code provided in this chapter was designed as an academic standalone tool and is not adequate for other purposes. The code was written in a way to assist the reader in gaining a better understanding of the theory. The code was not developed, nor is it intended to be used as part of an open-loop or a closed-loop simulation of any kind. The MATLAB code found in this textbook can be downloaded from this book's web page on the CRC Press website. Simply use your favorite web browser, go to *www.crcpress.com*, and search for keyword *"Mahafza"* to locate this book's web page.

MATLAB Function *"improv_fac.m"* Listing

```
function impr_of_np = improv_fac (np, pfa, pd)
% This function computes the non-coherent integration improvement
% factor using the empirical formula defind in Eq. (13.10)
% Inputs
    % np          == number of pulses
    % pfa         == probability of false alaram
    % pd          == probability of detection
%% Output
    % impr_of_np    == improvement factor for np pulses
fact1 = 1.0 + log10( 1.0 / pfa) / 46.6;
fact2 = 6.79 .* (1.0 + 0.253 .* pd);
fact3 = 1.0 - 0.14 .* log10(np) + 0.0183 .* (log10(np)).^2;
impr_of_np = fact1 .* fact2 .* fact3 .* log10(np);
end
```

MATLAB Program *"Fig13_2.m"* Listing

```
% This program is used to produce Fig. 13.2
% It uses the function "improv_fac".
clc
clear all
close all
Pfa = [1e-2, 1e-6, 1e-8, 1e-10];
Pd = [.5 .8 .95 .99];
np = linspace(1,1000,10000);
I(1,:) = improv_fac (np, Pfa(1), Pd(1));
I(2,:) = improv_fac (np, Pfa(2), Pd(2));
I(3,:) = improv_fac (np, Pfa(3), Pd(3));
I(4,:) = improv_fac (np, Pfa(4), Pd(4));
index = [1 2 3 4];
L(1,:) = 10.*log10(np) - I(1,:);
L(2,:) = 10.*log10(np) - I(2,:);
L(3,:) = 10.*log10(np) - I(3,:);
L(4,:) = 10.*log10(np) - I(4,:);
subplot(2,1,2)
semilogx (np, L(1,:), 'k:', np, L(2,:), 'k-.', np, L(3,:), 'k-.', np, L(4,:), 'k','linewidth',1.5)
xlabel ('\bfNumber of pulses');
ylabel ('\bfIntegration loss in dB')
axis tight
grid
subplot(2,1,1)
semilogx (np, I(1,:), 'k:', np, I(2,:), 'k-.', np, I(3,:), 'k--', np, I(4,:), 'k','linewidth',1.5)
```

```
%set (gca,'xtick',[1 2 3 4 5 6 7 8  10 20 30 100]);
xlabel ('\bfNumber of pulses');
ylabel ('\bfImprovement factor in dB')
legend   ('P_D=.5,   P_f_a=10^-^2','P_D=.8,   P_f_a=10^-^6','P_D=.95,   P_f_a=10^-^8','P_D=.99,
P_f_a=10^-^1^0');
grid
axis tight
```

MATLAB Function "threshold.m" Listing

```
function [pfa, vt] = threshold (nfa, np)
% This function calculates the threshold value from nfa and np.
% The newton-Raphson recursive formula
% This function uses "gammainc.m".
% Inputs
   % nfa          == number of false alarm
   % np           == number of pulses
%% Outputs
   % Pfa          == probability of false alarm
   % vt           == threshold
%
delta = eps;
pfa = np * log(2) / nfa;
sqrtpfa = sqrt(-log10(pfa));
sqrtnp = sqrt(np);
vt0 = np - sqrtnp + 2.3 * sqrtpfa * (sqrtpfa + sqrtnp - 1.0);
vt = vt0;
while (delta < (vt0/10000));
  igf = gammainc(vt0,np);
  num = 0.5^(np/nfa) - igf;
  deno = -exp(-vt0) * vt0^(np-1) /factorial(np-1);
  vt = vt0 - (num / (deno+eps));
  delta = abs(vt - vt0);
  vt0 = vt;
end
```

MATLAB Program "Fig13_4.m" Listing

```
% Use this program to reproduce Fig. 13.4 of text
clear all
close all
for n= 1: 1:10000
  [pfa1 y1(n)] = threshold(1e4,n);
  [pfa2 y3(n)] = threshold(1e8,n);
  [pfa3 y4(n)] = threshold(1e12,n);
end
n =1:1:10000;
loglog(n,y1,'k',n,y3,'k--',n,y4,'k-.','linewidth',1.5);
xlabel ('\bfNumber of pulses');
ylabel('\bfThreshold')
legend('nfa=10^1^2','nfa=10^8','nfa=10^8')
grid
```

MATLAB Function "pd_swerling5.m" Listing

```
function pd = pd_swerling5 (input1, indicator, np, snrbar)
% This function is used to calculate the probability of detection
% for Swerling 5 or 0 targets for np>1.
%
% Inputs
  % input1      == Pfa or nfa
  % indicator   == 1 when input1 = Pfa; 2 when input1 = nfa
  % np          == number of pulses
  % snrbar      == SNR
% Outputs
  % pd          == probability of detection
if(np == 1)
  'Stop, np must be greater than 1'
  return
end
format long
snrbar = 10.0.^(snrbar./10.);
eps = 0.00000001;
delmax = .00001;
delta =10000.;
% Calculate the threshold Vt
if (indicator ~=1)
  nfa = input1;
  pfa =  np * log(2) / nfa;
else
  pfa = input1;
  nfa = np * log(2) / pfa;
end
sqrtpfa = sqrt(-log10(pfa));
sqrtnp = sqrt(np);
vt0 = np - sqrtnp + 2.3 * sqrtpfa * (sqrtpfa + sqrtnp - 1.0);
vt = vt0;
while (delta < (vt0/10000));
  igf = gammainc(vt0,np);
  num = 0.5^(np/nfa) - igf;
  deno = -exp(-vt0) * vt0^(np-1) /factorial(np-1);
  vt = vt0 - (num / (deno+eps));
  delta = abs(vt - vt0);
  vt0 = vt;
end
% Calculate the Gram-Chrlier coefficients
temp1 = 2.0 .* snrbar + 1.0;
omegabar = sqrt(np .* temp1);
c3 = -(snrbar + 1.0 / 3.0) ./ (sqrt(np) .* temp1.^1.5);
c4 = (snrbar + 0.25) ./ (np .* temp1.^2.);
c6 = c3 .* c3 ./2.0;
V = (vt - np .* (1.0 + snrbar)) ./ omegabar;
Vsqr = V .*V;
val1 = exp(-Vsqr ./ 2.0) ./ sqrt( 2.0 * pi);
val2 = c3 .* (V.^2 -1.0) + c4 .* V .* (3.0 - V.^2) -...
  c6 .* V .* (V.^4 - 10. .* V.^2 + 15.0);
q = 0.5 .* erfc (V./sqrt(2.0)); pd =  q - val1 .* val2;
return
```

MATLAB Program "Fig13_5.m" Listing

```
% This program is used to produce Fig. 13.5
clc
close all
clear all
pfa = 1e-9;
nfa = log(2) / pfa;
b = sqrt(-2.0 * log(pfa));
index = 0;
for snr = 0:.1:20
   index = index +1;
   a = sqrt(2.0 * 10^(.1*snr));
   pro(index) = marcumsq(a,b);
   prob205(index) = pd_swerling5 (pfa, 1, 10, snr);
end
x = 0:.1:20;
plot(x, pro,'k',x,prob205,'k:','linewidth',1.5);
axis([0 20 0 1])
xlabel ('\bfSNR in dB')
ylabel ('\bfProbability of detection')
legend('n_p = 1','n_p = 10')
grid on
```

MATLAB Function "pd_swerling1.m" Listing

```
function pd = pd_swerling1 (nfa, np, snrbar)
% This function is used to calculate the probability of detection
% for Swerling 1 targets.
%
% Inputs
   % nfa          == Marcum's false alarm number
   % np           == number of integrated pulses
   % snrbar       == SNR
%
% outputs
   % pd           == probability of detection
format long
snrbar = 10.0^(snrbar/10.);
eps = 0.00000001;
delta = eps;
% Calculate the threshold Vt
pfa =  np * log(2) / nfa;
sqrtpfa = sqrt(-log10(pfa));
sqrtnp = sqrt(np);
vt0 = np - sqrtnp + 2.3 * sqrtpfa * (sqrtpfa + sqrtnp - 1.0);
vt = vt0;
while (delta < (vt0/10000));
   igf = gammainc(vt0,np);
   num = 0.5^(np/nfa) - igf;
   deno = -exp(-vt0) * vt0^(np-1) /factorial(np-1);
   vt = vt0 - (num / (deno+eps));
   delta = abs(vt - vt0);
   vt0 = vt;
end
```

```
if (np == 1)
  temp = -vt / (1.0 + snrbar);
  pd = exp(temp);
  return
end
  temp1 = 1.0 + np * snrbar;
  temp2 = 1.0 / (np *snrbar);
  temp = 1.0 + temp2;
  val1 = temp^(np-1.);
  igf1 = gammainc(vt,np-1);
  igf2 = gammainc(vt/temp,np-1);
  pd = 1.0 - igf1 + val1 * igf2 * exp(-vt/temp1);
  return
```

MATLAB Program "Fig13_6.m" Listing

```
% This program is used to reproduce Fig. 13.6
clc
close all
clear all
pfa = 1e-9;
nfa = log(2) / pfa;
b = sqrt(-2.0 * log(pfa));
index = 0;
for snr = 0:.01:22
   index = index +1;
   a = sqrt(2.0 * 10^(.1*snr));
   swer0(index) = marcumsq(a,b);
   swer1(index) = pd_swerling1 (nfa, 1, snr);
end
x = 0:.01:22;
%figure(10)
plot(x, swer0,'k',x,swer1,'k:','linewidth', 1.5);
axis([2 22 0 1])
xlabel ('\bfSNR in dB')
ylabel ('\bfProbability of detection')
legend('Swerling 0', 'Swerling I')
grid on
```

MATLAB Program "Fig13_7.m" Listing

```
% This program is used to produce Fig. 13.7
clc
clear all
close all
pfa = 1e-6;
nfa = log(2) / pfa;
index = 0;
for snr = -10:.5:30
   index = index +1;
   prob1(index) = pd_swerling1 (nfa, 15, snr);
   prob0(index) = pd_swerling5 (nfa, 2, 15, snr);
   end
x = -10:.5:30;
```

```
plot(x, prob1,'k',x,prob0,'k:','linewidth',1.5);
axis([-10 30 0 1])
xlabel ('\bfSNR in dB')
ylabel ('\bfProbability of detection')
legend('Swerling I','Swerling 0')
title('\bfP_f_a =10^-^6;  n_p=5')
grid
```

MATLAB Function "pd_swerling2.m" Listing

```
function pd = pd_swerling2 (nfa, np, snrbar)
% This function is used to calculate the probability of detection
% for Swerling 2 targets.
% Inputs
   % nfa           == number of fals alarm
   % np            == number of pulses
   % snrbar        == SNR
%
% Outputs
   % pd            == proability of detection
format long
snrbar = 10.0^(snrbar/10.);
eps = 0.00000001;
delta = eps;
% Calculate the threshold Vt
pfa =  np * log(2) / nfa;
sqrtpfa = sqrt(-log10(pfa));
sqrtnp = sqrt(np);
vt0 = np - sqrtnp + 2.3 * sqrtpfa * (sqrtpfa + sqrtnp - 1.0);
vt = vt0;
while (delta < (vt0/10000));
  igf = gammainc(vt0,np);
  num = 0.5^(np/nfa) - igf;
  deno = -exp(-vt0) * vt0^(np-1) /factorial(np-1);
  vt = vt0 - (num / (deno+eps));
  delta = abs(vt - vt0);
  vt0 = vt;
end
if (np <= 50)
  temp = vt / (1.0 + snrbar);
  pd = 1.0 - gammainc(temp,np);
  return
else
  temp1 = snrbar + 1.0;
  omegabar = sqrt(np) * temp1;
  c3 = -1.0 / sqrt(9.0 * np);
  c4 = 0.25 / np;
  c6 = c3 * c3 /2.0;
  V = (vt - np * temp1) / omegabar;
  Vsqr = V *V;
  val1 = exp(-Vsqr / 2.0) / sqrt( 2.0 * pi);
  val2 = c3 * (V^2 -1.0) + c4 * V * (3.0 - V^2) - ...
    c6 * V * (V^4 - 10. * V^2 + 15.0);
  q = 0.5 * erfc (V/sqrt(2.0));
```

```
    pd = q - val1 * val2;
end
return
```

MATLAB Program "Fig13_8.m" Listing

```
% This program is used to produce Fig. 13.8
clc
clear all
close all
pfa = 1e-7;
nfa = log(2) / pfa;
index = 0;
for snr = -10:.5:30
    index = index +1;
    prob1(index) = pd_swerling1 (nfa, 5, snr);
    prob0(index) = pd_swerling5 (nfa, 2, 5, snr);
    prob2(index) = pd_swerling2 (nfa, 5, snr);
end
x = -10:.5:30;
plot(x, prob0,'k',x,prob1,'k:',x,prob2,'k--','linewidth',1.5);
axis([-10 30 0 1])
xlabel ('\bfSNR in dB')
ylabel ('\bfProbability of detection')
legend('Swerling 0','Swerling I','Swerling II')
title('P_f_a =10^-^7;  n=5')
grid
```

MATLAB Program "Fig13_9.m" Listing

```
% This program is used to produce Fig. 13.9
clear all
close all
pfa = 1e-6;
nfa = log(2) / pfa;
index = 0;
b = sqrt(-2.0 * log(pfa));
for snr = -10:.5:30
    a = sqrt(2.0 * 10^(.1*snr));
    index = index +1;
    prob1(index) = pd_swerling1 (nfa, 2, snr);
    prob0(index) =  marcumsq(a,b);
    prob2(index) = pd_swerling2 (nfa, 2, snr);
end
x = -10:.5:30;
plot(x, prob0,'k',x,prob1,'k:',x,prob2,'k--','linewidth',1.5);
axis([-10 30 0 1])
xlabel ('\bfSNR in dB')
ylabel ('\bfProbability of detection')
legend('Swerling 0','Swerling I','Swerling II')
title('P_f_a =10^-^6;  n_p = 2')
grid on
```

MATLAB Function "pd_swerling3.m" Listing

```
function pd = pd_swerling3 (nfa, np, snrbar)
% This function is used to calculate the probability of detection
% for Swerling 2 targets.
% Inputs
  % nfa      == false alarm number
  % np       == number of pulses
  % snrbar   == SNR
% Outputs
  % pd       == probability of detection
format long
snrbar = 10.0^(snrbar/10.);
eps = 0.00000001;
delta = eps;
% Calculate the threshold Vt
pfa = np * log(2) / nfa;
sqrtpfa = sqrt(-log10(pfa));
sqrtnp = sqrt(np);
vt0 = np - sqrtnp + 2.3 * sqrtpfa * (sqrtpfa + sqrtnp - 1.0);
vt = vt0;
while (delta < (vt0/10000));
  igf = gammainc(vt0,np);
  num = 0.5^(np/nfa) - igf;
  deno = -exp(-vt0) * vt0^(np-1) /factorial(np-1);
  vt = vt0 - (num / (deno+eps));
  delta = abs(vt - vt0);
  vt0 = vt;
end
temp1 = vt / (1.0 + 0.5 * np *snrbar);
temp2 = 1.0 + 2.0 / (np * snrbar);
temp3 = 2.0 * (np - 2.0) / (np * snrbar);
ko = exp(-temp1) * temp2^(np-2.) * (1.0 + temp1 - temp3);
if (np <= 2)
  pd = ko;
  return
else
  ko = exp(-temp1) * temp2^(np-2.) * (1.0 + temp1 - temp3);
  temp4 = vt^(np-1.) * exp(-vt) / (temp1 * (factorial(np-2.)));
  temp5 = vt / (1.0 + 2.0 / (np *snrbar));
  pd = temp4 + 1.0 - gammainc(vt,np-1.) + ko * gammainc(temp5,np-1.);
end; return
```

MATLAB Program "Fig13_10.m" Listing

```
% This program is used to produce Fig. 13.10
clc
close all
clear all
pfa = 1e-9;
nfa = log(2) / pfa;
index = 0;
for snr = -10:.5:30
  index = index +1;
  prob1(index) = pd_swerling3 (nfa, 1, snr);
```

```
  prob10(index) = pd_swerling3 (nfa, 10, snr);
  prob50(index) = pd_swerling3(nfa, 50, snr);
  prob100(index) = pd_swerling3 (nfa, 100, snr);
end
x = -10:.5:30;
plot(x, prob1,'k',x,prob10,'k:',x,prob50,'k--', x, prob100,'k-.','linewidth',1.5);
axis([-10 30 0 1])
xlabel ('SNR in dB')
ylabel ('Probability of detection')
legend('np = 1','np = 10','np = 50','np = 100')
grid on
```

MATLAB Program "Fig13_11.m" Listing

```
% This program is used to produce Fig. 13.11
clc
clear all
close all
pfa = 1e-7;
nfa = log(2) / pfa;
index = 0;
for snr = -10:.5:30
  index = index +1;
  prob1(index) = pd_swerling1 (nfa, 5, snr);
  prob0(index) = pd_swerling5 (nfa, 2, 5, snr);
  prob2(index) = pd_swerling2 (nfa, 5, snr);
  prob3(index) = pd_swerling3 (nfa, 5, snr);
end
x = -10:.5:30;
plot(x, prob0,'k',x,prob1,'k:',x,prob2,'k--',x,prob3,'k-.','linewidth',1,'linewidth',1.5);
axis([-10 30 0 1])
xlabel ('\bfSNR in dB')
ylabel ('P\bfrobability of detection')
legend('Swerling 0','Swerling I','Swerling II', 'Swerling III')
title('P_f_a =10^-^7;  n=5')
grid on
```

MATLAB Function "pd_swerling4.m" Listing

```
function pd = pd_swerling4 (nfa, np, snrbar)
% This function is used to calculate the probability of detection
% for Swerling 4 targets.
% Inputs
  % nfa          == number of false alarm
  % np           == number of pulses
  % snrbar       == SNR
% Output
  % pd           == probability of detection
format long
snrbar = 10.0^(snrbar/10.);
eps = 0.00000001;
delta = eps;
% Calculate the threshold Vt
pfa =  np * log(2) / nfa;
```

```
sqrtpfa = sqrt(-log10(pfa));
sqrtnp = sqrt(np);
vt0 = np - sqrtnp + 2.3 * sqrtpfa * (sqrtpfa + sqrtnp - 1.0);
vt = vt0;
while (delta < (vt0/10000));
  igf = gammainc(vt0,np);
  num = 0.5^(np/nfa) - igf;
  deno = -exp(-vt0) * vt0^(np-1) /factorial(np-1);
  vt = vt0 - (num / (deno+eps));
  delta = abs(vt - vt0);
  vt0 = vt;
end
h8 = snrbar /2.0;
beta = 1.0 + h8;
beta2 = 2.0 * beta^2 - 1.0;
beta3 = 2.0 * beta^3;
if (np >= 50)
  temp1 = 2.0 * beta -1;
  omegabar = sqrt(np * temp1);
  c3 = (beta3 - 1.) / 3.0 / beta2 / omegabar;
  c4 = (beta3 * beta3 - 1.0) / 4. / np /beta2 /beta2;;
  c6 = c3 * c3 /2.0;
  V = (vt - np * (1.0 + snrbar)) / omegabar;
  Vsqr = V *V;
  val1 = exp(-Vsqr / 2.0) / sqrt( 2.0 * pi);
  val2 = c3 * (V^2 -1.0) + c4 * V * (3.0 - V^2) - ...
    c6 * V * (V^4 - 10. * V^2 + 15.0);
  q = 0.5 * erfc (V/sqrt(2.0));
  pd =  q - val1 * val2;
  return
else
  gamma0 = gammainc(vt/beta,np);
  a1 = (vt / beta)^np / (factorial(np) * exp(vt/beta));
  sum = gamma0;
  for i = 1:1:np
    temp1 = gamma0;
    if (i == 1)
      ai = a1;
    else
      ai = (vt / beta) * a1 / (np + i -1);
    end
    gammai = gamma0 - ai;
    gamma0 = gammai;
    a1 = ai;
    for ii = 1:1:i
      temp1 = temp1 * (np + 1 - ii);
    end
    term = (snrbar /2.0)^i * gammai * temp1 / (factorial(i));
    sum = sum + term;
  end
  pd = 1.0 - (sum / beta^np);
end
pd = max(pd,0.);
return
```

MATLAB Program "Fig13_12.m" Listing

```
% This program is used to produce Fig. 13.12 of text
clear all
close all
pfa = 1e-6;
nfa = log(2) / pfa;
index = 0;
for snr = -7:.15:10
  index = index +1;
  prob1(index) =  pd_swerling4 (nfa, 5, snr);
  prob10(index) =  pd_swerling4 (nfa, 10, snr);
  prob25(index) =  pd_swerling4(nfa, 25, snr);
  prob75(index) =  pd_swerling4 (nfa, 75, snr);
end
x = -7:.15:10;
plot(x, prob1,'k',x,prob10,'k.',x,prob25,'k:',x, prob75,'k-.','linewidth',1.5);
xlabel ('\bfSNR in dB')
ylabel ('\bfProbability of detection')
legend('np = 5','np = 10','np = 25','np = 75')
grid on; axis tight
```

MATLAB Function "fluct_loss.m" Listing

```
function [SNR] = fluct(pd, nfa, np, sw_case)
% This function calculates the SNR fluctuation loss for Swerling models
% A negative Lf value indicates SNR gain instead of loss
% Inputs
   % pd      == desired probability of detection
   % nfa     == desired number of false alarms
   % np      == number of pulses
   % sw_case  == 0, 1, 2, 3, or 4 depending on the desired Swerling case
% Output
   % SNR      == Resulting SNR
format long
% *************** Swerling 5 case ***************
% check to make sure that np>1
pfa =  np * log(2) / nfa;
if (sw_case == 0)
if (np ==1)
  nfa = 1/pfa;
  b = sqrt(-2.0 * log(pfa));
  Pd_Sw5 = 0.001;
  snr_inc = 0.1 - 0.005;
  while(Pd_Sw5 <= pd)
    snr_inc = snr_inc + 0.005;
    a = sqrt(2.0 * 10^(.1*snr_inc));
    Pd_Sw5 = marcumsq(a,b);
  end
  PD_SW5 = Pd_Sw5;
  SNR = snr_inc;
else
  % np > 1 use MATLAB function pd_swerling5.m
  snr_inc = 0.1 - 0.001;
  Pd_Sw5 = 0.001;
```

```
    while(Pd_Sw5 <= pd)
       snr_inc = snr_inc + 0.001;
       Pd_Sw5 = pd_swerling5(pfa, 1, np, snr_inc);
    end
    PD_SW5 = Pd_Sw5;
    SNR = snr_inc;
  end
end
% ************** End Swerling 5 case ***********
% ************** Swerling 1 case ***************
% compute the false alarm number
if (sw_case == 1)
  Pd_Sw1 = 0.001;
  snr_inc = 0.1 - 0.001;
  while(Pd_Sw1 <= pd)
     snr_inc = snr_inc + 0.001;
     Pd_Sw1 = pd_swerling1(nfa, np, snr_inc);
  end
  PD_SW1 = Pd_Sw1;
  SNR = snr_inc;
end
% ************** End Swerling 1 case ***********
% ************** Swerling 2 case ***************
if (sw_case == 2)
  Pd_Sw2 = 0.001;
  snr_inc = 0.1 - 0.001;
  while(Pd_Sw2 <= pd)
     snr_inc = snr_inc + 0.001;
     Pd_Sw2 = pd_swerling2(nfa, np, snr_inc);
  end
  PD_SW2 = Pd_Sw2;
  SNR = snr_inc;
end
% ************** End Swerling 2 case ***********
% ************** Swerling 3 case ***************
if (sw_case == 3)
  Pd_Sw3 = 0.001;
  snr_inc = 0.1 - 0.001;
  while(Pd_Sw3 <= pd)
     snr_inc = snr_inc + 0.001;
     Pd_Sw3 = pd_swerling3(nfa, np, snr_inc);
  end
  PD_SW3 = Pd_Sw3;
  SNR = snr_inc;
end
% ************** End Swerling 3 case ***********
% ************** Swerling 4 case ***************
if (sw_case == 4)
  Pd_Sw4 = 0.001;
  snr_inc = 0.1 - 0.001;
  while(Pd_Sw4 <= pd)
     snr_inc = snr_inc + 0.001;
     Pd_Sw4 = pd_swerling4(nfa, np, snr_inc);
  end
```

```
    PD_SW4 = Pd_Sw4;
    SNR = snr_inc;
end
% ************** End Swerling 4 case ************
return
```

MATLAB Program "Fig13_13.m" Listing

```
% Use this program to reproduce Fig. 13.13 of text
clear all
close all
index = 0.;
for pd = 0.01:.05:1
    index = index + 1;
    [Lf,Pd_Sw5] = fluct_loss(pd, 1e-7,1,1);
    Lf1(index) = Lf;
    [Lf,Pd_Sw5] = fluct_loss(pd, 1e-7,1,4);
    Lf4(index) = Lf;
end
pd = 0.01:.05:1;
figure (3)
plot(pd, Lf1, 'k',pd, Lf4,'K:','linewidth',1.5)
xlabel('\bfProbability of detection')
ylabel('\bfFluctuation loss - dB')
legend('Swerling I & II','Swerling III & IV')
title('P_f_a = 10^-^9, n_p = 1')
grid on
```

MATLAB Program "Fig13A_1.m" Listing

```
% This program can be used to reproduce Fig. 13A.1
clc
close all
clear all
x=linspace(0,20,200);
y1 = gammainc(x,1);
y2 = gammainc(x,3);
y3 = gammainc(x,5);
y4 = gammainc(x,8);
plot(x,y1,'k',x,y2,'k:',x,y3,'k--',x,y4,'k-.','linewidth',1.5)
legend('q = 1','q = 3','q = 5','q = 8')
xlabel('\bfx')
ylabel('\bfIncomplete Gamma function (x,q)')
grid
```

Part V

Radar Special
Topics

Chapter 14

Radar Cross Section (RCS)

This chapter was coauthored with Mr. Walton C. Gibson.[1]

14.1. RCS Definition

Electromagnetic waves, with any specified polarization, are normally diffracted or scattered in all directions when incident on a target. These scattered waves are broken down into two parts. The first part is made of waves that have the same polarization as the receiving antenna. The other portion of the scattered waves will have a different polarization to which the receiving antenna does not respond. The two polarizations are orthogonal and are referred to as the Principal Polarization (PP) and Orthogonal Polarization (OP), respectively. The intensity of the *backscattered* energy that has the same polarization as the radar's receiving antenna is used to define the target RCS. When a target is illuminated by RF energy, it acts like an antenna, and will have near and far scattered fields. Waves reflected and measured in the near field are, in general, spherical. Alternatively, in the far field the wavefronts are decomposed into a linear combination of plane waves.

Assume the power density of a wave incident on a target located at range R away from the radar is P_{Di}, as illustrated in Fig. 14.1. The amount of reflected power from the target is

$$P_r = \sigma P_{Di} \qquad \qquad \text{Eq. (14.1)}$$

where σ denotes the target cross section. Define P_{Dr} as the power density of the scattered waves at the receiving antenna. It follows that

$$P_{Dr} = P_r / (4\pi R^2). \qquad \qquad \text{Eq. (14.2)}$$

Equating Eqs. (14.1) and (14.2) yields

$$\sigma = 4\pi R^2 \left(\frac{P_{Dr}}{P_{Di}}\right) \qquad \qquad \text{Eq. (14.3)}$$

---

1. Mr. Gibson is associated with Tripoint Industries, Inc. in Huntsville, Alabama, *www.tripointindustries.com*.

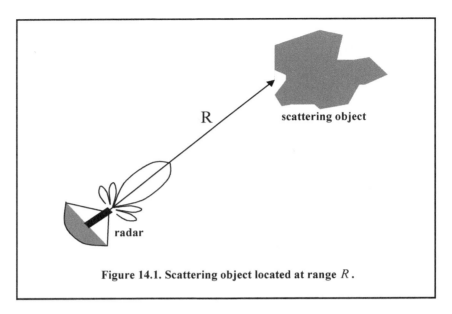

Figure 14.1. Scattering object located at range R.

and in order to ensure that the radar receiving antenna is in the far field (i.e., scattered waves received by the antenna are planar), Eq. (14.3) is modified to

$$\sigma = 4\pi R^2 \lim_{R \to \infty} \left(\frac{P_{Dr}}{P_{Di}}\right).$$ **Eq. (14.4)**

The RCS defined by Eq. (14.4) is often referred to as the monostatic RCS, the backscattered RCS, or simply the target RCS.

The backscattered RCS is measured from all waves scattered in the direction of the radar and has the same polarization as the receiving antenna. It represents a portion of the total scattered target RCS σ_t, where $\sigma_t > \sigma$. Assuming a spherical coordinate system defined by (ρ, θ, φ), then at range ρ, the target scattered cross section is a function of (θ, φ). Let the angles (θ_i, φ_i) define the direction of propagation of the incident waves. Also, let the angles (θ_s, φ_s) define the direction of propagation of the scattered waves. The special case, when $\theta_s = \theta_i$ and $\varphi_s = \varphi_i$, defines the monostatic RCS. The RCS measured by the radar at angles $\theta_s \neq \theta_i$ and $\varphi_s \neq \varphi_i$ is called the bistatic RCS. The total target scattered RCS is given by

$$\sigma_t = \frac{1}{4\pi} \int_{\varphi_s = 0}^{2\pi} \int_{\theta_s = 0}^{\pi} \sigma(\theta_s, \varphi_s) \sin\theta_s \ d\theta \ d\varphi_s.$$ **Eq. (14.5)**

The amount of backscattered waves from a target is proportional to the ratio of the target extent (size) to the wavelength, λ, of the incident waves. In fact, a radar will not be able to detect targets much smaller than its operating wavelength. For example, if weather radars use L-band frequency, rain drops become nearly invisible to the radar since they are much smaller than the wavelength. The frequency region, where the target extent and the wavelength are comparable, is referred to as the Rayleigh region. Alternatively, the frequency region where the target extent is much larger than the radar operating wavelength is referred to as the optical region. In practice, the majority of radar applications fall within the optical region.

The analysis presented in this book mainly assumes far field monostatic RCS measurements in the optical region. Near field RCS, bistatic RCS, and RCS measurements in the Rayleigh region will not be considered since their treatment falls beyond this book's intended scope. Additionally, RCS treatment in this chapter is mainly concerned with Narrow Band (NB) cases. In other words, the extent of the target under consideration falls within a single range bin of the radar. Wideband (WB) RCS measurements will be briefly addressed in a later section. Wideband radar range bins are small (typically 10 - 50 cm); hence, the target under consideration may cover many range bins. The RCS value in an individual range bin corresponds to the portion of the target falling within that bin.

14.2. RCS Dependency on Aspect Angle and Frequency

Radar cross section fluctuates as a function of radar aspect angle and frequency. For the purpose of illustration, isotropic point scatterers are considered. An isotropic scatterer is one that scatters incident waves equally in all directions. Consider the geometry shown in Fig. 14.2. In this case, two unity ($1 m^2$) isotropic scatterers are aligned and placed along the radar line of sight (zero aspect angle) at a far field range R. The spacing between the two scatterers is 1 meter. The radar aspect angle is then changed from zero to 180 degrees, and the composite RCS of the two scatterers measured by the radar is computed.

This composite RCS consists of the superposition of the two individual radar cross sections. At zero aspect angle, the composite RCS is $2m^2$. Taking scatterer-1 as a phase reference, when the aspect angle is varied, the composite RCS is modified by the phase that corresponds to the electrical spacing between the two scatterers. For example, at aspect angle $10°$, the electrical spacing between the two scatterers is

$$elec\text{–}spacing = \frac{2 \times (1.0 \times \cos(10°))}{\lambda}.$$

Eq. (14.6)

λ is the radar operating wavelength.

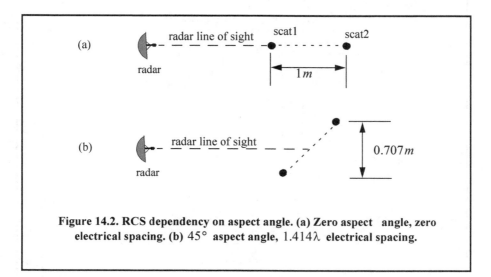

Figure 14.2. RCS dependency on aspect angle. (a) Zero aspect angle, zero electrical spacing. (b) $45°$ aspect angle, 1.414λ electrical spacing.

MATLAB Function "rcs_aspect.m"

The function *"rcs_aspect.m"* computes the RCS dependency on the aspect angle. Its syntax is as follows:

$$[rcs] = rcs\_aspect \ (scat\_spacing, freq)$$

where

Symbol	Description	Units	Status
scat_spacing	scatterer spacing	meters	input
freq	radar frequency	Hz	input
rcs	array of RCS versus aspect angle	dBsm	output

Figure 14.3 shows the composite RCS corresponding to this experiment. This plot can be reproduced using MATLAB program *"Fig.14.3.m"* listed in Appendix 14-A. As clearly indicated by Fig. 14.3, RCS is dependent on the radar aspect angle; thus, knowledge of this constructive and destructive interference between the individual scatterers can be very critical when a radar tries to extract the RCS of complex or maneuvering targets. This is true because of two reasons. First, the aspect angle may be continuously changing. Second, complex target RCS can be viewed as made up from contributions of many individual scattering points distributed on the target surface. These scattering points are often called scattering centers. Many approximate RCS prediction methods generate a set of scattering centers that define the back-scattering characteristics of such complex targets.

Next, to demonstrate RCS dependency on frequency, consider the experiment shown in Fig. 14.4. In this case, two far field unity isotropic scatterers are aligned with radar line of sight, and the composite RCS is measured by the radar as the frequency is varied from $8GHz$ to $12.5GHz$ (X-band).

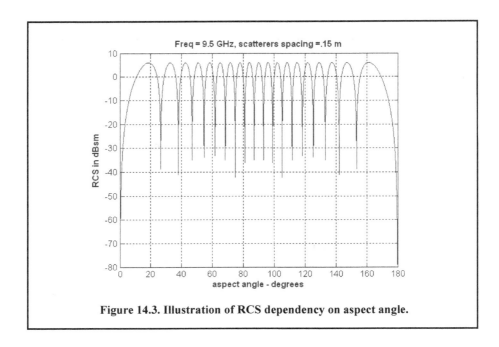

Figure 14.3. Illustration of RCS dependency on aspect angle.

MATLAB Function "rcs_frequency.m"

The function *"rcs_frequency.m"* computes the RCS dependency on frequency. Its syntax is as follows:

$$[rcs] = rcs\_frequency \ (scat\_spacing, frequ, freql)$$

where

Symbol	Description	Units	Status
scat_spacing	scatterer spacing	meters	input
freql, frequ	start and end of frequency band	Hz	input

Figures 14.5 and 14.6 show the composite RCS versus frequency for scatterer spacing of 0.25 and 0.75 meters. The plots shown in Figs. 14.5 and 14.6 can be reproduced using the MATLAB program *"Fig.14_5_6.m,"* listed in Appendix 14-A. From those two Figures, RCS fluctuation as a function of frequency is evident. A small frequency change can cause serious RCS fluctuation when the scatterer spacing is large. Alternatively, when scattering centers are relatively close, it requires more frequency variation to produce significant RCS fluctuation.

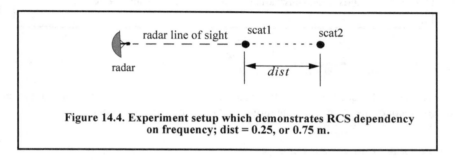

Figure 14.4. Experiment setup which demonstrates RCS dependency on frequency; dist = 0.25, or 0.75 m.

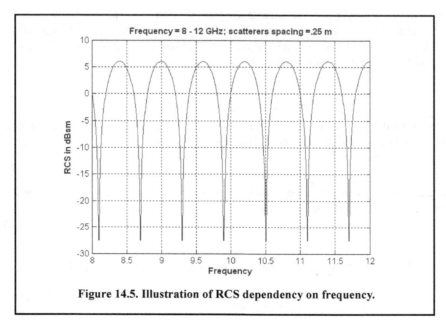

Figure 14.5. Illustration of RCS dependency on frequency.

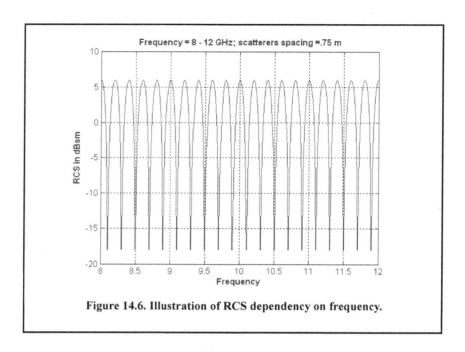

Figure 14.6. Illustration of RCS dependency on frequency.

14.3. RCS Dependency on Polarization

The material in this section covers two topics. First, a review of polarization fundamentals is presented. Second, the concept of the target scattering matrix is introduced.

14.3.1. Normalized Electric Field

In most radar simulations, it is desirable to obtain the complex-valued electric field scattered by the target at the radar. In such cases, it is useful to use a quantity called the normalized electric field. It is assumed that the incident electric field has a magnitude of unity, and is phase centered at a point at the target (usually the center of gravity). More precisely,

$$E_i = e^{jk(\vec{r_i} \cdot \vec{r})}$$

Eq. (14.7)

where $\vec{r_i}$ is the direction of incidence and $\vec{r}$ as a location at the target, each with respect to the phase center. The normalized scattered field is then given by

$$E_s = \sigma E_i$$

Eq. (14.8)

The quantity E_s is independent of radar and target location. It may be combined with an incident magnitude and phase.

14.3.2. Polarization

The x and y electric field components for a wave traveling along the positive z direction are given by

$$E_x = E_1 \sin(\omega t - kz)$$

Eq. (14.9)

$$E_y = E_2 \sin(\omega t - kz + \delta) \qquad \text{Eq. (14.10)}$$

where $k = 2\pi / \lambda$, ω is the wave frequency, the angle δ is the time phase angle at which E_y leads E_x, and finally, E_1 and E_2 are, respectively, the wave amplitudes along the x and y directions. When two or more electromagnetic waves combine, their electric fields are integrated vectorially at each point in space for any specified time. In general, the combined vector traces an ellipse when observed in the x-y plane. This is illustrated in Fig. 14.7.

The ratio of the major to the minor axes of the polarization ellipse is called the Axial Ratio (AR). When AR is unity, the polarization ellipse becomes a circle, and the resultant wave is then called circularly polarized. Alternatively, when $E_1 = 0$ and $AR = \infty$, the wave becomes linearly polarized.

Eqs. (14.9) and (14.10) can be combined to give the instantaneous total electric field,

$$\vec{E} = \hat{a}_x E_1 \sin(\omega t - kz) + \hat{a}_y E_2 \sin(\omega t - kz + \delta) \qquad \text{Eq. (14.11)}$$

where $\hat{a}_x$ and $\hat{a}_y$ are unit vectors along the x and y directions, respectively. At $z = 0$, $E_x = E_1 \sin(\omega t)$ and $E_y = E_2 \sin(\omega t + \delta)$, then by replacing $\sin(\omega t)$ by the ratio E_x / E_1 and by using trigonometry properties Eq. (14.11) can be rewritten as

$$\frac{E_x^2}{E_1^2} - \frac{2E_x E_y \cos\delta}{E_1 E_2} + \frac{E_y^2}{E_2^2} = (\sin\delta)^2 . \qquad \text{Eq. (14.12)}$$

Note that Eq. (14.12) has no dependency on ωt. In the most general case, the polarization ellipse may have any orientation, as illustrated in Fig. 14.8. The angle ξ is called the tilt angle of the ellipse. In this case, AR is given by

$$AR = \frac{OA}{OB} \qquad (1 \leq AR \leq \infty). \qquad \text{Eq. (14.13)}$$

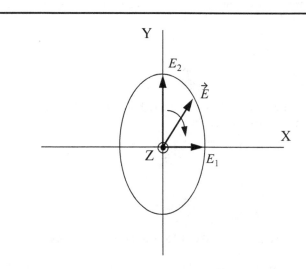

Figure 14.7. Electric field components along the *x* and *y* directions. The positive *z* direction is off of the page.

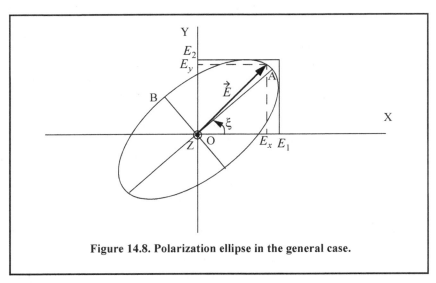

Figure 14.8. Polarization ellipse in the general case.

When $E_1 = 0$, the wave is said to be linearly polarized in the y direction, while if $E_2 = 0$, the wave is said to be linearly polarized in the x direction. Polarization can also be linear at an angle of 45° when $E_1 = E_2$ and $\xi = 45°$. When $E_1 = E_2$ and $\delta = 90°$, the wave is said to be Left Circularly Polarized (LCP), while if $\delta = -90°$ the wave is said to Right Circularly Polarized (RCP). It is a common notation to call the linear polarizations along the x and y directions by the names horizontal and vertical polarizations, respectively.

In general, an arbitrarily polarized electric field may be written as the sum of two circularly polarized fields. More precisely,

$$\vec{E} = \vec{E_R} + \vec{E_L}$$

Eq. (14.14)

where $\vec{E_R}$ and $\vec{E_L}$ are the RCP and LCP fields, respectively. Similarly, the RCP and LCP waves can be written as

$$\vec{E_R} = \vec{E_V} + j\vec{E_H}$$

Eq. (14.15)

$$\vec{E_L} = \vec{E_V} - j\vec{E_H}$$

Eq. (14.16)

where $\vec{E_V}$ and $\vec{E_H}$ are the fields with vertical and horizontal polarizations, respectively. Combining Eqs. (14.15) and (14.16) yields

$$E_R = \frac{E_H - jE_V}{\sqrt{2}}$$

Eq. (14.17)

$$E_L = \frac{E_H + jE_V}{\sqrt{2}}.$$

Eq. (14.18)

Using matrix notation, Eqs. (14.17) and (14.18) can be rewritten as

$$\begin{bmatrix} E_R \\ E_L \end{bmatrix} = \frac{1}{\sqrt{2}} \begin{bmatrix} 1 & -j \\ 1 & j \end{bmatrix} \begin{bmatrix} E_H \\ E_V \end{bmatrix} = [T] \begin{bmatrix} E_H \\ E_V \end{bmatrix}$$

Eq. (14.19)

$$\begin{bmatrix} E_H \\ E_V \end{bmatrix} = \frac{1}{\sqrt{2}} \begin{bmatrix} 1 & 1 \\ j & -j \end{bmatrix} \begin{bmatrix} E_R \\ E_L \end{bmatrix} = [T]^{-1} \begin{bmatrix} E_H \\ E_V \end{bmatrix}.$$ **Eq. (14.20)**

For many targets, the scattered waves will have different polarization than the incident waves. This phenomenon is known as depolarization or cross-polarization. However, perfect reflectors reflect waves in such a fashion that an incident wave with horizontal polarization remains horizontal, and an incident wave with vertical polarization remains vertical but is phase shifted $180°$. Additionally, an incident wave that is RCP becomes LCP when reflected, and a wave that is LCP becomes RCP after reflection from a perfect reflector. Therefore, when a radar uses LCP waves for transmission, the receiving antenna needs to be RCP polarized in order to capture the PP RCS, and LCR to measure the OP RCS.

14.3.3. Target Scattering Matrix

Target backscattered RCS is commonly described by a matrix known as the scattering matrix, and is denoted by $[S]$. When an arbitrarily linearly polarized wave is incident on a target, the backscattered field is then given by

$$\begin{bmatrix} E_1^s \\ E_2^s \end{bmatrix} = [S] \begin{bmatrix} E_1^i \\ E_2^i \end{bmatrix} = \begin{bmatrix} s_{11} & s_{12} \\ s_{21} & s_{22} \end{bmatrix} \begin{bmatrix} E_1^i \\ E_2^i \end{bmatrix}.$$ **Eq. (14.21)**

The superscripts i and s denote incident and scattered fields. The quantities s_{ij} are in general complex, and the subscripts 1 and 2 represent any combination of orthogonal polarizations. More precisely, $1 = H, R$, and $2 = V, L$. From Eq. (14.3), the backscattered RCS is related to the scattering matrix components by the following relation:

$$\begin{bmatrix} \sigma_{11} & \sigma_{12} \\ \sigma_{21} & \sigma_{22} \end{bmatrix} = 4\pi R^2 \begin{bmatrix} |s_{11}|^2 & |s_{12}|^2 \\ |s_{21}|^2 & |s_{22}|^2 \end{bmatrix}$$ **Eq. (14.22)**

It follows that once a scattering matrix is specified, the target backscattered RCS can be computed for any combination of transmitting and receiving polarizations. The reader is advised to see Ruck (1970) for ways to calculate the scattering matrix $[S]$.

Rewriting Eq. (14.22) in terms of the different possible orthogonal polarizations yields

$$\begin{bmatrix} E_H^s \\ E_V^s \end{bmatrix} = \begin{bmatrix} s_{HH} & s_{HV} \\ s_{VH} & s_{VV} \end{bmatrix} \begin{bmatrix} E_H^i \\ E_V^i \end{bmatrix}$$ **Eq. (14.23)**

$$\begin{bmatrix} E_R^s \\ E_L^s \end{bmatrix} = \begin{bmatrix} s_{RR} & s_{RL} \\ s_{LR} & s_{LL} \end{bmatrix} \begin{bmatrix} E_R^i \\ E_L^i \end{bmatrix}.$$ **Eq. (14.24)**

By using the transformation matrix $[T]$ in Eq. (14.19), the circular scattering elements can be computed from the linear scattering elements

$$\begin{bmatrix} s_{RR} & s_{RL} \\ s_{LR} & s_{LL} \end{bmatrix} = [T] \begin{bmatrix} s_{HH} & s_{HV} \\ s_{VH} & s_{VV} \end{bmatrix} \begin{bmatrix} 1 & 0 \\ 0 & -1 \end{bmatrix} [T]^{-1}$$ Eq. (14.25)

and the individual components are

$$s_{RR} = \frac{-s_{VV} + s_{HH} - j(s_{HV} + s_{VH})}{2}$$

$$s_{RL} = \frac{s_{VV} + s_{HH} + j(s_{HV} - s_{VH})}{2}$$

$$s_{LR} = \frac{s_{VV} + s_{HH} - j(s_{HV} - s_{VH})}{2}$$ Eq. (14.26)

$$s_{LL} = \frac{-s_{VV} + s_{HH} + j(s_{HV} + s_{VH})}{2}$$

Similarly, the linear scattering elements are given by

$$\begin{bmatrix} s_{HH} & s_{HV} \\ s_{VH} & s_{VV} \end{bmatrix} = [T]^{-1} \begin{bmatrix} s_{RR} & s_{RL} \\ s_{LR} & s_{LL} \end{bmatrix} \begin{bmatrix} 1 & 0 \\ 0 & -1 \end{bmatrix} [T]$$ Eq. (14.27)

and the individual components are

$$s_{HH} = \frac{s_{RR} - s_{RL} + s_{LR} - s_{LL}}{2}$$

$$s_{VH} = \frac{j(s_{RR} - s_{LR} - s_{RL} + s_{LL})}{2}$$

$$s_{HV} = \frac{-j(s_{RR} + s_{LR} + s_{RL} + s_{LL})}{2}$$ Eq. (14.28)

$$s_{VV} = \frac{s_{RR} + s_{LL} - s_{RL} - s_{LR}}{2}$$

14.4. RCS of Simple Objects

Electromagnetic wave scattering from simple objects has historically received a great amount of attention as analytic expressions since the scattered fields can often be derived. Among these are objects such as spheres and ellipsoids, and two-dimensional cylinders, half planes and wedges. The study of such shapes is of great value as they lend insight into the important scattering mechanisms inherent in wave interactions with real-world objects. These analytic scattering equations are also used to test and verify Computational Electromagnetic (CEM) software codes. Readers interested in these subjects should consider sources such as Bowman and Ruck, which summarize research into the scattering from such bodies.

This section presents a sample set of simple object radar cross section. Most of the expressions presented represent the radar cross section of the object when it is large compared to the wavelength. These are derived from analytic expressions, often series or complex-plane integrations, using asymptotic limits for wavelength or empirical fits to simplify their evaluation. These approximations are said to operate in the "high-frequency" or "optical" scattering regime. Computational methods will be discussed later in this chapter. In this case, other methods in the "low-frequency" or "resonance" regime are used to calculate the RCS.

This section presents examples of a backscattered radar cross section for a number of simple shape objects. In all cases, except for the perfectly conducting sphere, only optical region approximations are presented. Radar designers and RCS engineers consider the perfectly conducting sphere to be the simplest target to examine. Even in this case, the complexity of the exact solution, when compared to the optical region approximation, is overwhelming. Most formulas presented are Physical Optics (PO) approximation for the backscattered RCS measured by a far field radar in the direction (θ, φ), as illustrated in Fig. 14.9. In this section, it is assumed that the radar is always illuminating an object from the positive z-direction.

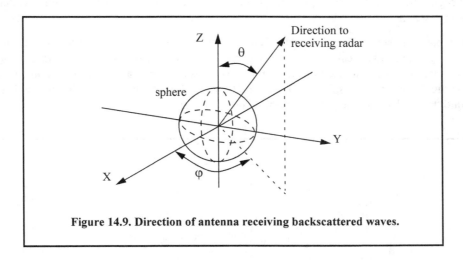

Figure 14.9. Direction of antenna receiving backscattered waves.

14.4.1. Sphere

Due to symmetry, waves scattered from a perfectly conducting sphere are co-polarized (have the same polarization) with the incident waves. This means that the cross-polarized backscattered waves are practically zero. For example, if the incident waves were Left Circularly Polarized (LCP), then the backscattered waves will also be LCP. However, because of the opposite direction of propagation of the backscattered waves, they are considered to be Right Circularly Polarized (RCP) by the receiving antenna. Therefore, the PP backscattered waves from a sphere are LCP, while the OP backscattered waves are negligible.

The normalized exact backscattered RCS for a perfectly conducting sphere is a Mie series given by

$$\frac{\sigma}{\pi r^2} = \left(\frac{j}{kr}\right) \sum_{n=1}^{\infty} (-1)^n (2n+1) \left[\left(\frac{kr J_n(kr) - n J_n(kr)}{kr H_{n-1}(kr) - n H_n^{(1)}(kr)}\right) - \left(\frac{J_n(kr)}{H_n^{(1)}(kr)}\right)\right] \qquad \text{Eq. (14.29)}$$

where r is the radius of the sphere, $k = 2\pi/\lambda$, λ is the wavelength, J_n is the spherical Bessel of the first kind of order n, and $H_n^{(1)}$ is the Hankel function of order n, and is given by

$$H_n^{(1)}(kr) = J_n(kr) + j Y_n(kr). \qquad \text{Eq. (14.30)}$$

Y_n is the spherical Bessel function of the second kind of order n. Plots of the normalized perfectly conducting sphere RCS as a function of its circumference in wavelength units are shown in Figs. 14.10a and 14.10b. These plots can be reproduced using the function *"rcs_sphere.m."* In Fig. 14.10, three regions are identified. First is the optical region (corresponds to a large sphere). In this case,

$$\sigma = \pi r^2 \qquad r \gg \lambda.$$

Eq. (14.31)

Second is the Rayleigh region (small sphere). In this case,

$$\sigma \approx 9\pi r^2 (kr)^4 \qquad r \ll \lambda.$$

Eq. (14.32)

The region between the optical and Rayleigh regions is oscillatory in nature and is called the Mie or resonance region.

The backscattered RCS for a perfectly conducting sphere is constant in the optical region. For this reason, radar designers typically use spheres of known cross sections to experimentally calibrate radar systems. For this purpose, spheres are flown attached to balloons. In order to obtain Doppler shift, spheres of known RCS are dropped out of an airplane and towed behind the airplane whose velocity is known to the radar.

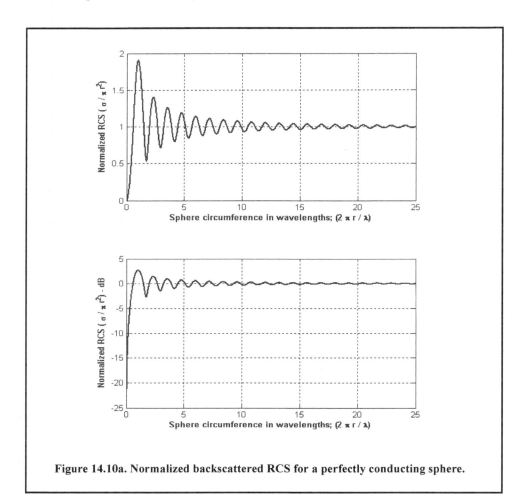

Figure 14.10a. Normalized backscattered RCS for a perfectly conducting sphere.

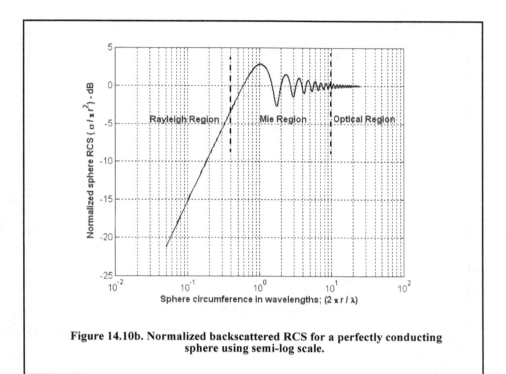

Figure 14.10b. Normalized backscattered RCS for a perfectly conducting sphere using semi-log scale.

14.4.2. Ellipsoid

An ellipsoid centered at (0,0,0) is shown in Fig. 14.11. It is defined by the following equation:

$$\left(\frac{x}{a}\right)^2 + \left(\frac{y}{b}\right)^2 + \left(\frac{z}{c}\right)^2 = 1.$$ **Eq. (14.33)**

One widely accepted approximation for the ellipsoid backscattered RCS is given by

$$\sigma = \frac{\pi a^2 b^2 c^2}{\left(a^2(\sin\theta)^2(\cos\varphi)^2 + b^2(\sin\theta)^2(\sin\varphi)^2 + c^2(\cos\theta)^2\right)^2}.$$ **Eq. (14.34)**

When $a = b$, the ellipsoid becomes roll symmetric. Thus, the RCS is independent of φ, and Eq. (14.34) is reduced to

$$\sigma = \frac{\pi b^4 c^2}{\left(a^2(\sin\theta)^2 + c^2(\cos\theta)^2\right)^2},$$ **Eq. (14.35)**

and for the case when $a = b = c$,

$$\sigma = \pi c^2.$$ **Eq. (14.36)**

Note that Eq. (14.36) defines the backscattered RCS of a sphere. This should be expected, since under the condition $a = b = c$ the ellipsoid becomes a sphere.

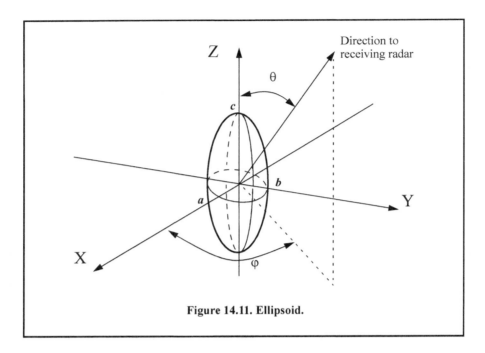

Figure 14.11. Ellipsoid.

MATLAB Function "rcs_ellipsoid.m"

The MATLAB function *"rcs_ellipsoid.m"* computes the RCS of an ellipsoid by implementing Eqs. (14.34) and (14.35). Its syntax is as follows:

$$[rcs] = rcs\_ellipsoid\ (a,\ b,\ c,\ phi)$$

where

Symbol	Description	Units	Status
a	ellipsoid a-radius	meters	input
b	ellipsoid b-radius	meters	input
c	ellipsoid c-radius	meters	input
phi	ellipsoid roll angle	degrees	input
rcs	array of RCS versus aspect angle	dBsm	output

Figure 14.12a shows the backscattered RCS for an ellipsoid versus the aspect angle θ for $\varphi = 0°$, $\varphi = 45°$, and $\theta = 90°$. Note that at normal incidence ($\theta = 90°$), the RCS corresponds to that of a sphere of radius c, and is often referred to as the broadside specular RCS value. This figure can be reproduced using MATLAB program *"Fig14_12a.m,"* listed in Appendix 14-A.

A MATLAB-based graphical user interface (GUI) was developed for this purpose. Figure 14.12b shows the GUI workspace associated with the function. To execute this GUI, first download its MATLAB code from this book's web page on the CRC Press website, then in the MATLAB command window, type *"rcs_ellipsoid_gui."*

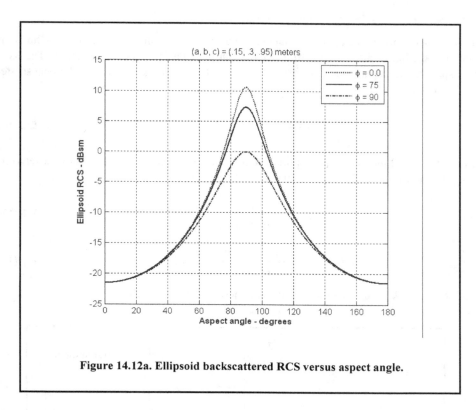

Figure 14.12a. Ellipsoid backscattered RCS versus aspect angle.

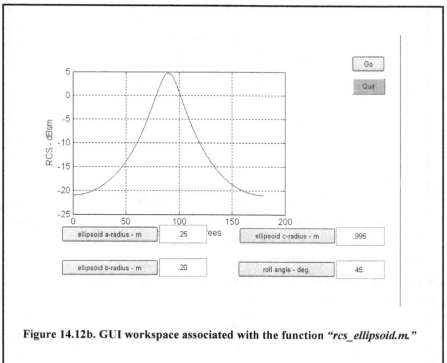

Figure 14.12b. GUI workspace associated with the function *"rcs_ellipsoid.m."*

14.4.3. Circular Flat Plate

Figure 14.13 shows a circular flat plate of radius r, centered at the origin. Due to the circular symmetry, the backscattered RCS of a circular flat plate has no dependency on φ. The RCS is only aspect angle dependent. For normal incidence (i.e., zero aspect angle), the backscattered RCS for a circular flat plate is

$$\sigma = \frac{4\pi^3 r^4}{\lambda^2} \qquad \theta = 0°. \qquad \text{Eq. (14.37)}$$

For non-normal incidence, two approximations for the circular flat plate backscattered RCS for any linearly polarized incident wave are

$$\sigma = \frac{\lambda r}{8\pi \sin\theta (\tan(\theta))^2} \qquad \text{Eq. (14.38)}$$

$$\sigma = \pi k^2 r^4 \left(\frac{2J_1(2kr\sin\theta)}{2kr\sin\theta}\right)^2 (\cos\theta)^2 \qquad \text{Eq. (14.39)}$$

where $k = 2\pi/\lambda$, and J_1 is the first-order spherical Bessel function.

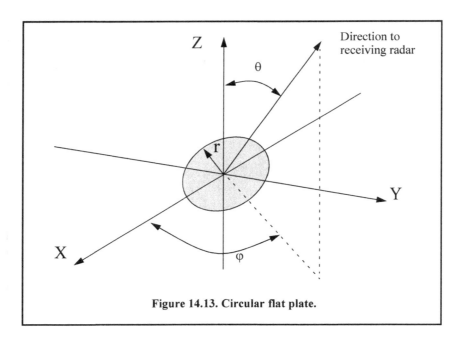

Figure 14.13. Circular flat plate.

MATLAB Function "rcs_circ_plate.m"

The function *"rcs_circ_plate.m"* calculates and plots the backscattered RCS from a circular plate. The syntax is as follows:

[rcs] = rcs_circ_plate (r, freq)

where

Symbol	Description	Units	Status
r	*radius of circular plate*	*meters*	*input*
freq	*frequency*	*Hz*	*input*
rcs	*array of RCS versus aspect angle*	*dBsm*	*output*

A MATLAB-based GUI was developed to implement this function. Figure 14.14 shows the GUI workspace associated with function and a typical output. To execute this GUI, first download its MATLAB code from this book's web page on the CRC Press website, then in the MATLAB command window type, *"rcs_circ_gui.m."*

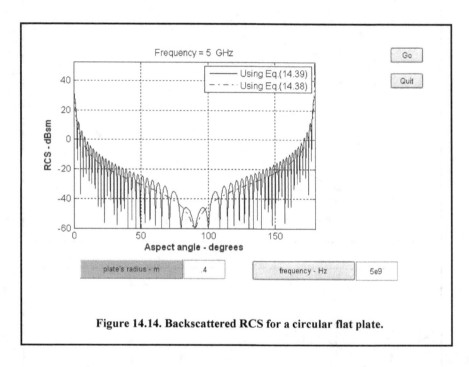

Figure 14.14. Backscattered RCS for a circular flat plate.

14.4.4. Truncated Cone (Frustum)

Figures 14.15 and 14.16 show the geometry associated with a frustum. The half cone angle α is given by

$$\tan\alpha = \frac{(r_2 - r_1)}{H} = \frac{r_2}{L}.$$ **Eq. (14.40)**

Define the aspect angle at normal incidence with respect to the frustum's surface (broadside) as θ_n. Thus, when a frustum is illuminated by a radar located at the same side as the cone's small end, the angle θ_n is

$$\theta_n = 90° - \alpha.$$ **Eq. (14.41)**

Alternatively, normal incidence occurs at

$$\theta_n = 90° + \alpha .$$ **Eq. (14.42)**

At normal incidence, one approximation for the backscattered RCS of a truncated cone due to a linearly polarized incident wave is

$$\sigma_{\theta_n} = \frac{8\pi(z_2^{3/2} - z_1^{3/2})^2}{9\lambda\sin\theta_n}\tan\alpha(\sin\theta_n - \cos\theta_n\tan\alpha)^2$$ **Eq. (14.43)**

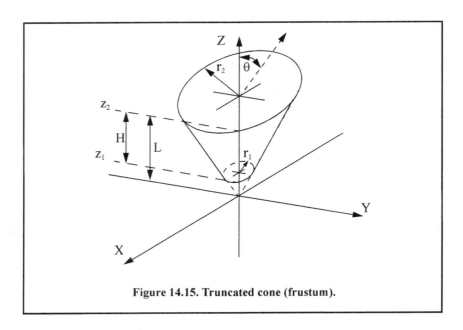

Figure 14.15. Truncated cone (frustum).

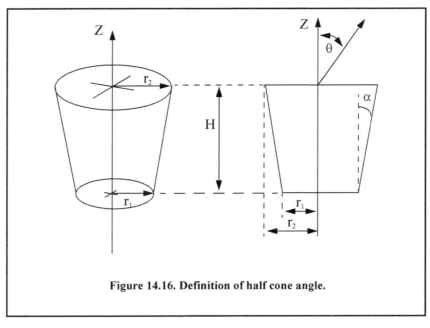

Figure 14.16. Definition of half cone angle.

$$\sigma_{\theta_n} = \frac{8\pi(z_2^{3/2} - z_1^{3/2})^2}{9\lambda} \frac{\sin\alpha}{(\cos\alpha)^4}.$$ **Eq. (14.44)**

For non-normal incidence, the backscattered RCS due to a linearly polarized incident wave is

$$\sigma = \frac{\lambda z \tan\alpha}{8\pi \sin\theta}\left(\frac{\sin\theta - \cos\theta \tan\alpha}{\sin\theta \tan\alpha + \cos\theta}\right)^2$$ **Eq. (14.45)**

where z is equal to either z_1 or z_2, depending on whether the RCS contribution is from the small or the large end of the cone. Again, using trigonometric identities Eq. (14.45) (assuming the radar illuminates the frustum starting from the large end) is reduced to

$$\sigma = \frac{\lambda z \tan\alpha}{8\pi \sin\theta}\ (\tan(\theta - \alpha))^2$$ **Eq. (14.46)**

where λ is the wavelength, and z_1, z_2 are defined in Fig. 14.15.

When the radar illuminates the frustum starting from the small end (i.e., the radar is in the negative z direction in Fig. 14.15), Eq. (14.46) should be modified to

$$\sigma = \frac{\lambda z \tan\alpha}{8\pi \sin\theta}\ (\tan(\theta + \alpha))^2.$$ **Eq. (14.47)**

MATLAB Function "rcs_frustum.m"

The function *"rcs_frustum.m"* computes and plots the backscattered RCS of a truncated conic section. The syntax is as follows:

[rcs] = rcs_frustum (r1, r2, freq, indicator)

where

Symbol	Description	Units	Status
r1	small end radius	meters	input
r2	large end radius	meters	input
freq	frequency	Hz	input
indicator	indicator = 1 when viewing from large end indicator = 0 when viewing from small end	none	input
rcs	array of RCS versus aspect angle	dBsm	output

For example, consider a frustum defined by $H = 20.945\,cm$, $r_1 = 2.057\,cm$, and $r_2 = 5.753\,cm$. It follows that the half cone angle is $10°$. Figure 14.17a shows a plot of its RCS when illuminated by a radar in the positive z direction. Figure 14.17b shows the same thing, except in this case, the radar is in the negative z direction. Note that for the first case, normal incidence occurs at $100°$, while for the second case it occurs at $80°$. A MATLAB-based GUI was developed to implement this function. To execute this GUI, first download the its MATLAB code from this book's web page on the CRC Press website, then in the MATLAB command window type, *"rcs_frustum_gui.m."*

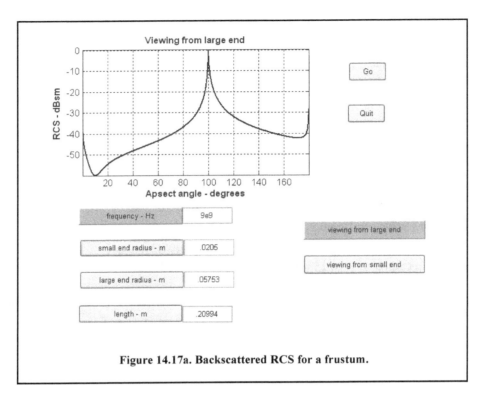

Figure 14.17a. Backscattered RCS for a frustum.

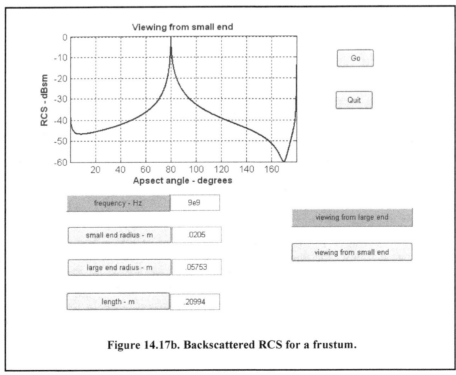

Figure 14.17b. Backscattered RCS for a frustum.

14.4.5. *Cylinder*

Figure 14.18 shows the geometry associated with a finite-length conducting cylinder. Two cases are presented: first, the general case of an elliptical cross section cylinder; second, the case of a circular cross section cylinder. The normal and non-normal incidence backscattered RCS due to a linearly polarized incident wave from an elliptical cylinder with minor and major radii being r_1 and r_2 are, respectively, given by

$$\sigma_{\theta_n} = \frac{2\pi H^2 r_2^2 r_1^2}{\lambda [r_1^2 (\cos\varphi)^2 + r_2^2 (\sin\varphi)^2]^{1.5}}$$ Eq. (14.48)

$$\sigma = \frac{\lambda r_2^2 r_1^2 \sin\theta}{8\pi \ (\cos\theta)^2 [r_1^2 (\cos\varphi)^2 + r_2^2 (\sin\varphi)^2]^{1.5}}$$ Eq. (14.49)

For a circular cylinder of radius r, due to roll symmetry, Eqs. (14.48) and (14.49), respectively, reduce to

$$\sigma_{\theta_n} = \frac{2\pi H^2 r}{\lambda}$$ Eq. (14.50)

$$\sigma = \frac{\lambda r \sin\theta}{8\pi (\cos\theta)^2}.$$ Eq. (14.51)

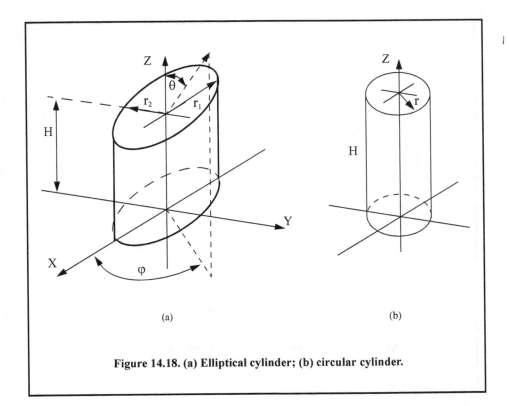

Figure 14.18. (a) Elliptical cylinder; (b) circular cylinder.

MATLAB Function "rcs_cylinder.m"

The function *"rcs_cylinder.m"* computes and plots the backscattered RCS of a cylinder. The syntax is as follows:

$$[rcs] = rcs\_cylinder(r1, r2, h, freq, phi, CylinderType)$$

where

Symbol	Description	Units	Status
r1	radius r1	meters	input
r2	radius r2	meters	input
h	length of cylinder	meters	input
freq	frequency	Hz	input
phi	roll viewing angle	degrees	input
Cylinder Type	"Circular," i.e., $r_1 = r_2$; "Elliptic," i.e., $r_1 \neq r_2$	none	input
rcs	array of RCS versus aspect angle	dBsm	output

Figure 14.19a shows a plot of the cylinder backscattered RCS for a symmetrical cylinder. Figure 14.19b shows the backscattered RCS for an elliptical cylinder. Figure 14.19 can be reproduced using the MATLAB program *"Fig14_19.m,"* listed in Appendix 14-A.

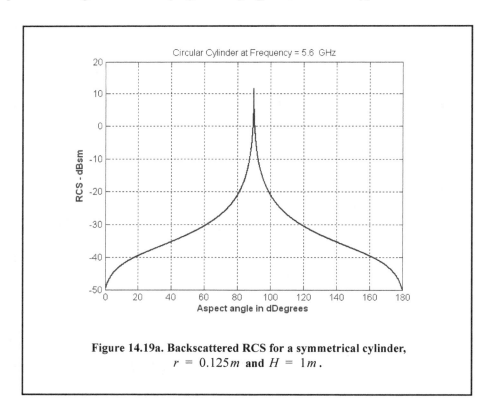

Figure 14.19a. Backscattered RCS for a symmetrical cylinder,
$r = 0.125m$ **and** $H = 1m$.

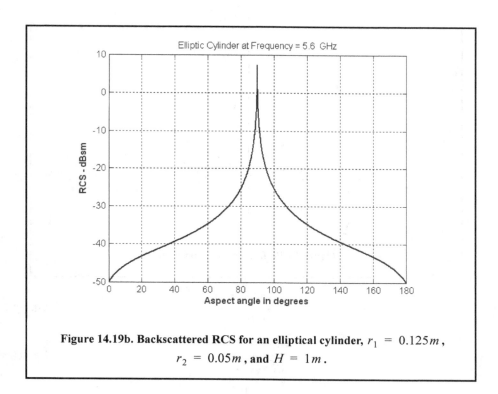

Figure 14.19b. Backscattered RCS for an elliptical cylinder, $r_1 = 0.125m$,
$r_2 = 0.05m$, **and** $H = 1m$.

14.4.6. Rectangular Flat Plate

Consider a perfectly conducting rectangular thin flat plate in the *x-y* plane as shown in Fig. 14.20. The two sides of the plate are denoted by $2a$ and $2b$. For a linearly polarized incident wave in the *x-z* plane, the horizontal and vertical backscattered RCS are, respectively, given by

$$\sigma_V = \frac{b^2}{\pi} \left| \sigma_{1V} - \sigma_{2V} \left[\frac{1}{\cos\theta} + \frac{\sigma_{2V}}{4}(\sigma_{3V} + \sigma_{4V}) \right] \sigma_{5V}^{-1} \right|^2 \qquad \text{Eq. (14.52)}$$

$$\sigma_H = \frac{b^2}{\pi} \left| \sigma_{1H} - \sigma_{2H} \left[\frac{1}{\cos\theta} - \frac{\sigma_{2H}}{4}(\sigma_{3H} + \sigma_{4H}) \right] \sigma_{5H}^{-1} \right|^2 \qquad \text{Eq. (14.53)}$$

where $k = 2\pi/\lambda$ and

$$\sigma_{1V} = \cos(ka\sin\theta) - j\frac{\sin(ka\sin\theta)}{\sin\theta} = (\sigma_{1H})^* \qquad \text{Eq. (14.54)}$$

$$\sigma_{2V} = \frac{e^{j(ka - \pi/4)}}{\sqrt{2\pi}(ka)^{3/2}} \qquad \text{Eq. (14.55)}$$

$$\sigma_{3V} = \frac{(1 + \sin\theta)e^{-jka\sin\theta}}{(1 - \sin\theta)^2} \qquad \text{Eq. (14.56)}$$

$$\sigma_{4V} = \frac{(1 - \sin\theta)e^{jka\sin\theta}}{(1 + \sin\theta)^2} \qquad \text{Eq. (14.57)}$$

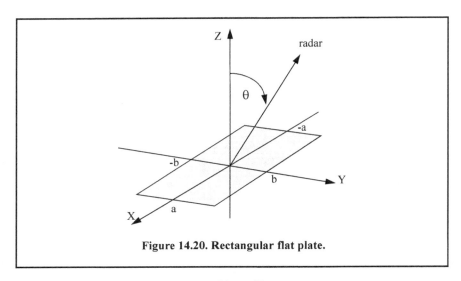

Figure 14.20. Rectangular flat plate.

$$\sigma_{5V} = 1 - \frac{e^{j(2ka - \pi/2)}}{8\pi(ka)^3} \qquad \textbf{Eq. (14.58)}$$

$$\sigma_{2H} = \frac{4e^{j(ka + \pi/4)}}{\sqrt{2\pi}(ka)^{1/2}} \qquad \textbf{Eq. (14.59)}$$

$$\sigma_{3H} = \frac{e^{-jk\,a\sin\theta}}{1 - \sin\theta} \qquad \textbf{Eq. (14.60)}$$

$$\sigma_{4H} = \frac{e^{jk\,a\sin\theta}}{1 + \sin\theta} \qquad \textbf{Eq. (14.61)}$$

$$\sigma_{5H} = 1 - \frac{e^{j(2ka + (\pi/2))}}{2\pi(ka)}. \qquad \textbf{Eq. (14.62)}$$

Equations (14.52) and (14.53) are valid and quite accurate for aspect angles $0° \le \theta \le 80$. For aspect angles near 90°, Ross[1] obtained, by extensive fitting of measured data, an empirical expression for the RCS. It is given by

$$\sigma_H \to 0$$

$$\sigma_V = \frac{ab^2}{\lambda}\left\{\left[1 + \frac{\pi}{2(2a/\lambda)^2}\right] + \left[1 - \frac{\pi}{2(2a/\lambda)^2}\right]\cos\left(2ka - \frac{3\pi}{5}\right)\right\}. \qquad \textbf{Eq. (14.63)}$$

The backscattered RCS for a perfectly conducting thin rectangular plate for incident waves at any θ, φ, can be approximated by

$$\sigma = \frac{4\pi a^2 b^2}{\lambda^2}\left(\frac{\sin(ak\sin\theta\cos\varphi)}{ak\sin\theta\cos\varphi}\frac{\sin(bk\sin\theta\sin\varphi)}{bk\sin\theta\sin\varphi}\right)^2(\cos\theta)^2 \qquad \textbf{Eq. (14.64)}$$

1. Ross, R. A., Radar Cross Section of Rectangular Flat Plate as a Function of Aspect Angle, *IEEE Trans.*, AP-14, 320, 1966.

Note that, Eq. (14.64) is independent of the polarization, and is only valid for aspect angles $\theta \leq 20°$.

MATLAB Function "rcs_rect_plate.m"

The function *"rcs_rect_plate.m"* calculates and plots the backscattered RCS of a rectangular flat plate. Its syntax is as follows:

$$[rcs] = rcs\_rect\_plate\ (a,\ b,\ freq)$$

where

Symbol	Description	Units	Status
a	*short side of plate*	*meters*	*input*
b	*long side of plate*	*meters*	*input*
freq	*frequency*	*Hz*	*input*
rcs	*array of RCS versus aspect angle*	*dBsm*	*output*

Figure 14.21 shows an example for the backscattered RCS of a rectangular flat plate, for both vertical (Fig. 14.21a) and horizontal (Fig. 14.21b) polarizations, using Eqs. (14.52), (14.53), and (14.64). In this example, $a = b = 10.16cm$ and wavelength $\lambda = 3.33cm$. This plot can be reproduced using MATLAB function *"rcs_rect_plate."* Figure 14.21c shows the GUI workspace associated with this function.

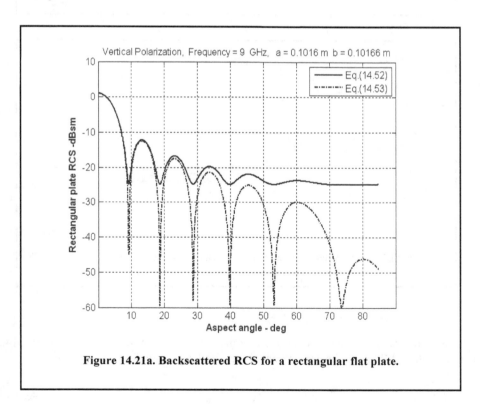

Figure 14.21a. Backscattered RCS for a rectangular flat plate.

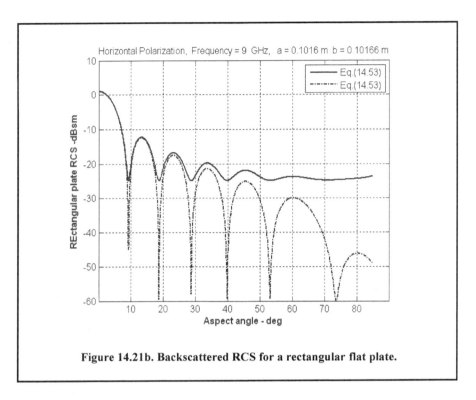

Figure 14.21b. Backscattered RCS for a rectangular flat plate.

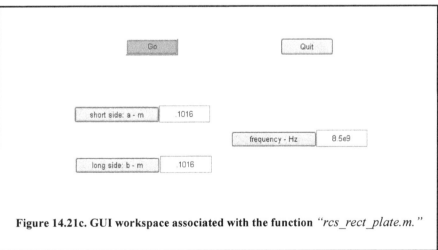

Figure 14.21c. GUI workspace associated with the function *"rcs_rect_plate.m."*

14.4.7. Triangular Flat Plate

Consider the triangular flat plate defined by the isosceles triangle as oriented in Fig. 14.22. The backscattered RCS can be approximated for small aspect angles ($\theta \leq 30°$) by

$$\sigma = \frac{4\pi A^2}{\lambda^2}(\cos\theta)^2 \sigma_0 \qquad\qquad \textbf{Eq. (14.65)}$$

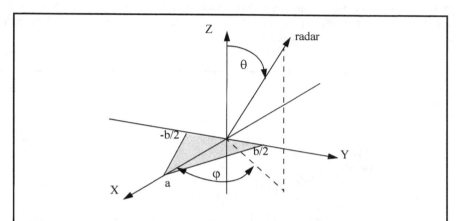

Figure 14.22. Coordinates for a perfectly conducting isosceles triangular plate.

$$\sigma_0 = \frac{[(\sin\alpha)^2 - (\sin(\beta/2))^2]^2 + \sigma_{01}}{\alpha^2 - (\beta/2)^2}$$

Eq. (14.66)

$$\sigma_{01} = 0.25(\sin\varphi)^2[(2a/b)\cos\varphi\sin\beta - \sin\varphi\sin2\alpha]^2$$

Eq. (14.67)

where $\alpha = ka\sin\theta\cos\varphi$, $\beta = kb\sin\theta\sin\varphi$, and $A = ab/2$. For waves incident in the plane $\varphi = 0$, the RCS reduces to

$$\sigma = \frac{4\pi A^2}{\lambda^2}(\cos\theta)^2\left[\frac{(\sin\alpha)^4}{\alpha^4} + \frac{(\sin2\alpha - 2\alpha)^2}{4\alpha^4}\right],$$

Eq. (14.68)

and for incidence in the plane $\varphi = \pi/2$,

$$\sigma = \frac{4\pi A^2}{\lambda^2}(\cos\theta)^2\left[\frac{(\sin(\beta/2))^4}{(\beta/2)^4}\right].$$

Eq. (14.69)

MATLAB Function "rcs_isosceles.m"

The function *"rcs_isosceles.m"* calculates and plots the backscattered RCS of a triangular flat plate. Its syntax is as follows:

[rcs] = rcs_isosceles (a, b, freq, phi)

where

Symbol	Description	Units	Status
a	*height of plate*	*meters*	*input*
b	*base of plate*	*meters*	*input*
freq	*frequency*	*Hz*	*input*
phi	*roll angle*	*degrees*	*input*
rcs	*array of RCS versus aspect angle*	*dBsm*	*output*

Figure 14.23 shows a plot for the normalized backscattered RCS from a perfectly conducting isosceles triangular flat plate. In this example $a = 0.2m$, $b = 0.75m$. This plot can be reproduced using MATLAB GUI *"rcs_isosceles_gui.m."*

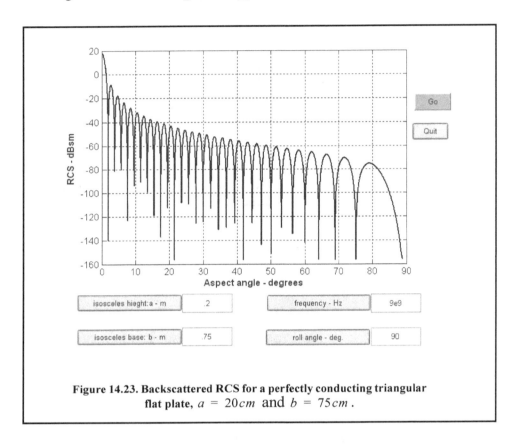

Figure 14.23. Backscattered RCS for a perfectly conducting triangular flat plate, $a = 20cm$ and $b = 75cm$.

14.5. RCS of Complex Objects

A complex target RCS is normally computed by coherently combining the cross sections of the simple shapes that make that target. In general, a complex target RCS can be modeled as a group of individual scattering centers distributed over the target. The scattering centers can be modeled as isotropic point scatterers (N-point model) or as simple shape scatterers (N-shape model). In any case, knowledge of the scattering centers' locations and strengths is critical in determining complex target RCS. This is true because as seen in Section 14.3, relative spacing and aspect angles of the individual scattering centers drastically influence the overall target RCS. Complex targets that can be modeled by many equal scattering centers are often called Swerling 1 or 2 targets. Alternatively, targets that have one dominant scattering center and many other smaller scattering centers are known as Swerling 3 or 4 targets.

In narrowband (NB) radar applications, contributions from all scattering centers combine coherently to produce a single value for the target RCS at every aspect angle. However, in wideband (WB) applications, a target may straddle many range bins. For each range bin, the average RCS extracted by the radar represents the contributions from all scattering centers that fall within that bin.

As an example, consider a circular cylinder with two perfectly conducting circular flat plates on both ends. Assume linear polarization and let $H = 1m$ and $r = 0.125m$. The backscattered RCS for this object versus aspect angle is shown in Fig. 14.24. Note that at aspect angles close to $0°$ and $180°$, the RCS is mainly dominated by the circular plate, while at aspect angles close to normal incidence, the RCS is dominated by the cylinder broadside specular return. The reader can reproduce this plot using the MATLAB program *"rcs_cylinder_complex.m,"* listed in Appendix 14-A.

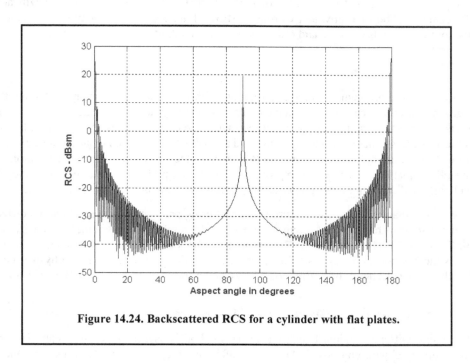

Figure 14.24. Backscattered RCS for a cylinder with flat plates.

14.6. RCS Prediction Methods

Before presenting the different RCS calculation methods, it is important to understand the significance of RCS prediction. Most radar systems use RCS as a means of discrimination. Therefore, accurate prediction of target RCS is critical in order to design and develop robust discrimination algorithms. Additionally, measuring and identifying the scattering centers (sources) for a given target aid in developing RCS reduction techniques. Another reason of lesser importance is that RCS calculations require broad and extensive technical knowledge; thus, many scientists and scholars find the subject challenging and intellectually motivating. Two categories of RCS prediction methods are available: exact and approximate.

Exact methods of RCS prediction are very complex, even for simple shape objects. This is because they require solving either differential or integral equations that describe the scattering problem under the proper set of boundary conditions. Such boundary conditions are governed by Maxwell's equations. Even when exact solutions are achievable, they are often difficult to interpret and to program using digital computers.

Due to the difficulties associated with the exact RCS prediction, approximate methods become the viable alternative. The majority of the approximate methods are valid in the optical region, and each has its own strengths and limitations. Most approximate methods can predict RCS within a few *dB* of the truth. In general, such a variation is quite acceptable to radar engineers and designers. Approximate methods are usually the main source for predicting the RCS of complex and extended targets such as aircrafts, ships, and missiles. When experimental results are available, they can be used to validate and verify the approximations.

Some of the most commonly used approximate methods are Geometrical Optics (GO), Physical Optics (PO), Geometrical Theory of Diffraction (GTD), Physical Theory of Diffraction (PTD), and Method of Equivalent Currents (MEC). Interested readers may consult Knott or Ruck for more details on these and other approximate methods.

14.6.1. Computational Electromagnetics

Most scattering problems involve radar targets with very complicated shapes. Among these are ground-based targets such as trucks, tanks, and artillery; air targets such as aircraft, helicopters, and missiles; and space-based targets, such as reentry vehicles and satellites. For such an object, there is generally no analytic method available to predict the radar cross section. The field of Computational Electromagnetics (CEM) uses the power of a computer to implement Maxwell's Equations and solve these problems. CEM has applications in other areas, too, such as antennas and waveguide design, wave propagation, and medical imaging.

There exist many CEM techniques to solve scattering problems, each employing a different numerical analysis technique. Among the most popular methods used are the Finite Difference Time Domain (FDTD) method, the Finite Element Method (FEM), integral equation methods such as the Method of Moments (MoM), and asymptotic techniques such as Physical Optics (PO), the Physical Theory of Diffraction (PTD), and Shooting and Bouncing Rays (SBR).

14.6.2. Finite Difference Time Domain Method

The Finite Difference Time Domain (FDTD) method is useful for solving scattering problems involving objects composed of complex, often inhomogeneous media. It uses a finite difference scheme to discretize Maxwell's equations in the time domain. This has the advantage of allowing waveforms with wide bandwidths to be used as an excitation.

The main drawbacks for the FDTD method include the requirements on the grid size and non-conformal grid shape, which often results in poor discretization of target geometry and high memory requirements, particularly in three-dimensional cases. The object and its adjacent region must be discretized, and an artificial absorbing layer used to truncate the grid to simulate an unbounded space. It is also challenging to create a purely planar wave in such simulations.

The FDTD method makes use of finite difference approximations to directly discretize Maxwell's equations in the time domain. Consider the "forward difference" approximation for the first derivative:

$$\dot{f}(x_o) \approx \frac{f(x_o + \Delta x) - f(x_o)}{\Delta x} .$$

Eq. (14.70)

The backward difference approximation is

$$\dot{f}(x_o) \approx \frac{f(x_o) - f(x_o - \Delta x)}{\Delta x} .$$

Eq. (14.71)

The central difference approximation is

$$\dot{f}(x_o) \approx \frac{f(x_o + \Delta x) - f(x_o - \Delta x)}{2\Delta x}.$$

Eq. (14.72)

Second derivatives can be approximated by a similar procedure

$$\ddot{f}(x_o) = \frac{[f(x_o + \Delta x) - f(x_o)] - [f(x_o) - f(x_o) - \Delta x]}{(\Delta x)^2},$$

Eq. (14.73)

where the second-order derivative makes use of forward and backward first derivatives.

Next consider an example of using FDTD to implement two-dimensional simulation. First, consider the time domain form of Maxwell's equations in a charge and conductive-free region

$$\nabla \times \vec{E} = -\mu \frac{\partial}{\partial t} \vec{H}$$

Eq. (14.74)

$$\nabla \times \vec{H} = -\varepsilon \frac{\partial}{\partial t} \vec{E} + \vec{J}$$

Eq. (14.75)

$$\nabla \cdot \vec{D} = 0$$

Eq. (14.76)

$$\nabla \cdot \vec{B} = 0$$

Eq. (14.77)

where $\vec{E}$ is the electric field intensity in volts/meter, $\vec{H}$ is the magnetic field intensity in ampere/meter^2, $\vec{J}$ is the current density in coulombs/meter^3, $\vec{D}$ is the displacement flux in coulombs/meter^2, $\vec{B}$ is the magnetic induction flux in Tesla or Weber/meter^2, μ is the permeability, and ε is the permittivity. Note that the region may comprise several homogeneous areas, each with its own μ and ε.

In rectangular coordinates, one can write these equations as

$$\mu \frac{\partial H_x}{\partial t} = \frac{\partial E_y}{\partial z} - \frac{\partial E_z}{\partial y}$$

Eq. (14.78)

$$\mu \frac{\partial H_y}{\partial t} = \frac{\partial E_z}{\partial x} - \frac{\partial E_x}{\partial z}$$

Eq. (14.79)

$$\mu \frac{\partial H_z}{\partial t} = \frac{\partial E_x}{\partial y} - \frac{\partial E_y}{\partial x}$$

Eq. (14.80)

$$\varepsilon \frac{\partial E_x}{\partial t} = \frac{\partial H_z}{\partial y} - \frac{\partial H_y}{\partial z} - \vec{J}_x$$

Eq. (14.81)

$$\varepsilon \frac{\partial E_y}{\partial t} = \frac{\partial H_x}{\partial z} - \frac{\partial H_z}{\partial x} - \vec{J}_y$$

Eq. (14.82)

$$\varepsilon \frac{\partial E_z}{\partial t} = \frac{\partial H_y}{\partial x} - \frac{\partial H_x}{\partial y} - \vec{J}_z.$$

Eq. (14.83)

Consider a filamentary current source $\vec{J}$ with direction $\hat{z}$, which excites TM-polarized (E_z) waves only. The above three equations then reduce to

$$\varepsilon \, \frac{\partial E_z}{\partial t} = \frac{\partial H_y}{\partial x} - \frac{\partial H_x}{\partial y} - \vec{J}_z \qquad\qquad \textbf{Eq. (14.84)}$$

$$\mu \, \frac{\partial H_x}{\partial t} = - \frac{\partial E_z}{\partial y} \qquad\qquad \textbf{Eq. (14.85)}$$

$$\mu \, \frac{\partial H_y}{\partial t} = - \frac{\partial E_z}{\partial x} \qquad\qquad \textbf{Eq. (14.86)}$$

These expressions can be discretized using a 2-D Yee Algorithm.[1] In this scheme, the electric and magnetic fields are arranged on grids that are a half point in distance and time away from each other.

First-order derivatives may be applied to the above equations to obtain values at these time and grid points. More precisely,

$$E_{z,i,j}^{t+1/2} = E_{z,i,j}^{t-1/2} + \frac{\Delta t}{\varepsilon_{i,j}} \left\{ \left(\frac{1}{\Delta x}[H_{y,i+1/2,j}^{t} - H_{y,i-1/2,j}^{t}] \right) \right. \qquad\qquad \textbf{Eq. (14.87)}$$

$$\left. -\frac{1}{\Delta y}[H_{x,i+1/2,j}^{t} - H_{x,i-1/2,j}^{t}] \right\} - J_{z,i,j}^{t}$$

$$H_{x,i,j+1/2}^{t+1} = H_{x,i,j+1/2}^{t} + \frac{\Delta t}{\mu_{i,j+1/2}\Delta y}[E_{z,i,j+1}^{t+1/2} - E_{z,i,j}^{t+1/2}] \qquad\qquad \textbf{Eq. (14.88)}$$

$$H_{y,i,j+1/2}^{t+1} = H_{y,i+1/2,j}^{t} + \frac{\Delta t}{\mu_{i+1,j}\Delta x}[E_{z,i+1,j}^{t+1/2} - E_{z,i,j}^{t+1/2}]. \qquad\qquad \textbf{Eq. (14.89)}$$

As an example, consider a two-dimensional box with width and height of 2 meters. In the center of the box, place a $\hat{z}$ directed current source and assign to it the excitation function,

$$J(t) = \left(4\left(\frac{t}{\tau}\right)^3 - \left(\frac{t}{\tau}\right)^4 \right) e^{-t/\tau} \qquad\qquad \textbf{Eq. (14.90)}$$

where $\tau = 1/(4\pi f_o)$ and $f_o = 600 MHz$. Place a 1×0.2 meter dielectric slab with $\varepsilon = 8\varepsilon_o$ at $(1.0, 0.7)$ to introduce an obstruction to the spreading wavefront. The current J_z is shown in Figs. 14.25 and 14.26 at different times. Note that the wave slows down inside the slab by a factor of $\sqrt{8}$, and hence the wavelength is compressed. The wavefront starts penetrating the slab at time $t = 2ns$. At time $t = 3.36ns$ the wave starts to leave the dielectric slab and the unobstructed wavefronts reach the walls of the box.

The MATLAB program *"fdtd.m"* was developed to simulate this example. It is listed in Appendix 14-A. Readers are strongly advised to run this program and observe how the wave front spreads through the dielectric slab.

---

1. Taflove, Allen, *Computational Electromagnetic: The Finite-Difference-Time-Domain Method*, Artech House, 1995.

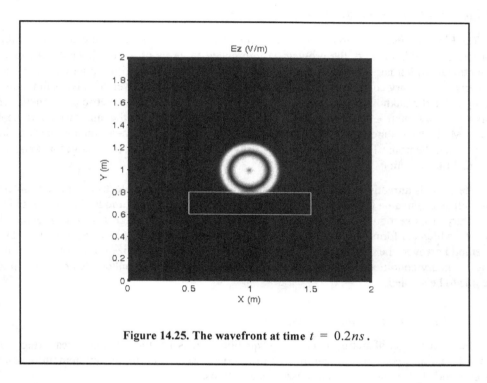

Figure 14.25. The wavefront at time $t = 0.2ns$.

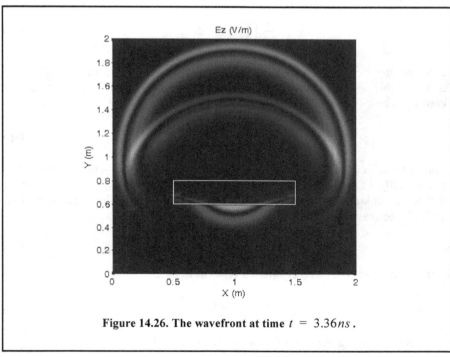

Figure 14.26. The wavefront at time $t = 3.36ns$.

14.6.3. Finite Element Method

The Finite Element Method (FEM) is a popular CEM technique for solving boundary-valued problems. In this method, the problem is formulated in terms of a variational expression, or functional, which has a minimum corresponding to the governing differential equation under the given boundary conditions. A trial function comprising a set of weighted basis functions is assigned to the unknown quantity in the region of study, typically the electric or magnetic field distribution. A matrix equation is then formed that can be solved for the unknown coefficients. In FEM, the basis functions are typically assigned to surface or volume elements, such as triangles or tetrahedrons. The coefficients are usually defined at the vertices (or edges in some vector FEM formulations).

The FEM is attractive for solving both static and time-harmonic electromagnetic problems, as well as eigenvalue problems such as determining the fundamental modes in a waveguide of arbitrary cross section. Most regions are easily discretized into triangular or quadrilateral elements, which conform very well to object boundaries and dielectric interfaces. Like the FDTD method however, the object and the adjoining space must be discretized, requiring an absorbing boundary condition imposed at the terminating boundary if an unbounded radiation problem is to be studied.

14.6.4. Integral Equations

There exists a set of auxiliary scattering equations that assists the solution of scattering problems in unbounded regions. One of the most widely used is the frequency-domain magnetic vector potential $\vec{A}$, derived from Maxwell's equations,

$$\vec{A} = \frac{\mu}{4\pi} \iint_S \vec{J} \, \frac{e^{-jkR}}{R} \, ds$$

Eq. (14.91)

where $\vec{J}$ is an electric surface current, and S is the surface in free space on which the current resides. Using this definition, the scattered electric field at all points in space is given by the well-known Electric Field Integral Equation (EFIE)

$$\vec{E}_s = -j\omega\vec{A} - \frac{j}{\varepsilon\mu\omega}\nabla\nabla \cdot \vec{A} \, .$$

Eq. (14.92)

This equation relates the scattered field $\vec{E}_s$ to a known current $\vec{J}$. In a general scattering problem, it is typically the incident electric field that is known, and the surface current and scattered field that are the unknowns. If we assume a conducting surface for the currents, the tangential electric field must vanish, producing

$$\vec{E}_i^{\tan} = -\vec{E}_s^{\tan} \, .$$

Eq. (14.93)

The EFIE can be rewritten using the known incident field $\vec{E}_i$

$$-\vec{E}_s^{\tan} = \left(-j\omega\vec{A} - \frac{j}{\varepsilon\mu\omega}\nabla\nabla \cdot \vec{A}\right)^{\tan} \, .$$

Eq. (14.94)

This represents an integral equation for the unknown current $\vec{J}$.

The Method of Moments (MoM) is a technique used to solve such integral equations, and has received much attention in the last 30 years. In using the MoM to solve the EFIE, Maxwell's equations are represented exactly, and the solution is described as "exact" or "full wave." This means that all electromagnetic effects and dominant scattering mechanisms are represented in the result. To solve the EFIE, the current is usually discretized according to

$$\vec{J} = \sum_{n=1}^{N} I_n \vec{f}_n(\vec{r})$$

Eq. (14.95)

where the $\vec{f}_n(\vec{r})$ are basis functions chosen to represent the behavior of the current, and the I_n are unknown coefficients. The target surface is typically broken up into small subdomains and a basis function assigned to each. Inserting Eq. (14.95) into Eq. (14.93) yields a single equation in N unknowns.

To create N equations in N unknowns, the EFIE is tested or enforced over all subdomains by employing an inner product of Eq. (14.93) by a set of testing functions. Most often, the basis functions $\vec{f}_n(\vec{r})$ are used (the Galerkin method). The resulting system may then be solved for the unknown coefficients by Gaussian elimination, or an iterative technique.

The MoM has been used extensively to solve scattering problems involving rotationally symmetric objects. The McDonnell Douglas[1] code (CICERO), solves the body-of-revolution scattering problem for objects that have various conducting and dielectric coated surfaces. The MoM has also been applied to three-dimensional bodies, and this is often done according to the method proposed by Rao, Wilton, and Glisson, who introduced a basis function suitable for use with surfaces described by connected triangular patches.

While the MoM achieves excellent accuracy, the size of the matrix system is proportional to the square of the radar wavelength. Until recently this has limited the maximum object size that could be stored in system memory, typically a few wavelengths at most for three-dimensional problems. Methods that approximate the system's Green's Function have been developed in recent years in an attempt to reduce the required memory. The Adaptive Integral Method (AIM) and the Fast Multipole Method (FMM) are two methods that were developed to alleviate this problem. The FMM has proved quite successful and is used in the Fast Illinois Solver Code (FISC) at the University of Illinois.

In the next few sections a brief discussion of asymptotic, or so, called high-frequency techniques is presented.

14.6.5. Geometrical Optics

The method of Geometrical Optics (GO) treats the radar energy as small ray tubes that propagate according to Fermat's Principle. The specular reflection points on the target are found and divergence and spreading of energy are accounted for by analyzing the radii of curvature at the reflection points. GO is limited by its applicability at caustic points, and it does not handle diffractions from tips and edges, or account for creeping waves. Keller introduced the Geometrical Theory of Diffraction (GTD) in an attempt to handle diffraction effects; however, GTD suffers from the same problem at caustics and shadow boundaries. The Uniform Theory of Diffraction (UTD) was introduced to further improve this method.

---

1. Medgyesi-Mitschang, Louis and Putnam, John, Electromagnetic Scattering from Axially Inhomogeneous Bodies of Revolution, *IEEE Trans. Antennas Propagation*, Vol. 32, pp.707-806, August 1984.

14.6.6. Physical Optics

Physical Optics (PO) is a technique that approximates that surface current $\vec{J}$ in the illuminated portion of the target by assuming that locally, the target's surface can be considered flat and planar. If at each point the surface is considered to be an infinite half plane, image theory allows the surface current to be written directly in terms of the incident magnetic field and the local surface normal,

$$\vec{J} = 2\hat{n} \times \vec{H}_i.$$

Eq. (14.96)

This is called the physical optics approximation. With the current known, the scattered electric field is obtained directly via Eq. (14.91).

While the PO method is used extensively in high-frequency CEM computations, it is limited in its accuracy to near-specular observations. It does not treat diffractions, traveling or creeping waves, multiple bounces, or other scattering phenomena. These mechanisms are often supplemented in the PO solution by other techniques, some of which are discussed later.

Rectangular Plate

Consider a rectangular plate of length b and width a in the xy-plane. The $\hat{\theta}$ polarized incident electric and magnetic fields on the plate are, respectively, given by

$$\vec{E}_i = \hat{\theta} e^{jk(\hat{r}_i \cdot \vec{r})}$$

Eq. (14.97)

$$\vec{H}_i = -\frac{\hat{r}_i}{z_o} \times \vec{E}_i = \left(-\frac{2}{z_o}\right)\hat{\phi} e^{jk(\hat{r}_i \cdot \vec{r})}$$

Eq. (14.98)

where $\hat{r}_i$ is the direction on incidence, and z_o is the free space impedance. This field generates the current

$$\vec{J} = \frac{2}{z_o} \hat{\phi} \times \hat{n} e^{jk(\hat{r}_i \cdot \vec{r})},$$

Eq. (14.99)

resulting in magnetic vector potential given by

$$\vec{A} = \frac{\mu}{2\pi z_o}\frac{e^{-jkR_s}}{R_s} \int_{-a/2}^{a/2} \int_{-b/2}^{b/2} \vec{J} e^{jk(\hat{r}_i \cdot \vec{r})} \, dx\,dy =$$

Eq. (14.100)

$$\frac{\mu}{2\pi z_o}\frac{e^{-jkR_s}}{R_s}\hat{\phi} \times \hat{n} \int_{-a/2}^{a/2} \int_{-b/2}^{b/2} e^{2jk(\hat{r}_i \cdot \vec{r})} \, dx\,dy$$

where $\vec{r} = x\hat{x} + y\hat{y}$, and for phase variation, the range R_s is approximated as

$$R_s - (\hat{r}_i \cdot \vec{r}).$$

Eq. (14.101)

Evaluating Eq. (14.100) analytically yields

$$\vec{A} = \frac{\mu}{2\pi z_o}\frac{e^{-jkR_s}}{R_s}\hat{\phi} \times \hat{n}ab\frac{\sin(k\sin\theta\cos\phi)}{ak\sin\theta\cos\phi}\frac{\sin(k\sin\theta\sin\phi)}{bk\sin\theta\sin\phi}.$$

Eq. (14.102)

In the far field, the $\hat{\theta}$ polarized scattered field is given by

$$\vec{E}_s = -j\omega(\hat{\theta} \cdot \vec{A}).$$

Eq. (14.103)

Figure 14.27 shows the backscattered RCS for a rectangular plate versus incident angle, using the technique presented in this section. This plot can be reproduced using the MATLAB program *"rectplate.m,"* listed in Appendix 14-A.

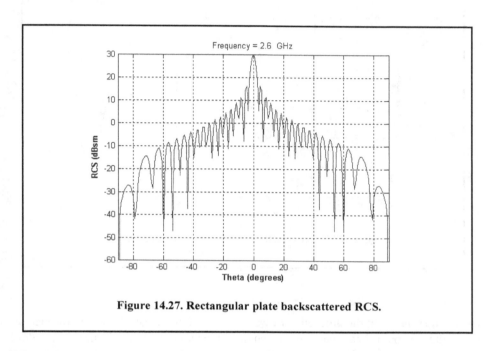

Figure 14.27. Rectangular plate backscattered RCS.

N-Sided Polygon

The integral in Eq. (14.100) has the form of a Fourier transform over the planar extent of the rectangular plate. In general, this 2-D Fourier transform is given by

$$S(u, v) = \int_x \int_y e^{j(ux + vy)} dx\, dy.$$

Eq. (14.104)

The expression for an arbitrarily N-sided polygon in a local coordinate system has been evaluated analytically as

$$S(u, v) = \sum_{n=1}^{N} e^{j(\vec{\omega} \cdot \vec{\gamma})} \left[\frac{\hat{n} \times \hat{\alpha}_n \cdot \hat{\alpha}_{n-1}}{(\vec{\omega} \cdot \hat{\alpha}_n)(\vec{\omega} \cdot \hat{\alpha}_{n-1})} \right]$$

Eq. (14.105)

where $\vec{\gamma}_n$ are the polygon vertices and α_n are the edge vectors given by

$$\hat{\alpha}_n = \frac{\vec{\gamma}_{n+1} - \vec{\gamma}_n}{|\vec{\gamma}_{n+1} - \vec{\gamma}_n|},$$

Eq. (14.106)

and $\vec{\omega} = u\hat{x} + v\hat{y}$. In the summation above, $\alpha_o = \alpha_N$. Figure 14. 28 shows a plot for the backscattered RCS for a N-sided polygon versus angle of incidence. This figure can be reproduced using the MATLAB program *"polygon.m,"* listed in Appendix 14-A.

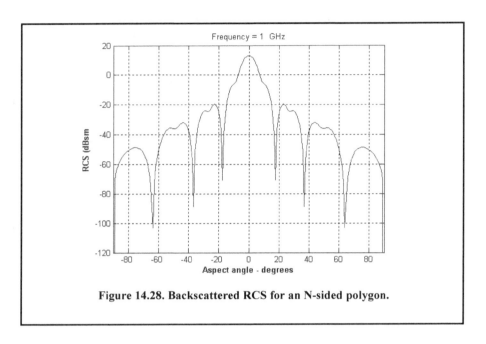

Figure 14.28. Backscattered RCS for an N-sided polygon.

14.6.7. Edge Diffraction

The PO does not treat the diffraction of waves at edges. In the late 1950s the Russian physicist, P. Ufimtsev, published a paper on a technique now known as the Physical Theory of Diffraction. In this paper, Ufimtsev introduced expressions for the edge diffraction at arbitrary incidence and scattering angles that complemented Physical Optics. This method was extended by K. Mitzer at Northrop who applied the PTD to incremental length edges in three dimensions. The PTD method was used in codes such as Northrop's (MISCAT), and was instrumental in the design of low cross-section aircraft such as Lockheed's F-117 Stealth fighter.

14.7. Multiple Bounce

Multiple reflections are a very important scattering mechanism in some complex targets. Many real-world targets have cavities or other concave areas where energy may be reflected and scattered several times. Examples are rocket boosters with nozzles and fuel tanks and aircraft with deep engine inlets. This type of scattering often results in high RCS at certain aspect angles, and significantly delayed returns that may cause the target to appear much longer in the downrange direction than it actually is.

A popular method for modeling these interactions is to treat the incident plane wave as a bundle of "ray tubes" as in GO theory, incorporating material effects and ray tube spreading and divergence. At the exit aperture of the ray bundle, a PO-type integral is performed over the ray tube footprint. This technique is known as Shooting and Bouncing Rays (SBR), and was developed at the University of Illinois in the late 1980s. This technique, as well as the PO and

PTD methods, are used in the well-known software called XPATCH, which has been used in high-frequency signature prediction for many years.

Problems

14.1. Design a cylindrical RCS calibration target such that its broadside RCS (cylinder) and end (flat plate) RCS are equal to $10m^2$ at $f = 9.5GHz$. The RCS for a flat plate of area A is $\sigma_{fp} = 4\pi f^2 A^2 / c^2$.

14.2. The following table is constructed from a radar cross-section measurement experiment. Calculate the mean and standard deviation of the radar cross section.

Number of samples	RCS, m²
2	55
6	67
12	73
16	90
20	98
24	110
26	117
19	126
13	133
8	139
5	144
3	150

14.3. Develop a MATLAB simulation to compute and plot the backscattered RCS for the following objects. Utilize the simple shape MATLAB functions developed in this chapter. Assume that the radar is located on the left side of the page and that its line of sight is aligned with the target body axis. Assume an X-band radar.

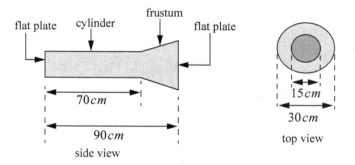

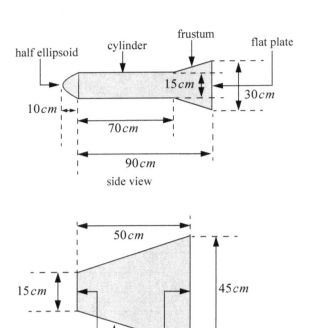

14.4. The backscattered RCS for a corner reflector is given by

$$\sigma = \left[\sqrt{\frac{16\pi a^4}{\lambda^2}(\sin\theta)^2} + \sqrt{\frac{4\pi a^4}{\lambda^2}\left(\frac{\sin\left(\frac{2\pi a}{\lambda}\sin\theta\right)}{\frac{2\pi a}{\lambda}\sin\theta}\right)^2} \right]^2 \qquad 0° \leq \theta \leq 45°.$$

This RCS is symmetric about the angle $\theta = 45°$. Develop a MATLAB program to compute and plot the RCS for a corner reflector. The RCS at the $\theta = 45°$ is

$$\sigma = \frac{8\pi a^2 b^2}{\lambda^2}.$$

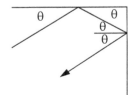 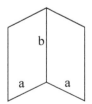

corner reflector

Appendix 14-A: Chapter 14 MATLAB Code Listings

The MATLAB code provided in this chapter was designed as an academic standalone tool and is not adequate for other purposes. The code was written in a way to assist the reader in gaining a better understanding of the theory. The code was not developed, nor is it intended to be used as part of an open-loop or a closed-loop simulation of any kind. The MATLAB code found in this textbook can be downloaded from this book's web page on the CRC Press website. Simply use your favorite web browser, go to *www.crcpress.com*, and search for keyword *"Mahafza"* to locate this book's web page.

MATLAB Function "rcs_aspect.m" Listing

```
function [rcs] = rcs_aspect(scat_spacing, freq)
% This function demonstrates the effect of aspect angle on RCS
% The default frequency is 3GHz. The radar is observing two unity
% point scatterers separated by 1.0 meters. Initially the two scatterers
% are aligned with radar line of sight. The aspect angle is changed from
% 0 degrees to 180 degress and the equivalant RCS is computed.
% The RCS as measured by the radar versus aspect angle is then plotted.
% Inputs
    % scat_spacing in meters
    % freq radar frequency in Hz
%
% Output
    % rcs in dBsm
% Users may vary frequency, and/or scatteres spacing to observe RCS variation
eps = 0.0001;
wavelength = 3.0e+8 / freq;
% Compute aspect angle vector
aspect_degrees = 0.:.05:180.;
aspect_radians = (pi/180) .* aspect_degrees;
% Compute electrical scatterer spacing vector in wavelength units
elec_spacing = (2.0 * scat_spacing / wavelength) .* cos(aspect_radians);
% Compute RCS (rcs = RCS_scat1 + RCS_scat2)
% Scat1 is taken as phase refernce point
rcs = abs(1.0 + cos((2.0 * pi) .* elec_spacing) ...
        + i * sin((2.0 * pi) .* elec_spacing));
rcs = rcs + eps;
rcs = 20.0*log10(rcs); % RCS in dBsm
end
```

MATLAB Program "Fig.14_3.m" Listing

```
% generates Fig.. 14.3 of text
clc
close all
clear all
% Enter scatterer spacing, in meters
distance = input('Enter scatterer spacing, in meters \n');
% Enter frequency
freq = input('Enter Enter frequency in Hz \n');
rcs = rcs_aspect(distance,freq);
Figure (1);
aspect_degrees = 0.:.05:180.;
```

```
plot(aspect_degrees,rcs);
grid;
xlabel('\bfaspect angle - degrees');
ylabel('\bfRCS in dBsm');
```

MATLAB Function "rcs_frequency.m" Listing

```
function [rcs] = rcs_frequency (scat_spacing, frequ, freql)
% This program demonstrates the dependency of RCS on wavelength
% The default assumes two unity point scatterers separated
% The radar line of sight is aligned with the two scatterers
% Inputs
    % scat_spacing in meters
    % freql lower frequency limit in Hz
    % frequ upper frequency limit in Hz
% Output
    % rcs in dBsm
eps = 0.0001;
freq_band = frequ - freql;
delfreq = freq_band / 500.;
index = 0;
for freq = freql: delfreq: frequ
    index = index +1;
    wavelength(index) = 3.0e+8 / freq;
end
% Compute electrical scatterer spacing vector in wavelength units
elec_spacing = 2.0 * scat_spacing ./ wavelength;
% Compute RCS (RCS = RCS_scat1 + RCS_scat2)
rcs = abs ( 1 + cos((2.0 * pi) .* elec_spacing)+ i * sin((2.0 * pi) .* elec_spacing));
rcs = rcs + eps;
rcs = 20.0*log10(rcs); % RCS ins dBsm
end
```

MATLAB Program "Fig.14_5_6.m" Listing

```
% Generates plot like Fig.. 14.5 and Fig. 14.6
% Enter scatterer spacing, in meters
clc
close all
clear all
scat_spacing = input('Enter scatterer spacing, in meters \n');
% Enter frequency band
freql = input('Enter lower frequency limit in Hz \n');
frequ = input('Enter upper frequency limit in Hz \n');
[rcs] = rcs_frequency (scat_spacing, frequ, freql);
N = size(rcs,2) ;
freq = linspace(freql,frequ,N)./1e9;
Figure (1);
plot(freq,rcs);
grid on;
xlabel('\bfFrequency');
ylabel('\bfRCS in dBsm');
```

MATLAB Program "Fig.14_10.m" Listing

```
% This program calculates the back-scattered RCS for a perfectly
% conducting sphere using Eq.(14.28), and produce plots similar to Fig.14.8
% Spherical Bessel functions are computed using series approximation and recursion.
clc
close all
clear all
eps  = 0.00001;
index = 0;
% kr limits are [0.05 - 15] ===> 300 points
for kr = 0.05:0.01:25
  index = index + 1;
  sphere_rcs  = 0. + 0.*i;
  f1   = 0. + 1.*i;
  f2   = 1. + 0.*i;
  m    = 1.;
  n    = 0.;
  q    = -1.;
  % initially set del to huge value
  del =100000+100000*i;
  while(abs(del) > eps)
    q  = -q;
    n  = n + 1;
    m  = m + 2;
    del = (2.*n-1) * f2 / kr-f1;
    f1 = f2;
    f2 = del;
    del = q * m /(f2 * (kr * f1 - n * f2));
    sphere_rcs = sphere_rcs + del;
  end
  rcs(index)  = abs(sphere_rcs);
  sphere_rcsdb(index) = 10. * log10(rcs(index));
  end
Figure(1);
n=0.05:.01:25;
subplot(2,1,1)
plot (n,rcs,'k','linewidth',1.5);
% set (gca,'xtick',[1 2 3 4 5 6 7 8 9 10 11 12 13 14 15]);
xlabel ('\bfSphere circumference in wavelengths; (2 \pi r / \lambda)');
ylabel ('\bf Normalized RCS ( \sigma / \pi r^2)');
grid on
subplot(2,1,2)
plot (n,sphere_rcsdb,'k','linewidth',1.5);
%set (gca,'xtick',[1 2 3 4 5 6 7 8 9 10 11 12 13 14 15]);
xlabel ('\bfSphere circumference in wavelengths; (2 \pi r / \lambda)');
ylabel ('\bf Normalized RCS ( \sigma / \pi r^2) - dB');
grid;
Figure (2);
semilogx (n,sphere_rcsdb,'k','linewidth',1.5);
xlabel ('\bfSphere circumference in wavelengths; (2 \pi r / \lambda)');
ylabel ('\bf Normalized sphere RCS ( \sigma / \pi r^2) - dB');
grid on
gtext('\bfRayleigh Region')
gtext('\bfMie Region')
```

gtext('\bfOptical Region')

MATLAB Function "rcs_ellipsoid.m" Listing

```
function [rcs_db] = rcs_ellipsoid (a, b, c, phi)
% This program computes the back-scattered RCS for an ellipsoid.
% The angle phi is fixed, while the angle theta is varied from 0-180 deg.
% Inputs
    % a          == ellipsoid a-radius in meters
    % b          == ellipsoid b-radius in meters
    % c          == ellipsoid c-radius in meters
    % phi        == ellipsoid roll angle in degrees
%Output
    % rcs        == ellipsoid rcs versus aspect angle in dBsm
eps = 0.00001;
sin_phi_s = sin(phi)^2;
cos_phi_s = cos(phi)^2;
% Generate aspect angle vector
theta = 0.:.05:180;
theta = (theta .* pi) ./ 180.;
if(a ~= b & a ~= c)
   rcs = (pi * a^2 * b^2 * c^2) ./ (a^2 * cos_phi_s .* (sin(theta).^2) + ...
   b^2 * sin_phi_s .* (sin(theta).^2) + ...
   c^2 .* (cos(theta).^2)).^2 ;
else
  if(a == b & a ~= c)
    rcs = (pi * b^4 * c^2) ./ ( b^2 .* (sin(theta).^2) + ...
      c^2 .* (cos(theta).^2)).^2 ;
  else
    if (a == b & a ==c)
      rcs = pi * c^2;
    end
  end
end
rcs_db = 10.0 * log10(rcs);
return
```

MATLAB Program "Fig14_12a.m" Listing

```
% generates Fig 14.12a of text
clc
close all
clear all
% Enter the ellpsiod a radius
a = .15;
% Enter the ellpsiod b radius
b = .3;
% Enter the ellpsiod c radius
c = .95;
% Enter the ellpsiod roll angle in degrees
phi = [0 75 90];
[rcs_db1] = rcs_ellipsoid (a, b, c, phi(1));
[rcs_db2] = rcs_ellipsoid (a, b, c, phi(2));
[rcs_db3] = rcs_ellipsoid (a, b, c, phi(3));
```

```
N = size(rcs_db1,2);
theta = linspace(0.0, pi, N);
theta = theta .* 180 ./ pi;
figure (1);
plot(theta,rcs_db1,'k:',theta,rcs_db2,'k',theta,rcs_db3,'k-.','linewidth',1.5);
xlabel ('\bfAspect angle - degrees');
ylabel ('\bfEllipsoid RCS - dBsm');
legend ('\phi = 0.0', '\phi = 75', '\phi = 90')
title('(a, b, c) = (.15, .3, .95) meters')
grid on;
```

MATLAB Function "rcs_circ_plate.m" Listing

```
function [rcsdb] = rcs_circ_plate (r, freq)
% This program calculates and plots the backscattered RCS of
% circular flat plate of radius r.
eps = 0.000001;
% Compute aspect angle vector
% Compute wavelength
lambda = 3.e+8 / freq; % X-Band
index = 0;
for aspect_deg = 0.:.1:180
  index = index +1;
  aspect = (pi /180.) * aspect_deg;
% Compute RCS using Eq. (2.37)
  if (aspect == 0 | aspect == pi)
    rcs_po(index) = (4.0 * pi^3 * r^4 / lambda^2) + eps;
    rcs_mu(index) = rcs_po(1);
  else
    x = (4. * pi * r / lambda) * sin(aspect);
    val1 = 4. * pi^3 * r^4 / lambda^2;
    val2 = 2. * besselj(1,x) / x;
    rcs_po(index) = val1 * (val2 * cos(aspect))^2 + eps;
% Compute RCS using Eq. (2.36)
    val1m = lambda * r;
    val2m = 8. * pi * sin(aspect) * (tan(aspect)^2);
    rcs_mu(index) = val1m / val2m + eps;
  end
 end
 % Compute RCS using Eq. (2.35) (theta=0,180)
rcsdb = 10. * log10(rcs_po);
rcsdb_mu = 10 * log10(rcs_mu);
angle = 0:.1:180;
plot(angle,rcsdb,'k',angle,rcsdb_mu,'k-.')
grid;
xlabel ('\bfAspect angle - degrees');
ylabel ('\bfRCS - dBsm');
axis tight
legend('Using Eq.(14.39)','Using Eq.(14.38)')
freqGH = num2str(freq*1.e-9);
title (['Frequency = ',[freqGH],' GHz']);
end
```

MATLAB Function "rcs_frustum.m" Listing

```
function [rcs] = rcs_frustum (r1, r2, h, freq, indicator)
% This program computes the monostatic RCS for a frustum.
% Incident linear Polarization is assumed.
% When viewing from the small end of the frustum
% normal incedence occurs at aspect pi/2 - half cone angle
% When viewing from the large end, normal incidence occur at
% pi/2 + half cone angle.
% RCS is computed using Eq. (14.43). This program assumes a geometry
% similar top Fig. 14.13
% Inputs
   % r1       == small end radius in meters
   % r2       == large end radius in meters
   % freq     == frequency in Hz
   % indicator == 1 when viewing from large end 0 when viewing from small end
% Output
   % rcs      == array of RCS versus aspect angle
format long
index = 0;
eps = 0.000001;
lambda = 3.0e+8 /freq;
% Enter frustum's small end radius
%r1 =.02057;
% Enter Frustum's large end radius
%r2 = .05753;
% Compute Frustum's length
%h = .20945;
% Comput half cone angle, alpha
alpha = atan(( r2 - r1)/h);
% Compute z1 and z2
z2 = r2 / tan(alpha);
z1 = r1 / tan(alpha);
delta = (z2^1.5 - z1^1.5)^2;
factor = (8. * pi * delta) / (9. * lambda);
%('enter 1 to view frustum from large end, 0 otherwise')
large_small_end = indicator;
if(large_small_end == 1)
  % Compute normal incidence, large end
  normal_incedence = (180./pi) * ((pi /2) + alpha)
  % Compute RCS from zero aspect to normal incidence
  for theta = 0.001:.1:normal_incedence-.5
    index = index +1;
    theta = theta * pi /180.;
    rcs(index) = (lambda * z1 * tan(alpha) *(tan(theta - alpha))^2) /...
      (8. * pi *sin(theta)) + eps;
  end
  %Compute broadside RCS
  index = index +1;
  rcs_normal = factor * sin(alpha) / ((cos(alpha))^4) + eps;
  rcs(index) = rcs_normal;
  % Compute RCS from broad side to 180 degrees
  for theta = normal_incedence+.5:.1:180
    index = index + 1;
    theta =  theta * pi / 180. ;
```

```
     rcs(index) = (lambda * z2 * tan(alpha) *(tan(theta - alpha))^2) / ...
       (8. * pi *sin(theta)) + eps;
   end
else
   % Compute normal incidence, small end
   normal_incidence = (180./pi) * ((pi /2) - alpha)
   % Compute RCS from zero aspect to normal incidence (large end of frustum)
   for theta = 0.001:.1:normal_incedence-.5
     index = index +1;
     theta = theta * pi /180.;
     rcs(index) = (lambda * z1 * tan(alpha) *(tan(theta + alpha))^2) / ...
       (8. * pi *sin(theta)) + eps;
   end
   %Compute broadside RCS
   index = index +1;
   rcs_normal = factor * sin(alpha) / ((cos(alpha))^4) + eps;
   rcs(index) = rcs_normal;
   % Compute RCS from broad side to 180 degrees (small end of frustum)
   for theta = normal_incedence+.5:.1:180
     index = index + 1;
     theta =  theta * pi / 180. ;
     rcs(index) = (lambda * z2 * tan(alpha) *(tan(theta + alpha))^2) / ...
       (8. * pi *sin(theta)) + eps;
   end
end
% Plot RCS versus aspect angle
delta = 180 /index;
angle = 0.001:delta:180;
plot (angle,10*log10(rcs),'k','linewidth',1.5);
grid;
xlabel ('\bfApsect angle - degrees');
ylabel ('\bfRCS - dBsm');
axis tight
if(indicator ==1)
   title ('\bfViewing from large end');
else
   title ('\bfViewing from small end');
end
```

MATLAB Function "rcs_cylinder.m" Listing

```
function [rcs] = rcs_cylinder(r1, r2, h, freq, phi, CylinderType)
% rcs_cylinder.m
% This program compute monostatic RCS for a finite length
% cylinder of either circular or elliptical cross-section.
% Plot of RCS versus aspect angle theta is generated at a specified
r = r1;         % radius of the circular cylinder
eps =0.00001;
dtr = pi/180;
phir = phi*dtr;
lambda = 3.0e+8 /freq;     % wavelength
% CylinderType= 'Elliptic';   % 'Elliptic' or 'Circular'

switch CylinderType
```

```
  case 'Circular'
    % Compute RCS from 0 to (90-.5)  degrees
    index = 0;
    for theta = 0.0:.1:90-.5
       index = index +1;
       thetar = theta * dtr;
       rcs(index) = (lambda * r * sin(thetar) / ...
          (8. * pi * (cos(thetar))^2)) + eps;
    end
    % Compute RCS for broadside specular at 90 degree
    thetar = pi/2;
    index = index +1;
    rcs(index) = (2. * pi * h^2 * r / lambda )+ eps;
    % Compute RCS from (90+.5) to 180 degrees
    for theta = 90+.5:.1:180.
       index = index + 1;
       thetar = theta * dtr;
       rcs(index) = ( lambda * r * sin(thetar) / ...
          (8. * pi * (cos(thetar))^2)) + eps;
    end
  case 'Elliptic'
   r12 = r1*r1;
   r22 = r2*r2;
   h2 = h*h;
    % Compute RCS from 0 to (90-.5)  degrees
    index = 0;
    for theta = 0.0:.1:90-.5
       index = index +1;
       thetar = theta * dtr;
       rcs(index) =  lambda * r12 * r22 * sin(thetar) / ...
          ( 8*pi* (cos(thetar)^2)* ( (r12*cos(phir)^2 + r22*sin(phir)^2)^1.5 ))+ eps;
    end
    % Compute RCS for broadside specular at 90 degree
    index = index +1;
    rcs(index) = 2. * pi * h2 * r12 * r22 / ...
          ( lambda*( (r12*cos(phir)^2 + r22*sin(phir)^2)^1.5 ))+ eps;
    % Compute RCS from (90+.5) to 180 degrees
    for theta = 90+.5:.1:180.
       index = index + 1;
       thetar = theta * dtr;
       rcs(index) =  lambda * r12 * r22 * sin(thetar) / ...
          ( 8*pi* cos(thetar)^2* ( (r12*cos(phir)^2 + r22*sin(phir)^2)^1.5 ))+ eps;
    end
end
end
```

MATLAB Program "Fig14_19.m" Listing

```
% generates Fig. 14.19 of text
clc
close all
clear all
r1 = .125;
r2 = 0.05;
```

```
h = 1;
phi = 45;
freq = 5.6e9;
freqGH = num2str(freq*1.e-9);
% Fig 14.19a
[rcs1] = rcs_cylinder(r1, r1, h, freq, phi,'Circular');
figure(1)
angle = linspace(0,180,size(rcs1,2));
plot(angle,10*log10(rcs1),'k','linewidth',1.5);
grid on;
xlabel ('\bfAspect angle in dDegrees');
ylabel ('\bfRCS - dBsm');
title  (['Circular Cylinder at Frequency = ',[freqGH],' GHz']);
% Fig. 14.19b
[rcs2] = rcs_cylinder(r1, r2, h, freq, phi,'Elliptic');
figure(2)
angle = linspace(0,180,size(rcs2,2));
plot(angle,10*log10(rcs2),'k','linewidth',1.5);
grid on;
xlabel ('\bfAspect angle in degrees');;
ylabel ('\bfRCS - dBsm');
title  (['Elliptic Cylinder at Frequency = ',[freqGH],' GHz']);
```

MATLAB Function "rcs_rect_plate.m" Listing

```
function [rcsdb_h,rcsdb_v] = rcs_rect_plate(a, b, freq)
% This program computes the backscattered RCS for a rectangular
% flat plate. The RCS is computed for vertical and horizontal
% polarization based on Eq.s(14.52)through (14.62). Also Physical
% Optics approximation Eq.(14.64) is computed.
% User may vary frequency, or the plate's dimensions.
% Default values are a=b=10.16cm; lambda=3.25cm.
eps = 0.000001;
% Enter a, b, and lambda
lambda = .0325;
ka = 2. * pi * a / lambda;
% Compute aspect angle vector
theta_deg = 0.05:0.1:85;
theta = (pi/180.) .* theta_deg;
sigma1v = cos(ka .*sin(theta)) - i .* sin(ka .*sin(theta)) ./ sin(theta);
sigma2v = exp(i * ka - (pi /4)) / (sqrt(2 * pi) *(ka)^1.5);
sigma3v = (1. + sin(theta)) .* exp(-i * ka .* sin(theta)) ./ ...
   (1. - sin(theta)).^2;
sigma4v = (1. - sin(theta)) .* exp(i * ka .* sin(theta)) ./ ...
   (1. + sin(theta)).^2;
sigma5v = 1. - (exp(i * 2. * ka - (pi / 2)) / (8. * pi * (ka)^3));
sigma1h = cos(ka .*sin(theta)) + i .* sin(ka .*sin(theta)) ./ sin(theta);
sigma2h = 4. * exp(i * ka * (pi / 4.)) / (sqrt(2 * pi * ka));
sigma3h =  exp(-i * ka .* sin(theta)) ./ (1. - sin(theta));
sigma4h = exp(i * ka * sin(theta)) ./ (1. + sin(theta));
sigma5h = 1. - (exp(j * 2. * ka + (pi / 4.)) / 2. * pi * ka);
% Compute vertical polarization RCS
rcs_v = (b^2 / pi) .* (abs(sigma1v - sigma2v .*((1. ./ cos(theta)) ...
   + .25 .* sigma2v .* (sigma3v + sigma4v)) .* (sigma5v).^-1)).^2 + eps;
```

```
% compute horizontal polarization RCS
rcs_h = (b^2 / pi) .* (abs(sigma1h - sigma2h .*((1. ./ cos(theta)) ...
    - .25 .* sigma2h .* (sigma3h + sigma4h)) .* (sigma5h).^-1)).^2 + eps;
% Compute RCS from Physical Optics, Eq.(2.62)
angle = ka .* sin(theta);
rcs_po = (4. * pi* a^2 * b^2 / lambda^2 ).* (cos(theta)).^2 .* ...
    ((sin(angle) ./ angle).^2) + eps;
rcsdb_v = 10. .*log10(rcs_v);
rcsdb_h = 10. .*log10(rcs_h);
rcsdb_po = 10. .*log10(rcs_po);
figure
plot (theta_deg, rcsdb_v,'k',theta_deg,rcsdb_po,'k -.','linewidth',1.5);
set(gca,'xtick',[10:10:85]);
freqGH = num2str(freq*1.e-9);
A = num2str(a);
B = num2str(b);
title (['Vertical Polarization,  ','Frequency = ',[freqGH],' GHz, ',' a = ', [A], ' m',' b = ',[B],' m']);
ylabel ('\bfRectangular plate RCS -dBsm');
xlabel ('\bfAspect angle - deg');
legend('Eq.(14.52)','Eq.(14.53)')
grid on
figure
plot (theta_deg, rcsdb_h,'k',theta_deg,rcsdb_po,'k -.','linewidth',1.5);
set(gca,'xtick',[10:10:85]);
title (['Horizontal Polarization,  ','Frequency = ',[freqGH],' GHz, ',' a = ', [A], ' m',' b = ',[B],' m']);
ylabel ('\bfREctangular plate RCS -dBsm');
xlabel ('\bfAspect angle - deg');
legend('Eq.(14.53)','Eq.(14.53)')
grid on
```

MATLAB Function "rcs_isosceles.m" Listing

```
function [rcs] = rcs_isosceles (a, b, freq, phi)
% This program calculates the backscattered RCS for a perfectly
% conducting triangular flat plate, using Eq.s (14.65) through (14.67)
% The default case is to assume phi = pi/2. These equations are
% valid for aspect angles less than 30 degrees
% Users may vary wavelength, or plate's dimensions
% Inputs
   % a    == height of plate in meters
   % b    == base of plate in meters
   % freq  == frequency in Hz
   % phi   == roll angle in degrees
% Output
   % rcs   == array of RCS versus aspect angle
A = a * b / 2.;
lambda = 3.e+8 / freq;
ka = 2. * pi / lambda;
kb = 2. *pi / lambda;
% Compute theta vector
theta_deg = 0.01:.05:89;
theta = (pi /180.) .* theta_deg;
alpha = ka * cos(phi) .* sin(theta);
beta = kb * sin(phi) .* sin(theta);
```

```
if (phi == pi / 2)
  rcs = (4. * pi * A^2 / lambda^2) .* cos(theta).^2 .* (sin(beta ./ 2)).^4 ...
    ./ (beta./2).^4 + eps;
end
if (phi == 0)
  rcs = (4. * pi * A^2 / lambda^2) .* cos(theta).^2 .* ...
    ((sin(alpha).^4 ./ alpha.^4) + (sin(2 .* alpha) - 2.*alpha).^2 ...
    ./ (4 .* alpha.^4)) + eps;
end
if (phi ~= 0 & phi ~= pi/2)
  sigmao1 = 0.25 *sin(phi)^2 .* ((2. * a / b) * cos(phi) .* ...
    sin(beta) - sin(phi) .* sin(2. .* alpha)).^2;
  fact1 = (alpha).^2 - (.5 .* beta).^2;
  fact2 = (sin(alpha).^2 - sin(.5 .* beta).^2).^2;
  sigmao = (fact2 + sigmao1) ./ fact1;
  rcs = (4. * pi * A^2 / lambda^2) .* cos(theta).^2 .* sigmao + eps;
end
rcsdb = 10. *log10(rcs);
plot(theta_deg,rcsdb,'k','linewidth', 1.5)
xlabel ('\bfAspect angle - degrees');
ylabel ('\bfRCS - dBsm')
grid on
```

MATLAB Program "rcs_cylinder_cmplx.m" Listing

```
clc
close
clear all
indes = 0;
eps =0.00001;
a1 =.125;
h = 1.;
lambda = 3.0e+8 /9.5e+9;
lambda = 0.00861;
index = 0;
for theta = 0.0:.1:90-.1
  index = index +1;
  theta = theta * pi /180.;
  rcs(index) = (lambda * a1 * sin(theta) / (8 * pi * (cos(theta))^2)) + eps;
end
theta*180/pi
theta = pi/2;
index = index +1
rcs(index) = (2 * pi * h^2 * a1 / lambda )+ eps;
for theta = 90+.1:.1:180.
  index = index + 1;
  theta = theta * pi / 180.;
  rcs(index) = ( lambda * a1 * sin(theta) / (8 * pi * (cos(theta))^2)) + eps;
end
%%%%%%%%%%%%%
r = a1;
index = 0;
for aspect_deg = 0.:.1:180
  index = index +1;
```

```
  aspect = (pi /180.) * aspect_deg;
% Compute RCS using Eq. (2.37)
  if (aspect == 0 | aspect == pi)
     rcs_po(index) = (4.0 * pi^3 * r^4 / lambda^2) + eps;
     rcs_mu(index) = rcs_po(1);
   else
     x = (4. * pi * r / lambda) * sin(aspect);
     val1 = 4. * pi^3 * r^4 / lambda^2;
     val2 = 2. * besselj(1,x) / x;
     rcs_po(index) = val1 * (val2 * cos(aspect))^2 + eps;
   end
 end
rcs_t =(rcs_po + rcs);
 %%%%%%%%%%%%%%
 angle = 0:.1:180;
 plot(angle,10*log10(rcs_t(1:1801)),'k');
 xlabel('\bfAspect angle in degrees')
 ylabel('\bfRCS - dBsm')
 grid
```

MATLAB Program "fdtd.m" Listing

```
clear all
%
mu_o = pi*4.0e-7;              % free space permeability
epsilon_o = 8.854e-12;          % free space permittivity
%
c = 1.0/sqrt(mu_o * epsilon_o);  % speed of light
%
length_x = 2.0;               % x-width of region
nx = 200;                     % number of x grid points
dx = length_x / (nx - 1);      % x grid size
%
x = linspace(0.0, length_x, nx); % x array
%
length_y = 2.0;               % y-width of region
ny = 200;                     % number of y grid points
dy = length_y / (ny - 1);      % y grid size
%
y = linspace(0.0, length_y, ny); % y array
%
max_timestep = c*sqrt(1.0/(dx*dx) + 1.0/(dy*dy));   % max tstep for FDTD
max_timestep = 1.0/max_timestep;
%
delta_t = 0.5*max_timestep;     % delta t a little less than max tstep
%
er = 8.0;                     % relative permittivity of slab
%
epsilon = epsilon_o*ones(ny, nx); % epsilon array
mu = mu_o*ones(ny - 1, nx - 1);   % mu array
%
a1 = [0.5 1.5 1.5 0.5 0.5];    % for drawing slab on plot
a2 = [0.6 0.6 0.8 0.8 0.6];    % for drawing slab on plot
%
```

```
x1 = fix(0.5/dx)+1;          % grid extents for slab
y1 = fix(0.6/dy);            % grid extents for slab
x2 = fix(1.5/dx)+1;          % grid extents for slab
y2 = fix(0.8/dy);            % grid extents for slab
%
epsilon(y1:y2,x1:x2) = er*epsilon_o;  % set epsilon inside slab
%
j_x = nx/2;                  % x location of current source
j_y = ny/2;                  % y location of current source
%
e_z_1 = zeros(ny, nx);       % initialize array. e_z at boundaries will remain 0
h_x_1 = zeros(ny - 1, nx - 1);  % initialize array
h_y_1 = zeros(ny - 1, nx - 1);  % initialize array
e_z_2 = zeros(ny, nx);       % initialize array. e_z at boundaries will remain 0
h_x_2 = zeros(ny - 1, nx - 1);  % initialize array
h_y_2 = zeros(ny - 1, nx - 1);  % initialize array
%
ntim = 300;                  % number of desired time points
f_o = 600e6;                 % base frequency for pulse
tau = 1.0/(4.0*pi*f_o);      % tau for pulse
%
for i_t = 1:ntim
%
  time(i_t) = i_t * delta_t;
%
   i_t
   time(i_t)
%
  if time(i_t) > 3.36e-9
     break
  end
%
  jz(i_t) = (4.0 * (time(i_t)/tau)^3 - (time(i_t)/tau)^4) * exp(-time(i_t)/tau);
%
   for i_x = 2:nx-1     % ez at boundaries remains zero
    for i_y = 2:ny-1  % ez at boundaries remains zero
%
      j = 0.0;
      if i_x == j_x
       if i_y == j_y
         j = jz(i_t);
       end
      end
%
      if rem(i_t, 2) == 1
        a = 1.0/dx*(h_y_1(i_y, i_x) - h_y_1(i_y, i_x - 1));
        b = 1.0/dy*(h_x_1(i_y, i_x) - h_x_1(i_y - 1, i_x));
        e_z_2(i_y, i_x) = e_z_1(i_y, i_x) + (delta_t/epsilon(i_y, i_x))*(a - b) - j;
      else
        a = 1.0/dx*(h_y_2(i_y, i_x) - h_y_2(i_y, i_x - 1));
        b = 1.0/dy*(h_x_2(i_y, i_x) - h_x_2(i_y - 1, i_x));
        e_z_1(i_y, i_x) = e_z_2(i_y, i_x) + (delta_t/epsilon(i_y, i_x))*(a - b) - j;
      end
%
```

```
     end
   end
%
  for i_x = 1:nx-1
    for i_y = 1:ny-1
%
       if rem(i_t, 2) == 1
         h_x_2(i_y, i_x) = h_x_1(i_y, i_x) - (delta_t/mu(i_y, i_x)/dy)*(e_z_2(i_y + 1, i_x) - e_z_2(i_y, i_x));
            h_y_2(i_y, i_x) = h_y_1(i_y, i_x) + (delta_t/mu(i_y, i_x)/dx)*(e_z_2(i_y, i_x + 1) - e_z_2(i_y,
i_x));
       else
         h_x_1(i_y, i_x) = h_x_2(i_y, i_x) - (delta_t/mu(i_y, i_x)/dy)*(e_z_1(i_y + 1, i_x) - e_z_1(i_y, i_x));
            h_y_1(i_y, i_x) = h_y_2(i_y, i_x) + (delta_t/mu(i_y, i_x)/dx)*(e_z_1(i_y, i_x + 1) - e_z_1(i_y,
i_x));
       end
     end
   end
%
  pcolor(x, y, abs(e_z_2))
  line(a1, a2, 'Linewidth', 1.0, 'Color', 'white');
  xlabel('X (m)')
  ylabel('Y (m)')
  title('Ez (V/m)')
  axis square
  shading interp
  %colormap gray
  caxis([0 .1])
  %axis([0 2 0 2 0 .1])
  fr(i_t) = getframe;
end
```

MATLAB Program "rectplate.m" Listing

```
close all
clear all
frequency = 2.6e9;              % desired radar frequency
freqGH = num2str(frequency*1.e-9);
c = 299795645.0;               % speed of light
w = 2.0*pi*frequency;           % radian frequency
wavenumber = w/c;               % free space wavenumber
mu = 4.0*pi*1.0e-7;             % free space permeability
z_o = 376.7343;                % free space wave impedance
l_x = 1.0;                % length of plate
l_y = 1.0;                % width of plate
normal_vect = [0 0 1];          % +z normal for x-y plane
theta_points = 180;             % number of points in theta
phi_points = 1;                 % number of points in phi
 theta = linspace(-0.5*pi, 0.5*pi, theta_points);
phi = linspace(0.0, 2.0*pi, phi_points);
for i_theta = 1:theta_points
    for i_phi = 1:phi_points
        theta_vect(1) = cos(theta(i_theta))*cos(phi(i_phi));
    theta_vect(2) = cos(theta(i_theta))*sin(phi(i_phi));
    theta_vect(3) = -sin(theta(i_theta));
```

```
        phi_vect(1) = -sin(phi(i_phi));
   phi_vect(2) = cos(phi(i_phi));
   phi_vect(3) = 0.0;
        u = sin(theta(i_theta))*cos(phi(i_phi));
   v = sin(theta(i_theta))*sin(phi(i_phi));
        vect_term = dot(theta_vect, cross(phi_vect, normal_vect));
                        es(i_theta, i_phi) = -j*w*mu/2.0/pi/z_o*vect_term*l_x*l_y*sinc(wavenum-
ber*u*l_x)*sinc(wavenumber*v*l_y);
   end
end
rcs = 20.0*log10(sqrt(4*pi)*abs(es));
plot(180*theta/pi, rcs)
axis([-90 90 -60 30])
xlabel('\bfTheta (degrees)')
ylabel('\bfRCS (dBsm')
grid on
title (['Frequency = ',[freqGH],' GHz']);
return
```

MATLAB Program "polygon.m" Listing

```
% this routine calculates the scattered electric field of an arbitrary
% N-sided polygon located in the x-y plane.
clc
clear all
close all
frequency = 1.0e9;              % desired radar frequency
freqGH = num2str(frequency*1.e-9);
c = 299795645.0;               % speed of light
w = 2.0*pi*frequency;            % radian frequency
wavenumber = w/c;               % free space wavenumber
mu = 4.0*pi*1.0e-7;             % free space permeability
z_o = 376.7343;                % free space wave impedance

nsides = 3;                    % number of polygon sides
vertices(1,:) = [0.0 0.0 0.0];      % vertexes of polygon (counterclockwise)
vertices(2,:) = [1.0 0.5 0.0];      % vertexes of polygon (counterclockwise)
vertices(3,:) = [1.5 0.0 0.0];      % vertexes of polygon (counterclockwise)

for n = 1:nsides
  if n == nsides
    alpha_n(nsides,1) = vertices(1,1) - vertices(nsides,1);
    alpha_n(nsides,2) = vertices(1,2) - vertices(nsides,2);
    alpha_n(nsides,3) = vertices(1,3) - vertices(nsides,3);
  else
    alpha_n(n,1) = vertices(n+1,1) - vertices(n,1);
    alpha_n(n,2) = vertices(n+1,2) - vertices(n,2);
    alpha_n(n,3) = vertices(n+1,3) - vertices(n,3);
  end
  alpha_n(n, 1:3) = alpha_n(n, 1:3)/norm(alpha_n(n, 1:3));
end

normal_vect = [0 0 1];          % +z normal for x-y plane
```

```
theta_points = 180;              % number of points in theta
phi_points = 1;                  % number of points in phi

theta = linspace(-0.5*pi, 0.5*pi, theta_points);
phi = linspace(0.0, 2.0*pi, phi_points);

for i_theta = 1:theta_points

  for i_phi = 1:phi_points

    theta_vect(1) = cos(theta(i_theta))*cos(phi(i_phi));
    theta_vect(2) = cos(theta(i_theta))*sin(phi(i_phi));
    theta_vect(3) = -sin(theta(i_theta));

    phi_vect(1) = -sin(phi(i_phi));
    phi_vect(2) = cos(phi(i_phi));
    phi_vect(3) = 0.0;

    w_vect(1) = 2*wavenumber*sin(theta(i_theta))*cos(phi(i_phi));
    w_vect(2) = 2*wavenumber*sin(theta(i_theta))*sin(phi(i_phi));
    w_vect(3) = 0.0;

    s_term = 0.0;

    for n = 1:nsides
       exterm = exp(i*dot(w_vect, vertices(n,1:3)));
       if n == 1
         num = dot(cross(normal_vect, alpha_n(n,1:3)), alpha_n(nsides,1:3));
         denom = dot(w_vect, alpha_n(n,1:3))*dot(w_vect, alpha_n(nsides,1:3));
       else
         num = dot(cross(normal_vect, alpha_n(n,1:3)), alpha_n(n-1,1:3));
         denom = dot(w_vect, alpha_n(n,1:3))*dot(w_vect, alpha_n(n-1,1:3));
       end
       s_term = s_term + num*exterm/denom;
    end

    vect_term = dot(theta_vect, cross(phi_vect, normal_vect));

    es(i_theta, i_phi) = -j*w*mu/2.0/pi/z_o*vect_term*s_term;

  end
end

rcs = 20.0*log10(sqrt(4*pi)*abs(es));
plot(180*theta/pi, rcs)
axis([-90 90 -120 20])
xlabel('\bfAspect angle - degrees')
ylabel('\bfRCS (dBsm')
grid on
title (['Frequency = ',[freqGH],' GHz']);
```

Chapter 15

Phased Array Antennas

15.1. Directivity, Power Gain, and Effective Aperture

Radar antennas can be characterized by the directive gain G_D, power gain G, and effective aperture A_e. Antenna gain is a term used to describe the ability of an antenna to concentrate the transmitted energy in a certain direction. Directive gain, or simply directivity, is more representative of the antenna radiation pattern, while power gain is normally used in the radar equation. Plots of the power gain and directivity, when normalized to unity, are called the *antenna radiation pattern*. The directivity of a transmitting antenna can be defined by

$$G_D = \frac{maximum \ \ radiation \ \ intensity}{average \ \ radiation \ \ intensity}.$$ **Eq. (15.1)**

The radiation intensity is the power-per-unit solid angle in the direction (θ, ϕ) and denoted by $P(\theta, \phi)$. The average radiation intensity over 4π radians (solid angle) is the total power divided by 4π. Hence, Eq. (15.1) can be written as

$$G_D = \frac{4\pi(maximum \ \ radiated \ \ power / unit \ \ solid \ \ angle)}{total \ \ radiated \ \ power}.$$ **Eq. (15.2)**

It follows that

$$G_D = 4\pi \ \frac{P(\theta, \phi)_{max}}{\int_0^\pi \int_0^{2\pi} P(\theta, \phi) d\theta \, d\phi}.$$ **Eq. (15.3)**

As an approximation, it is customary to rewrite Eq. (15.3) as

$$G_D \approx \frac{4\pi}{\theta_3 \phi_3}$$ **Eq. (15.4)**

where θ_3 and ϕ_3 are the antenna half-power (3-dB) beamwidths in either direction. The antenna power gain and its directivity are related by

$$G = \rho_r G_D$$ **Eq. (15.5)**

where ρ_r is the radiation efficiency factor. In this book, the antenna power gain will be denoted as *gain*. The radiation efficiency factor accounts for the ohmic losses associated with the antenna. Therefore, the definition for the antenna gain is also given in Eq. (15.1). The antenna effective aperture A_e is related to gain by

$$A_e = \frac{G\lambda^2}{4\pi}$$

Eq. (15.6)

where λ is the wavelength. The relationship between the antenna's effective aperture A_e and the physical aperture A is

$$A_e = \rho A$$
$$0 \le \rho \le 1.$$

Eq. (15.7)

ρ is referred to as the aperture efficiency, and good antennas require $\rho \to 1$ (in this book $\rho = 1$ is always assumed, i.e., $A_e = A$).

Using simple algebraic manipulations of Eqs. (15.4) through (15.6) (assuming that $\rho_r = 1$) yields

$$G = \frac{4\pi A_e}{\lambda^2} \approx \frac{4\pi}{\theta_3 \phi_3}$$

Eq. (15.8)

Consequently, the angular cross section of the beam is

$$\theta_3 \phi_3 \approx \frac{\lambda^2}{A_e}.$$

Eq. (15.9)

Eq. (15.9) indicates that the antenna beamwidth decreases as $\sqrt{A_e}$ increases. Thus, in surveillance operations, the number of beam positions an antenna will take on to cover a volume V is

$$N_{Beams} > \frac{V}{\theta_3 \phi_3}.$$

Eq. (15.10)

and when V represents the entire hemisphere, Eq. (15.10) is modified to

$$N_{Beams} > \frac{2\pi}{\theta_3 \phi_3} \approx \frac{2\pi A_e}{\lambda^2} \approx \frac{G}{2}.$$

Eq. (15.11)

15.2. Near and Far Fields

The electric field intensity generated from the energy emitted by an antenna is a function of the antenna physical aperture shape and the electric current amplitude and phase distribution across the aperture. Plots of the modulus of the electric field intensity of the emitted radiation, $|E(\theta, \phi)|$, are referred to as the *intensity pattern* of the antenna. Alternatively, plots of $|E(\theta, \phi)|^2$ are called the *power radiation pattern* (the same as $P(\theta, \phi)$).

Based on the distance from the face of the antenna, where the radiated electric field is measured, three distinct regions are identified. They are the near field, Fresnel, and the Fraunhofer regions. In the near field and the Fresnel regions, rays emitted from the antenna have spherical wavefronts (equiphase fronts). In the Fraunhofer region, the wavefronts can be locally represented by plane waves. The near field and the Fresnel regions are normally of little interest to most radar applications. Most radar systems operate in the Fraunhofer region, which is also known as the far field region. In the far field region, the electric field intensity can be computed from the aperture Fourier transform.

Construction of the far criterion can be developed with the help of Fig. 15.1. Consider a radiating source at point O that emits spherical waves. A receiving antenna of length d is at distance r away from the source. The phase difference between a spherical wave and a local plane wave at the receiving antenna can be expressed in terms of the distance δr. The distance δr is given by

$$\delta r = \overline{AO} - \overline{OB} = \sqrt{r^2 + \left(\frac{d}{2}\right)^2} - r,$$
Eq. (15.12)

and since in the far field $r \gg d$, Eq. (15.12) is approximated via binomial expansion by

$$\delta r = r\left(\sqrt{1 + \left(\frac{d}{2r}\right)^2} - 1\right) \approx \frac{d^2}{8r}.$$
Eq. (15.13)

It is customary to assume far field when the distance δr corresponds to less than $1/16$ of a wavelength (i.e., $22.5°$). More precisely, if

$$\delta r = d^2/8r \leq \lambda/16,$$
Eq. (15.14)

then a useful expression for far field is

$$r \geq 2d^2/\lambda.$$
Eq. (15.15)

Note that far field is a function of both the antenna size and the operating wavelength.

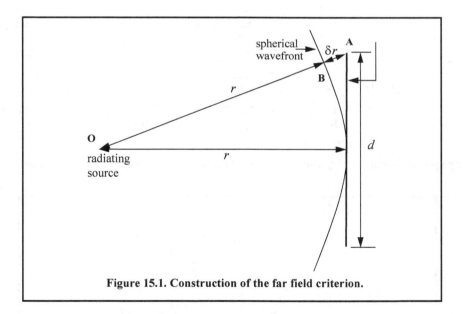

Figure 15.1. Construction of the far field criterion.

15.3. General Arrays

An array is a composite antenna formed from two or more basic radiators. Each radiator is denoted as an element. The elements forming an array could be dipoles, dish reflectors, slots in a wave guide, or any other type of radiator. Array antennas synthesize narrow directive beams that may be steered, mechanically or electronically, in many directions. Electronic steering is achieved by controlling the phase of the current feeding the array elements. Arrays with electronic beam steering capability are called phased arrays. Phased array antennas, when compared to other simple antennas such as dish reflectors, are costly and complicated to design. However, the inherent flexibility of phased array antennas to steer the beam electronically, and also the need for specialized multifunction radar systems, have made phased array antennas attractive for radar applications.

Figure 15.2 shows the geometrical fundamentals associated with this problem. In general, consider the radiation source located at (x_1, y_1, z_1) with respect to a phase reference at $(0, 0, 0)$. The electric field measured at far field point P is

$$E(\theta, \phi) = I_0 \frac{e^{-jkR_1}}{R_1} f(\theta, \phi)$$

Eq. (15.16)

where I_0 is the complex amplitude, $k = 2\pi/\lambda$ is the wave number, and $f(\theta, \phi)$ is the radiation pattern.

Now, consider the case where the radiation source is an array made of many elements, as shown in Fig. 15.3. The coordinates of each radiator with respect to the phase reference is (x_i, y_i, z_i), and the vector from the origin to the *ith* element is given by

$$\vec{r}_i = \hat{a}_x x_i + \hat{a}_y y_i + \hat{a}_z z_i.$$

Eq. (15.17)

The far field components that constitute the total electric field are

$$E_i(\theta, \phi) = I_i \frac{e^{-jkR_i}}{R_i} f(\theta_i, \phi_i)$$

Eq. (15.18)

where

$$R_i = |\vec{R}_i| = |\vec{r} - \vec{r}_i| = \sqrt{(x - x_i)^2 + (y - y_i)^2 + (z - z_i)^2}$$

$$= r\sqrt{1 + \frac{(x_i^2 + y_i^2 + z_i^2)}{r^2} - 2\frac{(xx_i + yy_i + zz_i)}{r^2}}.$$

Eq. (15.19)

Using spherical coordinates, where $x = r\sin\theta\cos\varphi$, $y = r\sin\theta\sin\varphi$, and $z = r\cos\theta$, yields

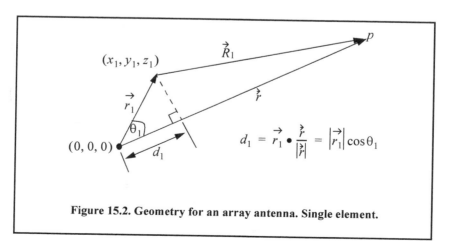

Figure 15.2. Geometry for an array antenna. Single element.

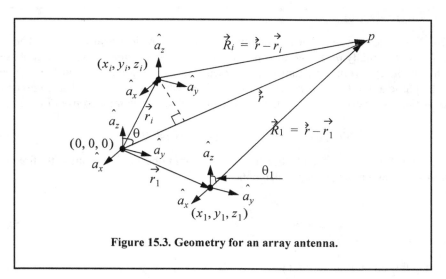

Figure 15.3. Geometry for an array antenna.

$$\frac{(x_i^2 + y_i^2 + z_i^2)}{r^2} = \frac{|\vec{r}_i|^2}{r^2} \ll 1.$$ Eq. (15.20)

Thus, a good approximation (using binomial expansion) for Eq. (15.19) is

$$R_i = r - r(x_i \sin\theta\cos\phi + y_i \sin\theta\sin\phi + z_i \cos\theta).$$ Eq. (15.21)

It follows that the phase contribution at the far field point from the *ith* radiator with respect to the phase reference is

$$e^{-jkR_i} = e^{-jkr}\, e^{jk(x_i \sin\theta\cos\phi + y_i \sin\theta\sin\phi + z_i \cos\theta)}.$$ Eq. (15.22)

Remember, however, that the unit vector $\vec{r}_0$ along the vector $\vec{r}$ is

$$\vec{r}_0 = \frac{\vec{r}}{|\vec{r}|} = \hat{a}_x \sin\theta\cos\phi + \hat{a}_y \sin\theta\sin\phi + \hat{a}_z \cos\theta.$$ Eq. (15.23)

Hence, we can rewrite Eq. (15.22) as

$$e^{-jkR_i} = e^{-jkr}\, e^{jk(\vec{r}_i \cdot \vec{r}_0)} = e^{-jkr} e^{j\Psi_i(\theta, \phi)}.$$ Eq. (15.24)

Finally, by virtue of superposition, the total electric field is

$$E(\theta, \phi) = \sum_{i=1}^{N} I_i e^{j\Psi_i(\theta, \phi)},$$ Eq. (15.25)

which is known as the array factor for an array antenna where the complex current for the *ith* element is I_i.

In general, an array can be fully characterized by its array factor. This is true since knowing the array factor provides the designer with knowledge of the array's (1) 3-dB beamwidth; (2) null-to-null beamwidth; (3) distance from the main peak to the first sidelobe; (4) height of the first sidelobe as compared to the main beam; (5) location of the nulls; (6) rate of decrease of the sidelobes; and (7) grating lobe locations.

15.4. Linear Arrays

Figure 15.4 shows a linear array antenna consisting of N identical elements. The element spacing is d (normally measured in wavelength units). Let element #1 serve as a phase reference for the array. From the geometry, it is clear that an outgoing wave at the nth element leads the phase at the $(n+1)th$ element by $kd\sin\theta$, where $k = 2\pi/\lambda$. The combined phase at the far field observation point P is independent of ϕ and is computed from Eq. (15.24) as

$$\Psi(\theta, \phi) = k(\vec{r}_n \bullet \vec{r}_0) = (n-1)kd\sin\theta .$$

Eq. (15.26)

Thus, from Eq. (15.25), the electric field at a far field observation point with direction-sine equal to $\sin\theta$ (assuming isotropic elements) is

$$E(\sin\theta) = \sum_{n=1}^{N} e^{j(n-1)(kd\sin\theta)} .$$

Eq. (15.27)

Expanding the summation in Eq. (15.27) yields

$$E(\sin\theta) = 1 + e^{jkd\sin\theta} + \ldots + e^{j(N-1)(kd\sin\theta)} .$$

Eq. (15.28)

The right-hand side of Eq. (15.28) is a geometric series, which can be expressed in the form

$$1 + a + a^2 + a^3 + \ldots + a^{(N-1)} = \frac{1-a^N}{1-a} .$$

Eq. (15.29)

Replacing a by $e^{jkd\sin\theta}$ yields

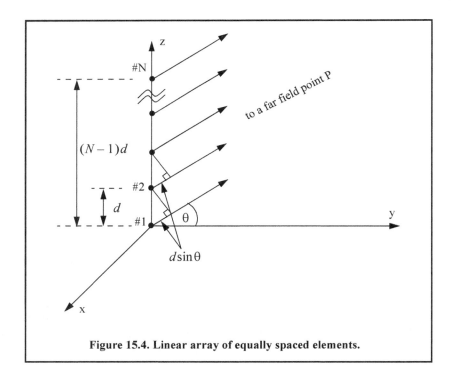

Figure 15.4. Linear array of equally spaced elements.

$$E(\sin\theta) = \frac{1 - e^{jNkd\sin\theta}}{1 - e^{jkd\sin\theta}} = \frac{1 - (\cos Nkd\sin\theta) - j(\sin Nkd\sin\theta)}{1 - (\cos kd\sin\theta) - j(\sin kd\sin\theta)}.$$ **Eq. (15.30)**

The far field array intensity pattern is then given by

$$|E(\sin\theta)| = \sqrt{E(\sin\theta)E^*(\sin\theta)}.$$ **Eq. (15.31)**

Substituting Eq. (15.30) into Eq. (15.31) and collecting terms yields

$$|E(\sin\theta)| = \sqrt{\frac{(1 - \cos Nkd\sin\theta)^2 + (\sin Nkd\sin\theta)^2}{(1 - \cos kd\sin\theta)^2 + (\sin kd\sin\theta)^2}} = \sqrt{\frac{1 - \cos Nkd\sin\theta}{1 - \cos kd\sin\theta}},$$ **Eq. (15.32)**

and using the trigonometric identity $1 - \cos\theta = 2(\sin\theta/2)^2$ yields

$$|E(\sin\theta)| = \left|\frac{\sin(Nkd\sin\theta/2)}{\sin(kd\sin\theta/2)}\right|,$$ **Eq. (15.33)**

which is a periodic function of $kd\sin\theta$, with a period equal to 2π.

The maximum value of $|E(\sin\theta)|$, which occurs at $\theta = 0$, is equal to N. It follows that the normalized intensity pattern is equal to

$$|E_n(\sin\theta)| = \frac{1}{N}\left|\frac{\sin((Nkd\sin\theta)/2)}{\sin((kd\sin\theta)/2)}\right|.$$ **Eq. (15.34)**

The normalized two-way array pattern (radiation pattern) is given by

$$G(\sin\theta) = |E_n(\sin\theta)|^2 = \frac{1}{N^2}\left(\frac{\sin((Nkd\sin\theta)/2)}{\sin((kd\sin\theta)/2)}\right)^2.$$ **Eq. (15.35)**

Figure 15.5 shows a plot of Eq. (15.35) versus $\sin\theta$ for $N = 8$. The pattern $G(\sin\theta)$ has cylindrical symmetry about its axis ($\sin\theta = 0$), and is independent of the azimuth angle. Thus, it is completely determined by its values within the interval $(0 < \theta < \pi)$. This figure can be reproduced using MATLAB program *"Fig15_5.m,"* listed in Appendix 15-A.

The main beam of an array can be steered electronically by varying the phase of the current applied to each array element. Steering the main beam into the direction-sine $\sin\theta_0$ is accomplished by making the phase difference between any two adjacent elements equal to $kd\sin\theta_0$. In this case, the normalized radiation pattern can be written as

$$G(\sin\theta) = \frac{1}{N^2}\left(\frac{\sin[(Nkd/2)(\sin\theta - \sin\theta_0)]}{\sin[(kd/2)(\sin\theta - \sin\theta_0)]}\right)^2$$ **Eq. (15.36)**

If $\theta_0 = 0$, then the main beam is perpendicular to the array axis, and the array is said to be a broadside array. Alternatively, the array is called an endfire array when the main beam points along the array axis.

The radiation pattern maxima are computed using L'Hopital's rule when both the denominator and numerator of Eq. (15.35) are zeros. More precisely,

$$\left(\frac{kd\sin\theta}{2} = \pm m\pi\right); \quad m = 0, 1, 2, \ldots$$ **Eq. (15.37)**

Solving for θ yields

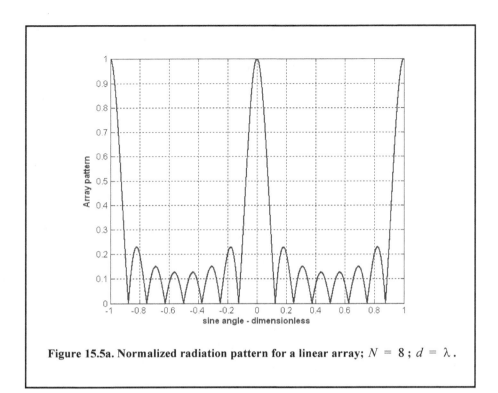

Figure 15.5a. Normalized radiation pattern for a linear array; $N = 8$; $d = \lambda$.

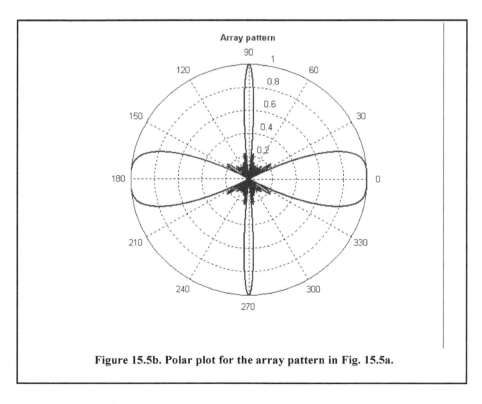

Figure 15.5b. Polar plot for the array pattern in Fig. 15.5a.

$$\theta_m = \text{asin}\left(\pm\frac{\lambda m}{d}\right); \quad m = 0, 1, 2, \dots \qquad \text{Eq. (15.38)}$$

where the subscript m is used as a maxima indicator. The first maximum occurs at $\theta_0 = 0$, and is denoted as the main beam (lobe). Other maxima occurring at $|m| \geq 1$ are called grating lobes. Grating lobes are undesirable and must be suppressed. The grating lobes occur at non-real angles when the absolute value of the arc-sine argument in Eq. (15.38) is greater than unity; it follows that $d < \lambda$. Under this condition, the main lobe is assumed to be at $\theta = 0$ (broadside array). Alternatively, when electronic beam steering is considered, the grating lobes occur at

$$|\sin\theta - \sin\theta_0| = \pm\frac{\lambda n}{d}; \quad n = 1, 2, \dots \qquad \text{Eq. (15.39)}$$

Thus, in order to prevent the grating lobes from occurring between $\pm 90°$, the element spacing should be $d < \lambda/2$.

The radiation pattern attains secondary maxima (sidelobes) when the numerator of Eq. (15.35) is maximum, or equivalently

$$\frac{Nkd\sin\theta}{2} = \pm(2l + 1)\frac{\pi}{2}; \quad l = 1, 2, \dots \qquad \text{Eq. (15.40)}$$

Solving for θ yields

$$\theta_l = \text{asin}\left(\pm\frac{\lambda}{2d}\frac{2l+1}{N}\right); \quad l = 1, 2, \dots \qquad \text{Eq. (15.41)}$$

where the subscript l is used as an indication of sidelobe maxima. The nulls of the radiation pattern occur when only the numerator of Eq. (15.35) is zero. More precisely,

$$\frac{N}{2}kd\sin\theta = \pm n\pi; \quad \begin{matrix} n = 1, 2, \dots \\ n \neq N, 2N, \dots \end{matrix} \qquad \text{Eq. (15.42)}$$

Again solving for θ yields

$$\theta_n = \text{asin}\left(\pm\frac{\lambda}{d}\frac{n}{N}\right); \quad \begin{matrix} n = 1, 2, \dots \\ n \neq N, 2N, \dots \end{matrix} \qquad \text{Eq. (15.43)}$$

where the subscript n is used as a null indicator. Define the angle that corresponds to the half power point as θ_h. It follows that the half power (3dB) beamwidth is $2|\theta_m - \theta_h|$. This occurs when

$$\frac{N}{2}kd\sin\theta_h = 1.391 \ radians \Rightarrow \theta_h = \text{asin}\left(\frac{\lambda}{2\pi d}\frac{2.782}{N}\right). \qquad \text{Eq. (15.44)}$$

15.4.1. Array Tapering

Figure 15.6 shows a normalized two-way radiation pattern of a uniformly excited linear array of size $N = 8$, element spacing $d = \lambda/2$. The first sidelobe is $13.46dB$ below the main lobe, and for most radar applications this may not be sufficient, particularly in the presence of a strong source of jamming or high levels of noise. Under such conditions, target detection in the main beam becomes rather challenging, since the SNR is reduced.

In order to reduce the sidelobe levels, the array must be designed to radiate more power toward the center, and much less at the edges. This can be achieved through tapering (windowing) the current distribution over the face of the array. There are many possible tapering sequences that can be used for this purpose. However, as known from spectral analysis, windowing reduces sidelobe levels at the expense of widening the main beam. Thus, for a given radar application, the choice of the tapering sequence must be based on the trade-off between sidelobe reduction and main-beam widening. The same type of windows discussed earlier in Chapter 3 can be used for array tapering. Table 15.1 summarizes the impact of most common windows on the array pattern in terms of main-beam widening and peak reduction. Note that the rectangular window is used as the baseline. This is also illustrated in Fig. 15.7, which can be reproduced using MATLAB program *"Fig15_7.m,"* listed in Appendix 15-A.

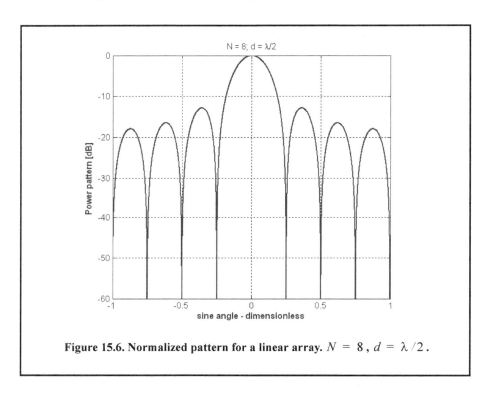

Figure 15.6. Normalized pattern for a linear array. $N = 8$, $d = \lambda / 2$.

TABLE 15.1. Common windows.

Window	Null-to-Null Beamwidth	Peak Reduction
Rectangular	*1*	*1*
Hamming	*2*	*0.73*
Hanning	*2*	*0.664*
Blackman	*6*	*0.577*
Kaiser ($\beta = 6$*)*	*2.76*	*0.683*
Kaiser ($\beta = 3$*)*	*1.75*	*0.882*

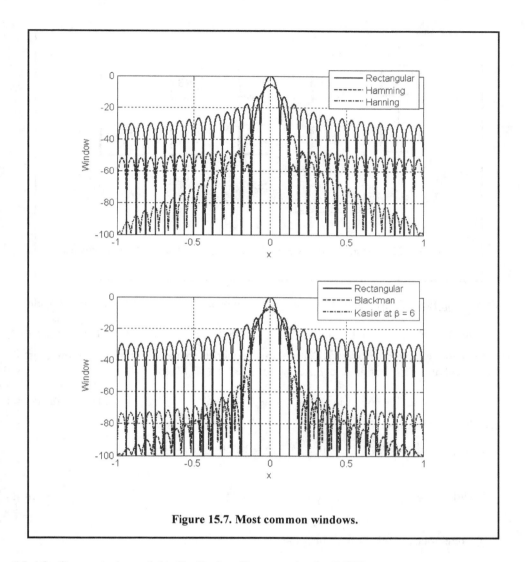

Figure 15.7. Most common windows.

15.4.2. Computation of the Radiation Pattern via the DFT

Figure 15.8 shows a linear array of size N, element spacing d, and wavelength λ. The radiators are circular dishes of diameter d. Let $w(n)$ and $\Phi(n)$, respectively, denote the tapering and phase shifting sequences. The normalized electric field at a far field point in the direction-sine $\sin\theta$ is

$$E(\sin\theta) = \sum_{n=0}^{N-1} w(n)e^{j\Delta\theta\left(n - \left(\frac{N-1}{2}\right)\right)}$$

Eq. (15.45)

where in this case the phase reference is taken as the physical center of the array, and

$$\Delta\theta = \frac{2\pi d}{\lambda}\sin\theta .$$

Eq. (15.46)

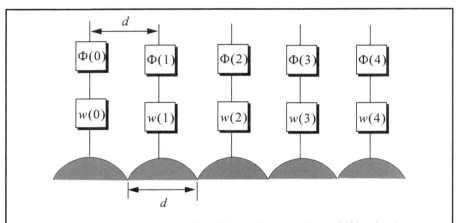

Figure 15.8. Linear array of size 5, with tapering and phase shifting hardware.

Expanding Eq. (15.45) and factoring the common phase term $\exp[j(N-1)\Delta\theta/2]$ yields

$$E(\sin\theta) = e^{j(N-1)\Delta\theta/2}\{w(0)e^{-j(N-1)\Delta\theta} + w(1)e^{-j(N-2)\Delta\theta} + \dots + w(N-1)\}. \qquad \textbf{Eq. (15.47)}$$

By using the symmetry property of a window sequence (remember that a window must be symmetrical about its central point), we can rewrite Eq. (15.47) as

$$E(\sin\theta) = e^{j\theta_0}\{w(N-1)e^{-j(N-1)\Delta\theta} + w(N-2)e^{-j(N-2)\Delta\theta} + \dots + w(0)\} \qquad \textbf{Eq. (15.48)}$$

where $\theta_0 = (N-1)\Delta\theta/2$.

Define $\{V_1^n = \exp(-jn\Delta\theta); n = 0, 1, \dots, N-1\}$. It follows that

$$E(\sin\theta) = e^{j\theta_0}[w(0) + w(1)V_1^1 + \dots + w(N-1)V_1^{N-1}] = e^{j\theta_0}\sum_{n=0}^{N-1} w(n)V_1^n. \qquad \textbf{Eq. (15.49)}$$

The discrete Fourier transform of the sequence $w(n)$ is defined as

$$W(q) = \sum_{n=0}^{N-1} w(n)e^{-\frac{(j2\pi nq)}{N}} \quad ; \quad q = 0, 1, \dots, N-1. \qquad \textbf{Eq. (15.50)}$$

The set $\{\sin\theta_q\}$ that makes V_1 equal to the DFT kernel is

$$\sin\theta_q = \frac{\lambda q}{Nd}; \quad q = 0, 1, \dots, N-1. \qquad \textbf{Eq. (15.51)}$$

Then, by using Eq. (15.51) in Eq. (15.50) yields

$$E(\sin\theta) = e^{j\phi_0}W(q). \qquad \textbf{Eq. (15.52)}$$

The one-way array pattern is computed as the modulus of Eq. (15.52). It follows that the one-way radiation pattern of a tapered linear array of circular dishes is

$$G(\sin\theta) = G_e \ |W(q)|$$ **Eq. (15.53)**

where G_e is the element pattern. In practice, phase shifters are normally implemented as part of the Transmit/Receive (TR) modules, using a finite number of bits. Consequently, due to the quantization error (difference between desired phase and actual quantized phase) the sidelobe levels are affected.

MATLAB Function "linear_array.m"

The function *"linear_array.m"* computes and plots the linear array gain pattern as a function of real sine-space. The syntax is as follows:

[theta, patternr, patterng] = linear_array(Nr, dolr, theta0, winid, win, nbits)

where

Symbol	Description	Units	Status
Nr	*number of elements in array*	*none*	*input*
dolr	*element spacing in lambda units*	*wavelengths*	*input*
theta0	*steering angle*	*degrees*	*input*
winid	*-1== No weighting; 1== weighting = user specified window*	*none*	*input*
win	*window for sidelobe control*	*none*	*input*
nbits	*negative #: perfect quantization* *positive #: use 2^{nbits} quantization levels*	*none*	*input*
theta	*real angle available for steering*	*degrees*	*output*
patternr	*array pattern*	*dB*	*output*
patterng	*gain pattern*	*dB*	*output*

The MATLAB-based graphical user interface (GUI) in Fig. 15.9 implements this function. This GUI was used to produce Figs. 15.10 through 15.18 assuming the following cases:

[theta, patternr, patterng] = linear_array(19, 0.5, 0, -1, -1, -3);

[theta, patternr, patterng] = linear_array(19, 0.5, 0, 1, 'hamming', -3);

[theta, patternr, patterng] = linear_array(19, 0.5, 5, -1, -1, 3);

[theta, patternr, patterng] = linear_array(19, 0.5, 5, 1, 'hamming', 3);

[theta, patternr, patterng] = linear_array(19, 0.5, 25, 1, 'hamming', 3);

[theta, patternr, patterng] = linear_array(19, 1.5, 48, -1, -1, -3);

[theta, patternr, patterng] = linear_array(19, 1.5, 48, 1, 'hamming', -3);

[theta, patternr, patterng] = linear_array(19, 1.5, -48, -1, -1, 3);

[theta, patternr, patterng] = linear_array(19, 1.5, -38, 1, 'hamming', 3);

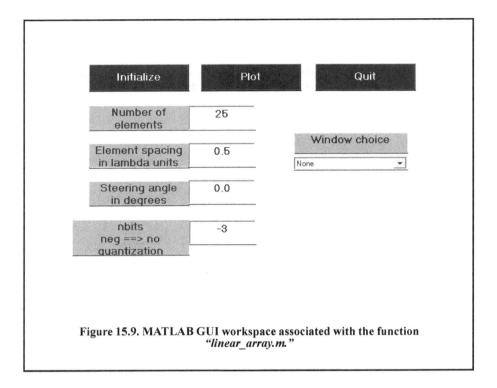

Figure 15.9. MATLAB GUI workspace associated with the function
"linear_array.m."

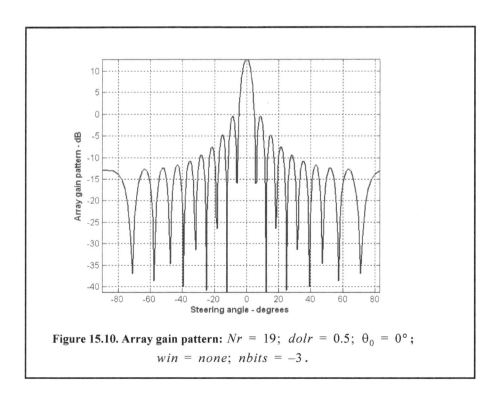

Figure 15.10. Array gain pattern: $Nr = 19$; $dolr = 0.5$; $\theta_0 = 0°$;
$win = none$; $nbits = -3$.

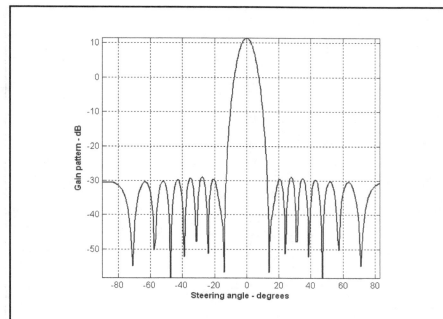

Figure 15.11. Array gain pattern: $Nr = 19$; $dolr = 0.5$; $\theta_0 = 0°$;
win = *Hamming*; *nbits* = –3

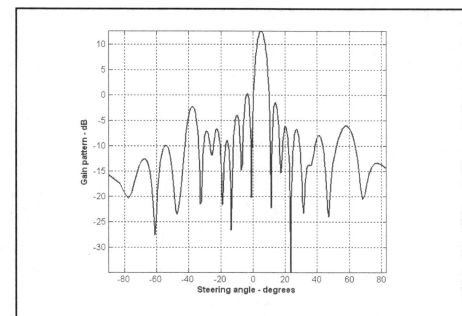

Figure 15.12. Array gain pattern: $Nr = 19$; $dolr = 0.5$; $\theta_0 = 5°$;
win = *none*; *nbits* = 3

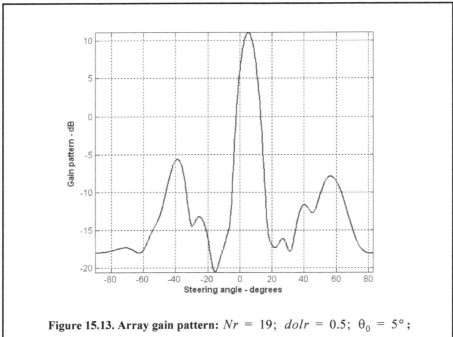

Figure 15.13. Array gain pattern: $Nr = 19$; $dolr = 0.5$; $\theta_0 = 5°$;
win = Hamming; *nbits = 3*

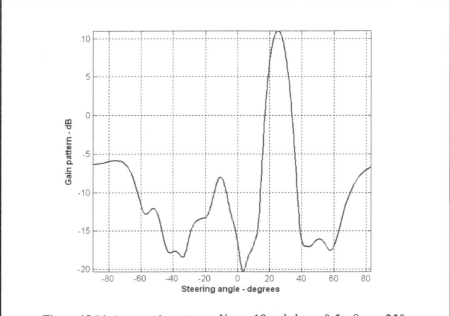

Figure 15.14. Array gain pattern: $Nr = 19$; $dolr = 0.5$; $\theta_0 = 25°$;
win = Hamming; *nbits = 3*

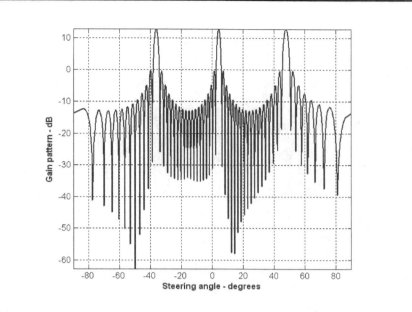

Figure 15.15. Array gain pattern: $Nr = 19$; $dolr = 1.5$; $\theta_0 = 48°$; $win = none$; $nbits = -3$

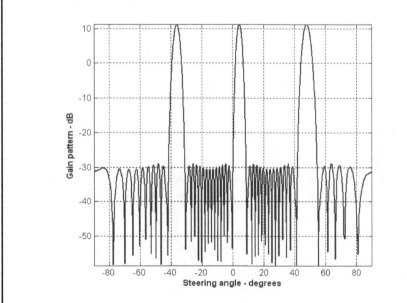

Figure 15.16. Array gain pattern: $Nr = 19$; $dolr = 1.5$; $\theta_0 = 48°$; $win = Hamming$; $nbits = -3$

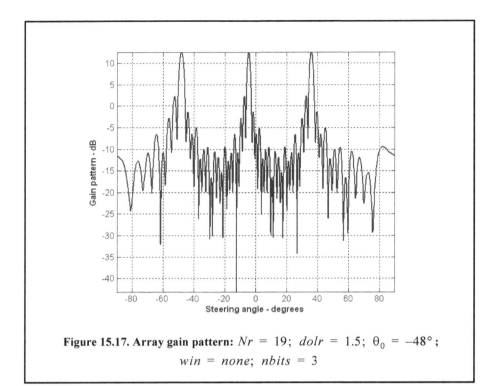

Figure 15.17. Array gain pattern: $Nr = 19$; $dolr = 1.5$; $\theta_0 = -48°$;
win = *none*; *nbits* = 3

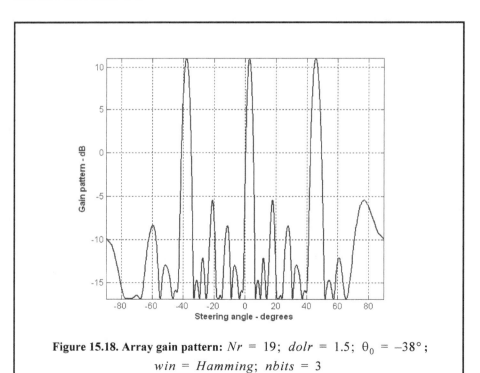

Figure 15.18. Array gain pattern: $Nr = 19$; $dolr = 1.5$; $\theta_0 = -38°$;
win = *Hamming*; *nbits* = 3

15.5. *Planar Arrays*

Planar arrays are a natural extension of linear arrays. Planar arrays can take on many configurations, depending on the element spacing and distribution defined by a "grid." Examples include rectangular, rectangular with circular boundary, hexagonal with circular boundary, circular, and concentric circular grids, as illustrated in Fig. 15.19.

Planar arrays can be steered in elevation and azimuth ((θ, ϕ), as illustrated in Fig. 15.20 for a rectangular grid array. The element spacing along the x- and y-directions are respectively denoted by d_x and d_y. The total electric field at a far field observation point for any planar array can be computed using Eqs. (15.24) and (15.25).

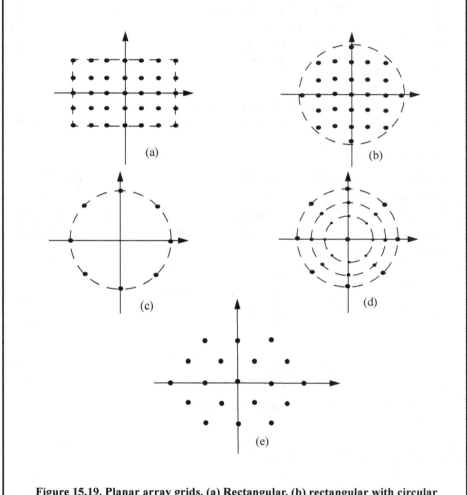

Figure 15.19. Planar array grids. (a) Rectangular, (b) rectangular with circular boundary, (c) circular, (d) concentric circular, and (e) hexagonal.

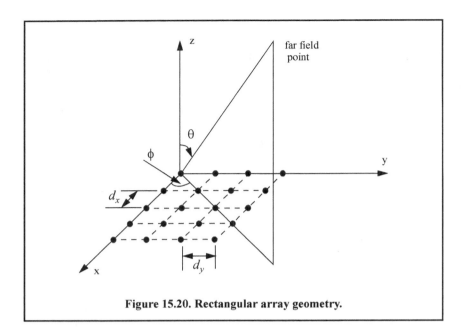

Figure 15.20. Rectangular array geometry.

15.5.1. Rectangular Grid Arrays

Consider the $N \times M$ rectangular grid as shown in Fig. 15.20. The dot product $\vec{r_i} \bullet \vec{r_0}$, where the vector $\vec{r_i}$ is the vector to the *ith* element in the array and $\vec{r_0}$ is the unit vector to the far field observation point, can be broken linearly into its *x*- and *y*-components. It follows that the electric field components due to the elements distributed along the x- and y-directions are, respectively,

$$E_x(\theta, \phi) = \sum_{n=1}^{N} I_{x_n} e^{j(n-1)kd_x \sin\theta \cos\phi} \qquad \text{Eq. (15.54)}$$

$$E_y(\theta, \phi) = \sum_{m=1}^{N} I_{y_m} e^{j(m-1)kd_y \sin\theta \sin\phi} . \qquad \text{Eq. (15.55)}$$

The total electric field at the far field observation point is then given by

$$E(\theta, \phi) = E_x(\theta, \phi) E_y(\theta, \phi) = \left(\sum_{m=1}^{N} I_{y_m} e^{j(m-1)kd_y \sin\theta \sin\phi} \right) \left(\sum_{n=1}^{N} I_{x_n} e^{j(n-1)kd_x \sin\theta \cos\phi} \right) \text{Eq. (15.56)}$$

Eq. (15.56) can be expressed in terms of the directional cosines

$$\begin{aligned} u &= \sin\theta \cos\phi \\ v &= \sin\theta \sin\phi \end{aligned} \qquad \text{Eq. (15.57)}$$

$$\phi = \text{atan}\left(\frac{u}{v}\right)$$

Eq. (15.58)

$$\theta = \text{asin}\sqrt{u^2 + v^2}$$

the visible region is then defined by

$$\sqrt{u^2 + v^2} \le 1.$$

Eq. (15.59)

It is very common to express a planar array's ability to steer the beam in space in terms of the U, V space instead of the angles θ, ϕ. Figure 15.21 shows how a beam steered in a certain θ, ϕ direction is translated into U, V space.

The rectangular array one-way intensity pattern is then equal to the product of the individual patterns. More precisely for a uniform excitation ($I_{y_m} = I_{x_n} = const$),

$$E(\theta, \phi) = \left| \frac{\sin\left(\dfrac{Nkd_x\sin\theta\cos\phi}{2}\right)}{\sin\left(\dfrac{kd_x\sin\theta\cos\phi}{2}\right)} \right| \left| \frac{\sin\left(\dfrac{Nkd_y\sin\theta\sin\phi}{2}\right)}{\sin\left(\dfrac{kd_y\sin\theta\sin\phi}{2}\right)} \right|.$$

Eq. (15.60)

The radiation pattern maxima, nulls, sidelobes, and grating lobes in both the x- and y-axes are computed in a similar fashion to the linear array case. Additionally, the same conditions for grating lobe control are applicable. Note the symmetry is about the angle ϕ.

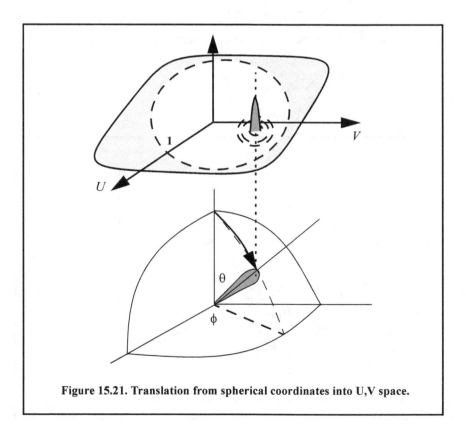

Figure 15.21. Translation from spherical coordinates into U,V space.

15.5.2. Circular Grid Arrays

The geometry of interest is shown in Fig. 15.19c. In this case, N elements are distributed equally on the outer circle whose radius is a. For this purpose, consider the geometry shown in Fig. 15.22. From the geometry

$$\Phi_n = \frac{2\pi}{N} n \qquad ; n = 1, 2, ..., N.$$

Eq. (15.61)

The coordinates of the *nth* element are

$$x_n = a \cos\Phi_n$$
$$y_n = a \sin\Phi_n.$$
$$z_n = 0$$

Eq. (15.62)

It follows that

$$k(\vec{r}_n \bullet \vec{r}_0) = \Psi_n = k(a\sin\theta\cos\phi\cos\Phi_n + a\sin\theta\sin\phi\sin\Phi_n + 0).$$

Eq. (15.63)

Equation (15.63) can be rearranged as

$$\Psi_n = ak\sin\theta(\cos\phi\cos\Phi_n + \sin\phi\sin\Phi_n).$$

Eq. (15.64)

Then, by using the identity $\cos(A - B) = \cos A \cos B + \sin A \sin B$, Eq.(15.63) collapses to

$$\Psi_n = ak\sin\theta\cos(\Phi_n - \phi).$$

Eq. (15.65)

Finally, by using Eq. (15.25), the far field electric field is then given by

$$E(\theta, \phi; a) = \sum_{n=1}^{N} I_n \exp\left\{ j\frac{2\pi a}{\lambda}\sin\theta\cos(\Phi_n - \phi) \right\}$$

Eq. (15.66)

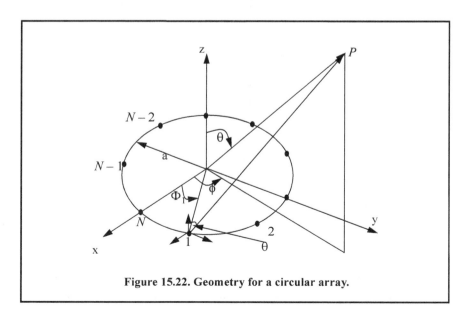

Figure 15.22. Geometry for a circular array.

where I_n represents the complex current distribution for the *nth* element. When the array main beam is directed in the (θ_0, ϕ_0), Eq. (15.65) takes on the following form

$$E(\theta, \phi; a) = \sum_{n=1}^{N} I_n \exp\left\{ j\frac{2\pi a}{\lambda}[\sin\theta\cos(\Phi_n - \phi) - \sin\theta_0\cos(\Phi_n - \phi_0)] \right\}. \qquad \textbf{Eq. (15.67)}$$

MATLAB program "circular_array.m"

The MATLAB program *"circular_array.m"* calculates and plots the rectangular and polar array patterns for a circular array versus θ and ϕ constant planes. The input parameters to this program are:

Symbol	Description	Units
a	*Circular array radius*	λ
N	*number of elements*	*none*
theta0	*main direction in* θ	*degrees*
phi0	*main direction in* ϕ	*degrees*
Variations	*'Theta'; or 'Phi'*	*none*
phid	*constant* ϕ *plane*	*degrees*
thetad	*constant* θ *plane*	*degrees*

As an example, consider the following two cases with inputs defined in Table 15.2:

Table 15.2. Parameters to be used in Figs. 15.23 through 15.32

Parameter	Case I	Case II
a	*1.*	*1.5*
N	*10*	*10*
θ_0	*45*	*45*
ϕ_0	*60*	*60*
variation	*'Theta'*	*'Phi'*
ϕ_d	*60*	*60*
θ_d	*45*	*45*

Figures 15.23 and 15.24 respectively show the array patterns in relative amplitude and the power patterns versus the angle θ corresponding to Case I parameters. Figures 15.25 and 15.26 are similar to Figs. 15.23 and 15.24, except in this case the patterns are plotted in polar coordinates. Figure 15.27 shows a plot of the normalized single element pattern (upper left corner), the normalized array factor (upper right corner), the total array pattern (lower left corner). Figures 15.28 through 15.32 are similar to those in Figs. 15.23 through 15.27, except in this case the input parameters correspond to Case II.

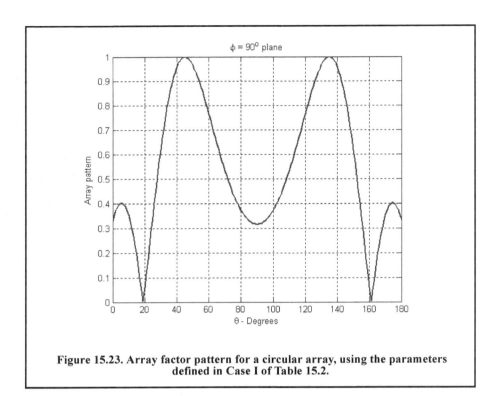

Figure 15.23. Array factor pattern for a circular array, using the parameters defined in Case I of Table 15.2.

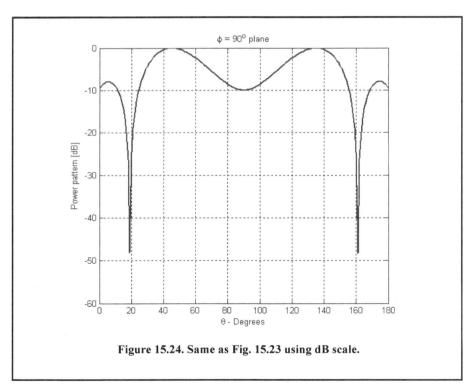

Figure 15.24. Same as Fig. 15.23 using dB scale.

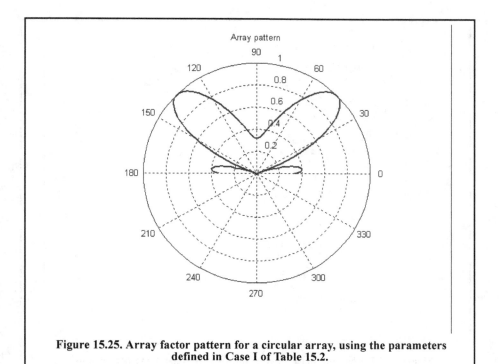

Figure 15.25. Array factor pattern for a circular array, using the parameters defined in Case I of Table 15.2.

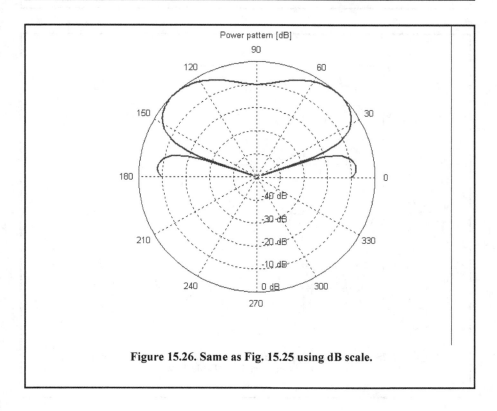

Figure 15.26. Same as Fig. 15.25 using dB scale.

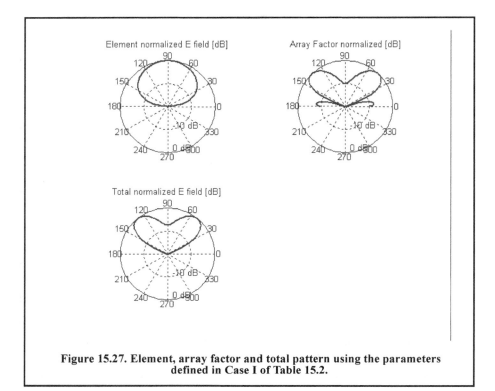

Figure 15.27. Element, array factor and total pattern using the parameters defined in Case I of Table 15.2.

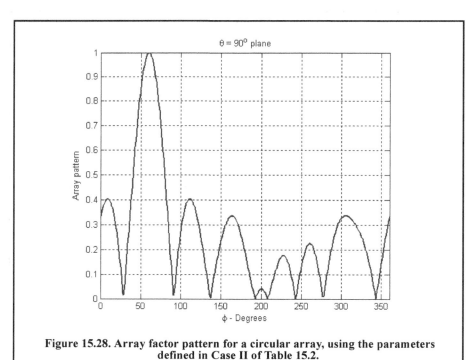

Figure 15.28. Array factor pattern for a circular array, using the parameters defined in Case II of Table 15.2.

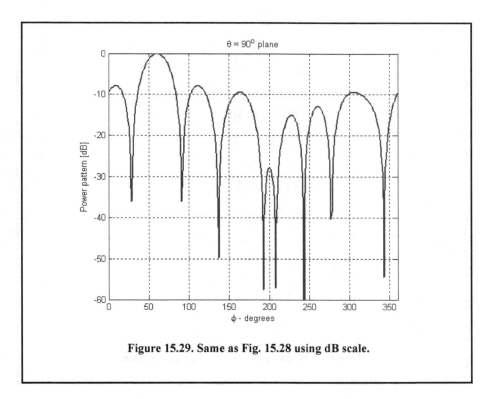

Figure 15.29. Same as Fig. 15.28 using dB scale.

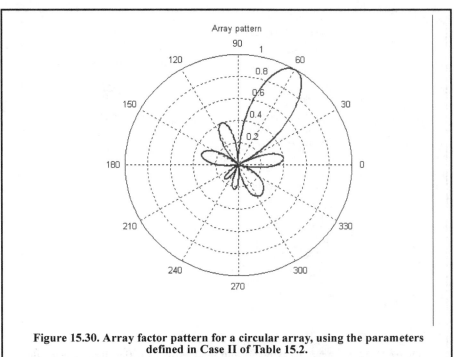

Figure 15.30. Array factor pattern for a circular array, using the parameters defined in Case II of Table 15.2.

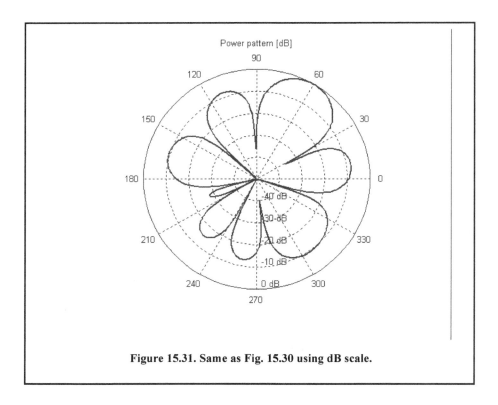

Figure 15.31. Same as Fig. 15.30 using dB scale.

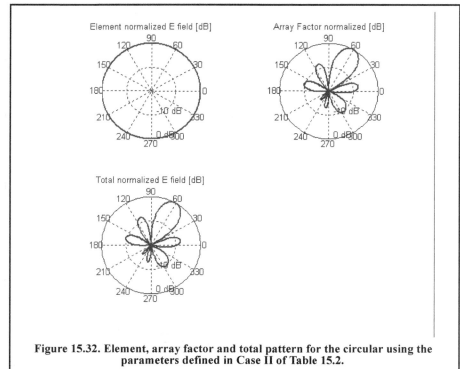

Figure 15.32. Element, array factor and total pattern for the circular using the parameters defined in Case II of Table 15.2.

15.5.3. Concentric Grid Circular Arrays

The geometry of interest is shown in Fig. 15.33. In this case, N_2 elements are distributed equally on the outer circle whose radius is a_2, while another N_1 elements are linearly distributed on the inner circle whose radius is a_1. The element located on the center of both circles is used as the phase reference. In this configuration, there are $N_1 + N_2 + 1$ total elements in the array.

The array factor is derived in two steps. First, the array factor corresponding to a linearly distributed circular array is computed. Second, the overall array factor corresponding to all elements will be the product of each individual circular array times the pattern of the central element. More precisely,

$$E(\theta, \phi) = E_1(\theta, \phi; a_1) E_2(\theta, \phi; a_2) E_0(\theta, \phi) \tag{8.68}$$

Fig. 15.34 shows a 3-D plot for a concentric circular array in the θ, ϕ space for the following parameters: $a_1 = 1\lambda$, $N_1 = 8 = N_2$, and $a_2 = 2\lambda$

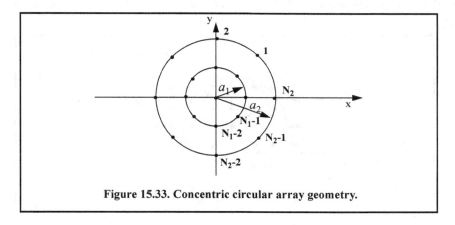

Figure 15.33. Concentric circular array geometry.

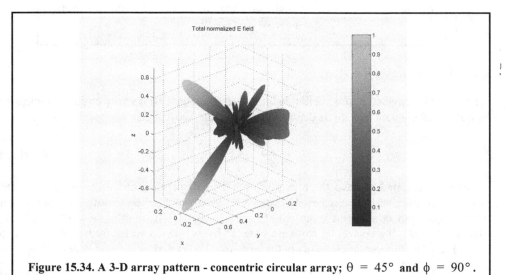

Figure 15.34. A 3-D array pattern - concentric circular array; $\theta = 45°$ and $\phi = 90°$.

15.5.4. Rectangular Grid with Circular Boundary Arrays

The far field electric field associated with this configuration can be easily obtained from that corresponding to a rectangular grid. In order to accomplish this task, follow these steps: First, select the desired maximum number of elements along the diameter of the circle and denote it by N_d. Also select the associated element spacings d_x, d_y. Define a rectangular array of size $N_d \times N_d$. Draw a circle centered at $(x, y) = (0, 0)$ with radius r_d where

$$r_d = \frac{N_d - 1}{2} + \Delta x \qquad \text{Eq. (15.69)}$$

and $\Delta x \leq d_x / 4$. Finally, modify the weighting function across the rectangular array by multiplying it with the two-dimensional sequence $a(m, n)$, where

$$a(m, n) = \begin{cases} 1 & , \; \textit{if dis to } (m, n)\textit{th element} < r_d \\ 0 & ; \; \textit{elsewhere} \end{cases} \qquad \text{Eq. (15.70)}$$

where distance, dis, is measured from the center of the circle. This is illustrated in Fig. 15.35.

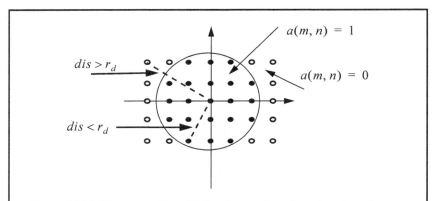

Figure 15.35. Elements with solid dots have $a(m, n) = 1$ **; other elements have** $a(m, n) = 0$ **.**

15.5.5. Hexagonal Grid Arrays

The analysis provided in this section is limited to hexagonal arrays with circular boundaries. The horizontal element spacing is denoted as d_x and the vertical element spacing is

$$d_y = \frac{\sqrt{3}}{2} \, d_x. \qquad \text{Eq. (15.71)}$$

The array is assumed to have the maximum number of identical elements along the x-axis ($y = 0$). This number is denoted by N_x, where N_x is an odd number, in order to obtain a symmetric array, where an element is present at $(x, y) = (0, 0)$. The number of rows in the array is denoted by M. The horizontal rows are indexed by m, which varies from $-(N_x - 1)/2$ to $(N_x - 1)/2$. The number of elements in the mth row is denoted by N_r and is defined by

$$N_r = N_x - |m|. \qquad \text{Eq. (15.72)}$$

The electric field at a far field observation point is computed using Eq. (15.24) and (15.25). The phase associated with $(m, n)th$ location is

$$\psi_{m, n} = \frac{2\pi d_x}{\lambda} \sin\theta \left[\left(m + \frac{n}{2} \right) \cos\phi + n\frac{\sqrt{3}}{2} \sin\phi \right].$$ **Eq. (15.73)**

MATLAB Function "rect_array.m"

The function *"rect_array.m"* computes and plots the rectangular antenna gain pattern in the visible U, V space. The syntax is as follows:

[pattern] = rect_array(Nxr, Nyr, dolxr, dolyr, theta0, phi0, winid, win, nbits)

where

Symbol	Description	Units	Status
Nxr	*number of elements along x*	*none*	*input*
Nyr	*number of elements along y*	*none*	*input*
dolxr	*element spacing in lambda units along x*	*wavelengths*	*input*
dolyr	*element spacing in lambda units along y*	*wavelengths*	*input*
theta0	*elevation steering angle*	*degrees*	*input*
phi0	*azimuth steering angle*	*degrees*	*input*
winid	*-1: No weighting is used* *1: Use weighting defined in win*	*none*	*input*
win	*window for sidelobe control*	*none*	*input*
nbits	*negative #: perfect quantization* *positive #: use 2^{nbits} quantization levels*	*none*	*input*
pattern	*gain pattern*	*dB*	*output*

A MATLAB-based GUI workspace called *"array.m"* was developed for this function. It is shown in Fig. 15.36. The user is advised to use this MATLAB GUI workspace to generate array gain patterns that match this requirement. Figures 15.37 through 15.42, respectively, show plots of the array gain pattern in the *U-V* space, for the following cases:

Case I: [pattern] = rect_array(15, 15, 0.5, 0.5, 0, 0, -1, -1, -3)

Case II: [pattern] = rect_array(15, 15, 0.5, 0.5, 20, 30, -1, -1, -3)

Case III: [pattern] = rect_array(15, 15, 0.5, 0.5, 45, 45, 1, 'Hamming', -3)

Case IV: [pattern] = rect_array(15, 15, 0.5, 0.5, 10, 20, -1, -1, 3)

Case V: [pattern] = rect_array(15, 15, 1, 0.5, 20, 25, -1, -1, -3)

Case VI: [pattern] = rect_array(15, 15, 1.25, 1.25, 0, 0, -1, -1, -3)

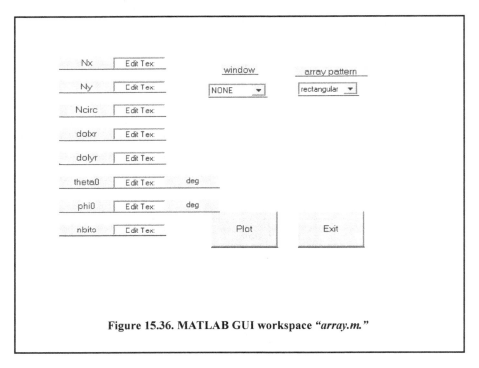

Figure 15.36. MATLAB GUI workspace *"array.m."*

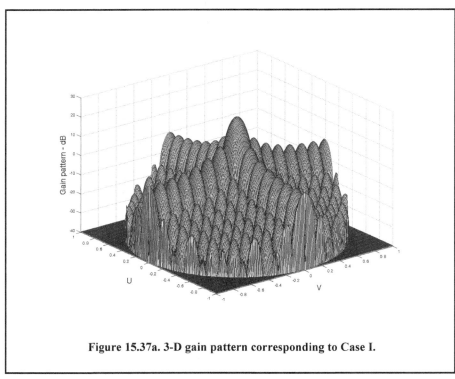

Figure 15.37a. 3-D gain pattern corresponding to Case I.

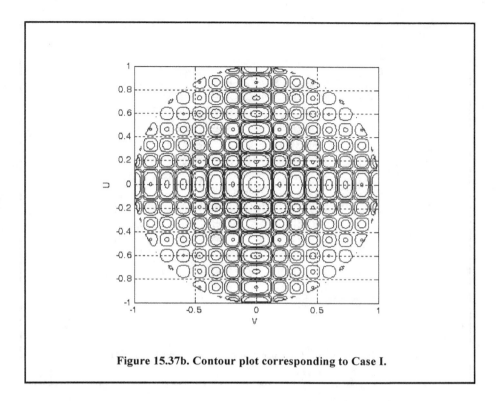

Figure 15.37b. Contour plot corresponding to Case I.

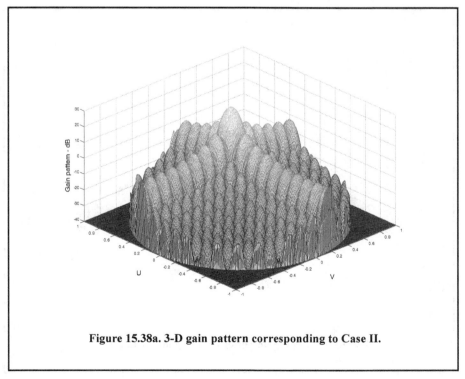

Figure 15.38a. 3-D gain pattern corresponding to Case II.

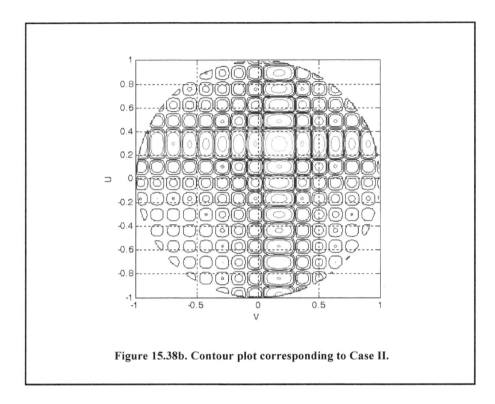

Figure 15.38b. Contour plot corresponding to Case II.

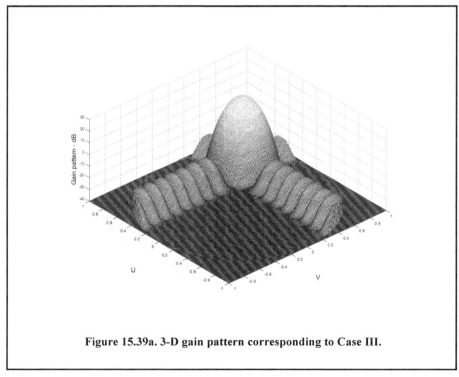

Figure 15.39a. 3-D gain pattern corresponding to Case III.

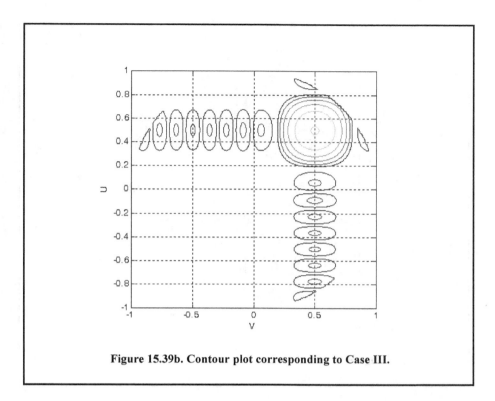

Figure 15.39b. Contour plot corresponding to Case III.

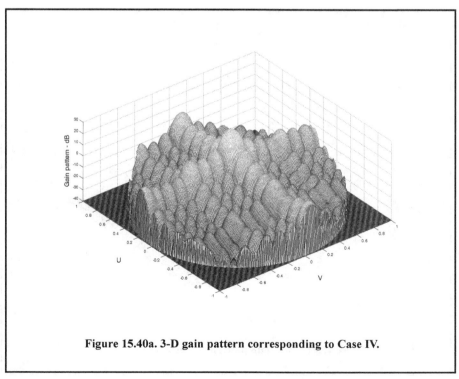

Figure 15.40a. 3-D gain pattern corresponding to Case IV.

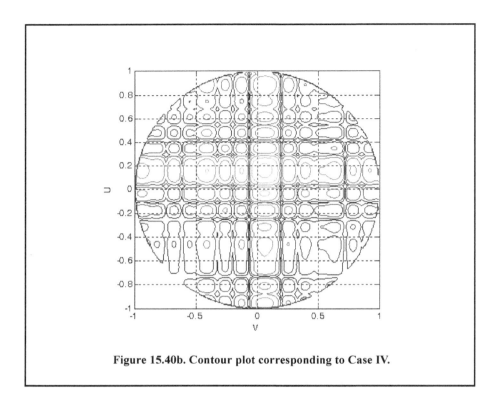

Figure 15.40b. Contour plot corresponding to Case IV.

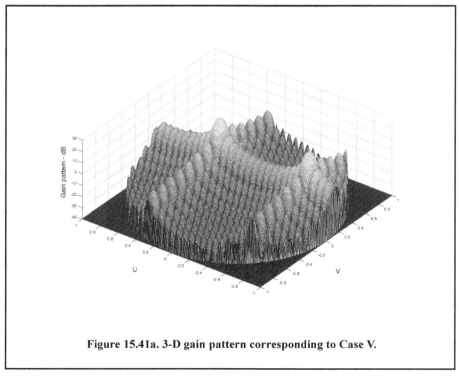

Figure 15.41a. 3-D gain pattern corresponding to Case V.

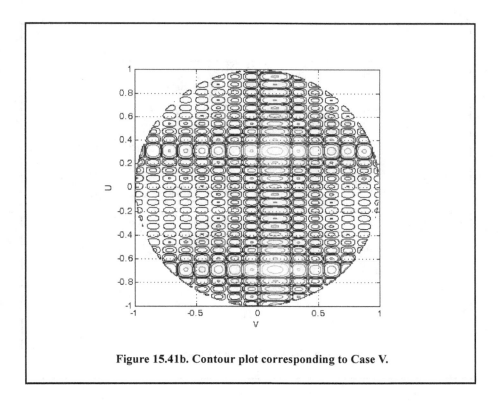

Figure 15.41b. Contour plot corresponding to Case V.

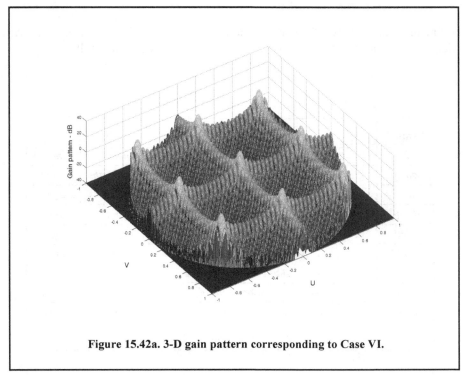

Figure 15.42a. 3-D gain pattern corresponding to Case VI.

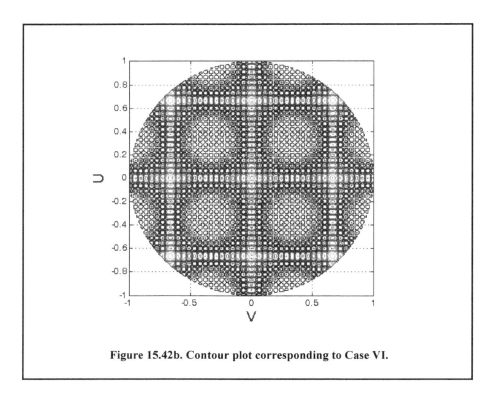

Figure 15.42b. Contour plot corresponding to Case VI.

MATLAB Function "circ_array.m"

The function *"circ_array.m"* computes and plots the rectangular grid with a circular array boundary antenna gain pattern in the visible *U,V* space. The syntax is as follows:

[pattern, amn] = circ_array(N, dolxr, dolyr, theta0, phi0, winid, win, nbits);

where

Symbol	Description	Units	Status
N	*number of elements along diameter*	*none*	*input*
dolxr	*element spacing in lambda units along x*	*wavelengths*	*input*
dolyr	*element spacing in lambda units along y*	*wavelengths*	*input*
theta0	*elevation steering angle*	*degrees*	*input*
phi0	*azimuth steering angle*	*degrees*	*input*
winid	*-1: No weighting is used* *1: Use weighting defined in win*	*none*	*input*
win	*window for sidelobe control*	*none*	*input*
nbits	*negative #: perfect quantization* *positive #: use 2^{nbits} quantization levels*	*none*	*input*

Symbol	Description	Units	Status
patterng	*gain pattern*	*dB*	*output*
amn	*a(m,n) sequence defined in Eq. (15.68)*	*none*	*output*

Figures 15.43 through 15.48, respectively, show plots of the array gain pattern versus steering for the following cases:

Case I: [pattern, amn] = circ_array(15, 0.5, 0.5, 0, 0, -1, -1, -3)

Case II: [pattern, amn] = circ_array(15, 0.5, 0.5, 20, 30, -1, -1, -3)

Case III: [pattern, amn] = circ_array(15, 0.5, 0.5, 30, 30, 1, 'Hamming', -3)

Case IV: [pattern, amn] = circ_array(15, 0.5, 0.5, 30, 30, -1, -1, 3)

Case V: [pattern, amn] = circ_array(15, 1, 0.5, 30, 30, -1, -1, -3)

Case VI: [pattern, amn] = circ_array(15, 1, 1, 0, 0, -1, -1, -3)

Note that the function *"circ_array.m"* uses the function *"rec_to_circ.m,"* which computes the array $a(m, n)$. Also note that the GUI workspace shown in Fig. 15.36 can be used in this case by applying the *"Ncirc"* option on the GUI, where *"Ncirc"* refers to the number of array elements along the diameter.

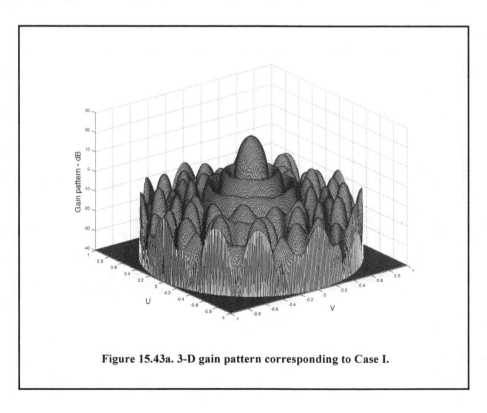

Figure 15.43a. 3-D gain pattern corresponding to Case I.

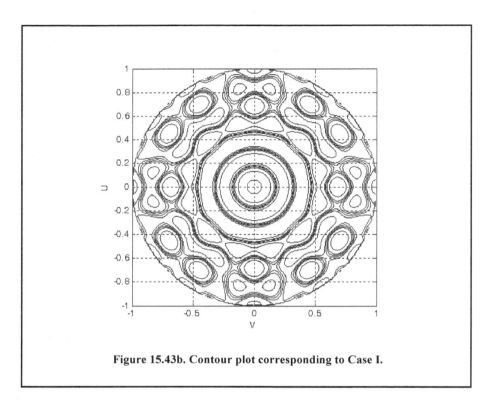

Figure 15.43b. Contour plot corresponding to Case I.

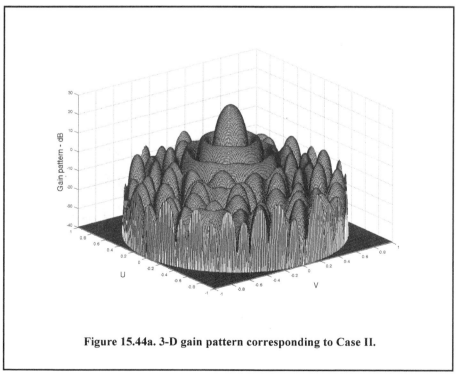

Figure 15.44a. 3-D gain pattern corresponding to Case II.

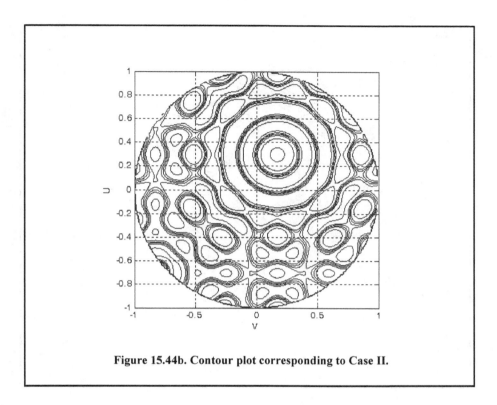

Figure 15.44b. Contour plot corresponding to Case II.

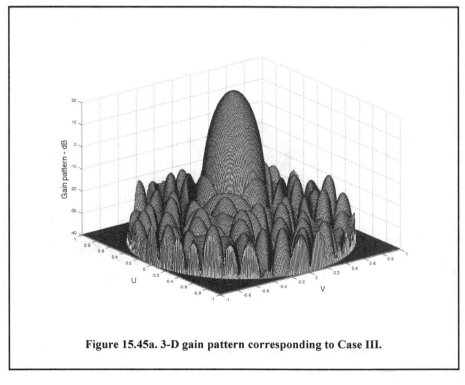

Figure 15.45a. 3-D gain pattern corresponding to Case III.

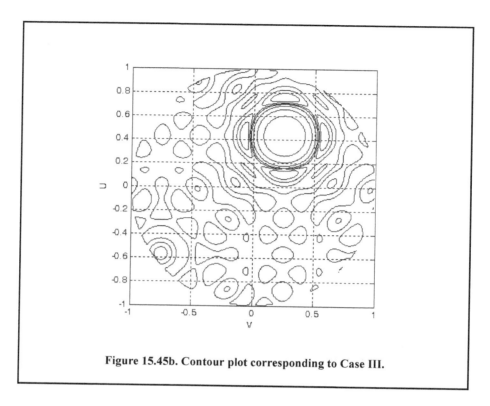

Figure 15.45b. Contour plot corresponding to Case III.

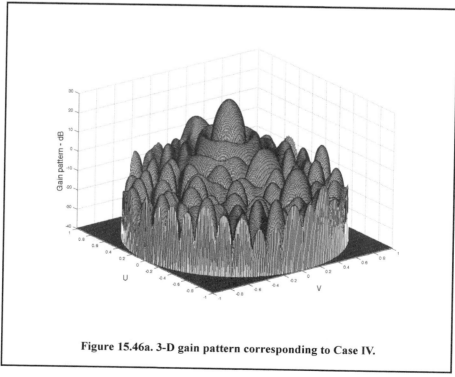

Figure 15.46a. 3-D gain pattern corresponding to Case IV.

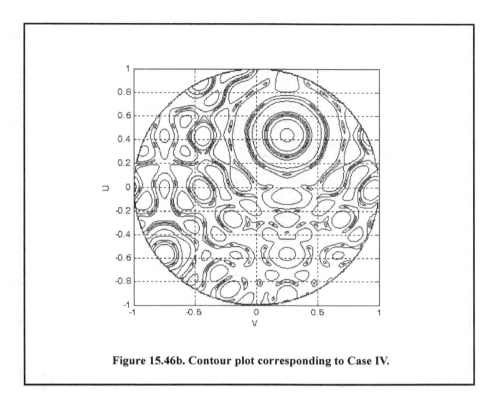

Figure 15.46b. Contour plot corresponding to Case IV.

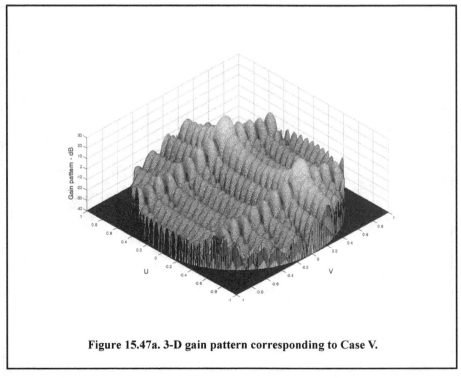

Figure 15.47a. 3-D gain pattern corresponding to Case V.

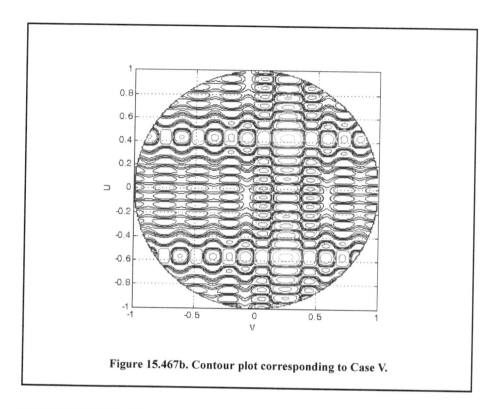

Figure 15.467b. Contour plot corresponding to Case V.

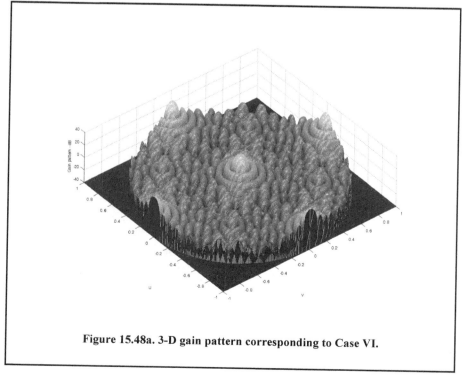

Figure 15.48a. 3-D gain pattern corresponding to Case VI.

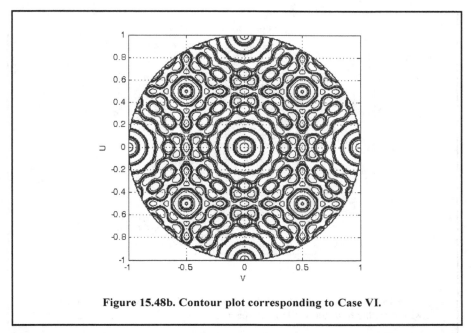

Figure 15.48b. Contour plot corresponding to Case VI.

The program *"array.m"* also plots the array's element spacing pattern. Figures 15.49a and 15.49b show two examples. The *"x's"* indicate the location of actual active array elements, while the *"o's"* indicate the location of dummy or virtual elements created merely for computational purposes. More precisely, Figure 15.49a shows a rectangular grid with circular boundary as defined in Eqs. (15.67) and (15.68) with $d_x = d_y = 0.5\lambda$ and $a = 0.35\lambda$. Figure 15.49b is similar, except for an element spacing $d_x = 1.5\lambda$ and $d_y = 0.5\lambda$.

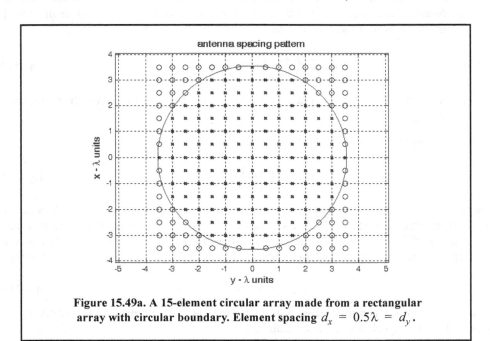

Figure 15.49a. A 15-element circular array made from a rectangular array with circular boundary. Element spacing $d_x = 0.5\lambda = d_y$.

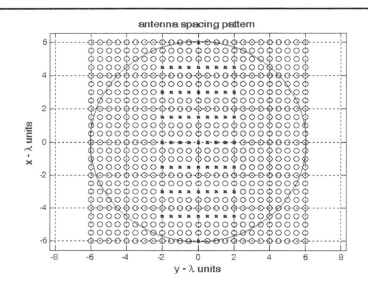

Figure 15.49b. A 15-element circular array made from a rectangular array with circular boundary. Element spacing $d_y = 0.5\lambda$ and $d_x = 1.5\lambda$.

15.6. Array Scan Loss

Phased arrays experience gain loss when the beam is steered away from the array boresight, or zenith (normal to the face of the array). This loss is due to the fact that the array effective aperture becomes smaller, and consequently the array beamwidth is broadened, as illustrated in Fig. 15.50. This loss in antenna gain is called scan loss, L_{scan}, where

$$L_{scan} = \left(\frac{A}{A_\theta}\right)^2 = \left(\frac{G}{G_\theta}\right)^2 .$$ **Eq. (15.74)**

A_θ is the effective aperture area at scan angle θ, and G_θ is the effective array gain at the same angle.

The beamwidth at scan angle θ is

$$\Theta_\theta = \frac{\Theta_{broadside}}{\cos\theta}$$ **Eq. (15.75)**

due to the increased scan loss at large scanning angles. In order to limit the scan loss to under some acceptable practical values, most arrays do not scan electronically beyond about $\theta = 60°$. Such arrays are called Full Field of View (FFOV) arrays. FFOV arrays employ element spacing of 0.6λ or less to avoid grating lobes. FFOV array scan loss is approximated by

$$L_{scan} \approx (\cos\theta)^{2.5} .$$ **Eq. (15.76)**

Arrays that limit electronic scanning to under $\theta = 60°$ are referred to as Limited Field of View (LFOV) arrays. In this case the scan loss is

$$L_{scan} = \left[\frac{\sin\left(\frac{\pi d}{\lambda}\sin\theta\right)}{\frac{\pi d}{\lambda}\sin\theta} \right]^{-4}.$$

Eq. (15.77)

Figure 15.51 shows a plot for scan loss versus scan angle. This figure can be reproduced using MATLAB program *"Fig15_50.m,"* listed in Appendix 15-A.

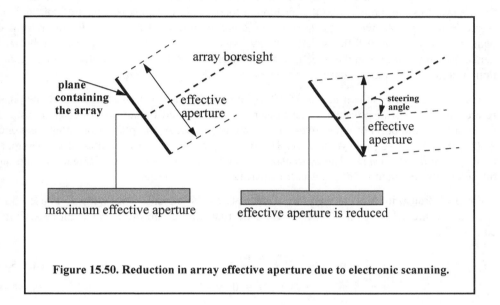

Figure 15.50. Reduction in array effective aperture due to electronic scanning.

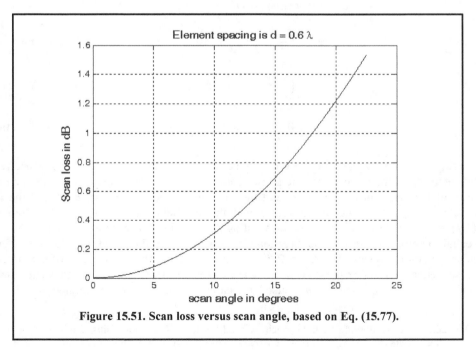

Figure 15.51. Scan loss versus scan angle, based on Eq. (15.77).

15.7. Multiple Input Multiple Output (MIMO) - Linear Array

In this section, a multiple input multiple output (MIMO) target detection technique is presented. This section is based on Mahafza[1] et al (1996). In this approach, each array element (or subarray; super-element) has its own receive channel as described in Fig. 15.52. The radar is assumed to transmit a burst of N pulses, where N is equal to the number of elements in the array. The phase reference of transmitted pulses is assumed to move linearly from the first element in the array when the first pulse in the burst is transmitted, to the last element in the array for the last pulse in the burst. In this fashion, a total of N^2 complex returns are collected and stored in memory. As will be explained later, there are a total of $(2N-1)$ distinct returns of equal two-way phase. It follows that an array twice as large as the actual one is synthesized; hence the term *synthetic* in the title. This synthetic array effectively doubles the angular resolution as compared to the standard operation and the SNR is greatly improved.

Consider the array shown in Fig. 15.52. A burst of N pulses is transmitted where the phase reference for n^{th} is the physical center of element n. The echo signals are collected and stored coherently on the basis of equal two-way geometric phase. A complex information sequence $\{b(m); m = 0, 2N-2\}$ is synthesized. The two-way array pattern is computed as the amplitude spectrum of $\{b(m)\}$. The synthesized sequence has natural triangular windowing, and the sidelobes are about -27dB, thus extra tapering may not be required.

For each transmitted pulse there are a total of N echo signals. The two-way phase corresponding to the echo signal for the ith element transmitting and the jth element receiving is computed as

$$b(n) = e^{jk(\mathbf{r}_i \cdot \mathbf{r}_0 + \mathbf{r}_j \cdot \mathbf{r}_0)} = e^{j\phi_i}e^{j\phi_j} \qquad \text{Eq. (15.78)}$$

$$m = i+j; \ (i,j) = 0, N-1 \qquad \text{Eq. (15.79)}$$

$$b(m) = p(i,j)\exp(j\phi_i)\exp(j\phi_j) \qquad \text{Eq. (15.80)}$$

$$\phi_i = \left(-\left(\frac{N-1}{2}\right) + i\right)\Delta\theta \qquad \text{Eq. (15.81)}$$

$$\phi_j = \left(-\left(\frac{N-1}{2}\right) + j\right)\Delta\theta \qquad \text{Eq. (15.82)}$$

$$\Delta\theta = \frac{2\pi d}{\lambda}\sin\theta \qquad \text{Eq. (15.83)}$$

where $\mathbf{r}_i \cdot \mathbf{r}_0$ and $\mathbf{r}_j \cdot \mathbf{r}_0$ are the dot products between the vectors $\mathbf{r}_0$ and $\mathbf{r}_i$, $\mathbf{r}_j$, respectively. The vector $\mathbf{r}_0$ is the ray between the phase reference point of the array, taken as the physical center of the array, in this analysis, and the far field target; the vectors $\mathbf{r}_i$, $\mathbf{r}_j$ are the rays between i^{th}, j^{th} elements of the array and the far field target, respectively. Note that $p(i,j) = 1$ if the path: i^{th} element transmitting and j^{th} element receiving exists, otherwise it is equal to zero, and $\sin\theta$ is the direction-sine toward which the radiation pattern is steered. The information sequence has $N_a = 2N-1$ distinct entries. The components of the information sequence have linear phase, and the phase increment between any two adjacent terms is equal to $\Delta\phi$. The sequence $\{b(m)\}$ also has triangular shape weighting, defined by

1. Mahafza, B. R., Heifner, L.A., and Gracchi, V. C., Multitarget Detection Using Synthetic Sampled Aperture Radars (SSAMAR), *IEEE - AES Trans.*, Vol. 31, No. 3, July 1995, pp. 1127-1132.

$$\{c(m);\ m = 0, 2N-2\} = \begin{cases} m+1;\ m = 0, N-2 \\ N;\ m = N-1 \\ 2N-1-m;\ m = N, 2N-2 \end{cases}.$$ **Eq. (15.84)**

Through zero padding, the sequence $\{b(m)\}$ is extended to the next power of 2. The two-way pattern in the direction $\sin\theta$, is computed as the modulus of the Discrete Fourier Transform (DFT) of the extended sequence $\{b(m)\}$.

Assume an incident plane wave defined by amplitude A_1 and direction-sine $\sin\theta_1$, and zero-mean, white additive noise w_n with variance σ^2. Then, the mth sample of the information sequence is

$$b(m) = A_1 s(m) + w_n(m);\ m = 0, N_a - 1$$ **Eq. (15.85)**

$$A_1 = G_e^2(\sin\theta_1)\left(\frac{R_o}{R}\right)^4 \rho_1$$ **Eq. (15.86)**

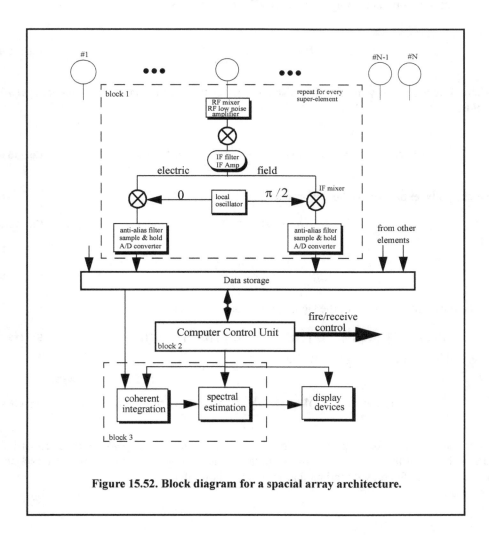

Figure 15.52. Block diagram for a spacial array architecture.

$$s(m) = c(m)\exp\{j[m - (N-1)]k_1\} \qquad \text{Eq. (15.87)}$$

where, G_e^2 represents the two-way element gain, R_0 is the reference range, and ρ_1 is the wave amplitude. It follows that if there are L incident plane waves defined by $\{\sin\theta_i; \ i = 1, L\}$, then the composite information sequence is

$$b(m) = \sum_{i=1}^{L} A_i s_i(m) + w_n(m); \ m = 0, N_a - 1, \qquad \text{Eq. (15.88)}$$

which can be written in vector notation as

$$\mathbf{b} = \sum_{i=1}^{L} A_i \mathbf{s}_i + \mathbf{w}_n. \qquad \text{Eq. (15.89)}$$

Assuming that the noise is spatially incoherent and is uncorrelated with the signal samples, the autocorrelation matrix for the field sensed by the array is

$$\Re = E[\mathbf{b}\mathbf{b}^\dagger] = \sigma^2 \mathbf{I} + \sum_{i=1}^{L} P_i \mathbf{s}\mathbf{s}^\dagger \qquad \text{Eq. (15.90)}$$

where $\mathbf{I}$ is the identity matrix. The discrete Fourier transformation of the sequence $\{b(m)\}$ yields,

$$B(q) = \sum_{m=0}^{N_a - 1} b(m)\exp\left(-j\frac{2\pi qm}{N_a}\right); \ q = 0, N_a - 1, \qquad \text{Eq. (15.91)}$$

which can be expressed as the dot product

$$B(q) = \mathbf{a}^\dagger(q) \cdot \mathbf{b} \qquad \text{Eq. (15.92)}$$

where

$$a(q;m) = \exp\left(j\frac{2\pi qm}{N_a}\right). \qquad \text{Eq. (15.93)}$$

The power at the output of the signal processor at frequency bin q is

$$P(q) = E[|B(q)|^2] = E[\{\mathbf{a}^\dagger(q)\mathbf{b}\}\{\mathbf{a}^\dagger(q)\mathbf{b}\}^\dagger] = \mathbf{a}^\dagger(q)\Re\mathbf{a}(q) \qquad \text{Eq. (15.94)}$$

where $\Re$ is defined in Eq. (15.90). Thus,

$$P(q) = \sigma_v^2 \mathbf{a}^\dagger(q)\mathbf{I}\mathbf{a}(q) + \sum_{i=1}^{L} P_1 \mathbf{a}^\dagger(q)\mathbf{s}_i \mathbf{s}_i^\dagger \mathbf{a}(q). \qquad \text{Eq. (15.95)}$$

After compensation for range attenuation and antenna gain, spectral peaks will be proportional to amplitudes of incident waves. For example, a peak at an arbitrary bin q_j will correspond to a plane wave defined by direction-sine $\sin\theta_j$. It follows that

$$P(q_j) = 2N\sigma^2 + N^4 P_j + \sum_{\substack{i = 1 \\ i \neq j}}^{L} P_i \mathbf{a}^\dagger(q_j) \mathbf{s}_i \mathbf{s}_i^\dagger \mathbf{a}(q_j).$$ Eq. (15.96)

The first term of the right-hand side of Eq. (15.96) represents the noise power at q_j. The last term corresponds to spectral leakage, while the signal power at q_j is given by the middle term. Note that the sequence $\{c(m)\}$ is the reason for the N^4 factor. More precisely,

$$N^4 = \sum_{m = 0}^{2N-2} [c(m)]^2.$$ Eq. (15.97)

Thus, the SNR is

$$SNR|_{q_j} = \left(\frac{N^3}{2}\right)\left(\frac{P_j}{\sigma_\upsilon^2}\right).$$ Eq. (15.98)

Recall that in conventional phased array radars the SNR is improved by a factor of N, where N is the size of the array. Examination of Eq. (15.98) indicates that the SNR improvement factor using this MIMO mode of operation over conventional array operation and signal processing is

$$I_{SNR} = (20 \log N - 3) dB.$$ Eq. (15.99)

Problems

15.1. Consider an antenna whose diameter is $d = 3m$. What is the far field requirement for an X-band or an L-band radar that is using this antenna?

15.2. Consider an antenna with electric field intensity in the xy-plane $E(\varsigma)$. This electric field is generated by a current distribution $D(y)$ in the yz-plane. The electric field intensity is computed using the integral

$$E(\varsigma) = \int_{-r/2}^{r/2} D(y) \exp\left(2\pi j \frac{y}{\lambda} \sin \varsigma\right) dy$$

where λ is the wavelength and r is the aperture. (a) Write an expression for $E(\varsigma)$ when $D(Y) = d_0$ (a constant). (b) Write an expression for the normalized power radiation pattern and plot it in dB.

15.3. A linear phased array consists of 50 elements with $\lambda/2$ element spacing. (a) Compute the $3dB$ beamwidth when the main-beam steering angle is $0°$ and $45°$. (b) Compute the electronic phase difference for any two consecutive elements for steering angle $60°$.

15.4. A linear phased array antenna consists of eight elements spaced with $d = \lambda$ element spacing. (a) Give an expression for the antenna gain pattern (assume no steering and uniform aperture weighting). (b) Sketch the gain pattern versus the sine of the off-boresight angle β. What problems do you see is using $d = \lambda$ rather than $d = \lambda/2$?

15.5. In Section 15.4.2 we showed how a DFT can be used to compute the radiation pattern of a linear phased array. Consider a linear phased array of 64 elements at half wavelength spacing, where an FFT of size 512 is used to compute the pattern. What are the FFT bins that correspond to steering angles $\beta = 30°, 45°$?

15.6. Derive Eq. (15.73).

15.7. Consider the two-element array shown in the figure below. If the composite array electric field is $E(\theta) = a_1 E_1(\theta) + a_2 E_2(\theta)$, where a_1, a_2 are constants (can be complex) and E_1, E_2 are the individual elements fields. Determine a_1, a_2 so that the electric field is maximum at θ_o. Plot the resulting array pattern.

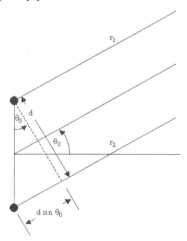

15.8. Use the FFT to compute the radiation pattern for an array of size 21 elements and element spacing (a) $d = 0.5\lambda$, and (b) $d = 0.8\lambda$. In each case, compute and plot the array pattern with and without using Hamming weights.

15.9. Modify the FFT routine developed in the previous problem to compute and plot the power gain pattern.

15.10. Repeat Problems 15.8 and 15.9, where in this case the array pattern can be steered in any off-boresight direction.

15.11. Why do the grating lobes appear when the array beam is steered to angles other than the boresight? Include reasonable plots to back up your argument.

Appendix 15-A: Chapter 15 MATLAB Code Listings

The MATLAB code provided in this chapter was designed as an academic standalone tool and is not adequate for other purposes. The code was written in a way to assist the reader in gaining a better understanding of the theory. The code was not developed, nor is it intended to be used as part of an open-loop or a closed-loop simulation of any kind. The MATLAB code found in this textbook can be downloaded from this book's web page on the CRC Press website. Simply use your favorite web browser, go to *www.crcpress.com*, and search for keyword *"Mahafza"* to locate this book's web page.

MATLAB Program "Fig15_5.m" Listing

```
% Use this code to produce Fig. 15.5a and 15.5b based on Eq.(15.35)
clc
clear all
close all
eps = 0.00001;
k = 2*pi;
theta = -pi : pi / 10791 : pi;
var = sin(theta);
nelements = 8;
d = 1;       % d = 1;
num = sin((nelements * k * d * 0.5) .* var);
%
if(abs(num) <= eps)
   num = eps;
end
den = sin((k* d * 0.5) .* var);
if(abs(den) <= eps)
   den = eps;
end
%
pattern = abs(num ./ den);
maxval = max(pattern);
pattern = pattern ./ maxval;
 %
figure(1)
plot(var,pattern,'linewidth', 1.5)
xlabel('\bfsine angle - dimensionless')
ylabel('\bfArray pattern')
grid
%
figure(2)
plot(var,20*log10(pattern),'linewidth', 1.5)
axis ([-1 1 -60 0])
xlabel('\bfsine angle - dimensionless')
ylabel('\bfPower pattern [dB]')
grid;
%
figure(3)
theta = theta +pi/2;
polar(theta,pattern)
title ('\bfArray pattern')
```

MATLAB Program "Fig15_7.m" Listing

```
% Use this program to reproduce Fig. 15.7 of text
clc
clear all
close all
eps =0.00001;
N = 32;
rect(1:32) = 1;
ham = hamming(32);
han = hanning(32);
blk = blackman(32);
k3 = kaiser(32,3);
k6 = kaiser(32,6);
RECT = 20*log10(abs(fftshift(fft(rect, 1024)))./32 +eps);
HAM = 20*log10(abs(fftshift(fft(ham, 1024)))./32 +eps);
HAN = 20*log10(abs(fftshift(fft(han, 1024)))./32+eps);
BLK = 20*log10(abs(fftshift(fft(blk, 1024)))./32+eps);
K6 = 20*log10(abs(fftshift(fft(k6, 1024)))./32+eps);
x = linspace(-1,1,1024);
figure
subplot(2,1,1)
plot(x,RECT,'k',x,HAM,'k--',x,HAN,'k-.','linewidth',1.5);
xlabel('x')
ylabel('Window')
grid
axis tight
legend('Rectangular','Hamming','Hanning')
subplot(2,1,2)
plot(x,RECT,'k',x,BLK,'k--',x,K6,'K-.','linewidth',1.5)
xlabel('x')
ylabel('Window')
legend('Rectangular','Blackman','Kasier at \beta = 6')
grid
axis tight
```

MATLAB Function "linear_array.m" Listing

```
function [theta,patternr,patterng] = linear_array(Nr,dolr,theta0,winid,win,nbits);
% This function computes and returns the gain radiation pattern for a linear array
% It uses the FFT to computes the pattern
%%%% *INPUTS ********** %%%%%%%%%%%%%%%%%%%%%%%
% Nr ==> number of elements; dolr ==> element spacing (d) in lambda units divided by lambda
% theta0 ==> steering angle in degrees; winid ==> use winid negative for no window, winid positive to
enter your window of size(Nr)
% win is input window, NOTE that win must be an NrX1 row vector; nbits ==> number of bits used in the
pahse shifters
% negative nbits mean no quantization is used
%%%% *OUTPUTS ********** %%%%%%%%%%%%%%%%%%%%%%%
% theta ==> real-space angle; patternr ==> array radiation pattern in dBs
% patterng ==> array directive gain pattern in dBs
%%%%%%%%% ********************** %%%%%%%%%%%%%%%%%%%%%%
eps = 0.00001;
n = 0:Nr-1;
i = sqrt(-1);
```

```
%if dolr is > 0.5 then; choose dol = 0.25 and compute new N
if(dolr <=0.5)
    dol = dolr;
    N = Nr;
else
    ratio = ceil(dolr/.25);
    N = Nr * ratio;
    dol = 0.25;
end
% choose proper size fft, for minimum value choose 256
Nrx = 10 * N;
nfft = 2^(ceil(log(Nrx)/log(2)));
if nfft < 256
    nfft = 256;
end
% convert steering angle into radians; and compute the sine of angle
theta0 = theta0 *pi /180.;
sintheta0 = sin(theta0);
% detrmine and comupte quantized steering angle
if nbits < 0
    phase0 = exp(i*2.0*pi .* n * dolr * sintheta0);
else
    % compute and add the phase shift terms (WITH nbits quantization)
    % Use formula thetal = (2*pi*n*dol) * sin(theta0) divided into 2^nbits
    % and rounded to the nearest qunatization level
    levels = 2^nbits;
    qlevels = 2.0 * pi / levels; % compute quantization levels
% compute the phase level and round it to the closest quantizatin level
    angleq = round(dolr .* n * sintheta0 * levels) .* qlevels; % vector of possible angles
    phase0 = exp(i*angleq);
end
% generate array of elements with or without window
if winid < 0
    wr(1:Nr) = 1;
else
    wr = win';
end
% add the phase shift terms
 wr = wr .* phase0;
 % determine if interpolation is needed (i.e N > Nr)
if N > Nr
    w(1:N) = 0;
    w(1:ratio:N) = wr(1:Nr);
else
    w = wr;
end
% compute the sine(theta) in real space sthat correspond to the FFT index
arg = [-nfft/2:(nfft/2)-1] ./ (nfft*dol);
idx = find(abs(arg) <= 1);
sinetheta = arg(idx);
theta = asin(sinetheta);
% convert angle into degrees
theta = theta .* (180.0 / pi);
% Compute fft of w (radiation pattern)
```

```
patternv = (abs(fftshift(fft(w,nfft)))).^2;
% convert raditiona pattern to dBs
patternr = 10*log10(patternv(idx) ./Nr  + eps);
% Compute directive gain pattern
rbarr  = 0.5 *sum(patternv(idx)) ./ (nfft * dol);
patterng = 10*log10(patternv(idx) + eps) - 10*log10(rbarr + eps);
return
```

MATLAB Program "circular_array.m" Listing

```
%Circular Array in the x-y plane
% Element is a short dipole antenna parallel to the z axis
% 2D Radiation Patterns for fixed phi or fixed theta
% dB polar plots uses the polardb.m file
%
%%%% Element expression needs to be modified if different
%%%% than a short dipole antenna along the z axis
%
clear all
clf
close all
%
% ====  Input Parameters  ====
a = 1.;        % radius of the circle
N = 10;          % number of Elements of the circular array
theta0 = 45;    % main beam Theta direction
phi0 = 60;      % main beam Phi direction
% Theta or Phi variations for the calculations of the far field pattern
Variations = 'Phi';  % Correct selections are  'Theta' or 'Phi'
phid = 60;       % constant phi plane for theta variations
thetad = 45;     % constant theta plane for phi variations
% ====  End of Input parameters section  ====
%
dtr = pi/180;          % conversion factors
rtd = 180/pi;
phi0r = phi0*dtr;
theta0r = theta0*dtr;
lambda = 1;
k = 2*pi/lambda;
ka = k*a;          % Wavenumber times the radius
jka = j*ka;
I(1:N) = 1;          % Elements excitation Amplitude and Phase
alpha(1:N) =0;
for n = 1:N          % Element positions Uniformly distributed along the circle
   phin(n) = 2*pi*n/N;
end
%
switch Variations
case 'Theta'
   phir = phid*dtr;    % Pattern in a constant Phi plane
   i = 0;
   for theta = 0.001:1:181
      i = i+1;
      thetar(i) = theta*dtr;
```

```
        angled(i) = theta;  angler(i) = thetar(i);
        Arrayfactor(i) = 0;
        for n = 1:N
            Arrayfactor(i) = Arrayfactor(i) + I(n)*exp(j*alpha(n)) ...
                    * exp( jka*(sin(thetar(i))*cos(phir -phin(n))) ...
                        -jka*(sin(theta0r  )*cos(phi0r-phin(n)))  );
        end
        Arrayfactor(i) = abs(Arrayfactor(i));
        Element(i) = abs(sin(thetar(i)+0*dtr));  % use the abs function to avoid
    end
case 'Phi'
   thetar = thetad*dtr;  % Pattern in a constant Theta plane
   i = 0;
   for phi = 0.001:1:361
      i = i+1;
      phir(i)   = phi*dtr;
      angled(i) = phi;  angler(i) = phir(i);
      Arrayfactor(i) = 0;
      for n = 1:N
          Arrayfactor(i) = Arrayfactor(i) +  I(n)*exp(j*alpha(n)) ...
                  * exp( jka*(sin(thetar )*cos(phir(i)-phin(n))) ...
                      -jka*(sin(theta0r)*cos(phi0r -phin(n)))  );
      end
      Arrayfactor(i) = abs(Arrayfactor(i));
      Element(i) = abs(sin(thetar+0*dtr));  % use the abs function to avoid
   end
end
angler = angled*dtr;
Element = Element/max(Element);
Array = Arrayfactor/max(Arrayfactor);
ArraydB = 20*log10(Array);
EtotalR =(Element.*Arrayfactor)/max(Element.*Arrayfactor);
%
figure(1)
plot(angled,Array,'linewidth',1.5)
ylabel('Array pattern')
grid
switch Variations
case 'Theta'
  axis ([0 180 0 1 ])
%  theta = theta +pi/2;
   xlabel('\theta - Degrees')
   title ( '\phi = 90^o plane')
case 'Phi'
axis ([0 360 0 1 ])
   xlabel('\phi - Degrees')
   title ( '\theta = 90^o plane')
end
%
figure(2)
plot(angled,ArraydB,'linewidth',1.5)
%axis ([-1 1 -60 0])
ylabel('Power pattern [dB]')
grid;
```

```
switch Variations
case 'Theta'
  axis ([0 180 -60 0 ])
   xlabel('Theta [Degrees]')
     title ( '\phi = 90^o plane')
case 'Phi'
axis ([0 360 -60 0 ])
   xlabel('\phi - degrees')
     title ( '\theta = 90^o plane')
end
%
figure(3)
polar(angler,Array)
title ('Array pattern')
%
figure(4)
polardb(angler,Array)
title ('Power pattern [dB]')
 %
% the plots provided above are for the array factor based on the circular
% array plots for other patterns such as those for the antenna element
% (Element)or the total pattern (Etotal based on Element*Arrayfactor) can
% also be displayed by the user as all these patterns are already computed
% above.
 %
figure(10)
subplot(2,2,1)
polardb (angler,Element,'b-'); % rectangular plot of element pattern
title('Element normalized E field [dB]')
subplot(2,2,2)
polardb(angler,Array,'b-')
title(' Array Factor normalized [dB]')
subplot(2,2,3)
polardb(angler,EtotalR,'b-');  % polar plot
title('Total normalized E field [dB]')
```

MATLAB Function "rect_array.m" Listing

```
function [pattern] = rect_array(Nxr,Nyr,dolxr,dolyr,theta0,phi0,winid,win,nbits);
%%%%% ************************* %%%%%%%%%%%%%%%%%%
% This function computes the 3-D directive gain patterns for a planar array
% This function uses the fft2 to compute its output
%%%%%%%%%% ************ INPUTS ************ %%%%%%%%%%%%%%
% Nxr ==> number of along x-aixs; Nyr ==> number of elemnts along y-axis
% dolxr ==> element spacing in x-direction; dolyr ==> element spacing in y-direction Both are in
lambda units
% theta0 ==> elevation steering angle in degrees, phi0 ==> azimuth steering angle in degrees
% winid ==> window identifier; winid negative ==> no window ; winid positive ==> use window given
by win
% win ==> input window function (2-D window) MUST be of size (Nxr X Nyr)
% nbits is the number of nbits used in phase quantization; nbits negative ==> NO quantization
%%%%%%%% ********** OUTPUTS ************ %%%%%%%%%%%%%
% pattern ==> directive gain pattern
%%%%%%%%%%%%%%%%%% ************************* %%%%%%%%%%%%
```

```
eps = 0.0001;
nx = 0:Nxr-1;
ny = 0:Nyr-1;
i = sqrt(-1);
% check that window size is the same as the array size
[nw,mw] = size(win);
if winid >0
  if nw ~= Nxr
  fprintf('STOP == Window size must be the same as the array')
  return
end
if mw ~= Nyr
  fprintf('STOP == Window size must be the same as the array')
  return
end
end
%if dol is > 0.5 then; choose dol = 0.5 and compute new N
if(dolxr <=0.5)
  ratiox = 1 ;
  dolx = dolxr ;
  Nx = Nxr ;
else
  ratiox = ceil(dolxr/.5) ;
  Nx = (Nxr -1 ) * ratiox + 1 ;
  dolx = 0.5 ;
end
if(dolyr <=0.5)
  ratioy = 1 ;
  doly = dolyr ;
  Ny = Nyr ;
else
  ratioy = ceil(dolyr/.5) ;
  Ny = (Nyr -1) * ratioy + 1 ;
  doly = 0.5 ;
end
% choose proper size fft, for minimum value choose 256X256
Nrx = 10 * Nx;
Nry = 10 * Ny;
nfftx = 2^(ceil(log(Nrx)/log(2)));
nffty = 2^(ceil(log(Nry)/log(2)));
if nfftx < 256
  nfftx = 256;
end
if nffty < 256
  nffty = 256;
end
% generate array of elements with or without window
if winid < 0
  array = ones(Nxr,Nyr);
else
  array = win;
end
% convert steering angles (theta0, phi0) to radians
theta0 = theta0 * pi / 180;
```

```
phi0 = phi0 * pi / 180;
% convert steering angles (theta0, phi0) to U-V sine-space
u0 = sin(theta0) * cos(phi0);
v0 = sin(theta0) * sin(phi0);
% Use formula theta1 = (2*pi*n*dol) * sin(theta0) divided into 2^m levels
% and rounded to the nearest qunatization level
if nbits < 0
   phasem = exp(i*2*pi*dolx*u0 .* nx *ratiox);
   phasen = exp(i*2*pi*doly*v0 .* ny *ratioy);
else
   levels = 2^nbits;
   qlevels = 2.0*pi / levels; % compute quantization levels
   sinthetaq = round(dolx .* nx * u0 * levels * ratiox) .* qlevels; % vector of possible angles
   sinphiq = round(doly .* ny * v0 * levels *ratioy) .* qlevels; % vector of possible angles
   phasem = exp(i*sinthetaq);
   phasen = exp(i*sinphiq);
end
 % add the phase shift terms
array = array .* (transpose(phasem) * phasen);
 % determine if interpolation is needed (i.e. N > Nr)
if (Nx > Nxr )| (Ny > Nyr)
  for xloop = 1 : Nxr
    temprow = array(xloop, :) ;
    w( (xloop-1)*ratiox+1, 1:ratioy:Ny) =  temprow ;
  end
  array = w;
else
  w = array ;
%   w(1:Nx, :) = array(1:N,:);
end
% Compute array pattern
arrayfft = abs(fftshift(fft2(w,nfftx,nffty))).^2 ;
%compute [su,sv] matrix
U = [-nfftx/2:(nfftx/2)-1] ./(dolx*nfftx);
indexx = find(abs(U) <= 1);
U = U(indexx);
V = [-nffty/2:(nffty/2)-1] ./(doly*nffty);
indexy = find(abs(V) <= 1);
V = V(indexy);
%Normalize to generate gain pattern
rbar=sum(sum(arrayfft(indexx,indexy))) / dolx/doly/4./nfftx/nffty;
arrayfft = arrayfft(indexx,indexy) ./rbar;
[SU,SV] = meshgrid(V,U);
indx = find((SU.^2 + SV.^2) >1);
arrayfft(indx) = eps/10;
pattern = 10*log10(arrayfft +eps);
figure(1)
mesh(V,U,pattern);
xlabel('V')
ylabel('U');
zlabel('Gain pattern - dB')
figure(2)
contour(V,U,pattern)
grid
```

```
axis image
xlabel('V')
ylabel('U');
axis([-1 1 -1 1])
figure(3)
x0 = (Nx+1)/2 ;
y0 = (Ny+1)/2 ;
radiusx = dolx*((Nx-1)/2) ;
radiusy = doly*((Ny-1)/2) ;
[xxx, yyy]=find(abs(array)>eps);
xxx = xxx-x0 ;
yyy = yyy-y0 ;
plot(yyy*doly, xxx*dolx,'rx')
hold on
axis([-radiusy-0.5 radiusy+0.5 -radiusx-0.5  radiusx+0.5]);
grid
title('antenna spacing pattern');
xlabel('y - \lambda units')
ylabel('x - \lambda units')
[xxx0, yyy0]=find(abs(array)<=eps);
xxx0 = xxx0-x0 ;
yyy0 = yyy0-y0 ;
plot(yyy0*doly, xxx0*dolx,'co')
axis([-radiusy-0.5 radiusy+0.5 -radiusx-0.5  radiusx+0.5]);
hold off
return
```

MATLAB Function "circ_array.m" Listing

```
function [pattern,amn] = circ_array(N,dolxr,dolyr,theta0,phi0,winid,win,nbits);
%%%%%% ********************** %%%%%%%%%%%%%%%%
% This function computes the 3-D directive gain patterns for a circular planar array
% This function uses the fft2 to compute its output. It assumes that there are the same number
% of elements along the major x- and y-axes
%%%%%%%%% *********** INPUTS *********** %%%%%%%%% N ==> number of ele-
ments along x-aixs or y-axis
% dolxr ==> element spacing in x-direction; dolyr ==> element spacing in y-direction. Both are in
lambda units
% theta0 ==> elevation steering angle in degrees, phi0 ==> azimuth steering angle in degrees
% This function uses the function (rec_to_circ) which computes the circular array from a square
% array (of size NXN) using the notation developed by ALLEN,J.L.,"The Theory of Array Antennas
% (with Emphasis on Radar Application)" MIT-LL Technical Report No. 323,July, 25 1965.
% winid ==> window identifier; winid negative ==> no window ; winid positive ==> use window given
by win
% win ==> input window function (2-D window) MUST be of size (Nxr X Nyr)
% nbits is the number of nbits used in phase quantization; nbits negative ==> NO quantization
%%%%%%%%%%% *********** OUTPUTS *********** %%%%%%%%%
% amn ==> array of ones and zeros; ones indicates true element location on the grid
% zeros mean no elements at that location; pattern ==> directive gain pattern
%%%%%%%%%%%%%%% *********************** %%%%%%%%%%%
eps = 0.0001;
nx = 0:N-1;
ny = 0:N-1;
i = sqrt(-1);
```

```
% check that window size is the same as the array size
[nw,mw] = size(win);
if winid >0
  if mw ~= N
    fprintf('STOP == Window size must be the same as the array')
    return
  end
  if nw ~= N
    fprintf('STOP == Window size must be the same as the array')
    return
  end
end
%if dol is > 0.5 then; choose dol = 0.5 and compute new N
if(dolxr <=0.5)
  ratiox = 1 ;
  dolx = dolxr ;
  Nx = N ;
else
  ratiox = ceil(dolxr/.5) ;
  Nx = (N-1) * ratiox + 1 ;
  dolx = 0.5 ;
end
if(dolyr <=0.5)
  ratioy = 1 ;
  doly = dolyr ;
  Ny = N ;
else
  ratioy = ceil(dolyr/.5);
  Ny = (N-1)*ratioy + 1 ;
  doly = 0.5 ;
end
% choose proper size fft, for minimum value choose 256X256
Nrx = 10 * Nx;
Nry = 10 * Ny;
nfftx = 2^(ceil(log(Nrx)/log(2)));
nffty = 2^(ceil(log(Nry)/log(2)));
if nfftx < 256
  nfftx = 256;
end
if nffty < 256
  nffty = 256;
end
% generate array of elements with or without window
if winid < 0
  array = ones(N,N);
else
  array = win;
end
% convert steering angles (theta0, phi0) to radians
theta0 = theta0 * pi / 180;
phi0 = phi0 * pi / 180;
% convert steering angles (theta0, phi0) to U-V sine-space
u0 = sin(theta0) * cos(phi0);
v0 = sin(theta0) * sin(phi0);
```

```
% Use formula theta1 = (2*pi*n*dol) * sin(theta0) divided into 2^m levels
% and rounded to the nearest quantization level
if nbits < 0
    phasem = exp(i*2*pi*dolx*u0 .* nx * ratiox);
    phasen = exp(i*2*pi*doly*v0 .* ny * ratioy);
else
    levels = 2^nbits;
    qlevels = 2.0*pi / levels; % compute quantization levels
    sinthetaq = round(dolx .* nx * u0 * levels * ratiox) .* qlevels; % vector of possible angles
    sinphiq = round(doly .* ny * v0 * levels *ratioy) .* qlevels; % vector of possible angles
    phasem = exp(i*sinthetaq);
    phasen = exp(i*sinphiq);
end
% add the phase shift terms
array = array .* (transpose(phasem) * phasen) ;
% determine if interpolation is needed (i.e N > Nr)
if (Nx > N )| (Ny > N)
    for xloop = 1 : N
        temprow = array(xloop, :) ;
        w( (xloop-1)*ratiox+1, 1:ratioy:Ny) =  temprow ;
    end
    array = w;
else
    w(1:Nx, :) = array(1:N,:);
end
% Convert rectangular array into circular using function rec_to_circ
[m,n] = size(w) ;
NC = max(m,n);  % Use Allens algorithm
if Nx == Ny
    temp_array = w;
else
    midpoint = (NC-1)/2 +1 ;
    midwm = (m-1)/2 ;
    midwn = (n-1)/2 ;
    temp_array = zeros(NC,NC);
    temp_array(midpoint-midwm:midpoint+midwm, midpoint-midwn:midpoint+midwn) = w ;
end
amn = rec_to_circ(NC);  % must be rectangular array (Nx=Ny)
amn = temp_array .* amn ;
% Compute array pattern
arrayfft = abs(fftshift(fft2(amn,nfftx,nffty))).^2 ;
%compute [su,sv] matrix
U = [-nfftx/2:(nfftx/2)-1] ./(dolx*nfftx);
indexx = find(abs(U) <= 1);
U = U(indexx);
V = [-nffty/2:(nffty/2)-1] ./(doly*nffty);
indexy = find(abs(V) <= 1);
V = V(indexy);
[SU,SV] = meshgrid(V,U);
indx = find((SU.^2 + SV.^2) >1);
arrayfft(indx) = eps/10;
%Normalize to generate gain pattern
rbar=sum(sum(arrayfft(indexx,indexy))) / dolx/doly/4./nfftx/nffty;
```

```
arrayfft = arrayfft(indexx,indexy) ./rbar;
[SU,SV] = meshgrid(V,U);
indx = find((SU.^2 + SV.^2) >1);
arrayfft(indx) = eps/10;
pattern = 10*log10(arrayfft +eps);
figure(1)
mesh(V,U,pattern);
xlabel('V')
ylabel('U');
zlabel('Gain pattern - dB')
figure(2)
contour(V,U,pattern)
axis image
grid
xlabel('V')
ylabel('U');
axis([-1 1 -1 1])
figure(3)
x0 = (NC+1)/2 ;
y0 = (NC+1)/2 ;
radiusx = dolx*((NC-1)/2 + 0.05/dolx) ;
radiusy = doly*((NC-1)/2 + 0.05/dolx) ;
theta = 5  ;
[xxx, yyy]=find(abs(amn)>0);
xxx = xxx-x0 ;
yyy = yyy-y0 ;
plot(yyy*doly, xxx*dolx,'rx')
axis equal
hold on
axis([-radiusy-0.5 radiusy+0.5 -radiusx-0.5  radiusx+0.5]);
grid
title('antenna spacing pattern');
xlabel('y - \lambda units')
ylabel('x - \lambda units')
[x, y]= makeellip( 0, 0, radiusx, radiusy, theta) ;
plot(y, x) ;
axis([-radiusy-0.5 radiusy+0.5 -radiusx-0.5  radiusx+0.5]);
[xxx0, yyy0]=find(abs(amn)<=0);
xxx0 = xxx0-x0 ;
yyy0 = yyy0-y0 ;
plot(yyy0*doly, xxx0*dolx,'co')
axis([-radiusy-0.5 radiusy+0.5 -radiusx-0.5  radiusx+0.5]);
axis equal
hold off ;
return
```

MATLAB Function "rect_to_circ.m" Listing

```
function amn = rec_to_circ(N)
midpoint = (N-1)/2 + 1;
amn = zeros(N);
array1(midpoint,midpoint) = N;
x0 = midpoint;
y0 = x0;
```

```
for i = 1:N
  for j = 1:N
      distance(i,j) = sqrt((x0-i)^2 + (y0-j)^2);
  end
end
idx = find(distance < (N-1)/2 + .025);
amn (idx) = 1;
return
```

MATLAB Program "Fig15_51.m" Listing

```
%Use this program to reproduce Fig. 15.51 of text
clear all
close all
d = 0.6; % element spacing in lambda units
betadeg = linspace(0,22.5,1000);
beta = betadeg .*pi ./180;
den = pi*d .* sin(beta);
numarg = den;
num = sin(numarg);
lscan = (num./den).^-4;
LSCAN = 10*log10(lscan+eps);
figure (1)
plot(betadeg,LSCAN,'linewidth',1.5)
xlabel('\bfscan angle in degrees')
ylabel('\bfScan loss in dB')
grid
title('Element spacing is d = 0.6 \lambda ')
```

Chapter 16

Adaptive Signal
Processing

The emphasis in this chapter is on adaptive signal processing to include adaptive array processing and Space Time Adaptive processing (STAP). Adaptive arrays employ phased array antennas to adaptively sense and eliminate unwanted signals entering the radar's Field of View (FOV) while enhancing reception about the desired target returns. For this purpose, adaptive arrays utilize a rather complicated combination of hardware and require demanding levels of software implementation. Through feedback networks, a proper set of complex weights is computed and applied to each channel of the array.

STAP processing refers to the ability to simultaneously process spatial sensor and temporal (time dependent) input data. For this purpose, phased arrays (spatial component) along with time delay units (temporal component) are used to optimally detect targets in the presence of high clutter or interference environment.

16.1. Nonadaptive Beamforming

In adaptive beamforming the beam of interest is formed (generated) by continuously changing a set of weights through feedback circuits to minimize an output error signal. Nonadaptive or conventional beamformers do the same thing in the sense that the beam of interest is generated using a set of unique weights. Except in this case, these weights are determined a priori so that interference from a specific angle of arrival is minimized or eliminated. Different sets of weights will produce nulls in different directions in the array's field of view.

Consider a linear array of N equally spaced elements, and a plane wave $\exp(j2\pi f_0 t))$ incident on the aperture with direction-sine $\sin\theta$, as shown in Fig. 16.1. The weights w_i, $i = 0, 1, ...N-1$ are, in general, complex constants. The output of the beamformer is

$$y(t) = \sum_{n=0}^{N-1} w_n x_n(t - \tau_n)$$

Eq. (16.1)

$$\tau_n = n\frac{d}{c}\sin\theta; \quad n = 0, 1, ..., (N-1)$$

Eq. (16.2)

where d is the element spacing and c is the speed of light. Fourier transformation of Eq. (16.1) yields

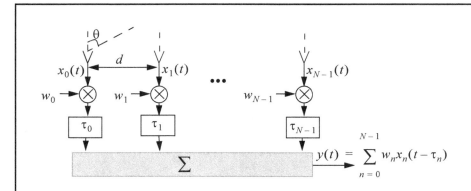

Figure 16.1. A linear array of size N, element spacing d, and an incident plane wave defined by $\sin\theta$.

$$Y(\omega) = \sum_{n=0}^{N-1} w_n X_n(\omega) exp(-j\omega\tau_n) = \sum_{n=0}^{N-1} w_n X_n(\omega) e^{-jn\Delta\theta}.$$ **Eq. (16.3)**

The phase term $\Delta\theta$ is defined as

$$\Delta\theta = 2\pi f_0 \frac{d}{c}\sin\theta = \frac{2\pi}{\lambda}d\sin\theta,$$ **Eq. (16.4)**

$\omega = 2\pi f_0$ and $f_0/c = 1/\lambda$. Eq. (16.3) can be written in vector form as

$$\mathbf{Y} = \mathbf{s}^\dagger\mathbf{x}$$ **Eq. (16.5)**

$$\mathbf{s}^\dagger = \begin{bmatrix} 1 & e^{j\Delta\theta} & ... & e^{j(N-1)\Delta\theta} \end{bmatrix}$$ **Eq. (16.6)**

$$\mathbf{x}^\dagger = \begin{bmatrix} w_0 X_o & w_1 X_1 & ... & ...w_{N-1}X_{N-1} \end{bmatrix}^*$$ **Eq. (16.7)**

where the superscripts $*$ and $\dagger$, respectively, indicate complex conjugate and complex conjugate transpose.

Let A_1 be the amplitude of the wavefront defined by $\sin\theta_1$; it follows that the vector $\mathbf{x}$ is given by

$$\mathbf{x} = A_1\mathbf{s}_1^*$$ **Eq. (16.8)**

where $\mathbf{s}_1$ is a steering vector and can be written as,

$$\mathbf{s}^\dagger_1 = \begin{bmatrix} w_0 & w_1 e^{-j\Delta\theta_1} & ...w_{N-1}e^{-j(N-1)\Delta\theta_1} \end{bmatrix} \quad ; \quad \Delta\theta_1 = \frac{2\pi d}{\lambda}\cdot\sin\theta_1.$$ **Eq. (16.9)**

Using this notation, Eq. (16.5) can be expressed in the form

$$\mathbf{Y} = \mathbf{s}^\dagger \mathbf{x} = A_1 \mathbf{s}^\dagger \mathbf{s}^*_1 . \qquad \text{Eq. (16.10)}$$

The array pattern of the beam steered at θ_1 is computed as the expected value of $\mathbf{Y}$. In other words, the power spectrum density for the beamformer output is given by

$$S(k) = E[\mathbf{Y}\mathbf{Y}^\dagger] = P_1 \mathbf{s}^\dagger \Re \mathbf{s} \qquad \text{Eq. (16.11)}$$

where $P_1 = E[|A_1|^2]$ and $\Re$ is the correlation matrix given by

$$\Re = E\{\mathbf{s}_1 \mathbf{s}^\dagger_1\} . \qquad \text{Eq. (16.12)}$$

Consider L incident plane waves with directions of arrival defined by

$$\Delta\theta_i = \frac{2\pi d}{\lambda}\sin\theta_i; \quad i = 1, L . \qquad \text{Eq. (16.13)}$$

The *nth* sample at the output of the *mth* sensor is

$$y_m(n) = \upsilon(n) + \sum_{i=1}^{L} A_i(n)exp(-jm\Delta\theta_i); \quad m = 0, N-1 \qquad \text{Eq. (16.14)}$$

where $A_i(n)$ is the amplitude of the *ith* plane wave and $\upsilon(n)$ is white, zero-mean noise with variance σ_υ^2, and it is assumed to be uncorrelated with the signals. Equation (16.44) can be written in vector notation as

$$\mathbf{y}(n) = \upsilon(n) + \sum_{i=1}^{L} A_i(n)\mathbf{s}_i^* . \qquad \text{Eq. (16.15)}$$

A set of L steering vectors is needed to simultaneously form L beams. Define the steering matrix $\aleph$ as

$$\aleph = \begin{bmatrix} \mathbf{s}_1 & \mathbf{s}_2 & \dots & \mathbf{s}_L \end{bmatrix} . \qquad \text{Eq. (16.16)}$$

Then the autocorrelation matrix of the field measured by the array is

$$\Re = E\{\mathbf{y}_m(n)\mathbf{y}_m^\dagger(n)\} = \sigma_\upsilon^2\mathbf{I} + \aleph\mathbf{C}\aleph^\dagger \qquad \text{Eq. (16.17)}$$

where $\mathbf{C} = dig\begin{bmatrix} P_1 & P_2 & \dots & P_L \end{bmatrix}$, and $\mathbf{I}$ is the identity matrix.

For example, consider the case depicted in Fig. 16.2, where an interfering signal is located at angle $\theta_i = \pi/6$ off the antenna boresight. The desired signal is at $\theta_t = 0°$. The desired output should contain only the signal $s(t)$. From Eq. (16.3) and Eq. (16.4), the desired output is

$$y_d(t) = \sum_{n=0}^{1} w_n x_n(t - \tau_{n_t}) = w_0 x_0 + w_1 x_1 e^{-j\frac{2\pi}{\lambda}d\sin\theta_t} . \qquad \text{Eq. (16.18)}$$

Since the angle $\theta_t = 0°$, it follows that

$$y_d(t) = \{Ae^{j2\pi f_0 t}\}\{(w_{0R} + jw_{0I}) + (w_{1R} + jw_{1I})\} \qquad \text{Eq. (16.19)}$$

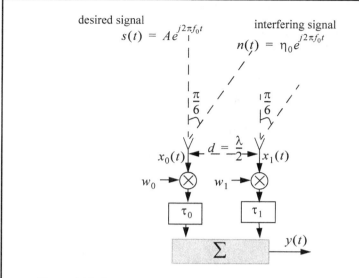

Figure 16.2. Two-element array with an interfering signal at $\theta_i = \pi/6$.

$$w_0 = w_{0R} + jw_{0I}$$
$$w_1 = w_{1R} + jw_{1I} .$$

Eq. (16.20)

Thus, in order to produce the desired signal, $s(t)$, at the output of the beamformer, it is required that

$$w_{0R} + w_{1R} = 1 \Rightarrow w_{0R} = 1 - w_{1R}$$
$$w_{0I} + w_{1I} = 0 \Rightarrow w_{0I} = -w_{1I} .$$

Eq. (16.21)

Next, the output due to the interfering signal is

$$y_i(t) = \sum_{n=0}^{1} w_n x_n(t - \tau_{n_i}) = w_0 x_0 + w_1 x_1 e^{-j\frac{2\pi}{\lambda}d\sin\theta_i}$$

Eq. (16.22)

Since the angle $\theta_i = \pi/6$, it follows that

$$y_i(t) = \{\eta_0 e^{j2\pi f_0 t}\}\{(w_{0R} + jw_{0I}) - j(w_{1R} + jw_{1I})\} ,$$

Eq. (16.23)

and in order to eliminate the interference signal from the output of the beamformer, it is required that

$$w_{0R} + w_{1I} = 0 \Rightarrow w_{0R} = -w_{1I}$$
$$w_{0I} - w_{1R} = 0 \Rightarrow w_{0I} = w_{1R} .$$

Eq. (16.24)

Solving Eq. (16.21) and Eq. (16.24) yields

$$w_{0R} = \frac{1}{2}; w_{0I} = \frac{1}{2}; w_{1R} = \frac{1}{2}; w_{1I} = \frac{-1}{2} .$$

Eq. (16.25)

Using the weights given in Eq. (16.25) will allow the desired signal to get through the beam-former unaffected; however, the interference signal will be completely eliminated from the output.

16.2. Adaptive Signal Processing Using Least Mean Square (LMS)

Adaptive signal processing evolved as a natural evolution from adaptive control techniques of time-varying systems. Advances in digital processing computation techniques and associated hardware have facilitated maturing adaptive processing techniques and algorithms.

Consider the basic adaptive digital system shown in Fig. 16.3. The system input is the sequence $x[k]$ and its output is the sequence $y[k]$. What differentiates adaptive from non-adaptive systems is that in adaptive systems the transfer function $H_k(z)$ is now time varying. The arrow through the transfer function box is used to indicate adaptive processing (or time varying transfer function). The sequence $d[k]$ is referred to as the *desired* response sequence. The error sequence is the difference between the desired response and the actual response. Remember that the desired sequence is not completely known; otherwise, if it were completely known, one would not need any adaptive processing to compute it. The definition of this desired response is dependent on the system-specific requirements.

Many different techniques and algorithms have been developed to minimize the error sequence. Using one technique over another depends heavily on the operating environment under consideration. For example, if the input sequence is a stationary random process, then minimizing the error signal is nothing more than solving the least mean squares problem. However, in most adaptive processing systems, the input signal is a non-stationary process. In this section, the least mean squares technique is examined.

The least mean squares (LMS) algorithm is the most commonly utilized algorithm in adaptive processing, primary because of its simplicity. The time-varying transfer function of order L can be written as a Finite Impulse Response (FIR) filter defined by

$$H_k(z) = b_0 + b_1 z^{-1} + \dots + b_L z^{-L}.$$

Eq. (16.26)

The input output relationship is given by the discrete convolution

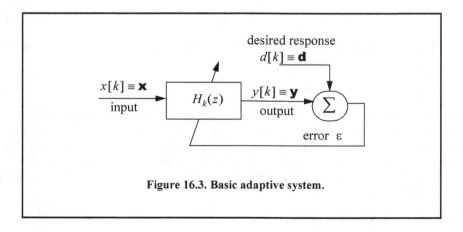

Figure 16.3. Basic adaptive system.

$$y(k) = \sum_{n=0}^{L} b_n(k)x(k-n).$$ **Eq. (16.27)**

The goal of the adaptive LMS process is to adjust the filter coefficients toward an optimum minimum mean square error (MMSE). The most common approach to achieving this MMSE utilizes the method of steepest descent. For this purpose, define the filter coefficients in vector notation as

$$\mathbf{b}_k = \begin{bmatrix} b_o(k) & b_1(k) \ldots & b_L(k) \end{bmatrix}^\dagger,$$ **Eq. (16.28)**

then

$$\mathbf{b}_{k+1} = \mathbf{b}_k - \mu \nabla_k$$ **Eq. (16.29)**

where μ is a parameter that controls how fast the error converges to the desired MMSE value, and the gradient vector ∇_k is defined by

$$\nabla_k = \frac{\partial}{\partial \mathbf{b}_k} E[\varepsilon_k^2] = \begin{bmatrix} \frac{\partial}{\partial \mathbf{b}_0(k)} E[\varepsilon_k^2] & \cdots & \frac{\partial}{\partial \mathbf{b}_L(k)} E[\varepsilon_k^2] \end{bmatrix}^\dagger.$$ **Eq. (16.30)**

As clearly indicated by Eq. (16.29), the adaptive filter coefficients update rate is proportional to the negative gradient; thus, if the gradient is known at each step of the adaptive process, then better computation of the coefficient is obtained. In other words, the MMSE decreases from step k to step $k+1$. Of course, once the solution is found, the gradient becomes zero and the coefficient will not change any more.

When the gradient is not known, estimates of the gradient are used based only on the instantaneous squared error. These estimates are defined by

$$\hat{\nabla}_k = \frac{\partial}{\partial \mathbf{b}_k}[\varepsilon_k^2] = 2\varepsilon_k \frac{\partial}{\partial \mathbf{b}_k}(d_k - y_k).$$ **Eq. (16.31)**

Since the desired sequence $d[k]$ is independent from the output $y[k]$, Eq. (16.31) can be written as

$$\hat{\nabla}_k = -2\varepsilon_k \mathbf{x}_k$$ **Eq. (16.32)**

where the vector $\mathbf{x}_k$ is the input signal sequence. Substituting Eq. (16.32) into Eq. (16.29) yields

$$\mathbf{b}_{k+1} = \mathbf{b}_k + 2\varepsilon_k \mu \mathbf{x}_k.$$ **Eq. (16.33)**

The choice of the convergence parameter μ plays a significant role in determining the system performance. This is clear because as indicated by Eq. (16.33), a successful implementation of the LMS algorithm depends on the input signal, the choice of the desired signal, and the convergence parameter. Much research and effort has been devoted to selecting the optimal value for μ. Nonetheless, no universal value has been found. However, a range for this parameter has been determined to be $0 < \mu < 1$.

Often, a normalized value for the convergence parameter μ_N can be used instead of its absolute value. That is,

$$\mu_N = \frac{\mu}{(L+1)\sigma^2}$$

Eq. (16.34)

where L is the order of the adaptive FIR filter and σ^2 is the variance (power) of the input signal. When the input signal is not stationary and its variance is varying with time, a time-varying estimate of σ^2 is used. That is

$$\hat{\sigma}_k^2 = \alpha x_k^2 + (1-\alpha)\hat{\sigma}_{k-1}^2$$

Eq. (16.35)

where α is a factor selected such that $0 < \alpha < 1$. Finally, Eq. (16.33) can be written as

$$\mathbf{b}_{k+1} = \mathbf{b}_k + \frac{2\varepsilon_k \mu \mathbf{x}_k}{(L+1)\hat{\sigma}_k^2}$$

Eq. (16.36)

MATLAB Function "LMS.m"

The MATLAB function "LMS.m" implements Eq. (16.36). Its syntax is as follows:

```
Xout = LMS(Xin, D, B, mu, sigma, alpha)
```

where

Symbol	Description	Status
X	input data sequence - corrupted	input
D	desired signal sequence	input
B	adaptive coefficient	input
mu	convergence parameters	input
sigma	input signal power estimate	input
alpha	forgetting factor, see Eq. (16.35)	input
Xout	predicted out sequence	output

As an example and in reference to Fig. 16.3, let the input and desired signals be defined as

$$x[k] = \sqrt{2}\ \sin\left(\frac{2\pi k}{20}\right) + n[k] \qquad ;\ k = 0, 1, ..., 500$$

Eq. (16.37)

$$d[k] = \sqrt{2}\ \sin\left(\frac{2\pi k}{20}\right) \qquad ;\ k = 0, 1, ..., 500$$

Eq. (16.38)

where $n[k]$ is additive white noise with zero mean and variance $\sigma_n^2 = 2$. Figure 16.4 shows the output of the LMS algorithm defined in Eq. (16.36) when $\mu = 0.1$ and $\alpha = 0$. Figure 16.5 is similar to Fig. 16.4 except in this case, $\mu = 0.01$ and $\alpha = 0.1$.

Note that in Fig. 16.5, the rate of convergence is reduced since μ is smaller than that used in Fig. 16.4; however, the filter's output is less noise because α is greater than zero, which allows for more accurate updates of the noise variance as defined in Eq. (16.35). These plots can be reproduced using the MATLAB program *"Fig16_4_5.m,"* listed in Appendix 16-A.

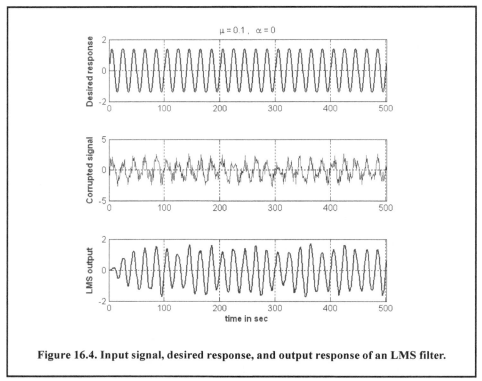

Figure 16.4. Input signal, desired response, and output response of an LMS filter.

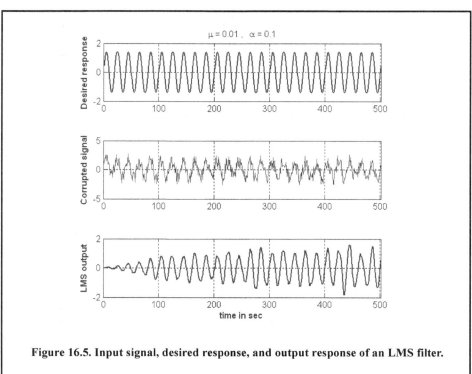

Figure 16.5. Input signal, desired response, and output response of an LMS filter.

16.3. The LMS Adaptive Array Processing

Consider the LMS adaptive array shown in Fig. 16.6. The difference between the reference signal and the array output constitutes an error signal. The error signal is then used to adaptively calculate the complex weights, using a predetermined convergence algorithm. The reference signal is assumed to be an accurate approximation of the desired signal (or desired array response). This reference signal can be computed using a training sequence or spreading code, which is supposed to be known at the radar receiver. The format of this reference signal will vary from one application to another. But in all cases, the reference signal is assumed to be correlated with the desired signal. An increased amount of this correlation significantly enhances the accuracy and speed of the convergence algorithm being used. In this section, the LMS algorithm is assumed.

In general, the complex envelope of a bandpass signal and its corresponding analytical (pre-envelope) signal can be written using the quadrature components pair $(x_I(t), x_Q(t))$. Recall that the quadrature components are related using the Hilbert transform as follows:

$$x_Q(t) = \hat{x}_I(t) \; ; \text{ and } \hat{x}_Q = -x_I \qquad \text{Eq. (16.39)}$$

where $\hat{x}_I$ and $\hat{x}_Q$ are, respectively, the Hilbert transforms of x_I and x_Q. A bandpass signal $x(t)$ can be expressed as follows (visit Chapter 3 for a refresher):

$$x(t) = x_I(t)\cos 2\pi f_0 t - x_Q(t)\sin 2\pi f_0 t \qquad \text{Eq. (16.40)}$$

$$\psi_x(t) = x(t) + j\hat{x}(t) \equiv \tilde{x}(t)e^{j2\pi f_0 t} \qquad \text{Eq. (16.41)}$$

$$\tilde{x}(t) = x_I(t) + jx_Q(t) \qquad \text{Eq. (16.42)}$$

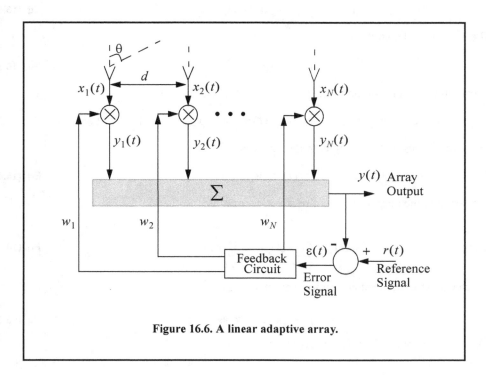

Figure 16.6. A linear adaptive array.

where $\psi(t)$ is the pre-envelope and $\tilde{x}(t)$ is the complex envelope. Equation (16.42) can be written using Eq. (16.39) as

$$\tilde{x}(t) = x_I(t) + jx_Q(t) = x_I(t) + j\hat{x}_I(t).$$ Eq. (16.43)

Using this notation, the adaptive array output signal, its reference signal, and the error signal can also be written using the same notation as

$$y(t) = y(t) + j\hat{y}(t)$$ Eq. (16.44)

$$\tilde{r}(t) = r(t) + j\hat{r}(t)$$ Eq. (16.45)

$$\tilde{\varepsilon}(t) = \varepsilon(t) + j\hat{\varepsilon}(t).$$ Eq. (16.46)

Referencing Fig. 16.6, denote the output of the nth array input signal as $y_n(t)$ and assume complex weights given by

$$w_n = w_{nI} - jw_{nQ}.$$ Eq. (16.47)

It follows that

$$y_n(t) = w_{nI}\, x_{nI}(t) + w_{nQ}\, x_{nQ}(t).$$ Eq. (16.48)

Taking the Hilbert transform of Eq. (16.48) yields

$$\hat{y}_n(t) = w_{nI}\, \hat{x}_{nI}(t) + w_{nQ}\, \hat{x}_{nQ}(t).$$ Eq. (16.49)

By using Eq. (16.39) into Eq. (16.49), one gets

$$\hat{y}_n(t) = w_{nI}\, x_{nQ}(t) - w_{nQ}\, x_{nI}(t).$$ Eq. (16.50)

The nth channel analytic signal is

$$\psi_{y_n}(t) = y_n(t) + j\hat{y}_n(t).$$ Eq. (16.51)

Substituting Eq. (16.48) and Eq. (16.49) into Eq. (16.50) gives

$$\psi_{y_n}(t) = w_{nI}\, x_{nI}(t) + w_{nQ}\, x_{nQ}(t) + j[w_{nI}\, x_{nQ}(t) - w_{nQ}\, x_{nI}(t)].$$ Eq. (16.52)

Collecting terms yields, using complex notation,

$$\psi_{y_n}(t) = w_n\tilde{x}_n(t).$$ Eq. (16.53)

Therefore, the output of the entire adaptive array is

$$\tilde{y}(t) = \sum_{n=1}^{N} \tilde{y}_n(t) = \sum_{n=1}^{N} w_n\tilde{x}_n(t),$$ Eq. (16.54)

which can be written using vector notation as

$$\tilde{\mathbf{y}} = \mathbf{w}^t\tilde{\mathbf{x}} = \tilde{\mathbf{x}}^t\mathbf{w}$$ Eq. (16.55)

where the vectors **x** and **w** are given by

$$\tilde{\mathbf{x}} = \begin{bmatrix} \tilde{x}_1(t) & \tilde{x}_2(t) & \dots \tilde{x}_N(t) \end{bmatrix}^t \qquad \text{Eq. (16.56)}$$

$$\mathbf{w} = \begin{bmatrix} w_1 & w_2 & \dots w_N \end{bmatrix}^t. \qquad \text{Eq. (16.57)}$$

The superscript $\{ ^t \}$ indicates the transpose operation.

As discussed earlier, one common technique to achieving the MMSE of an LMS algorithm is to use the steepest descent. Thus, the complex weights in the LMS adaptive array are related as defined in Eq. (16.29). That is,

$$\mathbf{w}_{k+1} = \mathbf{w}_k - \mu \nabla_k \qquad \text{Eq. (16.58)}$$

where again, μ is the convergence parameter. The subscript k indicates time samples. In this case, the gradient vector ∇_k is defined by

$$\nabla_k = \frac{\partial}{\partial \mathbf{w}_k} E[\tilde{\varepsilon}_k^2] = \begin{bmatrix} \frac{\partial}{\partial \mathbf{w}_0(k)} E[\tilde{\varepsilon}_k^2] & \dots & \frac{\partial}{\partial \mathbf{w}_N(k)} E[\tilde{\varepsilon}_k^2] \end{bmatrix}^t. \qquad \text{Eq. (16.59)}$$

Rearranging Eq. (16.58) so that the rate of change between consecutive estimates of the complex weights is on one side of the equation yields

$$\mathbf{w}_{k+1} - \mathbf{w}_k = -\mu \frac{\partial}{\partial \mathbf{w}_k} (E[\tilde{\varepsilon}_k^2]) \qquad \text{Eq. (16.60)}$$

where the middle portion of Eq. (16.59) was also substituted for the gradient vector. In this format, the left-hand side of Eq. (16.60) represents the rate of change of the complex weights with respect to time (i.e., the derivative of the weights with respect to time). It follows that

$$\frac{d}{dt} \mathbf{w} = -\mu \frac{\partial}{\partial \mathbf{w}} (E[\tilde{\varepsilon}^2(t)]). \qquad \text{Eq. (16.61)}$$

However, see from Fig. 16.5, that the error signal complex envelope is

$$\tilde{\varepsilon}(t) = \tilde{r}(t) - \sum_{n=1}^{N} w_n \tilde{x}_n(t) \Rightarrow \tilde{\varepsilon}(t) = \tilde{r}(t) - \tilde{\mathbf{x}}^t \mathbf{w}. \qquad \text{Eq. (16.62)}$$

It can be shown (see Problem 16.1) that

$$\frac{\partial}{\partial \mathbf{w}} E[\tilde{\varepsilon}^2(t)] = -E[\tilde{\mathbf{x}}^* \tilde{\varepsilon}(t)]. \qquad \text{Eq. (16.63)}$$

Therefore, Eq. (16.61) can be written as

$$\frac{d}{dt} \mathbf{w} = \mu E[\tilde{\mathbf{x}}^* \tilde{\varepsilon}(t)]. \qquad \text{Eq. (16.64)}$$

Substituting Eq. (16.62) into Eq. (16.64) gives

$$\frac{d}{dt} \mathbf{w} = \mu E[\tilde{\mathbf{x}}^* (\tilde{r}(t) - \tilde{\mathbf{x}}^t \mathbf{w})]. \qquad \text{Eq. (16.65)}$$

Equivalently,

$$\frac{d}{dt}\mathbf{w} + \mu E[\tilde{\mathbf{x}}^*\tilde{\mathbf{x}}^t]\mathbf{w} = \mu E[\tilde{\mathbf{x}}^*\tilde{r}(t)].$$ Eq. (16.66)

The covariance matrix is by definition

$$\mathbf{C} = E[\tilde{\mathbf{x}}^*\tilde{\mathbf{x}}^t] = \begin{bmatrix} \tilde{x}_1{}^*\tilde{x}_1 & \tilde{x}_1{}^*\tilde{x}_2 & \cdots \\ \tilde{x}_2{}^*\tilde{x}_1 & \tilde{x}_2{}^*\tilde{x}_2 & \cdots \\ \cdots & \cdots & \end{bmatrix},$$ Eq. (16.67)

and the reference signal correlation vector **s** is

$$\mathbf{s} = E[\tilde{\mathbf{x}}^*\tilde{r}(t)] = E\begin{bmatrix} \tilde{x}_1{}^*\tilde{r} & \tilde{x}_2{}^*\tilde{r} & \cdots \end{bmatrix}^t.$$ Eq. (16.68)

Using Eq. (16.68) and Eq. (16.67), one can rewrite the differential equation (DE) given Eq. (16.66) as

$$\frac{d}{dt}\mathbf{w} + \mu\mathbf{C}\mathbf{w} = \mu\mathbf{s}.$$ Eq. (16.69)

The steady state solution for the DE defined in Eq. (16.69) (provided that the covariance matrix is not singular) is

$$\mathbf{w} = \mathbf{C}^{-1}\mathbf{s}.$$ Eq. (16.70)

As the size of the covariance matrix increases (i.e., number of channels in the adaptive array), so does the complexity associated with computing the adaptive weights in real time. This is true because computing the inverse of large matrices in real time can be extremely challenging and demands a significant amount of computing power. Consequently, the effectiveness of adaptive arrays has been limited to small-sized arrays, where only a few interfering signals can be eliminated (cancelled). Additionally, computing of a good estimate of the covariance matrix in real time is also difficult in practical applications. In order to mitigate that effect, a reasonable estimate for $E\{x_i x_j{}^*\}$ (the i,j element of the covariance matrix) is derived by averaging m independent samples of data from the same distribution. This approach can be extended to the entire covariance matrix by collecting M independent "snapshots" of data from N channels. Thus, the estimate of the covariance matrix can be given as,

$$\tilde{\mathbf{C}} \approx (\tilde{\mathbf{x}}^\dagger\tilde{\mathbf{x}})/M.$$ Eq. (16.71)

The transient solution of Eq. (16.69) (see Problem 16.2) is

$$\mathbf{w}(t) = \sum_{n=1}^{N} \mathbf{p}_i e^{-\mu\lambda_i t}$$ Eq. (16.72)

where the vectors $\mathbf{p}_i$ are constants that depend on the initial value of $\mathbf{w}(t)$, and λ_i are the eigenvalues of the matrix $\mathbf{C}$. It follows that the complete solution of Eq. (16.69) is

$$\mathbf{w}(t) = \sum_{n=1}^{N} \mathbf{p}_i e^{-\mu\lambda_i t} + \mathbf{C}^{-1}\mathbf{s}.$$ Eq. (16.73)

A very common measure of effectiveness of an adaptive array is the ratio of the total output interference power, S_o, to the internal noise power, S_n.

Example:

Consider the two-element array in Section 16.2. Assume the desired signal is at directional-sine $\sin(\theta_r)$ and the interference signal is at $\sin(\theta_i)$. Calculate the adaptive weights so that the interference signal is cancelled.

Solution:

From Fig. 16.6

$$\tilde{x}_1(t) = \tilde{d}_1(t) + \tilde{n}_1(t) + \tilde{I}_1(t)$$

$$\tilde{x}_2(t) = \tilde{d}_2(t) + \tilde{n}_2(t) + \tilde{I}_2(t)$$

where d is the desired response, n is the noise, signal, and I is the interference signal. The noise signal is spatially incoherent, more specifically

$$E[\tilde{n}_i{}^*(t)\tilde{n}_j(t)] = \begin{cases} 0 & i \neq j \\ \sigma_n^2 & i = j \end{cases}.$$

Also

$$E[\tilde{d}_i{}^*(t)\tilde{n}_i(t)] = 0 \qquad \text{for all } (i,j) .$$

The desired signal is

$$\tilde{d}(t) = \tilde{d}_1(t) + \tilde{d}_2(t) = A_d e^{j2\pi f_0 t} e^{j\Theta_d} + A_d e^{j2\pi f_0 t} e^{j\Theta_d} e^{-j\pi \sin\theta_d}$$

where Θ_d is a uniform random variable. The interference signal is

$$\tilde{I}(t) = \tilde{I}_1(t) + \tilde{I}_2(t) = A_i e^{j2\pi f_0 t} e^{j\Theta_i} + A_i e^{j2\pi f_0 t} e^{j\Theta_i} e^{-j\pi \sin\theta_i}$$

where Θ_i is a uniform random variable. Of course the random variables Θ_d and Θ_i are assumed to be statistically independent. In vector format,

$$\tilde{\mathbf{x}}_d = A_d e^{j2\pi f_0 t} e^{j\Theta_d} \begin{bmatrix} 1 \\ e^{-j\pi \sin\theta_d} \end{bmatrix}$$

$$\tilde{\mathbf{x}}_i = A_i e^{j2\pi f_0 t} e^{j\Theta_i} \begin{bmatrix} 1 \\ e^{-j\pi \sin\theta_i} \end{bmatrix} .$$

Of course, the noise vector is

$$\tilde{\mathbf{x}}_n = \begin{bmatrix} \tilde{n}_1(t) \\ \tilde{n}_2(t) \end{bmatrix},$$

and the reference signal is (this is an assumption so that the desired and reference signal are correlated)

$$\tilde{r}(t) = A_r e^{j2\pi f_0 t} e^{j\Theta_d}.$$

Note that the input SNR is

$$SNR_d = A_d^2 / \sigma_n^2$$

and the interference to noise ratio is

$$SNR_i = A_i^2 / \sigma_n^2.$$

The input signal can be written using vector notation as

$$\tilde{\mathbf{x}} = \tilde{\mathbf{x}}_d + \tilde{\mathbf{x}}_i + \tilde{\mathbf{x}}_n.$$

The covariance matrix is computed from Eq. (16.67) as

$$\mathbf{C} = E[\tilde{\mathbf{x}}_d^* \tilde{\mathbf{x}}_d^t] = \begin{bmatrix} A_d^2 + A_i^2 + \sigma_n^2 & A_d^2 e^{-j\pi \sin\theta_d} + A_i^2 e^{-j\pi \sin\theta_i} \\ A_d^2 e^{j\pi \sin\theta_d} + A_i^2 e^{j\pi \sin\theta_i} & A_d^2 + A_i^2 + \sigma_n^2 \end{bmatrix}.$$

In order to compute the covariance matrix eigenvalue, one needs to compute the determinant first

$$|\mathbf{C}| = 4A_d^2 A_i^2 \left(\sin\left(\frac{\theta_d + \theta_i}{s}\right) \right)^2 + 2A_d^2 \sigma_n^2 + 2A_i^2 \sigma_n^2 + \sigma_n^4.$$

Thus,

$$\mathbf{C}^{-1} = \frac{1}{|\mathbf{C}|} \begin{bmatrix} A_d^2 + A_i^2 + \sigma_n^2 & -A_d^2 e^{-j\pi \sin\theta_d} - A_i^2 e^{-j\pi \sin\theta_i} \\ -A_d^2 e^{j\pi \sin\theta_d} - A_i^2 e^{j\pi \sin\theta_i} & A_d^2 + A_i^2 + \sigma_n^2 \end{bmatrix}.$$

The reference correlation vector is

$$\mathbf{s} = E[\tilde{\mathbf{x}}^* \tilde{r}(t)] = A_d A_r \begin{bmatrix} 1 \\ e^{j\pi \sin\theta_d} \end{bmatrix}.$$

It follows that the weights are

$$\mathbf{w} = \frac{A_d A_r}{|\mathbf{C}|} \begin{bmatrix} A_i^2 + \sigma_n^2 - A_i^2 e^{j\pi(\sin\theta_d - \sin\theta_i)} \\ e^{j\pi \sin\theta_d} \{ A_i^2 + \sigma_n^2 - A_i^2 e^{j\pi(\sin\theta_i - \sin\theta_d)} \} \end{bmatrix}.$$

MATLAB Function "adaptive_array_lms.m"

The MATLAB function *"adaptive_array_lms.m"* implements the LMS adaptive array processing described in this section. Its syntax is as follows:

```
adaptive_array_lms(N, dol,tagt_angle, jam_angle)
```

where

Symbol	Description	Units	Status
N	*array size*	*none*	*input*
dol	*array element spacing*	*lambda*	*input*
tagt_angle	*desired beam spatial location*	*degrees*	*input*
jam_angle	*jammer spatial location*	*degrees*	*input*

The output of this function is a plot of the normalized array response in dB versus scan before and after adaptive processing is applied. Figure 16.7 shows an example using the following MATLAB call:

adaptive_array_lms(19, 0.5, 0, 35)

Note that the quality of the null (how deep and how narrow) heavily depends on the accuracy of the covariance matrix. In the *"adaptive_array_lms.m"* code, the MATLAB function *"mvn-rnd"* was employed to estimate the noise vector used in computing the covariance matrix. It follows that each time the code is executed, a different covariance matrix is calculated, and hence it is very likely that the adaptive null will differ in appearance from one run to another. This is illustrated in Figs. 16.8a and 16.8b. In this case, the main beam is steered to $\theta = -10°$, while the jammer is located at $\theta = 25°$.

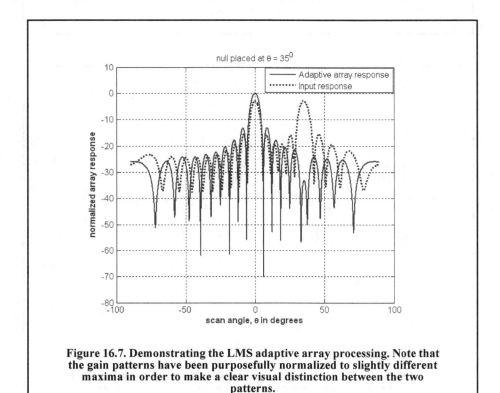

Figure 16.7. Demonstrating the LMS adaptive array processing. Note that the gain patterns have been purposefully normalized to slightly different maxima in order to make a clear visual distinction between the two patterns.

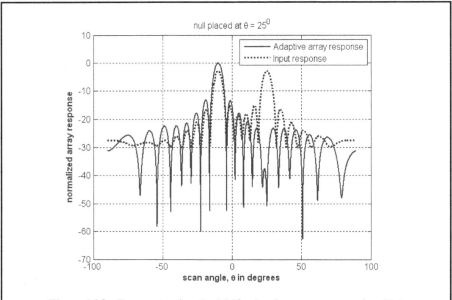

Figure 16.8a. Demonstrating the LMS adaptive array processing. Note that the gain patterns have been purposefully normalized to slightly different maxima in order to make a clear visual distinction between the two patterns.

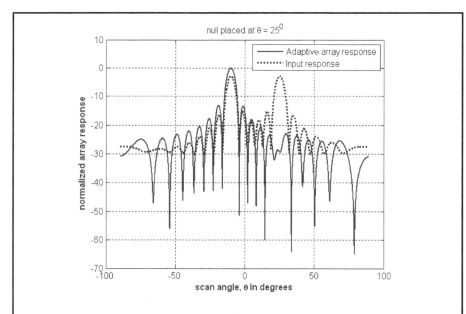

Figure 16.8b. Demonstrating the LMS adaptive array processing. Note that the gain patterns have been purposefully normalized to slightly different maxima in order to make a clear visual distinction between the two patterns.

16.4. Sidelobe Cancelers (SLC)

Sidelobe cancelers typically consist of a main antenna (which can be a phased array or a single element) and one or more auxiliary antennas. The main antenna is referred to as the main channel; it is assumed to be highly directional and is pointed toward the desired signal angular location. The interfering signal is assumed to be located somewhere off the main antenna boresight (in the sidelobes). Because of this configuration, the main channel receives returns from both the desired and the interfering signals. However, returns from the interfering signal in the main channel are weak because of the low main antenna sidelobe gain in the direction of the interfering signal. Also the auxiliary antenna returns are primarily from the interfering signal. This is illustrated in Fig. 16.9.

Referring to Fig. 16.9, $\tilde{s}(t)$ is the desired signal, $\tilde{n}(t)$ is the main channel noise signal, which is primarily from the interfering signal, while $\tilde{n}'(t)$ is the interfering signal in the auxiliary array. It is assumed that the signals $\tilde{s}(t)$ and $\tilde{n}(t)$ are uncorrelated. It is also assumed that the interfering signal is highly correlated with the noise signal in the main channel. The basic idea behind the SLC is to have the adaptive auxiliary channel produce an accurate estimate of the noise signal first, then to subtract that estimate from the main channel signal so that the output signal is mainly the desired signal.

The error signal is

$$\tilde{\varepsilon} = \tilde{\mathbf{d}} - \mathbf{w}^t \tilde{\mathbf{x}}$$

Eq. (16.74)

where $\mathbf{x}$ is the vector of the auxiliary array signal, and $\mathbf{w}$ is the adapted weights. The vector $\tilde{\mathbf{d}}$ of size M. The residual power is

$$P_{res} = E[\tilde{\varepsilon}\tilde{\varepsilon}^\dagger]$$

Eq. (16.75)

$$P_{res} = E[(\tilde{\mathbf{d}} - \mathbf{w}^t \tilde{\mathbf{x}})(\tilde{\mathbf{d}}^* - \tilde{\mathbf{x}}^\dagger \mathbf{w}^*)] .$$

Eq. (16.76)

It follows that

$$P_{res} = E[|\tilde{\mathbf{d}}|^2] - E[\tilde{\mathbf{d}}\tilde{\mathbf{x}}^\dagger \mathbf{w}^*] - E[\tilde{\mathbf{d}}^* \mathbf{w}^t \tilde{\mathbf{x}}] - \mathbf{w}^t E[\tilde{\mathbf{x}}\tilde{\mathbf{x}}^\dagger]\mathbf{w}^* .$$

Eq. (16.77)

Differentiate the residual power with respect to $\mathbf{w}$ and setting the answer equal to zero (to compute the optimal weights that minimize the power residual) yields

$$\frac{\partial P_{res}}{\partial \mathbf{w}} = \mathbf{0} = -\tilde{\mathbf{x}}\tilde{\mathbf{d}} + \mathbf{C}_a \mathbf{w}$$

Eq. (16.78)

where $\mathbf{C}_a$ is the covariance matrix of the auxiliary channel. Finally, the optimal weights are given by

$$\mathbf{w} = \mathbf{C}_a^{-1} \tilde{\mathbf{x}}\tilde{\mathbf{d}} .$$

Eq. (16.79)

Note that the vector $\tilde{\mathbf{x}}\tilde{\mathbf{d}}$ represents the components that are common to both main and auxiliary channels. Note that Eq. (16.79) makes intuitive sense where the objective is to isolate the components in the data which are common to the main and auxiliary channels, and we then wish to give them some heavy attenuation (which comes from inverting $\mathbf{C}_a$).

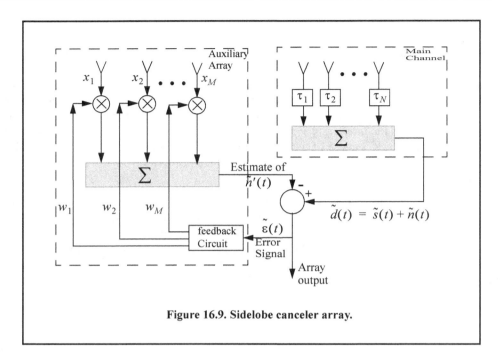

Figure 16.9. Sidelobe canceler array.

16.5. Space Time Adaptive Processing (STAP)

Space time adaptive processing (STAP) is the term used to describe adaptive arrays that simultaneously process spatial and temporal data. The spatial components of the signal are collected using the array sensors (same as in any array operation) while the temporal components of the signal are generated using time-delay units of equal intervals behind each array sensor. For this purpose, an array of size N will have N sub-channels (one behind each senor); within each sub-channel the signal from the jth range bin comprises M pulses interleaved by the radar pulse repetition interval ($T = 1/f_r$) where f_r is the PRF. The outputs from all M delayed responses are then summed coherently, then all N channels are coherently summed to generate the composite array response. The array input is assumed to be made of target returns, clutter returns, and interfering signals (e.g., jammers) returns.

The material in this section is presented in the following sequence: First, the concept of space time beamforming is introduced; then the analysis is extended to encompass space time adaptive processing.

16.5.1. Space Time Processing

The configuration of a space time beamformer is illustrated in Fig. 16.10. In this case, an array of N sensors and M pulses (interleaved by the radar PRI) comprise the beamformer output for each range bin. The signal output of the nth array sensor corresponding to the mth pulse and j^{th} range bin is

$$y_i(t_j - mT) \quad ; \begin{cases} i = 1, ..., N \\ j = 1, ..., J \\ m = 0, ..., M-1 \end{cases} \qquad \text{Eq. (16.80)}$$

where N is the number of sensors in the array, M is the number of pulses, and J is the number of range bins being processed.

Using this notation, the jth range bin return signal from all pulses is given by

$$\mathbf{Y}_j = \begin{bmatrix} \mathbf{y}_{1j} \\ \mathbf{y}_{2j} \\ \\ \mathbf{y}_{Nj} \end{bmatrix} = \begin{bmatrix} y_1(t_j) & y_1(t_j - T) & y_1(t_j - 2T) & \cdots & y_1(t_j - (M-1)T) \\ y_2(t_j) & y_2(t_j - T) & y_2(t_j - 2T) & & y_2(t_j - (M-1)T) \\ & & \vdots & & \\ y_N(t_j) & y_N(t_j - T) & y_N(t_j - 2T) & \cdots & y_N(t_j - (M-1)T) \end{bmatrix} . \qquad \text{Eq. (16.81)}$$

In this manner, the space time beamformer receives a series of M pulses from each of the N array elements for each of the J range bins. Hence, a data cube of returns is generated, as illustrated in Fig. 16.11. For this purpose, the data received from the jth range bin is made of $MN \times 1$ space (or time snapshots).

By taking element-1 the array phase reference, then the signal received by the nth array element (or sensor) at time t_j from a far field target whose angle of arrival is θ can be computed with the help of Eq. (16.1) as

$$y_n(t_j) = x(t_j)e^{-j\Delta\theta_n} \qquad \text{Eq. (16.82)}$$

where

$$\Delta\theta_n = \frac{2\pi d}{\lambda}(n-1)\sin\theta \qquad ; n = 1, 2, ..., N \qquad \text{Eq. (16.83)}$$

where, in general, the signal $x(t)$ is

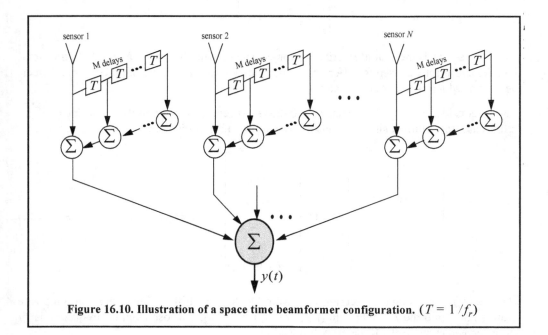

Figure 16.10. Illustration of a space time beamformer configuration. $(T = 1/f_r)$

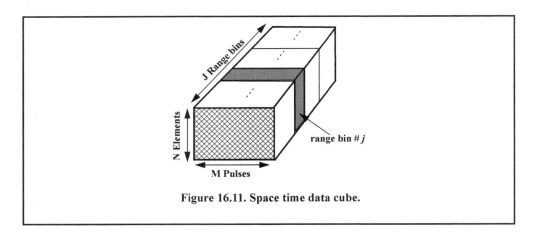

Figure 16.11. Space time data cube.

$$x(t) = e^{j2\pi f_0 t}.$$ **Eq. (16.84)**

f_0 is the radar operating frequency. It follows that

$$\mathbf{Y}_j = \begin{bmatrix} \mathbf{y}_{1j} \\ \mathbf{y}_{2j} \\ \\ \mathbf{y}_{Nj} \end{bmatrix} = x(t_j) \begin{bmatrix} 1 \\ e^{\left(-j\frac{2\pi d}{\lambda}\right)\sin\theta} \\ e^{\left(-j\frac{2\pi d}{\lambda}\right)2\sin\theta} \\ \\ e^{\left(-j\frac{2\pi d}{\lambda}\right)(N-1)\sin\theta} \end{bmatrix} = x(t_j)\mathbf{s}_s(\theta)$$ **Eq. (16.85)**

where the $\mathbf{s}_s(\theta)$ is the spatial steering vector associated with the arrival angle θ. In this notation, the subscript s is used to differentiate the spatial steering vector from the temporal steering vector, which will be defined later.

Next, consider the Doppler effects due to the target relative motion to the radar line of sight. In this case, the returned signal at sensor-1 due to M pulses is given by

$$\begin{bmatrix} y_1(t) \\ y_1(t-T) \\ \\ y_1(t-(M-1)T) \end{bmatrix} = x(t) \begin{bmatrix} 1 \\ e^{-j2\pi(f_d)} \\ e^{-j2\pi(2f_d)} \\ \\ e^{-j2\pi((M-1)f_d)} \end{bmatrix} = x(t)\mathbf{s}_t(f_d)$$ **Eq. (16.86)**

where $\mathbf{s}_t(f_d)$ is the temporal steering vector for the Doppler shift f_d. Therefore, the composite return signal from the jth range bin (i.e., time t_j) for a target whose Doppler frequency is f_d and is located at angle θ off the array boresight is

$$\mathbf{Y}_j = x(t_j)\{\mathbf{s}_s(\theta) \otimes \mathbf{s}_t(f_d)\} = x(t_j)\mathbf{s}_t \qquad \text{Eq. (16.87)}$$

where the symbol $\otimes$ indicates the Kronecker product and $\mathbf{s}_t = \mathbf{s}_s(\theta) \otimes \mathbf{s}_t(f_d)$.

16.5.2. Space Time Adaptive Processing

The space time adaptive beamformer is shown in Fig. 16.12. The output of the STAP beamformer is now given by,

$$\mathbf{Z}_j = \mathbf{W}\mathbf{Y}_j \qquad \text{Eq. (16.88)}$$

where $\mathbf{Y}_j$ was defined in the previous section and $\mathbf{W}$ is the adaptive weights matrix (see Fig. 16.11). As before, the input signal to the array is assumed to be made of the target returned signal, clutter and interference retuned signal, and thermal noise.

The total power output of the STAP beamformer is

$$P_{out} = E[\mathbf{Z}_j^*\mathbf{Z}_j] = P_{tgt}|\mathbf{W}^*\mathbf{s}_t|^2 + \mathbf{W}^*\mathbf{C}\mathbf{W} \qquad \text{Eq. (16.89)}$$

where $\mathbf{C}$ is the composite input signal covariance matrix and P_{tgt} is the desired target signal power. Note that the covariance matrix represents the combined target, clutter, noise, and interference signals; it is of size $MN \times MN$. It follows that the signal-to-interference plus noise ratio is

$$SINR_{out} = \frac{P_{tgt}|\mathbf{W}^*\mathbf{s}_t|^2}{\mathbf{W}^*\mathbf{C}\mathbf{W}}. \qquad \text{Eq. (16.90)}$$

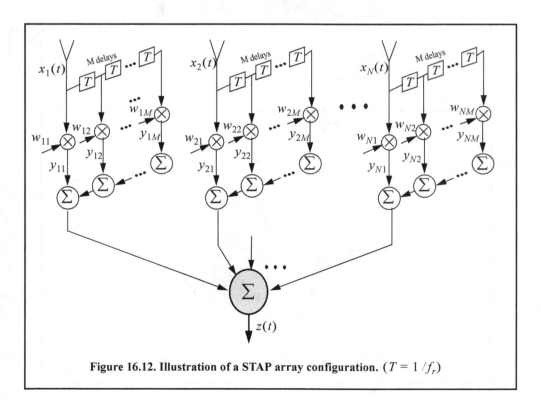

Figure 16.12. Illustration of a STAP array configuration. ($T = 1/f_r$)

Therefore, the optimal set of weights that maximize the ratio given in Eq. (16.90) is

$$\mathbf{W}_{opt} = \mathbf{C}^{-1}\mathbf{s}_t.$$

Eq. (16.91)

Figures 16.13 through 16.18 demonstrate STAP processing. In these examples, one target and one jammer are present. Figures 16.13 and 16.14 show the combined target, jammer, and clutter returns. Figures 16.15 and 16.16 show the target and jammer return after removing the clutter ridge. Figures 16.17 and 16.18 show the target return after removing the jammer and clutter ridge returns. These figures can be reproduced using the MATLAB program *"run_stap.m"* and its associated MATLAB functions, which are listed in Appendix 16-A.

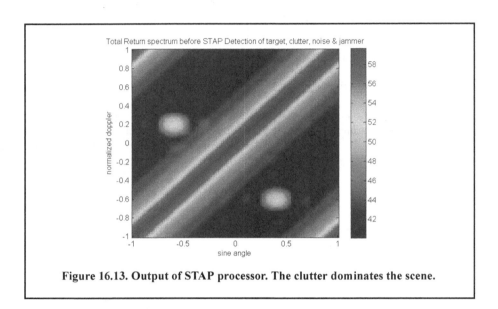

Figure 16.13. Output of STAP processor. The clutter dominates the scene.

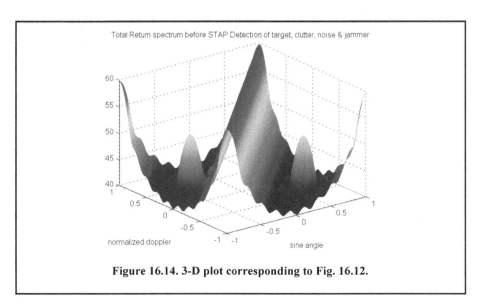

Figure 16.14. 3-D plot corresponding to Fig. 16.12.

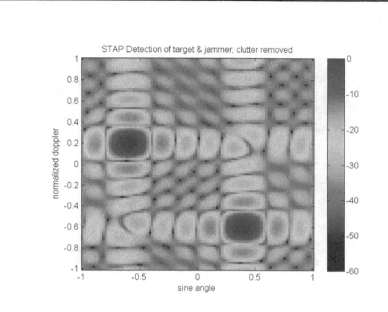

Figure 16.15. Output of STAP processor. Target and jammer returns; clutter ridge has been removed.

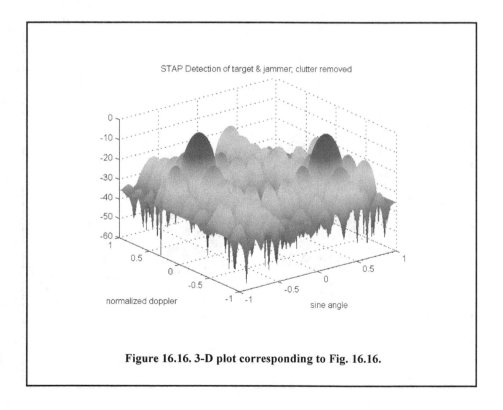

Figure 16.16. 3-D plot corresponding to Fig. 16.16.

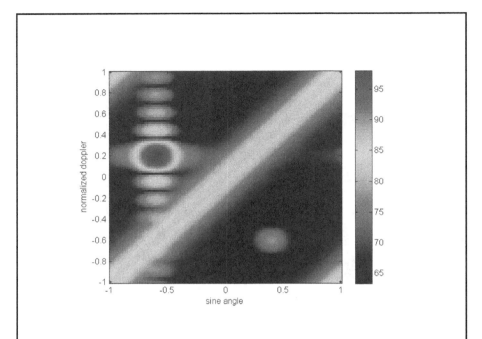

Figure 16.17. Output of STAP processor. Target only; jammer and clutter ridge returns have been removed.

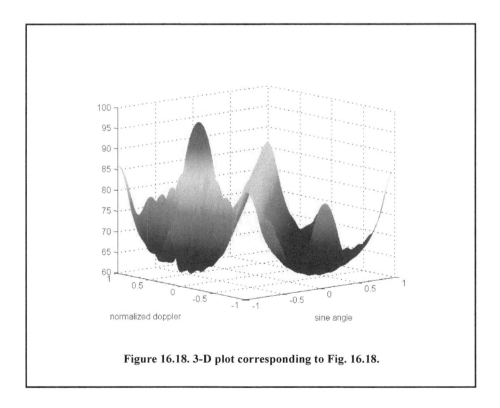

Figure 16.18. 3-D plot corresponding to Fig. 16.18.

Problems

16.1. Starting with Eq. (16.62), derive Eq. (16.63).

16.2. Compute the transient solution of the DE defined in Eq. (16.69).

16.3. Repeat the example in Section 16.1 for angle $\pi/4$ instead of $\pi/6$.

16.4. In Section 16.3, the MATLAB function *"adaptive_array_lms.m"* was developed to illustrate how linear arrays can adaptively place a null anywhere within the array's field of view. This code, however, assumed a single target (desired beam) and a single jammer (null). Extend this code (or develop your own) to account for multiple simultaneous desired beams (up to $N/2$) and multiple jammers (up to $N/2\text{-}1$) where N is the size of the array.

16.5. Building on the previous problem, in Chapter 15, the effect of having a limited number of bits to steer the main beam was demonstrated. Modify the code of the previous problem to include the effects of having a limited number of bits for phase shifting.

16.6. Figures 16.8a and 16.8b clearly demonstrate how the estimate of the covariance matrix impacts the quality of the adaptive null. In Section 16.3 (see Eq. (16.71)), a technique was described for estimating and improving the quality of the covariance matrix. Modify the MATLAB code *"adaptive_array_lms.m"* or develop your own code to implement Eq. (16.71). Briefly discuss how the quality of the adaptive null has been improved.

16.7. Develop a MATLAB code to implement the SLC canceler.

16.8. The MATLAB code *"run_stap.m"* used hard-coded (pre-determined) values for the SNR, CNR, and JSR power ratios. Modify this code, or write your own, to allow the user to change these values. Make a few runs with different combinations of these values and discuss your results.

16.9. Repeat the previous problem, where in this case, you will allow the number of elements in the array, N, to become a user-controlled variable. Run two cases one with low value for N (less than *10*) and one with a large value (more than *20*), briefly discuss your results.

Appendix 16-A: Chapter 16 MATLAB Code Listings

The MATLAB code provided in this chapter was designed as an academic standalone tool and is not adequate for other purposes. The code was written in a way to assist the reader in gaining a better understanding of the theory. The code was not developed, nor is it intended to be used as part of an open-loop or a closed-loop simulation of any kind. The MATLAB code found in this textbook can be downloaded from this book's web page on the CRC Press website. Simply use your favorite web browser, go to *www.crcpress.com*, and search for keyword *"Mahafza"* to locate this book's web page.

MATLAB Function "LMS.m" Listing

```
function Xout = LMS(Xin, D, B, mu, sigma, alpha)
%   This program was written by Stephen Robinson a senior radar
%   engineer at deciBel Research, Inc. in Huntsville, AL
%   Xin = data vector ; size = 1 x N
%   D = desired signal vector; size = 1 x N
%   N = number of data samples and of adaptive iterations
%   B = adaptive coefficients of Lht order fFIRfilter; size = 1 x L
%   L = order of adaptive system
%   mu = convergence parameter
%   sigma = input signal power estimate
%   alpha = exponential forgetting factor
N = size(Xin,2)
L = size(B,2)-1
px = B;
for k = 1:N
   px(1) = Xin(k);
   Xout(k) = sum(B.*px);
   E = D(k) - Xout(k);
   sigma = alpha*(px(1)^2) + (1 - alpha)*sigma;
   tmp = 2*mu/((L+1)*sigma);
   B = B + tmp*E*px;
   px(L+1:-1:2) = px(L:-1:1);
end; return
```

MATLAB Program "Fig16_4_5.m" Listing

```
% Figures 16.4 and 16.5
clc; close all;  clear all
N = 501;
mu = 0.1; % convergence parameter
Mu = num2str(mu);
L = 20; % FIR filter order
B = zeros(1,L+1); % FIR coefficients
sigma = 2; %Initial estimate for noise power
alpha = .100; % forgetting factor
Alpha =num2str(alpha);
k = 1:N;
noise = rand(1, length(k)) - .5; % Random noise
D = sqrt(2)*sin(2*pi*k/20);
X = D + sqrt(7)*noise;
Y = LMS(X, D, B, mu, sigma, alpha);
subplot(3,1,1)
```

```
plot(D,'linewidth',1.5);
xlim([0 501]); grid on;
ylabel('\bfDesired response');
title(['\mu = ',[Mu], ',   \alpha = ',[Alpha]])
subplot(3,1,2)
plot(X,'linewidth',1);
xlim([0 501]); grid on;
ylabel('\bfCorrupted signal')
subplot(3,1,3)
plot(Y,'linewidth',1.5);
xlim([0 501]);
grid on; xlabel('\bftime in sec'); ylabel('\bfLMS output')
```

MATLAB Function "adaptive_array_lms.m" Listing

```
function adaptive_array_lms(N, dol,tagt_angle, jam_angle)
% This function implements the adaptive array LMS algorithm described in
% Section 16.6 of text.
% This function calls two other function
   % la_sampled_wave  and
   % linear_array_FFT
% Inputs
   % N       == size of linear array
   % dol     == array element spacing in lambda units
   % tgt_angle == targte angle (desired signal) in degrees
   % jam_angle == jammer angle (desired location of null) in degrees
% Outputs
   % This function will display the before and after normalized array
   % response in dB versus scan angle in degrees
clc; close all
mu = [0 0]; % noise mean value
sigma = [.21 .21; .21 .210]; % noise variance
% N = 19;
% dol = 0.5;
% tgt_angle = 0;
% jam_angle = 40;
sine_tgt_angle = sin(tagt_angle *pi/180);
sine_jam_angle = sin(jam_angle*pi/180);
al = la_sampled_wave(N, dol, sine_tgt_angle);
jl  = la_sampled_wave(N, dol, sine_jam_angle);
x = al + jl;
n = mvnrnd(mu,sigma,N);
jl = jl + complex(n(:,1),n(:,2));
Xl = jl * jl';
Cl = cov(Xl) + eye(N);
Wl = inv(Cl) * al;
[G, R, u, theta] = linear_array_FFT(Wl, dol);
[G1, R1, u, theta] = linear_array_FFT(x, dol);
u_deg = asin(u) *180/pi;
plot(u_deg, 10*log10((G/max(G)+eps))','linewidth', 1.5);
grid on
hold on
plot(u_deg, 10*log10((G1/max(G1)+eps)),'k:','linewidth', 2);
xlabel('\bfscan angle, \theta in degrees')
```

```
ylabel('\bf normalized array response')
legend('Adaptive array response','Input response')
JAM = num2str(jam_angle); title (['null placed at \theta = ',[JAM],'^0'])
```

MATLAB Function "la_sampled_wave.m" Listing

```
function s = la_sampled_wave(N, dol, sinbeta)
 k = 2*pi * dol * sinbeta;
for m = 1: N,
   s(m) = exp(j*(m-1)*k);
end
% Return a column vector, not a row vector
if size(s,1)==1,
   s = s.';
end
```

MATLAB Function "Linear_array_FFT.m" Listing

```
function [G, R, u, theta] = linear_array_FFT(a, dol);

Nelt = length(a);
ratio = 1;
if dol<=0.5,
   Nr = Nelt;
   dolr = dol;
else
   ratio = ceil(dol/0.5);
   Nr = Nelt * ratio;
   dolr = dol/ratio;
   atemp = a;
   a = zeros(1, Nr);
   a(1: ratio: Nr) = atemp(1: Nelt);
end
% use a value for NFFT that is at least 10 times that of N
% I borrowed this piece of code
nfft = 2^(ceil(log(10*Nr)/log(2)));
nfft = 65536;
A = fftshift(fft(a, nfft));
% Compute u = sin(theta)
u = [-nfft/2 : nfft/2-1] * (1/dolr/nfft);
% 'k' gives us the bounds of visible space
k = find(abs(u)<=1);
R = (abs(A(k))).^2;
u = u(k);
theta = asin(u);
% Gain patterns
Rbar = 0.5 * sum(R) / (nfft*dol);
G = R / Rbar;
```

MATLAB Program "run_stap.m" Listing

```
clear all; close all
sintheta_t1 = .4;
wd_t1 =-.6;
sintheta_t2 = -.6;
```

```
wd_t2 = .2;
[LL, sintheta, wd] = stap_std(sintheta_t1, wd_t1, sintheta_t2, wd_t2);
LL = LL / max(max(abs(LL)));
LL = max(LL, 1e-6);
figure (3)
imagesc(sintheta, wd, 10*log10(abs(LL)))
colorbar
title('STAP Detection of target & jammer; clutter removed');
set(gca,'ydir','normal'), xlabel('sine angle'), ylabel('normalized doppler')
figure (4)
surf(sintheta, wd, 10*log10(abs(LL)))
shading interp
title('STAP Detection of target & jammer; clutter removed');
set(gca,'ydir','normal'), xlabel('sine angle'), ylabel('normalized doppler')
[LL, sintheta, wd] = stap_smaa(sintheta_t1, wd_t1, sintheta_t2, wd_t2);
LL = LL / max(max(abs(LL)));
LL = max(LL, 1e-6);
figure
imagesc(sintheta, wd, 10*log10(abs(LL)))
colorbar
set(gca,'ydir','normal'), xlabel('sine angle'), ylabel('normalized doppler')
title('STAP Detection of target; jammer & clutter removed');
figure
surf(sintheta, wd, 10*log10(abs(LL)))
shading interp
set(gca,'ydir','normal'), xlabel('sine angle'), ylabel('normalized doppler')
title('SNR after SMAA STAP Detection of target, clutter, noise & jammer');
```

MATLAB Function "stap_std.m" Listing

```
function [LL, sintheta, wd] = stap_std(sintheta_t1, wd_t1, sintheta_t2, wd_t2);
do_plot = 1;
N = 10;        % Sensors
M = 12;        % Pulses
No = 250;       % k-th clutter bins (refers to fig. 5)
beta = 1;       % The way the clutter fills the angle Doppler
dol = 0.5;      % d over lambda
CNR = 30;   % dB Clutter to Noise Ratio
SNR = 10;    % dB Signal to Noise Ratio
JSR = 0;   % dB Jammer to Signal Ratio
% Set the noise power
sigma2_n = 1;
% Clutter power
sigma2_c = sigma2_n * 10^(CNR/10);
sigma_c = sqrt(sigma2_c);
% Target 1 power
sigma2_t1 = sigma2_n * 10^(SNR/10);
sigma_t1 = sqrt(sigma2_t1);
% Target 2 (Jammer) power
sigma2_t2 = sigma2_t1 * 10^(JSR/10);
sigma_t2 = sqrt(sigma2_t2);
% Ground clutter is the primary source of interference
sintheta = linspace(-1, 1, No);
phi = 2 * dol * sintheta;
```

```
wd = beta * phi;
Rc = zeros(N*M);
ac_all = zeros(N*M,1);
for k = 1: length(phi),
   ac = sigma_c * st_steering_vector(phi(k), N, beta*phi(k), M);  % Xc
   Rc = Rc + ac * ac';    % covarience matrix of target "1" ,  "'" --> conjugate transpose
   ac_all = ac_all + ac;  % "w" not optimized yet
end
Rc = Rc / length(phi);
% Noise signals decorrelate from pulse-to-pulse
% With this assumption, noise covariance matrix is
Rn = sigma2_n * eye(M*N);
% Target 1 covariance matrix
% at1 = st_steering_vector(sintheta_t1, N, wd_t1, M);
% Rt1 = sigma2_t1 * at1 * at1';
at1 = sigma_t1 * st_steering_vector(sintheta_t1, N, wd_t1, M); % Xj1
Rt1 = at1 * at1';   % covarience matrix of target "1"
at2 = sigma_t2 * st_steering_vector(sintheta_t2, N, wd_t2, M);
Rt2 = at2 * at2';  % covarience matrix of target "2" == jammer
% Total covariance matrix
R = Rc + Rn + Rt1 + Rt2;
% Unweighted spectrum of the total return from the beamformer
sintheta = linspace(-1, 1);
wd = beta * sintheta;
Pb = zeros(length(wd), length(sintheta));
for nn = 1: length(sintheta),
  for mm = 1: length(wd),
    a = st_steering_vector(sintheta(nn), N, wd(mm), M);
    Pb(mm, nn) = a' * R * a;
  end
end
if do_plot,
  % Display the total return spectrum
  figure (1)
  imagesc(sintheta, wd, 10*log10(abs(Pb)))
  colorbar
  title('Total Return spectrum before STAP Detection of target, clutter, noise & jammer');
  set(gca,'ydir','normal'), xlabel('sine angle'), ylabel('normalized doppler')
  figure (2)
  surf(sintheta, wd, 10*log10(abs(Pb)))
  shading interp, , xlabel('sine angle'), ylabel('normalized doppler')
  title('Total Return spectrum before STAP Detection of target, clutter, noise & jammer');
end
% Total covariance matrix
R = Rc + Rn + Rt1 + Rt2;
% Calculate optimal weights
Rc = (ac_all * ac_all') / length(phi);
Rinv = inv(Rc + Rn );  n
wopt = Rinv' * (at1 + at2 );
% Log-Likelihood Function
% Calculating the SNR and switching to the run_stap.m to execute the log part
sintheta = linspace(-1, 1);
wd = beta * sintheta;
LL = zeros(length(wd), length(sintheta));
```

```
for nn = 1: length(sintheta),
   for mm = 1: length(wd),
     a = st_steering_vector(sintheta(nn), N, wd(mm), M);
   %   LL(mm,nn) = abs( a' * Rinv * (at1+at2+ac_all) )^2 / ( a' * Rinv * a ); % Original by Keith
     LL(mm,nn) = abs(a' * Rinv * (at1+at2+ac_all) )^2 / ( a' * (Rc + Rn ) * a ); % our expectation
   end
end
disp(size(a))
disp (size(Rinv))
```

MATLAB Function "stap_smaa.m" Listing

```
function [LL, sintheta, wd] = stap_smaa(sintheta_t1, wd_t1, sintheta_t2, wd_t2);
do_plot = 1;
N = 10; Na = 2*N-1;
M = 12;
No = 250;
beta = 1;
dol = 0.5;
CNR = 20; % dB
SNR = 0; % dB
JSR = 20; % dB
% Set the noise power
sigma2_n = 1;
% Clutter power
sigma2_c = sigma2_n * 10^(CNR/10);
sigma_c = sqrt(sigma2_c);
% Target 1 power
sigma2_t1 = sigma2_n * 10^(SNR/10);
sigma_t1 = sqrt(sigma2_t1);
% Target 2 (Jammer) power
sigma2_t2 = sigma2_t1 * 10^(JSR/10);
sigma_t2 = sqrt(sigma2_t2);
% Ground clutter is the primary source of interference
sintheta = linspace(-1, 1, No);
phi = 2 * dol * sintheta;
wd = beta * phi;
Rc = zeros(Na*M);
ac_all = zeros(Na*M,1);
for k = 1: length(phi),
   ac = sigma_c * smaa_st_steering_vector(phi(k), N, beta*phi(k), M);
   Rc = Rc + ac * ac';
   ac_all = ac_all + ac;
end
Rc = Rc / length(phi);
% Noise signals decorrelate from pulse-to-pulse
% With this assumption, noise covariance matrix is
Rn = sigma2_n * eye(M*Na);
% Target 1 covariance matrix
% at1 = smaa_st_steering_vector(sintheta_t1, N, wd_t1, M);
% Rt1 = sigma2_t1 * at1 * at1';
at1 = sigma_t1 * smaa_st_steering_vector(sintheta_t1, N, wd_t1, M);
Rt1 = at1 * at1';
% Target 1 covariance matrix
```

```
% at2 = smaa_st_steering_vector(sintheta_t2, N, wd_t2, M);
% Rt2 = sigma2_t2 * at2 * at2';
at2 = sigma_t2 * smaa_st_steering_vector(sintheta_t2, N, wd_t2, M);
Rt2 = at2 * at2';
% Total covariance matrix
R = Rc + Rn + Rt1 + Rt2;
% Unweighted spectrum of the total return from the beamformer
sintheta = linspace(-1, 1);
wd = beta * sintheta;
Pb = zeros(length(wd), length(sintheta));
for nn = 1: length(sintheta),
   for mm = 1: length(wd),
      a = smaa_st_steering_vector(sintheta(nn), N, wd(mm), M);
      Pb(mm, nn) = a' * R * a;
   end
end
if do_plot,
   % Display the total return spectrum
   figure (5)
   imagesc(sintheta, wd, 10*log10(abs(Pb)))
   set(gca,'ydir','normal'), xlabel('sine angle'), ylabel('normalized doppler')
   figure (6)
   surf(sintheta, wd, 10*log10(abs(Pb)))
   shading interp, xlabel('sine angle'), ylabel('normalized doppler')
end
% Calculate optimal weights
Rc = (ac_all * ac_all') / length(phi);
Rinv = inv(Rc + Rn);
wopt = Rinv * (at1 + at2);
% Log-Likelihood Function
sintheta = linspace(-1, 1);
wd = beta * sintheta;
LL = zeros(length(wd), length(sintheta));
for nn = 1: length(sintheta),
   for mm = 1: length(wd),
      a = smaa_st_steering_vector(sintheta(nn), N, wd(mm), M);
      LL(mm,nn) = abs( a' * Rinv * (at1+at2+ac_all) )^2 / ( a' * Rinv * a );
   end
end
```

MATLAB Function "st_steering_vector.m" Listing

```
function a = st_steering_vector(sintheta, N, wd, M)
a_N = exp(-j*pi*sintheta*[0:N-1]');
b_M = exp(-j*pi*wd     *[0:M-1]');
a = kron(b_M, a_N);
```

MATLAB Function "smaa_st_steering_vector.m" Listing

```
function a = smaa_st_steering_vector(sintheta, N, wd, M)
a_N = exp(-j*pi*sintheta*[-(N-1):+(N-1)]');
b_M = exp(-j*pi*wd     *[0:M-1]');
a_N = a_N .* ts_weighting(N);
a = kron(b_M, a_N);
```

Chapter 17

Target Tracking

Single Target Tracking

Tracking radar systems are used to measure the target's relative position in range, azimuth angle, elevation angle, and velocity. Then, by using and keeping track of these measured parameters the radar can predict their future values. Target tracking is important to military radars as well as to most civilian radars. In military radars, tracking is responsible for fire control and missile guidance; in fact, missile guidance is almost impossible without proper target tracking. Commercial radar systems, such as civilian airport traffic control radars, may utilize tracking as a means of controlling incoming and departing airplanes.

Tracking techniques can be divided into range/velocity tracking and angle tracking. It is also customary to distinguish between continuous single-target tracking radars and multi-target track-while-scan (TWS) radars. Tracking radars utilize pencil beam (very narrow) antenna patterns. It is for this reason that a separate search radar is needed to facilitate target acquisition by the tracker. Still, the tracking radar has to search the volume where the target's presence is suspected. For this purpose, tracking radars use special search patterns, such as helical, T.V. raster, cluster, and spiral patterns, to name a few.

17.1. Angle Tracking

Angle tracking is concerned with generating continuous measurements of the target's angular position in the azimuth and elevation coordinates. The accuracy of early-generation angle tracking radars depended heavily on the size of the pencil beam employed. Most modern radar systems achieve very fine angular measurements by utilizing monopulse tracking techniques.

Tracking radars use the angular deviation from the antenna main axis of the target within the beam to generate an error signal. This deviation is normally measured from the antenna's main axis. The resultant error signal describes how much the target has deviated from the beam main axis. Then, the beam position is continuously changed in an attempt to produce a zero error signal. If the radar beam is normal to the target (maximum gain), then the target angular position would be the same as that of the beam. In practice, this is rarely the case.

In order to be able to quickly change the beam position, the error signal needs to be a linear function of the deviation angle. It can be shown that this condition requires the beam's axis to be squinted by some angle (squint angle) off the antenna's main axis.

17.1.1. Sequential Lobing

Sequential lobing is one of the first tracking techniques that was utilized by the early generation of radar systems. Sequential lobing is often referred to as lobe switching or sequential switching. It has a tracking accuracy that is limited by the pencil beamwidth used and by the noise caused by either mechanical or electronic switching mechanisms. However, it is very simple to implement. The pencil beam used in sequential lobing must be symmetrical (equal azimuth and elevation beamwidths).

Tracking is achieved (in one coordinate) by continuously switching the pencil beam between two pre-determined symmetrical positions around the antenna's Line of Sight (LOS) axis. Hence, the name sequential lobing is adopted. The LOS is called the radar tracking axis, as illustrated in Fig. 17.1.

As the beam is switched between the two positions, the radar measures the returned signal levels. The difference between the two measured signal levels is used to compute the angular error signal. For example, when the target is tracked on the tracking axis, as the case in Fig. 17.1a, the voltage difference is zero. However, when the target is off the tracking axis, as in Fig. 17.1b, a nonzero error signal is produced. The sign of the voltage difference determines the direction in which the antenna must be moved. Keep in mind, the goal here is to make the voltage difference be equal to zero.

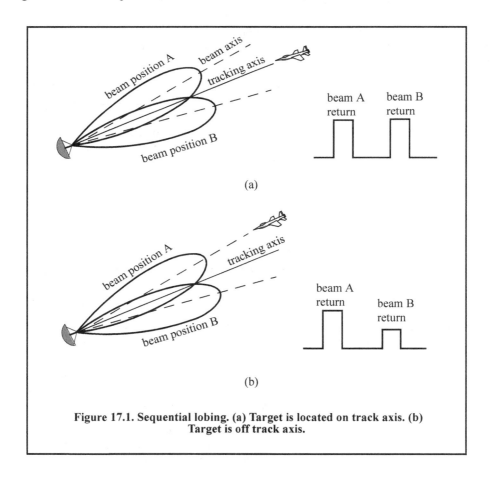

Figure 17.1. Sequential lobing. (a) Target is located on track axis. (b) Target is off track axis.

In order to obtain the angular error in the orthogonal coordinate, two more switching positions are required for that coordinate. Thus, tracking in two coordinates can be accomplished by using a cluster of four antennas (two for each coordinate) or by a cluster of five antennas. In the latter case, the middle antenna is used to transmit, while the other four are used to receive.

17.1.2. Conical Scan

Conical scan is a logical extension of sequential lobing where, in this case, the antenna is continuously rotated at an offset angle, or has a feed that is rotated about the antenna's main axis. Figure 17.2 shows a typical conical scan beam. The beam scan frequency, in radians per second, is denoted as ω_s. The angle between the antenna's LOS and the rotation axis is the squint angle φ. The antenna's beam position is continuously changed so that the target will always be on the tracking axis.

Figure 17.3 shows a simplified conical scan radar system. The envelope detector is used to extract the return signal amplitude, and the Automatic Gain Control (AGC) tries to hold the receiver output to a constant value. Since the AGC operates on large time constants, it can hold the average signal level constant and still preserve the signal rapid scan variation. It follows that the tracking error signals (azimuth and elevation) are functions of the target's RCS; they are functions of its angular position off the main beam axis.

In order to illustrate how conical scan tracking is achieved, we will first consider the case shown in Fig. 17.4. In this case, as the antenna rotates around the tracking axis, all target returns have the same amplitude (zero error signal). Thus, no further action is required.

Next, consider the case depicted by Fig. 17.5. Here, when the beam is at position B, returns from the target will have maximum amplitude, and when the antenna is at position A, returns from the target have minimum amplitude. Between those two positions, the amplitude of the target returns will vary between the maximum value at position B, and the minimum value at position A. In other words, Amplitude Modulation (AM) exists on top of the returned signal. This AM envelope corresponds to the relative position of the target within the beam. Thus, the extracted AM envelope can be used to derive a servo-control system in order to position the target on the tracking axis.

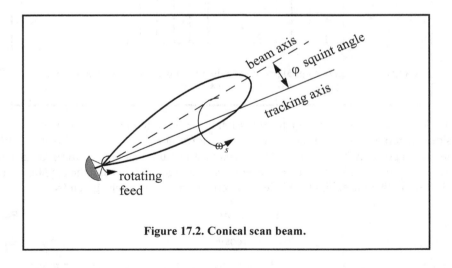

Figure 17.2. Conical scan beam.

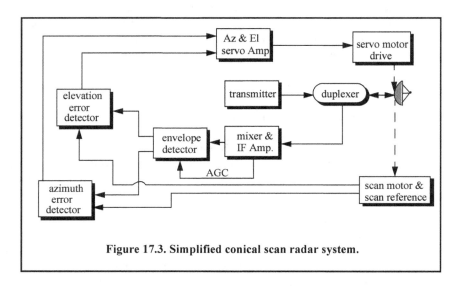

Figure 17.3. Simplified conical scan radar system.

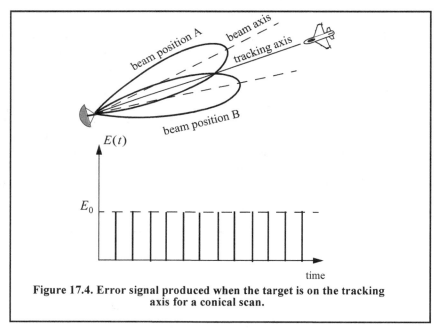

Figure 17.4. Error signal produced when the target is on the tracking axis for a conical scan.

Now, let us derive the error signal expression that is used to drive the servo-control system. Consider the top view of the beam axis location shown in Fig. 17.6. Assume that $t = 0$ is the starting beam position. The locations for maximum and minimum target returns are also identified. The quantity ε defines the distance between the target location and the antenna's tracking axis. It follows that the azimuth and elevation errors are, respectively, given by

$$\varepsilon_a = \varepsilon \sin\varphi \qquad\qquad\qquad \textbf{Eq. (17.1)}$$

$$\varepsilon_e = \varepsilon \cos\varphi . \qquad\qquad\qquad \textbf{Eq. (17.2)}$$

These are the error signals that the radar uses to align the tracking axis on the target.

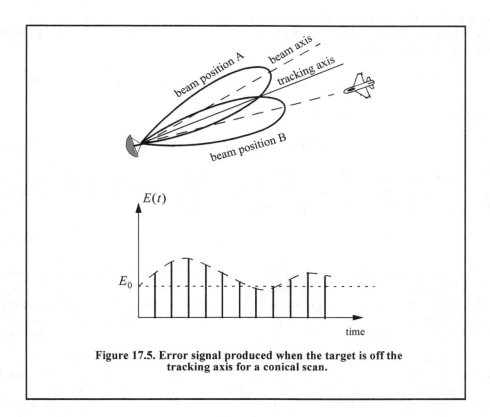

Figure 17.5. Error signal produced when the target is off the tracking axis for a conical scan.

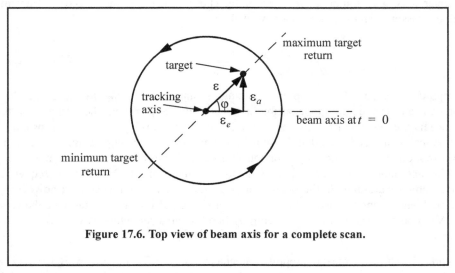

Figure 17.6. Top view of beam axis for a complete scan.

The AM signal $E(t)$ can then be written as

$$E(t) = E_0\cos(\omega_s t - \varphi) = E_0\varepsilon_e\cos\omega_s t + E_0\varepsilon_a\sin\omega_s t \qquad \textbf{Eq. (17.3)}$$

where E_0 is a constant called the error slope, ω_s is the scan frequency in radians per second, and φ is the angle already defined. The scan reference is the signal that the radar generates to keep track of the antenna's position around a complete path (scan). The elevation error signal is obtained by mixing the signal $E(t)$ with $\cos\omega_s t$ (the reference signal) followed by lowpass filtering. More precisely,

$$E_e(t) = E_0\cos(\omega_s t - \varphi)\cos\omega_s t = -\frac{1}{2}E_0\cos\varphi + \frac{1}{2}\cos(2\omega_s t - \varphi),$$ **Eq. (17.4)**

and after lowpass filtering we get

$$E_e(t) = -\frac{1}{2}E_0\cos\varphi.$$ **Eq. (17.5)**

Negative elevation error drives the antenna beam downward, while positive elevation error drives the antenna beam upward. Similarly, the azimuth error signal is obtained by multiplying $E(t)$ by $\sin\omega_s t$ followed by lowpass filtering. It follows that

$$E_a(t) = \frac{1}{2}E_0\sin\varphi.$$ **Eq. (17.6)**

The antenna scan rate is limited by the scanning mechanism (mechanical or electronic), where electronic scanning is much faster and more accurate than mechanical scanning. In either case, the radar needs at least four target returns to be able to determine the target azimuth and elevation coordinates (two returns per coordinate). Therefore, the maximum conical scan rate is equal to one fourth of the PRF. Rates as high as 30 scans per second are commonly used.

The conical scan squint angle needs to be large enough so that a good error signal can be measured. However, due to the squint angle, the antenna gain in the direction of the tracking axis is less than maximum. Thus, when the target is in track (located on the tracking axis), the SNR suffers a loss equal to the drop in the antenna gain. This loss is known as the squint or crossover loss. The squint angle is normally chosen such that the two-way (transmit and receive) crossover loss is less than a few decibels.

17.2. Amplitude Comparison Monopulse

Amplitude comparison monopulse tracking is similar to lobing in the sense that four squinted beams are required to measure the target's angular position. The difference is that the four beams are generated simultaneously rather than sequentially. For this purpose, a special antenna feed is utilized such that the four beams are produced using a single pulse, hence the name "monopulse." Additionally, monopulse tracking is more accurate and is not susceptible to lobing anomalies, such as AM jamming and gain inversion ECM. Finally, in sequential and conical lobing, variations in the radar echoes degrade the tracking accuracy; however, this is not a problem for monopulse techniques since a single pulse is used to produce the error signals. Monopulse tracking radars can employ both antenna reflectors as well as phased array antennas.

Figure 17.7 show a typical monopulse antenna pattern. The four beams A, B, C, and D represent the four conical scan beam positions. Four feeds, mainly horns, are used to produce the monopulse antenna pattern. Amplitude monopulse processing requires that the four signals have the same phase and different amplitudes.

A good way to explain the concept of amplitude monopulse technique is to represent the target echo signal by a circle centered at the antenna's tracking axis, as illustrated by Fig. 17.8a, where the four quadrants represent the four beams. In this case, the four horns receive an equal amount of energy, which indicates that the target is located on the antenna's tracking axis. However, when the target is off the tracking axis (Figs. 17.8b-d), an imbalance of energy occurs in the different beams. This imbalance of energy is used to generate an error signal that drives the servo-control system. Monopulse processing consists of computing a sum Σ and two difference Δ (azimuth and elevation) antenna patterns. Then by dividing a Δ channel voltage by the Σ channel voltage, the angle of the signal can be determined.

The radar continuously compares the amplitudes and phases of all beam returns to sense the amount of target displacement off the tracking axis. It is critical that the phases of the four signals be constant in both transmit and receive modes. For this purpose, either digital networks or microwave comparator circuitry are utilized. Figure 17.9 shows a block diagram for a typical microwave comparator, where the three receiver channels are declared as the sum channel, elevation angle difference channel, and azimuth angle difference channel.

To generate the elevation difference beam, one can use the beam difference (A−D) or (B−C). However, by first forming the sum patterns (A+B) and (D+C) and then computing the difference (A+B)−(D+C), we achieve a stronger elevation difference signal, Δ_{el}. Similarly, by first forming the sum patterns (A+D) and (B+C) and then computing the difference (A+D)−(B+C), a stronger azimuth difference signal, Δ_{az}, is produced.

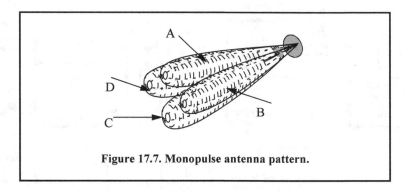

Figure 17.7. Monopulse antenna pattern.

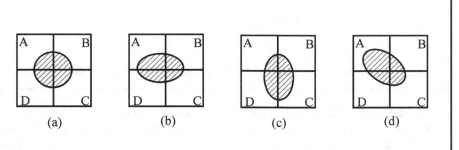

(a)   (b)   (c)   (d)

Figure 17.8. Illustration of monopulse concept. (a) Target is on the tracking axis. (b) - (d) Target is off the tracking axis.

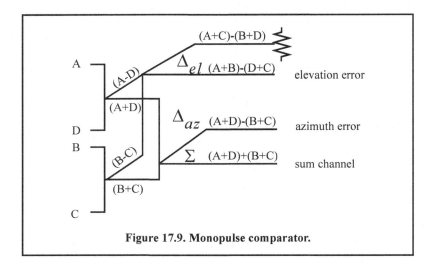

Figure 17.9. Monopulse comparator.

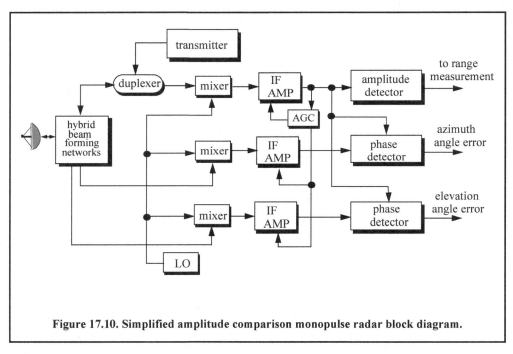

Figure 17.10. Simplified amplitude comparison monopulse radar block diagram.

A simplified monopulse radar block diagram is shown in Fig. 17.10. The sum channel is used for both transmit and receive. In the receive mode, the sum channel provides the phase reference for the other two difference channels. Range measurements can also be obtained from the sum channel. In order to illustrate how the sum and difference antenna patterns are formed, we will assume a $\sin\varphi/\varphi$ single-element antenna pattern and squint angle φ_0. The sum signal in one coordinate (azimuth or elevation) is then given by

$$\Sigma(\varphi) = \frac{\sin(\varphi - \varphi_0)}{(\varphi - \varphi_0)} + \frac{\sin(\varphi + \varphi_0)}{(\varphi + \varphi_0)}, \qquad \textbf{Eq. (17.7)}$$

and a difference signal in the same coordinate is

$$\Delta(\varphi) = \frac{\sin(\varphi - \varphi_0)}{(\varphi - \varphi_0)} - \frac{\sin(\varphi + \varphi_0)}{(\varphi + \varphi_0)}.$$ **Eq. (17.8)**

MATLAB Function "mono_pulse.m"

The function "mono_pulse.m" implements Eqs. (17.7) and (17.8). Its output includes plots of the sum and difference antenna patterns as well as the difference-to-sum ratio. The syntax is as follows:

mono_pulse (phi0)

where *phi0* is the squint angle in radians.

Figure 17.11 (a-c) shows the corresponding plots for the sum and difference patterns for $\varphi_0 = 0.15$ radians. Fig. 17.12 (a-c) is similar to Fig. 17.11, except in this case $\varphi_0 = 0.75$ radians. Clearly, the sum and difference patterns depend heavily on the squint angle. Using a relatively small squint angle produces a better sum pattern than that resulting from a larger angle. Additionally, the difference pattern slope is steeper for the small squint angle.

The difference channels give us an indication of whether the target is on or off the tracking axis. However, this signal amplitude depends not only on the target angular position, but also on the target's range and RCS. For this reason, the ratio Δ / Σ (delta over sum) can be used to accurately estimate the error angle that only depends on the target's angular position.

Let us now address how the error signals are computed. First, consider the azimuth error signal. Define the signals S_1 and S_2 as

$$S_1 = A + D$$ **Eq. (17.9)**

$$S_2 = B + C.$$ **Eq. (17.10)**

The sum signal is $\Sigma = S_1 + S_2$, and the azimuth difference signal is $\Delta_{az} = S_1 - S_2$. If $S_1 \geq S_2$, then both channels have the same phase $0°$ (since the sum channel is used for phase reference). Alternatively, if $S_1 < S_2$, then the two channels are $180°$ out of phase. Similar analysis can be done for the elevation channel, where in this case $S_1 = A + B$ and $S_2 = D + C$. Thus, the error signal output is

$$\varepsilon_\varphi = \frac{|\Delta|}{|\Sigma|} \cos\xi$$ **Eq. (17.11)**

where ξ is the phase angle between the sum and difference channels and it is equal to $0°$ or $180°$. More precisely, if $\xi = 0$, then the target is on the tracking axis; otherwise it is off the tracking axis. Figure 17.13 (a,b) shows a plot for the ratio Δ / Σ for the monopulse radar whose sum and difference patterns are in Figs. 17.11 and 17.12.

17.3. Phase Comparison Monopulse

Phase comparison monopulse is similar to amplitude comparison monopulse in the sense that the target angular coordinates are extracted from one sum and two difference channels. The main difference is that the four signals produced in amplitude comparison monopulse will have similar phases but different amplitudes; however, in phase comparison monopulse, the signals have the same amplitude and different phases. Phase comparison monopulse tracking radars use a minimum of a two-element array antenna for each coordinate (azimuth and elevation), as illustrated in Fig. 17.14. A phase error signal (for each coordinate) is computed from the phase difference between the signals generated in the antenna elements.

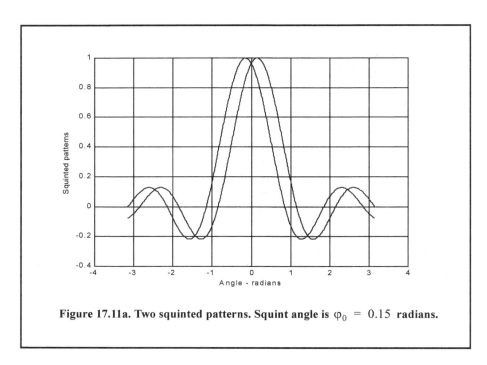

Figure 17.11a. Two squinted patterns. Squint angle is $\varphi_0 = 0.15$ radians.

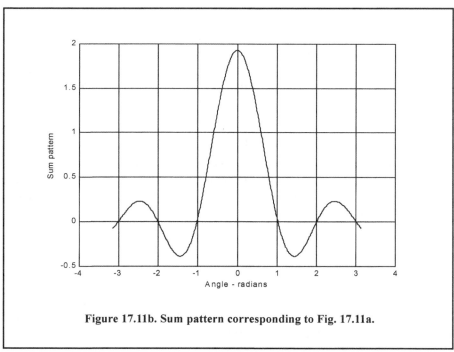

Figure 17.11b. Sum pattern corresponding to Fig. 17.11a.

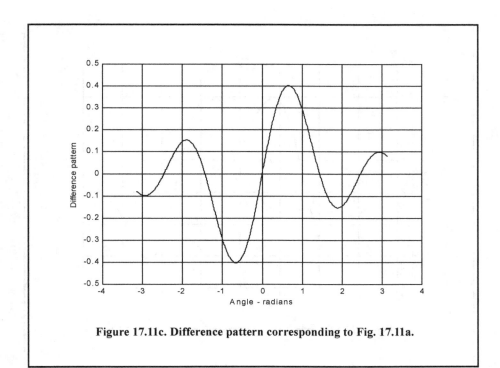

Figure 17.11c. Difference pattern corresponding to Fig. 17.11a.

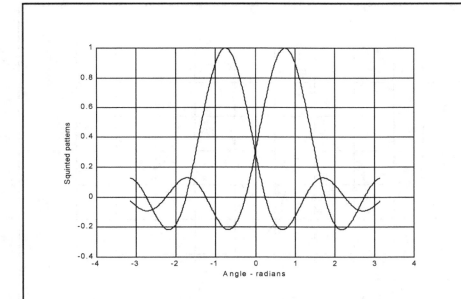

Figure 17.12a. Two squinted patterns. Squint angle is $\varphi_0 = 0.75$ **radians.**

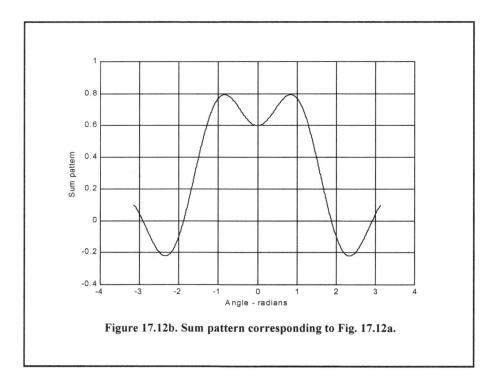

Figure 17.12b. Sum pattern corresponding to Fig. 17.12a.

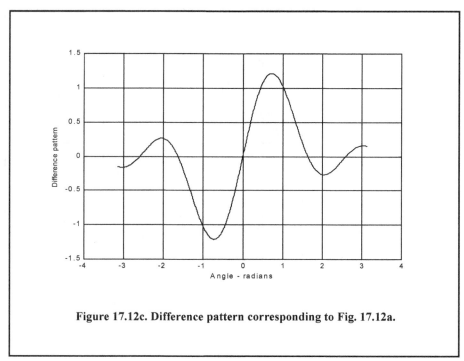

Figure 17.12c. Difference pattern corresponding to Fig. 17.12a.

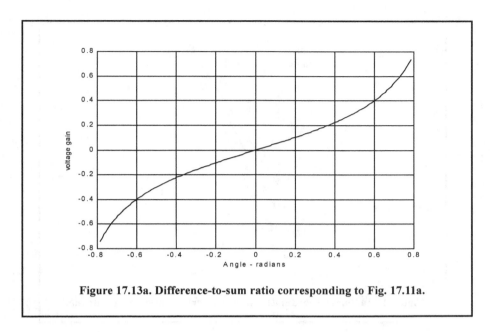

Figure 17.13a. Difference-to-sum ratio corresponding to Fig. 17.11a.

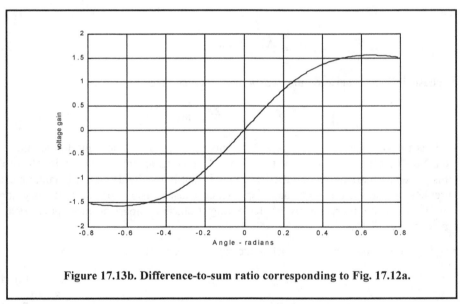

Figure 17.13b. Difference-to-sum ratio corresponding to Fig. 17.12a.

Consider Fig. 17.14; since the angle α is equal to $\varphi + \pi/2$, it follows that

$$R_1^2 = R^2 + \left(\frac{d}{2}\right)^2 - 2\frac{d}{2}R\cos\left(\varphi + \frac{\pi}{2}\right) = R^2 + \frac{d^2}{4} - dR\sin\varphi, \qquad \textbf{Eq. (17.12)}$$

and since $d \ll R$, we can use the binomial series expansion to get

$$R_1 \approx R\left(1 + \frac{d}{2R}\sin\varphi\right). \qquad \textbf{Eq. (17.13)}$$

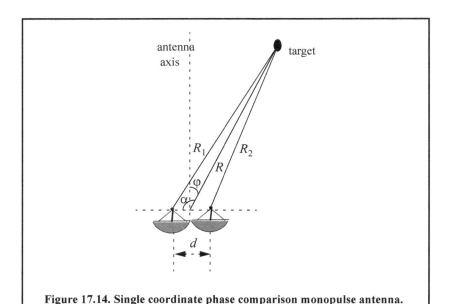

Figure 17.14. Single coordinate phase comparison monopulse antenna.

Similarly,

$$R_2 \approx R\left(1 - \frac{d}{2R}\sin\varphi\right).$$ **Eq. (17.14)**

The phase difference between the two elements is then given by

$$\phi = \frac{2\pi}{\lambda}(R_1 - R_2) = \frac{2\pi}{\lambda}d\sin\varphi$$ **Eq. (17.15)**

where λ is the wavelength. The phase difference ϕ is used to determine the angular target location. Note that if $\phi = 0$, then the target would be on the antenna's main axis. The problem with this phase comparison monopulse technique is that it is quite difficult to maintain a stable measurement of the off-boresight angle φ, which causes serious performance degradation. This problem can be overcome by implementing a phase comparison monopulse system as illustrated in Fig. 17.15.

The (single coordinate) sum and difference signals are, respectively, given by

$$\Sigma(\varphi) = S_1 + S_2$$ **Eq. (17.16)**

$$\Delta(\varphi) = S_1 - S_2$$ **Eq. (17.17)**

where the S_1 and S_2 are the signals in the two elements. Now, since S_1 and S_2 have similar amplitude and are different in phase by ϕ, we can write

$$S_1 = S_2 e^{-j\phi}.$$ **Eq. (17.18)**

It follows that

$$\Delta(\varphi) = S_2(1 - e^{-j\phi})$$ **Eq. (17.19)**

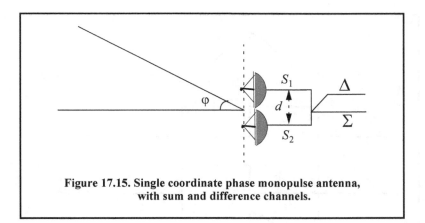

Figure 17.15. Single coordinate phase monopulse antenna, with sum and difference channels.

$$\Sigma(\varphi) = S_2(1 + e^{-j\phi}).$$

Eq. (17.20)

The phase error signal is computed from the ratio Δ / Σ. More precisely,

$$\frac{\Delta}{\Sigma} = \frac{1 - e^{-j\phi}}{1 + e^{-j\phi}} = j\tan\left(\frac{\phi}{2}\right),$$

Eq. (17.21)

which is purely imaginary. The modulus of the error signal is then given by

$$\frac{|\Delta|}{|\Sigma|} = \tan\left(\frac{\phi}{2}\right).$$

Eq. (17.22)

This kind of phase comparison monopulse tracker is often called the half-angle tracker.

17.4. Range Tracking

Target range is measured by estimating the round-trip delay of the transmitted pulses. The process of continuously estimating the range of a moving target is known as range tracking. Since the range to a moving target is changing with time, the range tracker must be constantly adjusted to keep the target locked in range. This can be accomplished using a split gate system, where two range gates (early and late) are utilized. The concept of split gate tracking is illustrated in Fig. 17.16, where a sketch of a typical pulsed radar echo is shown in the figure. The early gate opens at the anticipated starting time of the radar echo and lasts for half its duration. The late gate opens at the center and closes at the end of the echo signal. For this purpose, good estimates of the echo duration and the pulse center time must be reported to the range tracker so that the early and late gates can be placed properly at the start and center times of the expected echo. This reporting process is widely known as the "designation process."

The early gate produces positive voltage output while the late gate produces negative voltage output. The outputs of the early and late gates are subtracted, and the difference signal is fed into an integrator to generate an error signal. If both gates are placed properly in time, the integrator output will be equal to zero. Alternatively, when the gates are not timed properly, the integrator output is not zero, which gives an indication that the gates must be moved in time, left or right, depending on the sign of the integrator output.

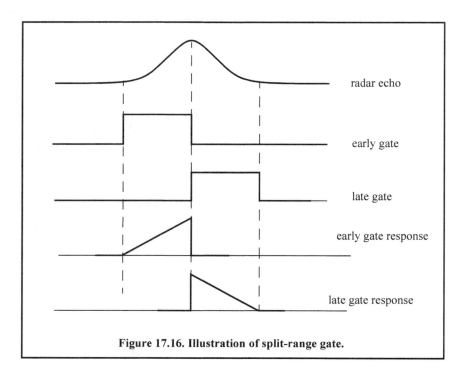

Figure 17.16. Illustration of split-range gate.

Multiple Target Tracking

Track-while-scan radar systems sample each target once per scan interval, and use sophisticated smoothing and prediction filters to estimate the target parameters between scans. To this end, the Kalman filter and the Alpha-Beta-Gamma ($\alpha\beta\gamma$) filter are commonly used. Once a particular target is detected, the radar may transmit up to a few pulses to verify the target parameters, before it establishes a track file for that target. Target position, velocity, and acceleration comprise the major components of the data maintained by a track file.

The principles of recursive tracking and prediction filters are presented in this part. First, an overview of state representation for Linear Time Invariant (LTI) systems is discussed. Then, second- and third-order one-dimensional fixed-gain polynomial filter trackers are developed. These filters are, respectively, known as the $\alpha\beta$ and $\alpha\beta\gamma$ filters (also known as the g-h and g-h-k filters). Finally, the equations for an n-dimensional multi-state Kalman filter is introduced and analyzed. As a matter of notation, lower case letters, with an underbar, are used.

17.5. Track-While-Scan (TWS)

Modern radar systems are designed to perform multi-function operations, such as detection, tracking, and discrimination. With the aid of sophisticated computer systems, multi-function radars are capable of simultaneously tracking many targets. In this case, each target is sampled once (mainly range and angular position) during a dwell interval (scan). Then, by using smoothing and prediction techniques, future samples can be estimated. Radar systems that can perform multi-tasking and multi-target tracking are known as Track-While-Scan (TWS) radars.

Once a TWS radar detects a new target, it initiates a separate track file for that detection; this ensures that sequential detections from that target are processed together to estimate the target's future parameters. Position, velocity, and acceleration comprise the main components of the track file. Typically, at least one other confirmation detection (verify detection) is required before the track file is established.

Unlike single target tracking systems, TWS radars must decide whether each detection (observation) belongs to a new target or belongs to a target that has been detected in earlier scans. And in order to accomplish this task, TWS radar systems utilize correlation and association algorithms. In the correlation process, each new detection is correlated with all previous detections in order to avoid establishing redundant tracks. If a certain detection correlates with more than one track, then a pre-determined set of association rules are exercised so that the detection is assigned to the proper track. A simplified TWS data processing block diagram is shown in Fig. 17.17.

Choosing a suitable tracking coordinate system is the first problem a TWS radar has to confront. It is desirable that a fixed reference of an inertial coordinate system be adopted. The radar measurements consist of target range, velocity, azimuth angle, and elevation angle. The TWS system places a gate around the target position and attempts to track the signal within this gate. The gate dimensions are normally azimuth, elevation, and range. Because of the uncertainty associated with the exact target position during the initial detections, a gate has to be large enough so that targets do not move appreciably from scan to scan; more precisely, targets must stay within the gate boundary during successive scans. After the target has been observed for several scans, the size of the gate is reduced considerably.

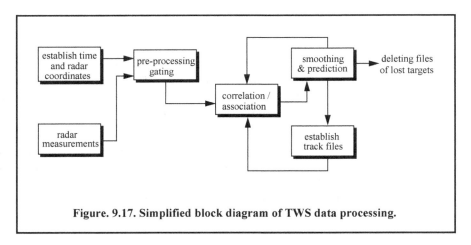

Figure. 9.17. Simplified block diagram of TWS data processing.

Gating is used to decide whether an observation is assigned to an existing track file, or to a new track file (new detection). Gating algorithms are normally based on computing a statistical error distance between a measured and an estimated radar observation. For each track file, an upper bound for this error distance is normally set. If the computed difference for a certain radar observation is less than the maximum error distance of a given track file, then the observation is assigned to that track.

All observations that have an error distance less than the maximum distance of a given track are said to correlate with that track. For each observation that does not correlate with any existing tracks, a new track file is established accordingly. Since new detections (measurements) are compared to all existing track files, a track file may then correlate with no observations or with one or more observations. The correlation between observations and all existing track files is identified using a correlation matrix. Rows of the correlation matrix represent radar observations, while columns represent track files. In cases where several observations correlate with more than one track file, a set of pre-determined association rules can be utilized so that a single observation is assigned to a single track file.

17.6. State Variable Representation of an LTI System

A linear time invariant system (continuous or discrete) can be described mathematically using three variables. They are the input, output, and the state variables. In this representation, any LTI system has observable or measurable objects (abstracts). For example, in the case of a radar system, range may be an object measured or observed by the radar tracking filter. States can be derived in many different ways. For the scope of this book, states of an object or an abstract are the components of the vector that contains the object and its time derivatives. For example, a third-order one-dimensional (in this case range) state vector representing range can be given by

$$\mathbf{x} = \begin{bmatrix} R \\ \dot{R} \\ \ddot{R} \end{bmatrix}$$

Eq. (17.23)

where R, $\dot{R}$, and $\ddot{R}$ are, respectively, the range measurement, range rate (velocity), and acceleration. The state vector defined in Eq. (17.23) can be representative of continuous or discrete states. In this book, the emphasis is on discrete time representation, since most radar signal processing is executed using digital computers. For this purpose, an n-dimensional state vector has the following form:

$$\mathbf{x} = \begin{bmatrix} x_1 & \dot{x}_1 & \dots & x_2 & \dot{x}_2 & \dots & x_n & \dot{x}_n & \dots \end{bmatrix}^t \qquad \text{Eq. (17.24)}$$

where the superscript indicates the transpose operation.

The LTI system of interest can be represented using the following state equations:

$$\dot{\mathbf{x}}(t) = \mathbf{A}\mathbf{x}(t) + \mathbf{B}\mathbf{w}(t) \qquad \text{Eq. (17.25)}$$

$$\mathbf{y}(t) = \mathbf{C}\mathbf{x}(t) + \mathbf{D}\mathbf{w}(t) \qquad \text{Eq. (17.26)}$$

where $\dot{\mathbf{x}}$ is the value of the $n \times 1$ state vector; $\mathbf{y}$ is the value of the $p \times 1$ output vector; $\mathbf{w}$ is the value of the $m \times 1$ input vector; $\mathbf{A}$ is an $n \times n$ matrix; $\mathbf{B}$ is an $n \times m$ matrix; $\mathbf{C}$ is $p \times n$ matrix; and $\mathbf{D}$ is an $p \times m$ matrix. The homogeneous solution (i.e., $\mathbf{w} = \mathbf{0}$) to this linear system, assuming known initial condition $\mathbf{x}(0)$ at time t_0, has the form

$$\mathbf{x}(t) = \Phi(t - t_0)\mathbf{x}(t - t_0). \qquad \text{Eq. (17.27)}$$

The matrix Φ is known as the state transition matrix, or fundamental matrix, and is equal to

$$\Phi(t - t_0) = e^{\mathbf{A}(t - t_0)}. \qquad \text{Eq. (17.28)}$$

Eq. (17.28) can be expressed in series format as

$$\Phi(t - t_0)\Big|_{t_0 = 0} = e^{\mathbf{A}(t)} = \mathbf{I} + \mathbf{A}t + \mathbf{A}^2\frac{t^2}{2!} + \dots = \sum_{k=0}^{\infty} \mathbf{A}^k \frac{t^k}{k!} \qquad \text{Eq. (17.29)}$$

where $\mathbf{I}$ is the identity matrix.

Example:

Compute the state transition matrix for an LTI system when

$$\mathbf{A} = \begin{bmatrix} 0 & 1 \\ -0.5 & -1 \end{bmatrix}.$$

Solution:

The state transition matrix can be computed using Eq. (17.29). For this purpose, compute $\mathbf{A}^2$ *and* $\mathbf{A}^3$.... *It follows*

$$\mathbf{A}^2 = \begin{bmatrix} -\dfrac{1}{2} & -1 \\ \dfrac{1}{2} & \dfrac{1}{2} \end{bmatrix} \qquad \mathbf{A}^3 = \begin{bmatrix} \dfrac{1}{2} & \dfrac{1}{2} \\ -\dfrac{1}{4} & 0 \end{bmatrix} \qquad \dots$$

Therefore,

$$
\Phi = \begin{bmatrix} 1 + 0t - \dfrac{\frac{1}{2}t^2}{2!} + \dfrac{\frac{1}{2}t^3}{3!} + \ldots & 0 + t - \dfrac{t^2}{2!} + \dfrac{\frac{1}{2}t^3}{3!} + \ldots \\[2em] 0 - \dfrac{1}{2}t + \dfrac{\frac{1}{2}t^2}{2!} - \dfrac{\frac{1}{4}t^3}{3!} + \ldots & 1 - t + \dfrac{\frac{1}{2}t^2}{2!} + \dfrac{0t^3}{3!} + \ldots \end{bmatrix}.
$$

The state transition matrix has the following properties (the proof is left as an exercise):

1. *Derivative property*

$$
\frac{\partial}{\partial t}\Phi(t - t_0) = \mathbf{A}\Phi(t - t_0) \tag{Eq. (17.30)}
$$

2. *Identity property*

$$
\Phi(t_0 - t_0) = \Phi(0) = \mathbf{I} \tag{Eq. (17.31)}
$$

3. *Initial value property*

$$
\left. \frac{\partial}{\partial t}\Phi(t - t_0) \right|_{t = t_0} = \mathbf{A} \tag{Eq. (17.32)}
$$

4. *Transition property*

$$
\Phi(t_2 - t_0) = \Phi(t_2 - t_1)\Phi(t_1 - t_0) \qquad ; \; t_0 \le t_1 \le t_2 \tag{Eq. (17.33)}
$$

5. *Inverse property*

$$
\Phi(t_0 - t_1) = \Phi^{-1}(t_1 - t_0) \tag{Eq. (17.34)}
$$

6. *Separation property*

$$
\Phi(t_1 - t_0) = \Phi(t_1)\Phi^{-1}(t_0) \tag{Eq. (17.35)}
$$

The general solution to the system defined in Eq. (17.25) can be written as

$$
\mathbf{x}(t) = \Phi(t - t_0)\mathbf{x}(t_0) + \int_{t_0}^{t} \Phi(t - \tau)\mathbf{B}\mathbf{w}(\tau)d\tau . \tag{Eq. (17.36)}
$$

The first term of the right-hand side of Eq. (17.36) represents the contribution from the system response to the initial condition. The second term is the contribution due to the driving force $\underline{w}$. By combining Eqs. (17.26) and (17.36), an expression for the output is computed as

$$
\mathbf{y}(t) = \mathbf{C}e^{\mathbf{A}(t - t_0)}\mathbf{x}(t_0) + \int_{t_0}^{t} [\mathbf{C}e^{\mathbf{A}(t - \tau)}\mathbf{B} - \mathbf{D}\delta(t - \tau)]\mathbf{w}(\tau)d\tau . \tag{Eq. (17.37)}
$$

Note that the system impulse response is equal to $\mathbf{C}e^{\mathbf{A}t}\mathbf{B} - \mathbf{D}\delta(t)$.

The difference equations describing a discrete time system, equivalent to Eqs. (17.25) and (17.26), are

$$
\mathbf{x}(n + 1) = \mathbf{A}\mathbf{x}(n) + \mathbf{B}\mathbf{w}(n) \tag{Eq. (17.38)}
$$

$$\mathbf{y}(n) = \mathbf{C}\ \mathbf{x}(n) + \mathbf{D}\mathbf{w}(n) \qquad\qquad \text{Eq. (17.39)}$$

where n defines the discrete time nT and T is the sampling interval. All other vectors and matrices were defined earlier. The homogeneous solution to the system defined in Eq. (17.38), with initial condition $\mathbf{x}(n_0)$, is

$$\mathbf{x}(n) = \mathbf{A}^{n - n_0}\mathbf{x}(n_0). \qquad\qquad \text{Eq. (17.40)}$$

In this case, the state transition matrix is an $n \times n$ matrix given by

$$\Phi(n, n_0) = \Phi(n - n_0) = \mathbf{A}^{n - n_0}. \qquad\qquad \text{Eq. (17.41)}$$

The following is the list of properties associated with the discrete transition matrix:

$$\Phi(n + 1 - n_0) = \mathbf{A}\Phi(n - n_0) \qquad\qquad \text{Eq. (17.42)}$$

$$\Phi(n_0 - n_0) = \Phi(0) = \mathbf{I} \qquad\qquad \text{Eq. (17.43)}$$

$$\Phi(n_0 + 1 - n_0) = \Phi(1) = \mathbf{A} \qquad\qquad \text{Eq. (17.44)}$$

$$\Phi(n_2 - n_0) = \Phi(n_2 - n_1)\Phi(n_1 - n_0) \qquad\qquad \text{Eq. (17.45)}$$

$$\Phi(n_0 - n_1) = \Phi^{-1}(n_1 - n_0) \qquad\qquad \text{Eq. (17.46)}$$

$$\Phi(n_1 - n_0) = \Phi(n_1)\Phi^{-1}(n_0) \qquad\qquad \text{Eq. (17.47)}$$

The solution to the general case (i.e., non-homogeneous system) is given by

$$\mathbf{x}(n) = \Phi(n - n_0)\mathbf{x}(n_0) + \sum_{m = n_0}^{n - 1} \Phi(n - m - 1)\mathbf{B}\mathbf{w}(m). \qquad\qquad \text{Eq. (17.48)}$$

It follows that the output is given by

$$\mathbf{y}(n) = \mathbf{C}\Phi(n - n_0)\mathbf{x}(n_0) + \sum_{m = n_0}^{n - 1} \mathbf{C}\ \Phi(n - m - 1)\mathbf{B}\mathbf{w}(m) + \mathbf{D}\mathbf{w}(n) \qquad\qquad \text{Eq. (17.49)}$$

where the system impulse response is given by

$$\mathbf{h}(n) = \sum_{m = n_0}^{n - 1} \mathbf{C}\ \Phi(n - m - 1)\mathbf{B}\underline{\delta}(m) + \mathbf{D}\underline{\delta}(n) \qquad\qquad \text{Eq. (17.50)}$$

where $\underline{\delta}$ is a vector.

Taking the Z-transform for Eqs. (17.38) and (17.39) yields

$$z\mathbf{x}(z) = \mathbf{A}\mathbf{x}(z) + \mathbf{B}\mathbf{w}(z) + z\mathbf{x}(0) \qquad\qquad \text{Eq. (17.51)}$$

$$\mathbf{y}(z) = \mathbf{C}\mathbf{x}(z) + \mathbf{D}\mathbf{w}(z). \qquad\qquad \text{Eq. (17.52)}$$

Manipulating Eqs. (17.51) and (17.52) yields

$$\mathbf{x}(z) = [z\mathbf{I} - \mathbf{A}]^{-1}\underline{B}\mathbf{w}(z) + [z\mathbf{I} - \mathbf{A}]^{-1}z\mathbf{x}(0) \qquad \text{Eq. (17.53)}$$

$$\mathbf{y}(z) = \{\mathbf{C}[z\mathbf{I} - \mathbf{A}]^{-1}\mathbf{B} + \mathbf{D}\}\mathbf{w}(z) + \mathbf{C}[z\mathbf{I} - \mathbf{A}]^{-1}z\mathbf{x}(0) . \qquad \text{Eq. (17.54)}$$

It follows that the state transition matrix is

$$\Phi(z) = z[z\mathbf{I} - \mathbf{A}]^{-1} = [\mathbf{I} - z^{-1}\mathbf{A}]^{-1} , \qquad \text{Eq. (17.55)}$$

and the system impulse response in the z-domain is

$$\mathbf{h}(z) = \mathbf{C}\Phi(z)z^{-1}\mathbf{B} + \mathbf{D} . \qquad \text{Eq. (17.56)}$$

17.7. The LTI System of Interest

For the purpose of establishing the framework necessary for the Kalman filter development, consider the LTI system shown in Fig. 17.18. This system (which is a special case of the system described in the previous section) can be described by the following first-order differential vector equations

$$\dot{\mathbf{x}}(t) = \mathbf{A}\ \mathbf{x}(t) + \mathbf{u}(t) \qquad \text{Eq. (17.57)}$$

$$\mathbf{y}(t) = \mathbf{G}\ \mathbf{x}(t) + \mathbf{v}(t) \qquad \text{Eq. (17.58)}$$

where $\mathbf{y}$ is the observable part of the system (i.e., output), $\mathbf{u}$ is a driving force, and $\mathbf{v}$ is the measurement noise. The matrices $\mathbf{A}$ and $\mathbf{G}$ vary depending on the system. The noise observation $\mathbf{v}$ is assumed to be uncorrelated. If the initial condition vector is $\mathbf{x}(t_0)$, then from Eq. (17.36) we get

$$\mathbf{x}(t) = \Phi(t - t_0)\mathbf{x}(t_0) + \int_{t_0}^{t}\Phi(t - \tau)\mathbf{u}(\tau)d\tau . \qquad \text{Eq. (17.59)}$$

The object (abstract) is observed only at discrete times determined by the system. These observation times are declared by discrete time nT where T is the sampling interval. Using the same notation adopted in the previous section, the discrete time representations of Eqs. (17.57) and (17.58) are

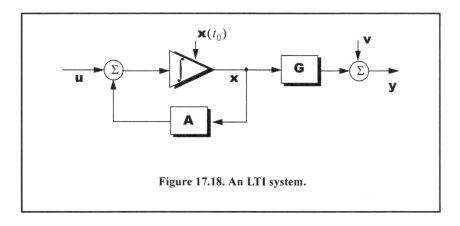

Figure 17.18. An LTI system.

$$\mathbf{x}(n) = \mathbf{A}\ \mathbf{x}(n-1) + \mathbf{u}(n) \qquad \text{Eq. (17.60)}$$

$$\mathbf{y}(n) = \mathbf{G}\mathbf{x}(n) + \mathbf{v}(n) \ . \qquad \text{Eq. (17.61)}$$

The homogeneous solution to this system is given in Eq. (17.27) for continuous time, and in Eq. (17.40) for discrete time.

The state transition matrix corresponding to this system can be obtained using Taylor series expansion of the vector $\mathbf{x}$. More precisely,

$$x = x + T\dot{x} + \frac{T^2}{2!}\ddot{x} + \dots$$
$$\dot{x} = \dot{x} + T\ddot{x} + \dots \qquad \text{Eq. (17.62)}$$
$$\ddot{x} = \ddot{x} + \dots$$

It follows that the elements of the state transition matrix are defined by

$$\Phi[ij] = \left\{ \begin{array}{ll} T^{j-i} \div (j-i)! & 1 \le i, j \le n \\ 0 & j < i \end{array} \right\}. \qquad \text{Eq. (17.63)}$$

Using matrix notation, the state transition matrix is then given by

$$\Phi = \begin{bmatrix} 1 & T & \frac{T^2}{2!} & \dots \\ 0 & 1 & T & \dots \\ 0 & 0 & 1 & \dots \\ \dots & \dots & \dots & \dots \end{bmatrix}. \qquad \text{Eq. (17.64)}$$

The matrix given in Eq. (17.64) is often called the Newtonian matrix.

17.8. Fixed-Gain Tracking Filters

This class of filters (or estimators) is also known as "Fixed-Coefficient" filters. The most common examples of this class of filters are the $\alpha\beta$ and $\alpha\beta\gamma$ filters and their variations. The $\alpha\beta$ and $\alpha\beta\gamma$ trackers are one-dimensional second- and third-order filters, respectively. They are equivalent to special cases of the one-dimensional Kalman filter. The general structure of this class of estimators is similar to that of the Kalman filter.

The standard $\alpha\beta\gamma$ filter provides smoothed and predicted data for target position, velocity (Doppler), and acceleration. It is a polynomial predictor/corrector linear recursive filter. This filter can reconstruct position, velocity, and constant acceleration based on position measurements. The $\alpha\beta\gamma$ filter can also provide a smoothed (corrected) estimate of the present position, which can be used in guidance and fire control operations.

Notation:

For the purpose of the discussion presented in the remainder of this chapter, the following notation is adopted: $x(n|m)$ represents the estimate during the nth sampling interval, using all data up to and including the mth sampling interval; y_n is the nth measured value; and e_n is the nth residual (error).

The fixed-gain filter equation is given by

$$\mathbf{x}(n|n) = \Phi\underline{x}(n-1|n-1) + \mathbf{K}[y_n - \mathbf{G}\Phi\mathbf{x}(n-1|n-1)].$$

Eq. (17.65)

Since the transition matrix assists in predicting the next state,

$$\mathbf{x}(n+1|n) = \Phi\underline{x}(n|n).$$

Eq. (17.66)

Substituting Eq. (17.66) into Eq. (17.65) yields

$$\mathbf{x}(n|n) = \mathbf{x}(n|n-1) + \mathbf{K}[y_n - \mathbf{G}\mathbf{x}(n|n-1)].$$

Eq. (17.67)

The term enclosed within the brackets on the right-hand side of Eq. (17.67) is often called the residual (error), which is the difference between the measured input and predicted output. Eq. (17.67) means that the estimate of $\mathbf{x}(n)$ is the sum of the prediction and the weighted residual. The term $\mathbf{G}\mathbf{x}(n|n-1)$ represents the prediction state. In the case of the $\alpha\beta\gamma$ estimator, $\mathbf{G}$ is the row vector given by

$$\mathbf{G} = \begin{bmatrix} 1 & 0 & 0 & ... \end{bmatrix},$$

Eq. (17.68)

and the gain matrix $\mathbf{K}$ is given by

$$\mathbf{K} = \begin{bmatrix} \alpha \\ \beta/T \\ \gamma/T^2 \end{bmatrix}.$$

Eq. (17.69)

One of the main objectives of a tracking filter is to decrease the effect of the noise observation on the measurement. For this purpose, the noise covariance matrix is calculated. More precisely, the noise covariance matrix is

$$\mathbf{C}(n|n) = E[(\mathbf{x}(n|n)\ )\mathbf{x}^t(n|n)] \qquad ;\ y_n = v_n$$

Eq. (17.70)

where E indicates the expected value operator. Noise is assumed to be a zero mean random process with variance equal to σ_v^2. Additionally, noise measurements are assumed to be uncorrelated,

$$E[v_n v_m] = \begin{cases} \delta\sigma_v^2 & n = m \\ 0 & n \neq m \end{cases}.$$

Eq. (17.71)

Eq. (17.65) can be written as

$$\mathbf{x}(n|n) = \mathbf{A}\mathbf{x}(n-1|n-1) + \mathbf{K}y_n$$

Eq. (17.72)

where

$$\mathbf{A} = (\mathbf{I} - \mathbf{K}\mathbf{G})\Phi.$$

Eq. (17.73)

Substituting Eqs. (17.72) and (17.73) into Eq. (17.70) yields

$$\mathbf{C}(n|n) = E[(\mathbf{A}\mathbf{x}(n-1|n-1) + \mathbf{K}y_n)(\mathbf{A}\mathbf{x}(n-1|n-1) + \mathbf{K}y_n)'].$$

Eq. (17.74)

Expanding the right-hand side of Eq. (17.74), and using Eq. (17.71), gives

$$\mathbf{C}(n|n) = \mathbf{AC}(n-1|n-1)\mathbf{A}^t + \mathbf{K}\sigma_v^2\mathbf{K}^t.$$ Eq. (17.75)

Under the steady-state condition, Eq. (17.75) collapses to

$$\mathbf{C}(n|n) = \mathbf{ACA}^t + \mathbf{K}\sigma_v^2\mathbf{K}^t$$ Eq. (17.76)

where $\mathbf{C}$ is the steady-state noise covariance matrix. In the steady-state,

$$\mathbf{C}(n|n) = \mathbf{C}(n-1|n-1) = \mathbf{C} \quad \textit{for any } n$$ Eq. (17.77)

Several criteria can be used to establish the performance of the fixed-gain tracking filter. The most commonly used technique is to compute the Variance Reduction Ratio (VRR). The VRR is defined only when the input to the tracker is noise measurements. It follows that in the steady-state case, the VRR is the steady-state ratio of the output variance (auto-covariance) to the input measurement variance.

In order to determine the stability of the tracker under consideration, consider the Z-transform for Eq. (17.72),

$$\mathbf{x}(z) = \mathbf{A}z^{-1}\mathbf{x}(z) + \mathbf{K}y_n(z).$$ Eq. (17.78)

Rearranging Eq. (17.78) yields the following system transfer functions:

$$\mathbf{h}(z) = \frac{\mathbf{x}(z)}{y_n(z)} = (\mathbf{I} - \mathbf{A}z^{-1})^{-1}\mathbf{K}$$ Eq. (17.79)

where $(\mathbf{I} - \mathbf{A}z^{-1})$ is called the characteristic matrix. Note that the system transfer functions can exist only when the characteristic matrix is a non-singular matrix. Additionally, the system is stable if and only if the roots of the characteristic equation are within the unit circle in the z-plane,

$$|(\mathbf{I} - \mathbf{A}z^{-1})| = 0.$$ Eq. (17.80)

The filter's steady-state errors can be determined with the help of Fig. 17.19. The error transfer function is

$$\mathbf{e}(z) = \frac{\mathbf{y}(z)}{1 + \mathbf{h}(z)},$$ Eq. (17.81)

and by using Abel's theorem, the steady-state error is

$$\mathbf{e}_\infty = \lim_{t \to \infty} \mathbf{e}(t) = \lim_{z \to 1}\left(\frac{z-1}{z}\right)\mathbf{e}(z).$$ Eq. (17.82)

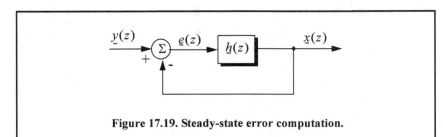

Figure 17.19. Steady-state error computation.

Substituting Eq. (17.82) into (17.81) yields

$$\mathbf{e}_\infty = \lim_{z \to 1} \frac{z-1}{z} \frac{\mathbf{y}(z)}{1 + \mathbf{h}(z)} .$$
Eq. (17.83)

17.8.1. The $\alpha\beta$ *Filter*

The $\alpha\beta$ tracker produces, on the *nth* observation, smoothed estimates for position and velocity, and a predicted position for the $(n + 1)th$ observation. Figure 17.20 shows an implementation of this filter. Note that the subscripts "*p*" and "*s*" are used to indicate, respectively, the predicated and smoothed values. The $\alpha\beta$ tracker can follow an input ramp (constant velocity) with no steady-state errors. However, a steady-state error will accumulate when constant acceleration is present in the input. Smoothing is done to reduce errors in the predicted position through adding a weighted difference between the measured and predicted values to the predicted position, as follows:

$$x_s(n) = x(n|n) = x_p(n) + \alpha(x_0(n) - x_p(n))$$
Eq. (17.84)

$$\dot{x}_s(n) = x'(n|n) = \dot{x}_s(n-1) + \frac{\beta}{T} (x_0(n) - x_p(n)) .$$
Eq. (17.85)

x_0 is the position input samples. The predicted position is given by

$$x_p(n) = x_s(n|n-1) = x_s(n-1) + T\dot{x}_s(n-1) .$$
Eq. (17.86)

The initialization process is defined by

$$x_s(1) = x_p(2) = x_0(1)$$

$$\dot{x}_s(1) = 0$$

$$\dot{x}_s(2) = \frac{x_0(2) - x_0(1)}{T} .$$

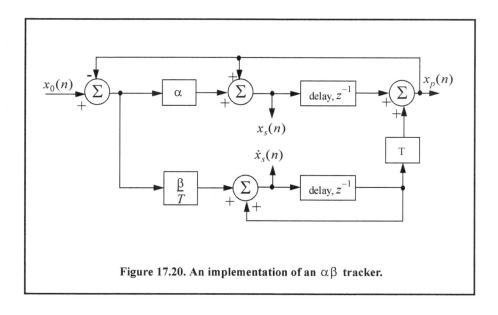

Figure 17.20. An implementation of an $\alpha\beta$ **tracker.**

A general form for the covariance matrix was developed in the previous section, and is given in Eq. (17.75). In general, a second-order one-dimensional covariance matrix (in the context of the $\alpha\beta$ filter) can be written as

$$\mathbf{C}(n|n) = \begin{bmatrix} C_{xx} & C_{x\dot{x}} \\ C_{\dot{x}x} & C_{\dot{x}\dot{x}} \end{bmatrix} \qquad \text{Eq. (17.87)}$$

where, in general, C_{xy} is

$$C_{xy} = E\{xy^t\}. \qquad \text{Eq. (17.88)}$$

By inspection, the $\alpha\beta$ filter has

$$\mathbf{A} = \begin{bmatrix} 1-\alpha & (1-\alpha)T \\ -\beta/T & (1-\beta) \end{bmatrix} \qquad \text{Eq. (17.89)}$$

$$\mathbf{K} = \begin{bmatrix} \alpha \\ \beta/T \end{bmatrix} \qquad \text{Eq. (17.90)}$$

$$\mathbf{G} = \begin{bmatrix} 1 & 0 \end{bmatrix} \qquad \text{Eq. (17.91)}$$

$$\Phi = \begin{bmatrix} 1 & T \\ 0 & 1 \end{bmatrix}. \qquad \text{Eq. (17.92)}$$

Finally, using Eqs. (17.89) through (17.92) in Eq. (17.72) yields the steady-state noise covariance matrix,

$$\mathbf{C} = \frac{\sigma_v^2}{\alpha(4-2\alpha-\beta)} \begin{bmatrix} 2\alpha^2 - 3\alpha\beta + 2\beta & \dfrac{\beta(2\alpha-\beta)}{T} \\ \dfrac{\beta(2\alpha-\beta)}{T} & \dfrac{2\beta^2}{T^2} \end{bmatrix}. \qquad \text{Eq. (17.93)}$$

It follows that the position and velocity VRR ratios are, respectively, given by

$$(VRR)_x = C_{xx}/\sigma_v^2 = \frac{2\alpha^2 - 3\alpha\beta + 2\beta}{\alpha(4-2\alpha-\beta)} \qquad \text{Eq. (17.94)}$$

$$(VRR)_{\dot{x}} = C_{\dot{x}\dot{x}}/\sigma_v^2 = \frac{1}{T^2} \frac{2\beta^2}{\alpha(4-2\alpha-\beta)}. \qquad \text{Eq. (17.95)}$$

The stability of the $\alpha\beta$ filter is determined from its system transfer functions. For this purpose, compute the roots for Eq. (17.80) with $\mathbf{A}$ from Eq. (17.89),

$$|\mathbf{I} - \mathbf{A}z^{-1}| = 1 - (2-\alpha-\beta)z^{-1} + (1-\alpha)z^{-2} = 0. \qquad \text{Eq. (17.96)}$$

Solving Eq. (17.96) for z yields

$$z_{1,2} = 1 - \frac{\alpha+\beta}{2} \pm \frac{1}{2}\sqrt{(\alpha-\beta)^2 - 4\beta}, \qquad \text{Eq. (17.97)}$$

and in order to guarantee stability,

$$|z_{1,2}| < 1.$$ **Eq. (17.98)**

Two cases are analyzed. First, $z_{1,2}$ are real. In this case (the details are left as an exercise),

$$\beta > 0 \qquad ; \ \alpha > -\beta.$$ **Eq. (17.99)**

The second case is when the roots are complex; in this case we find

$$\alpha > 0.$$ **Eq. (17.100)**

The system transfer functions can be derived by using Eqs. (17.79), (17.89), and (17.90),

$$\begin{bmatrix} h_x(z) \\ h_{\dot{x}}(z) \end{bmatrix} = \frac{1}{z^2 - z(2 - \alpha - \beta) + (1 - \alpha)} \begin{bmatrix} \alpha z\left(z - \dfrac{(\alpha - \beta)}{\alpha}\right) \\ \dfrac{\beta z(z - 1)}{T} \end{bmatrix}.$$ **Eq. (17.101)**

Up to this point all relevant relations concerning the $\alpha\beta$ filter were made with no regard to how to choose the gain coefficients (α and β). Before considering the methodology of selecting these coefficients, consider the main objective behind using this filter. The twofold purpose of the $\alpha\beta$ tracker can be described as follows:

1. *The tracker must reduce the measurement noise as much as possible.*

2. *The filter must be able to track maneuvering targets, with as little residual (tracking error) as possible.*

The reduction of measurement noise is normally determined by the VRR ratios. However, the maneuverability performance of the filter depends heavily on the choice of the parameters α and β.

A special variation of the $\alpha\beta$ filter was developed by Benedict and Bordner[1] and is often referred to as the Benedict-Bordner filter. The main advantage of the Benedict-Bordner is reducing the transient errors associated with the $\alpha\beta$ tracker. This filter uses both the position and velocity VRR ratios as measures of performance. It computes the sum of the squared differences between the input (position) and the output when the input has a unit step velocity at time zero. Additionally, it computes the squared differences between the real velocity and the velocity output when the input is as described earlier. Both error differences are minimized when

$$\beta = \frac{\alpha^2}{2 - \alpha}.$$ **Eq. (17.102)**

In this case, the position and velocity VRR ratios are, respectively, given by

$$(VRR)_x = \frac{\alpha(6 - 5\alpha)}{\alpha^2 - 8\alpha + 8}$$ **Eq. (17.103)**

$$(VRR)_{\dot{x}} = \frac{2}{T^2} \frac{\alpha^3/(2 - \alpha)}{\alpha^2 - 8\alpha + 8}.$$ **Eq. (17.104)**

1. Benedict, T. R. and Bordner, G. W., Synthesis of an Optimal Set of Radar Track-While-Scan Smoothing Equations. *IRE Transaction on Automatic Control, AC-7.* July 1962, pp. 27-32.

Another important sub-class of the $\alpha\beta$ tracker is the critically damped filter, often called the fading memory filter. In this case, the filter coefficients are chosen on the basis of a smoothing factor ξ, where $0 \leq \xi \leq 1$. The gain coefficients are given by

$$\alpha = 1 - \xi^2$$

Eq. (17.105)

$$\beta = (1 - \xi)^2.$$

Eq. (17.106)

Heavy smoothing means $\xi \to 1$ and little smoothing means $\xi \to 0$. The elements of the covariance matrix for a fading memory filter are

$$C_{xx} = \frac{1-\xi}{(1+\xi)^3} (1 + 4\xi + 5\xi^2) \sigma_v^2$$

Eq. (17.107)

$$C_{x\dot{x}} = C_{\dot{x}x} = \frac{1}{T} \frac{1-\xi}{(1+\xi)^3} (1 + 2\xi + 3\xi^2) \sigma_v^2$$

Eq. (17.108)

$$C_{\dot{x}\dot{x}} = \frac{2}{T^2} \frac{1-\xi}{(1+\xi)^3} (1-\xi)^2 \sigma_v^2.$$

Eq. (17.109)

17.8.2. The $\alpha\beta\gamma$ Filter

The $\alpha\beta\gamma$ tracker produces, for the *nth* observation, smoothed estimates of position, velocity, and acceleration. It also produces the predicted position and velocity for the $(n+1)th$ observation. An implementation of the $\alpha\beta\gamma$ tracker is shown in Fig. 17.21.

The $\alpha\beta\gamma$ tracker will follow an input whose acceleration is constant with no steady-state errors. Again, in order to reduce the error at the output of the tracker, a weighted difference between the measured and predicted values is used in estimating the smoothed position, velocity, and acceleration as follows:

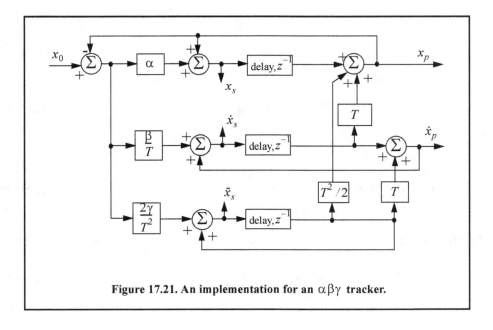

Figure 17.21. An implementation for an $\alpha\beta\gamma$ tracker.

$$x_s(n) = x_p(n) + \alpha(x_0(n) - x_p(n))$$
Eq. (17.110)

$$\dot{x}_s(n) = \dot{x}_s(n-1) + T\ddot{x}_s(n-1) + \frac{\beta}{T}(x_0(n) - x_p(n))$$
Eq. (17.111)

$$\ddot{x}_s(n) = \ddot{x}_s(n-1) + \frac{2\gamma}{T^2}(x_0(n) - x_p(n))$$
Eq. (17.112)

$$x_p(n+1) = x_s(n) + T\dot{x}_s(n) + \frac{T^2}{2}\ddot{x}_s(n) \ .$$
Eq. (17.113)

and the initialization process is

$$x_s(1) = x_p(2) = x_0(1)$$

$$\dot{x}_s(1) = \ddot{x}_s(1) = \ddot{x}_s(2) = 0$$

$$\dot{x}_s(2) = \frac{x_0(2) - x_0(1)}{T}$$

$$\ddot{x}_s(3) = \frac{x_0(3) + x_0(1) - 2x_0(2)}{T^2} \ .$$

Using Eq. (17.63), the state transition matrix for the $\alpha\beta\gamma$ filter is

$$\Phi = \begin{bmatrix} 1 & T & \dfrac{T^2}{2} \\ 0 & 1 & T \\ 0 & 0 & 1 \end{bmatrix}.$$
Eq. (17.114)

The covariance matrix (which is symmetric) can be computed from Eq. (17.76). For this purpose, note that

$$\mathbf{K} = \begin{bmatrix} \alpha \\ \beta/T \\ \gamma/T^2 \end{bmatrix}$$
Eq. (17.115)

$$\mathbf{G} = \begin{bmatrix} 1 & 0 & 0 \end{bmatrix}$$
Eq. (17.116)

and

$$\mathbf{A} = (\mathbf{I} - \mathbf{KG})\Phi = \begin{bmatrix} 1-\alpha & (1-\alpha)T & (1-\alpha)T^2/2 \\ -\beta/T & -\beta+1 & (1-\beta/2)T \\ -2\gamma/T^2 & -2\gamma/T & (1-\gamma) \end{bmatrix}.$$
Eq. (17.117)

Substituting Eq. (17.117) into (17.76) and collecting terms, the VRR ratios are computed as

$$(VRR)_x = \frac{2\beta(2\alpha^2 + 2\beta - 3\alpha\beta) - \alpha\gamma(4 - 2\alpha - \beta)}{(4 - 2\alpha - \beta)(2\alpha\beta + \alpha\gamma - 2\gamma)}$$
Eq. (17.118)

$$(VRR)_{\dot{x}} = \frac{4\beta^3 - 4\beta^2\gamma + 2\gamma^2(2-\alpha)}{T^2(4-2\alpha-\beta)(2\alpha\beta+\alpha\gamma-2\gamma)} \qquad \text{Eq. (17.119)}$$

$$(VRR)_{\ddot{x}} = \frac{4\beta\gamma^2}{T^4(4-2\alpha-\beta)(2\alpha\beta+\alpha\gamma-2\gamma)}. \qquad \text{Eq. (17.120)}$$

As in the case of any discrete time system, this filter will be stable if and only if all of its poles fall within the unit circle in the z-plane.

The $\alpha\beta\gamma$ characteristic equation is computed by setting

$$\left| \mathbf{I} - \mathbf{A}z^{-1} \right| = 0. \qquad \text{Eq. (17.121)}$$

Substituting Eq. (17.117) into (17.121) and collecting terms yields the following characteristic function:

$$f(z) = z^3 + (-3\alpha + \beta + \gamma)z^2 + (3 - \beta - 2\alpha + \gamma)z - (1-\alpha). \qquad \text{Eq. (17.122)}$$

The $\alpha\beta\gamma$ becomes a Benedict-Bordner filter when

$$2\beta - \alpha\left(\alpha + \beta + \frac{\gamma}{2}\right) = 0. \qquad \text{Eq. (17.123)}$$

Note that for $\gamma = 0$, Eq. (17.123) reduces to Eq. (17.102). For a critically damped filter the gain coefficients are

$$\alpha = 1 - \xi^3 \qquad \text{Eq. (17.124)}$$

$$\beta = 1.5(1-\xi^2)(1-\xi) = 1.5(1-\xi)^2(1+\xi) \qquad \text{Eq. (17.125)}$$

$$\gamma = (1-\xi)^3. \qquad \text{Eq. (17.126)}$$

Note that heavy smoothing takes place when $\xi \to 1$, while $\xi = 0$ means that no smoothing is present.

MATLAB Function "ghk_tracker.m"

The function *"ghk_tracker.m"* implements the steady-state $\alpha\beta\gamma$ filter. The syntax is as follows:

[residual, estimate] = ghk_tracker (X0, smoocof, inp, npts, T, nvar)

where

Symbol	Description	Status
X0	*initial state vector*	*input*
smoocof	*desired smoothing coefficient*	*input*
inp	*array of position measurements*	*input*
npts	*number of points in input position*	*input*
T	*sampling interval*	*input*
nvar	*desired noise variance*	*input*

Symbol	Description	Status
residual	*array of position error (residual)*	*output*
estimate	*array of predicted position*	*output*

Note that *"ghk_tracker.m"* uses MATLAB's function *"normrnd.m"* to generate zero mean Gaussian noise, which is part of MATLAB's Statistics Toolbox. If this toolbox is not available to the user, then *"ghk_tracker.m"* function-call must be modified to

[residual, estimate] = ghk_tracker1 (X0, smoocof, inp, npts, T)

In this case, noise measurements are either considered to be unavailable or are part of the position input array.

To illustrate how to use the functions *"ghk_tracker.m"* and *"ghk_tracker1.m,"* consider the inputs shown in Figs. 17.22 and 17.23. Figure 17.22 assumes an input with lazy maneuvering, while Figure 17.23 assumes an aggressive maneuvering case. These figures can be reproduced using MATLAB program *"Fig17_20s.m,"* listed in Appendix 17-A.

Figures 17.24 and 17.25 show the residual error and predicted position corresponding to Fig. 17.22 assuming the cases: heavy smoothing and little smoothing with and without noise. The noise is white Gaussian with zero mean and variance of $\sigma_v^2 = 0.05$. Figures 17. 26 and 17.27 show the residual error and predicted position corresponding to Fig. 17.23 with and without noise.

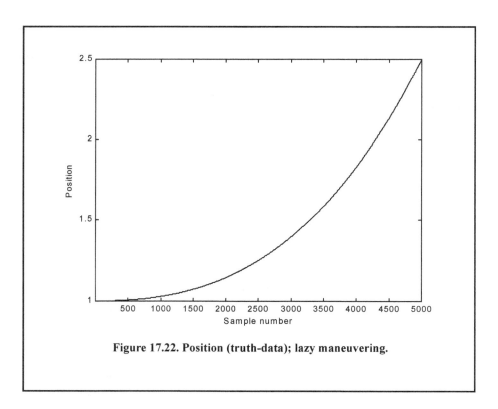

Figure 17.22. Position (truth-data); lazy maneuvering.

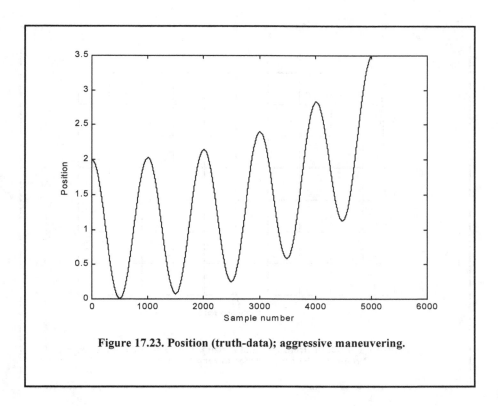

Figure 17.23. Position (truth-data); aggressive maneuvering.

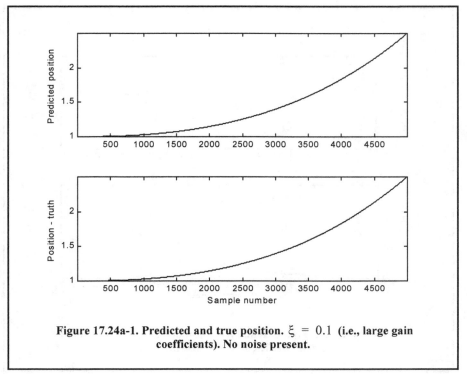

Figure 17.24a-1. Predicted and true position. $\xi = 0.1$ (i.e., large gain coefficients). No noise present.

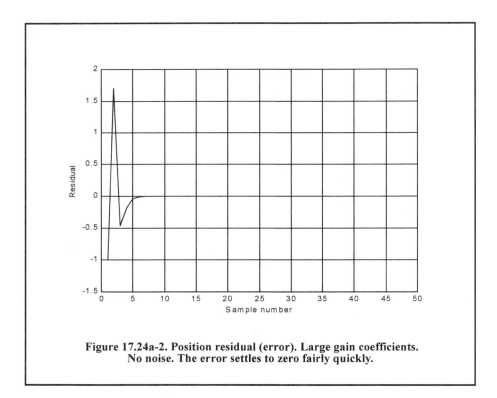

Figure 17.24a-2. Position residual (error). Large gain coefficients. No noise. The error settles to zero fairly quickly.

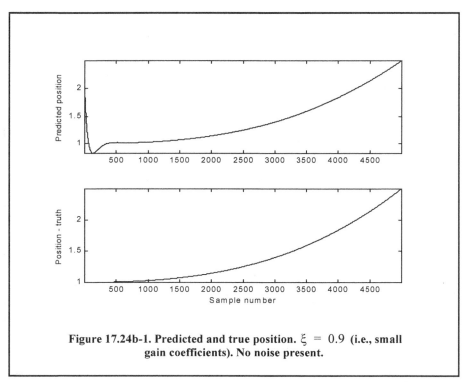

Figure 17.24b-1. Predicted and true position. $\xi = 0.9$ **(i.e., small gain coefficients). No noise present.**

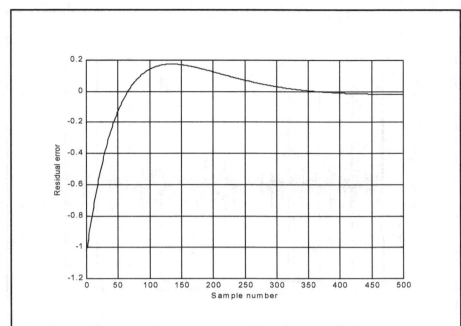

Figure 17.24b-2. Position residual (error). Small gain coefficients. No noise. It takes the filter a longer time for the error to settle down.

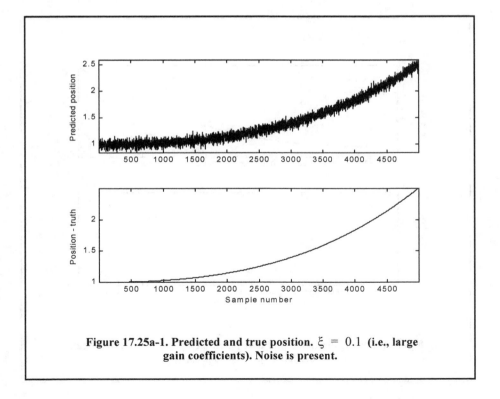

Figure 17.25a-1. Predicted and true position. $\xi = 0.1$ (i.e., large gain coefficients). Noise is present.

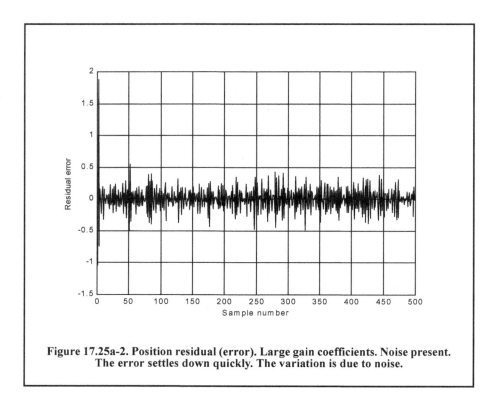

**Figure 17.25a-2. Position residual (error). Large gain coefficients. Noise present.
The error settles down quickly. The variation is due to noise.**

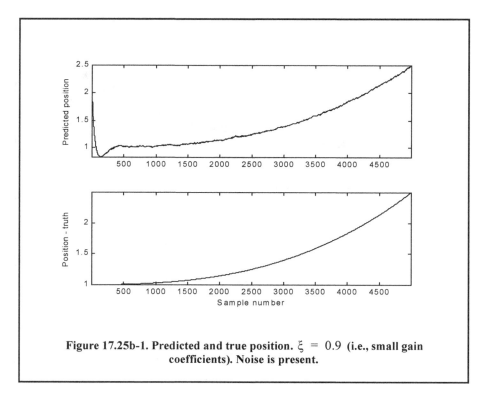

**Figure 17.25b-1. Predicted and true position. $\xi = 0.9$ (i.e., small gain
coefficients). Noise is present.**

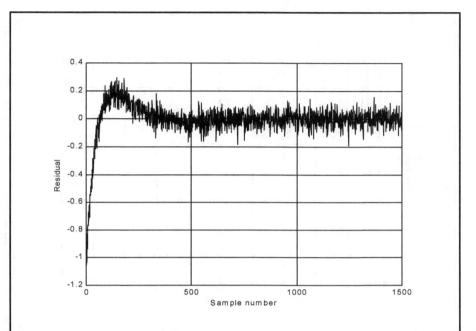

Figure 17.25b-2. Position residual (error). Small gain coefficients. Noise present. The error requires more time before settling down. The variation is due to noise.

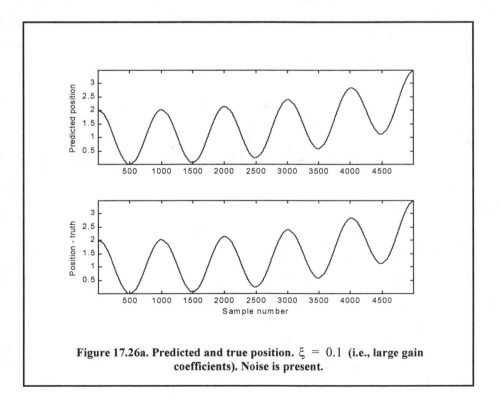

Figure 17.26a. Predicted and true position. $\xi = 0.1$ (i.e., large gain coefficients). Noise is present.

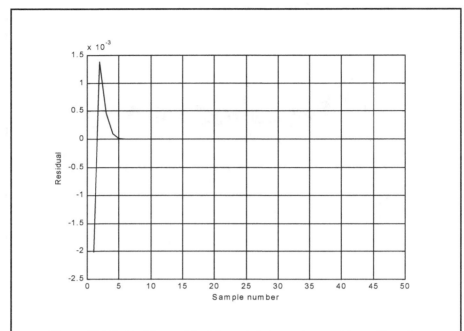

Figure 17.26b. Position residual (error). Large gain coefficients. No noise. The error settles down quickly.

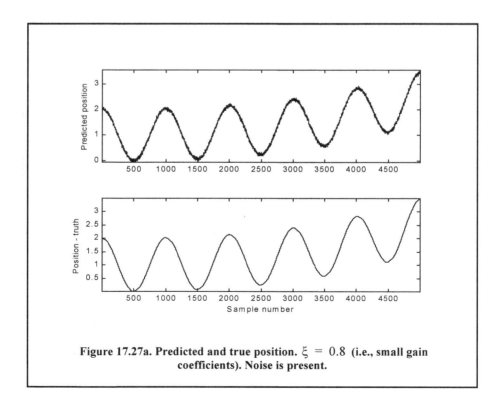

Figure 17.27a. Predicted and true position. $\xi = 0.8$ (i.e., small gain coefficients). Noise is present.

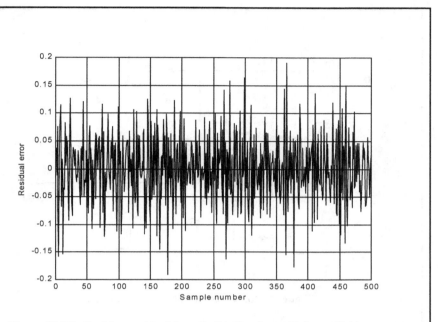

Figure 17.27b. Position residual (error). Small gain coefficients. Noise present. The error stays fairly large; however, its average is around zero. The variation is due to noise.

17.9. The Kalman Filter

The Kalman filter is a linear estimator that minimizes the mean squared error as long as the target dynamics are modeled accurately. All other recursive filters, such as the $\alpha\beta\gamma$ and the Benedict-Bordner filters, are special cases of the general solution provided by the Kalman filter for the mean squared estimation problem. Additionally, the Kalman filter has the following advantages:

1. *The gain coefficients are computed dynamically. This means that the same filter can be used for a variety of maneuvering target environments.*

2. *The Kalman filter gain computation adapts to varying detection histories, including missed detections.*

3. *The Kalman filter provides an accurate measure of the covariance matrix. This allows for better implementation of the gating and association processes.*

4. *The Kalman filter makes it possible to partially compensate for the effects of mis-correlation and mis-association.*

Many derivations of the Kalman filter exist in the literature; only results are provided in this chapter. Figure17.28 shows a block diagram for the Kalman filter. The Kalman filter equations can be deduced from Fig. 17.28. The filtering equation is

$$\mathbf{x}(n|n) = \mathbf{x}_s(n) = \mathbf{x}(n|n-1) + \mathbf{K}(n)[\mathbf{y}(n) - \mathbf{G}\mathbf{x}(n|n-1)].$$
\hfill **Eq. (17.127)**

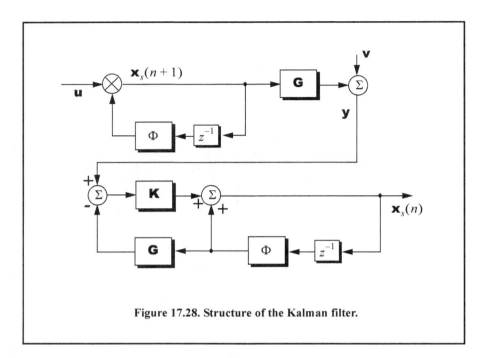

Figure 17.28. Structure of the Kalman filter.

The measurement vector is

$$\mathbf{y}(n) = \mathbf{G}\mathbf{x}(n) + \mathbf{v}(n)$$ Eq. (17.128)

where $\mathbf{y}(n)$ is zero mean, white Gaussian noise with covariance $\Re_c$,

$$\Re_c = E\{\mathbf{y}(n)\ \mathbf{y}^t(n)\}.$$ Eq. (17.129)

The gain (weight) vector is dynamically computed as

$$\mathbf{K}(n) = \mathbf{P}(n|n-1)\mathbf{G}^t[\mathbf{G}\mathbf{P}(n|n-1)\mathbf{G}^t + \Re_c]^{-1}$$ Eq. (17.130)

where the measurement noise matrix $\mathbf{P}$ represents the predictor covariance matrix, and is equal to

$$\mathbf{P}(n+1|n) = E\{\mathbf{x}_s(n+1)\mathbf{x}^*_s(n)\} = \Phi\mathbf{P}(n|n)\Phi^t + \mathbf{Q}$$ Eq. (17.131)

where $\mathbf{Q}$ is the covariance matrix for the input $\mathbf{u}$,

$$\mathbf{Q} = E\{\mathbf{u}(n)\ \mathbf{u}^t(n)\}.$$ Eq. (17.132)

The corrector equation (covariance of the smoothed estimate) is

$$\mathbf{P}(n|n) = [\mathbf{I} - \mathbf{K}(n)\mathbf{G}]\mathbf{P}(n|n-1).$$ Eq. (17.133)

Finally, the predictor equation is

$$\mathbf{x}(n+1|n) = \Phi\mathbf{x}(n|n).$$ Eq. (17.134)

17.9.1. The Singer $\alpha\beta\gamma$-Kalman Filter

The Singer[1] filter is a special case of the Kalman, where the filter is governed by a specified target dynamic model whose acceleration is a random process with autocorrelation function given by

$$E\{\ddot{x}(t) \ \ddot{x}(t+t_1)\} = \sigma_a^2 \ e^{-\frac{|t_1|}{\tau_m}} \qquad \text{Eq. (17.135)}$$

where τ_m is the correlation time of the acceleration due to target maneuvering or atmospheric turbulence. The correlation time τ_m may vary from as low as 10 seconds for aggressive maneuvering to as large as 60 seconds for lazy maneuvering cases.

Singer defined the random target acceleration model by a first-order Markov process given by

$$\ddot{x}(n+1) = \rho_m \ \ddot{x}(n) + \sqrt{1-\rho_m^2} \ \sigma_m \ w(n) \qquad \text{Eq. (17.136)}$$

where $w(n)$ is a zero mean, Gaussian random variable with unity variance, σ_m is the maneuver standard deviation, and the maneuvering correlation coefficient ρ_m is given by

$$\rho_m = e^{-\frac{T}{\tau_m}}. \qquad \text{Eq. (17.137)}$$

The continuous time domain system that corresponds to these conditions is the same as the Wiener-Kolmogorov whitening filter, which is defined by the differential equation

$$\frac{d}{dt}v(t) = -\beta_m v(t) + w(t) \qquad \text{Eq. (17.138)}$$

where β_m is equal to $1/\tau_m$. The maneuvering variance using Singer's model is given by

$$\sigma_m^2 = \frac{A_{max}^2}{3}[1 + 4P_{max} - P_0]. \qquad \text{Eq. (17.139)}$$

A_{max} is the maximum target acceleration with probability P_{max}, and the term P_0 defines the probability that the target has no acceleration.

The transition matrix that corresponds to the Singer filter is given by

$$\Phi = \begin{bmatrix} 1 & T & \frac{1}{\beta_m^2}(-1+\beta_m T+\rho_m) \\ 0 & 1 & \frac{1}{\beta_m}(1-\rho_m) \\ 0 & 0 & \rho_m \end{bmatrix}. \qquad \text{Eq. (17.140)}$$

Note that when $T\beta_m = T/\tau_m$ is small (the target has constant acceleration), then Eq. (17.140) reduces to Eq. (17.114). Typically, the sampling interval T is much less than the maneuvering

---

1. Singer, R. A., Estimating Optimal Tracking Filter Performance for Manned Maneuvering Targets, *IEEE Transaction on Aerospace and Electronics, AES-5*, July, 1970. pp. 473-483.

time constant τ_m; hence, Eq. (17.140) can be accurately replaced by its second-order approximation. More precisely,

$$\Phi = \begin{bmatrix} 1 & T & T^2/2 \\ 0 & 1 & T(1 - T/2\tau_m) \\ 0 & 0 & \rho_m \end{bmatrix}.$$

Eq. (17.141)

The covariance matrix was derived by Singer, and it is equal to

$$\mathbf{C} = \frac{2\sigma_m^2}{\tau_m} \begin{bmatrix} C_{11} & C_{12} & C_{13} \\ C_{21} & C_{22} & C_{23} \\ C_{31} & C_{32} & C_{33} \end{bmatrix}$$

Eq. (17.142)

where

$$C_{11} = \sigma_x^2 = \frac{1}{2\beta_m^5}\left[1 - e^{-2\beta_m T} + 2\beta_m T + \frac{2\beta_m^3 T^3}{3} - 2\beta_m^2 T^2 - 4\beta_m Te^{-\beta_m T}\right]$$

Eq. (17.143)

$$C_{12} = C_{21} = \frac{1}{2\beta_m^4}[e^{-2\beta_m T} + 1 - 2e^{-\beta_m T} + 2\beta_m Te^{-\beta_m T} - 2\beta_m T + \beta_m^2 T^2]$$

Eq. (17.144)

$$C_{13} = C_{31} = \frac{1}{2\beta_m^3}[1 - e^{-2\beta_m T} - 2\beta_m Te^{-\beta_m T}]$$

Eq. (17.145)

$$C_{22} = \frac{1}{2\beta_m^3}[4e^{-\beta_m T} - 3 - e^{-2\beta_m T} + 2\beta_m T]$$

Eq. (17.146)

$$C_{23} = C_{32} = \frac{1}{2\beta_m^2}[e^{-2\beta_m T} + 1 - 2e^{-\beta_m T}]$$

Eq. (17.147)

$$C_{33} = \frac{1}{2\beta_m}[1 - e^{-2\beta_m T}].$$

Eq. (17.148)

Two limiting cases are of interest:

1. *The short sampling interval case* $(T \ll \tau_m)$,

$$\lim_{\beta_m T \to 0} \mathbf{C} = \frac{2\sigma_m^2}{\tau_m} \begin{bmatrix} T^5/20 & T^4/8 & T^3/6 \\ T^4/8 & T^3/3 & T^2/2 \\ T^3/6 & T^2/2 & T \end{bmatrix}$$

Eq. (17.149)

and the state transition matrix is computed from Eq. (17.141) as

$$\lim_{\beta_m T \to 0} \Phi = \begin{bmatrix} 1 & T & T^2/2 \\ 0 & 1 & T \\ 0 & 0 & 1 \end{bmatrix},$$

Eq. (17.150)

which is the same as the case for the $\alpha\beta\gamma$ filter (constant acceleration).

2. *The long sampling interval ($T \gg \tau_m$). This condition represents the case when acceleration is a white noise process. The corresponding covariance and transition matrices are, respectively, given by*

$$\lim_{\beta_m T \to \infty} \mathbf{C} = \sigma_m^2 \begin{bmatrix} \dfrac{2T^3\tau_m}{3} & T^2\tau_m & \tau_m^2 \\ T^2\tau_m & 2T\tau_m & \tau_m \\ \tau_m^2 & \tau_m & 1 \end{bmatrix} \qquad \text{Eq. (17.151)}$$

$$\lim_{\beta_m T \to \infty} \Phi = \begin{bmatrix} 1 & T & T\tau_m \\ 0 & 1 & \tau_m \\ 0 & 0 & 0 \end{bmatrix}. \qquad \text{Eq. (17.152)}$$

Note that under the condition that $T \gg \tau_m$, the cross correlation terms C_{13} and C_{23} become very small. It follows that estimates of acceleration are no longer available, and thus a two-state filter model can be used to replace the three-state model. In this case,

$$\mathbf{C} = 2\sigma_m^2\tau_m \begin{bmatrix} T^3/3 & T^2/2 \\ T^2/2 & T \end{bmatrix} \qquad \text{Eq. (17.153)}$$

$$\Phi = \begin{bmatrix} 1 & T \\ 0 & 1 \end{bmatrix}. \qquad \text{Eq. (17.154)}$$

17.9.2. Relationship between Kalman and $\alpha\beta\gamma$ Filters

The relationship between the Kalman filter and the $\alpha\beta\gamma$ filters can be easily obtained by using the appropriate state transition matrix $\underline{\Phi}$, and gain vector $\underline{K}$ corresponding to the $\alpha\beta\gamma$ in Eq. (17.127). Thus,

$$\begin{bmatrix} x(n|n) \\ \dot{x}(n|n) \\ \ddot{x}(n|n) \end{bmatrix} = \begin{bmatrix} x(n|n-1) \\ \dot{x}(n|n-1) \\ \ddot{x}(n|n-1) \end{bmatrix} + \begin{bmatrix} k_1(n) \\ k_2(n) \\ k_3(n) \end{bmatrix} [x_0(n) - x(n|n-1)] \qquad \text{Eq. (17.155)}$$

with (see Fig. 17.21)

$$x(n|n-1) = x_s(n-1) + T\ \dot{x}_s(n-1) + \frac{T^2}{2}\ \ddot{x}_s(n-1) \qquad \text{Eq. (17.156)}$$

$$\dot{x}(n|n-1) = \dot{x}_s(n-1) + T\ \ddot{x}_s(n-1) \qquad \text{Eq. (17.157)}$$

$$\ddot{x}(n|n-1) = \ddot{x}_s(n-1). \qquad \text{Eq. (17.158)}$$

Comparing the previous three equations with the $\alpha\beta\gamma$ filter equations yields,

$$\begin{bmatrix} \alpha \\ \beta \\ T \\ \dfrac{\gamma}{T^2} \end{bmatrix} = \begin{bmatrix} k_1 \\ k_2 \\ k_3 \end{bmatrix}.$$

Eq. (17.159)

Additionally, the covariance matrix elements are related to the gain coefficients by

$$\begin{bmatrix} k_1 \\ k_2 \\ k_3 \end{bmatrix} = \frac{1}{C_{11} + \sigma_v^2} \begin{bmatrix} C_{11} \\ C_{12} \\ C_{13} \end{bmatrix}.$$

Eq. (17.160)

Eq. (17.160) indicates that the first gain coefficient depends on the estimation error variance of the total residual variance, while the other two gain coefficients are calculated through the covariances between the second and third states and the first observed state.

MATLAB Function *"kalman_filter.m"*

The function *"kalman_filter.m"* implements a state Singer-$\alpha\beta\gamma$ Kalman filter. The syntax is as follows:

[residual, estimate] = kalman_filter(npts, T, X0, inp, R, nvar)

where

Symbol	Description	Status
npts	*number of points in input position*	*input*
T	*sampling interval*	*input*
X0	*initial state vector*	*input*
inp	*input array*	*input*
R	*noise variance see Eq. (10-129)*	*input*
nvar	*desired state noise variance*	*input*
residual	*array of position error (residual)*	*output*
estimate	*array of predicted position*	*output*

Note that *"kalman_filter.m"* uses MATLAB's function *"normrnd.m"* to generate zero mean Gaussian noise, which is part of MATLAB's Statistics Toolbox.

To illustrate how to use the functions *"kalman_filter.m,"* consider the inputs shown in Figs. 17.22 and 17.23. Figures 17.29 and 17.30 show the residual error and predicted position corresponding to Figures 17.22 and 17.23. These plots can be reproduced using the MATLAM program *"Fig17_28.m,"* listed in Appendix 17-A.

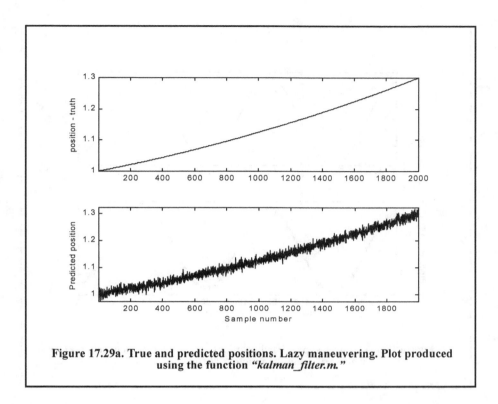

Figure 17.29a. True and predicted positions. Lazy maneuvering. Plot produced using the function *"kalman_filter.m."*

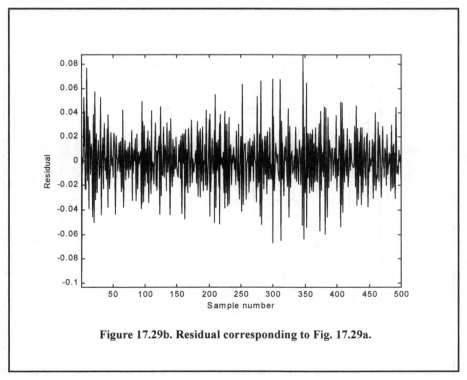

Figure 17.29b. Residual corresponding to Fig. 17.29a.

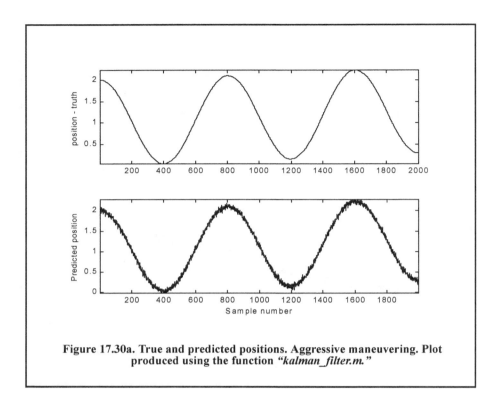

Figure 17.30a. True and predicted positions. Aggressive maneuvering. Plot produced using the function *"kalman_filter.m."*

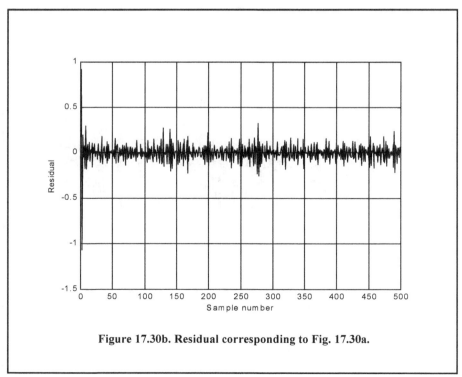

Figure 17.30b. Residual corresponding to Fig. 17.30a.

17.10. MATLAB Kalman Filter Simulation

For this purpose, the MATLAB GUI workspace entitled *"kalman_gui.m"* was developed. It is shown in Fig. 17.31. In this design, the inputs can be initialized to correspond to two target type kinematics (aircraft and missile). For example, when you click on the button *"ResetMissile,"* the initial *x-*, *y-*, and *z*-detection coordinates for the missile are loaded into the *"Starting Location"* field. The corresponding target velocity is also loaded in the *"velocity in x direction"* field. Finally, all other fields associated with the Kalman filter are also loaded using default values that are appropriate for this design case study. Note that the user can alter these entries as appropriate.

This program generates a fictitious default trajectory for the selected target type. This is accomplished using the function *"maketraj.m,"* listed in Appendix 17-A. Users can either use this program with its default trajectories, or import their own specific trajectory files. The function *"maketraj.m"* assumes constant altitude, and generates a maneuvering trajectory in the *x-y* plane, as shown in Fig. 17.32. This trajectory can be changed using the different fields in the *"trajectory Parameter"* fields.

Next the program corrupts the trajectory by adding white Gaussian noise. This is accomplished by the function *"addnoise.m,"* which is listed in Appendix 17-A. A six-state Kalman filter named *"kalfilt.m"* is then utilized to perform the tracking task. This function is also listed in Appendix 17-A.

The azimuth, elevation, and range errors are input to the program using their corresponding fields on the GUI. In the example used in this chapter, these entries are assumed constant throughout the simulation. In practice, this is not true and these values will change. They are calculated by the radar signal processor on a "per-processing-interval" basis and then are input into the tracker. For example, the standard deviation of the error in the range measurement is

$$\sigma_R = \frac{\Delta R}{\sqrt{2 \times SNR}} = \frac{c}{2B\sqrt{2 \times SNR}} \qquad \text{Eq. (17.161)}$$

where ΔR is the range resolution, c is the speed of light, B is the bandwidth, and SNR is the measurement SNR.

The standard deviation of the error in the velocity measurement is

$$\sigma_v = \frac{\lambda}{2\tau\sqrt{2 \times SNR}} \qquad \text{Eq. (17.162)}$$

where λ is the wavelength and τ is the uncompressed pulse width. The standard deviation of the error in the angle measurement is

$$\sigma_a = \frac{\Theta}{1.6\sqrt{2 \times SNR}} \qquad \text{Eq. (17.163)}$$

where Θ is the antenna beamwidth of the angular coordinate of the measurement (azimuth and elevation).

Table 17.1 lists the type of plots generated by this simulation. Figures 17.32 through Fig. 17.42 show typical outputs produced using this simulation, assuming the missile case, during any given run.

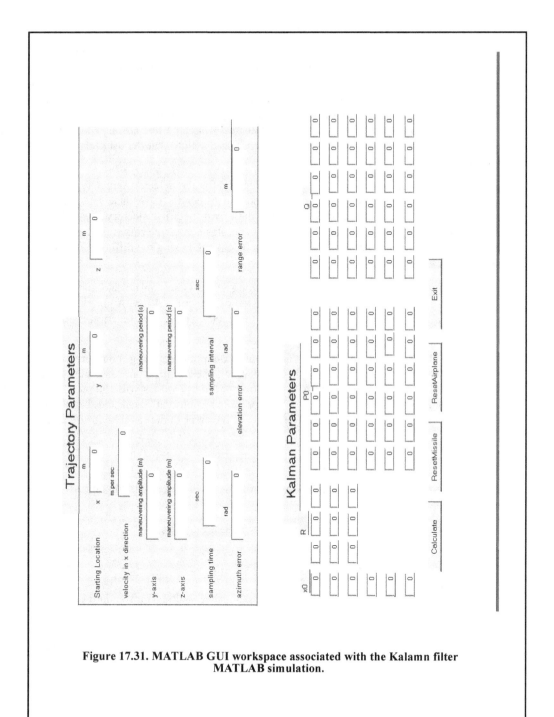

Figure 17.31. MATLAB GUI workspace associated with the Kalamn filter MATLAB simulation.

TABLE 17.1. Output list generated by the *"kalman_gui.m"* simulation

Figure #	Description
1	*uncorrupted input trajectory*
2	*corrupted input trajectory*
3	*corrupted and uncorrupted x-position*
4	*corrupted and uncorrupted y-position*
5	*corrupted and uncorrupted z-position*
6	*corrupted and filtered x-, y- and z- positions*
7	*predicted x-, y- and z- velocities*
8	*position residuals*
9	*velocity residuals*
10	*covariance matrix components versus time*
11	*Kalman filter gains versus time*

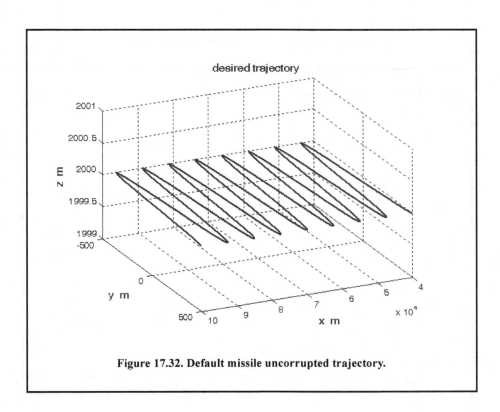

Figure 17.32. Default missile uncorrupted trajectory.

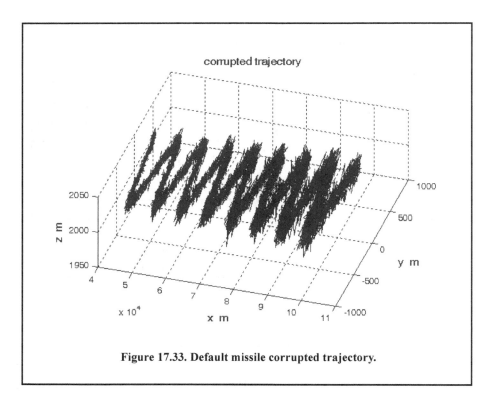

Figure 17.33. Default missile corrupted trajectory.

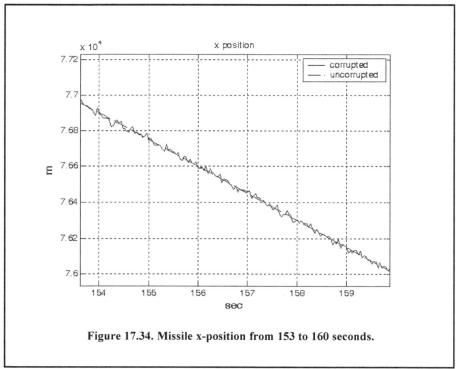

Figure 17.34. Missile x-position from 153 to 160 seconds.

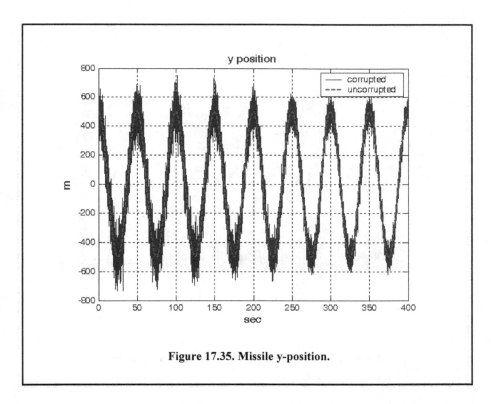

Figure 17.35. Missile y-position.

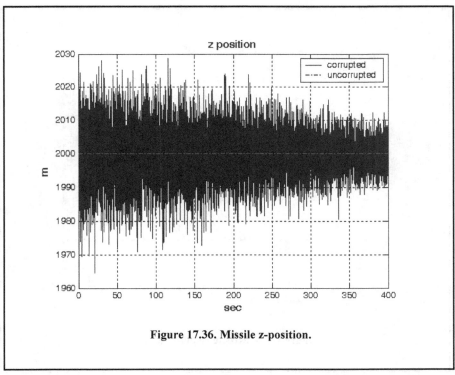

Figure 17.36. Missile z-position.

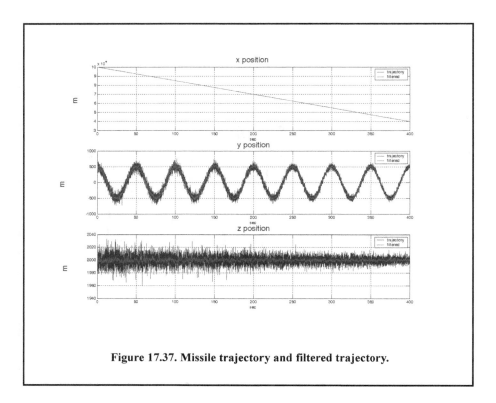

Figure 17.37. Missile trajectory and filtered trajectory.

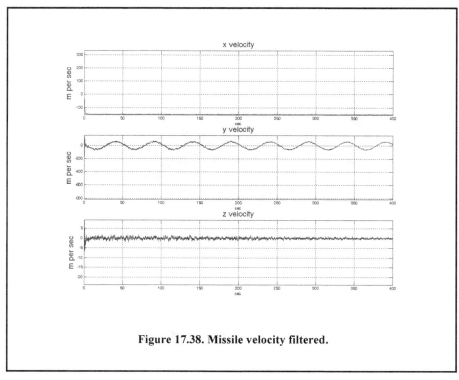

Figure 17.38. Missile velocity filtered.

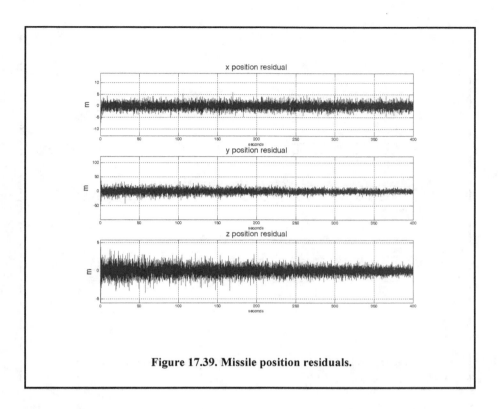

Figure 17.39. Missile position residuals.

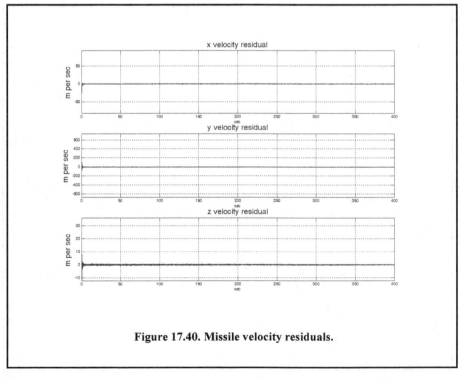

Figure 17.40. Missile velocity residuals.

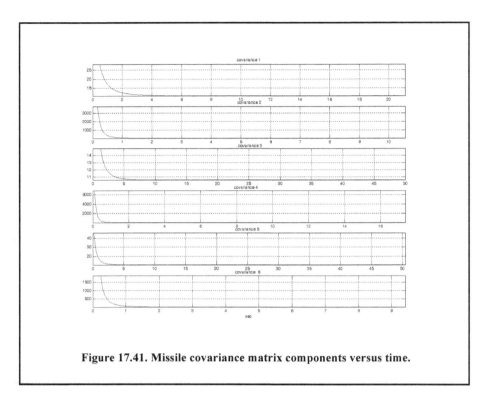

Figure 17.41. Missile covariance matrix components versus time.

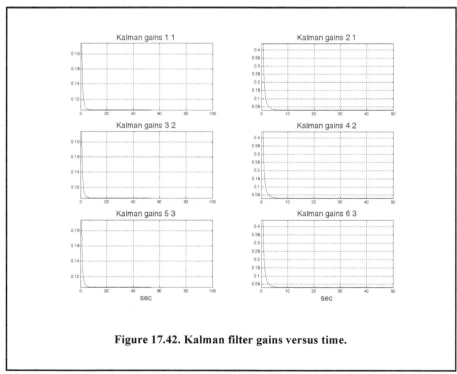

Figure 17.42. Kalman filter gains versus time.

Problems

17.1. Show that in order to be able to quickly achieve changing the beam position, the error signal needs to be a linear function of the deviation angle.

17.2. Prepare a short report on the vulnerability of conical scan to amplitude modulation jamming. In particular, consider the self-protecting technique called "Gain Inversion."

17.3. Consider a conical scan radar. The pulse repetition interval is $10 \mu s$. Calculate the scan rate so that at least ten pulses are emitted within one scan.

17.4. Consider a conical scan antenna whose rotation around the tracking axis is completed in 4 seconds. If during this time 20 pulses are emitted and received, calculate the radar PRF and the unambiguous range.

17.5. Reproduce Fig. 17.11 for $\varphi_0 = 0.05, 0.1$ and $\varphi_0 = 0.15$ radians.

17.6. Reproduce Fig. 17.13 for the squint angles defined in the previous problem.

17.7. Derive Eq. (17.33) and Eq. (17.34).

17.8. Consider a monopulse radar where the input signal is comprised of both target return and additive white Gaussian noise. Develop an expression for the complex ratio Σ / Δ.

17.9. To generate the sum and difference patterns for a linear array of size N, follow this algorithm: To form the difference pattern, multiply the first $N/2$ elements by -1 and the second $N/2$ elements by +1. Plot the sum and difference patterns for a linear array of size 60.

17.10. Generate the delta/sum patterns for a 21-element linear array using the form

$$\frac{\Delta}{\Sigma} = j \frac{V_\Delta}{\sqrt{|V_\Delta|^2 + |V_\Sigma|^2}}$$

where V_Δ is the difference voltage pattern and V_Σ is the sum voltage pattern.

17.11. Consider the sum and difference signals defined in Eqs. (17.7) and (17.8). What is the squint angle φ_0 that maximizes $\Sigma(\varphi = 0)$?

17.12. A certain system is defined by the following difference equation:

$$y(n) + 4y(n-1) + 2y(n-2) = w(n)$$

Find the solution to this system for $n > 0$ and $w = \delta$.

17.13. Prove the state transition matrix properties (i.e., Eqs. (17.30) through (17.36)).

17.14. Suppose that the state equations for a certain discrete time LTI system are

$$\begin{bmatrix} x_1(n+1) \\ x_2(n+1) \end{bmatrix} = \begin{bmatrix} 0 & 1 \\ -2 & -3 \end{bmatrix} \begin{bmatrix} x_1(n) \\ x_2(n) \end{bmatrix} + \begin{bmatrix} 0 \\ 1 \end{bmatrix} w(n).$$

If $y(0) = y(1) = 1$, find $y(n)$ when the input is a step function.

17.15. Derive Eq. (17.55).

17.16. Derive Eq. (17.75).

17.17. Using Eq. (17.83), compute a general expression (in terms of the transfer function) for the steady-state errors when the input sequence is:

$$u1 = \{0, 1, 1, 1, 1, ...\}$$

$$u2 = \{0, 1, 2, 3, ...\}$$

$$u3 = \{0, 1^2, 2^2, 3^2, ...\}$$

$$u4 = \{0, 1^3, 2^3, 3^3, ...\}$$

17.18. Verify the results in Eqs. (17.99) and (17.100).

17.19. Develop an expression for the steady-state error transfer function for an $\alpha\beta$ tracker.

17.20. Using the result of the previous problem and Eq. (17.83), compute the steady-state errors for the $\alpha\beta$ tracker with the inputs defined in Problem 17.13.

17.21. Design a critically damped $\alpha\beta$, when the measurement noise variance associated with position is $\sigma_v^2 = 50m$ and when the desired standard deviation of the filter prediction error is $5.5m$.

17.22. Derive Eqs. (17.118) through (17.120).

17.23. Derive Eq. (17.122).

17.24. Consider a $\alpha\beta\gamma$ filter. We can define six transfer functions: $H_1(z)$, $H_2(z)$, $H_3(z)$, $H_4(z)$, $H_5(z)$, and $H_6(z)$ (predicted position, predicted velocity, predicted acceleration, smoothed position, smoothed velocity, and smoothed acceleration). Each transfer function has the form

$$H(z) = \frac{a_3 + a_2 z^{-1} + a_1 z^{-2}}{1 + b_2 z^{-1} + b_1 z^{-2} + b_0 z^{-3}}.$$

The denominator remains the same for all six transfer functions. Compute all the relevant coefficients for each transfer function.

17.25. Verify the results obtained for the two limiting cases of the Singer-Kalman filter.

17.26. Verify Eq. (17.160).

Appendix 17-A: Chapter 17 MATLAB Code Listings

The MATLAB code provided in this chapter was designed as an academic standalone tool and is not adequate for other purposes. The code was written in a way to assist the reader in gaining a better understanding of the theory. The code was not developed, nor is it intended to be used as part of an open-loop or a closed-loop simulation of any kind. The MATLAB code found in this textbook can be downloaded from this book's web page on the CRC Press website. Simply use your favorite web browser, go to *www.crcpress.com*, and search for keyword *"Mahafza"* to locate this book's web page.

MATLAB Function "mono_pulse.m" Listing

```
function mono_pulse(phi0)
eps = 0.0000001;
angle = -pi:0.01:pi;
y1 = sinc(angle + phi0);
y2 = sinc((angle - phi0));
ysum = y1 + y2;
ydif = -y1 + y2;
figure (1)
plot (angle,y1,'k',angle,y2,'k');
grid;
xlabel ('Angle - radians')
ylabel ('Squinted patterns')
figure (2)
plot(angle,ysum,'k');
grid;
xlabel ('Angle - radians')
ylabel ('Sum pattern')
figure (3)
plot (angle,ydif,'k');
grid;
xlabel ('Angle - radians')
ylabel ('Difference pattern')
angle = -pi/4:0.01:pi/4;
y1 = sinc(angle + phi0);
y2 = sinc((angle - phi0));
ydif = -y1 + y2;
ysum = y1 + y2;
dovrs = ydif ./ ysum;
figure(4)
plot (angle,dovrs,'k');
grid;
xlabel ('Angle - radians')
ylabel ('voltage gain')
```

MATLAB Function "ghk_tracker.m" Listing

```
function [residual, estimate] = ghk_tracker (X0, smoocof, inp, npts, T, nvar)
rn = 1.;
% read the initial estimate for the state vector
X = X0;
theta = smoocof;
%compute values for alpha, beta, gamma
```

```
w1 = 1. - (theta^3);
w2 = 1.5 * (1. + theta) * ((1. - theta)^2) / T;
w3 = ((1. - theta)^3) / (T^2);
% setup the transition matrix PHI
PHI = [1. T (T^2)/2.;0. 1. T;0. 0. 1.];
while rn < npts ;
   %use the transition matrix to predict the next state
   XN = PHI * X;
   error = (inp(rn) + normrnd(0,nvar)) - XN(1);
   residual(rn) = error;
   tmp1 = w1 * error;
   tmp2 = w2 * error;
   tmp3 = w3 * error;
   % compute the next state
   X(1) = XN(1) + tmp1;
   X(2) = XN(2) + tmp2;
   X(3) = XN(3) + tmp3;
   estimate(rn) = X(1);
   rn = rn + 1.;
end
return
```

MATLAB Function "ghk_tracker1.m" Listing

```
function [residual, estimate] = ghk_tracker1 (X0, smoocof, inp, npts, T)
rn = 1.;
% read the initial estimate for the state vector
X = X0;
theta = smoocof;
%compute values for alpha, beta, gamma
w1 = 1. - (theta^3);
w2 = 1.5 * (1. + theta) * ((1. - theta)^2) / T;
w3 = ((1. - theta)^3) / (T^2);
% setup the transition matrix PHI
PHI = [1. T (T^2)/2.;0. 1. T;0. 0. 1.];
while rn < npts ;
   %use the transition matrix to predict the next state
   XN = PHI * X;
   error = inp(rn)  - XN(1);
   residual(rn) = error;
   tmp1 = w1 * error;
   tmp2 = w2 * error;
   tmp3 = w3 * error;
   % compute the next state
   X(1) = XN(1) + tmp1;
   X(2) = XN(2) + tmp2;
   X(3) = XN(3) + tmp3;
   estimate(rn) = X(1);
   rn = rn + 1.;
end
return
```

MATLAB Program "Fig17_20s.m" Listing

```
clear all
eps = 0.0000001;
npts = 5000;
del = 1./ 5000.;
t = 0. : del : 1.;
% generate input sequence
inp = 1.+ t.^3 + .5 .*t.^2 + cos(2.*pi*10 .* t) ;
% read the initial estimate for the state vector
X0 = [2,.1,.01]';
% this is the update interval in seconds
T = 100. * del;
% this is the value of the smoothing coefficient
xi = .91;
[residual, estimate] = ghk_tracker (X0, xi, inp, npts, T, .01);
figure(1)
plot (residual(1:500))
xlabel ('Sample number')
ylabel ('Residual error')
grid
figure(2)
NN = 4999.;
n = 1:NN;
plot (n,estimate(1:NN),'b',n,inp(1:NN),'r')
xlabel ('Sample number')
ylabel ('Position')
legend ('Estimated','Input')
```

MATLAB Function "kalman_filter.m" Listing

```
function [residual, estimate] = kalman_filter(npts, T, X0, inp, R, nvar)
N = npts;
rn=1;
% read the initial estimate for the state vector
X = X0;
% it is assumed that the measurement vector H=[1,0,0]
% this is the state noise variance
VAR = nvar;
% setup the initial value for the prediction covariance.
S = [1. 1. 1.; 1. 1. 1.; 1. 1. 1.];
% setup the transition matrix PHI
PHI = [1. T (T^2)/2.; 0. 1. T; 0. 0. 1.];
% setup the state noise covariance matrix
Q(1,1) = (VAR * (T^5)) / 20.;
Q(1,2) = (VAR * (T^4)) / 8.;
Q(1,3) = (VAR * (T^3)) / 6.;
Q(2,1) = Q(1,2);
Q(2,2) = (VAR * (T^3)) / 3.;
Q(2,3) = (VAR * (T^2)) / 2.;
Q(3,1) = Q(1,3);
Q(3,2) = Q(2,3);
Q(3,3) = VAR * T;
while rn < N ;
   %use the transition matrix to predict the next state
```

```
XN = PHI * X;
% Perform error covariance extrapolation
S = PHI * S * PHI' + Q;
% compute the Kalman gains
ak(1) = S(1,1) / (S(1,1) + R);
ak(2) = S(1,2) / (S(1,1) + R);
ak(3) = S(1,3) / (S(1,1) + R);
%perform state estimate update:
error = inp(rn) + normrnd(0,R) - XN(1);
residual(rn) = error;
tmp1 = ak(1) * error;
tmp2 = ak(2) * error;
tmp3 = ak(3) * error;
X(1) = XN(1) + tmp1;
X(2) = XN(2) + tmp2;
X(3) = XN(3) + tmp3;
estimate(rn) = X(1);
% update the error covariance
S(1,1) = S(1,1) * (1. -ak(1));
S(1,2) = S(1,2) * (1. -ak(1));
S(1,3) = S(1,3) * (1. -ak(1));
S(2,1) = S(1,2);
S(2,2) = -ak(2) * S(1,2) + S(2,2);
S(2,3) = -ak(2) * S(1,3) + S(2,3);
S(3,1) = S(1,3);
S(3,3) = -ak(3) * S(1,3) + S(3,3);
 rn = rn + 1.;
end
```

MATLAB Program "Fig17-29.m" Listing

```
% generates Fig 17.29
clc
close ll
clear all
npts = 2000;
del = 1/2000;
t = 0:del:1;
inp = (1+.2 .* t + .1 .*t.^2);% + cos(2. * pi * 2.5 .* t);
X0 = [1,.1,.01]';
% it is assumed that the measurmeny vector H=[1,0,0]
% this is the update interval in seconds
T = 1.;
% enter the mesurement noise variance
R = .01;
% this is the state noise variance
nvar = .18;
[residual, estimate] = kalman_filter(npts, T, X0, inp, R, nvar);
figure(1)
plot(residual(1:500),'k')
xlabel ('Sample number')
ylabel ('Residual')
figure(2)
subplot(2,1,1)
```

```
plot(inp,'k')
axis tight
ylabel ('position - truth')
subplot(2,1,2)
plot(estimate,'k')
axis tight
xlabel ('Sample number')
ylabel ('Predicted position')
```

MATLAB Program "Fig17_30.m" Listing

```
% generates Fig 17.30 of text
clc
close all
clear all
npts = 2000;
del = 1/2000;
t = 0:del:1;
inp = (1+.2 .* t + .1 .*t.^2) + cos(2. * pi * 2.5 .* t);
X0 = [1,.1,.01]';
% it is assumed that the measurement vector H=[1,0,0]
% this is the update interval in seconds
T = 1.;
% enter the mesurement noise variance
R = .035;
% this is the state noise variance
nvar = .5;
[residual, estimate] = kalman_filter(npts, T, X0, inp, R, nvar);
figure(1)
plot(residual,'k')
xlabel ('Sample number')
ylabel ('Residual')
figure(2)
subplot(2,1,1)
plot(inp,'k')
axis tight
ylabel ('position - truth')
subplot(2,1,2)
plot(estimate,'k')
axis tight
xlabel ('Sample number')
ylabel ('Predicted position')
```

MATLAB Function "maketraj.m" Listing

```
function [times , trajectory] = maketraj(start_loc, xvelocity, yamp, yperiod, zamp, zperiod, samplingtime,
deltat)
% maketraj.m
% USAGE:  [times , trajectory] = maketraj(start_loc, xvelocity, yamp, yperiod, zamp, zperiod, sampling-
time, deltat)
% NOTE: all coordinates are in radar reference coordinates.
% INPUTS
% name       dimension explanation                    units
%------      ------    --------------              -------
```

```
% start_loc    3 X 1    starting location of target        m
% xvelocity    1        velocity of target                m/s
% yamp         1        amplitude of oscillation y direction   m
% yperiod      1        period of oscillation y direction      m
% zamp         1        amplitude of oscillation z direction   m
% zperiod      1        period of oscillation z direction      m
% samplingtime 1        length of interval of trajectory      sec
% deltat       1        time between samples                  sec
%
% OUTPUTS
%
% name         dimension          explanation          units
%------        ----------         ---------------      ------
% times        1 X samplingtime/deltat vector of times
%                              corresponding to samples sec
% trajectory   3 X samplingtime/deltat trajectory x,y,z       m
%
times = 0: deltat: samplingtime ;
x = start_loc(1)+xvelocity.*times ;
if yperiod~=0
  y = start_loc(2)+yamp*cos(2*pi*(1/yperiod).*times) ;
else
  y = ones(1, length(times))*start_loc(2) ;
end
if zperiod~=0
  z = start_loc(3)+zamp*cos(2*pi*(1/zperiod).*times)  ;
else
  z = ones(1, length(times))*start_loc(3) ;
end
trajectory = [x ; y  ; z] ;
```

MATLAB Function "addnoise.m" Listing

```
function [noisytraj ] = addnoise(trajectory, sigmaaz, sigmael, sigmarange )
% addnoise.m
% USAGE: [noisytraj ] = addnoise(trajectory, sigmaaz, sigmael, sigmarange )
% INPUTS
% name         dimension  explanation                  units
%------        ------     ---------------              -------
% trajectory   3 X POINTS trajectory in radar reference coords  [m;m;m]
% sigmaaz      1          standard deviation of azimuth error    radians
% sigmael      1          standard deviation of elevation error  radians
% sigmarange   1          standard deviation of range error      m
%
% OUTPUTS
% name         dimension  explanation                  units
%------        ------     ---------------              -------
% noisytraj    3 X POINTS noisy trajectory             [m;m;m]
noisytraj = zeros(3, size(trajectory,2)) ;

for loop = 1 : size(trajectory,2)
  x = trajectory(1,loop);
  y = trajectory(2,loop);
  z = trajectory(3,loop);
```

```
azimuth_corrupted =  atan2(y,x) + sigmaaz*randn(1) ;
elevation_corrupted = atan2(z, sqrt(x^2+y^2)) + sigmael*randn(1) ;
range_corrupted = sqrt(x^2+y^2+z^2)  + sigmarange*randn(1) ;
x_corrupted = range_corrupted*cos(elevation_corrupted)*cos(azimuth_corrupted) ;
y_corrupted = range_corrupted*cos(elevation_corrupted)*sin(azimuth_corrupted) ;
z_corrupted = range_corrupted*sin(elevation_corrupted) ;
noisytraj(:,loop) = [x_corrupted ; y_corrupted; z_corrupted ] ;
end % next loop
```

MATLAB Function "kalfilt.m" Listing

```
function [filtered, residuals , covariances, kalmgains] = kalfilt(trajectory, x0, P0, phi, R, Q )
% kalfilt.m
% USAGE: [filtered, residuals , covariances, kalmgains] = kalfilt(trajectory, x0, P0, phi, R, Q)
%
% INPUTS
% name        dimension              explanation                      units
%------      ------      ---------------      -------
% trajectory     NUMMEASUREMENTS  X  NUMPOINTS   trajectory in radar reference coords
[m;m;m]
% x0        NUMSTATES X 1          initial estimate of state vector       m, m/s
% P0        NUMSTATES X NUMSTATES    initial estimate of covariance matrix     m, m/s
% phi        NUMSTATES X NUMSTATES     state transition matrix       -
% R        NUMMEASUREMENTS X NUMMEASUREMENTS   measurement error covariance matrix
m
% Q        NUMSTATES X NUMSTATES     state error covariance matrix        m, m/s
%
% OUTPUTS
% name        dimension              explanation                      units
%------      ------      ---------------      -------
%filtered    NUMSTATES X NUMPOINTS      filtered trajectory x,y,z pos, vel   [m; m/s; m; m/s; m; m/s]
% residuals   NUMSTATES X NUMPOINTS      residuals of filtering        [m;m;m]
% covariances  NUMSTATES X NUMPOINTS       diagonal of covariance matrix      [m;m;m]
% kalmgains    (NUMSTATES X NUMMEASUREMENTS)
%          X NUMPOINTS        Kalman gain matrix        -
NUMSTATES = 6 ;
NUMMEASUREMENTS = 3 ;
NUMPOINTS = size(trajectory, 2) ;
% initialize output matrices
filtered = zeros(NUMSTATES, NUMPOINTS) ;
residuals = zeros(NUMSTATES, NUMPOINTS) ;
covariances = zeros(NUMSTATES, NUMPOINTS) ;
kalmgains = zeros(NUMSTATES*NUMMEASUREMENTS, NUMPOINTS) ;
% set matrix relating measurements to states
H = [1 0 0 0 0 0 ; 0 0 1 0 0 0 ; 0 0 0 0 1 0];
xhatminus = x0 ;
Pminus = P0 ;
for loop = 1: NUMPOINTS
  % compute the Kalman gain
  K = Pminus*H'*inv(H*Pminus*H' + R) ;
  kalmgains(:,loop) = reshape(K, NUMSTATES*NUMMEASUREMENTS, 1) ;
  % update the estimate with the measurement z
  z = trajectory(:,loop) ;
  xhat = xhatminus + K*(z - H*xhatminus) ;
```

```
   filtered(:,loop) = xhat ;
   residuals(:,loop) = xhat - xhatminus ;
   % update the error covariance for the updated estimate
   P = ( eye(NUMSTATES, NUMSTATES) - K*H)*Pminus ;
   covariances(:,loop) = diag(P) ;  % only save diagonal of covariance matrix
   % project ahead
   xhatminus_next = phi*xhat ;
   Pminus_next = phi*P*phi' + Q ;
    xhatminus = xhatminus_next ;
    Pminus = Pminus_next ;
end
```

Chapter 18

Tactical Synthetic Aperture Radars

This chapter was coauthored with Brian J. Smith.[1]

This chapter provides an introduction to Tactical Synthetic Aperture Radar (TSAR). The purpose of this chapter is to further develop the readers' understanding of SAR by taking a closer look at high resolution spotlight SAR image formation algorithms, motion compensation techniques, autofocus algorithms, and performance metrics.

18.1. Introduction

Modern airborne radar systems are designed to perform a large number of functions which range from detection and discrimination of targets to mapping large areas of ground terrain. This mapping can be performed by the Synthetic Aperture Radar (SAR). Through illuminating the ground with coherent radiation and measuring the echo signals, SAR can produce high resolution two-dimensional (and in some cases three-dimensional) imagery of the ground surface. The quality of ground maps generated by SAR is determined by the size of the resolution cell. A resolution cell is specified by both range and azimuth resolutions of the system. Other factors affecting the size of the resolution cells are (1) size of the processed map and the amount of signal processing involved; (2) cost consideration; and (3) size of the objects that need to be resolved in the map. For example, mapping gross features of cities and coastlines does not require as much resolution when compared to resolving houses, vehicles, and streets.

SAR systems can produce maps of reflectivity versus range and Doppler (cross range). Range resolution is accomplished through range gating. Fine range resolution can be accomplished by using pulse compression techniques. The azimuth resolution depends on antenna size and radar wavelength. Fine azimuth resolution is enhanced by taking advantage of the radar motion in order to synthesize a larger antenna aperture. Let N_r denote the number of range bins and let N_a denote the number of azimuth cells. It follows that the total number of resolution cells in the map is $N_r N_a$. SAR systems that are generally concerned with improving azimuth resolution are often referred to as Doppler Beam-Sharpening (DBS) SARs. In this case, each range bin is processed to resolve targets in Doppler which corresponds to azimuth. This chapter is presented in the context of DBS.

---

1. Dr. Brian J. Smith is with the US Army Aviation and Missile Command (AMCOM), Redstone Arsenal, Alabama.

Due to the large amount of signal processing required in SAR imagery, the early SAR designs implemented optical processing techniques. Although such optical processors can produce high-quality radar images, they have several shortcomings. They can be very costly and are, in general, limited to making strip maps. Motion compensation is not easy to implement for radars that utilize optical processors. With the recent advances in solid state electronics and Very Large Scale Integration (VLSI) technologies, digital signal processing in real time has been made possible in SAR systems.

18.1.1. Side Looking SAR Geometry

Fig. 18.1 shows the geometry of the standard side looking SAR. We will assume that the platform carrying the radar maintains both fixed altitude h and velocity v. The antenna $3\,dB$ beamwidth is θ, and the elevation angle (measured from the z-axis to the antenna axis) is β. The intersection of the antenna beam with the ground defines a footprint. As the platform moves, the footprint scans a swath on the ground.

The radar position with respect to the absolute origin $\vec{O} = (0, 0, 0)$, at any time is the vector $\vec{a}(t)$. The velocity vector $\vec{a}'(t)$ is

$$\vec{a}'(t) = 0 \times \hat{a}_x + v \times \hat{a}_y + 0 \times \hat{a}_z .$$ **Eq. (18.1)**

The Line of Sight (LOS) for the current footprint centered at $\vec{q}(t_c)$ is defined by the vector $\vec{R}(t_c)$, where t_c denotes the central time of the observation interval T_{ob} (coherent integration interval). More precisely,

$$(t = t_a + t_c) \;\; ; \;\; -\frac{T_{ob}}{2} \leq t \leq \frac{T_{ob}}{2}$$ **Eq. (18.2)**

where t_a and t are the absolute and relative times, respectively. The vector $\vec{m}_g$ defines the ground projection of the antenna at central time. The minimum slant range to the swath is R_{min}, and the maximum range is denoted R_{max}, as illustrated by Fig. 18.2. It follows that

$$R_{min} = h / \cos(\beta - \theta / 2)$$
$$R_{max} = h / \cos(\beta + \theta / 2) .$$ **Eq. (18.3)**
$$\left| \vec{R}(t_c) \right| = h / \cos\beta$$

Notice that the elevation angle β is equal to

$$\beta = 90 - \psi_g$$ **Eq. (18.4)**

where ψ_g is the grazing angle. The size of the footprint is a function of the grazing angle and the antenna beamwidth, as illustrated in Fig. 18.3. The SAR geometry described in this section is referred to as SAR "strip mode" of operation. Another SAR mode of operation, which will not be discussed in this chapter, is called "spot-light mode," where the antenna is steered (mechanically or electronically) to continuously illuminate one spot (footprint) on the ground. In this case, one high resolution image of the current footprint is generated during an observation interval.

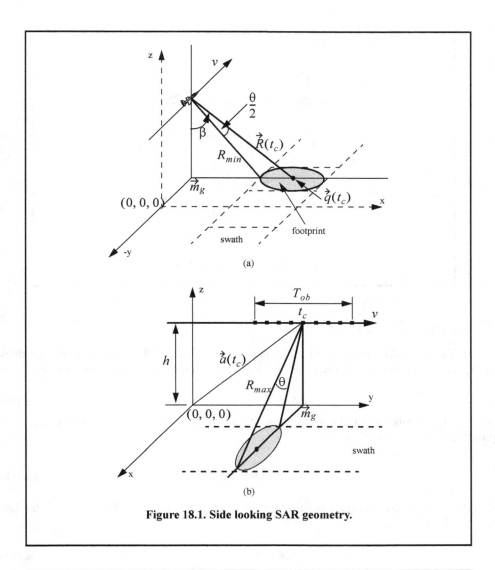

Figure 18.1. Side looking SAR geometry.

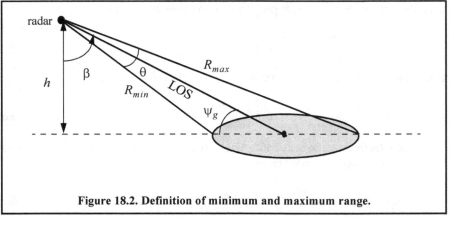

Figure 18.2. Definition of minimum and maximum range.

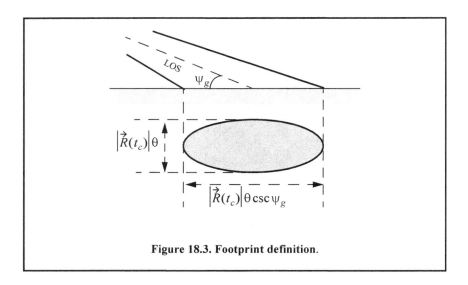

Figure 18.3. Footprint definition.

18.2. SAR Design Considerations

The quality of SAR images is heavily dependent on the size of the map resolution cell shown in Fig. 18.4. The range resolution, ΔR, is computed on the beam LOS, and is given by

$$\Delta R = (c\tau)/2 \qquad\qquad \textbf{Eq. (18.5)}$$

where τ is the pulse width. From the geometry in Fig. 18.5, the extent of the range cell ground projection ΔR_g is computed as

$$\Delta R_g = \frac{c\tau}{2}\sec\psi_g . \qquad\qquad \textbf{Eq. (18.6)}$$

The azimuth or cross range resolution for a real antenna with a $3\,dB$ beamwidth θ (radians) at range R is

$$\Delta A = \theta R . \qquad\qquad \textbf{Eq. (18.7)}$$

However, the antenna beamwidth is proportional to the aperture size,

$$\theta \approx \frac{\lambda}{L} \qquad\qquad \textbf{Eq. (18.8)}$$

where λ is the wavelength and L is the aperture length. It follows that

$$\Delta A = \frac{\lambda R}{L} . \qquad\qquad \textbf{Eq. (18.9)}$$

And since the effective synthetic aperture size is twice that of a real array, the azimuth resolution for a synthetic array is then given by

$$\Delta A = \frac{\lambda R}{2L} . \qquad\qquad \textbf{Eq. (18.10)}$$

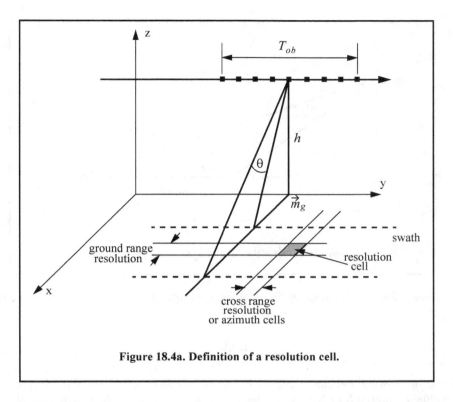

Figure 18.4a. Definition of a resolution cell.

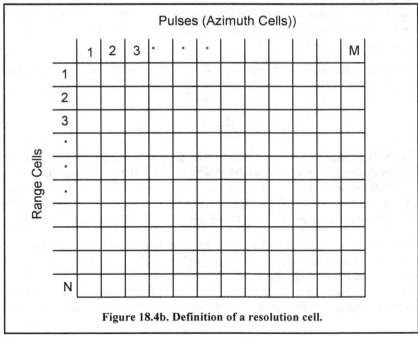

Figure 18.4b. Definition of a resolution cell.

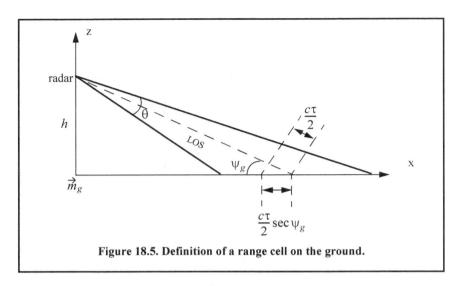

Figure 18.5. Definition of a range cell on the ground.

Furthermore, since the synthetic aperture length L is equal to vT_{ob}, Eq. (18.10) can be rewritten as

$$\Delta A = \frac{\lambda R}{2vT_{ob}}.$$

Eq. (18.11)

The azimuth resolution can be greatly improved by taking advantage of the Doppler variation within a footprint (or a beam). As the radar travels along its flight path, the radial velocity to a ground scatterer (point target) within a footprint varies as a function of the radar radial velocity in the direction of that scatterer. The variation of Doppler frequency for a certain scatterer is called the "Doppler history."

Let $R(t)$ denote the range to a scatterer at time t, and v_r be the corresponding radial velocity; thus the Doppler shift is

$$f_d = -\frac{2\dot{R}(t)}{\lambda} = \frac{2v_r}{\lambda}$$

Eq. (18.12)

where $\dot{R}(t)$ is the range rate to the scatterer. Let t_1 and t_2 be the times when the scatterer enters and leaves the radar beam, respectively, and t_c be the time that corresponds to minimum range. Fig. 18.6 shows a sketch of the corresponding $R(t)$. Since the radial velocity can be computed as the derivative of $R(t)$ with respect to time, one can clearly see that Doppler frequency is maximum at t_1, zero at t_c, and minimum at t_2, as illustrated in Fig. 18.7.

In general, the radar maximum PRF, $f_{r_{max}}$, must be low enough to avoid range ambiguity. Alternatively, the minimum PRF, $f_{r_{min}}$, must be high enough to avoid Doppler ambiguity. SAR unambiguous range must be at least as wide as the extent of a footprint. More precisely, since target returns from maximum range due to the current pulse must be received by the radar before the next pulse is transmitted, it follows that SAR unambiguous range is given by

$$R_u = R_{max} - R_{min}.$$

Eq. (18.13)

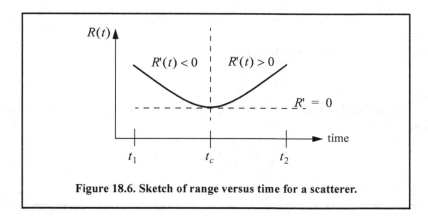

Figure 18.6. Sketch of range versus time for a scatterer.

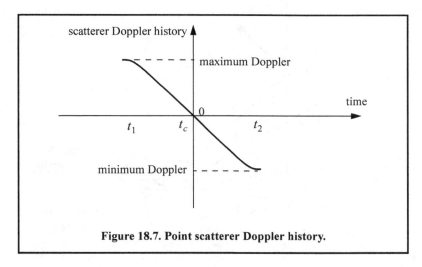

Figure 18.7. Point scatterer Doppler history.

An expression for unambiguous range was derived in Chapter 1, and is repeated here as Eq. (18.14),

$$R_u = \frac{c}{2f_r}.$$

Eq. (18.14)

Combining Eq. (18.14) and Eq. (18.13) yields

$$f_{r_{max}} \leq \frac{c}{2(R_{max} - R_{min})}.$$

Eq. (18.15)

SAR minimum PRF, $f_{r_{min}}$, is selected so that Doppler ambiguity is avoided. In other words, $f_{r_{min}}$ must be greater than the maximum expected Doppler spread within a footprint. From the geometry of Fig. 18.8, the maximum and minimum Doppler frequencies are, respectively, given by

$$f_{d_{max}} = \frac{2v}{\lambda} \, \sin\left(\frac{\theta}{2}\right) \sin\beta \; ; \; at \; t_1$$

Eq. (18.16)

$$f_{d_{min}} = -\frac{2v}{\lambda} \; \sin\left(\frac{\theta}{2}\right) \sin\beta \; ; \; at \; t_2 \, .$$

Eq. (18.17)

It follows that the maximum Doppler spread is

$$\Delta f_d = f_{d_{max}} - f_{d_{min}} \, .$$

Eq. (18.18)

Substituting Eqs. (18.16) and (18.17) into Eq. (18.18) and applying the proper trigonometric identities yield

$$\Delta f_d = \frac{4v}{\lambda} \; \sin\frac{\theta}{2} \; \sin\beta \, .$$

Eq. (18.19)

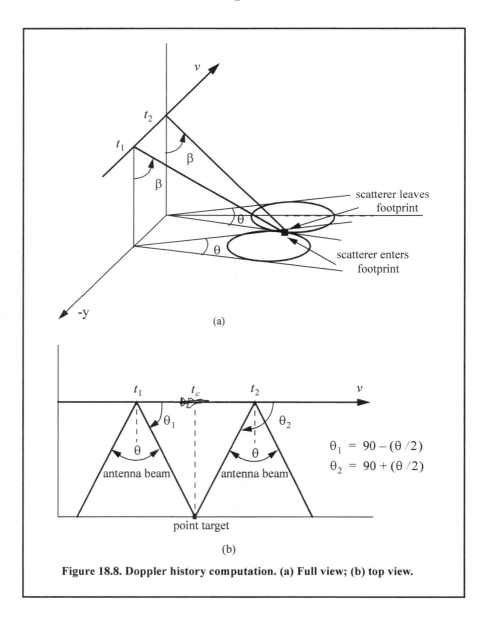

Figure 18.8. Doppler history computation. (a) Full view; (b) top view.

Finally, by using the small angle approximation we get

$$\Delta f_d \approx \frac{4v}{\lambda} \frac{\theta}{2} \sin\beta = \frac{2v}{\lambda} \theta \sin\beta.$$

Eq. (18.20)

Therefore, the minimum PRF is

$$f_{r_{min}} \geq \frac{2v}{\lambda} \theta \sin\beta.$$

Eq. (18.21)

Combining Eqs. (18.15) and (18.21) we get

$$\frac{c}{2(R_{max} - R_{min})} \geq f_r \geq \frac{2v}{\lambda} \theta \sin\beta.$$

Eq. (18.22)

It is possible to resolve adjacent scatterers at the same range within a footprint based only on the difference of their Doppler histories. For this purpose, assume that the two scatterers are within the *kth* range bin.

Denote their angular displacement as $\Delta\theta$, and let $\Delta f_{d_{min}}$ be the minimum Doppler spread between the two scatterers such that they will appear in two distinct Doppler filters. Using the same methodology that led to Eq. (18.20), we get

$$\Delta f_{d_{min}} = \frac{2v}{\lambda} \Delta\theta \sin\beta_k$$

Eq. (18.23)

where β_k is the elevation angle corresponding to the *kth* range bin.

The bandwidth of the individual Doppler filters must be equal to the inverse of the coherent integration interval T_{ob} (i.e., $\Delta f_{d_{min}} = 1/T_{ob}$). It follows that

$$\Delta\theta = \frac{\lambda}{2vT_{ob}\sin\beta_k}.$$

Eq. (18.24)

Substituting L for vT_{ob} yields

$$\Delta\theta = \frac{\lambda}{2L\sin\beta_k}.$$

Eq. (18.25)

Therefore, the SAR azimuth resolution (within the *kth* range bin) is

$$\Delta A_g = \Delta\theta R_k = R_k \frac{\lambda}{2L\sin\beta_k}.$$

Eq. (18.26)

Note that when $\beta_k = 90°$, Eq. (18.26) is identical to Eq. (18.10).

18.3. SAR Radar Equation

The single-pulse radar equation was derived in Chapter 2, and is repeated here as Eq. (18.27),

$$SNR = \frac{P_t G^2 \lambda^2 \sigma}{(4\pi)^3 R_k^4 kT_0 B L_{Loss}}$$

Eq. (18.27)

where P_t is peak power, G is antenna gain, λ is wavelength, σ is radar cross section, R_k is radar slant range to the *kth* range bin, k is Boltzman's constant, T_0 is receiver noise temperature, B is receiver bandwidth, and L_{Loss} is radar losses. The radar cross section is a function of the radar resolution cell and terrain reflectivity. More precisely,

$$\sigma = \sigma^0 \Delta R_g \Delta A_g = \sigma^0 \Delta A_g \frac{c\tau}{2} \sec \psi_g \qquad\qquad \textbf{Eq. (18.28)}$$

where σ^0 is the clutter scattering coefficient, ΔA_g is the azimuth resolution, and Eq. (18.6) was used to replace the ground range resolution. The number of coherently integrated pulses within an observation interval is

$$n = f_r T_{ob} = \frac{f_r L}{v} \qquad\qquad \textbf{Eq. (18.29)}$$

where L is the synthetic aperture size. Using Eq. (18.26) in Eq. (18.29) and rearranging terms yield

$$n = \frac{\lambda R f_r}{2 \Delta A_g v} \csc \beta_k. \qquad\qquad \textbf{Eq. (18.30)}$$

The radar average power over the observation interval is

$$P_{av} = (P_t / B) f_r. \qquad\qquad \textbf{Eq. (18.31)}$$

The SNR for n coherently integrated pulses is then

$$(SNR)_n = nSNR = n \; \frac{P_t G^2 \lambda^2 \sigma}{(4\pi)^3 R_k^4 k T_0 B L_{Loss}}. \qquad\qquad \textbf{Eq. (18.32)}$$

Substituting Eqs. (18.31), (18.30), and (18.28) into Eq. (18.32) and performing some algebraic manipulations give the SAR radar equation,

$$(SNR)_n = \frac{P_{av} G^2 \lambda^3 \sigma^0}{(4\pi)^3 R_k^3 k T_0 L_{Loss}} \; \frac{\Delta R_g}{2v} \; \csc \beta_k. \qquad\qquad \textbf{Eq. (18.33)}$$

Eq. (18.33) leads to the conclusion that in SAR systems, the SNR is (1) inversely proportional to the third power of range; (2) independent of azimuth resolution; (3) a function of the ground range resolution; (4) inversely proportional to the velocity v; and (5) proportional to the third power of wavelength.

18.4. SAR Signal Processing

There are two signal processing techniques to sequentially produce a SAR map or image; they are line-by-line processing and Doppler processing. The concept of SAR line-by-line processing is as follows: Through the radar linear motion, a synthetic array is formed, where the elements of the current synthetic array correspond to the position of the antenna transmissions during the last observation interval. Azimuth resolution is obtained by forming narrow synthetic beams through combinations of the last observation interval returns. Fine range resolution is accomplished in real time by utilizing range gating and pulse compression. For each range bin and each of the transmitted pulses during the last observation interval, the returns are recorded in a two-dimensional array of data that is updated for every pulse. Denote the two-dimensional array of data as MAP.

To further illustrate the concept of line-by-line processing, consider the case where a map of size $N_a \times N_r$ is to be produced, where N_a is the number of azimuth cells and N_r is the number of range bins. Hence, MAP is of size $N_a \times N_r$, where the columns refer to range bins, and the rows refer to azimuth cells. For each transmitted pulse, the echoes from consecutive range bins are recorded sequentially in the first row of MAP. Once the first row is completely filled (i.e., returns from all range bins have been received), all data (in all rows) are shifted downward one row before the next pulse is transmitted. Thus, one row of MAP is generated for every transmitted pulse. Consequently, for the current observation interval, returns from the first transmitted pulse will be located in the bottom row of MAP, and returns from the last transmitted pulse will be in the first row of MAP.

In SAR Doppler processing, the array MAP is updated once every N pulses so that a block of N columns is generated simultaneously. In this case, N refers to the number of transmissions during an observation interval (i.e., size of the synthetic array). From an antenna point of view, this is equivalent to having N adjacent synthetic beams formed in parallel through electronic steering.

18.5. Side Looking SAR Doppler Processing

Consider the geometry shown in Fig. 18.9, and assume that the scatterer C_i is located within the kth range bin. The scatterer azimuth and elevation angles are μ_i and β_i, respectively. The scatterer elevation angle β_i is assumed to be equal to β_k, the range bin elevation angle. This assumption is true if the ground range resolution, ΔR_g, is small; otherwise, $\beta_i = \beta_k + \varepsilon_i$ for some small ε_i; in this chapter $\varepsilon_i = 0$.

The normalized transmitted signal can be represented by

$$s(t) = \cos(2\pi f_0 t - \xi_0) \qquad \text{Eq. (18.34)}$$

where f_0 is the radar operating frequency, and ξ_0 denotes the transmitter phase. The returned radar signal from C_i is then equal to

$$s_i(t, \mu_i) = A_i \cos[2\pi f_0(t - \tau_i(t, \mu_i)) - \xi_0] \qquad \text{Eq. (18.35)}$$

where $\tau_i(t, \mu_i)$ is the round-trip delay to the scatterer, and A_i includes scatterer strength, range attenuation, and antenna gain. The round-trip delay is

$$\tau_i(t, \mu_i) = \frac{2r_i(t, \mu_i)}{c} \qquad \text{Eq. (18.36)}$$

where c is the speed of light and $r_i(t, \mu_i)$ is the scatterer slant range. From the geometry in Fig. 18.9, one can write the expression for the slant range to the ith scatterer within the kth range bin as

$$r_i(t, \mu_i) = \frac{h}{\cos\beta_i} \sqrt{1 - \frac{2vt}{h}\cos\beta_i \cos\mu_i \sin\beta_i + \left(\frac{vt}{h}\cos\beta_i\right)^2}. \qquad \text{Eq. (18.37)}$$

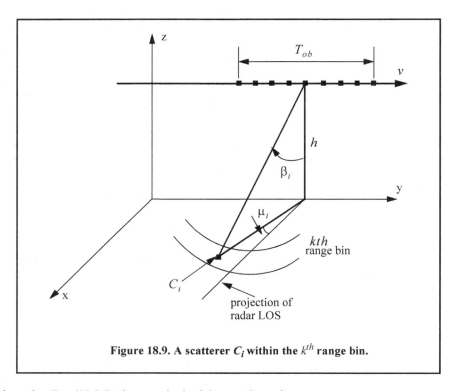

Figure 18.9. A scatterer C_i within the kth range bin.

And by using Eq. (18.36), the round-trip delay can be written as

$$\tau_i(t, \mu_i) = \frac{2}{c} \frac{h}{\cos \beta_i} \sqrt{1 - \frac{2vt}{h} \cos \beta_i \cos \mu_i \sin \beta_i + \left(\frac{vt}{h} \cos \beta_i\right)^2}.$$ **Eq. (18.38)**

The round-trip delay can be approximated using a two-dimensional second-order Taylor series expansion about the reference state $(t, \mu) = (0, 0)$. Performing this Taylor series expansion yields

$$\tau_i(t, \mu_i) \approx \bar{\tau} + \bar{\tau}_{t\mu} \ \mu_i t + \bar{\tau}_{tt} \ \frac{t^2}{2}$$ **Eq. (18.39)**

where the over-bar indicates evaluation at the state $(0, 0)$, and the subscripts denote partial derivatives. For example, $\bar{\tau}_{t\mu}$ means

$$\bar{\tau}_{t\mu} = \frac{\partial^2}{\partial t \partial \mu} \tau_i(t, \mu_i)\Big|_{(t, \mu) = (0, 0)}.$$ **Eq. (18.40)**

The Taylor series coefficients are

$$\bar{\tau} = \left(\frac{2h}{c}\right) \frac{1}{\cos \beta_i}$$ **Eq. (18.41)**

$$\bar{\tau}_{t\mu} = \left(\frac{2v}{c}\right) \sin \beta_i$$ **Eq. (18.42)**

$$\bar{\tau}_{tt} = \left(\frac{2v^2}{hc}\right)\cos\beta_i.$$

Eq. (18.43)

Note that other Taylor series coefficients are either zeros or very small. Hence, they are neglected. Finally, we can rewrite the returned radar signal as

$$s_i(t, \mu_i) = A_i\cos[\psi_i(t, \mu_i) - \xi_0]$$

$$\hat{\psi}_i(t, \mu_i) = 2\pi f_0\left[(1 - \bar{\tau}_{t\mu}\mu_i)t - \bar{\tau} - \bar{\tau}_{tt}\frac{t^2}{2}\right].$$

Eq. (18.44)

Observation of Eq. (18.44) indicates that the instantaneous frequency for the *ith* scatterer varies as a linear function of time due to the second-order phase term $2\pi f_0(\bar{\tau}_{tt}t^2/2)$ (this confirms the result we concluded about a scatterer Doppler history). Furthermore, since this phase term is range-bin dependent and not scatterer dependent, all scatterers within the same range bin produce this exact second-order phase term. It follows that scatterers within a range bin have identical Doppler histories. These Doppler histories are separated by the time delay required to fly between them, as illustrated in Fig. 18.10.

Suppose that there are *I* scatterers within the *kth* range bin. In this case, the combined returns for this cell are the sum of the individual returns due to each scatterer as defined by Eq. (18.44). In other words, superposition holds, and the overall echo signal is

$$s_r(t) = \sum_{i=1}^{I} s_i(t, \mu_i).$$

Eq. (18.45)

A signal processing block diagram for the *kth* range bin is illustrated in Fig. 18.11. It consists of the following steps. First, heterodyning with the carrier frequency is performed to extract the quadrature components.

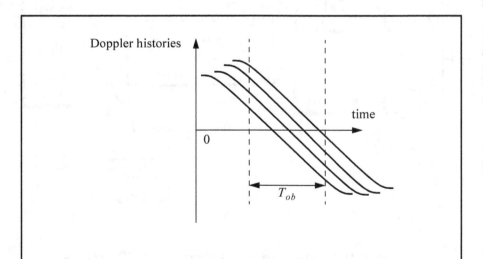

Figure 18.10. Doppler histories for several scatterers within the same range bin.

This is followed by LP filtering and A/D conversion. Next, deramping or focusing to remove the second-order phase term of the quadrature components is carried out using a phase rotation matrix. The last stage of the processing includes windowing, performing an FFT on the windowed quadrature components, and scaling the amplitude spectrum to account for range attenuation and antenna gain.

The discrete quadrature components are

$$\tilde{x}_I(t_n) = \tilde{x}_I(n) = A_i \cos[\tilde{\psi}_i(t_n, \mu_i) - \xi_0]$$
$$\tilde{x}_Q(t_n) = \tilde{x}_Q(n) = A_i \sin[\tilde{\psi}_i(t_n, \mu_i) - \xi_0]$$

$$\text{Eq. (18.46)}$$

$$\tilde{\psi}_i(t_n, \mu_i) = \hat{\psi}_i(t_n, \mu_i) - 2\pi f_0 t_n$$

$$\text{Eq. (18.47)}$$

and t_n denotes the *nth* sampling time (remember that $-T_{ob}/2 \le t_n \le T_{ob}/2$). The quadrature components after deramping (i.e., removal of the phase $\psi = -\pi f_0 \tilde{\tau}_{tt} t_n^2$) are given by

$$\begin{bmatrix} x_I(n) \\ x_Q(n) \end{bmatrix} = \begin{bmatrix} \cos\psi & -\sin\psi \\ \sin\psi & \cos\psi \end{bmatrix} \begin{bmatrix} \tilde{x}_I(n) \\ \tilde{x}_Q(n) \end{bmatrix}.$$

$$\text{Eq. (18.48)}$$

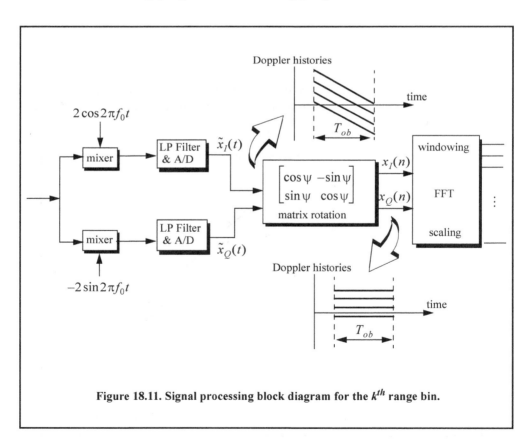

Figure 18.11. Signal processing block diagram for the kth range bin.

18.6. SAR Imaging Using Doppler Processing

It was mentioned earlier that SAR imaging is performed using two orthogonal dimensions (range and azimuth). Range resolution is controlled by the receiver bandwidth and pulse compression. Azimuth resolution is limited by the antenna beamwidth. A one-to-one correspondence between the FFT bins and the azimuth resolution cells can be established by utilizing the signal model described in the previous section. Therefore, the problem of target detection is transformed into a spectral analysis problem, where detection is based on the amplitude spectrum of the returned signal. The FFT frequency resolution Δf is equal to the inverse of the observation interval T_{ob}. It follows that a peak in the amplitude spectrum at $k_1 \Delta f$ indicates the presence of a scatterer at frequency $f_{d1} = k_1 \Delta f$.

For an example, consider the scatterer C_i within the kth range bin. The instantaneous frequency f_{di} corresponding to this scatterer is

$$f_{di} = \frac{1}{2\pi} \frac{d\psi}{dt} = f_0 \bar{\tau}_{t\mu} \mu_i = \frac{2v}{\lambda} \sin \beta_i \mu_i.$$

Eq. (18.49)

This is the same result derived in Eq. (18.23), with $\mu_i = \Delta\theta$. Therefore, the scatterers separated in Doppler by more than Δf can then be resolved.

Fig. 18.12 shows a two-dimensional SAR image for three point scatterers located $10Km$ downrange. In this case, the azimuth and range resolutions are equal to $1m$ and the operating frequency is $35GHz$. Fig. 18.13 is similar to Fig. 18.12, except in this case the resolution cell is equal to 6 inches. One can clearly see the blurring that occurs in the image. Figures 12.12 and 12.13 can be reproduced using the program *"Fig18_12_13.m,"* listed in Appendix 18-A.

18.7. Range Walk

As shown earlier, SAR Doppler processing is achieved in two steps: first, range gating and second, azimuth compression within each bin at the end of the observation interval. For this purpose, azimuth compression assumes that each scatterer remains within the same range bin during the observation interval. However, since the range gates are defined with respect to a radar that is moving, the range gate grid is also moving relative to the ground. As a result, a scatterer appears to be moving within its range bin. This phenomenon is known as range walk. A small amount of range walk does not bother Doppler processing as long as the scatterer remains within the same range bin. However, range walk over several range bins can constitute serious problems, where in this case Doppler processing is meaningless.

18.8. A Three-Dimensional SAR Imaging Technique

This section presents a new three-dimensional (3-D) Synthetic Aperture Radar (SAR) imaging based on Mahafza[1] et al. It utilizes a linear array in transverse motion to synthesize a two-dimensional (2-D) synthetic array. Elements of the linear array are fired sequentially (one element at a time), while all elements receive in parallel. A 2-D information sequence is computed from the equiphase two-way signal returns.

---

1. Mahafza, B. R. and Sajjadi, M., Three-Dimensional SAR Imaging Using a Linear Array in Transverse Motion, *IEEE - AES Trans.*, Vol. 32, No. 1, January 1996, pp. 499-510.

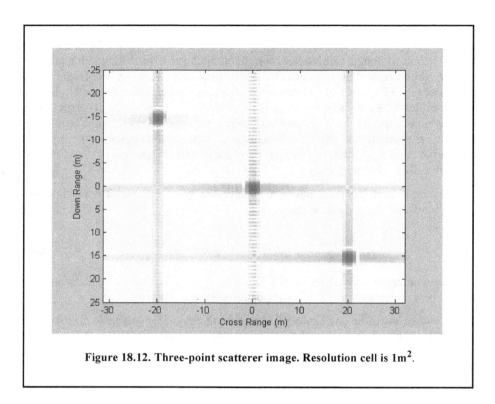

Figure 18.12. Three-point scatterer image. Resolution cell is 1m².

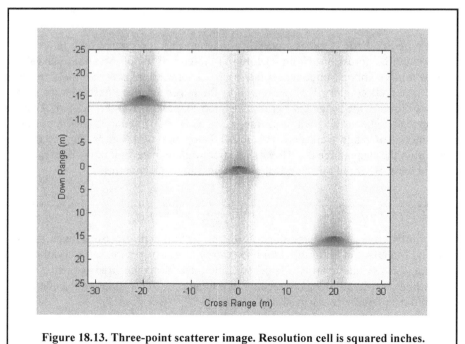

Figure 18.13. Three-point scatterer image. Resolution cell is squared inches.

A signal model based on a third-order Taylor series expansion about incremental relative time, azimuth, elevation, and target height is used. Scatterers are detected as peaks in the amplitude spectrum of the information sequence. Detection is performed in two stages. First, all scatterers within a footprint are detected using an incomplete signal model where target height is set to zero. Then, processing using the complete signal model is performed only on range bins containing significant scatterer returns. The difference between the two images is used to measure target height. Computer simulation shows that this technique is accurate and virtually impulse invariant.

18.8.1. Background

Standard Synthetic Aperture Radar (SAR) imaging systems are generally used to generate high resolution two-dimensional (2-D) images of ground terrain. Range gating determines resolution along the first dimension. Pulse compression techniques are usually used to achieve fine range resolution. Such techniques require the use of wideband receiver and display devices in order to resolve the time structure in the returned signals. The width of azimuth cells provides resolution along the other dimension. Azimuth resolution is limited by the duration of the observation interval.

This section presents a three-dimensional (3-D) SAR imaging technique based on Discrete Fourier Transform (DFT) processing of equiphase data collected in sequential mode (DFTSQM). It uses a linear array in transverse motion to synthesize a 2-D synthetic array. A 2-D information sequence is computed from the equiphase two-way signal returns. To this end, a new signal model based on a third-order Taylor series expansion about incremental relative time, azimuth, elevation, and target height is introduced. Standard SAR imaging can be achieved using an incomplete signal model where target height is set to zero. Detection is performed in two stages. First, all scatterers within a footprint are detected using an incomplete signal model, where target height is set to zero. Then, processing using the complete signal model is performed only on range bins containing significant scatterer returns. The difference between the two images is used as an indication of target height. Computer simulation shows that this technique is accurate and virtually impulse invariant.

18.8.2. DFTSQM Operation and Signal Processing

Linear Arrays

Consider a linear array of size N, uniform element spacing d, and wavelength λ. Assume a far field scatterer P located at direction-sine $\sin\theta_l$. DFTSQM operation for this array can be described as follows. The elements are fired sequentially, one at a time, while all elements receive in parallel. The echoes are collected and integrated coherently on the basis of equal phase to compute a complex information sequence $\{b(m); m = 0, 2N-1\}$. The x-coordinates, in d-units, of the x_n^{th} element with respect to the center of the array is

$$x_n = \left(-\frac{N-1}{2} + n\right); n = 0, \ldots N-1.$$ **Eq. (18.50)**

The electric field received by the x_2^{th} element due to the firing of the x_1^{th}, and reflection by the lth far field scatterer P is

$$E(x_1, x_2; s_l) = G^2(s_l)\left(\frac{R_0}{R}\right)^4 \sqrt{\sigma_l} \ exp(j\phi(x_1, x_2; s_l)) \qquad \text{Eq. (18.51)}$$

$$\phi(x_1, x_2; s_l) = \frac{2\pi}{\lambda}(x_1 + x_2)(s_l) \qquad \text{Eq. (18.52)}$$

$$s_l = \sin\theta_l \qquad \text{Eq. (18.53)}$$

where $\sqrt{\sigma_l}$ is the target cross section, $G^2(s_l)$ is the two-way element gain, and $(R_0/R)^4$ is the range attenuation with respect to reference range R_0. The scatterer phase is assumed to be zero, however it could be easily included. Assuming multiple scatterers in the array's FOV, the cumulative electric field in the path $x_1 \Rightarrow x_2$ due to reflections from all scatterers is

$$E(x_1, x_2) = \sum_{all \ l} [E_I(x_1, x_2; s_l) + jE_Q(x_1, x_2; s_l)] \qquad \text{Eq. (18.54)}$$

where the subscripts (I, Q) denote the quadrature components. Note that the variable part of the phase given in Eq. (18.52) is proportional to the integers resulting from the sums $\{(x_{n1} + x_{n2}); (n1, n2) = 0, \ldots N - 1\}$. In the far field operation, there are a total of $(2N - 1)$ distinct $(x_{n1} + x_{n2})$ sums. Therefore, the electric fields with paths of the same $(x_{n1} + x_{n2})$ sums can be collected coherently. In this manner, the information sequence $\{b(m); m = 0, 2N - 1\}$ is computed, where $b(2N - 1)$ is set to zero. At the same time, one forms the sequence $\{c(m); m = 0, \ldots 2N - 2\}$, which keeps track of the number of returns that have the same $(x_{n1} + x_{n2})$ sum. More precisely, for $m = n1 + n2; (n1, n2) = \ldots 0, N - 1$

$$b(m) = b(m) + E(x_{n1}, x_{n2}) \qquad \text{Eq. (18.55)}$$

$$c(m) = c(m) + 1. \qquad \text{Eq. (18.56)}$$

It follows that

$$\{c(m); m = 0, \ldots 2N - 2\} = \begin{cases} m + 1 \ ; \ m = 0, \ldots N - 2 \\ N \ ; \ m = N - 1 \\ 2N - 1 - m \quad m = N, \ldots 2N - 2 \end{cases} \qquad \text{Eq. (18.57)}$$

which is a triangular shape sequence.

The processing of the sequence $\{b(m)\}$ is performed as follows: (1) the weighting takes the sequence $\{c(m)\}$ into account; (2) the complex sequence $\{b(m)\}$ is extended to size N_F, a power integer of two, by zero padding; (3) the DFT of the extended sequence $\{b'(m); m = 0, N_F - 1\}$ is computed,

$$B(q) = \sum_{m=0}^{N_F - 1} b'(m) \cdot exp\left(-j\frac{2\pi qm}{N_F}\right); q = 0, \ldots N_F - 1; \qquad \text{Eq. (18.58)}$$

and, (4) after compensation for antenna gain and range attenuation, scatterers are detected as peaks in the amplitude spectrum $|B(q)|$. Note that step (4) is true only when

$$\sin\theta_q = \frac{\lambda q}{2Nd}; q = 0, \ldots 2N-1,$$

Eq. (18.59)

where $\sin\theta_q$ denotes the direction-sine of the qth scatterer, and $N_F = 2N$ is implied in Eq. (18.59).

The classical approach to multiple target detection is to use a phased array antenna with phase shifting and tapering hardware. The array beamwidth is proportional to (λ/Nd), and the first sidelobe is at about $-13dB$. On the other hand, multiple target detection using DFTSQM provides a beamwidth proportional to $(\lambda/2Nd)$ as indicated by (Eq. (18.59)), which has the effect of doubling the array's resolution. The first sidelobe is at about $-27dB$ due to the triangular sequence $\{c(m)\}$. Additionally, no phase shifting hardware is required for detection of targets within a single-element field of view.

Rectangular Arrays

DFTSQM operation and signal processing for 2-D arrays can be described as follows. Consider an $N_x \times N_y$ rectangular array. All N_xN_y elements are fired sequentially, one at a time. After each firing, all the N_xN_y array elements receive in parallel. Thus, N_xN_y samples of the quadrature components are collected after each firing, and a total of $(N_xN_y)^2$ samples will be collected. However, in the far field operation, there are only $(2N_x - 1) \times (2N_y - 1)$ distinct equiphase returns. Therefore, the collected data can be added coherently to form a 2-D information array of size $(2N_x - 1) \times (2N_y - 1)$. The two-way radiation pattern is computed as the modulus of the 2-D amplitude spectrum of the information array. The processing includes 2-D windowing, 2-D Discrete Fourier Transformation, antenna gain and range attenuation compensation. The field of view of the 2-D array is determined by the $3dB$ pattern of a single element. All the scatterers within this field will be detected simultaneously as peaks in the amplitude spectrum.

Consider a rectangular array of size $N \times N$, with uniform element spacing $d_x = d_y = d$, and wavelength λ. The coordinates of the nth element, in d-units, are

$$x_n = \left(-\frac{N-1}{2} + n\right) \quad ; n = 0, \ldots N-1$$

Eq. (18.60)

$$y_n = \left(-\frac{N-1}{2} + n\right) \quad ; n = 0, \ldots N-1.$$

Eq. (18.61)

Assume a far field point P defined by the azimuth and elevation angles (α, β). In this case, the one-way geometric phase for an element is

$$\varphi'(x, y) = \frac{2\pi}{\lambda}[x\sin\beta\cos\alpha + y\sin\beta\sin\alpha].$$

Eq. (18.62)

Therefore, the two-way geometric phase between the (x_1, y_1) and (x_2, y_2) elements is

$$\varphi(x_1, y_1, x_2, y_2) = \frac{2\pi}{\lambda}\sin\beta[(x_1 + x_2)\cos\alpha + (y_1 + y_2)\sin\alpha].$$

Eq. (18.63)

The two-way electric field for the lth scatterer at (α_l, β_l) is

$$E(x_1, x_2, y_1, y_2; \alpha_l, \beta_l) = G^2(\beta_l) \left(\frac{R_0}{R}\right)^4 \sqrt{\sigma_l} \ exp[j(\varphi(x_1, y_1, x_2, y_2))].$$

Eq. (18.64)

Assuming multiple scatterers within the array's FOV, then the cumulative electric field for the two-way path $(x_1, y_1) \Rightarrow (x_2, y_2)$ is given by

$$E(x_1, x_2, y_1, y_2) = \sum_{all\ scatterers} E(x_1, x_2, y_1, y_2; \alpha_l, \beta_l).$$

Eq. (18.65)

All formulas for the 2-D case reduce to those of a linear array case by setting $N_y = 1$ and $\alpha = 0$.

The variable part of the phase given in Eq. (18.63) is proportional to the integers $(x_1 + x_2)$ and $(y_1 + y_2)$. Therefore, after completion of the sequential firing, electric fields with paths of the same (i, j) sums, where

$$\{i = x_{n1} + x_{n2}; i = -(N-1), ...(N-1)\}$$

Eq. (18.66)

$$\{j = y_{n1} + y_{n2}; j = -(N-1), ...(N-1)\}$$

Eq. (18.67)

can be collected coherently. In this manner, the 2-D information array $\{b(m_x, m_y); (m_x, m_y) = 0, ...2N-1\}$ is computed. The coefficient sequence $\{c(m_x, m_y); (m_x, m_y) = 0, ...2N-2\}$ is also computed. More precisely,

$$for \ m_x = n1 + n2 \ and \ m_y = n1 + n2;$$
$$n1 = 0, ...N-1 \ , and \ n2 = 0, ...N-1$$

Eq. (18.68)

$$b(m_x, m_y) = b(m_x, m_y) + E(x_{n1}, y_{n1}, x_{n2}, y_{n2}).$$

Eq. (18.69)

It follows that

$$c(m_x, m_y) = (N_x - |m_x - (N_x - 1)|) \times (N_y - |m_y - (N_y - 1)|).$$

Eq. (18.70)

The processing of the complex 2-D information array $\{b(m_x, m_y)\}$ is similar to that of the linear case with the exception that one should use a 2-D DFT. After antenna gain and range attenuation compensation, scatterers are detected as peaks in the 2-D amplitude spectrum of the information array. A scatterer located at angles (α_l, β_l) will produce a peak in the amplitude spectrum at DFT indexes (p_l, q_l), where

$$\alpha_l = atan\left(\frac{q_l}{p_l}\right)$$

Eq. (18.71)

$$\sin \beta_l = \frac{\lambda p_l}{2Nd\cos \alpha_l} = \frac{\lambda q_l}{2Nd\sin \alpha_l}.$$

Eq. (18.72)

Derivation of Eq. (18.71) is in Section 12.9.7.

18.8.3. Geometry for DFTSQM SAR Imaging

Fig. 18.14 shows the geometry of the DFTSQM SAR imaging system. In this case, t_c denotes the central time of the observation interval, D_{ob}. The aircraft maintains both constant velocity v, and height h. The origin for the relative system of coordinates is denoted as $\vec{O}$. The vector $\vec{O}M$ defines the radar location at time t_c. The transmitting antenna consists of a linear real array operating in the sequential mode. The real array is of size N, element spacing d, and the radiators are circular dishes of diameter $D = d$. Assuming that the aircraft scans M transmitting locations along the flight path, then a rectangular array of size $N \times M$ is synthesized, as illustrated in Fig. 18.15.

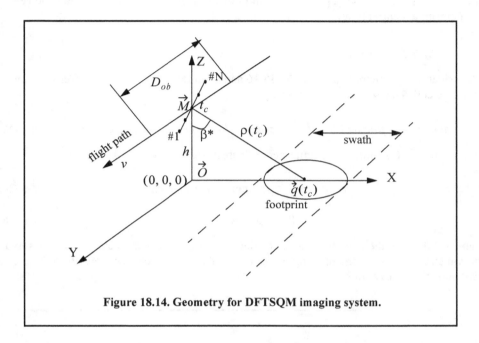

Figure 18.14. Geometry for DFTSQM imaging system.

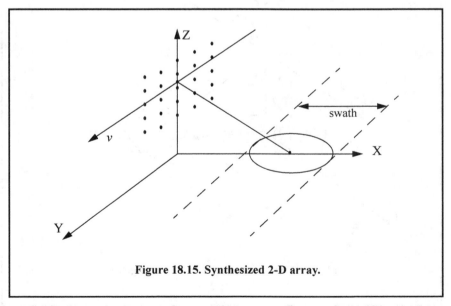

Figure 18.15. Synthesized 2-D array.

The vector $\vec{q}(t_c)$ defines the center of the $3dB$ footprint at time t_c. The center of the array coincides with the flight path, and it is assumed to be perpendicular to both the flight path and the line of sight $\rho(t_c)$. The unit vector $\vec{a}$, along the real array is

$$\vec{a} = \cos\beta^* \vec{a}_x + \sin\beta^* \vec{a}_z \qquad \text{Eq. (18.73)}$$

where β^* is the elevation angle, or the compliment of the depression angle, for the center of the footprint at central time t_c.

18.8.4. Slant Range Equation

Consider the geometry shown in Fig. 18.16 and assume that there is a scatterer $\vec{C}_i$ within the kth range cell. This scatterer is defined by

$$\{amplitiude, phase, elevation, azimuth, height\} = \{a_i, \phi_i, \beta_i, \mu_i, \tilde{h}_i\} . \qquad \text{Eq. (18.74)}$$

the scatterer $\vec{C}_i$ (assuming rectangular coordinates) is given by

$$\vec{C}_i = h\tan\beta_i\cos\mu_i\vec{a}_x + h\tan\beta_i\sin\mu_i\vec{a}_y + \tilde{h}_i\vec{a}_z \qquad \text{Eq. (18.75)}$$

$$\beta_i = \beta_k + \varepsilon \qquad \text{Eq. (18.76)}$$

where β_k denotes the elevation angle for the kth range cell at the center of the observation interval and ε is an incremental angle. Let $\vec{O}e_n$ refer to the vector between the nth array element and the point $\vec{O}$, then

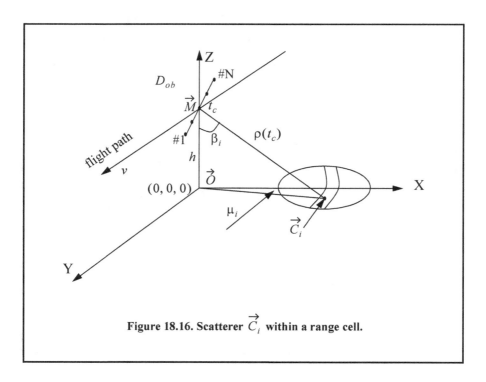

Figure 18.16. Scatterer $\vec{C}_i$ within a range cell.

$$\eth e_n = D_n \cos\beta^* \hat{a}_x + vt\hat{a}_y + (D_n \sin\beta^* + h)\hat{a}_z$$

Eq. (18.77)

$$D_n = \left(\frac{1-N}{2} + n\right)d \quad ;n = 0, \dots N-1.$$

Eq. (18.78)

The range between a scatterer $\tilde{C}$ within the kth range cell, and the nth element of the real array is

$$r_n^2(t, \varepsilon, \mu, \tilde{h}; D_n) = D_n^2 + v^2 t^2 + (h - \tilde{h})^2 + 2D_n \sin\beta^*(h - \tilde{h}) +$$
$$h\tan(\beta_k + \varepsilon)[h\tan(\beta_k + \varepsilon) - 2D_n \cos\beta^* \cos\mu - 2vt\sin\mu]$$

Eq. (18.79)

It is more practical to use the scatterer's elevation and azimuth direction-sines rather than the corresponding increments. Therefore, define the scatterer's azimuth and elevation direction-sines as

$$s = \sin\mu$$

Eq. (18.80)

$$u = \sin\varepsilon.$$

Eq. (18.81)

Then, one can rewrite Eq. (18.79) as

$$r_n^2(t, s, u, \tilde{h}; D_n) = D_n^2 + v^2 t^2 + (h - \tilde{h})^2 + h^2 f^2(u) +$$
$$2D_n \sin\beta^*(h - \tilde{h}) - (2D_n h \cos\beta^* f(u)\sqrt{1 - s^2} - 2vhtf(u)s)$$

Eq. (18.82)

$$f(u) = \tan(\beta_k + a\sin u)$$

Eq. (18.83)

Expanding r_n as a third-order Taylor series expansion about incremental $(t, s, u, \tilde{h})$ yields

$$r(t, s, u, \tilde{h}; D_n) = \bar{r} + \bar{r}_{\tilde{h}}\tilde{h} + \bar{r}_u u + \bar{r}_{\tilde{h}\tilde{h}}\frac{\tilde{h}^2}{2} + \bar{r}_{\tilde{h}u}\tilde{h}u + \bar{r}_{ss}\frac{s^2}{2} + \bar{r}_{st}st +$$

Eq. (18.84)

$$\bar{r}_{tt}\frac{t^2}{2} + \bar{r}_{uu}\frac{u^2}{2} + \bar{r}_{\tilde{h}\tilde{h}\tilde{h}}\frac{\tilde{h}^3}{6} + \bar{r}_{\tilde{h}\tilde{h}u}\frac{\tilde{h}^2 u}{2} + \bar{r}_{\tilde{h}st}\tilde{h}st + \bar{r}_{\tilde{h}uu}\frac{\tilde{h}u^2}{2} +$$

$$\bar{r}_{\tilde{h}ss}\frac{\tilde{h}s^2}{2} + \bar{r}_{uss}\frac{us^2}{2} + \bar{r}_{stu}stu + \bar{r}_{suu}\frac{su^2}{2} + \bar{r}_{\tilde{h}\tilde{h}t}\frac{t\tilde{h}^2}{2} + \bar{r}_{utt}\frac{ut^2}{2} + \bar{r}_{uuu}\frac{u^3}{6}$$

where subscripts denote partial derivations, and the over-bar indicates evaluation at the state $(t, s, u, \tilde{h}) = (0, 0, 0, 0)$. Note that

$$\{\bar{r}_s = \bar{r}_t = \bar{r}_{\tilde{h}s} = \bar{r}_{\tilde{h}t} = \bar{r}_{su} = \bar{r}_{tu} = \bar{r}_{\tilde{h}\tilde{h}s} = \bar{r}_{\tilde{h}\tilde{h}t} = \bar{r}_{\tilde{h}su} = \bar{r}_{\tilde{h}tu} = $$
$$\bar{r}_{sss} = \bar{r}_{sst} = \bar{r}_{stt} = \bar{r}_{ttt} = \bar{r}_{tsu} = 0\}$$

Eq. (18.85)

Section 12.9.8 has detailed expressions of all non-zero Taylor series coefficients for the kth range cell.

Even at the maximum increments $t_{mx}, s_{mx}, u_{mx}, \tilde{h}_{mx}$, the terms

$$\bar{r}_{\tilde{h}\tilde{h}\tilde{h}}\frac{\tilde{h}^3}{6}, \bar{r}_{\tilde{h}\tilde{h}u}\frac{\tilde{h}^2 u}{2}, \bar{r}_{\tilde{h}uu}\frac{\tilde{h}u^2}{2}, \bar{r}_{\tilde{h}ss}\frac{\tilde{h}s^2}{2}, \bar{r}_{uss}\frac{us^2}{2}, \bar{r}_{stu}stu, \bar{r}_{suu}\frac{su^2}{2}, \bar{r}_{\tilde{h}\tilde{h}t}\frac{t\tilde{h}^2}{2}, \bar{r}_{utt}\frac{ut^2}{2}, \bar{r}_{uuu}\frac{u^3}{6}$$

Eq. (18.86)

are small and can be neglected. Thus, the range r_n is approximated by

$$r(t, s, u, \tilde{h}; D_n) = \bar{r} + \bar{r}_{\tilde{h}}\tilde{h} + \bar{r}_u u + \bar{r}_{\tilde{h}\tilde{h}}\frac{\tilde{h}^2}{2} + \bar{r}_{\tilde{h}u}\tilde{h}u + \quad .$$ Eq. (18.87)

$$\bar{r}_{ss}\frac{s^2}{2} + \bar{r}_{st}st + \bar{r}_{tt}\frac{t^2}{2} + \bar{r}_{uu}\frac{u^2}{2} + \bar{r}_{\tilde{h}st}\tilde{h}st$$

Consider the following two-way path: the n_1^{th} element transmitting, scatterer $\vec{C}_i$ reflecting, and the n_2^{th} element receiving. It follows that the round-trip delay corresponding to this two-way path is

$$\tau_{n_1 n_2} = \frac{1}{c}(r_{n_1}(t, s, u, \tilde{h}; D_{n_1}) + r_{n_2}(t, s, u, \tilde{h}; D_{n_2}))$$ Eq. (18.88)

where c is the speed of light.

18.8.5. Signal Synthesis

The observation interval is divided into M subintervals of width $\Delta t = (D_{ob} \div M)$. During each subinterval, the real array is operated in sequential mode, and an array length of $2N$ is synthesized. The number of subintervals M is computed such that Δt is large enough to allow sequential transmission for the real array without causing range ambiguities. In other words, if the maximum range is denoted as R_{mx}, then

$$\Delta t > N\frac{2R_{mx}}{c} .$$ Eq. (18.89)

Each subinterval is then partitioned into N sampling subintervals of width $2R_{mx}/c$. The location t_{mn} represents the sampling time at which the nth element is transmitting during the mth subinterval.

The normalized transmitted signal during the mth subinterval for the nth element is defined as

$$s_n(t_{mn}) = \cos(2\pi f_o t_{mn} + \zeta)$$ Eq. (18.90)

where ζ denotes the transmitter phase, and f_o is the system operating frequency. Assume that there is only one scatterer, $\vec{C}_i$ within the kth range cell defined by $(a_i, \phi_i, s_i, u_i, \tilde{h}_i)$. The returned signal at the n_2^{th} element due to firing from the n_1^{th} element and reflection from the $\vec{C}_i$ scatterer is

$$s_i(n_1, n_2; t_{mn_1}) = a_i G^2(\sin\beta_i)(\rho_k(t_c)/\rho(t_c))^4 \cos[2\pi f_o(t_{mn_1} - \tau_{n_1 n_2}) + \zeta - \phi_i]$$ Eq. (18.91)

where G^2 represents the two-way antenna gain, and the term $(\rho_k(t_c)/\rho(t_c))^4$ denotes the range attenuation at the kth range cell. The analysis in this paper will assume that ζ and ϕ_i are both equal to zeroes.

Suppose that there are N_o scatterers within the kth range cell, with angular locations given by

$$\{(a_i, \phi_i, s_i, u_i, \tilde{h}_i); i = 1, \dots N_o\} .$$ Eq. (18.92)

The composite returned signal at time t_{mn_1} within this range cell due to the path $(n_1 \Rightarrow all\ \vec{C}_i \Rightarrow n_2)$ is

$$s(n_1, n_2; t_{mn_1}) = \sum_{i=1}^{N_o} s_i(n_1, n_2; t_{mn_1}).$$

Eq. (18.93)

The platform motion synthesizes a rectangular array of size $N \times M$, where only one column of N elements exists at a time. However, if $M = 2N$ and the real array is operated in the sequential mode, a square planar array of size $2N \times 2N$ is synthesized. The element spacing along the flight path is $d_y = vD_{ob}/M$.

Consider the kth range bin. The corresponding two-dimensional information sequence $\{b_k(n, m); (n, m) = 0, \ldots 2N-2\}$ consists of $2N$ similar vectors. The mth vector represents the returns due to the sequential firing of all N elements during the mth subinterval. Each vector has $(2N - 1)$ rows, and it is extended, by adding zeroes, to the next power of two. For example, consider the mth subinterval, and let $M = 2N = 4$. Then, the elements of the extended column $\{b_k(n, m)\}$ are

$$\{b_k(0, m), b_k(1, m), b_k(2, m), b_k(3, m), b_k(4, m), b_k(5, m),$$

Eq. (18.94)

$$b_k(6, m), b_k(7, m)\} = \{s(0, 0; t_{mn_0}), s(0, 1; t_{mn_0}) + s(1, 0; t_{mn_1}),$$
$$s(0, 2; t_{mn_0}) + s(1, 1; t_{mn_1}) + s(2, 0; t_{mn_2}), s(0, 3; t_{mn_0}) + s(1, 2; t_{mn_1}) +$$
$$s(2, 1; t_{mn_2}) + s(3, 0; t_{mn_3}), s(1, 3; t_{mn_1}) + s(2, 2; t_{mn_2}) +$$
$$s(3, 1; t_{mn_3}), s(2, 3; t_{mn_2}) + s(3, 2; t_{mn_3}), s(3, 3; t_{mn_3}), 0\}$$

18.8.6. Electronic Processing

Consider again the kth range cell during the mth subinterval, and the two-way path: n_1^{th} element transmitting and n_2^{th} element receiving. The analog quadrature components corresponding to this two-way path are

$$s_I^{\perp}(n_1, n_2; t) = B\cos\psi^{\perp}$$

Eq. (18.95)

$$s_Q^{\perp}(n_1, n_2; t) = B\sin\psi^{\perp}$$

Eq. (18.96)

$$\psi^{\perp} = 2\pi f_0 \left\{ t - \frac{1}{c} \left[2\bar{r} + (\bar{r}_{\tilde{h}}(D_{n_1}) + \bar{r}_{\tilde{h}}(D_{n_2}))\tilde{h} + (\bar{r}_u(D_{n_1}) + \bar{r}_u(D_{n_2}))u + \right. \right.$$

Eq. (18.97)

$$(\bar{r}_{\tilde{h}\tilde{h}}(D_{n_1}) + \bar{r}_{\tilde{h}\tilde{h}}(D_{n_2}))\frac{\tilde{h}^2}{2} + (\bar{r}_{\tilde{h}u}(D_{n_1}) + \bar{r}_{\tilde{h}u}(D_{n_2}))\tilde{h}u +$$

$$(\bar{r}_{ss}(D_{n_1}) + \bar{r}_{ss}(D_{n_2}))\frac{s^2}{2} + 2\bar{r}_{st}st + 2\bar{r}_{tt}\frac{t^2}{2} +$$

$$(\bar{r}_{uu}(D_{n_1}) + \bar{r}_{uu}(D_{n_2}))\frac{u^2}{2} + (\bar{r}_{\tilde{h}st}(D_{n_1}) + \bar{r}_{\tilde{h}st}(D_{n_2}))\tilde{h}st \left] \right\}$$

where B denotes antenna gain, range attenuation, and scatterers' strengths. The subscripts for t have been dropped for notation simplicity. Rearranging Eq. (18.97) and collecting terms yields

$$\psi^\perp = \frac{2\pi f_0}{c} \Bigg\{ \{ tc - [2\bar{r}_{st}s + (\bar{r}_{\tilde{h}st}(D_{n_1}) + \bar{r}_{\tilde{h}st}(D_{n_2}))\tilde{h}s]t - \bar{r}_{tt}t^2 \} -$$

$$\Big[2\bar{r} + (\bar{r}_{\tilde{h}}(D_{n_1}) + \bar{r}_{\tilde{h}}(D_{n_2}))\tilde{h} + (\bar{r}_u(D_{n_1}) + \bar{r}_u(D_{n_2}))u +$$

$$(\bar{r}_{uu}(D_{n_1}) + \bar{r}_{uu}(D_{n_2}))\frac{u^2}{2} + (\bar{r}_{\tilde{h}\tilde{h}}(D_{n_1}) + \bar{r}_{\tilde{h}\tilde{h}}(D_{n_2}))\frac{\tilde{h}^2}{2} +$$

$$(\bar{r}_{\tilde{h}u}(D_{n_1}) + \bar{r}_{\tilde{h}u}(D_{n_2}))\tilde{h}u + (\bar{r}_{ss}(D_{n_1}) + \bar{r}_{ss}(D_{n_2}))\frac{s^2}{2} \Big] \Bigg\}$$

Eq. (18.98)

After analog-to-digital (A/D) conversion, deramping of the quadrature components to cancel the quadratic phase $(-2\pi f_0 \bar{r}_{tt}t^2/c)$ is performed. Then, the digital quadrature components are

$$s_I(n_1, n_2; t) = B\cos\psi \qquad\qquad \textbf{Eq. (18.99)}$$

$$s_Q(n_1, n_2; t) = B\sin\psi \qquad\qquad \textbf{Eq. (18.100)}$$

$$\psi = \psi^\perp - 2\pi f_0 t + 2\pi f_0 \bar{r}_{tt}\frac{t^2}{c}. \qquad\qquad \textbf{Eq. (18.101)}$$

The instantaneous frequency for the ith scatterer within the kth range cell is computed as

$$f_{di} = \frac{1}{2\pi}\frac{d\psi}{dt} = -\frac{f_0}{c}[2\bar{r}_{st}s + (\bar{r}_{\tilde{h}st}(D_{n_1}) + \bar{r}_{\tilde{h}st}(D_{n_2}))\tilde{h}s]. \qquad \textbf{Eq. (18.102)}$$

Substituting the actual values for $\bar{r}_{st}$, $\bar{r}_{\tilde{h}st}(D_{n_1})$, $\bar{r}_{\tilde{h}st}(D_{n_2})$ and collecting terms yields

$$f_{di} = -\left(\frac{2v\sin\beta_k}{\lambda}\right)\left(\frac{\tilde{h}s}{\rho_k^2(t_c)}(h + (D_{n_1} + D_{n_2})\sin\beta^*) - s\right). \qquad \textbf{Eq. (18.103)}$$

Note that if $\tilde{h} = 0$, then

$$f_{di} = \frac{2v}{\lambda}\sin\beta_k\sin\mu, \qquad\qquad \textbf{Eq. (18.104)}$$

which is the Doppler value corresponding to a ground patch (see Eq. (18.49)).

The last stage of the processing consists of three steps: (1) two-dimensional windowing; (2) performing a two-dimensional DFT on the windowed quadrature components; and (3) scaling to compensate for antenna gain and range attenuation.

18.8.7. Derivation of Eq. (18.71)

Consider a rectangular array of size $N \times N$, with uniform element spacing $d_x = d_y = d$, and wavelength λ. Assume sequential mode operation where elements are fired sequentially, one at a time, while all elements receive in parallel. Assume far field observation defined by azimuth and elevation angles (α, β). The unit vector $\hat{n}$ on the line of sight, with respect to $\vec{O}$, is given by

$$\hat{u} = \sin\beta\cos\alpha \ \hat{a}_x + \sin\beta\sin\alpha \ \hat{a}_y + \cos\beta \ \hat{a}_z. \qquad \text{Eq. (18.105)}$$

The $(n_x, n_y)^{th}$ element of the array can be defined by the vector

$$\hat{e}(n_x, n_y) = \left(n_x - \frac{N-1}{2}\right)d \ \hat{a}_x + \left(n_y - \frac{N-1}{2}\right)d \ \hat{a}_y \qquad \text{Eq. (18.106)}$$

where $(n_x, n_y = 0, \dots N-1)$. The one-way geometric phase for this element is

$$\varphi'(n_x, n_y) = k(\hat{u} \bullet \hat{e}(n_x, n_y)) \qquad \text{Eq. (18.107)}$$

where $k = 2\pi/\lambda$ is the wavenumber, and the operator ($\bullet$) indicates dot product. Therefore, the two-way geometric phase between the (n_{x1}, n_{y1}) and (n_{x2}, n_{y2}) elements is

$$\varphi(n_{x1}, n_{y1}, n_{x2}, n_{y2}) = k[\hat{u} \bullet \{\hat{e}(n_{x1}, n_{y1}) + \hat{e}(n_{x2}, n_{y2})\}]. \qquad \text{Eq. (18.108)}$$

The cumulative two-way normalized electric field due to all transmissions in the direction (α, β) is

$$E(\hat{u}) = E_t(\hat{u})E_r(\hat{u}) \qquad \text{Eq. (18.109)}$$

where the subscripts t and r, respectively refer to the transmitted and received electric fields. More precisely,

$$E_t(\hat{u}) = \sum_{n_{xt}=0}^{N-1} \sum_{n_{yt}=0}^{N-1} w(n_{xt}, n_{yt})exp[jk\{\hat{u} \bullet \hat{e}(n_{xt}, n_{yt})\}] \qquad \text{Eq. (18.110)}$$

$$E_r(\hat{u}) = \sum_{n_{xr}=0}^{N-1} \sum_{n_{yr}=0}^{N-1} w(n_{xr}, n_{yr})exp[jk\{\hat{u} \bullet \hat{e}(n_{xr}, n_{yr})\}]. \qquad \text{Eq. (18.111)}$$

In this case, $w(n_x, n_y)$ denotes the tapering sequence. Substituting Eqs. (18.108), (18.110), and (18.111) into Eq. (18.109) and grouping all fields with the same two-way geometric phase yields

$$E(\hat{u}) = e^{j\delta} \sum_{m=0}^{N_a-1} \sum_{n=0}^{N_a-1} w'(m, n)exp[jkd\sin\beta(m\cos\alpha + n\sin\alpha)] \qquad \text{Eq. (18.112)}$$

$$N_a = 2N-1 \qquad \text{Eq. (18.113)}$$

$$m = n_{xt} + n_{xr}; m = 0, \dots 2N-2 \qquad \text{Eq. (18.114)}$$

$$n = n_{yt} + n_{yr}; n = 0, \dots 2N-2 \qquad \text{Eq. (18.115)}$$

$$\delta = \left(\frac{-d\sin\beta}{2}\right)(N-1)(\cos\alpha + \sin\alpha). \qquad \text{Eq. (18.116)}$$

The two-way array pattern is then computed as

$$|E(\vec{n})| = \left| \sum_{m=0}^{N_a-1} \sum_{n=0}^{N_a-1} w'(m,n) exp[jkd\sin\beta(m\cos\alpha + n\sin\alpha)] \right| .$$ Eq. (18.117)

Consider the two-dimensional DFT transform, $W'(p,q)$, of the array $w'(n_x, n_y)$

$$W'(p,q) = \sum_{m=0}^{N_a-1} \sum_{n=0}^{N_a-1} w'(m,n) exp\left(-j\frac{2\pi}{N_a}(pm+qn)\right); p,q = 0, ...N_a-1 .$$ Eq. (18.118)

Comparison of Eqs. (18.117) and Eq. (18.118) indicates that $|E(\vec{n})|$ is equal to $|W'(p,q)|$ if

$$-\left(\frac{2\pi}{N_a}\right)p = \frac{2\pi}{\lambda}d\sin\beta\cos\alpha$$ Eq. (18.119)

$$-\left(\frac{2\pi}{N_a}\right)q = \frac{2\pi}{\lambda}d\sin\beta\sin\alpha .$$ Eq. (18.120)

It follows that

$$\alpha = \tan^{-1}\left(\frac{q}{p}\right) .$$ Eq. (18.121)

18.8.8. Non-Zero Taylor Series Coefficients for the kth Range Cell

$$\bar{r} = \sqrt{D_n^2 + h^2(1 + \tan\beta_k) + 2hD_n\sin\beta^* - 2hD_n\cos\beta^*\tan\beta_k} = \rho_k(t_c)$$ Eq. (18.122)

$$\bar{r}_{\tilde{h}} = \left(\frac{-1}{\bar{r}}\right)(h + D_n\sin\beta^*)$$ Eq. (18.123)

$$\bar{r}_u = \left(\frac{h}{\bar{r}\cos^2\beta_k}\right)(h\tan\beta_k - D_n\cos\beta^*)$$ Eq. (18.124)

$$\bar{r}_{\tilde{h}\tilde{h}} = \left(\frac{1}{\bar{r}}\right) - \left(\frac{1}{\bar{r}^3}\right)(h + D_n\sin\beta^*)$$ Eq. (18.125)

$$\bar{r}_{\tilde{h}u} = \left(\frac{1}{\bar{r}^3}\right)\left(\frac{h}{\cos^2\beta_k}\right)(h + D_n\tan\beta^*)(h\tan\beta_k - D_n\cos\beta^*)$$ Eq. (18.126)

$$\bar{r}_{ss} = \left(\frac{-1}{4\bar{r}^3}\right) + \left(\frac{1}{\bar{r}}\right)(h\tan\beta_k - D_n\cos\beta^*)$$ Eq. (18.127)

$$\bar{r}_{st} = \left(\frac{-1}{\bar{r}}\right)hv\tan\beta_k$$ Eq. (18.128)

$$\bar{r}_{tt} = \frac{v^2}{\bar{r}}$$

Eq. (18.129)

$$\bar{r}_{uu} = \left(\frac{h}{\bar{r}\cos^3\beta_k}\right)\left\{\left(\frac{h}{\bar{r}^2\cos\beta_k}\right)(h\tan\beta_k - D_n\cos\beta^*) + \right.$$

Eq. (18.130)

$$\left. h\left(\left(\frac{1}{\cos\beta_k}\right) + 2\tan\beta_k\sin\beta_k\right) - 2\sin\beta_k D_n\cos\beta^*\right\}$$

$$\bar{r}_{\tilde{h}\tilde{h}\tilde{h}} = \left(\frac{3}{\bar{r}^3}\right)(h + D_n\sin\beta^*)\left[\left(\frac{1}{\bar{r}^2}\right)(h + D_n\sin\beta^*)^2 - 1\right]$$

Eq. (18.131)

$$\bar{r}_{\tilde{h}\tilde{h}u} = \left(\frac{h}{\bar{r}^3\cos^2\beta_k}\right)(h\tan\beta_k - D_n\cos\beta^*)\left[\left(\frac{-3}{\bar{r}^2}\right)(h + D_n\sin\beta^*)^2 + 1\right]$$

Eq. (18.132)

$$\bar{r}_{\tilde{h}st} = \left(\frac{hv\tan\beta_k}{\bar{r}^3}\right)(h + D_n\sin\beta^*)$$

Eq. (18.133)

$$\bar{r}_{\tilde{h}uu} = \left(\frac{-3}{\bar{r}^5}\right)\left(\frac{h^2}{\cos^4\beta_k}\right)(h + D_n\sin\beta^*)(h\tan\beta_k - D_n\cos\beta^*)$$

Eq. (18.134)

$$\bar{r}_{\tilde{h}ss} = \left(\frac{-1}{\bar{r}^3}\right)(h\tan\beta_k - D_n\cos\beta^*)(h + D_n\sin\beta^*)$$

Eq. (18.135)

$$\bar{r}_{uss} = \left(\frac{h}{\bar{r}\cos^2\beta_k}\right)(D_n\cos\beta^*)\left[\left(\frac{1}{\bar{r}^2}\right)(h\tan\beta_k - D_n\cos\beta^*)(h\tan\beta_k) + 1\right]$$

Eq. (18.136)

$$\bar{r}_{stu} = \left(\frac{-h\tan\beta_k}{\bar{r}^3\cos^2\beta_k}\right)(h\tan\beta_k - D_n\cos\beta^*)$$

Eq. (18.137)

$$\bar{r}_{suu} = \left(\frac{hD_n\cos\beta^*}{\bar{r}\cos^2\beta_k}\right)\left[\left(\frac{h\tan\beta_k}{\bar{r}^2}\right)(h\tan\beta_k - D_n\cos\beta^*) + 1\right]$$

Eq. (18.138)

$$\bar{r}_{\tilde{h}tt} = \left(\frac{v^2 h}{\bar{r}^3\cos^2\beta_k}\right)(h\tan\beta_k - D_n\cos\beta^*)$$

Eq. (18.139)

$$\bar{r}_{uuu} =$$

Eq. (18.140)

$$\left(\frac{h}{\bar{r}\cos^4\beta_k}\right)[8h\tan\beta_k + \sin^2\beta_k(h - D_n\cos\beta^*) - 2D_n\cos\beta^*] +$$

$$\left(\frac{3h^2}{\bar{r}^3\cos^5\beta_k}\right)(h\tan\beta_k - D_n\cos\beta^*) + \left[\left(\frac{3h^2}{\bar{r}^3\cos^5\beta_k}\right)(h\tan\beta_k - D_n\cos\beta^*)\right.$$

$$\left.\left(\frac{1}{2\cos\beta_k} + (h\tan\beta_k - D_n\cos\beta^*)\right)\right] + \left(\frac{3h^3}{\bar{r}^5\cos^6\beta_k}\right)(h\tan\beta_k - D_n\cos\beta^*)$$

Problems

18.1. A side looking SAR is traveling at an altitude of $15Km$; the elevation angle is $\beta = 15°$. If the aperture length is $L = 5m$, the pulse width is $\tau = 20\mu s$ and the wavelength is $\lambda = 3.5cm$, (a) calculate the azimuth resolution, (b) calculate the range and ground range resolutions.

18.2. An MMW side looking SAR has the following specifications: radar velocity $v = 70m/s$, elevation angle $\beta = 35°$, operating frequency $f_0 = 94GHz$, and antenna $3dB$ beamwidth $\theta_{3dB} = 65mrad$. (a) Calculate the footprint dimensions. (b) Compute the minimum and maximum ranges. (c) Compute the Doppler frequency span across the footprint. (d) Calculate the minimum and maximum PRFs.

18.3. A side looking SAR takes on eight positions within an observation interval. In each position, the radar transmits and receives one pulse. Let the distance between any two consecutive antenna positions be d, and define $\delta = 2\pi\frac{d}{\lambda}(\sin\beta - \sin\beta_0)$ to be the one-way phase difference for a beam steered at angle β_0. (a) In each of the eight positions a sample of the phase pattern is obtained after heterodyning. List the phase samples. (b) How will you process the sequence of samples using an FFT (do not forget windowing)? (c) Give a formula for the angle between the grating lobes.

18.4. Consider a synthetic aperture radar. You are given the following Doppler history for a scatterer: $\{1000Hz, 0, -1000HZ\}$, which corresponds to times $\{-10ms, 0, 10ms\}$. Assume that the observation interval is $T_{ob} = 20ms$, and a platform velocity $v = 200m/s$. (a) Show the Doppler history for another scatterer which is identical to the first one except that it is located in azimuth $1m$ earlier. (b) How will you perform deramping on the quadrature components (show only the general approach)? (c) Show the Doppler history for both scatterers after deramping.

18.5. You want to design a side looking synthetic aperture ultrasonic radar operating at $f_0 = 60KHz$ and peak power $P_t = 2W$. The antenna beam is conical with $3dB$ beamwidth $\theta_{3dB} = 5°$. The maximum gain is 16. The radar is at a constant altitude $h = 15m$ and is moving at a velocity of $10m/s$. The elevation angle defining the footprint is $\beta = 45°$. (a) Give an expression for the antenna gain assuming a Gaussian pattern. (b) Compute the pulse width corresponding to range resolution of $10mm$. (c) What are the footprint dimensions? (d) Compute and plot the Doppler history for a scatterer located on the central range bin. (e) Calculate the minimum and maximum PRFs. Do you need to use more than one PRF? (f) How will you design the system in order to achieve an azimuth resolution of $10mm$?

18.6. Validate Eq. (18.46).

18.7. In Section 18.7 we assumed the elevation angle increment ε is equal to zero. Develop an equivalent to Eq. (18.43) for the case when $\varepsilon \neq 0$. You need to use a third-order three-dimensional Taylor series expansion about the state $(t, \mu, \varepsilon) = (0, 0, 0)$ in order to compute the new round-trip delay expression.

Appendix 18-A: Chapter 18 MATLAB Code Listings

The MATLAB code provided in this chapter was designed as an academic standalone tool and is not adequate for other purposes. The code was written in a way to assist the reader in gaining a better understanding of the theory. The code was not developed, nor is it intended to be used as part of an open-loop or a closed-loop simulation of any kind. The MATLAB code found in this textbook can be downloaded from this book's web page on the CRC Press website. Simply use your favorite web browser, go to *www.crcpress.com*, and search for keyword *"Mahafza"* to locate this book's web page.

MATLAB Program "Fig18_12-13.m" Listing

```
%                Figures 18.12 and 18.13
%    Program to do Spotlight SAR using the rectangular format and
%    HRR for range compression.
%                13 June 2003
%                Dr. Brian J. Smith
clear all;
%%%%%%%%% SAR Image Resolution %%%%
dr = .50;
da = .10;
% dr = 6*2.54/100;
% da = 6*2.54/100;
%%%%%%%%%% Scatter Locations %%%%%%%%
xn = [10000 10015 9985];  % Scatter Location, x-axis
yn = [0 -20 20];          % Scatter Location, y-axis
Num_Scatter = 3;          % Number of Scatters
Rnom = 10000;
%%%%%%%%%% Radar Parameters %%%%%%%%%%
f_0 =  35.0e9;   % Lowest Freq. in the HRR Waveform
df =   3.0e6;    % Freq. step size for HRR, Hz
c =      3e8;    % Speed of light, m/s
Kr = 1.33;
Num_Pulse = 2^(round(log2(Kr*c/(2*dr*df))));
Lambda = c/(f_0 + Num_Pulse*df/2);
%%%%%%%%%% Synthetic Array Parameters %%%%%%%%%
du = 0.2;
L = round(Kr*Lambda*Rnom/(2*da));
U = -(L/2):du:(L/2);
Num_du = length(U);
%%%%%%%%%% This section generates the target returns %%%%%%%
Num_U = round(L/du);
I_Temp = 0;
Q_Temp = 0;
for I = 1:Num_U
   for J = 1:Num_Pulse
      for K = 1:Num_Scatter
         Yr = yn(K) - ((I-1)*du - (L/2));
         Rt = sqrt(xn(K)^2 + Yr^2);
         F_ci = f_0 + (J-1)*df;
         PHI = -4*pi*Rt*F_ci/c;
         I_Temp = cos(PHI) + I_Temp;
         Q_Temp = sin(PHI) + Q_Temp;
```

```
    end;
    IQ_Raw(J,I) = I_Temp + i*Q_Temp;
    I_Temp = 0.0;
    Q_Temp = 0.0;
  end;
end;
%%%%%%%%%%% End target return section %%%%%
%%%%%%%%%%% Range Compression %%%%%%%%%%%%%%%%
Num_RB = 2*Num_Pulse;
WR = hamming(Num_Pulse);
for I = 1:Num_U
  Range_Compressed(:,I) = fftshift(ifft(IQ_Raw(:,I).*WR,Num_RB));
end;
%%%%%%%%%%% Focus Range Compressed Data %%%%%
dn = (1:Num_U)*du - L/2;
PHI_Focus = -2*pi*(dn.^2)/(Lambda*xn(1));
for I = 1:Num_RB
  Temp = angle(Range_Compressed(I,:)) - PHI_Focus;
  Focused(I,:) = abs(Range_Compressed(I,:)).*exp(i*Temp);
end;
%Focused = Range_Compressed;
%%%%%%%%%%% Azimuth Compression %%%%%%%%%%%%%%
WA = hamming(Num_U);
for I = 1:Num_RB
  AZ_Compressed(I,:) = fftshift(ifft(Focused(I,:).*WA'));
end;
 SAR_Map = 10*log10(abs(AZ_Compressed));
 Y_Temp = (1:Num_RB)*(c/(2*Num_RB*df));
 Y = Y_Temp - max(Y_Temp)/2;
 X_Temp = (1:length(IQ_Raw))*(Lambda*xn(1)/(2*L));
 X = X_Temp - max(X_Temp)/2;
 image(X,Y,20-SAR_Map);  %
 %image(X,Y,5-SAR_Map);  %
 axis([-25 25 -25 25]); axis equal; colormap(gray(64));
 xlabel('Cross Range (m)'); ylabel('Down Range (m)');
 grid
 %print -djpeg .jpg
```

Bibliography

[1]   Abramowitz, M., and Stegun, I. A., Editors, *Handbook of Mathematical Functions, with Formulas, Graphs, and Mathematical Tables*, Dover Publications, New York, 1970.

[2]   Balanis, C. A., *Antenna Theory, Analysis and Design*, Harper & Row, New York, 1982.

[3]   Barkat, M., *Signal Detection and Estimation*, Artech House, Norwood, MA, 1991.

[4]   Barton, D. K., *Modern Radar System Analysis*, Artech House, Norwood, MA, 1988.

[5]   Bean, B. R., and Abbott, R., Oxygen and Water-Vapor Absorption of Radio Waves in the Atmosphere, *Review Geofisica Pura E Applicata (Milano)* 37:127, 1957.

[6]   Benedict, T., and Bordner, G., Synthesis of an Optimal Set of Radar Track-While-Scan Smoothing Equations, *IRE Transaction on Automatic Control, Ac-7*, July 1962, pp. 27-32.

[7]   Berkowitz, R. S., *Modern Radar: Analysis, Evaluation, and System Design*, John Wiley & Sons, Inc, New York, 1965.

[8]   Beyer, W. H., *CRC Standard Mathematical Tables*, 26th Edition, CRC Press, Boca Raton, FL, 1981.

[9]   Billetter, D. R., *Multifunction Array Radar*, Artech House, Norwood, MA, 1989.

[10]   Blackman, S. S., *Multiple-Target Tracking with Radar Application*, Artech House, Norwood, MA, 1986.

[11]   Blake, L. V., *A Guide to Basic Pulse-Radar Maximum Range Calculation. Part-1: Equations, Definitions, and Aids to Calculation*, Naval Res. Lab. Report 5868, 1969.

[12]   Blake, L. V., *Curves of Atmospheric-Absorption Loss for Use in Radar Range Calculation*, NRL Report 5601, March 23, 1961.

[13]   Blake, L. V., *Radar / Radio Tropospheric Absorption and Noise Temperature*, NRL Report Ad-753 197, October 1972. Distributed by the National Technical Information Service (NTIS).

[14]   Blake, L. V., *Radar-Range Performance Analysis*, Lexington Books, Lexington, MA, 1980.

[15]   Boothe, R. R., *A Digital Computer Program for Determining the Performance of an Acquisition Radar through Application of Radar Detection Probability Theory*, U.S. Army Missile Command, Report No. RD-TR-64-2, Redstone Arsenal, Alabama, 1964.

[16]   Boothe, R. R., *The Weibull Distribution Applied to the Ground Clutter Backscatter Coefficient*, U.S. Army Missile Command, Report No. RE-TR-69-15, Redstone Arsenal, Alabama, 1969.

[17]   Bowman, J. J., Piergiorgio, L. U., and Senior, T. B., *Electromagnetic and Acoustic Scattering by Simple Shapes*, North-Holland Pub. Co, Amsterdam, 1969.

[18]   Brookner, E., Editor, *Aspects of Modern Radar*, Artech House, Norwood, MA, 1988.

[19]   Brookner, E., Editor, *Practical Phased Array Antenna System*, Artech House, Norwood, MA, 1991.

[20]   Brookner, E., *Radar Technology*, Lexington Books, Lexington, MA, 1996.

[21]   Burdic, W. S., *Radar Signal Analysis*, Prentice Hall, Englewood Cliffs, NJ, 1968.

[22]   Brookner, E., *Tracking and Kalman Filtering Made Easy*, John Wiley & Sons, New York, 1998.

[23]   Cadzow, J. A., *Discrete-Time Systems, An Introduction with Interdisciplinary Applications*, Prentice Hall, Englewood Cliffs, NJ, 1973.

[24]   Carlson, A. B., *Communication Systems, An Introduction to Signals and Noise in Electrical Communication*, 3rd Edition, McGraw-Hill, New York, 1986.

[25]   Carpentier, M. H., *Principles of Modern Radar Systems*, Artech House, Norwood, MA, 1988.

[26]   Compton, R. T., *Adaptive Antennas*, Prentice Hall, Englewood Cliffs, NJ, 1988.

[27]   Cook, E. C., and Bernfeld, M., *Radar Signals: An Introduction to Theory and Application*, Artech House, Norwood, MA, 1993.

[28]   Costas, J. P., A Study of a Class of Detection Waveforms Having Nearly Ideal Range-Doppler Ambiguity Properties, *Proc. IEEE 72*, 1984, pp. 996-1009.

[29]   Crispin, J. W. Jr., and Siegel, K. M, Editors, *Methods of Radar Cross-Section Analysis*, Academic Press, New York, 1968.

[30]   Curry, G. R., *Radar System Performance Modeling*, Artech House, Norwood, MA, 2001.

[31]   DiFranco, J. V. and Rubin, W. L., *Radar Detection*. Artech House, Norwood, MA, 1980.

[32]   Dillard, R. A. and Dillard, G. M., *Detectability of Spread-Spectrum Signals*, Artech House, Norwood, MA, 1989.

[33]   Edde, B., *Radar Principles, Technology, Applications*, Prentice Hall, Englewood Cliffs, NJ, 1993.

[34]   Elsherbeni, A., Inman, M. J., and Riley, C., Antenna Design and Radiation Pattern Visualization, *The 19th Annual Review of Progress in Applied Computational Electromagnetics*, ACES'03, Monterey, CA, March 2003.

[35]   Fehlner, L. F., *Marcum's and Swerling's Data on Target Detection by a Pulsed Radar*, Johns Hopkins University, Applied Physics Lab. Rpt. # TG451, July 2, 1962, and Rpt. # TG451A, September 1964.

[36]   Fielding, J. E., and Reynolds, G. D., *VCCALC: Vertical Coverage Calculation Software and Users Manual*, Artech House, Norwood, MA, 1988.

[37]   Gabriel, W. F., Spectral Analysis and Adaptive Array Superresolution Techniques, *Proc. IEEE*, Vol. 68, June 1980, pp. 654-666.

[38]   Gelb, A., Editor, *Applied Optimal Estimation*, MIT Press, Cambridge, MA, 1974.

[39] Goldman, S. J., *Phase Noise Analysis in Radar Systems, Using Personal Computers*, John Wiley & Sons, New York, NY, 1989.

[40] Grewal, M. S., and Andrews, A. P., *Kalman Filtering: Theory and Practice Using MATLAB*, 2nd Edition, Wiley & Sons Inc., New York, 2001.

[41] Hamming, R. W., *Digital Filters*, 2nd Edition, Prentice Hall, Englewood Cliffs, NJ, 1983.

[42] Hanselman, D., and Littlefield, B., *Mastering MATLAB 5: A Complete Tutorial and Reference*, MATLAB Curriculum Series, Prentice Hall, Englewood Cliffs, NJ, 1998.

[43] Hirsch, H. L., and Grove, D. C., *Practical Simulation of Radar Antennas and Radomes*, Artech House, Norwood, MA, 1987.

[44] Hovanessian, S. A., *Radar System Design and Analysis*, Artech House, Norwood, MA, 1984.

[45] James, D. A., *Radar Homing Guidance for Tactical Missiles*, John Wiley & Sons, New York, 1986.

[46] Jin, J., *The Finite Element Method in Electromagnetics*, John Wiley & Sons, New York, 2002.

[47] Kanter, I., Exact Detection Probability for Partially Correlated Rayleigh Targets, *IEEE Trans, AES-22*, March 1986, pp. 184-196.

[48] Kay, S. M., *Fundamentals of Statistical Signal Processing: Estimation Theory*, Volume I, Prentice Hall Signal Processing Series, Englewood Cliffs, NJ, 1993.

[49] Kay, S. M., *Fundamentals of Statistical Signal Processing: Detection Theory*, Volume II, Prentice Hall Signal Processing Series, Englewood Cliffs, NJ, 1993.

[50] Keller, J. B., Geometrical Theory of Diffraction, *Journal Opt. Soc. Amer.*, Vol. 52, February 1962, pp. 116-130.

[51] Klauder, J. R., Price, A. C., Darlington, S., and Albershiem, W. J., The Theory and Design of Chirp Radars, *The Bell System Technical Journal*, Vol. 39, No. 4, 1960.

[52] Klemm, R., *Principles of Space-Time Adaptive Processing*, 3rd Edition, IET, London, UK, 2006.

[53] Knott, E. F., Shaeffer, J. F., and Tuley, M. T., *Radar Cross Section*, 2nd Edition, Artech House, Norwood, MA, 1993.

[54] Lativa, J., Low-Angle Tracking Using Multifrequency Sampled Aperture Radar, *IEEE-AES Trans.*, Vol. 27, No. 5, September 1991, pp. 797-805.

[55] Lee, S. W., and Mittra, R., Fourier Transform of a Polygonal Shape Function and Its Application in Electromagnetics, *IEEE Trans. Antennas and Propagation*, Vol. 31, January 1983, pp. 99-103.

[56] LeFande, R. A., *Attenuation of Microwave Radiation for Paths through the Atmosphere*, NRL Report 6766, Nov. 1968.

[57] Levanon, N., *Radar Principles*, John Wiley & Sons, New York, 1988.

[58] Levanon, N., and Mozeson, E., Nullifying ACF Grating Lobes in Stepped-Frequency Train of LFM Pulses, *IEEE-AES Trans.*, Vol. 39, No. 2, April 2003, pp. 694-703.

[59] Levanon, N., and Mozeson, E., *Radar Signals*, John Wiley-Interscience, Hoboken, NJ, 2004.

[60] Lewis, B. L., Kretschmer, Jr., F. F., and Shelton, W. W., *Aspects of Radar Signal Processing*, Artech House, Norwood, MA, 1986.

[61] Li, J., and Stoica, P., Editors, *MIMO Radar Signal Processing*, John Wiley & Sons Inc., New York, 2009.

[62] Long, M. W., *Radar Reflectivity of Land and Sea*, Artech House, Norwood, MA, 1983.

[63] Lothes, R. N., Szymanski, M. B., and Wiley, R. G., *Radar Vulnerability to Jamming*, Artech House, Norwood, MA, 1990.

[64] Maffett, A. L., *Topics for a Statistical Description of Radar Cross Section*, John Wiley & Sons, New York, 1989.

[65] Mahafza, B. R., *Introduction to Radar Analysis*, CRC Press, Boca Raton, FL, 1998.

[66] Mahafza, B. R., *Radar Systems Analysis and Design Using MATLAB*, 2nd Edition, Taylor & Francis, Boca Raton, FL, 2005.

[67] Mahafza, B. R., *Radar Signal Analysis and Signal Processing Using MATLAB*, Chapman and Hall/CRC, Boca Raton, FL, 2008.

[68] Mahafza, B. R., and Polge, R. J., Multiple Target Detection Through DFT Processing in a Sequential Mode Operation of Real Two-Dimensional Arrays, *Proc. of the IEEE Southeast Conf. '90*, New Orleans, LA, April 1990, pp. 168-170.

[69] Mahafza, B. R., Heifner, L.A., and Gracchi, V. C., Multitarget Detection Using Synthetic Sampled Aperture Radars (SSAMAR), *IEEE-AES Trans.*, Vol. 31, No. 3, July 1995, pp. 1127-1132.

[70] Mahafza, B. R., and Sajjadi, M., Three-Dimensional SAR Imaging Using a Linear Array in Transverse Motion, *IEEE-AES Trans.*, Vol. 32, No. 1, January 1996, pp. 499-510.

[71] Marchand, P., *Graphics and GUIs with MATLAB*, 2nd Edition, CRC Press, Boca Raton, FL, 1999.

[72] Marcum, J. I., A Statistical Theory of Target Detection by Pulsed Radar, Mathematical Appendix, *IRE Trans.*, Vol. IT-6, April 1960, pp. 259-267.

[73] Medgyesi-Mitschang, L. N., and Putnam, J. M., Electromagnetic Scattering from Axially Inhomogenous Bodies of Revolution, *IEEE Trans. Antennas and Propagation.*, Vol. 32, August 1984, pp. 797-806.

[74] Meeks, M. L., *Radar Propagation at Low Altitudes*, Artech House, Norwood, MA, 1982.

[75] Melsa, J. L., and Cohn, D. L., *Decision and Estimation Theory*, McGraw-Hill, New York, 1978.

[76] Mensa, D. L., *High Resolution Radar Imaging*, Artech House, Norwood, MA, 1984.

[77] Meyer, D. P., and Mayer, H. A., *Radar Target Detection: Handbook of Theory and Practice*, Academic Press, New York, 1973.

[78] Monzingo, R. A., and Miller, T. W., *Introduction to Adaptive Arrays,* John Wiley & Sons, New York, 1980.

[79] Morchin, W., *Radar Engineer's Sourcebook*, Artech House, Norwood, MA, 1993.

[80] Morris, G. V., *Airborne Pulsed Doppler Radar*, Artech House, Norwood, MA, 1988.

[81] Nathanson, F. E., *Radar Design Principles*, 2nd Edition, McGraw-Hill, New York, 1991.

[82] Navarro, Jr., A. M., *General Properties of Alpha Beta and Alpha Beta Gamma Tracking Filters*, Physics Laboratory of the National Defense Research Organization TNO, Report PHL 1977-92, January 1977.

[83] North, D. O., An Analysis of the Factors Which Determine Signal/Noise Discrimination in Pulsed Carrier Systems, *Proc. IEEE 51*, No. 7, July 1963, pp. 1015-1027.

[84] Oppenheim, A. V., and Schafer, R. W., *Discrete-Time Signal Processing*, Prentice Hall, Englewood Cliffs, NJ, 1989.

[85] Oppenheim, A. V., Willsky, A. S., and Young, I. T., *Signals and Systems*, Prentice Hall, Englewood Cliffs, NJ, 1983.

[86] Orfanidis, S. J., *Optimum Signal Processing, an Introduction*, 2nd Edition, McGraw-Hill, New York, 1988.

[87] Papoulis, A., *Probability, Random Variables, and Stochastic Processes*, 2nd Edition, McGraw-Hill, New York, 1984.

[88] Parl, S. A., New Method of Calculating the Generalized Q Function, *IEEE Trans. Information Theory,* Vol. IT-26, No. 1, January 1980, pp. 121-124.

[89] Peebles, P. Z., Jr., *Probability, Random Variables, and Random Signal Principles*, McGraw-Hill, New York, 1987.

[90] Peebles, P. Z., Jr., *Radar Principles*, John Wiley & Sons, New York, 1998.

[91] Pettit, R. H., *ECM and ECCM Techniques for Digital Communication Systems*, Lifetime Learning Publications, New York, 1982.

[92] Polge, R. J., Mahafza, B. R., and Kim, J. G., *Extension and Updating of the Computer Simulation of Range Relative Doppler Processing for MM Wave Seekers*, Interim Technical Report, Vol. I, prepared for the U.S. Army Missile Command, Redstone Arsenal, Alabama, January 1989.

[93] Polge, R. J., Mahafza, B. R., and Kim, J. G., Multiple Target Detection through DFT Processing in a Sequential Mode Operation of Real or Synthetic Arrays, *IEEE 21st Southeastern Symposium on System Theory*, Tallahassee, FL, 1989, pp. 264-267.

[94] Poularikas, A., *Signals and Systems Primer with MATLAB*, Taylor & Francis, Boca Raton, FL, 2007.

[95] Poularikas, A., and Ramadan, Z. M., *Adaptive Filtering Primer with MATLAB*, Taylor & Francis, Boca Raton, FL, 2006.

[96] Poularikas, A., and Seely, S., *Signals and Systems*, PWS Publishers, Boston, MA, 1984.

[97] Putnam, J. N., and Gerdera, M. B., CARLOS TM: A General-Purpose Three-Dimensional Method of Moments Scattering Code, *IEEE Trans. Antennas and Propagation*, Vol. 35, April 1993, pp. 69-71

[98] Reed, H. R. and Russell, C. M., *Ultra High Frequency Propagation*, Boston Technical Publishers, Inc., Lexington, MA, 1964.

[99] Resnick, J. B., *High Resolution Waveforms Suitable for a Multiple Target Environment*, MS thesis, MIT, Cambridge, MA, June 1962.

[100] Richards, M. A., *Fundamentals of Radar Signal Processing*, McGraw-Hill, New York, 2005.

[101] Rihaczek, A. W., *Principles of High Resolution Radars*, McGraw-Hill, New York, 1969.

[102] Robertson, G. H., Operating Characteristics for a Linear Detector of CW Signals in Narrow-Band Gaussian Noise, *Bell Sys. Tech. Journal*, Vol. 46 April 1967, pp. 755-774.

[103] Rosenbaum, B., *A Programmed Mathematical Model to Simulate the Bending of Radio Waves in the Atmospheric Propagation*, Goddard Space Flight Center Report X-551-68-367, Greenbelt, Maryland, 1968.

[104] Ross, R. A., Radar Cross Section of Rectangular Flat Plate as a Function of Aspect Angle, *IEEE Trans*. AP-14, 1966, p. 320.

[105] Ruck, G. T., Barrick, D. E., Stuart, W. D., and Krichbaum, C. K., *Radar Cross Section Handbook*, Volume 1, Plenum Press, New York, 1970.

[106] Ruck, G. T., Barrick, D. E., Stuart, W. D., and Krichbaum, C. K., *Radar Cross Section Handbook*, Volume 2, Plenum Press, New York, 1970.

[107] Rulf, B., and Robertshaw, G. A., *Understanding Antennas for Radar, Communications, and Avionics*, Van Nostrand Reinhold, 1987.

[108] Scanlan, M. J., Editor, *Modern Radar Techniques*, Macmillan, New York, 1987.

[109] Scheer, J. A., and Kurtz, J. L., Editor, *Coherent Radar Performance Estimation*, Artech House, Norwood, MA, 1993.

[110] Schelher, D. C., *MTI and Pulsed Doppler Radar with MATLAM*, 2nd Edition, Artech House, Norwood MA. 2010.

[111] Shanmugan, K. S., and Breipohl, A. M., *Random Signals: Detection, Estimation and Data Analysis*, John Wiley & Sons, New York, 1988.

[112] Shatz, M. P., and Polychronopoulos, G. H., *An Algorithm for Evaluation of Radar Propagation in the Spherical Earth Diffraction Region*. IEEE Transactions on Antenna and Propagation, VoL. 38, No.8, August 1990, pp. 1249-1252.

[113] Sherman, S. M., *Monopulse Principles and Techniques*, Artech House, Norwood, MA.

[114] Singer, R. A., Estimating Optimal Tracking Filter Performance for Manned Maneuvering Targets, *IEEE Transaction on Aerospace and Electronics, AES-5*, July 1970, pp. 473-483.

[115] Skillman, W. A., *DETPROB: Probability of Detection Calculation Software and User's Manual*, Artech House, Norwood, MA, 1991.

[116] Skolnik, M. I., *Introduction to Radar Systems*, McGraw-Hill, New York, 1982.

[117] Skolnik, M. I., Editor, *Radar Handbook*, 2nd Edition, McGraw-Hill, New York, 1990.

[118] Song, J. M., Lu, C. C., Chew, W. C., and Lee, S. W., Fast Illinois Solver Code (FISC), *IEEE Trans. Antennas and Propagation*, Vol. 40, June 1998, pp. 27-34.

[119] Stearns, S. D., and David, R. A., *Signal Processing Algorithms*, Prentice Hall, Englewood Cliffs, NJ, 1988.

[120] Stimson, G. W., *Introduction to Airborne Radar*, Hughes Aircraft Company, El Segundo, CA, 1983.

[121] Stratton, J. A., *Electromagnetic Theory*, McGraw-Hill, New York, 1941.

[122] Stremler, F. G., *Introduction to Communication Systems*, 3rd Edition, Addison-Wesley, New York, 1990.

[123] Stutzman, G. E., Estimating Directivity and Gain of Antennas, *IEEE Antennas and Propagation Magazine 40*, August 1998, pp. 7-11.

[124] Swerling, P., Probability of Detection for Fluctuating Targets, *IRE Transaction on Information Theory*, Vol. IT-6, April 1960, pp. 269-308.

[125] Taflove, A., *Computational Electromagnetics: The Finite-Difference Time-Domain Method*, Artech House, Norwood, MA, 1995.

[126] Trunck, G. V., *Automatic Detection, Tracking, and Senor Integration*, NRL Report 9110, June, 1988.

[127] Van Trees, H. L., *Detection, Estimation, and Modeling Theory*, Part I, Wiley & Sons, Inc., New York, 2001.

[128] Van Trees, H. L., *Detection, Estimation, and Modeling Theory*, Part III, Wiley & Sons, New York, 2001.

[129] Van Trees, H. L., *Optimum Array Processing*, Part IV of *Detection, Estimation, and Modeling Theory*, Wiley & Sons, New York, 2002.

[130] Tzannes, N. S., *Communication and Radar Systems*, Prentice Hall, Englewood Cliffs, NJ, 1985.

[131] Urkowitz, H., *Decision and Detection Theory*, Unpublished Lecture Notes, Lockheed Martin Co., Moorestown, NJ.

[132] Urkowtiz, H., *Signal Theory and Random Processes*, Artech House, Norwood, MA, 1983.

[133] Van Vleck, J. H., The Absorption of Microwaves by Oxygen, *Physical Review*, Vol. 71, 1947.

[134] Van Vleck, J. H., The Absorption of Microwaves by Uncondensed Water Vapor, *Physical Review*, Vol. 71, 1947.

[135] Vaughn, C. R., Birds and Insects as Radar Targets: A Review, *Proc. IEEE*, Vol. 73, No. 2, February 1985, pp. 205-227.

[136] Wehner, D. R., *High Resolution Radar*, Artech House, Norwood, MA, 1987.

[137] Weiner, M. M., Editor, *Adaptive Antennas and Receivers*, Taylor & Francis, Boca Raton, FL, 2006.

[138] White, J. E., Mueller, D. D., and Bate, R. R., *Fundamentals of Astrodynamics*, Dover Publications, New York, 1971.

[139] Ziemer, R. E., and Tranter, W. H., *Principles of Communications, Systems, Modulation, and Noise,* 2nd Edition, Houghton Mifflin, Boston, MA, 1985.

[140] Zierler, N., *Several Binary-Sequence Generators*, MIT Technical Report No. 95, Sept. 1955.

Index